CLINICAL DECISION SUPPORT

CLINICAL DECISION SUPPORT

THE ROAD AHEAD

Edited by

ROBERT A. GREENES, M.D., Ph.D.
Harvard Medical School and Brigham & Women's Hospital
Boston, Massachusetts
USA

AMSTERDAM • BOSTON • HEIDELBERG • LONDON
NEW YORK • OXFORD • PARIS • SAN DIEGO
SAN FRANCISCO • SINGAPORE • SYDNEY • TOKYO

Academic Press is an imprint of Elsevier

30 Corporate Drive, Suite 400, Burlington, MA 01803, USA
525 B Street, Suite 1900, San Diego, California 92101-4495, USA
84 Theobald's Road, London WC1X 8RR, UK

This book is printed on acid-free paper. ♾

Library of Congress Cataloging-in-Publication Data
Application Submitted

British Library Cataloguing-in-Publication Data
A catalogue record for this book is available from the British Library.

ISBN 13: 978-0-12-369377-8
ISBN 10: 0-12-369377-2

For information on all Academic Press publications
visit our Web site at www.books.elsevier.com

Printed in the United States of America
06 07 08 09 10 9 8 7 6 5 4 3 2 1

CONTENTS

CONTRIBUTORS

Numbers in parentheses indicate the pages on which the authors' contributions begin.

Joan S. Ash (385) Oregon Health and Science University, Department of Medical Informatics and Clinical Epidemiology School of Medicine, 3181 SW Sam Jackson Park Road, Portland, OR

David W. Bates (127) Brigham and Women's Hospital, General Medicine Division, 1620 Tremont Street, 3rd Floor, BC3-2M, Boston, MA

Paul Biondich (111) Indiana University School of Medicine, Regenstrief Institute, 1050 Wishard Boulevard RG5, Indianapolis, IN

Richard L. Bradshaw (469) Intermountain Healthcare, Department of Medical Informatics, 4646 West Lake Park Boulevard, West Valley City, UT

James J. Cimino (345) Columbia University, Department of Biomedical Informatics, 622 West 168th Street, VC-5, New York, NY

Guilherme del Fiol (345) Intermountain Healthcare and University of Utah, 4646 Lake Park Boulevard, West Valley City, UT

R. Scott Evans (143) LDS Hospital/Intermountain Healthcare, Department of Medical Informatics, 8th Avenue and C Street, Salt Lake City, UT

Kent Gale (169) Klas Enterprises, LLC, 630 E. Technology Avenue, Orem, UT

John Glaser (403) Partners HealthCare, 800 Boylston Street, Boston, MA

Robert A. Greenes (3, 31, 79, 195, 373, 541) Harvard Medical School, Brigham and Women's Hospital, L1 Mezzanine, 75 Francis Street, Boston, MA

Jason Hess (169) Klas Enterprises, LLC, 630 E. Technology Avenue, Orem, UT

Tonya Hongsermeier (403) Partners HealthCare System, Inc., Clinical Informatics Research and Development, 93 Worcester Street, PO Box 81905, Wellesley, MA

Stanley M. Huff (307) Intermountain Health Care, 4646 Lake Park Boulevard, Salt Lake City, UT

Nathan C. Hulse (307) Intermountain Health Care, 4646 Lake Park Boulevard, Salt Lake City, UT

Robert A. Jenders (267) Cedars-Sinai Medical Center, UCLA, 8700 Beverly Boulevard, SSB-309, Los Angeles, CA

Vipul Kashyap (447) Partners HealthCare System, Inc., Clinical Informatics Research and Development, 93 Worcester Street, PO Box 81905, Wellesley, MA

Kensaku Kawamoto (503) Duke University Medical Center, Division of Clinical Informatics, Department of Community and Family Medicine, Box 2914, Durham, NC

Joseph Lau (249) Tufts–New England Medical Center, Institute for Clinical Research and Health Policy Studies, 750 Washington Street, Box 63, Boston, MA

Helen G. Lo (127) Brigham and Women's Hospital, General Medicine Division, 1620 Tremont Street, 3rd Floor, BC3-2M, Boston, MA

Burke Mamlin (111) Indiana University School of Medicine, Regenstrief Institute, 1050 Wishard Boulevard RG5, Indianapolis, IN

Michael E. Matheny (227) Decision Systems Group, Brigham and Women's Hospital, 75 Francis Street, Harvard Medical School, Boston, MA

Clement McDonald (111) Indiana University School of Medicine, Regenstrief Institute, 1050 Wishard Boulevard RG5, Indianapolis, IN

Randolph A. Miller (423) Vanderbilt University Medical Center, Room B003C, Eskind Biomedical Library, 2209 Garland Avenue, Nashville, TN

Sarah Miller (423) Peterhouse, Cambridge University, Cambridge, UK CB2 1RD

Lucila Ohno-Machado (227) Decision Systems Group, Brigham and Women's Hospital, 75 Francis Street, Harvard Medical School, Boston, MA

Marc Overhage (111) Indiana University School of Medicine, Regenstrief Institute, 1050 Wishard Boulevard RG5, Indianapolis, IN

Matvey B. Palchuk (325) Partners HealthCare System, Inc., Clinical Informatics Research and Development, 93 Worcester Street, PO Box 81905, Wellesley, MA

Vimla L. Patel (207) Departments of Biomedical Informatics and Psychiatry, Laboratory of Decision Making and Cognition, Columbia University, Vanderbilt Clinic, Fifth Floor, 622 West 168th Street, New York, NY

Mor Peleg (281) University of Haifa, Department of Management Information Systems, 31905, Israel

Roberto A. Rocha (469) Intermountain Health Care, Information Systems – Medical Informatics, 4646 West Lake Park Boulevard, West Valley City, UT

Beatriz H.S.C. Rocha (469) Intermountain Healthcare, Department of Medical Informatics, 4646 West Lake Park Boulevard, West Valley City, UT

Edward H. Shortliffe (207) Columbia University Medical Center, 622 West 168th Street, VC-550, New York, NY

Margarita Sordo (325) Decision Systems Group, Brigham and Women's Hospital; 75 Francis Street, Harvard Medical School, Boston, MA

William Tierney (111) Indiana University School of Medicine, Wishard Hospital, 1001 West 10th Street, OPW-0200, Indianapolis, IN

PREFACE

"Good ideas are not adopted automatically. They must be driven into practice with courageous impatience. Once implemented they can be easily overturned or subverted through apathy or lack of follow-up, so a continuous effort is required."

Admiral Hyman G. Rickover

This sentiment was expressed by Admiral Hyman G. Rickover in a speech delivered at Columbia University in 1982, and, I believe, serves as an appropriate introduction to this book. Several citations of this quote on the Internet present a curious discrepancy, in that the word "impatience" is replaced with "patience." I find this variation intriguing and suitable, as it seems to me that both *courageous patience* and *courageous impatience* must be manifest for the greatest progress to occur.

My idea for this book was based on the observation that, although computer-based clinical decision support (CDS) has the potential to be truly transformative in health care, and despite considerable creativity and experimentation by enthusiasts over more than four decades, as well as convincing demonstration of effectiveness in particular settings, the adoption of CDS has proceeded at a snail's pace. This slow progress has not accelerated significantly even with major recent national and regional efforts in a number of nations to promote the use of the electronic health record (EHR), computer-based physician order entry (CPOE), electronic prescribing, and the personal health record (PHR), all of which are important substrates on which CDS can operate (and for which the prospect of CDS itself is a major driver). Some capabilities have made their way into commercial health information system products. Examples include advice and warnings during CPOE to ensure proper doses, avoid harmful interactions, or warn about allergies, the provision of alerts to providers when an abnormal laboratory result is found, and the use of order sets, or groupings of orders, for specific clinical problems and settings, such as coronary care unit admission or post-operative care after a hip replacement. Nonetheless, CDS usage remains spotty at best, most prevalent but by no means ubiquitous at academic medical centers, less so in community hospitals, and almost non-existent in office practice. Although now the public frequently turns to the Internet for medical knowledge, automated decision support oriented to patients and consumers in terms of reminders, alerts, and patient-specific advice is similarly lacking.

What's wrong with this picture? Is CDS perhaps not really a good idea? Are the requirements for wide dissemination and use beyond reach? Or are there initiatives that can be undertaken that can change the dynamic, and significantly boost adoption? This book is an effort to address this conundrum. Our goal is to examine CDS in detail from many perspectives–its history and motivations, enabling technologies, psychological and human factors concerning its deployment and use, sociological, organizational, and management considerations, financial and economic drivers and constraints, marketplace and business opportunities, and lastly, the role of communal and top-down initiatives such as standardization and creation of infrastructure and sharable resources.

The premise is that if CDS is truly a good idea, then the sluggish progress in adoption and use to date can only mean that we are in need of a new approach. But to develop a new approach to a complex, multifaceted problem, one that would have a better chance of success than current incremental, uncoordinated efforts, the effort will require the participation of many stakeholders representing a range of perspectives. It is much easier in such situations to preserve the status quo, or to introduce minor tweaks than to take concerted action. Only by finding a way for these participants to come together with a commonality of purpose can this good idea be driven forward with the requisite duality of courageous patience and impatience.

Thus, this book is an effort to develop a common ground for addressing this challenge. It should be of most interest to health care organization managers, policy makers, and other senior leadership, payers, government funding agencies, and foundations focused on health care delivery, medical informatics researchers, and students, information technology development managers, information systems, and knowledge product and service vendors, and clinical investigators and health care providers more generally who have interest in the issues of health care quality, safety, and cost-effectiveness.

Let's look a bit more at the motivations for CDS and the challenge of aligning them. In this book we adopt a view of CDS as decision support aimed at individual patient-specific health care. It is advice and guidance offered by computers (more properly, information and communication technology) to aid the problem solving and decision making of health care providers, patients, and the public (i.e., including those not currently patients). CDS is in most views not only a *good idea* but an *essential* one. The most compelling reasons for CDS are to help practitioners avoid errors, optimize quality, and improve efficiency in health care. Many pressures fuel the need: an explosion of biomedical knowledge over the past several decades; the multiplicity of diagnostic and therapeutic choices available for patient care; specialization and fragmentation of care; time constraints on practitioners; regulatory and compliance demands; malpractice concerns; increasing prevalence of multi-system diseases as the population ages; rising costs of health care; enhanced activism and involvement of individuals in their own health care; and the emerging capbilities for "personalized medicine" enabled by genomics, biomarkers, and increasingly structured clinical phenotype data.

Although the descriptors used to characterize the above trends have a kind of desperate urgency to them that collectively suggest hyperbole, the fact is that they reflect the reality of health care today. Because the trends and their consequences impact on different stakeholders in various ways, however, the combined extent of the need perhaps has not been appreciated to its full degree by individuals, as a result of which there has been little impetus for a broad-based effort to address it.

It is important to understand the differences in perspective of the various stakeholders, and to recognize their motivations and needs. For practitioners, despite the benefits offered, there are a number of reasons why CDS is not unequivocally endorsed. Providers often become quite expert in their own particular subject domains, keep up with the literature, and don't feel a compelling need for computers to make recommendations. Despite this, they generally accept the value of CDS to monitor their actions, especially where the aim is to help avoid accidental errors, appreciate alerts for unexpected lab results (if false positives are kept to a minimum), find timely reminders for schedulable actions useful, and take advantage of predefined order sets for frequent clinical situations they encounter. In general, and not surprisingly, the satisfaction with CDS by practitioners seems mostly related to the degree to which it is supportive, patient-specific, relevant, and provided in a way that doesn't interfere with care or require inordinate effort and time.

In some circumstances, however, the use of CDS not only requires extra time and effort by the practitioner, but the benefits of its use aren't seen as accruing to the practitioner or the patient. Examples are applications of CDS aimed at limiting orders for expensive tests and treatments. This purpose has been manifested in adoption of drug formularies, substitution of generic for brand name drugs, utilization review and management, and requirements for preapproval/prior authorization from payers for imaging procedures, specialty referrals, or surgery as a function of clinical indication. Such applications of CDS may be tolerated by practitioners as necessary evils but they are regarded primarily as interference with medical judgment and are hardly embraced. Patients also exhibit disdain for and distrust of such applications. So, although a societal net benefit may be at play, it is difficult to align support of institutions, payers, providers, and patients in cost containment circumstances.

Similar difficulties have been experienced in introducing many information system innovations in health care. Given the frequently tenuous acceptance or tolerance of such technologies by those who must interact with them directly, instances in which their implementation has been poorly executed have been quick to be seized on by unhappy users and critics as examples of why the adoption of such systems should be resisted. Before an information system innovation is introduced, careful thought and experience with testbeds and pilot implementations are needed. Planning, both locally and at an organizational or enterprise level, needs to be directed at issues of how to motivate the participants, and how to make the innovation function effectively, including considerations of process and work flow, responsibilities and prerogatives of the users, ease of use, and perception of benefit by them.

With respect to the deployment and support of CDS, it also appears that a major barrier to progress is lack of appreciation of the difficulty of the problem. On the surface, most CDS does not appear to be very complicated to implement. Methods for provision of CDS have been the subject of study throughout its long history, and many useful approaches have been identified, explored, demonstrated, and evaluated. As I've noted, some of these methods have become highly successful at achieving the intended goals in operational settings. Although some forms of CDS are quite complex, by and large, the most effective approaches have indeed been relatively simple, such as the use of *if ... then* rules for applications such as mentioned earlier for determining appropriateness of an action such as a medication order, for recognizing an abnormality of a laboratory result, or for generating a reminder for a test or procedure. Another straightforward form of CDS is the establishment of groups of orders into order sets for particular clinical situations. As also noted, these approaches have made their way into a number of commercial products.

The point that is often overlooked, however, is that robust, sustainable use of CDS is not at all simple, even for *if ... then* rules or order sets, when one considers it not with respect to a single point in time but from a long-term maintenance and update perspective. This Sustainability is difficult even in a single site with a limited focus, but the complexity of managing CDS, and the knowledge embedded in it, increases dramatically when one considers its deployment and frequent update on an enterprise-wide basis, driven in large part by the continual efforts in most health care organizations to improve patient safety, quality and cost-effectiveness. The knowledge assets underlying CDS are time-consuming and expensive to generate, voluminous, and subject to change, so sharing and reuse of them, once created, would be highly advantageous. The knowledge derived from multiple research studies and analyses must be collected, validated, and refined. The knowledge considered most useful for CDS must then be assembled, curated, and represented formally and unambiguously. Knowledge should be represented in standard form, so that it can be disseminated and used widely, and needs to be updated on a regular basis. As sites then seek to use such knowledge in clinical applications, the tools for doing so should be shared and leveraged. The knowledge must be adapted to local requirements and constraints, integrated into CDS, and interfaced with clinical applications. The experiences with these activities, both positive and negative, should be made available broadly as well.

These tasks may be considered to comprise three interacting but separate lifecycle processes – knowledge generation and validation, knowledge management and dissemination, and CDS implementation and evaluation. Currently much of the work involved in these tasks tends to occur haphazardly or without explicit delineation, and without a formalized infrastructure. The provision of such capabilities on a sharable, communal basis is an area in which effort has barely begun, and where alignment of stakeholders is essential for substantive progress. I believe that the lack of such capabilities is one of the primary impediments to driving widespread CDS adoption and use. A further key obstacle has been the slow progress in development of standards for representing knowledge for decision support. The lack of such standards limits the ability to use knowledge in CDS, except after extensive adaptation and recoding for specific platforms. This results in perpetuation of multiple incompatible proprietary system implementations and contributes greatly to the difficulty in maintaining and updating both the CDS and its knowledge content.

In the first two sections of this book, we introduce CDS in terms of its various purposes, design, motivations for use, and experiences over the years with implementation, in both academic settings and commercial products.

In subsequent sections, we consider the issues underlying knowledge generation, knowledge management, and CDS deployment, and current approaches to formalizing these processes. After examining efforts in various organizations, we consider the prospect of mustering forces on a national or international scale in order to move ahead more rapidly. We conclude with suggestions about how such a communal process might be initiated on a modest scale in order to gain experience and build support and momentum for larger scale efforts.

In summary, the widespread adoption of CDS presents a very complex, multi-faceted challenge that must be attacked on many fronts. My hope is that this book will enable readers with a variety of backgrounds to gain an appreciation of both the nature of the challenge and the benefits of tackling it. Given that this odyssey will require the shared vision and cooperation of many stakeholders as noted earlier, the intent is that parts of the book will address issues relevant to their various perspectives.

In writing this book, I am grateful for the participation of extraordinarily gifted and experienced colleagues who have contributed a number of excellent chapters. While I take responsibility for the overall vision of the road ahead espoused here, I know that all of the contributors have a commitment to widespread implementation of clinical decision support and have a passion for its general goals. I am proud and honored to have such wonderful colleagues, whom I count as friends and fellow travellers on this important journey.

Robert A. Greenes
Boston, Massachusetts

I

COMPUTER-BASED CLINICAL DECISION SUPPORT: CONCEPTS AND ORIGINS

1 DEFINITION, SCOPE, AND CHALLENGES

ROBERT A. GREENES

1.1 INTRODUCTION

The application of computers in health care is already an "old" pursuit, when assessed by the yardsticks of the electronics age where product life cycles and the rise and fall of industries happens in months and years rather than decades. Efforts to automate aspects of health care began in earnest as far back as the early 1960s—some 45 ago. Yet the rate of adoption and degree of impact of computers and information technology remain low in comparison to the extent to which they have become primary, even driving, forces in other fields such as engineering, physics, finance, and even personal communication.

In the earliest days of computer use in health care, the prospect that computers could play an active role in helping to solve problems and make decisions stimulated much interest and excitement, and was among the primary motivators for pursuing the use of computers. This was manifested both in terms of research and development as well as public imagination and attention. This hope for the computer in health care has continued, and in recent years has grown to a plea bordering on desperation. Health care practitioners continually confront a wide range of challenges—seeking to make difficult diagnoses, avoid errors, ensure highest quality, maximize efficacy, and save money—all at the same time! Patients and the public have many questions and needs in evaluating their health and in making decisions that also require help. The array of choices, the tradeoffs, and the other mitigating factors that affect decisions are more complex and require more detailed knowledge than ever before. Computers, with their speed, vast memories, and stored knowledge, can surely help us do these things.

A large number of computer-based clinical decision aids have been developed and their usefulness evaluated over the past 40+ years. Many of these have involved simple types of decision support like recognizing that a laboratory test result is out of normal range, or that a medication being ordered has a dangerous interaction with another one that a patient is taking, or determining that a patient is now due for a flu shot. In numerous studies, such checks, warnings, and reminders have been demonstrated to be effective. In Section II, we examine numerous examples of successful decision support at three leading academic centers, which have been shown in published studies to reduce errors, encourage best practices, reduce costs, and provide a variety of other benefits. Work also has progressed on more complex forms of decision support such as that involved in constructing a differential diagnosis or selecting an optimal treatment strategy.

These positive developments have been dramatic and encouraging; therefore, it is quite surprising to note that the innovations and successful approaches, by and large, have not led to broad dissemination and widespread adoption of such approaches. To be sure, as we also review in Section II, although much of the initial work has been in academic settings, some of it has found its way into commercial products, as part of clinical information systems, or as components that can be added onto or integrated with clinical systems. Nonetheless, availability has been decidedly spotty, and limited, not just for complex forms of decision support, but even for the more simple aids like alerts and reminders or drug interaction checks.

Despite all the promise and eager anticipation, the prospect of using computers in decision support has turned out to be a much harder problem than generally is appreciated. Even for the simplest forms of decision support, it takes a large scale-up of effort to go from an initial implementation, aimed at showing that clinical decision support is effective in a particular application setting, to having the ability to provide ongoing management of decision support in the same setting. A further leap is required to move from that capability to wider deployment beyond a single application, even within a single institution or across a single enterprise. This becomes not just a big leap but a giant one, if the goal expands so as to address the possibilities of regional or national adoption of accepted clinical practices and guidelines—even for limited aspects of health care, such as appropriate utilization of imaging procedures or prescribing of high-cost medications. Challenges that are manageable with some effort in a single environment become much more difficult in a multi-institutional setting. These relate to maintenance and updating of the knowledge underlying decision support; managing the corpus of knowledge, in terms of conflicts, overlaps, and gaps; determining the best ways to deploy various forms of decision support, in terms of their integration with practice and impact on efficiency and workflow; and disseminating knowledge that has been well established so that it can be reused in multiple sites, making such knowledge platform-independent. Addressing this last challenge, in particular, is essential to leveraging and making the effort involved in knowledge management economically feasible on a broad scale.

The story that has unfolded over the years is one about a provocative, tantalizing, and yet frustrating relationship between the computer and the practitioners and recipients of health care regarding decision support—one that has seen great moments, and strong proponents, but which has not caught fire, not reached a "tipping point" (Gladwell 2000) where it is considered a necessary component of the health care process (see Table 1-1). Although the relationship between the computer and health care practitioners and recipients has languished for some four and a half decades, nonetheless, there are signs that it may now be coming out of this period of dormancy—older, wiser, and better equipped to enter into a mature union. This book is a tale of that relationship—the early allure (technical and medical benefits), the roots of attraction (the characteristics of the parties), the realities of the long courtship (the cultural, social, and organizational milieu that have both encouraged and held back the relationship), the

TABLE 1-1 Relationship between computers as source of clinical decision support and providers and recipients of health care.

Phase of relationship	Duration (date ranges approximate)	Hallmarks
A long infatuation	1960–1985	Enthusiasm for clinical decision support, research, new ideas
A troubled courtship	1985–1998	Successful implementations, evaluations showing benefit, but limited dissemination
Renewed passions	1998–2003	Knowledge explosion, safety and quality agendas
Building the foundations for a lasting relationship	2003–	National agendas, call to action, roll out of electronic health records (EHRs), computer-based provider order entry (CPOE), electronic prescribing (eRx), personal health records (PHRs)
A new party to the relationship	2004–	Recognizing knowledge management as a necessary infrastructure

growth and adaptation that have occurred (improved understanding of requirements and limitations), and the underpinnings that are now being recognized as necessary for the relationship to thrive.

Our goal for this book is to address the question of how to achieve broad impact of clinical decision support on patient safety, health care quality, and health care cost-effectiveness. This involves three primary foci:

1. We need to understand the issues involved in identifying what kinds of decision support are useful for these purposes, as shown in implementations and evaluation studies.
2. We need to understand the problems and challenges that must be addressed in order to broadly disseminate and replicate these successes, as well as to extend successful approaches to other settings so that the long-anticipated benefits can be realized.
3. Ultimately, for that to occur, we need to identify how various stakeholders will need to participate, and the resources, commitments, and coordinated, sustained effort that will need to be marshaled.

Two recent publications are helpful in conjunction with this book. A HIMSS 2005 publication, *Improving Outcomes with Clinical Decision Support: An Implementer's Guide* (Osheroff, Pifer et al. 2005) provides a set of practical activities that can be taken to prepare an environment for receptivity and success in deploying CDS. A white paper commissioned by the U.S. Office of the National Coordinator for Health Information Technology (ONC) on *A Roadmap for National Action on Clinical Decision Support* (Osheroff, Teich et al. 2006) calls for the creation of an organized national effort in the United States to develop and deploy needed infrastructure and alignment of the multiple stakeholders involved.

This book is aimed at providing an appreciation of the whole landscape of clinical decision support, which we believe will be useful (a) to strategic planners seeking to develop consensus and infrastructure on a national scale, (b) to managers and implementers of clinical decision support in medical centers and practices, (c) to knowledge content providers in identifying potential opportunities for providing their content and services, and (d) to clinical information system vendors in terms of the ways in which they can facilitate the development of the needed interfaces and application capabilities.

1.2 DEFINITION OF COMPUTER-BASED CLINICAL DECISION SUPPORT

Computer-based clinical decision support (CDS) can be defined as the use of the computer to bring relevant knowledge to bear on the health care and well being of a patient.

There are several key words in the phrase "computer-based clinical decision support" and in the preceding definition. Here, the term *computer* is really shorthand for *information and communication technologies*, collectively. Clinical decision support can, of course, be provided by textbooks, teaching, manual feedback, and a variety of other methods; by *computer-based*, we mean that our focus is specifically on use of information and communication technology as the *basis* for providing it. By *clinical decisions*, we mean those that bear on the management of health and health care of an *individual* person (the patient). By *support*, we mean the *aiding* of rather than the *making* of decisions. By *relevant knowledge* we mean the selection of knowledge that is directly pertinent to the specific patient.

1.3 FEATURES OF CDS

CDS has a number of characteristics that apply to most, if not all, of the many ways it can occur. We will discuss these features in some detail in this and the following chapters in this section.

1. The general aim of CDS can be one or both of the following:
   - To make data about a patient easier to assess by, or more apparent to, a human.
   - To foster optimal problem-solving, decision-making, and action by the human. The exact nature of a particular form of CDS depends on its specific purpose.
2. The decision support is provided to a user—who may be a physician, a nurse, a laboratory technologist, a pharmacist, a patient, or other individual with a need for it. In some instances, the user may be a computer program rather than a human user. Many possible settings can give rise to this need, such as a problem arising in clinical practice, a health maintenance/preventive care question of a patient, or a training/educational exercise.
3. A primary task of the computer is to select knowledge that is pertinent, and/or to process data to create the pertinent knowledge. To the extent that the computer can make the selection based on patient-specific data, the relevance of the CDS to the individual patient is enhanced.
4. The selection of knowledge and processing of data involve carrying out some sort of inferencing process, algorithm, rule, or association method.
5. The result of CDS is to perform some action, usually to make a recommendation.

1.4 THE TALE OF A RELATIONSHIP

1.4.1 A Long Infatuation

Many attractive scenarios have been explored for use of CDS over its approximately 45-year history. Consider a doctor seeing a patient with a skin rash. The physician could benefit from a variety of forms of CDS in this setting; for example, he or she could be presented with a set of similar eruptions, their differential diagnosis, and information about the next steps for evaluating them. This could be in the form of an atlas of images from a Web site, a textbook reference, a differential diagnosis interactive software program, or a clinical practice guideline for skin rashes. The interactive software program and practice guideline, if used together such that guideline steps are selected based on patient-specific information entered by the patient or provider, would allow the decision support to be customized for that patient. It could be used to directly make recommendations for treatment or to trigger the display of a set of potential orders for a prescription medication, topical remedies, dietary recommendations, and other associated activities; it could perhaps also identify educational materials to be printed and made available to the patient.

For decision support resources such as those referred to in the preceding example to work optimally, they should be able to access high-quality, evidence-based medical knowledge. Further, items of knowledge should be automatically selected or derived based on the clinical context and the particular findings of the patient so that the CDS is as relevant as possible.

1.4.1.1 Simple and Complex

The roles and proper uses of CDS have intrigued investigators from the outset (Blois 1980; Collen 1986; Diamond, Pollock et al. 1987). The preponderant applications of CDS have been of the more *straightforward* variety we have alluded to already. Computers can be used for information retrieval, by providing search capabilities to find answers to specific clinical questions. They can do very basic error checks, enabling them to be guardians of safety—to detect problems when they occur or to prevent them altogether. A particularly valuable yet simple task is to perform data entry validation, as in the checking of a requested dose in a physician-entered medication order against predefined limits. Another practical and uncomplicated function is to continuously monitor new test results in a clinical laboratory, to identify conditions such as a critically low potassium level that require notification of the patient's physician. Yet another is to identify conditions that trigger reminders such as for scheduling an annual mammogram in a woman over 50 or for giving a flu shot to an elderly patient in the winter flu season.

More complex uses are also of value. Among these, the idea of putting the computer to work to help make difficult diagnoses has been especially intriguing from the earliest days of computer use. In fact, from those earliest days up to the present, if one were to ask a layperson how a computer could be most useful for decision support in medicine, chances are that the person would say that it would be for making diagnoses. One of my own first exposures to CDS in clinical medicine was the seminal paper in *Science* by Ledley and Lusted, published in 1959, entitled "Reasoning Foundations of Medical Diagnosis" (Ledley and Lusted 1959). This manuscript explored a combination of logical manipulation and probability, in particular, Bayes theorem, to identify most likely diagnoses given a particular set of findings. Over the ensuing four and a half decades, multiple applications and extensions of the approach have occurred, as well as development and evaluation of a number of alternative models for differential diagnosis (Caceres 1963; Pipberger, Stallmann et al. 1963; Warner, Toronto et al. 1964; Lodwick 1965; Gorry and Barnett 1968; deDombal 1975). (It is interesting to note that among these activities, some of the earliest developments were in the area of electrocardiographic diagnosis (Caceres 1963; Pipberger, Stallmann et al. 1963), which involved signal processing and analysis of the ECG tracing, whereas the other efforts all dealt with clinical diagnosis requiring entry of findings by a user. As we will discuss further in subsection 1.4.2.2, the singularity of focus of ECG analysis and lack of need for human data entry likely contributed to the wider adoption and use of computer-based ECG interpretation today than other clinical diagnostic applications.)

Beyond diagnosis, the computer can support a variety of other complex decision-making tasks. It can help determine optimal workup strategy (Greenes, Tarabar et al. 1989) in sequencing of tests and procedures for evaluating a clinical problem (e.g., staging of colon cancer in an elderly man or evaluation of a breast lump in a young woman). It can assist in selecting treatment (Shortliffe, Davis et al. 1975), or in evaluating alternative treatment strategies (Kassirer, Moskowitz et al. 1987) in order to select an optimal one for those conditions. It can be used to perform detailed treatment plans, in terms of dose calculations for chemotherapy (Knaup, Wiedemann et al. 2002) or detailed 3D modeling and dosimetry calculations for radiation therapy (Ten Haken, Fraass et al. 1995). It can provide estimates of prognosis and risk of complications for alternative treatments (Resnic, Popma et al. 2000; Inza, Merino et al. 2001).

In complex decision-making problem areas such as workup, diagnosis, treatment, and long-term management, just the ability to organize and coordinate the sequence of steps for performing various actions, evaluating results, and making choices of next steps is valuable. Thus, decision support in the form of clinical practice guidelines is of interest (Ohno-Machado, Gennari et al. 1998; Shiffman, Liaw et al. 1999; Miller, Frawley et al. 2000; Peleg, Boxwala et al. 2000; Greenes, Peleg et al. 2001). Guidelines also can be used to embody best practices, with the hope that their use will improve health care quality, reduce variation, and improve efficiency and workflow.

1.4.1.2 Evidence of Usefulness

The usefulness of CDS for the kinds of applications just mentioned, as well as many others, has been demonstrated in a number of evaluation studies. The experiences of three academic institutions that have pioneered the practical use of CDS in clinical systems and systematically evaluated them are detailed in Section II, primarily with respect to the use of alerts and reminders, and drug interaction and dosing checks in computer-based provider order entry (CPOE). Comparative evaluation studies of various types of systems providing CDS also have been carried out, for example, for differential diagnosis (deDombal 1975; Berner, Webster et al. 1994; Friedman, Elstein et al. 1998), information retrieval for clinical questions (Haynes, McKibbon et al. 2005), and clinical guidelines (Sintchenko, Coiera et al. 2004), and for CDS in general

(Garg, Adhikari et al. 2005; Kawamoto, Houlihan et al. 2005; Sittig, Krall et al. 2006). Success factors for CDS appear to relate to the degree of patient-specificity (with appropriate true positive vs. false positive rate), degree of integration with clinical practice workflow (without excessive demand for additional or extraneous effort by practitioners), and delivery at the point of need (in the appropriate context and at the time it is maximally useful or able to be acted upon).

1.4.2 A Troubled Courtship

Demonstration of success does not always translate into widespread dissemination and adoption. The first three chapters in Section II describe case histories of three leading academic health care delivery organizations that have integrated CDS capabilities into their clinical information systems and have studied them extensively to determine what works and what doesn't, and have evolved their approaches over the years. Some of these approaches have made their way into commercial offerings, but many haven't. Chapter 7 explores the commercial marketplace as of the present, in terms of available CDS capabilities in deployed clinical information systems. As that chapter shows, the penetration of CDS in clinical systems is decidedly spotty although growing.

Reasons for lack of widespread dissemination and adoption of CDS are both technical and nontechnical, and appear to relate to the complexity of providing CDS. This is true not only for *inherently* complex types of CDS such as differential diagnosis and treatment selection, but even for more simple forms such as alerts and reminders. In fact, a central thesis of this book is that the difficulty in deploying and disseminating CDS is in large part due to the lack of recognition of how hard the job is and lack of availability of tools and resources to make this job easier. The many reasons for lack of penetration and the requirements for breaking down barriers in order to move ahead and realize the potential for CDS are examined in later chapters.

1.4.2.1 Why CDS Is a Hard Problem

As we have noted, the prospect of providing simple forms of CDS appears deceptively easy; for example, the incorporation of an *if...then* rule for checking whether a laboratory test result exceeds a threshold for abnormality. And, in fact, deploying such a rule in a *single* computer system is relatively straightforward to do.

The problems relate to everything *else* surrounding this act—the necessity of considering not just the single use of the logic in an application at a point in time, but how the rule is intended to interact with users, whether the use of the rule is cost effective, how to tune it so that there are not too many false positives yet the appropriate numbers of true positives, how the knowledge underlying the rule will be maintained and updated, how the rule will be encoded or interfaced with the application that will use it, how it relates to other rules, how it can be deployed in other applications or in other system platforms (with different programming languages and architectures and developer conventions), and how knowledge and CDS approaches that are found to be effective can be disseminated and used more broadly, in terms of whether there are sustainable, viable mechanisms for doing this that are supportable by commercial or other means.

These requirements pertain to all forms of CDS that we shall examine, ranging from simple *if... then* rules to the more complex forms of CDS. Going from a single instance of use with demonstration of effectiveness, to continued use over time, to update, to deployment in more than one setting, and to possible adaptation for other uses are all challenging tasks that must be addressed if the desire is to make CDS broadly available.

One of the challenges for CDS cited earlier is its dependency on the computer environment in which it is to be deployed. This relates to more than just hardware and software platform, programming language, and architecture. As noted in subsection 1.4.1.2, to be optimally effective, CDS needs to be highly patient-specific and delivered at the point in patient care where it is most appropriate or most likely to be needed. Thus CDS depends on the existence of computer-based electronic health records (EHRs), and on the suite of applications that are available in which CDS can be incorporated. For example, use of CDS for identifying potential drug–drug interactions is most effective if interfaced with CPOE, but if that capability is not available, an alternative way to introduce the CDS may need to be sought, such as incorporating it into an application based in the pharmacy. In this case, dangerous interactions are identified before the order is filled rather than at the time of entry of the order; although not optimal, the overall effect may still be positive.

1.4.2.2 The Technology of CDS

To understand in more detail why CDS deployment on a broad scale is difficult, we will begin by examining differential diagnosis, in terms of aspects that must be addressed for it to be successful. We will then show how these aspects must be addressed even for simpler forms of decision support.

Diagnostic Decision Support as an Example

As noted earlier in this chapter, computer-based differential diagnosis is a well-studied problem, and efforts have been made over the years to deploy some systems, yet they have had limited penetration. Analysis of the reasons for this is thus helpful in understanding the inherent complexities of deploying CDS.

The core of the methodology proposed by Ledley and Lusted (1959), and further refined by many subsequent investigators, is a *knowledge base* of the various diseases under consideration and the set of findings that might occur in patients with those diseases. Note that we use the term "knowledge base" to mean a distillation of collected data or experience into a statement of facts or relationships. The knowledge base Ledley and Lusted used had both logical and probabilistic aspects. The logical part of the knowledge base encoded relationships pertaining to absolute indications of whether a particular value of a finding was required to be present or absent for a specific disease to be diagnosed, such as *female* as a necessary value of the finding *gender* in the condition ("disease") *pregnancy* or *cancer of the cervix*. The probabilistic part of the knowledge base consisted of (a) the *a priori* probabilities of the various diseases under consideration, and (b) the conditional probability of each possible combination of findings given each of the diseases.

TABLE 1-2 Components of clinical decision support.

CDS Component	Description
Purpose	The task or process of clinical care for which the CDS is intended
Structure	The components specifying the way CDS is to be carried out
• Decision model	The method of organizing or analyzing data and knowledge to arrive at a recommendation
• Knowledge base	The knowledge content used by CDS
• Information model	The manner of representing and naming the clinical and decision support parameters used by the inferencing method to arrive at a recommendation
• Result specification	The output of the decision model
• Application environment	The manner in which the CDS interacts with host applications and users to obtain data, communicate results, and enable the host application to make recommendations or perform actions

Constructing a differential diagnosis consists of the following steps:

1. Ascertain values for the various findings in a patient. Although the method for obtaining the data was not addressed by Ledley and Lusted, these data would need to be obtained from the user or from an electronic health record. In the simplest application of Bayes theorem, each finding is indicated in binary form, that is, as only being present or absent.
2. Logically eliminate those diseases for which finding(s) are incompatible with the presence of the disease. For pathognomonic findings, exclude all diseases that do not have the finding(s).
3. For those diseases remaining under consideration (except in the case of diseases already established by the presence of pathognomonic finding(s)), perform a Bayes theorem calculation of the conditional probability of each disease, given the set of findings (more detailed discussion is given in Chapter 2).

This early method demonstrated the importance of several aspects of CDS (see Table 1-2):

- It has a focus or purpose, that of differential diagnosis.
- It has a structure that includes, either explicitly or implicitly, five components:
  1. A **decision model** or representation of the problem, in this case, a combination of logical and Bayesian manipulation.
  2. A **knowledge base**, consisting of the logical constraints, prior probabilities of disease, and conditional probabilities of findings given disease, for all findings and diseases under consideration.
  3. An **information model** for referring to the data elements needed for calculations, logic manipulations, and such.
  4. A **result specification**, that is, a kind of output or means of representing the results of the operation of the model.
  5. An **application environment**, that is, a method of interacting with a user or an information system, to obtain necessary inputs, and communicate results, to enable a host application to interpret the results in the form of recommendations or actions to be performed.

These components of CDS will be discussed in more detail in Chapter 3.

Two early applications of the methodology proposed by Ledley and Lusted (1959) were in the domains of: (1) congenital heart disease diagnosis, developed by Warner et al. (1964), and (2) bone tumor diagnosis developed by Lodwick (1965). An assumption was made in these models, for simplification, that each finding was conditionally independent of all other findings, given a particular disease, and that the set of diseases was mutually exclusive (non-overlapping) and exhaustive (covering all possible diseases). Without the conditional independence simplification, it would be necessary to determine estimates of conditional probabilities of all possible combinations of values of findings, an exponentially difficult task as numbers of findings increase). The mutually exclusive and exhaustive simplification means that the sum of the probabilities of the diseases adds up to 1.

Gorry and Barnett (1968) developed an approach to sequential Bayes diagnosis, which structured the decision problem as a series of steps requiring less than the total number of findings, and which decided, based on results at each step, what additional findings would best contribute to a diagnosis. Work by deDombal and colleagues in the 1970s and 1980s (deDombal 1975) resulted in one of the most widely used diagnosis programs, a Bayesian algorithm for diagnosis of abdominal pain.

Computer-aided diagnosis has been a major pursuit by a number of investigators over the years, with methods that have been based on Bayes theorem as well as on a number of other approaches that have evolved or been developed since the late 1950s, as we will review in Chapter 2. Later approaches included artificial intelligence models (rule-based, frame-based, and heuristic-reasoning-based) (Shortliffe, Davis et al. 1975; Pauker, Gorry et al. 1976; Miller, Pople et al. 1982; Miller, Masarie et al. 1986; Barnett, Cimino et al. 1987), and other Bayesian implementations (Warner 1989). Today, of those general differential diagnosis programs, DXplain (Barnett, Cimino et al. 1987) is one of the few that remains widely available (along with some newer programs such as Isabel (Ramnarayan, Tomlinson et al. 2004)).

Despite more than 40 years of interest and activity, computer-aided diagnosis is rarely actually used in practice, including DXplain. This bespeaks several points:

- Diagnostic challenges do not arise all that often in medicine. Many of the problems with which physicians deal in clinical practice have more to do with determining optimal workup strategy, choosing or modifying treatments, and assessing prognosis and response to treatment.
- When they do arise, diagnostic problems require analysis of many detailed data items. A number of these are not readily available in the electronic health record, or if present, are not sufficiently structured to be directly used by differential diagnosis programs, but rather must be encoded or mapped to the format required by the programs (e.g., as present/absent or high/medium/low or other categorical classification scheme).
- Use of differential diagnosis tools thus requires considerable manual entry of data, which is time-consuming and cumbersome.
- Finally, differential diagnosis CDS typically is available only as a separate capability that must be actively invoked by the physician when needed, rather than being smoothly integrated into operational applications

of the clinical information system. Given the other obstacles, it tends to be used only when the physician is faced with a real "stumper" of a clinical problem.

Many of these points indicate that, in seeking to provide differential diagnosis in a clinical environment, one of the major obstacles has been the lack of a good application environment (component 5 of the structure of a CDS capability, as described earlier) in which it could be incorporated, so that it could be smoothly and effectively used without undue effort. The much greater adoption of computer-based ECG analysis, as mentioned in subsection 1.4.1.1, is likely due to the fact that it can be performed, to a large extent, solely by analyzing the ECG signal itself, without the need for clinical data entry by a user. More sophisticated ECG analysis, of course, needs to take into account patient medications, presence of an electronic pacemaker, and other clinical factors, but the basic measurements and analyses can be done without this information. This emphasizes that a major goal that CDS must achieve is to have a perceived usefulness to physicians commensurate with or greater than the effort required by them to utilize it.

Components of Other Kinds of Decision Support

We now turn our attention to simpler kinds of decision support that have been more widely implemented and shown to have considerable impact, such as:

- The performance of calculations when needed (e.g., computing creatinine clearance or adjusting a drug dose in an infant, elderly patient, or patient with renal failure)
- Evaluation of simple conditional expressions in order to provide immediate feedback (e.g., in CPOE, a message to the physician when exceeding a recommended drug dose limit or attempting to order a medication in the presence of an interaction with another medication, or an allergy)
- Triggering the evaluation of a conditional expression to generate an alert or reminder that is communicated to a provider (or patient) (e.g., the presence of a critical abnormal lab result)

Even these forms of CDS have all the elements described earlier for diagnostic decision support, although considerably simpler. First, they have specific purposes, whether it is to provide alerts of critical values, remind physicians about timely actions, or perform useful calculations. They also all have the five components required: a decision model (e.g., evaluation of a formula or a Boolean conditional expression), a knowledge base (e.g., the formulas or conditional expressions themselves), an information model (the data elements needed), a result specification (e.g., the format of the calculated result of a formula or the possible truth values for conditional expressions), and an application environment, which determines how the CDS will be triggered or invoked, how needed data will be obtained from a host information system or from a user, and how the CDS will interact with the user.

Thus a common theme for both simple and complex types of CDS has been the set of challenges of providing formal specification for each of these components. Other underlying requirements are an infrastructure to maintain and update the knowledge base content resources, and a suitable mode of integration into clinical IT system environments, as well as tools for making it easier to carry out this integration in diverse environments.

1.4.2.3 Where the Focus Has Been in Computer Applications to Health Care

Some of the major advances in computer use in health care during the past four and a half decades have dealt with relatively mundane matters such as approaches to capturing and storing information, communicating it, retrieving it, and producing and distributing reports. These capabilities have greatly reduced transcription errors, improved legibility of reports, eliminated redundancy, facilitated billing and financial functions, and provided a wide variety of other benefits, which indirectly do, of course, affect patient safety, health care quality, and efficacy.

The main thrust of information technology (IT) efforts recently has been to encourage broad adoption of EHRs, and clinical systems and applications based on the EHR that provide various kinds of functionality. Adoption rate is high among academic medical centers, but much less so at community hospitals and in office practices. Estimates of adoption vary, depending on what one considers to represent an EHR (ranging from the ability to review laboratory and radiology results or other limited functionality to that of systems that include CPOE), and with adoption in the United States thus inexactly estimated to be from below 10 percent to close to 40 percent (Ash and Bates 2005; Bates 2005b; Berner, Detmer et al. 2005; Middleton, Hammond et al. 2005). Rates are higher in some other countries, particularly those with national health care systems, and adoption is accelerating in the United States due to several new initiatives led by the federal government, industry consortia, payer groups, and other stakeholders. Among key functionalities of current interest are CPOE, electronic prescription writing and communication (electronic prescribing, e-prescribing, or eRx), and the Personal Health Record (PHR).

- CPOE is the use of the computer to enable a physician or other provider to enter orders (for medications, procedures, or other actions), ensuring that they are legible, so they can be unambiguously stored in the EHR, communicated to pharmacies or other entities responsible for carrying them out, monitored for completion, and billed. CPOE is a valuable foundational platform on which to incorporate a wide variety of CDS capabilities, as we shall explore, but the basic CPOE functionality does not itself include it. The rate of adoption of CPOE as a foundation for CDS is reviewed in Chapter 7 (see also (Kaushal, Shojania et al. 2003; Ash, Gorman et al. 2004)) but lags that of EHR by a considerable degree, and with the lack of adoption again greatest in office and community practice.
- eRx has some features shared with CPOE, but focuses on the entry of prescriptions into a computer, and may include the ability to communicate prescription information to pharmacy benefit management (PBM) systems for approval of insurance coverage, and to pharmacies for filling of the prescriptions, as well as storing the information in an EHR if available, and printing of hard copy (Schectman, Schorling et al. 2005; Teich, Osheroff et al. 2005; Wang, Marken et al. 2005). Thus e-prescribing is also a potentially valuable platform for CDS.
- PHRs represent a growing capability for enabling access by patients and the health care public to maintain their own health care records (Kimmel, Greenes et al. 2005; Tang and Lansky 2005). This can be done by extracting information from institution-based or provider practice-based EHRs, providing patient-oriented views into EHRs, and enabling

the recording of other health-related data directly by patients. Some PHRs are thus "tethered" to a provider-based EHR system, although others are offered by third parties and are "untethered," but must interface with various EHRs to obtain updates of clinical information from them. PHRs are valuable in that they potentially provide a single place for viewing all the various aspects of a patient's health and health care, integrating across various providers and sources, as well as a longitudinal record over a patient's lifetime. Thus a PHR can also be an important substrate for CDS functionality.

As we note from this, the CDS that can be delivered depends to a large extent on the existence of the EHR and/or on the various applications such as CPOE, eRx, and the PHR that depend on or interact with the EHR. The smallest community hospitals and office practices have relied on computers primarily for financial and billing purposes. As the studies cited earlier have shown, acceptance of the EHR and of clinical IT systems in smaller hospitals and office practices has been limited by concerns about cost of implementation and of ongoing support, the technical experience needed, the extra time and effort required to use it, the confusion in the marketplace due to a plethora of nonstandard systems offerings by many vendors, and lack of positive incentives for introducing these systems.

This is significant in that the majority of patient encounters occur in such smaller office and community settings. Estimates of the magnitude of the opportunity are difficult to obtain, but it stands to reason that, in the community setting, though acuteness of the problems is typically less, the potential for optimizing care, preventing disease, and avoiding acute problems requiring medical center admission is potentially quite large. A recent study (Johnston, Pan et al. 2003) has projected that ambulatory CPOE in the United States, for example, could result in the avoidance of 2 million adverse drug events and 190,000 hospitalizations, and savings of $44 billion per year. In the United States, new legislative proposals, and national and regional campaigns and programs sponsored by professional specialty organizations, health insurers, and employer-based consortia such as the Leapfrog group (http://www.leapfroggroup.org) currently are promoting adoption of the EHR, CPOE, and related initiatives. The Office of the National Coordinator for Health Information Technology (ONC) was established by the U.S. government in April 2004, with the aims of aligning the various stakeholders to provide incentives and support for widespread adoption of the EHR, for establishing a National Health Information Network (NHIN), and for the systems that depend on it, such as CPOE, eRx, and the PHR, and for building national consensus on standards for interoperability of clinical systems.

1.4.3 Renewed Passions

Against the backdrop of slow but now accelerating advances in clinical IT automation in health care, specific computer-based support for doctors, nurses, and other providers, and for patients directly, to help them with clinical problem solving and decision making, until recently, has remained distinctly in the background. During most of the past 45 years, CDS has been more a research pursuit than a practical one.

While the lack of broad availability of EHRs and CPOE has clearly held back CDS deployment, a number of economic, social, and cultural pressures have raised its importance, especially recently. Consequently, it is the prospect of being able to provide CDS that—just as in the earliest days of computer application to health care—is paradoxically again most often cited, in the campaigns promoting adoption of EHRs, CPOE, eRx, PHRs, and other IT functionality, as the primary reason for doing so.

Many factors have contributed to this resurgence of interest, as we shall examine in Chapter 3, but root causes appear to stem from the relentlessly increasing complexity, costs, and constraints on the delivery of health care that are occurring. The stresses and strains include steadily growing demand for medical services generally, especially as a result of continued aging of our population, with attendant increased frequency and multiplicity of chronic diseases. With the growing complexity of care of patients, more specialists are involved, resulting in an ever-increasing fragmentation of the health care process. The range of diagnostic and therapeutic technologies and medications available is increasing, and becoming more expensive, contributing additional financial strain to a system already under pressure. Doctors are seeing more patients with decreased time available for each patient, and with more paperwork, including that associated with increasing regulatory compliance.

The explosion of biomedical knowledge in the genomic and post-genomic era, the growth of the Internet, a better informed and demanding public, the increase in health care malpractice awards and cost of insurance, and the highly publicized reports, both in professional books and journals—notably the landmark report from the Institute of Medicine (IOM), *To Err is Human* (Kohn, Corrigan et al. 1999)—and in the lay press, regarding the frequency of preventable medical error, have all contributed to a growing sense of desperation that something needs to be done. Added to this is the recognition of variations in the quality of medical practice—again brought to the forefront by an IOM report, *Crossing the Quality Chasm* (IOM 2001)—and of the lack of availability of even basic care to large segments of our population. Another notable initiative is the "100K Lives Campaign" begun in 2005 by the Institute for Healthcare Improvement (http://www.ihi.org (Gosfield and Reinertsen 2005), which calls for concerted effort by hospital and other health care organizations to adopt six specific approaches to improving health care safety and quality aimed at saving 100,000 lives over an 18-month period (and every 18 months thereafter). The approaches, which target specific problems including acute myocardial infarction, adverse drug events, central line infections, surgical site infections, and ventilator-associated pneumonias, and a more generic initiative for rapid response teams at the first sign of patient decline, all rely on the ability to apply best practices relying on specific knowledge at the point of care.

Some of the most important kinds of CDS that can occur in a CPOE setting are those that directly address ways of avoiding or reducing the occurrence of medication errors—one of the leading causes of iatrogenic death as well as increased morbidity and cost—by ensuring that dosages are appropriate, drug interactions are avoided, and contraindications such as allergies are recognized before prescribing a medication. Besides decision support built into CPOE, we have highlighted earlier a number of other kinds of applications, such as reminders, alerts, calculations, guidelines, and prognostic aids.

1.4.4 Building the Foundation for a Lasting Relationship

The slow process of adoption and relative lack of impact of CDS have taught us that, despite new urgency and enthusiasm, we must not now rush into CDS without providing a suitable foundation. As we have noted, providing even "simple" CDS is a hard problem, if it is to be done well, accepted by users, integrated into practice, and capable of being maintained over time.

Among the five structural components of CDS (see Table 1-2), the component relating to the application environment is most critical. Successful deployment of CDS requires effective coupling and interaction of all the components with the functions and operations of clinical practice. The clinical data needed must be either entered by a user or obtained through access to an EHR. The actions to be carried out as a result of CDS need to be communicated to the appropriate entities: if in the form of recommendations, the users must be notified; if in the form of tasks to be performed by the IT system, the target applications must be notified.

For these kinds of interactions to work, the specification of the data elements needed by CDS must be compatible with those in the IT system, and the actions that CDS determines should be performed must be capable of being carried out by the IT system. This means that either the CDS specification must be highly specific to the host IT environment, in order to ensure compatibility, or that an agreed upon means of mapping of clinical IT data to those data parameters needed by CDS must exist, as well as an agreed upon mapping between a recommendation or other action determined necessary by CDS and the particular functions the IT system must perform to carry it out.

For acceptance and adoption to be successful, the coupling and interaction of the CDS method's operation and clinical IT systems ideally should be done in a way that does not disrupt the workflow and practice patterns of the intended users, and the CDS capability should be perceived as helpful. Three recent publications, demonstrating negative consequences resulting from the implementation of CPOE in leading medical centers, highlight the difficulties in implementing any intervention that requires that providers devote effort to it, the need to understand the nature of their activities, processes and workflow patterns, and the importance of ensuring that the intervention is perceived by them as having a net benefit. The implementation of CPOE at Cedars-Sinai Medical Center, in Los Angeles in 2002, was perceived as too slow and triggered a physician revolt causing the system to be taken down (Shabot 2004); an implementation at the Hospital of the University of Pennsylvania, in Philadelphia, was found, paradoxically, to increase medication errors (Koppel, Metlay et al. 2005); and an implementation at the Children's Hospital at the University of Pittsburgh was found to coincide with an increased incidence of mortality in a pediatric critical care unit (Han, Carcillo et al. 2005). These reports, and the considerable discussions that ensued within the clinical informatics community, both as blogs and published articles, illuminate many reasons why undesirable results can occur, and suggestions about how they might have been avoided (see, for example, published comments on the Koppel et al. report and the authors' response to them (Bates 2005a; Horsky, Zhang et al. 2005; Koppel, Localio et al. 2005; Nemeth and Cook 2005)). However, the unfortunate fact is that very little has been published in terms of scientific and rigorous approaches to identifying the right versus wrong ways to carry out such implementations. Although the focus of these three experiences was CPOE, the same issues pertain to CDS, as some of the experiences in the case studies reported in Section II of this book illustrate. The organizational and management issues that need to be considered in order to create a positive and constructive approach to implementation are discussed in Section IV.

1.4.5 Knowledge Management—A New Party to the Relationship

As we have noted earlier, it is relatively easy to implement a CDS capability directly into a single host IT environment. Nonetheless, if the organization has multiple host environments (commercial or home-grown), the CDS implementation must be redone for every system in which it is to be deployed. If external knowledge bases are to be used or adapted locally, they must be rendered into a form that is compatible with the local host IT environment(s). If the same knowledge can be used to provide different kinds of CDS in various applications (e.g., a medication dose that should be adjusted based on a lab result, which might be of importance to order entry as well as to a lab result alerting application), it needs to be recoded or invoked in each of those contexts. If the knowledge is embedded in the applications, it is of course not readily transparent or viewable by subject experts, and is difficult to review, maintain, or update when necessary.

As we look at the reasons why the example systems described in Section II, despite their success in local settings in which they had been developed, have not enjoyed widespread penetration and usage, we recognize a number of infrastructural elements that must exist for such widespread use, but which largely do not. Thus a major theme we will explore is the infrastructure required for successful CDS deployment and long-term management. This is a relatively new area of focus for clinical informatics, in which a few institutions are taking the lead, but which has not been widely recognized or pursued. I believe that properly addressing this is an essential element necessary to move CDS to a higher level of activity and focus. The infrastructure elements that we will examine deal with what has come to be known as "knowledge management" (KM). KM has been of interest in the business community for many years (Gupta, LS et al. 2000), yet it has rarely been addressed with respect to clinical knowledge. Earliest references are in the late 1980s and early 1990s (Greenes, Tarabar et al. 1989; Chute, Cesnik et al. 1994). It is now mentioned more frequently in a clinical context (Bali, Dwivedi et al. 2005; Bali, Feng et al. 2005; Dieng-Kuntz, Minier et al. 2005; Ghosh and Scott 2005; Hussain and Abidi 2005; Gray 2006), but still tends to mean a variety of things to different people.

The most comprehensive exposition on the topic of KM in health care is a recent book by Bali (2005). Citing Gupta et al. (2000), Bali points out that "the cornerstone of any KM project is to transform tacit knowledge to explicit knowledge so as to allow its effective dissemination." We explore how two large academic health care delivery organizations tackle this challenge in Section VI of this book, where the focus is on the infrastructure needed to support the whole process from knowledge generation to validation, obtaining consensus on it, documentation, organization, representation, dissemination, and update. How these functions are done, both on a local scale as well as a larger scale involving professional specialty organizations, and national or even international entities, is critical to the ability to create robust repositories of knowledge.

The separation of a knowledge base from the engine that delivers it is also highly desirable, because it allows the two to be separately maintained and refined. A separation of the CDS execution engine from the clinical application environment may also be desirable in some circumstances, in that it could facilitate dissemination and installation of generic execution engines that are able to communicate through message protocols or Web service interfaces to the clinical information systems with which they interact. The benefits of doing this are explored also in Section VI of this book.

1.4.5.1 Three Intersecting and Interacting Life Cycles

A primary goal of this book is to consider the range of issues and challenges involved in moving from the general concept of CDS to implementation, maintenance, dissemination, and update. This involves understanding of the "life cycles" of CDS. Notice that we use the plural here. Three aspects of CDS that we have touched on earlier appear to progress as if they have their own life cycles:

- Knowledge generation, refinement, and update
- Knowledge management and dissemination
- Clinical decision support method development and refinement

These three life cycles are somewhat interdependent (see Fig. 1-1), but they evolve at different paces, have their own constituencies, and involve separate processes.

Knowledge generation and validation. The knowledge underlying CDS can be generated in a variety of possible ways. These are examined in Section III of this book. In general, the knowledge is initially unstructured and unassembled, or even only implicit, and must be extracted (from experts, from databases, or from the literature), organized and synthesized, analyzed for consistency and accuracy, and represented in an unambiguous form that can be computer-interpretable and acted upon. There may be gaps or overlaps with existing knowledge, calling for studies to refine the knowledge. Any synthesis of knowledge about a topic should have an appropriate expiration date, at which time the

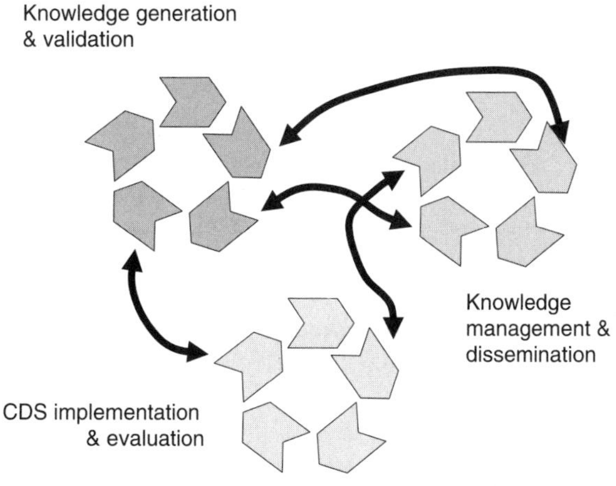

FIGURE 1-1 Three intersecting and interacting life cycles underlying clinical decision support systems technology.

sources should be re-reviewed and the knowledge updated if necessary. Thus each item of knowledge must go through a continuous life cycle process (see Fig. 1-2).

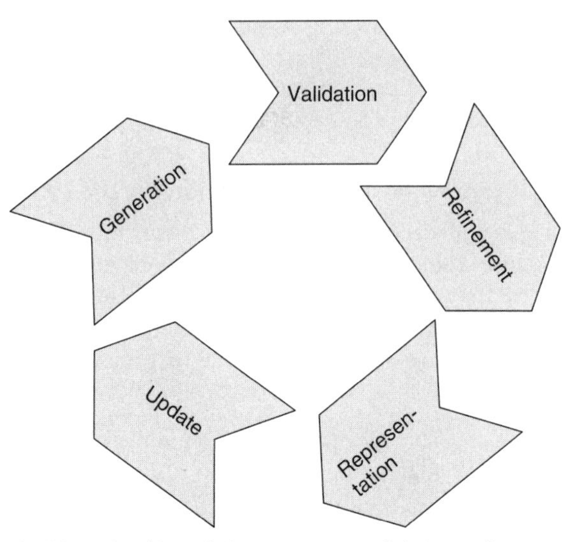

FIGURE 1-2 The life cycle of knowledge generation, validation, refinement, and update.

Knowledge management and dissemination. Although we have considered the life cycle of individual knowledge content resources earlier, there is another task related to the corpus of knowledge in use and other knowledge that is being prepared for use. Imagine all the knowledge resources incorporated in or invoked by various applications throughout an enterprise. A subject expert in diabetes now wants to have the institution provide a set of checks and reminders for compliance with quality guidelines, such as periodic testing of a patient's HbA1c, eye examination, and foot examination. It is important not only to decide what the guiding knowledge should be, in terms of rules logic, order sets, and structured documentation templates, but also how this relates to similar knowledge resources that may already be implemented. What is needed is a means of curation of knowledge resources, to identify those existing items of knowledge pertaining to a topic of interest that are already in use, as an aid to the subject expert in creating new knowledge or refining existing knowledge, to avoid redundancy, to ensure consistency and avoid contradictions, and to recognize gaps where the opportunity for additional CDS may be needed. It may be useful to have a formal editorial process, with panels of experts, peer review and approval mechanisms in place, in order to accept new knowledge into a system. In short, a resource is needed to facilitate content management and collaborative authoring and review. Once knowledge is implemented in applications, it is necessary to keep track of where it is used, to be able to identify those instances when updates are required.

On a broader scale, it may be desirable to maintain common repositories of computer-interpretable, unambiguous knowledge content (e.g., guidelines, or decision rules) for use across an enterprise that has multiple information system platforms. A still more ambitious goal would be to have regional, or national, or even international repositories of knowledge that are maintained and supported by government agencies, payers, or professional specialty or disease-focused organizations. For this to be useful, such knowledge resources need to be made available in a common format that is capable of being adapted to different platforms. This requires development and refinement of standards for representation of the knowledge. The notion of reuse would also benefit greatly from tools or standard approaches for adapting the content for different platforms, adapting to local customs or work processes, interfacing it to host patient databases, and invoking CDS through external services interfaces. Figure 1-3 depicts this life cycle process.

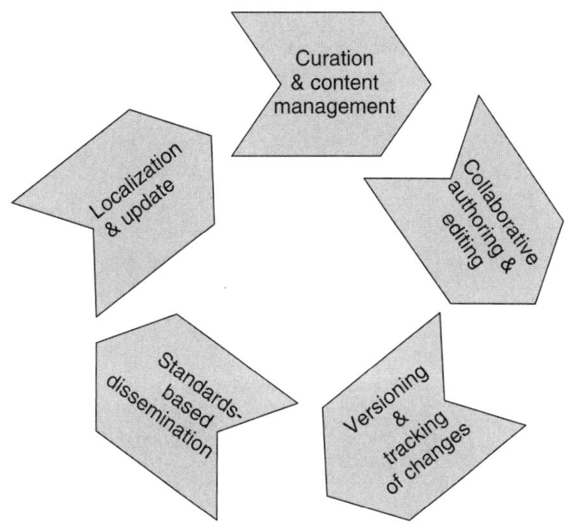

FIGURE 1-3 The life cycle of knowledge content management and dissemination.

Clinical decision support implementation and evaluation. This process is a very complicated one. A life cycle is involved in developing the method for providing CDS (see Fig. 1-4), in terms of the decision model and intended decision support delivery approach. The decision model and its detailed methodology may evolve over time as nuances of the decision problem are recognized that require more or different parameters, or more complex computational or logical manipulations, or that require it to reformulate its advice for delivery in different ways to accommodate new needs. The optimal application environments in which the CDS is to be used must be determined, often by experimentation and pilot use, feedback, and refinement. The goals are to learn how best to integrate the CDS into specific clinical settings, and to evaluate effectiveness. This process must continuously iterate. The representation scheme and mappings and modes of integration of CDS into host IT systems may evolve over time and must be updated. The methods for delivering CDS will depend on the availability and adoption of capabilities for integrating it into applications. For example, CPOE is most desirable as a platform for performing drug–drug, drug–lab, and drug–allergy checks to provide real-time feedback to physicians. Alerts and reminders require some kind of event or time trigger to cause the logic to be evaluated. Use of interactive data checks

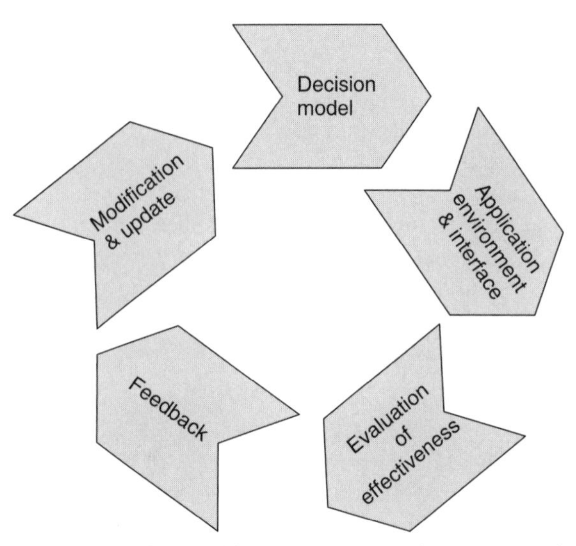

FIGURE 1-4 The life cycle of clinical decision support implementation and evaluation.

in structured data entry, groupings of knowledge into structured data entry forms and order sets, and methods for process and workflow optimization require appropriate application environments in which to provide their capabilities. All these methods of CDS need to be continually revised and updated as experience is gained with approaches that are successful versus those that are not.

Support for these three kinds of capabilities and their evolution through life cycle processes is a major part of the organizational, financial, and societal commitment that will be needed to make broad use of high-quality CDS a reality.

1.5 SCOPE AND PLAN OF THIS BOOK

For the introduction of CDS into a health care environment, two kinds of factors need to be considered:

1. *The kinds of CDS that are possible, the circumstances in which they are most useful, and technical aspects involved in providing them.* We are concerned here with the questions of what should be realistic goals for CDS, in what settings CDS would be most useful, and what forms of CDS are appropriate in each of the settings.
2. *The practical, organizational, and management approaches that are needed to make the deployment of CDS successful.* Each type of CDS requires a particular mode of integration with existing human processes, as well as with IT systems that provide access to data and synchronization, coordination, and communication resources that the CDS will need to utilize.

In Section I—this chapter and the next two—I introduce and elaborate on these two perspectives. It is important to determine not only technical feasibility, but relative importance or priority, practical ease of implementation, and prospects for success, for various types of CDS. Subsequent sections of the book will deal with both the technical, as well as the practical, organizational, and management aspects of CDS deployment.

So far, I have discussed the prospects for widespread adoption of CDS in terms of the ***relationship*** between the computer (actually, information systems and communication technology) and the provider, as that relationship has developed and become better understood. At the risk of shifting metaphors, the remainder of this book will discuss these prospects for widespread adoption of CDS in terms of a ***journey***—from theory, development, and experimentation, to demonstration of success, to methods of dissemination, adoption, and improvement, and to the development of robust infrastructure for supporting these processes and for knowledge management. Progress on this journey depends on establishment of the three life cycle processes we have identified, and on their being well supported as they iterate and improve, and as they interact with each other. We are still in the middle phases of this journey, where the processes are still being established, where the stakeholders and commitments required are still being identified and formed, and where experience with making the processes robust has not yet occurred.

As we consider this journey, the focus in this book will be on the elements needed not only to travel along the road ahead, but in fact to build the road that is needed, and the capabilities that must exist at the destination. My belief is that unless we understand these elements and the path that must be pursued, the resources needed won't be marshaled and progress will likely continue to languish. Hopefully this book will help to illuminate these requirements.

What We Do Not Cover

There have been and continue to be inroads in many exciting areas. We only touch on some of them in this book. Our main focus is on capabilities needed for broad deployment of CDS to impact on health care safety, quality, and efficacy. There are a variety of other important kinds of decision support that we will not cover in depth, because they are outside of that scope or take us in directions that would be too diverse to cover adequately here. These include:

1. Image and signal analysis methods for interpreting clinical data in such diverse modalities as radiographs, cytology smears, immunoglobulin assays, electrocardiograms, and electroencephalograms. Such methods can recognize patterns, extract findings, and make diagnostic assessments or measurements.
2. Treatment planning and image-guided intervention, including radiation dosimetry, three-dimensional modeling, and virtual reality applications. The latter may include, for example, use of heads-up display for integrating images from MRI in real time with a surgeon's view of the operating field.
3. Support for the information-seeking and problem-solving tasks of the patient or health care consumer. Many of the kinds of CDS we discuss here are useful for patients and consumers as well as providers, such as information retrieval, alerts, reminders, guidelines, and structured forms for documentation. Yet consumer decision support has a number of unique challenges, which we won't address here, related to adapting to the health literacy requirements of the user, and goals of fostering increased involvement and participation of the individual in

his/her own health, adoption and adherence to healthy lifestyles, and use of decision tools and approaches to shared decision-making between doctor and patient.

4. Public health approaches to identifying early outbreaks of disease, toxic conditions, or other hazards, improving compliance with preventive care measures, and promotion of cost-effective health care policy.

A key point with respect to the first two of these categories is that they tend to be *niche* applications; that is, those that interact only with a limited part of the health care system, such as an image processing method for a CT scanner or for use in a Picture Archiving and Communication System (PACS) workstation, or a signal processing method for ECG interpretation. Niche applications are more likely to be disseminated and adopted widely, when shown to be successful, than innovations, even simple ones, that impact on broader portions of the health care system. This suggests that the primary impediments to adoption are often not the technologies, but the difficulties of adaptation of the business, organizational, and cultural aspects of health care practice in order to integrate the new technologies into them smoothly and effectively. The latter two applications relate to clinical information systems, but in a less direct way. Although these are both rapid areas of growth, they are beyond our scope, since we deal here only with CDS as directly incorporated in or interacting with clinical information systems.

In Section II, we examine case histories of three institutions that have implemented CDS, both in terms of their positive experiences as well as lessons learned. We also review the current state of availability of CDS capabilities through the commercial marketplace. We then consider the issues and obstacles that have prevented more widespread adoption and examine the processes and approaches that need to be embraced if more widespread adoption is to occur.

To address the problems identified in Section II, Sections III and IV deal with the technical challenges of implementing CDS. In Section III, we focus on the knowledge embedded in CDS, and examine how it is generated and validated. There are many possible ways to obtain knowledge, which we classify into three main categories of methods for deriving knowledge: 1) human-intensive techniques; that is, developed either by watching and analyzing human experts, or by debriefing and extracting knowledge from experts systematically; 2) data-intensive techniques; that is, developed through data mining and analysis, and involving models that perform a task via methods that do not necessarily correspond to human approaches but which can be measured objectively against data; and 3) literature-derived; that is, developed by meta-analysis and extraction of information from published literature.

In Section IV, we consider how the knowledge, once derived, can be represented in standardized form so that it can be used in a variety of proprietary and nonproprietary settings, facilitating wider dissemination and reuse of successful methods. Formalization and standardization are complicated processes, which involve both technical and political and economic considerations, and standardization initiatives tend to be quite slow, typically proceeding in fits and starts. Benefits of standardization of CDS (i.e., reuse and easier implementation across platforms and settings), depend on not only an agreed-upon method for representing the knowledge, but also an information model for the data elements referred to in the knowledge representation, a vocabulary or taxonomy of terms used to denote these data elements, and methods for communicating between a CDS tool and the clinical IT system for obtaining the necessary data elements and communicating the results of the evaluations by the CDS tool in terms of recommendations are actions that need to be performed.

We come back to the challenges of integrating CDS into operational settings in Section V, which deals with organizational and business issues involved in implementing decision support. These issues involve consideration of change management and of the need for creating a culture that is supportive of CDS, the costs for development and maintenance of CDS capabilities, the business rationales and drivers for implementing CDS, and the liabilities and regulatory issues involved in providing CDS (or not providing it).

Organizational and business aspects also involve establishment of formal approaches for knowledge management. This includes methods for generating the knowledge, standardizing its representation, supporting authoring and editing, and otherwise managing knowledge resources and integrating them into applications. These issues need to be considered both on an enterprise scale, as well as with respect to the challenges of larger national or even international approaches to dissemination and maintenance and update of knowledge resources.

Section VI addresses organizational strategies and technical approaches for implementing and maintaining CDS, in terms of the knowledge management infrastructure needed. Knowledge management must support all three of the overlapping life cycles described in subsection 1.4.5.1.

Finally in Section VII we consider the requirements for coordinated national/international efforts to manage knowledge and deploy CDS widely. This will demand broad understanding of the issues, detailed planning, and the sustained commitment of resources by multiple stakeholders to be effective.

By the end of the book it should become apparent that the development, delivery, and support of CDS are substantial undertakings that require concerted efforts on many fronts in order to be successful. It is very easy to underestimate the problems of CDS, and then to become frustrated by the apparent lack of progress that one sees. By delineating the issues, it is hoped that the book can serve as a kind of call to action, so as to marshal the resources needed to systematically tackle the complexities involved. The challenges have long been dealt with on an ad hoc basis by various groups, a number of which have been academic, who have been particularly motivated to work in this area. However, these efforts typically have been done without either the platform or the support for extending their approaches to the wider community. There has been essentially no concerted effort to look at this problem systematically, to develop a roadmap and a plan for establishing a suitable infrastructure. As noted in subsection 1.1, there is now national interest in the United States for a roadmap for CDS, with a white paper suggesting a possible approach (Osheroff, Teich et al. 2006). This book should be a useful resource for such planning activities.

I hope that this book can play a role in enunciating requirements and stimulating a process of systematic design and development of needed infrastructure. National priorities for such goals as increasing patient safety, reducing unnecessary health care expenditures, reducing practice variation, and encouraging best practices make beautiful slogans, but they are difficult to achieve without the kind of infrastructure we discuss. Rather, the race to implement EHRs, CPOE, as well as various kinds of decision support on top of them, without the systematic attention to the problems and issues we have identified, may have unfortunate consequences, as illustrated by the recent experiences of three different CPOE implementations cited in subsection 1.4.4.

Efforts to make a science out of the process of implementation that pay attention to possible reasons for failure and the corrective actions that are necessary are badly needed. Negative experiences, such as those with the recent CPOE implementations, can set the field back substantially. There are likely to be many such experiences in the absence of a rigorous approach, but unfortunately there has been little recognition of the need for that. The bottom-line message is that the investment and effort for creating knowledge management infrastructure for delivering CDS are essential. The methods needed are not fully known, although the fact that substantial progress has occurred is illustrated by the approaches discussed in Section VI. This is an area that should receive considerable future attention as a focus for further development and refinement. We have a long road to travel from where we are now, so let the journey begin.

REFERENCES

Ash, J. S. and Bates, D. W. (2005). Factors and forces affecting EHR system adoption: report of a 2004 ACMI discussion. *J Am Med Inform Assoc* **12**(1): 8–12.

Ash, J. S., Gorman, P. N., Seshadri, V. and Hersh, W. R. (2004). Computerized physician order entry in U.S. hospitals: results of a 2002 survey. *J Am Med Inform Assoc* **11**(2): 95–99.

Bali, R. (2005). *Clinical Knowledge Management: Opportunities and Challenges*. Hershey, PA, Idea Group Publishing.

Bali, R. K., Dwivedi, A. and Naguib, R. N. (2005). The efficacy of knowledge management for personalised healthcare. *Stud Health Technol Inform* **117**: 104–107.

Bali, R. K., Feng, D. D., Burstein, F. and Dwivedi, A. N. (2005). Introduction to the special issue on advances in clinical and health-care knowledge management. *IEEE Trans Inf Technol Biomed* **9**(2): 157–161.

Barnett, G. O., Cimino, J. J., Hupp, J. A. and Hoffer, E. P. (1987). DXplain. An evolving diagnostic decision-support system. *JAMA* **258**(1): 67–74.

Bates, D. W. (2005a). Computerized physician order entry and medication errors: finding a balance. *J Biomed Inform* **38**(4): 259–261.

Bates, D. W. (2005b). Physicians and ambulatory electronic health records. U.S. Physicians are ready to make the transition to EHRs—which is clearly overdue, given the rest of the world's experience. *Health Aff (Millwood)* **24**(5): 1180–1189.

Berner, E. S., Detmer, D. E. and Simborg, D. (2005). Will the wave finally break? A brief view of the adoption of electronic medical records in the United States. *J Am Med Inform Assoc* **12**(1): 3–7.

Berner, E. S., Webster, G. D., Shugerman, A. A., Jackson, J. R., Algina, J., Baker, A. L., et al. (1994). Performance of four computer-based diagnostic systems. *N Engl J Med* **330**(25): 1792–1796.

Blois, M. S. (1980). Clinical judgment and computers. *N Engl J Med* **303**(4): 192–197.

Caceres, C. A. (1963). Electrocardiographic analysis by a computer system. *Arch Intern Med* **111**: 196–202.

Chute, C. G., Cesnik, B. and van Bemmel, J. H. (1994). Medical data and knowledge management by integrated medical workstations: summary and recommendations. *Int J Biomed Comput* **34**(1–4): 175–183.

Collen, M. F. (1986). Origins of medical informatics. *West J Med* **145**(6): 778–785.

deDombal, F. T. (1975). Computer-aided diagnosis and decision-making in the acute abdomen. *J R Coll Physicians Lond* **9**(3): 211–218.

Diamond, G. A., Pollock, B. H. and Work, J. W. (1987). Clinician decisions and computers. *J Am Coll Cardiol* **9**(6): 1385–1396.

Dieng-Kuntz, R., Minier, D., Ruzicka, M., Corby, F., Corby, O. and Alamarguy, L. (2005). Building and using a medical ontology for knowledge management and cooperative work in a health care network. *Comput Biol Med.*

Friedman, C., Elstein, A., Wolf, F., Murphy, G., Franz, T., Fine, P., et al. (1998). Measuring the quality of diagnostic hypothesis sets for studies of decision support. *Medinfo* **9** (Lt 2): 864–868.

Garg, A. X., Adhikari, N. K., McDonald, H., Rosas-Arellano, M. P., Devereaux, P. J., Beyene, J., et al. (2005). Effects of computerized clinical decision support systems on practitioner performance and patient outcomes: a systematic review. *JAMA* **293**(10): 1223–1238.

Ghosh, B. and Scott, J. E. (2005). Comparing knowledge management in health-care and technical support organizations. *IEEE Trans Inf Technol Biomed* **9**(2): 162–168.

Gladwell, M. (2000). *The Tipping Point.* New York, Little, Brown and Co.

Gorry, G. A. and Barnett, G. O. (1968). Experience with a model of sequential diagnosis. *Comput Biomed Res* **1**(5): 490–507.

Gosfield, A. and Reinertsen, J. (2005). The 100,000 Lives Campaign: Crystallizing standards of care for hospitals. *Health Affairs* **24**(6): 1560–1570.

Gray, S. M. (2006). Knowledge management: a core skill for surgeons who manage. *Surg Clin North Am* **86**(1): 17–39, vii–viii.

Greenes, R. A., Peleg, M., Boxwala, A., Tu, S., Patel, V. and Shortliffe, E. H. (2001). Sharable computer-based clinical practice guidelines: rationale, obstacles, approaches, and prospects. *Medinfo* **10**(Pt 1): 201–205.

Greenes, R. A., Tarabar, D. B., Krauss, M., Anderson, G., Wolnik, W. J., Cope, L., et al. (1989). Knowledge management as a decision support method: a diagnostic workup strategy application. *Comput Biomed Res* **22**(2): 113–135.

Gupta, B., LS, I. and Aronson, J. (2000). Knowledge management: Practices and challenges. *Industrial Management & Data Systems* **100**(1): 17–21.

Han, Y. Y., Carcillo, J. A., Venkataraman, S. T., Clark, R. S., Watson, R. S., Nguyen, T. C., et al. (2005). Unexpected increased mortality after implementation of a commercially sold computerized physician order entry system. *Pediatrics* **116**(6): 1506–1512.

Haynes, R., McKibbon, K., Wilczynski, N., Walter, S. and Werre, S. (2005). Optimal search strategies for retrieving scientifically strong studies of treatment from MEDLIN. **330**(May 21): 1179–1182.

Horsky, J., Zhang, J. and Patel, V. L. (2005). To err is not entirely human: complex technology and user cognition. *J Biomed Inform* **38**(4): 264–266.

Hussain, F. and Abidi, S. S. (2005). A knowledge management framework to morph clinical cases with clinical practice guidelines. *Stud Health Technol Inform* **116**: 731–736.

Inza, I., Merino, M., Larranaga, P., Quiroga, J., Sierra, B. and Girala, M. (2001). Feature subset selection by genetic algorithms and estimation of distribution algorithms. A case study in the survival of cirrhotic patients treated with TIPS. *Artif Intell Med* **23**(2): 187–205.

IOM (2001). Crossing the quality chasm: a new health system for the 21st century. Institute of Medicine. Washington, D.C., National Academy Press.

Johnston, D., Pan, E., Middleton, B., Walker, J. Bates, D. and (2003). The Value of Computerized Provider Order Entry in Ambulatory Settings. Report. Boston, MA, Center for Information Technology Leadership.

Kassirer, J. P., Moskowitz, A. J., Lau, J. and Pauker, S. G. (1987). Decision analysis: a progress report. *Ann Intern Med* **106**(2): 275–291.

Kaushal, R., Shojania, K. G. and Bates, D. W. (2003). Effects of computerized physician order entry and clinical decision support systems on medication safety: a systematic review. *Arch Intern Med* **163**(12): 1409–1416.

Kawamoto, K., Houlihan, C. A., Balas, E. A. and Lobach, D. F. (2005). Improving clinical practice using clinical decision support systems: a systematic review of trials to identify features critical to success. *BMJ* **330**(7494): 765.

Kimmel, Z., Greenes, R. A. and Liederman, E. (2005). Personal health records. *J Med Pract Manage* **21**(3): 147–152.

Knaup, P., Wiedemann, T., Bachert, A., Creutzig, U., Haux, R. and Schilling, F. (2002). Efficiency and safety of chemotherapy plans for children: CATIPO—a nationwide approach. *Artif Intell Med* **24**(3): 229–242.

Kohn, L., Corrigan, J., Donaldson, M. E. and Committee on Quality of Health Care in America, I. O. M. (1999). *To Err Is Human: Building a Safer Health System.* Washington, D.C., National Academies Press.

Koppel, R., Localio, A. R., Cohen, A. and Strom, B. L. (2005). Neither panacea nor black box: responding to three Journal of Biomedical Informatics papers on computerized physician order entry systems. *J Biomed Inform* **38**(4): 267–269.

Koppel, R., Metlay, J. P., Cohen, A., Abaluck, B., Localio, A. R., Kimmel, S. E., et al. (2005). Role of computerized physician order entry systems in facilitating medication errors. *JAMA* **293**(10): 1197–1203.

Ledley, R. S. and Lusted, L. B. (1959). Reasoning foundations of medical diagnosis; symbolic logic, probability, and value theory aid our understanding of how physicians reason. *Science* **130**(3366): 9–21.

Lodwick, G. S. (1965). A probabilistic approach to the diagnosis of bone tumors. *Radiol Clin North Am* 3(3): 487–497.

Middleton, B., Hammond, W. E., Brennan, P. F. and Cooper, G. F. (2005). Accelerating U.S. EHR adoption: how to get there from here. Recommendations based on the 2004 ACMI retreat. *J Am Med Inform Assoc* **12**(1): 13–19.

Miller, P. L., Frawley, S. J. and Sayward, F. G. (2000). Informatics issues in the national dissemination of a computer-based clinical guideline: a case study in childhood immunization. *Proc AMIA Symp*: 580–584.

Miller, R., Masarie, F. E. and Myers, J. D. (1986). Quick medical reference (QMR) for diagnostic assistance. *MD Comput* 3(5): 34–48.

Miller, R. A., Pople, H. E., Jr. and Myers, J. D. (1982). Internist-1, an experimental computer-based diagnostic consultant for general internal medicine. *N Engl J Med* **307**(8): 468–476.

Nemeth, C. and Cook, R. (2005). Hiding in plain sight: what Koppel et al. tell us about healthcare IT. *J Biomed Inform* **38**(4): 262–263.

Ohno-Machado, L., Gennari, J. H., Murphy, S. N., Jain, N. L., Tu, S. W., Oliver, D. E., et al. (1998). The guideline interchange format: a model for representing guidelines. *J Am Med Inform Assoc* **5**(4): 357–372.

Osheroff, J., Pifer, E., Teich, J., Sittig, D. and Jenders, R. (2005). *Improving Outcomes with Clinical Decision Support: An Implementer's Guide.* Chicago, IL, Healthcare Information and Management Systems Society (HIMSS).

Osheroff, J., Teich, J., Middleton, B., Steen, E., Wright, A. and Detmer, D. (2006). A Roadmap for National Action on Clinical Decision Support. Report. Bethesda, MD, American Medical Informatics Association.

Pauker, S. G., Gorry, G. A., Kassirer, J. P. and Schwartz, W. B. (1976). Towards the simulation of clinical cognition. Taking a present illness by computer. *Am J Med* **60**(7): 981–996.

Peleg, M., Boxwala, A. A., Ogunyemi, O., Zeng, Q., Tu, S., Lacson, R., et al. (2000). GLIF3: the evolution of a guideline representation format. *Proc AMIA Symp*: 645–649.

Pipberger, H. V., Stallmann, F. W., Yano, K. and Draper, H. W. (1963). Digital computer analysis of the normal and abnormal electrocardiogram. *Prog Cardiovasc Dis* **5**: 378–392.

Ramnarayan, P., Tomlinson, A., Kulkarni, G., Rao, A. and Britto, J. (2004). A novel diagnostic aid (ISABEL): development and preliminary evaluation of clinical performance. *Medinfo* **11**(Pt 2): 1091–1095.

Resnic, F. S., Popma, J. J. and Ohno-Machado, L. (2000). Development and evaluation of models to predict death and myocardial infarction following coronary angioplasty and stenting. *Proc AMIA Symp*: 690–693.

Schectman, J. M., Schorling, J. B., Nadkarni, M. M. and Voss, J. D. (2005). Determinants of physician use of an ambulatory prescription expert system. *Int J Med Inform* **74**(9): 711–717.

Shabot, M. M. (2004). Ten commandments for implementing clinical information systems. *Proc (Bayl Univ Med Cent)* **17**(3): 265–269.

Shiffman, R. N., Liaw, Y., Brandt, C. A. and Corb, G. J. (1999). Computer-based guideline implementation systems: a systematic review of functionality and effectiveness. *J Am Med Inform Assoc* **6**(2): 104–114.

Shortliffe, E. H., Davis, R., Axline, S. G., Buchanan, B. G., Green, C. C. and Cohen, S. N. (1975). Computer-based consultations in clinical therapeutics: explanation and rule acquisition capabilities of the MYCIN system. *Comput Biomed Res* **8**(4): 303–320.

Sintchenko, V., Coiera, E., Iredell, J. R. and Gilbert, G. L. (2004). Comparative impact of guidelines, clinical data, and decision support on prescribing decisions: an interactive web experiment with simulated cases. *J Am Med Inform Assoc* **11**(1): 71–77.

Sittig, D. F., Krall, M. A., Dykstra, R. H., Russell, A. and Chin, H. L. (2006). A survey of factors affecting clinician acceptance of clinical decision support. *BMC Med Inform Decis Mak* **6**: 6.

Tang, P. C. and Lansky, D. (2005). The missing link: bridging the patient-provider health information gap. Electronic personal health records could transform the patient-provider relationship in the twenty-first century. *Health Aff (Millwood)* **24**(5): 1290–1295.

Teich, J. M., Osheroff, J. A., Pifer, E. A., Sittig, D. F. and Jenders, R. A. (2005). Clinical decision support in electronic prescribing: recommendations and an action plan: report of the joint clinical decision support workgroup. *J Am Med Inform Assoc* **12**(4): 365–376.

Ten Haken, R. K., Fraass, B. A., Kessler, M. L. and McShan, D. L. (1995). Aspects of enhanced three-dimensional radiotherapy treatment planning. *Bull Cancer* **82** (**Suppl** 5): 592s–600s.

Wang, C. J., Marken, R. S., Meili, R. C., Straus, J. B., Landman, A. B. and Bell, D. S. (2005). Functional characteristics of commercial ambulatory electronic prescribing systems: a field study. *J Am Med Inform Assoc* **12**(3): 346–356.

Warner, H. R., Jr. (1989). Iliad: moving medical decision-making into new frontiers. *Methods Inf Med* **28**(4): 370–372.

Warner, H. R., Toronto, A. F. and Veasy, L. G. (1964). Experience With Baye's Theorem For Computer Diagnosis Of Congenital Heart Disease. *Ann N Y Acad Sci* **115**: 558–567.

2 A BRIEF HISTORY OF CLINICAL DECISION SUPPORT: TECHNICAL, SOCIAL, CULTURAL, ECONOMIC, AND GOVERNMENTAL PERSPECTIVES

ROBERT A. GREENES

In this chapter we seek to provide a historical context for the current high degree of interest in clinical decision support systems. We focus on two aspects of history:

1. The development and evolution of the scientific and technical basis for the field, in terms of the primary research methodologies that have been proposed, tested, refined, extended, and in some cases deployed and evaluated in operational settings. Some of these themes represent key dimensions of activity in CDS currently, whereas others are related more to the underlying research to create the databases or knowledge used in CDS, and still others are of interest mainly in terms of their historical influence on current approaches.
2. Major social, cultural, economic, management, and governmental influences on health care providers, health care delivery organizations, and the health care public that have contributed to the current climate of enthusiasm and eagerness for wide deployment and usage of CDS, but also to some of the barriers to success.

2.1 PRIMARY RESEARCH METHODOLOGIES THAT HAVE BEEN PURSUED AND EXTENDED

Many different research approaches have been explored over the years to deliver clinical decision support. We discuss these here from the point of view of the methodology and technology used. An orthogonal classification of research in clinical decision support, in terms of purpose—for example, for diagnosis, treatment selection, or prognosis assessment—is presented in Chapter 3. But our objective here is to show how various methodologies have

Clinical Decision Support: The Road Ahead

TABLE 2-1 Principal methodologies for clinical decision support, their uses, and key developments.

Methodology	Major uses	Key developments
Information retrieval	Finding information, answering questions	Taxonomies, ontologies, text-based methods, patient-specific context keys, automatic invocation
Evaluation of logical conditions	Alerts, reminders, constraints, inferencing systems	Decision tables, event-condition-action rules, production rules
Probabilistic and data-driven classification or prediction	Diagnosis, technology assessment, treatment selection, classification and prediction, prognosis estimation, evidence-based medicine	Bayes theorem, decision theory, ROC analysis, data mining, logistic regression, artificial neural networks, belief networks, meta-analysis
Heuristic modeling and expert systems	Diagnostic and therapeutic reasoning, capturing nuances of human expertise	Rule-based systems, frame-based reasoning
Calculations, algorithms, and multistep processes	Execution of computational processes, flow-chart-based guidelines and consultations, interactive dialogue control, biomedical image and signal processing	Process flow and workflow modeling, guideline formalisms and modeling languages
Associative groupings of elements	Structured data entry, structured reports, order sets, other specialized presentations and data views	Report generators and document construction tools, document architectures, templates, markup languages, ontology tools, ontology languages

evolved over the 45-year history of the field to date. A number of these topics are explored in more depth in Sections III and IV of this book, so our intent here is to give an overview of the various methodologies and how they relate to one another (see Table 2-1) rather than attempting to be comprehensive.

2.1.1 Information Retrieval

The ability to find information relevant to a problem is a basic form of CDS. Determining whether a clinical laboratory test result is abnormal is, in its simplest form, a retrieval task—that of searching for a document that defines the range of normality in a population of interest. This could be a report from the primary literature or a document produced by the clinical laboratory itself regarding the normal limits of tests that it performs. Building on this basic capability, a clinical information system can transition increasingly into providing direct decision support about laboratory test results, by implementing means for increasing the patient-specificity and the integration with other applications.

2.1.1.1 User-Initiated

The simplest form of information retrieval is initiated by a user, through utilization of a search tool. MEDLINE is the classic example of an information retrieval resource. MEDLINE, the history of which is reviewed by Smith (2005), became available in 1964. That first incarnation of MEDLINE was a bibliographic retrieval system for the biomedical literature intended for librarians, called MEDLARS, developed by the National Library of Medicine (NLM). This evolved to the MEDLINE online resource in the mid-to-late 1970s, which now is accessed in its most convenient and widely used form through the Web via PubMed, operated by the National Center for Biotechnology Information of the NLM. Many other online bibliographic resources, databases, and knowledge bases have been added to MEDLINE and to PubMed since the latter's beginning, and there are today, of course, many other online reference sources. Over the years, a growing array of textbooks, handbooks, other reference materials, and interactive medical knowledge resources has also become available online through subscriptions and other arrangements.

There are two main ways of doing basic information retrieval as illustrated in Figure 2-1, taxonomy-based or ontology-based search, or text-based search.

Taxonomy-based or Ontology-based Search

The classic way of retrieving information resources relies on the use of a controlled vocabulary, consisting of terms (or *keywords*) by which the content has been preindexed. The Medical Subject Headings (MeSH), the heading terms used to index articles stored in MEDLINE (Coletti and Bleich 2001), have been in use for more than 40 years, since the early days of the precursor printed resource known as Index Medicus. The terms in MeSH are arranged hierarchically into trees covering a number of topic axes, and many terms actually occur in several MeSH subtrees. MeSH terms can be suggested by content authors, but the indexing for MEDLINE actually is done by specialists who choose the terms, in order to introduce consistency. This process is aided by automated tools for suggesting terms (Yang 1996; Aronson, Mork et al. 2004; Joubert, Peretti et al. 2005).

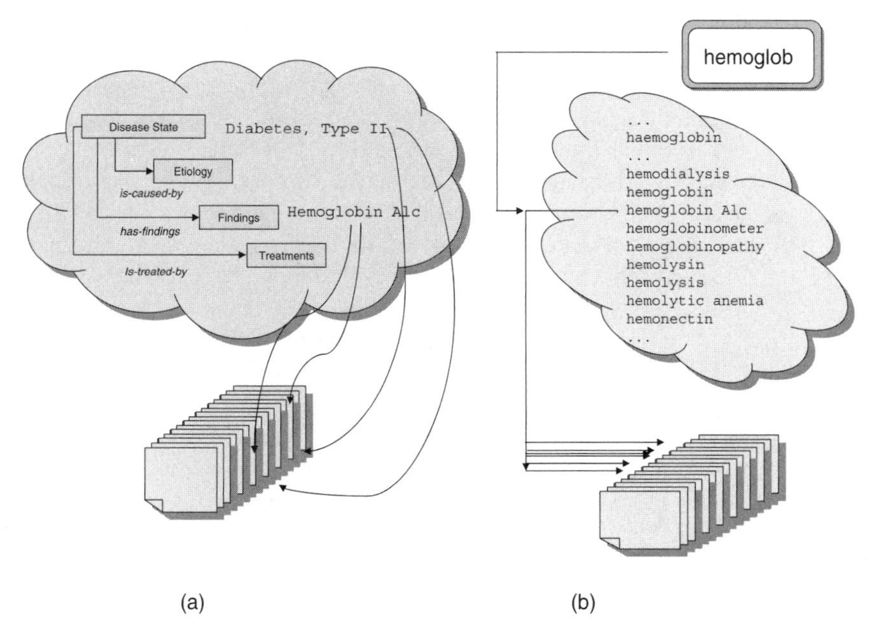

FIGURE 2-1 (a) Ontology-based retrieval. User can navigate terms and relations to select those wanted. Child terms can be included. (b) Text-based retrieval. Text entered by user is looked up in text index to find occurrences.

The user doing retrieval from a term-indexed resource must use the same resource to identify appropriate terms describing the topic of interest, and typically may create Boolean expressions comprised of logical combinations of terms, to do more complicated retrieval. For example, one might browse the MeSH tree to find terms for retrieving articles from PubMed, or browse therapeutic categories in a drug index, to find anti-hypertensive medications from a drug formulary, perhaps further indexed by mechanism of action such as diuretics, ACE inhibitors, and the like before selecting a particular medication for retrieval of its details. Since the MeSH tree is hierarchical, a term at a higher level can be used to connote the desire to "explode" the subtree at that point to be able to search for articles containing any lower-level terms as well. Use of MeSH terms was the primary way to retrieve information from MEDLINE in the past, and is one of the most precise ways to find relevant items, although free-text search using PubMed is now more widely used. A disadvantage of term-based searching is that, since it relies on indexing by an expert (or automatic indexing), the choice of terms used to index may not correspond to the particular interest or focus of a user.

The term *taxonomy* refers to the use of a hierarchical classification of controlled terms to provide a conceptual framework for a domain of interest, as an aid to organization, analysis, or information retrieval. The hierarchical relations are various types of parent-child, such as *is-a*, *is-member-of*, or *is-part-of*. An *ontology* is similar to a taxonomy in that it is also a means of describing a knowledge domain. An ontology uses a controlled vocabulary to formally represent concepts that describe objects and the relations among them. Ontologies typically use richer semantic relationships than taxonomies, to describe concepts and their attributes, and have a set of formal rules and constraints about how terms are defined and relations are specified. Because of the potential richness of an ontology, it can be thought of as a knowledge representation, rather than just as a method to control the terms used for the domain's description. Thus, using an ontology can allow one to actually reason about a domain.

The MeSH tree is a taxonomic system. The terms used are controlled, but, as noted earlier, they do not represent unique concepts, in that the same term can occur in multiple subtrees of the MeSH hierarchy. Thus tagging entities with MeSH terms themselves leaves some ambiguity. Using the controlled terms of an ontology would be a more precise way to do this, since each concept represented by the controlled term is unique. An alternative would be to specify the actual tree positions ("MeSH Tree Numbers") of the particular MeSH terms used, so that alternative uses of those terms are not considered.

If a taxonomy or ontology is present, we can traverse its classes and relations to find categories of interest. The Unified Medical Language System (UMLS) Metathesaurus (Lindberg, Humphreys et al. 1993), developed by the NLM, is the most comprehensive collection of terms from a variety of terminological systems. In the UMLS Metathesaurus, these various terms are mapped to unique UMLS concepts, with their own Concept Unique Identifiers (CUIs). The UMLS concepts are classified into semantic types using the UMLS Semantic Network (McCray and Nelson 1995), and semantic relations among types in the Semantic Network facilitate navigation of the UMLS. The UMLS, having incorporated and interrelated terms from many leading taxonomies, including ICD9/10, SNOMED, MeSH, and LOINC, provides a means for concept-based information retrieval rather than that based simply on using terms. Such an approach also has the advantage that it can identify synonyms and alternative forms of the selected concepts to improve retrieval. An example of concept-based retrieval is the SAPHIRE system (Hersh and Greenes 1990; Hersh and Hickam 1993).

Free Text Search

Direct text-based searching has become the most common way to search, since most Web-based search engines (such as Google® and Yahoo®), rely on this method. This is also now the most popular way to search PubMed, even though MeSH-heading-based retrieval also continues to be available. An obvious advantage of free text search is that it is the easiest to do, because it requires little effort by the user and almost no learning curve. If the search fails to provide appropriate results, then alternative terms can be tried.

A disadvantage is that one does not know how many potential items were not retrieved due to the wrong choice of term for free text search, when there are multiple possible synonyms. *Recall* is defined as the fraction of the true relevant documents that were actually retrieved, analogous to sensitivity or true positive rate. Another disadvantage is that the retrieval may include false positives when the chosen search term has alternative meanings. *Precision* is defined as the fraction of true relevant documents in the set of all documents retrieved, analogous to a positive predictive value. It is possible to combine elements of taxonomy/ontology-based search and free text search (Hersh and Hickam 1993), although in practice this is rarely done. The approach would be to map free-text search terms entered by a user to concepts and then search for terms that are variants of the concept or subsumed by it.

2.1.1.2 Semi-automated or Automated

Retrieval can be made more patient-specific and context/situation-specific, if additional conditions are used to constrain results (e.g., information about the application being invoked, the work setting, the user characteristics, and data about the patient). A user-initiated search in an application context could be modified automatically by adding terms related to such conditions, for example. If the retrieval is initiated by an application, the parameters for these conditions can be generated automatically by the application, thus enabling information retrieval to be more tightly integrated and even automated. The *infobutton* (Cimino, Li et al. 2002) is an example of this—a visual icon that can be placed in display screens at various points indicating the presence of information relevant to the adjacent displayed items. Retrieval can be implemented by predefined links to specific resources or can be generated dynamically by passing search parameters to a retrieval engine. This is discussed further in Chapter 16, which describes a proposed standard being developed for implementing infobutton capabilities. Of course, hard links to content also can be incorporated in applications at specific points, either in the user interface or for automatic invocation, where particular fixed content is desired, but the advantage that is intended from the infobutton approach is its ability to dynamically find appropriate resources at execution time (see Fig. 2-2).

In general, a search is designed to provide information resources for human perusal rather than automated decision support. But as with a clinical laboratory abnormal result, as mentioned earlier and discussed in Chapter 3, there is a continuum of degrees of CDS that can be provided, with respect to

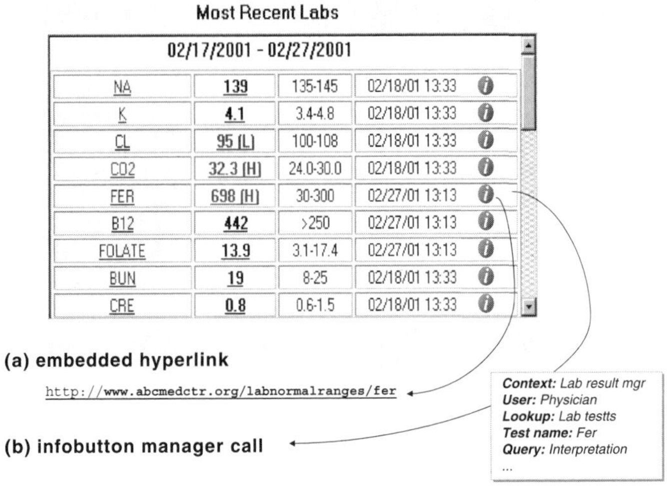

Most Recent Labs

02/17/2001 - 02/27/2001			
NA	139	135-145	02/18/01 13:33
K	4.1	3.4-4.8	02/18/01 13:33
CL	95 (L)	100-108	02/18/01 13:33
CO2	32.3 (H)	24.0-30.0	02/18/01 13:33
FER	698 (H)	30-300	02/27/01 13:13
B12	442	>250	02/27/01 13:13
FOLATE	13.9	3.1-17.4	02/27/01 13:13
BUN	19	8-25	02/18/01 13:33
CRE	0.8	0.6-1.5	02/18/01 13:33

(a) embedded hyperlink

http://www.abcmedctr.org/labnormalranges/fer

(b) infobutton manager call

Context: Lab result mgr
User: Physician
Lookup: Lab testts
Test name: Fer
Query: Interpretation
...

FIGURE 2-2 Automated or semi-automated retrieval from within an application by (a) embedded predefined hyperlinks, or (b) infobutton manager invocation with context-specific parameters.

the extent of integration with care and patient-specificity, and depending on system capabilities. Among such methods are those that automatically label results as abnormal, and those that automatically trigger alerts (see the next section).

2.1.2 Evaluation of Logical Conditions

Logical conditions are among the most widely used forms of clinical decision support, used in a wide variety of settings. Many different ways to represent logical conditions have been explored.

2.1.2.1 Decision Tables

One of the first themes to be pursued in clinical decision support was the idea of using logic as a way of refining and reducing the numbers of diagnostic possibilities. Let us consider n possible diseases D_i, where $i = 1, \ldots, n$. We also consider m possible findings f_j, where $j = 1, \ldots, m$.

For purposes of illustration here we make the simplifying assumption that a finding f_j can be only either positive, negative, or unspecified, denoted as $f_j =$ 1 or 2 or "–" (unspecified) for each j. For a finding to be unspecified in a definition, this means that its value is irrelevant or immaterial to the disease.

If all the diseases under consideration can be characterized by their findings, then a vector F_i can be constructed consisting of all the values of individual findings f_j for disease D_i.

Now consider arranging the findings characterizing all the diseases in a decision table, as depicted in Table 2-2. Since each of the columns of the table represents a different disease, there are n such columns. Each of the rows represents a different finding, so there are m such rows.

The diseases can be sorted (i.e., the columns rearranged), based on the value of any finding. Table 2-2 is sorted by the first finding, f_1, so all the diseases for which f_1 has a value of 1 are in columns to the left of those for which its value is 2. All the diseases for which the finding doesn't matter have

TABLE 2-2 An example of a decision table for *n* diseases and *m* possible findings. For a disease, a finding can be positive, negative, or immaterial, symbolized by 1, 2, and –, respectively.

Finding \ Disease	D_1	D_2	D_3	...	D_n
f_1	1	1	2		–
f_2	1	2	2		1
f_3	2	2	–		1
...					
f_m	–	1	1		2

a "–" in the cell for that finding, and appear to the right of those with either 1 or 2. (For clarity, the subscripts of the D_is are labeled from 1 to n in the table, but after the sort, this order will, of course, likely be different.)

Assume, for instance, that f_1 represents the sex of the patient, and 1 means *female*, and 2 means *male*; then this sorting has separated all diseases for which the sex of the patient must be female from those for which the patient must be male, and from those for which the disease can occur in either sex.

One can do similar manipulations for combinations of findings by doing subsorts of the sorted columns based on the values of other findings. In this example, if some diseases can occur only in infant females, and finding f_2 corresponds to 1: $age \leq 2$ and 2: $age > 2$, then a subsort of the table that is already sorted by sex will subgroup columns by this age criterion, as shown in the table. This shows by inspection that only D_1 has this combination of findings values. Note that if a particular finding value (or combination of findings values) is pathognomonic for a disease, then the particular value(s) should be present in only one column for that row (or combination of rows).

In the classic 1959 *Science* article by Ledley and Lusted (1959) and in Ledley's subsequent book (Ledley 1965), a similar discussion of the role of logic manipulation is presented as a prelude to other methodologies. Although Ledley and Lusted were particularly interested in probabilistic manipulation of findings to construct a differential diagnosis using Bayes theorem, they used logical manipulation to first reduce the range of possibilities that needed to be considered.

The representation of logic can be done in multiple ways. Decision tables similar to those just illustrated are one way to do that. An advantage of decision tables is the ability to sort and group columns with similar values in their rows. Another example of the way in which a decision table can be used is for representation of clinical practice guidelines, which are discussed further in subsection 2.1.3. This use, pursued by Shiffman et al. in the 1990s (Shiffman and Greenes 1991; Shiffman and Greenes 1992; Shiffman, Leape et al. 1993), involves constructing a findings-action matrix in which the upper set of rows corresponds to individual findings, and the lower set of rows corresponds to possible therapeutic actions (see Table 2-3). Individual columns are used to represent each of the possible combinations of findings that could occur (actual diseases are not explicitly considered here, just findings). So if we have two possible findings that can be present or absent, four columns are needed to represent the possible combinations of the findings. If there are n findings, each of which can be present or absent, then there will need to be 2^n columns to represent the various combinations.

TABLE 2-3 A guideline represented as a decision table. There are three findings that can be present (indicated by 1) or absent (indicated by 2), so there are 3^2 possible combinations, represented by the 8 columns, C_i. The recommended actions for each C_i are indicated by an X in the rows corresponding to those actions, for that column.

Findings	C_1	C_3	C_3	C_4	C_5	C_6	C_7	C_8
f_1	1	1	1	1	2	2	2	2
f_2	1	1	2	2	1	1	2	2
f_3	1	2	1	2	1	2	1	2
Actions								
a_1	X		X	X			X	
a_2		X	X	X	X		X	
a_3	X			X	X			X

The cells corresponding to the action rows in a given column indicate the treatment actions that are to be carried out given the particular findings in that column. (Note that more than one treatment may be appropriate.) Thus each column represents a step in a clinical guideline: the conditions for being at that particular step (i.e., the eligibility or applicability criteria for a set of recommended actions) correspond to the combination of findings indicated in the findings section of that column; the actions that are appropriate given those findings are indicated in the action part of that column. An X in an action cell means that the particular treatment should be carried out, and a blank means it should not. We can use this kind of decision table to look up any finding complex, and to determine what treatments or other actions are applicable. Also, if we desire to determine under what conditions a particular treatment should be done, we can use this table simply to sort by treatments, and find all the applicable findings complexes.

We can also use this approach to help design clinical guidelines and to check them for consistency and completeness. For example, we can simplify the table by eliminating redundant columns: if we inspect all the columns for which a particular treatment regimen is recommended versus those for which other treatment regimens are recommended, we can determine if particular findings are unique to that regimen. If the value of a finding does not distinguish between those columns for which the treatment regimen is recommended and those for which it is not, then that finding is irrelevant to the decision regarding that treatment regimen. Thus the various columns representing alternative values for that finding can be collapsed into a single column with a "don't care" symbol (–) in the cell for that finding. For example, in Table 2-3, columns C_3 and C_7 have the same action specification, and their findings differ only with respect to f_1, so f_1 is irrelevant to the decision to carry out this treatment regimen, and the two columns can be replaced by a single column with "–" in the cell corresponding to f_1. Other manipulations can be done to identify inconsistencies, for example, columns with identical findings specifications with different action recommendations. Columns with combinations of conditions that do not have actions associated with them (e.g., C_6 in the table) can be considered as either omissions that need to be completed, or impossible or irrelevant combinations for which the columns can be eliminated.

This discussion oversimplifies the process of using decision tables to represent clinical guidelines, because it does not consider sequences of tests, the performance characteristics of the tests, or the costs or risks of them and does not distinguish among the relative costs and utilities of various treatment options. Extensions of decision tables to create so-called "augmented decision tables" were pursued by Shiffman and Greenes (1992) to deal with such issues.

2.1.2.2 Venn Diagrams

Another well-known approach to representing clinical logic is by the use of Venn diagrams—shapes such as circles used to enclose groups of entities representing particular individual characteristics. The overlap of these shapes indicates the possible subgroups that can occur with combinations of these characteristics. If we consider a circle to represent those entities having attribute *A*, and another circle to represent those having attribute *B*, then the entities corresponding to the overlap of these two circles have both attribute *A* and attribute *B*. The area outside of either circle has neither attribute *A* nor attribute *B*. We can introduce a third circle representing attribute *C*, creating three more areas of overlap. Feinstein (1967; 1994) pursued the use of Venn diagrams such as this for thinking about and teaching clinical judgment, as shown in Figure 2-3(a) for three symptoms in acute rheumatic fever and (b) for four aspects of cancer, as adapted from his 1967 book. The process gets much more complex, however, when dealing with combinations of more than three or four logical entities, at which point visual display via Venn diagrams generally is considered to be impractical.

2.1.2.3 Logical Expressions

The usual mode for representing logical conditions is by *logical expressions* composed of Boolean combinations of terms. Individual terms are of the form *parameter operator value*, where *parameter* denotes the entity being evaluated, *operator* denotes a comparison operation (e.g., =, >, *or* <) to be performed, and *value* is the object of the logical operation and may be another parameter or a literal. The evaluation of the term yields a result of *true* or *false*. Terms are combined into *expressions* of arbitrary complexity by Boolean logical operators (e.g., *and*, *or*, *not*; note that *not* is a unary operator on a term).

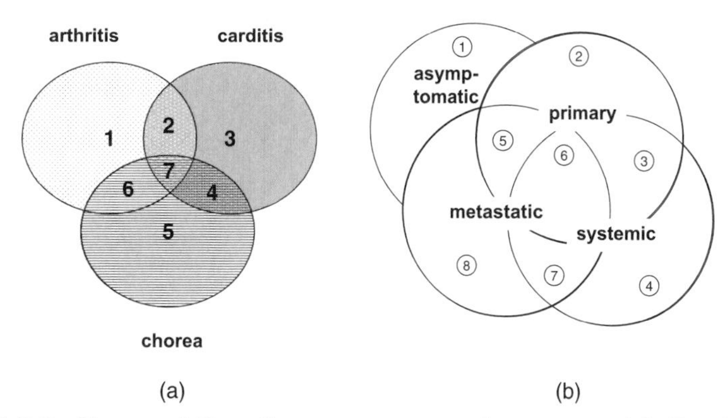

FIGURE 2-3 The use of Venn diagrams to represent the spectrum of findings in disease, adapted from Feinstein, *Clinical Judgment*, 1967. (a) The spectrum of rheumatic fever. (b) A simplified representation of the clinical spectrum of cancer.

One common approach to CDS is to build *expert systems* based on the capture and representation of knowledge from human experts. This topic is discussed further in subsection 2.4 and in Chapter 9, but for our discussion here of logical approaches to CDS, we point out that a major form of representation of expert human-derived knowledge is as rule-based systems. A rule-based system consists of a set of statements called *production rules* of the form *If (condition) then (action)*, where *condition* is a Boolean logical expression, and an inferencing method (possibly as a separate inference engine) for processing sequences of rules in a knowledge base to arrive at a conclusion. For each production rule, if the logical expression evaluates to *true*, then the *action* is performed.

In the business community, rules of the *if... then* form, known as *event-condition-action (ECA)* rules, of the form:

On	*event*
If	*condition*
Then	*action*

have been used to trigger rule logic evaluation in response to events. ECA rules automatically perform actions when triggered if the stated conditions hold. The ECA formalism originated in the active database field (where an active database system is one that has mechanisms to be able to respond automatically to events either inside or outside the database system) (Chakravarthy 1995). Bailey et al. (2002) review the history of ECA rules in the business community, noting that they also are used in workflow management, network management, personalization and publish/subscribe technology, and specifying and implementing business processes.

Similar methods to those of ECA rules have been used in medicine, in alerts and reminders, as well as in various kinds of conditions embedded in applications.

Alerts

One of the useful applications of CDS is to monitor inputs and to evaluate conditions triggered by those inputs for inappropriate values, or for notification of the presence of a situation that needs attention. In online interactive applications such as CPOE, an inappropriate dose of a medication or the presence of a drug interaction or contraindication such as allergy may be the conditions that are being evaluated. However, some alerts may be triggered by background events such as the identification in the clinical laboratory of an abnormal test result that requires attention (McDonald 1976; Haug, Gardner et al. 1994; Kuperman, Teich et al. 1996).

Once triggered by an event, a rule evaluator examines each rule that has identified the triggering event as a reason for invocation, and proceeds to determine whether its conditions evaluate to true, then carries out the specified action if the result is true. The action is typically to notify a physician (by pager, e-mail, or telephone).

This capability was first developed in the CARE system at Regrenstrief Institute in Indianapolis (see Chapter 4) and has also been a central part of the BICS system at Brigham and Women's Hospital (see Chapter 5) and the HELP system at LDS Hospital in Salt Lake City (see Chapter 6). Arden Syntax is a formal representation scheme, explicitly based on the ECA formalism, for specifying alerts and reminders, in terms of their evoking conditions, the data elements referenced, the logic, and the action to be performed, and which has become an HL7 standard (Hripcsak 1991; Hripcsak, Ludemann et al. 1994; Jenders, Huang et al. 1998). IT and successor approaches for standardizing the representation of logic in conditional expressions are discussed in Chapter 12.

Reminders

Reminders are similar to alerts; generally they are not triggered by inappropriate data entries or abnormal findings, but rather by the existence of conditions or the passage of time that makes specific actions desirable. Examples are a reminder for an HbA1c test that is generated at the time of a clinic visit of a diabetic patient who has not had the test in the past six months, or a reminder in an office to schedule an annual mammogram for a woman over 50. The representation approach for reminders is also generally in the form of a production rule.

Embedded Conditions/Constraints

Logical conditions are used in many types of CDS discussed later, where constraints or alternative actions need to be specified. The usual representation is in the form of a Boolean logical expression, although decision tables or other formalisms may be used in some circumstances.

A guideline with branching pathways has conditional expressions at each branch point, specifying the conditions for each alternative branch. Similarly, interactive dialogues of consultations might be designed such that conditional expressions determine the sequences of questions based on previous answers. In this manner, they function much like guidelines.

Data entry forms may have conditional expressions for validation of entered data as being within defined ranges or corresponding to a specific format (e.g., legitimate telephone number) or from a predefined list (e.g., abbreviations for U.S. state names, or drug name in a formulary). Conditional expressions may be used to determine whether certain data elements are to be included on the form, and to dynamically tailor the form. For example, if a patient is female and of child-bearing age, a form for entering a patient's past medical history might include entries for prior pregnancies and current pregnancy status.

2.1.2.4 Other Logic Models

Other variation on logical manipulations address the fact that binary (true/false) logic is often insufficient to capture the nuances of medical knowledge. A data item may be present, or absent, but if it is absent, this may be because it is unknown, unavailable, or not obtained. Multivalued logic may be more appropriate to deal with such situations (Gensler 2002); yet a review of the literature reveals that, although multivalued logic is widely used in engineering, particularly in circuit design, it has not been adopted in medicine. Another circumstance arises when the value of a data item may have an inherent uncertainty about it, reflected in a range of possible values. Because of the arbitrariness of asserting categories for some classifications, such as blood pressure characterized as *high*, *normal*, or *low*, fuzzy logic has been used to create qualitative degrees of membership to such categories (Zadeh 1965). The use of multivalued and fuzzy logic is discussed further in Chapter 10. These notions also bear upon the need to incorporate probabilistic reasoning and uncertainty into CDS, as discussed next.

2.1.3 Probabilistic and Data-driven Classification or Prediction

Most clinical judgments are not deterministic, and CDS needs to recognize the inherent variability of medical data, the imprecision of tests and measurements, and the fact that many principles of practice are based on limited evidence or just on expert opinion.

2.1.3.1 Updating Probabilities Based on Evidence

As we have noted in Chapter 1, the *Science* paper, "Reasoning Foundations of Medical Diagnosis," by Ledley and Lusted (1959), first introduced the idea of applying Bayes theorem to medical diagnosis. The core of Bayes theorem is the formula,

$$P'(D_i|F_j) = \frac{P(D_i)P(F_j|D_i)}{\sum_{k=1}^{n} P(D_k)P(F_j|D_k)}$$

where:

D_i is a particular disease of n mutually exclusive and exhaustive diseases
F_j is a set of findings
$P(D_i)$ is the prior probability of D_i
$P(F_j|D_i)$ is conditional probability of F_j given D_i
$P'(D_i|F_j)$ is the posterior probability of D_i, i.e., the probability of D_i given F_j

As we discussed in subsection 2.1.1, the Ledley and Lusted method actually used a prior step before applying Bayes theorem, in that they first used logical manipulations to look for logical combinations of findings that would absolutely rule in certain diseases. In the preceding formulation, the diseases D_i are mutually exclusive and exhaustive; that is, their probabilities sum to 1. Ledley and Lusted introduced a more general formulation in which combinations of diseases could occur, but in many cases this simplification is reasonable, or can be augmented by new "diseases" representing likely combinations of individual diseases. This makes it easier to develop the necessary set of estimates of prior probabilities of each disease.

Bayes theorem can be used to produce a posterior probability for each of the diseases under consideration, with the rank order of these thus corresponding to the relative magnitudes of the posterior probabilities. Since all possible diseases are considered, the sum of the posterior disease probabilities is equal to 1.0.

Bayes theorem is defined on findings complexes (the vector F_j for each possible combination of individual findings); this means that to be used operationally, conditional probabilities must be known for each such finding complex given each possible disease under consideration. Obtaining such joint probabilities is a monumental task, and databases containing such information are not usually available except in rare circumstances where only a few possible findings are under consideration. As a result of this, two other simplifying assumptions are typically made:

1. Each finding is considered to be conditionally independent of every other finding for a particular disease. This allows the probability of a combination of findings, given a disease, to be computed by multiplying the conditional probabilities of each individual finding given the disease.
2. The values for a finding are grouped into limited categories or bins, with the simplest categorization being binary (e.g., present, absent; male, female; age ≤ 20, age > 20).

Assumption (1) is the most tenuous, in that it often does not hold in practice; as a result, some combinations of findings that are not actually conditionally independent tend to be over-supported when the independence assumption is used.

In differential diagnosis it is often desirable to do some initial testing to narrow the range of diagnostic possibilities before proceeding to more definitive, and perhaps more invasive and expensive, tests. In the late 1960s, Gorry and Barnett (1968) explored the idea of a sequential Bayesian approach, in which certain tests were selected first, and based on the resulting diagnostic rankings, picked additional tests considered likely to contribute pertinent information based on various heuristics (e.g., reduction of entropy, or cost). The Iliad program, developed by H. Warner, Jr., and colleagues (Warner 1989; Guo, Lincoln et al. 1991) in the mid to late 1980s, used Bayes theorem confined to smaller subdomains of findings in its diagnostic model, but was used more as an aid to teaching and quality assurance than for direct CDS.

2.1.3.2 Decision Analysis

As early as Ledley and Lusted's 1959 article, the need to decide among treatment options by some sort of "value theory"-based rating of the alternatives was recognized. The formal methodology of statistical decision theory was just beginning to be developed at that time, as a successor to value theory. One of the pioneers of this was Howard Raiffa, a professor at Harvard Business School. His classic book on this topic, first published in 1970, has had a number of subsequent editions (Raiffa 1997). It wasn't until the mid-to-late 1970s, however, that the application of these methods for clinical decision making began to be done in earnest (Schwartz, Gorry et al. 1973; Pauker 1976). The essence of formal decision analysis is to develop a decision tree, in which a decision problem is structured in terms of branching sequences of decision nodes and chance nodes (Kassirer 1976; Pauker and Kassirer 1978).

The initial node of the tree is often a decision, such as whether to treat a patient now, test further, or do nothing (see Fig. 2-4). Decision nodes are indicated by square boxes in the figure, and chance nodes by circles. An example would be a young boy with abdominal pain, where the question is whether he has appendicitis or nonspecific abdominal pain. If surgery is done immediately, there is a possibility that a normal appendix will be found. There is also some risk associated with surgery, along with cost and morbidity. If one decides to test further (e.g., perform a CT scan), there is still a possibility that the test will be wrong, but if one uses the test as a basis for deciding on surgery or doing nothing, different probabilities will be associated with the outcomes, and the desirability of the endpoints needs to account for the cost of the test. If one decides to do nothing, surgery may not be required, and expenditure, discomfort, and risk of surgery have been avoided. However, there is now a possibility that the appendix may rupture, requiring more urgent and risky surgery.

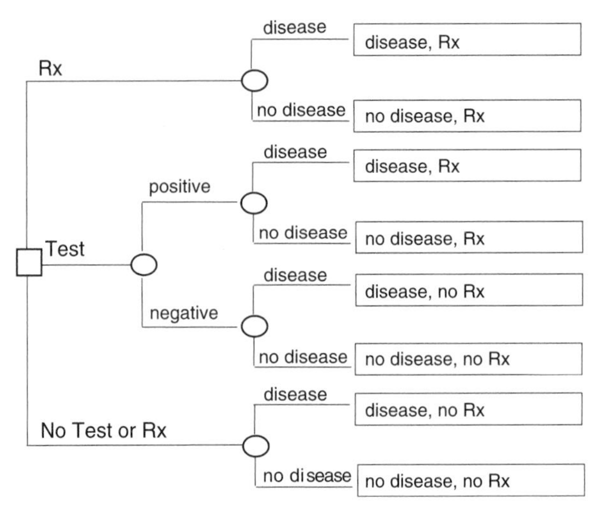

FIGURE 2-4 Decision tree for a generic treat now vs. test further vs. do nothing decision problem. Square boxes indicate decision nodes; circles indicate chance nodes.

One of the main purposes of a decision tree is to lay out the sequences of decisions and possible outcomes at each step, so that the decision maker can focus on the critical variables (e.g., the probabilities of various branches, the costs or risks of the treatments or the performance characteristics of the tests). The tree is expanded until all branches reach points that can be considered endpoints from the point of view of the analysis. The desirability of the endpoints is assessed, by assigning *utilities* to them (Plante, Kassirer et al. 1986). All branches from a chance node are assigned conditional probabilities (the probability of that branch occurring given the state of the patient at the immediately preceding chance node). Since the decision tree is intended to determine what the initial decision should be, the solution proceeds by a method known as *fold-back analysis*. The process begins by looking at the distal branches of the tree. If the immediately preceding node of an endpoint is a chance node, then an *Expected Utility (EU)* is computed for the chance node as the sum, over all the branches arising from that node, of the product of the utility of each of the endpoints times the conditional probability associated with the branch leading to it. If the immediately preceding node of an endpoint is not a chance node but a decision node, the process is different. Since the optimal decision generally is considered to be one that maximizes EU, the EU of the decision node is determined to be the maximum EU of the branches extending from it, rather than the sum of the EUs of the branches. The chance node or decision node thus assigned an EU is now considered an endpoint, and the process continues backward in the tree to the next more proximal node, iteratively, until the initial decision node of the tree is reached. At that point, the optimal choice, the subtree with the maximum EU, is determined.

An important feature of the decision tree is that the conclusion about optimal choice will differ as a function of estimates of probability and utility. Thus it is important to do a *sensitivity analysis* on the various estimated parameters, to see how robust the decision is over reasonable ranges of key parameters, and to determine the *threshold* for a parameter (Pauker and Kassirer 1980), such as a probability estimate or a utility (depending on one's focus), which would cause the optimal choice to change. In the example of the young boy with abdominal pain, it is likely that over some range of probabilities of nonspecific causes, nonsurgical management would be optimal, whereas over some lower range of likelihood of nonspecific causes (with higher likelihood of appendicitis), the surgical option would be optimal, if only to avoid the infrequent complications of surgery for a ruptured appendix. There is thus some threshold probability at which the decision choice would change. Likewise if the risk of surgery of either the simple type or that for ruptured appendix had a higher or lower frequency of undesirable outcomes (greater or lesser likelihood of complications or death), this would affect the point at which the choice to do surgery becomes optimal.

Multiple parameters can go into assessing a utility—for example, quality of life, length of life, and cost of treatment—in which case it may be necessary to rank the utility of each independently and then find a way to balance one against another or combine them into a single metric. Quality adjusted life years (QALYs) (Miyamoto and Eraker 1985; Smith 1987; Weinstein 1988) is a metric that has been used widely as a way to assess long-term consequences of various treatment options (controversial in that it can be associated with policy decisions about rationing of health care resources).

Other notable methods in decision analysis include the use of Markov modeling (Beck and Pauker 1983; Pauker and Kassirer 1987) for chance processes that may occur at unknown or multiple times in the future, and the declining exponential assessment of life expectancy (DEALE) (Beck, Pauker et al. 1982), which enables the physician to collate various survival data with information on morbidity to determine a quality-adjusted expected survival for a potential management plan.

Although decision analysis is sometimes used as a bedside decision support method, its primary use is for policy analysis (Weinstein 1980; 1989) and the formulation of guidelines and rules. It is through these that the fruits of decision analysis are usually actually realized as CDS. However, another use of decision analysis is in the assessment of patient preferences (Pauker and McNeil 1981; Eraker and Politser 1982; Fortin, Hirota et al. 2001), and in shared patient-doctor decision making (Col, Eckman et al. 1997; Fortin, Hirota et al. 2001).

2.1.3.3 Bayesian Belief Networks

Recognizing the difficulties in developing conditional probabilities for Bayesian analysis, and seeking to model causality and the probabilistic dependencies of events on one another, Pearl, working at Stanford in the 1970s (Pearl 1988), developed the notion of Bayesian belief networks that depict explicitly the various dependencies in the form of an acyclic directed graph. The first application to medical problems was the work of Cooper in the 1980s (Cooper 1986). This permits probabilities that are known to be explicitly entered into the network, others estimated where possible, and the remaining ones derived by inference from those that have been entered. Figure 2-5 is an example of a Bayesian network being explored for breast cancer risk prediction. See Chapter 10 for further discussion on the role of Bayesian networks in CDS.

2.1.3.4 Technology Assessment

Probabilistic decision support has been further aided by advances in technology assessment including receiver operating curve (ROC) analysis of performance characteristics of diagnostic technologies. Although not decision support

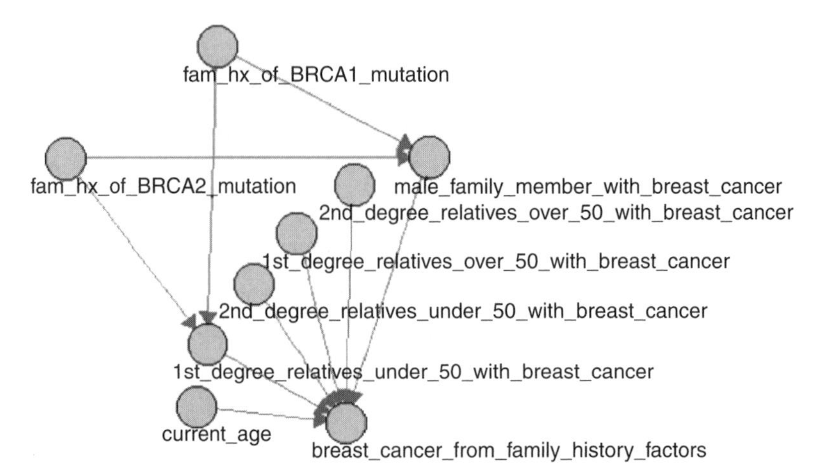

FIGURE 2-5 Bayesian network for breast cancer risk prediction, adapted from poster by Ogunyemi Chlebowski et al. (2004) reproduced by permission of the author.

methods themselves, technology assessment techniques have been invaluable in improving Bayesian probability estimation, by providing systematic approaches to quantifying diagnostic procedures in terms of the conditional probabilities of test results given specific disease conditions (Metz 1978; Swets 1979; Begg and Greenes 1983; McNeil and Hanley 1984; Greenes and Begg 1985; Hanley 1989).

2.1.3.5 Database Prediction: Data Mining and Machine Learning

Probabilistic decision making is critically dependent on the nature of the data on which it is based. The ideal situation is one in which large databases containing well-structured data are available that allow precise retrieval of patients similar to a current patient. Analysis of the responses of those patients to various treatments could then be used to decide upon the best treatment for the current patient. Two early approaches in the 1970s demonstrated the power of database prediction, that of Fries et al. (Dannenberg, Shapiro et al. 1979; Fries 1984; Bruce and Fries 2005) in rheumatological disease with a system known as ARAMIS, and that of Rosati et al. (Starmer, Rosati et al. 1974; Rosati, McNeer et al. 1975) in evaluating patients with coronary disease.

Much more often, however, databases are either not so well structured, or contain so many parameters and dependencies that considerable effort must be made to extract meaningful information from them. Despite these limitations, interest in database prediction has exploded due to the massive growth in the sizes and numbers of variables in data repositories now available, stimulated in large part by the advances in molecular biology, and by genomics, in particular. Statistical techniques such as regression and nearest neighbor, as well as newer nonlinear techniques and use of fuzzy logic, have been refined and improved over many years, and are mainstays of database prediction. But statistical techniques are largely hypothesis-driven, with the aim of proving or disproving hypotheses, a tedious and time-consuming process. With the huge numbers of variables in some databases, the desirability of shifting the paradigm to one that focuses on examining a large number of features to *discover interesting hypotheses* has led to the pursuit of data mining approaches.

Data mining is a blend of statistical methods, artificial intelligence (AI) techniques, and database design/retrieval approaches (Hill and Lewicki 2006). Progress in *data mining* and *machine learning*, or "knowledge discovery from databases" (KDD) has been stimulated by development of AI methods such as *artificial neural networks* (ANNs). The principal tasks are to identify key features that are important for the classification or prediction problem, and to determine the way in which these features should be combined to create an output variable representing the classification or prediction.

The history of ANNs is a tortuous one. The notion of an ANN can be traced back to the work by McCulloch and Pitts (1943) to model the human brain and cognitive processes by simulating neuronal pathways. As with their early work, subsequent research has continued to involve collaborations among computer scientists and neuroscientists, engineers, and psychologists. There was much excitement in 1958 when Rosenblatt reported on his work with *perceptrons* (Rosenblatt 1958). Using an input and output layer, and a middle "association layer," a perceptron could learn to associate specific inputs to particular output units. However, in a 1969 book by Minsky and Papert (1969), the authors pointed out significant theoretical limitations in single layer perceptrons. Although the authors didn't generalize this limitation to multilayered systems, as discussed further in Chapter 10, algorithms for estimating weights associated with the nodes for such systems did not exist at the time. Their result had so much impact on the field that the enthusiasm for machine learning cooled substantially. Funding for neural network simulation dried up for many years, and work in this area was considered a waste of time. A two-decade period of disenchantment followed, with only minimal work in neuroscience-oriented research on pattern recognition occurring. Despite this, progress was slowly made. In 1974 Werbos developed a learning method based on a back-propagation method for his Harvard PhD thesis (Werbos 1974), and a different threshold function for the artificial neuron than that used in the single-layer perceptron. Progress in ANNs slowly continued over the next two decades. Stimulated by the need for such techniques by the tremendous growth in size of available databases, and demonstrations of success in recent years, the field once again has become robust.

Two general approaches are used in machine learning, *supervised* and *unsupervised learning*. In supervised learning, the model specifies that one set of features known as inputs will have an effect on another set of features known as outputs; that is, they are presumed to be connected by a causal chain, although other mediating variables may occur between them. In unsupervised learning, the features are all assumed to be at the end of the causal chain, and the task is to discover latent variables that predict them. It is possible, though, for input features and latent variables to be considered in combination as causes of the output features. In supervised learning, the goal is to find the connection between two sets of observations, whereas in unsupervised learning, it is possible to develop larger and more complex models, as well as to deal with the situation in which there is a large causal gap. As the model is built up, it may become easier, at higher levels of abstraction, to bridge the gap.

Besides ANNs, a variety of statistical and AI methods have been explored in various combinations for data mining and machine learning, including linear discriminant analysis, k-nearest neighbor, logistic regression, classification and regression trees, genetic algorithms, and support vector machines.

Chapter 10 reviews these with particular reference to those that have proven most useful for generating knowledge for the purpose of CDS.

An approach that lent itself to ready integration into clinical practice was the creation of a clinical prediction rule, derived from logistic regression techniques, but which may be reduced to a simple linear scoring rule (Wasson, Sox et al. 1985). A well-known example of this is the "Goldman rule" for evaluation of patients with chest pain in an emergency department to determine whether they should be admitted to the coronary intensive care unit (Lee, Juarez et al. 1991). This was particularly useful in the days when computing the score by hand was necessary; in these days of ubiquitous computers and PDAs, of course, the full logistic regression model can be used instead of the simple scoring rule derived from it.

2.1.3.6 Evidence-based Medicine

Any discussion of probabilistic methods would not be complete without recognizing the movement toward evidence-based medicine (EBM), which basically seeks to annotate any clinical action with appropriate justification in the literature. The term has been used since the early 1990s in a series of JAMA articles from the Evidence-Based Medicine Working Group, based at McMaster University (Guyatt, Sackett et al. 1993; Oxman, Sackett et al. 1993). EBM attempts to develop objective ways to ensure high quality and safety of medical practice, aims to reduce health care costs, and seeks to speed the transfer of clinical research into practice. Although the goals are worthwhile and have been broadly espoused, some concerns are that clinical trial settings may be dissimilar to situations encountered in practice, and that the role of clinical experience and judgment cannot be minimized.

A major activity in EBM is the Cochrane Collaboration, the formation of EBM centers for systematic review of clinical trials to make the results of the analyses widely available. The idea was proposed by A. Cochrane, a British epidemiologist, and the first Cochrane Center was established at Oxford University in 1992 (Herxheimer 1993). A number of Cochrane Centers exist around the world, as well as evidence-based practice centers in the United States and Canada, funded by the U.S. Agency for Healthcare Quality (AHRQ). In 2005 AHRQ also funded the John M. Eisenberg Clinical Decisions and Communications Science Center in Portland, Oregon, directed by D. Hickham, to focus on applying evidence about treatment effectiveness for creating decision aids to foster patient participation in clinical decision making, as well as other information tools for patients, providers, and policy makers.

Often there are no randomized clinical trials pertinent to a particular clinical question, or the studies that are available differ in some respects from one another, and are insufficient in size to resolve the question. Thus, a major function of EBM is the use of meta-analysis of multiple studies to arrive at its conclusions. Methodologies for EBM and meta-analysis are described in further detail in Chapter 10.

2.1.4 Heuristic Modeling and Expert Systems

An alternative approach to modeling based on data is to develop models based on an attempt to emulate human expertise and reasoning processes. This is particularly needed in settings where there are insufficient data to be able to derive the needed estimates for probabilistic approaches, but where decisions nonetheless need to be made. An underlying motivation is the belief that it is useful to be able to understand cognitive processes of the human, and both to capture such knowledge and understand better what its limitations are. A secondary benefit is that, unlike probabilistic systems, heuristic systems can explain their conclusions, since their reasoning processes in fact are based on heuristics understandable to humans. These approaches are discussed in detail in Chapter 9.

2.1.4.1 Rule-based Systems

Rule-based systems are an example of a heuristic approach, in which individual logical statements in the form of production rules (as described in subsection 2.1.2) are obtained by observing human experts, or interviewing and debriefing them, and then combined in an attempt to emulate the reasoning processes of experts. A production rule may be of the form, Rule 7: *if x then diagnose y*, or Rule 17: *if m then assert n*. A rule-based system running in top-down (*backward chaining*) mode might start with a goal of seeking to diagnose *y* (see Fig. 2-6). That goal is the action part of a top-level rule. In order to determine if that rule can fire, the inference engine must establish that the antecedent condition *x* is true. Since *x* is typically a Boolean combination of terms, the engine must find data or other rules whose action parts, if asserted, would satisfy terms in condition *x*. To determine if those rules can fire, the inference engine then proceeds to try to evaluate their antecedents conditions in the same way. This process continues recursively until data are found or not found that satisfy the conditions of a rule. If a rule invoked recursively cannot be satisfied, then that chain of reasoning is abandoned. If a rule's condition is established, higher level rules that depend on the rule's action part can then be evaluated, and the process proceeds up the chain seeking to establish the goal. If satisfying conditions are not found, the goal fails to be established, or data are requested that may be able to subsequently establish the goal. This was the approach used in the MYCIN system of Shortliffe et al. (Shortliffe, Davis et al. 1975), which was the first medical application of a production rule system, with the goal of choosing appropriate antimicrobial therapy for a patient. Note that this system did not use purely logical rules, but also developed a model for dealing with uncertainty, which they called "certainty factors" (Shortliffe 1976). The work on MYCIN led to a number of extensions, including a shell known as EMYCIN (van Melle, Shortliffe et al. 1984); GUIDON, an intelligent tutoring application (Clancey 1987); and TEIRESIAS, an explanation facility (Davis and Lenat 1982).

A rule-based system can also be executed in bottom-up (*forward chaining*) mode (Bartels and Hiessl 1989). This is a useful approach in applications where data are arriving on a regular basis, as in patient monitoring (Rudowski, Frostell et al. 1989). A set of data is provided, and all production rules for which antecedent conditions refer to those data are evaluated. The results that are asserted from the rules that evaluate to true can serve as conditions for other rules that are then able to assert additional results, and the process continues until a rule is evaluated that establishes a goal, or until no more rules are able to be evaluated, and the process fails to reach a goal. Forward chaining is less efficient than backward chaining, in that many rules that are evaluated don't lead to actions.

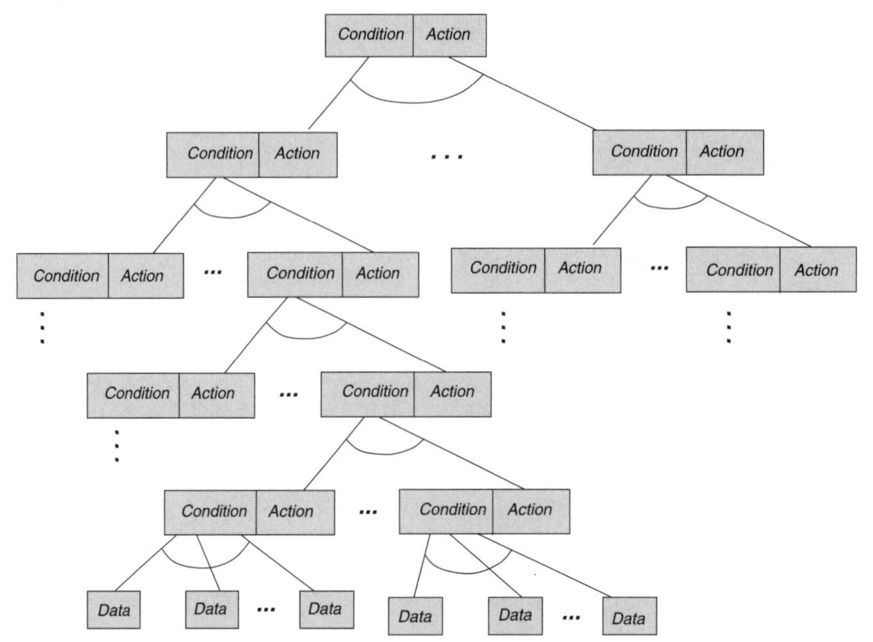

FIGURE 2-6 Rule-based systems. In backward-chaining mode, a top-level goal is to perform an action indicated by the action part of the primary rule (e.g., make a diagnosis, select a treatment). This is possible if the condition part of the rule is satisfied. To do so, the inference engine looks for rules whose action parts in Boolean combination (collectively indicated by the arc) satisfy the condition. Rules are invoked recursively in this manner until data are found or not found to satisfy a rule. If all the conditions in a chain are ultimately satisfied by data, the goal is established, otherwise not. In forward-chaining mode, rules are fired if data exist that satisfy their conditions. Higher level rules whose condition parts depend on the actions are evaluated, recursively, until a top-level goal is established. There may be more than one top-level goal, and none, one, or more can be satisfied by the data.

2.1.4.2 Other Heuristic Modeling Approaches

Another modeling approach that generated considerable interest is the use of *frames*. As defined by Minsky (1975), a frame is a data structure for representing a model of a stereotyped situation, represented as a network of nodes and relations, that pertain to various aspects of the situation, such as how to use the frame, expectations for the situation, and what to do about these expectations when they are satisfied. In medicine, we can think of frames as collecting attributes about diseases, patients, or other entities. We can then create relations among frames, such as ones portraying descriptive, causal, temporal, part/whole, or other relationships, and carry out reasoning processes by traversing the relations.

A frame-based representation was used in the 1970s in the Present Illness Program of Pauker et al. (1976) and the Internist-1 program of Miller et al. (1982), later refined and further developed by Miller et al. (1986) as QMR. The general reasoning approach was to rank differential diagnoses by attempting to match a patient's findings with those of stored profiles of diseases. A heuristic scoring rule was used, which weighted the frequency of a particular finding in a given disease, the evoking strength of a finding (i.e., how strongly its presence should raise suspicion of the disease's presence), and the importance of the disease. The first measure is similar, in a qualitative way, to the conditional probability of a finding given a disease, and the second is analogous to a posterior probability of disease, given a finding, although developers of both systems explicitly avoided considering them as probabilities. The third factor is intended to reduce the likelihood of missing serious diagnoses that otherwise might be ranked low. Internist-1 did a better job when more than one disease coexists, by using a partitioning heuristic that allowed it to develop a set of competing hypotheses.

DXplain, the diagnostic system by Barnett et al. (1987) also uses heuristic scoring rules to develop its differential diagnoses. CASNET/Glaucoma, developed beginning in the early 1970s by Kulikowski and colleagues (1982) used a causal-association network as a basis for its reasoning. This involved a combination of statistical pattern recognition, inference networks with probabilistic scoring of hypotheses, a conceptual structure to represent disease processes, and a normative set of rules for inferring pathophysiological state from observed patterns of findings, and preferring a treatment based on observed findings. AI/Rheum, developed in the 1980s by Kingsland and Lindberg (1983) used a knowledge representation system known as *criteria tables*. Findings or observations that could be entered by users were classified as Major or Minor decision elements for each disease in the knowledge base. The criteria for diagnosing any disease at particular levels of certainty are represented in a criteria table that indicates which observations could be present for the disease, are mandatory, or should cause the diagnosis to be excluded.

Other approaches attempted to develop causal models through deep reasoning; for example, for electrolyte and acid/base disorders (Patil 1987), or qualitative simulations (e.g., for nephritic syndrome (Kuipers and Kassirer 1984)). Ramoni and Riva (1997) have argued that with new deep knowledge from basic science, the possibilities for such approaches are now enhanced. A critiquing approach was pursued by Miller and colleagues, in hypertension (Miller and Black 1984) and in anesthesia management (Miller 1983), in which clinical actions were proposed by users and then analyzed by computer.

A recent approach to reasoning is the work of Fox et al. whose *PROforma* system includes a set of modular components for modeling a decision process such as a guideline and uses a formalism for "argumentation"; that is, for combining arguments for and against a hypothesis, as a way of determining whether to accept the hypothesis (Huang, Fox et al. 1993; Glasspool, Fox et al. 2001). This has some similarities to the criteria table approach used in AI/Rheum, described earlier.

A new diagnostic decision support system, known as Isabel (Ramnarayan, Tomlinson et al. 2003), employs pattern recognition methods to extract key concepts from textual descriptions to create a knowledge base that can be used to provide a differential diagnosis. The system currently concentrates on pediatrics, and reportedly has over 25,000 users worldwide, accessed through a stand-alone Web interface (although it can be incorporated into an EHR). Isabel contains medical content on over 6,500 diagnoses and heuristic rules such as applicability in particular age or gender groups. Clinical features entered by a user are matched with the knowledge base and filtered by the heuristic rules, and the disease entities thus matched are presented to the user for consideration. The system relies on a commercial concept matching software package known as Autonomy (Autonomy Corp, Cambridge, UK), which uses nonlinear adaptive digital signal processing derived from Bayesian inference and information theory to estimate the probability that a particular document is about a specific subject.

2.1.5 Calculations, Algorithms, and Multistep Processes

Many CDS approaches, such as those in subsection 2.1.4, involve multistep processes that may include logical or computational processes, probability assessments, heuristic methods, or other methods, and often can be decomposed into those components. The added element that ties them together is an implicit or explicit "execution semantics" that governs the flow of control from one step to another.

In some cases, we can use the *flow chart* as a familiar and convenient representation paradigm for the execution semantics of multistep processes. In a flow chart, the execution is considered to be sequential from one step to the next, except at points at which decision steps indicate alternative choices, where the choice made is based on evaluation of a condition. To enable a flow chart representation to fulfill this role, we posit three features of the flow chart: 1) the processes that can occur at each step can be arbitrary; and 2) the model for choice at a decision step can be arbitrary as well, that is, it can involve any combination of logical, probabilistic and heuristic processes that are appropriate for the decision-making task; and 3) any step can be decomposed into subprocesses, represented by their own flowcharts. Other elements of flowcharts are also needed to cover the semantics, such as iteration and stopping conditions, branching to two or more concurrent steps, and synchronization steps that wait for completion of concurrent steps.

As we consider various types of multistep processes in this section, it should be noted that an explicit representation of the execution semantics, whether by flowchart or other means, is often lacking. The flow of control is embedded in the application providing CDS, and the underlying flow can be inferred only through detailed inspection.

2.1.5.1 Interactive Dialogue and Structured Data Entry Control

The simple flowchart with sequential execution except at decision steps is used much more frequently than may be initially recognized. This model of execution, in fact, underlies (either explicitly or implicitly) many, if not all, interactive dialogues between a user and a computer, in which questions are asked of the user, the responses are evaluated by the computer, and based on logical conditions associated with alternative answers, the computer's next output to the user is determined. This process continues, with further inputs by the user, and evaluations of next steps by the computer, based on logical conditions that determine them, until the computer is able to make a recommendation or the user ends the session.

The earliest examples include the interactive history taking program developed in the 1960s by Slack and colleagues (1966) for interviewing patients about their asthma history, and an editor/driver developed by Swedlow and colleagues for creating and administering branching question-answer dialogues to gather patient histories from patients (Grossman, Barnett et al. 1971; Swedlow, Barnett et al. 1972). Many developments have subsequently focused on the use of question/answer dialogue and form editors for managing human-computer interaction, validating responses, evaluating them to assess conditions for subsequent branching, and thus controlling the flow of execution from the current question/answer element or form to subsequent ones (Bell, Greenes et al. 1992; Poon and Fagan 1994; Kahn 1997; van Mulligen, Stam et al. 1998). These approaches usually require specification, by a designer, of the

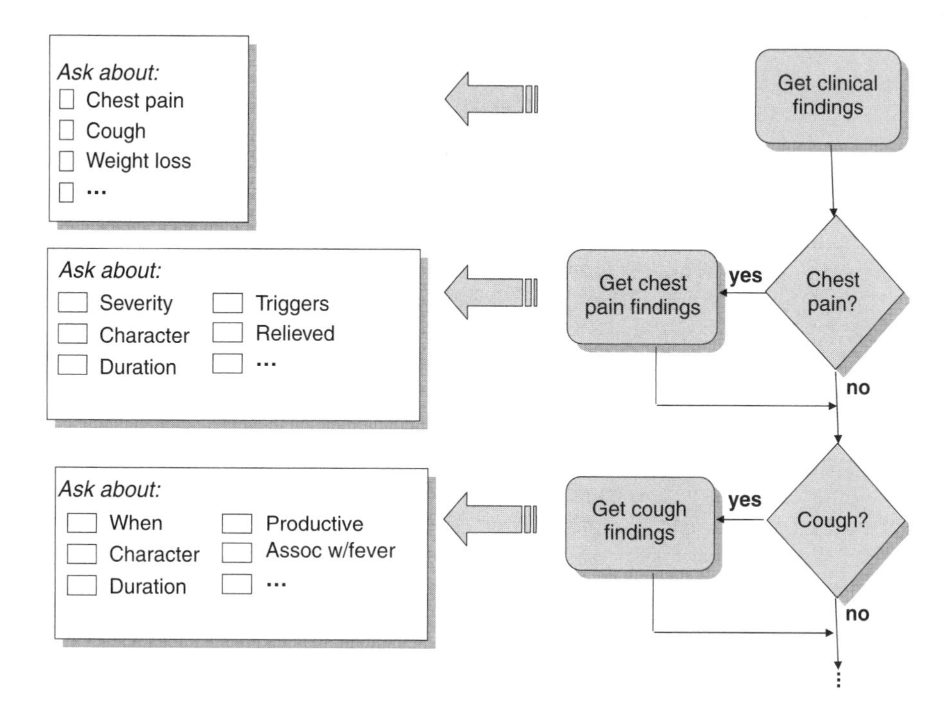

FIGURE 2-7 Interactive dialogue guided by an implicit or explicit underlying flowchart or guideline.

underlying control structure, but typically this is done only on a per-step basis, in terms of conditions for branching at each step. Rarely is the linkage of the entire dialogue to an underlying flowchart made explicit, the exception being when a clinical practice guideline is used to drive a user-computer interaction, an example of which was Shiffman's use of a guideline for pediatric outpatient asthma care (Shiffman 1994). Figure 2-7 depicts an explicit or implicit flowchart that represents the process flow underlying an interactive dialogue.

The problem-oriented medical record developed by Weed (1968), later implemented in computer form as a system called PROMIS (Schultz 1986), and the system for structured data entry for progress notes by Greenes et al. (1970) and radiology reports by Bauman et al. (1972) are also examples of structured environments for user data entry under implicit control of a system for managing process flow based on evaluation of inputs.

2.1.5.2 Computer-based Consultations

Algorithms for performing complex tasks involving multiple steps also follow this model. An early acid-base/electrolyte disorder consultation program developed by Bleich (1969; 1971) carried out an interactive dialogue with a physician, collecting data about a patient's electrolyte status, evaluating the data and performing calculations, and making recommendations regarding electrolyte replacement requirements. Although the software incorporated both the algorithm and the user interaction in a single program, the two components can be thought of as separate.

Formal algorithms and heuristic combinations of rules in multistep processes have also been used in many applications aimed at medication dose calculation and adjustment (Walton, Dovey et al. 1999; Walton, Harvey et al. 2001). For example, Swartout developed a digitalis therapy advisor with explanation capability (Swartout 1977). Other programs are used to calculate adjustment of medication dosage for children, the elderly, or those with renal failure (Chertow, Lee et al. 2001), or to use pharamacokinetic models to determine optimal dosage regimens (Jelliffe, Schumitzky et al. 1998). Another application of algorithms, heuristics, and multistep processes is in the control of devices, for example, for intravenous infusion of medications (Larsen, Parker et al. 2005) or ventilator management (Fagan, Shortliffe et al. 1979; Rutledge, Thomsen et al. 1991; Uckun 1994).

2.1.5.3 Clinical Practice Guidelines

The use of clinical guidelines as an embodiment of clinical practice can be traced back to the work of Komaroff, Sherman, and Reiffen (Komaroff, Black et al. 1974; Sherman and Komaroff 1976) at Beth Israel Hospital in Boston in the early 1970s, during which period a number of structured clinical protocols were developed to guide ambulatory practice, particularly for physician assistants. Another early setting in which clinical practice guidelines were deployed was for monitoring hospital inpatient utilization based on Diagnosis Related Groups (DRGs), in terms of specifying optimal patient care plans or clinical pathways, identifying deviations from these norms, and reviewing them (Huertas-Portocarrero, Ruiz et al. 1988; Tan, McCormick et al. 1993).

In the 1980s and early 1990s development of formal methods for specification of clinical practice guidelines was pursued by Margolis et al. (Margolis 1983; Gottlieb, Margolis et al. 1990) and Abendroth and Greenes (Abendroth and Greenes 1989), using a flowchart paradigm, and by Shiffman and Greenes (1991; 1992) in terms of the use of decision tables as described in subsection 2.1.2. Formal guideline modeling languages with execution semantics were developed in the mid-1990s and subsequently, by many investigators, as reviewed in more detail in Chapter 13. These included work by Lobach et al. (1997), Fox et al. (1997), Johnson, Purves, et al. (2000), Shahar et al. (1996); and Tu and Musen (2001) who pursued the idea of modeling clinical guidelines through the construction of formal modular elements that represented the different kinds of tasks and decision processes that a guideline could carry out. The GUIDE system (Ciccarese, Caffi et al. 2005) has explored the use of guidelines as a way of specifying and managing workflow in a clinical environment. Other work has been focused on mark up of narrative guidelines to extract computable elements, notably the Guideline Elements Markup (GEM) approach developed by Shiffman et al. (2000). Still other work has focused on the specification of clinical trial protocols, which are essentially similar to clinical practice guidelines but are more prescriptive and include a randomization component that assigns patients to one or another arm of the flow chart (Hickam, Shortliffe et al. 1985; Greenes, Tu et al. 2002).

The InterMed project in the late 1990s was a joint effort by biomedical informatics groups at Harvard, Columbia, and Stanford to develop a common formalism incorporating the best features of other available modeling languages for clinical guidelines, which could become a basis for interchange and sharing of guidelines. That effort resulted in the Guideline Interchange Format (GLIF), version 2 of which was published in 1998 (Ohno-Machado, Gennari et al. 1998), and version 3, in 2003–2004 (Peleg, Tu et al. 2003; Boxwala, Peleg et al. 2004).

Since that time activities aimed at seeking to create a common guideline modeling formalism have been pursued by the Clinical Guidelines Special Interest Group of the Health Level 7 (HL7) standards development organization. The issues of clinical practice guideline modeling and standardization are described in Chapter 13. Peleg et al. (2003) have characterized a number of common features of many guideline languages, in an effort to focus efforts at convergence on a common model. In addition to concerns about interchange and dissemination of guidelines, other foci have been on the issues of integrating guidelines into clinical environments and applications, such as the ATHENA system developed in collaboration between Stanford and the Palo Alto Veterans Administration Hospital (Goldstein, Coleman et al. 2004) and the SAGE project, which was a collaboration between IDX Corporation and Stanford University, Mayo Clinic, University of Nebraska, and Intermountain Healthcare (Tu, Musen et al. 2004).

2.1.5.4 Biomedical Signal and Image Processing

Another category of application of multistep processes is the analysis of biomedical signals and images. We will not explore this important but specialized category of decision support in this book. As we have noted in subsection 1.5.1, our reason for this relates to the highly specialized nature of these applications and their niche areas of focus—even though such specialized use has paradoxically enabled them to have had considerable impact (within those niches). Suffice it to say, however, that signal and image processing has had a very long history. One of the earliest applications was in the realm of ECG interpretation, which focused in particular on approaches to signal analysis of the ECG tracing. Another area of long interest that has become an increasingly important topic has been the pattern recognition and extraction of features from biomedical images (e.g., CT and MRI studies), volumetric modeling, dynamic imaging and computation of flow or metabolic activity or neuronal activation, and use of imaging interactively to guide surgery. Using computers for treatment planning, for example in radiation dosimetry, has also long been of interest and has involved considerable image analysis. These applications may involve a combination of algorithmic and logical operations, probabilistic methods, and heuristic reasoning.

2.1.6 Associative Groupings of Elements

A final category we address is the use of strategies for organizing data for presentation. This can occur in a variety of applications, in data entry forms, screens for selecting or customizing an order set, structured reports, dashboards, flow sheets, and other forms of input or output. We have already discussed in subsection 2.1.5 the use of an explicit or implicit underlying guideline to control the sequencing of interaction in human-computer dialogues such as for structured data entry or consultation. Here we concentrate on the strategies for combining data elements into *groups* so that they are presented, viewed, and considered together by a user. In all of these uses, the grouping is guided by an underlying reason, such as similarity of topic, similarity of action required, or priority. For example, the elements in particular sections of a clinical history input form relate to family history of cancer, the cardiovascular findings, or the response to medications. An order set groups orders that relate to a particular situation (e.g., admission to the ICU after surgery), and includes orders that have particular intents or purposes, such as vital sign check, pain control, diet, activity restrictions, intravenous fluids, and oxygen, in addition to procedure-specific orders.

Although grouping is generally considered desirable in terms of the convenience it offers in entering data or making selections from a set of choices when arranged in a logical structure, or the clarity of presentation offered by reports organized in such a manner, it is not often thought of as a decision support method. Yet we maintain that CDS is actually one of the more important benefits that can be achieved by the wise use of grouping. We refer to this as *associative grouping (AG)*, because of its purpose—to represent important associations and to elicit them in the mind of the user. AG can play a subtle but effective role in encouraging best practices simply by reminding the user to think about or include presented elements. The subtlety lies in the fact that the grouping of elements is not explicitly directive. The effectiveness derives from the consequences of using it: grouping elements together makes the desired behavior, that is, consideration of those elements (for either data entry or review purposes), the most convenient one to do. This is why we consider AG as a form of CDS.

The importance of AG as a CDS method has come to the fore partly as a result of the convergence of three recent trends:

1. Recognition of the large size and scope of effort required by an enterprise to manage its knowledge assets. This has been driven by new imperatives to deliver CDS (which we review in the next subsection of this chapter), and has revealed how much of the knowledge in the enterprise is embedded in the way forms and reports are designed and used.
2. Recognition of the limitations on reuse of executable knowledge content without more precise definition of the data elements used. This has stimulated increased efforts in structured data capture, and thus also in methods for designing forms for that purpose.
3. Development of tools and resources for structuring of documents for the Web, and ontology languages and representation schemes aimed at managing the metadata associated with such documents. The advent of the Web and its potential for sharing knowledge content has both stimulated interest and highlighted the limitations in actually achieving reuse. Capabilities provided by new and emerging tools offer the opportunity to organize components of documents to facilitate reuse.

Groupings of data elements for presentation typically are described by *templates*. The way the specification of such groupings and associations is embodied in the structure of documents and the status of efforts to develop standards for document architectures and for templates are described in Chapter 15. The knowledge management associated with the organization and categorization of grouped content elements to facilitate best practices and encourage reuse is a relatively new activity, also described in that chapter, for which standardization efforts have not yet begun.

Conditional logic expressions are used in data entry applications for specifying constraints on allowable data types, formats, and ranges, and for both data entry and output applications, for specifying conditions for inclusion, omission or modification of certain elements in the presence of others (e.g., omission of pregnancy history items in a male patient). Thus this aspect of document specification also draws upon approaches described in subsections 1.2.1.1 and 1.2.1.5.

The origin of document structuring approaches can be traced back to some of the earliest efforts to create forms for structured data entry and programs for report generation from data, and such capabilities are associated with most commercial database systems. In medicine, of course, the capture of structured data and its presentation in reports has faced the difficulty that much of the content does not lend itself to a high degree of structure, and the effort to capture it in structured form is either so unnatural to users or so time-consuming that efforts to do this have not been successful except in limited settings. Work in this area was briefly reviewed in subsection 2.1.5, in the discussion on interactive dialogues and structured data entry control.

A compromise that often has been adopted is to divide the form into sections relating to various categories of information. Just that simple device provides a context for the data contained within each section that can be used as an aid to retrieval, analysis, and interpretation. The degree of structure can, of course, range from minimal top-level headings to much more detailed divisions into subheadings and sections within them, depending on the application. This is true for both input forms and output documents—there is often a tradeoff between the IT perspective seeking structure down to the level of specific data elements and the user perspective seeking ease of use, for data entry, in terms of time required, and for reports, in terms of conciseness and clarity of presentation.

Reports and other documents produced from structured or semistructured input typically retain the structure, although the content may be rearranged and presented differently. Sometimes form-based data are presented in reports in tabular or flow sheet format, but the data alternatively may be rendered as prose narrative. The decision regarding mode of presentation should ideally reflect user preferences; for example, the study by Bell and Greenes (1994), which demonstrated physician preference to outline mode vs. narrative prose output of ultrasound procedure reports. The structure of the underlying data or sections of it facilitates some degree of manipulation for presentation purposes, and the ability to recast information in different views or for different media or form factors (e.g., printed report vs. display on a desktop monitor vs. PDA output).

The most important capability for describing documents in a way that allows identification of its structure and its contained elements is the use of *markup languages*; that is, conventions for organizing and tagging sections or elements of a document. This is a complex topic beyond the scope of this book, but since some aspects of its history are relevant, they are briefly highlighted. *Standard Generalized Markup Language (SGML)*, which was developed and standardized in 1986 by the International Organization for Standards (ISO) (ISO 1986), is widely used to facilitate management of large documents that are revised often and need to be printed in a variety of formats. A first version of *Hypertext Markup Language (HTML)* (Berners-Lee and Connolly 1993), used to describe pages in the World Wide Web, was specified in 1993 as a way of defining and interpreting tags consistent with SGML rules, although not a strict subset of SGML, and has gone through several updates.

Because the focus of HTML is just on the formatting and appearance of Web pages and not on their content, other capabilities were needed "to meet the needs of large-scale Web content providers for industry-specific markup, vendor-neutral data exchange, media-independent publishing, one-on-one marketing, workflow management in collaborative authoring environments, and the processing of Web documents by intelligent clients." The preceding statement was part of a 1997 press release (W3C 1997) for the first specification of the *Extensible Markup Language (XML)* by the W3 Consortium. XML (W3C 2006a) is the universal format for structured documents and data on the Web, which is characterized as a subset of SGML. (Of note, the next version of HTML will be reformulated as an application of XML.)

Despite the benefits of XML, composition and formatting of information (as data) provided by relational and XML databases are not sufficient to meet the goal of semantic structuring for which ontologies are needed. In the vision of the *Semantic Web* (W3C 2004), interaction among disparate applications or agents can be made unambiguous through shared reference to ontologies available on the network. The subject of ontologies and how to build them is itself a broad topic, but a domain-specific ontology needs to represent a consensual view of a domain.

A number of languages for building and managing ontologies have been developed, including general logic programming languages like Prolog, as well as the Open Knowledge Base Connectivity (OKBC) model and languages like KIF and its successor CL. An approach known as description logics is the basis for the *Web Ontology Language (OWL)* standard (W3C 2006b), and tools such as the open source Protégé (Musen 1992; Noy, Crubezy et al. 2003) and commercial products provide ontology editing capabilities. The wide array of information residing on the Web has given ontology use an impetus, and ontology languages increasingly rely on W3C technologies like RDF Schema as a language layer, XML Schema for data typing, and RDF to assert data.

Based on this massive effort focused on the Web, we are seeing a new generation of report generators, document builders, and form builders, based on Web technologies, replacing the old generation of such tools that were tied primarily to database systems. Thus we are now entering a stage where cataloguing the components of documents can be practically done and managed by ontology tools. This is further explored in Chapters 15 and 21, in terms of its potential value for knowledge management. However, a note of caution to the reader is that this is still largely conjecture, as the scope of efforts required to do this is just beginning to be explored.

2.2 DRIVING FORCES FOR CDS

The preceding section should convey the richness of the field of CDS in terms of the range of models, methods, and applications of computers and information technology that have been explored over the past four or more decades. In Chapter 1, we noted that, despite such research and development activity, CDS has failed to become a mainstream capability, and that initiatives for deployment on a large scale have been largely dormant. However, we are seeing a rekindling of the excitement about prospects for CDS. The reinvigoration of the impetus for CDS has its roots in a number of developments and trends on the political, sociocultural front, which we review here. All in all, we identify 17 driving forces for CDS, as listed in Table 2-4, and discussed in the remainder of this chapter.

2.2.1 The Technology Imperative

Clearly, progress in computer science, cognitive science, artificial intelligence, statistics, and the computer and communication technologies have created a

TABLE 2-4 Driving forces for CDS adoption.

Factor	Importance to CDS
Technology imperative	Opening up possibilities for CDS and stimulating development
Knowledge explosion	Need to find patient-specific, context-specific resources of high quality, cope with "information anxiety"
New technologies for diagnosis and treatment	Need to compare to existing approaches, assess relative costs and benefits, determine appropriate use
Assimilating discovery and knowledge	Need to digest literature, determine best practices, create evidence-based knowledge repositories and other resources
The Internet society	Can find information much more rapidly, can answer many more questions on the spot, changed expectations as a result
Empowerment of patient and consumer	Better informed public, need for more bidirectional communication and shared decision-making
Medical errors	Recognition of importance of proactive support to ensure safety
Variability in quality, access, and adoption of best practices	Need for reminders, alerts, recommendations and other approaches to foster highest quality
Spread of EHRs	Providing the substrate needed for CDS
Aging of population and increased complexity of disease	Causing increased stress on health care system, time and resource demands, need for CDS to help manage effectively and efficiently
The no-win proposition: decreasing time and increasing pressure on doctors	Again, need for CDS to help manage effectively and efficiently
Fragmentation and difficulty coordinating care	Need for improving problem-specific transfer of information to/from specialists, preapproval support
Defensive medicine	Need to set standards of care, make it easier to meet those standards, reduce liability concerns
Health care costs	Need for standards of care, to be able to determine appropriate tests and treatments independent of (dis)incentives imposed by reimbursement plan
Pay for performance	Need for CDS to attain the rewards for achieving practice performance goals
Demonstrated benefits	Successes of CDS in some settings as driver for adoption more broadly
Top-down initiatives	Creating a focus on the need for standards, interoperability, and baseline levels of functionality, as a substrate for effective CDS

momentum that has driven a broad spectrum of societal change. Part of the drive to use computers for CDS can be ascribed to the so-called "technology imperative." This is the view that technological developments, once under way, are unstoppable, inevitable, and irreversible. In other words, simply because something can be done, it will be attempted. If we can do something, let's see if it will be useful.

The prospect of being able to use computers to help make decisions has fascinated investigators both in medicine and in other fields from the earliest days of computing. Luger's textbook of AI (Luger 2005) traces the history to early philosophy and literature, predating computers by millennia, and to the relatively recent pursuits in the 1950s and 1960s of psychology, philosophy, neurophysiology, and in particular, cognitive science, to determine the extent to which human intelligence and reasoning ability could be understood sufficiently to be embodied in a computer.

The complexity of medicine has offered fertile ground for investigation and development and refinement of reasoning and problem-solving models, for diagnosis, treatment planning, and estimation of prognosis, as exemplified by the work reviewed in the preceding section of this chapter. Understanding the cognitive processes underlying diagnosis and decision-making by humans, such as the hypothetico-deductive model of diagnostic reasoning proposed by Elstein et al. (Elstein, Shulman et al. 1978), and the heuristics and biases identified by Kahneman and Tversky (Kahneman and Tversky 1982) have also stimulated development of various strategies and models.

In addition to the challenges of cognitive modeling and emulation of the human, another major intellectual stimulus came from the other major branch of AI—that focused on understanding and modeling perceptual and neurophysiological processes and function. Early work in the 1950s and 1960s, also reviewed in Luger's textbook (Luger 2005), included study of self-organizing systems and cybernetics. Advances in this area, together with statistical methodology development, have given rise to feature extraction and pattern recognition capabilities that underlie signal and image processing, artificial neural networks and other data mining methods, and robotics.

Major theoretical advances addressing a number of problems in medicine also have had impact. Examples, to name a very select few, are methods for technology assessment of diagnostic tests, such as Receiver Operating Characteristic curve analysis (Swets 1979), decision theory (Raiffa 1997), the variety of extensions to decision analysis for assessing utilities, survival analysis, and Markov modeling reviewed in subsection 2.1.3, and pharmacokinetic approaches to modeling of drug distribution and metabolism (Jelliffe, Schumitzky et al. 1998).

2.2.2 Knowledge Explosion

In 2004 NLM indexed more than 575,000 articles from 4,800 journals (Kotzin 2005) and the numbers of both articles and journals is increasing every year, including a growing number of journals that appear only in electronic form (reported to be 10% in 2004). A report in *Nature Genetics* in 1998 (Ermolaeva, Rastogi et al. 1998) showed the pace of discovery of DNA sequences to be outstripping the cumulative growth of the biology and genetics literature subset of MEDLINE at an exponentially increasing rate (see Fig. 2-8). We can extrapolate this to the present day as likely continuing to occur given the progress in genomics, proteomics, and pharmacotherapeutics. Many of these discoveries find their way into databases such as GenBank or the many other databases accessed via Entrez, the search interface maintained by the National Center for Biotechnology Information of the NLM, and don't end up being reported in the literature. Even within the literature, a new active area of research is that of mining the literature for discoveries that may be found on the basis of unrecognized connections among separate reports (de Bruijn and Martin 2002; Chaussabel 2004).

In the practice of medicine, the revered highly skilled intuitive clinician who could analyze subtle findings and integrate current knowledge at the bedside to come up with the correct diagnosis when all others have been unable to do so, has long ago become an anachronism. It's not that clinical

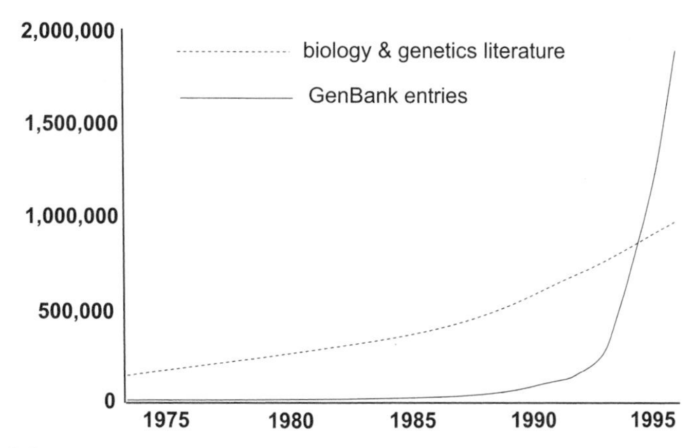

FIGURE 2-8 Cumulative growth in DNA sequences in GenBank vs. growth in biology and genetics literature subset in MEDLINE for period from 1975–1995, adapted from (Ermolaeva, Rastogi et al. 1998) by permission from Nature publishing Group.

judgment is becoming obsolete, of course, but the ability to have the entire needed knowledge base in one's head is long gone. It is virtually impossible to keep up with the literature except in very highly specialized niche fields. This has contributed to a growing sense of "information anxiety" by practitioners, particularly those in primary care, who are expected to be sufficiently conversant in all the aspects of medicine to know either how to manage a case directly or when it is necessary to refer the patient to a specialist.

2.2.3 New Technologies for Diagnosis and Treatment

Practitioners are confronted with a continuous stream of new products and reports of their efficacy for diagnosis, treatment, or prevention. As voluminous as the stream of new products are, and the difficulty in sorting out the claims and determining how they relate to existing approaches, these represent only a fraction of the discoveries that are reported in the medical literature. An illuminating study by Balas and Boren (2000) that is frequently cited showed that it takes an average of 17 years for scientific discoveries to make their way into products, and that only 14% of those discoveries actually become products that are used. Yet, even given that such throttling down process occurs, the pace of development of new medications and diagnostic and therapeutic technologies is relentless and overwhelming.

In the United States, the indications and optimal use of medications and health care technology have been largely identified, as well as their performance characteristics, costs, contraindications, and expected side effects, as part of the regulatory filings and approval process required by the Food and Drug Administration and its Center for Devices and Radiological Health (http://www.fda.gov). Nonetheless many diagnostic devices have overlapping indications and uses, as do many therapies. Clinical trials have not always clearly established optimal uses, superiority of one versus another among alternatives, or appropriate sequences. Guidelines, if they insist, may be conflicting. Given the slowness of dissemination of knowledge about best practices, and the complexities involved in evaluating alternatives, there is thus considerable unevenness in the adoption of optimal diagnostic and therapeutic technologies.

2.2.4 Assimilating Discovery and Knowledge

Elsewhere we described the growth of EBM. This has arisen in large part because of the complexities in making the judgments noted in subsection 2.2.3. As clinical trials and the literature proliferate, the need for summarizing data from these experiences expands. EBM and its principal methodology, meta-analysis, are discussed in detail in Chapter 11. Another important capability is the assembly of databases of clinical trials, as in the Web site http://www.clinicaltrials.gov (McCray 2000), and the movement spearheaded by leading journal editors to compel registration of clinical trials as a precondition for reports about them to be published (DeAngelis, Drazen et al. 2004).

2.2.5 The Internet Society

Although much of the information that is needed is available via the Internet, the sheer volume of Internet-based content presents its own set of problems. A simple Google® query will often retrieve hundreds, if not thousands, of related Web pages. Fortunately its page rank algorithm (and those of other search engines) tend to be very good at selecting those pages most likely to contain the requested information. Google® is often the first place one goes to for health care-related information retrieval. Reportedly, PubMed searches increasingly occur as a result of an initial Google® retrieval (although I could find no firm documentation of this trend). Because of the sheer volume of content, however, it is very difficult to assess the quality and source or provenance of the information retrieved, or to distinguish among multiple sources to determine which is most credible. Also, Web pages often do not contain a date, and it is difficult to determine how current the information retrieved is.

In addition, information is available not only to practitioners but to the health care public, not just on the Internet but via traditional media, which creates an increasing burden on the practitioner to have adequate background knowledge and be able to respond appropriately to suggestions or requests from patients.

Not only do practitioners not have time to sift through the many resources that are retrieved even from the simplest of queries, but they are ill-equipped to assess much of that information. This creates an urgent need for high-quality CDS tools to present predigested, certified, authoritative information when critical decisions need to be made.

2.2.6 Empowerment of Patients and Consumers

The "flattening of the world" that Friedman has written about (Friedman 2005) is as true in the realm of health as it is in other spheres. Along with increased availability of information to the health care consumer and patient has been a growing sense of empowerment. The health care public are showing a growing willingness to take responsibility for their own health, and to make their own decisions, or to at least participate more actively in the decision-making process (Deber, Kraetschmer et al. 1996; Coulter 1997; Cahill 1998).

Patients also increasingly have access to their own health care data through access to Personal Health Record (PHR) systems, either managed by third parties or provided through gateways to health care institutional EHR systems (Kimmel, Greenes et al. 2005; Tang, Ash et al. 2006).

These trends have the potential to cause a considerable shift in the nature of the doctor-patient interaction, with patients less likely to accept physician recommendations at face value. Physicians now need to take more time to go over and respond to patients' concerns, or to evaluate requests from patients based on information that the patients have obtained elsewhere.

An active area of research is the development of decision support software aimed at facilitating joint patient–doctor evaluation of alternatives for therapy, and engagement in a shared decision-making process (Fortin, Goldberg et al. 2003; Ruland 2004).

2.2.7 Medical Errors

Ever since the Institute of Medicine produced its landmark report in 1999, "To Err is Human" (Kohn, Corrigan et al. 1999), which called attention to medical errors as the eighth leading cause of death in the United States in that year, this grim statistic has sent shock waves through the health care community.

Release of the IOM report is probably the most pivotal event in the surge of activity in health care IT that has occurred in recent years. The goal of preventing medical errors has triggered a wide variety of responses aimed at improving the safety of patient care, ranging from analysis and improvement of procedures in individual health care institutions and practices, to education and training to avoid errors, development of clinical information systems capabilities, particularly adoption of the EHR, CPOE, e-prescribing, positive-identification medication administration systems (e.g., through the use of bar codes), and the use of CDS to detect and prevent errors. It has also resulted in the call for legislation and regulations aimed at improving patient safety through the use of IT (Bell and Friedman 2005; Middleton 2005).

2.2.8 Variability in Quality, Access, and Adoption of Best Practices

Health disparities have been widely recognized as a problem throughout the world; for example see Vidyasagar (2006). In countries with advanced health care such as the United States, the disparities owing to socioeconomic status differences and uneven health insurance coverage and access to care are striking (Smedley, Stith et al. 2003; Kirby, Taliaferro et al. 2006). Even within a single socioeconomic category, however, there are wide variations in the adoption of best practices and in the quality of results achieved for similar problems by different practitioners (Asch, Kerr et al. 2006).

The follow-up to the IOM's report on medical errors by its 2001 report on "Bridging the Quality Chasm" (IOM 2001) pointed out the frequency of nonoptimal patient care, the wide variations in practice, and the inefficiencies, dangers, and inequalities that have resulted. That report also called attention to six attributes that health care systems needed to have for high-quality care to be achieved: safety, effectiveness, patient-centered emphasis, timeliness, efficiency, and equitability.

This report further stimulated recognition of the centrality of IT systems and CDS to the reduction of errors and the achievement of quality. The goals of such systems are clearly recognized as 1) detecting inappropriate practices before they are carried out in order to suggest alternatives, 2) monitoring and providing feedback on existing practices, and 3) carrying out studies aimed at EBM; that is, determining optimal practices for situations where the optimal approach is not clearly defined at present. The emphasis on care being patient-centered also has stimulated interest in consumer health information sites, health appraisal tools, guidelines, scorecards, and information about performance of various medical centers and practitioners, and the growth of PHRs.

2.2.9 Spread of EHRs

The presence of an electronic health record facilitates the implementation of CDS capabilities, particularly when the operation of the CDS application requires access to stored patient data or use of communication and interaction mechanisms of the clinical IT system that hosts the EHR. In addition, the prospect of being able to provide CDS may itself be a major driver for the adoption of EHRs due to considerations such as those described earlier.

As a consequence, we are seeing a number of programs in the United States, at federal, state, and local levels, as well as through professional societies, aimed at establishing legislation and regulations, as well as providing funds to encourage the adoption of EHRs, CPOE, and CDS. The slowest adoption is in small hospitals and office practices, but even their adoption rates are slowly rising (Gans, Kralewski et al. 2005; Middleton, Hammond et al. 2005). Similar initiatives are also under way in a number of other countries (Haux 2006).

2.2.10 Aging of Population and Increased Complexity of Disease

The age distribution of populations in developed countries continues to shift toward the older age part of the spectrum, owing in part to the increased success of modern health care, as well as of preventive measures, to prolong life. Another factor in the United States is that a large percent of the population represented by members of the post-World War II "baby boomer" generation are just now beginning to enter their 60s. Along with increased age, of course, come the plagues of old age—among them, hypertension, diabetes, congestive heart failure, renal failure, chronic pulmonary disease, and neurological deterioration. Patients not only have these chronic diseases but frequently have more than one of them. These patients require multiple treatments, mainly in the form of medications, which have side effects and cause adverse events. Thus, treatment by medication, as with other therapy, requires balancing in terms of risks and benefits, but also in terms of interactions with medications required for other concurrent conditions.

Because of the combination of increased age of the population, increased frequency of chronic diseases, and the multiple treatments involved, it is estimated that chronic disease consumed close to 80 percent of all U.S. health care expenditures in 2000 (Ray, Collin et al. 2000). This situation will only have become more pronounced now and continue to be exacerbated in the future, further adding to the load on our health care system. Concern about this trend has stimulated the need for patient-specific decision support, especially tailored to the management of chronic disease, and to "disease management" programs and other strategies aimed at focusing on optimization procedures that can be put in place for patients with various chronic diseases (Bodenheimer 2003; Seroussi, Bouaud et al. 2004; Dorr, Wilcox et al. 2006).

2.2.11 The No-Win Proposition: Decreasing Time and Increasing Pressure on Doctors

Because of the relentless growth of health care expenditures, due to inefficiencies, increased cost of new technologies, and sheer numbers of patients needing care, there are ever-increasing pressures on practitioners in terms of cost containment measures, efforts to increase efficiency, and therefore decreased time available to see each patient.

This pressure on time and emphasis on efficiency are occurring at the same time as the complexity is actually demanding more time per patient. Doctors also face increased requirements for documentation of care, regulatory compliance, and other paperwork demands. Preauthorization has long been required in the United States by payers for certain costly procedures and specialty referrals. The Medicare Part D prescription drug benefit introduced in the United States in 2005 includes the need for prior authorization for many medications. Such preapprovals require time-consuming paperwork or telephone calls to obtain approval for care.

Under the stressful conditions of current medical practice, CDS has the potential for keeping physicians from making errors and overlooking actions that need to be done. CDS could also save time by making it possible for payer preapprovals to be done (automatically or semiautomatically, if an EHR is in use, based on a combination of stored and entered data); at least one study has shown this to be welcomed by practitioners (Kerr, Mittman et al. 2000).

2.2.12 Fragmentation and Difficulty Coordinating Care

Another consequence of the complexity of care, the optimal management of which often extends beyond the capabilities of the primary care practitioner, is the reliance on specialists for various aspects of the patient's care. Because of the difficulties in maintaining communication in both directions following a specialist referral, and the usual lack of a common EHR shared between primary care practitioner and specialist, referrals tend to increase the fragmentation of health care.

Among the problems caused by reliance on multiple specialists are poor coordination and communication. This results in lack of availability of needed test results, so tests often are repeated. Information on the patient's medications, concurrent conditions, and the actions of other providers for the same patient are often unknown to the specialist, resulting in inefficiencies and errors. Further, the gatekeeper function that primary care physicians often are required to perform to control referrals is a cause of dissatisfaction by many physicians (Weinstein 2001; Sturm 2002).

CDS can potentially be useful in overcoming the fragmentation problem, by helping to assess when referrals are justified (Hongsermeier 1997). It has been stated that primary care physicians should be coordinators rather than gatekeepers of care (Bodenheimer, Lo et al. 1999); CDS could help to facilitate this transformation by organizing and assembling the information needed for referral based on the clinical indication for it, and at the specialist end, helping to ensure that needed information for the problem to be assessed is assembled and that duplicate test ordering is avoided.

2.2.13 Defensive Medicine

Over the past few decades, the cost of malpractice insurance in the United States has risen dramatically. The reasons for this may be attributed to multiple factors, but among these are increasing recognition of the large numbers of medical errors, the relatively poor systems and safeguards in place to prevent them, the increasing vulnerability of physicians to errors because of the pressures and constraints cited earlier, and the increasing awareness and sophistication of the public regarding optimal health care practice, as a result of which a physician's decisions are more apt to be questioned.

One consequence of the threat of malpractice litigation is a tendency for physicians to over-utilize tests and aggressive evaluations, to avoid even the rarest possibility of missing a diagnosis. Another tendency is to avoid procedures with any degree of risk, even when justified, for fear of a bad outcome. Although these may be useful coping strategies for an individual practitioner, widespread practice of defensive medicine only further drives up the costs of health care, and may lead to worse outcomes for the patient (Kessler and McClellan 2002; Studdert, Mello et al. 2005).

The importance of CDS, particularly guidelines that reflect the accepted standards of care, is that it can both bolster a physician's decision to take a specific course of action as well as provide protection against liability should a bad outcome occur.

2.2.14 Health Care Costs

Financing of health care is a multifaceted topic that we won't attempt to do justice to here, but a key point that should be recognized is that sound clinical judgment needs to be kept separated as much as possible from financial considerations. This is, of course, an impossible goal, because resources are ultimately limited. But unchecked, most financing systems end up having undesirable consequences.

The financing system for health care in the United States has been chaotic, and has seen various attempts over the years to control costs, and to make the delivery of care more rational and accountable. Traditionally, health care was reimbursed on a fee-for-service basis. But because of a lack of constraints on spending in a pure fee-for-service model, and the growth of health care costs that was experienced particularly in the last 40 years of the twentieth century, a number of approaches were introduced over the years to chip away at this model, if not upend it altogether (Rappoport and Jacobs 2004). Some approaches focused on controlling reimbursement, such as reducing the fee schedules paid for physician and hospital services, use of Diagnosis-Related-Groups (DRGs) to determine reimbursement to hospitals for inpatient care for specific classes of diagnoses and severity, utilization review and utilization management to identify overuse of hospital resources, and use of Relative-Value-Unit (RVU)-based schedules to determine reimbursement for procedures such as surgery and radiology interpretation. Other approaches focused on discouraging expenditures by requiring preauthorization before allowing certain expensive tests, surgeries, or other high-cost procedures.

The managed care movement of the 1980s and 1990s was in large part a response to the difficulties in controlling fee-for-service care, where incentives are naturally oriented toward increasing volume (numbers of visits, hospitalization days, and procedures), by moving in the other direction. Managed care introduces incentives for reducing the amount of care provided within a prepaid health care context. Schemes such as relying on the primary care provider as gatekeeper for specialty referrals, and capitation, which provide incentives for reducing referrals or reliance on expensive tests, have been introduced. The public and providers are generally unhappy with managed care, and with capitation in particular, however, seeing it as fostering inferior care because the emphasis is so much in the direction of cutting costs. As a result, the United States is still largely in a fee-for-service environment, albeit with many more constraints such as fee schedules, and requirements such as preapproval to help control costs. Around the world, many other models, including national health coverage, have had varying success, but have often had their own share of problems, either by constraining resources, so that access to services is limited, or by fostering two-tiered systems where the wealthier patients go outside of the national plan.

Controlling costs in either a fee-for-service or a managed care environment requires heavy reliance on computer systems to track actual practices, compile statistics, provide practitioners with feedback about how they compare with their peers, implement feedback about undesirable practices and outcomes, and make recommendations through CDS to encourage appropriate practices.

2.2.15 Pay for Performance

Given the inadequacy of both fee-for-service medicine, with its relatively unchecked expenditures, and of managed care and capitation, with their preoccupation with controlling costs, the latest approaches that are being pursued focus on what is known as "Pay for Performance" (P4P), schemes for rewarding physicians for medical care practices that are considered to be optimal, and penalizing them for suboptimal practices (Beaulieu and Horrigan 2005; Schaeffer and McMurtry 2005). This approach is not without its skeptics, however, who believe that either the measures being checked are too simplistic, or that it won't change behavior as much as reward those whose performance is already high at baseline (Rosenthal, Frank et al. 2005).

It is clear that to achieve a goal of tracking performance and rewarding optimal practices requires extensive use of EHRs. Although simple measures such as tracking the frequency of ordering HbA1c tests in diabetics can be obtained from chart audits, more sophisticated ways of encouraging and measuring improvement of quality require CDS, such as computer-based rules and advice driven by patient-specific information in the EHR. In fact, some policy analysts recommend the use of quality incentives as a way to drive adoption of health care IT (Hackbarth and Milgate 2005).

2.2.16 Demonstrated Benefits

A primary reason for interest in the adoption of CDS has been the successful demonstration of capabilities in selected settings. These results have been reported in journal articles and described in professional meetings, and have come in large measure from academic medical centers. The first three chapters of Section II review the experiences at three such centers.

We have pointed out, though, that the results of those evaluations have been well known for at least ten years, yet many of the good ideas demonstrated in those systems have not yet been successfully replicated to a significant extent beyond the centers in which they have been developed. This may be due in part to the delays inherent to the whole process of technology transfer from research to practice cited earlier (see subsection 2.2.3 and (Balas and Boren 2000)). We explore a number of other possible reasons for this lack of dissemination in Chapter 8.

2.2.17 Top-down Initiatives

The final factor in the resurgence in demand for CDS is the articulation of the need at the highest levels of the health care establishment, government, and other arenas. The many factors we have cited already have contributed to a general sense of need for such capabilities, even desperation, but without a clearly identified path to realize their establishment. The stakeholders have been physicians, payers, professional specialty organizations, medical centers, government, and the ultimate stakeholder—the health care consumer/patient. For real action to occur, each of these participant groups needs to take concerted action, both within its own sphere of influence, and working together. Prodded by the urgency of need, as well as the growth of technology capability for meeting the need, we are now seeing that beginning to occur on a number of fronts, in many parts of the world.

We cited several examples of such activity in the United States in subsection 1.4.2.3. It is unclear how much influence any of these initiatives will ultimately have. The action front is largely political, and political winds shift as priorities and available resources change. A number of legislative bills have been proposed and some enacted, to establish programs for funding, developing, and evaluating IT approaches to health care safety and quality. The President of the United States has included a call for action in his State of the Union message in both 2005 and 2006, with particular reference to EHR adoption. In the United Kingdom, the Connecting for Health project of the National Health Service (NHS 2006) is devoting several billion pounds to a 10-year initiative to use modern computer systems to improve patient care and services, one of the largest initiatives of its kind in the world. A number of other nations are also pursuing modernization programs to varying degrees (FCW 2006).

The U.S. Office of the National Coordinator for Health Information Technology (ONC) has sought to encourage forging of multi-stakeholder partnerships among public agencies, private foundations, public and private health care delivery organizations, professional societies, payers, insurers, IT system vendors, and standards development organizations. Major goals are for universal adoption of standards, interoperability, and baseline functionality. A challenge is to find achievable goals around which incentives can be aligned. Another is to stimulate further efforts by carrying out a series of demonstration projects aimed at showing the value of IT for achieving goals of increased safety, quality, and efficiency in health care.

The ultimate goal of establishing a National Health Information Network (NHIN) is being pursued by funding a number of Regional Health Information Organizations (RHIOs), each of which forms the nidus of activities that are regionally driven, but with the goal of embracing common standards for interoperability (Overhage, Evans et al. 2005). The differing approaches of the various RHIOs are intended to encourage experimentation and innovation based on local needs and motivations. Legislative approval of budgets for ONC and for funding of RHIOs and other health care information technology initiatives has increased in recent years, although not to levels that have been sought by ONC and its advisory groups. This activity is still in its early days, having been initiated in 2004, with a variety of models and early experiences in the many RHIOs that have been formed (Bartschat, Burrington-Brown et al. 2006; Blair 2006).

Intrinsic to the goals for an NHIN are the universal availability of EHRs, widespread adoption of CPOE, and the introduction of various forms of CDS that can foster patient safety, health care quality, and cost-effectiveness. Major foci of RHIO activity are expected to be aimed at these initiatives, and to achieving adoption of standards that foster interoperability among systems for functions of the RHIOs and the NHIN that require interaction and communication.

2.3 CONCLUSION

In this chapter, we have reviewed a wide variety of technical approaches that have been explored for developing CDS over almost a half century. Some of the approaches have been focused on generating and representing knowledge, others more on how to deliver and use it. We then considered the many factors that have contributed to a current sense of urgency for robust, high-quality CDS. The first of those factors is a technology imperative, born of the rich history of research and development that has made CDS possible, and in a sense inevitable—*it is possible, so it will be done*. The remaining factors that have become particularly compelling in the most recent few years constitute a sort of "sociocultural imperative"—*it is needed, so it will be done.*

Given a sociocultural imperative to implement robust, widespread CDS to foster health care safety, quality, and efficacy, yet recognizing the slow progress to date, we next set about to appreciate what will be necessary to accomplish this. In the next chapter, we explore the many uses of CDS and the components that comprise it. Understanding of these features will prepare us to more effectively design both CDS capabilities and health care IT environments so that the two can be harmoniously integrated. At the same time, given the need for separate maintenance and update of CDS, this understanding will help to identify the needed infrastructure to support its life cycle processes effectively.

REFERENCES

Abendroth, T. W. and Greenes, R. A. (1989). Computer presentation of clinical algorithms. *MD Comput* 6(5): 295–299.

Aronson, A. R., Mork, J. G., Gay, C. W., Humphrey, S. M. and Rogers, W. J. (2004). The NLM Indexing Initiative's Medical Text Indexer. *Medinfo* 11(Pt 1): 268–272.

Asch, S. M., Kerr, E. A., Keesey, J., Adams, J. L., Setodji, C. M., Malik, S., et al. (2006). Who is at greatest risk for receiving poor-quality health care? *N Engl J Med* **354**(11): 1147–1156.

Bailey, J., Poulovassilis, A. and Wood, P. (2002). An Event-Condition-Action language for XML. *Proc. WWW2002*, Hawaii.

Balas E. A. and Boren, S. A. (2000). Managing clinical knowledge for health care improvement. *Yearbook of Medical Informatics*. Van Bemmel, J. and McCray, A. T. Stuttgart, Schattauer Verlagsgesellschaft mbH: 65–70.

Barnett, G. O., Cimino, J. J., Hupp, J. A. and Hoffer, E. P. (1987). DXplain. An evolving diagnostic decision-support system. *JAMA* **258**(1): 67–74.

Bartels, P. H. and Hiessl, H. (1989). Expert systems in histopathology. II. Knowledge representation and rule-based systems. *Anal Quant Cytol Histol* **11**(3): 147–153.

Bartschat, W., Burrington-Brown, J., Carey, S., Chen, J., Deming, S., Durkin, S., et al. (2006). Surveying the RHIO landscape. A description of current RHIO models, with a focus on patient identification. *J Ahima* 77(1): 64A–64D.

Bauman, R., Pendergrass, H., Greenes, R. and Kalayan, R. (1972). Further development of an on-line computer system for radiology reporting. *roc Conf on Computer Applications in Radiol*, DHEW Publication no. (FDA)73-8018.

Beaulieu, N. D. and Horrigan, D. R. (2005). Putting smart money to work for quality improvement. *Health Serv Res* **40**(5 Pt 1): 1318–1334.

Beck, J. R. and Pauker, S. G. (1983). The Markov process in medical prognosis. *Med Decis Making* 3(4): 419–458.

Beck, J. R., Pauker, S. G., Gottlieb, J. E., Klein, K. and Kassirer, J. P. (1982). A convenient approximation of life expectancy (the "DEALE"). II. Use in medical decision-making. *Am J Med* 73(6): 889–897.

Begg, C. B. and Greenes, R. A. (1983). Assessment of diagnostic tests when disease verification is subject to selection bias. *Biometrics* **39**(1): 207–215.

Bell, D. S. and Friedman, M. A. (2005). E-prescribing and the medicare modernization act of 2003. Paving the on-ramp to fully integrated health information technology? *Health Aff (Millwood)* **24**(5): 1159–1169.

Bell, D. S. and Greenes, R. A. (1994). Evaluation of UltraSTAR: performance of a collaborative structured data entry system. *Proc Annu Symp Comput Appl Med Care*: 216–222.

Bell, D. S., Greenes, R. A. and Doubilet, P. (1992). Form-based clinical input from a structured vocabulary: initial application in ultrasound reporting. *Proc Annu Symp Comput Appl Med Care*: 789–790.

Berners-Lee, T. and Connolly, D. (1993). Hypertext Markup Language (HTML): A Repre sentation of Textual Information and MetaInformation for Retrieval and Interchange, from http://www.w3.org/MarkUp/draft-ietf-iiir-html-01.txt.

Blair, R. (2006). RHIO nation. *Health Manag Technol* **27**(2): 56–62.

Bleich, H. L. (1969). Computer evaluation of acid-base disorders. *J Clin Invest* **48**(9): 1689–1696.

Bleich, H. L. (1971). The computer as a consultant. *N Engl J Med* **284**(3): 141–147.

Bodenheimer, T. (2003). Interventions to improve chronic illness care: evaluating their effectiveness. *Dis Manag* **6**(2): 63–71.

Bodenheimer, T., Lo, B. and Casalino, L. (1999). Primary care physicians should be coordinators, not gatekeepers. *JAMA* **281**(21): 2045–2049.

Boxwala, A. A., Peleg, M., Tu, S., Ogunyemi, O., Zeng, Q. T., Wang, D., et al. (2004). GLIF3: a representation format for sharable computer-interpretable clinical practice guidelines. *J Biomed Inform* **37**(3): 147–161.

Bruce, B. and Fries, J. F. (2005). The Arthritis, Rheumatism and Aging Medical Information System (ARAMIS): still young at 30 years. *Clin Exp Rheumatol* **23**(5 Suppl 39): S163–167.

Cahill, J. (1998). Patient participation—a review of the literature. *J Clin Nurs* 7(2): 119–128.

Chakravarthy, S. (1995). Early active database efforts: a capsule summary. *IEEE Transactions on Knowledge and Data Engineering* 7(6): 1008–1010.

Chaussabel, D. (2004). Biomedical literature mining: challenges and solutions in the 'omics' era. *Am J Pharmacogenomics* **4**(6): 383–393.

Chertow, G. M., Lee, J., Kuperman, G. J., Burdick, E., Horsky, J., Seger, D. L., et al. (2001). Guided medication dosing for inpatients with renal insufficiency. *JAMA* **286**(22): 2839–2844.

Ciccarese, P., Caffi, E., Quaglini, S. and Stefanelli, M. (2005). Architectures and tools for innovative Health Information Systems: the Guide Project. *Int J Med Inform* **74**(7–8): 553–562.

Cimino, J. J., Li, J., Bakken, S. and Patel, V. L. (2002). Theoretical, empirical and practical approaches to resolving the unmet information needs of clinical information system users. *Proc AMIA Symp*: 170–174.

Clancey, W. (1987). *Knowledge-Based Tutoring: The GUIDON Program. Series in Artificial Intelligence*. Cambridge, MA, MIT Press.

Col, N. F., Eckman, M. H., Karas, R. H., Pauker, S. G., Goldberg, R. J., Ross, E. M., et al. (1997). Patient-specific decisions about hormone replacement therapy in postmenopausal women. *JAMA* **277**(14): 1140–1147.

Coletti, M. H. and Bleich, H. L. (2001). Medical subject headings used to search the biomedical literature. *J Am Med Inform Assoc* **8**(4): 317–323.

Cooper, G. F. (1986). A diagnostic method that uses causal knowledge and linear programming in the application of Bayes' formula. *Comput Methods Programs Biomed* **22**(2): 223–237.

Coulter, A. (1997). Partnerships with patients: the pros and cons of shared clinical decision-making. *J Health Serv Res Policy* **2**(2): 112–121.

Dannenberg, A. L., Shapiro, A. R. and Fries, J. F. (1979). Enhancement of clinical predictive ability by computer consultation. *Methods Inf Med* **18**(1): 10–14.

Davis, R. and Lenat, D. (1982). *Knowledge-Based Systems in Artificial Intelligence: AM and TEIRESIAS*. New York, McGraw-Hill.

de Bruijn, B. and Martin, J. (2002). Getting to the (c)ore of knowledge: mining biomedical literature. *Int J Med Inform* **67**(1–3): 7–18.

DeAngelis, C. D., Drazen, J. M., Frizelle, F. A., Haug, C., Hoey, J., Horton, R., et al. (2004). Clinical trial registration: a statement from the International Committee of Medical Journal Editors. *JAMA* **292**(11): 1363–1364.

Deber, R. B., Kraetschmer, N. and Irvine, J. (1996). What role do patients wish to play in treatment decision making? *Arch Intern Med* **156**(13): 1414–1420.

Dorr, D. A., Wilcox, A., Burns, L., Brunker, C. P., Narus, S. P. and Clayton, P. D. (2006). Implementing a multidisease chronic care model in primary care using people and technology. *Dis Manag* **9**(1): 1–15.

Elstein, A. S., Shulman, L. S. and Sprafka, S. A. (1978). *Medical Problem Solving: An Analysis of Clinical Reasoning*. Cambridge, MA, Harvard University Press.

Eraker, S. A. and Politser, P. (1982). How decisions are reached: physician and patient. *Ann Intern Med* **97**(2): 262–268.

Ermolaeva, O., Rastogi, M., Pruitt, K. D., Schuler, G. D., Bittner, M. L., Chen, Y., et al. (1998). Data management and analysis for gene expression arrays. *Nat Genet* **20**(1): 19–23.

Fagan, L., Shortliffe, E. and Buchanan, B. (1979). Computer-based medical decision making: from MYCIN to VM. *Stanford heuristic programming project*. Report. Stanford, CA, Stanford University.

FCW. (2006). A guide to public policy and its applications in health IT: International Sites. *Government Health IT*; FCW Media Group, from http://www.govhealthit.com/resources/international.asp

Feinstein, A. (1967). *Clinical Judgment*. Baltimore, Williams & Wilkins Co.

Feinstein, A. R. (1994). "Clinical Judgment" revisited: the distraction of quantitative models. *Ann Intern Med* **120**(9): 799–805.

Fortin, J. M., Goldberg, R. J., Kaplan, S., Chuo, J., O'Connor, A. M. and Col, N. F. (2003). Impact of a personalized decision support aid on menopausal women—results from a randomized controlled trial. *AMIA Annu Symp Proc*: 843.

Fortin, J. M., Hirota, L. K., Bond, B. E., O'Connor, A. M. and Col, N. F. (2001). Identifying patient preferences for communicating risk estimates: a descriptive pilot study. *BMC Med Inform Decis Mak* **1**: 2.

Fox, J., Johns, N., Lyons, C., Rahmanzadeh, A., Thomson, R. and Wilson, P. (1997). PROforma: a general technology for clinical decision support systems. *Comput Methods Programs Biomed* **54**(1–2): 59–67.

Friedman, T. L. (2005). *The World Is Flat: A Brief History of the Twenty-first Century*. New York, Farrar, Straus, and Giroux.

Fries, J. F. (1984). The chronic disease data bank: first principles to future directions. *J Med Philos* **9**(2): 161–180.

Gans, D., Kralewski, J., Hammons, T. and Dowd, B. (2005). Medical groups' adoption of electronic health records and information systems. Practices are encountering greater-than-expected barriers to adopting an EHR system, but the adoption rate continues to rise. *Health Aff (Millwood)* **24**(5): 1323–1333.

Gensler, H. (2002). *Introduction to Logic*. New York, Routledge.

Glasspool, D. W., Fox, J., Coulson, A. S. and Emery, J. (2001). Risk assessment in genetics: a semi-quantitative approach. *Medinfo* **10**(Pt 1): 459–463.

Goldstein, M. K., Coleman, R. W., Tu, S. W., Shankar, R. D., O'Connor, M. J., Musen, M. A., et al. (2004). Translating research into practice: organizational issues in implementing automated decision support for hypertension in three medical centers. *J Am Med Inform Assoc* **11**(5): 368–376.

Gorry, G. A. and Barnett, G. O. (1968). Experience with a model of sequential diagnosis. *Comput Biomed Res* **1**(5): 490–507.

Gottlieb, L. K., Margolis, C. Z. and Schoenbaum, S. C. (1990). Clinical practice guidelines at an HMO: development and implementation in a quality improvement model. *QRB Qual Rev Bull* **16**(2): 80–86.

Greenes, R. A., Barnett, G. O., Klein, S. W., Robbins, A. and Prior, R. E. (1970). Recording, retrieval and review of medical data by physician-computer interaction. *N Engl J Med* **282**(6): 307–315.

Greenes, R. A. and Begg, C. B. (1985). Assessment of diagnostic technologies. Methodology for unbiased estimation from samples of selectively verified patients. *Invest Radiol* **20**(7): 751–756.

Greenes, R. A., Tu, S., Boxwala, A., Peleg, M. and Shortliffe, E. H. (2002). Toward a shared representation of clinical trial protocols: application of the GLIF guideline modeling framework. *Cancer Informatics: Essential Technologies for Clinical Trials.* Silva, J., Ball, M., Chute, C. G. et al. New York, Springer-Verlag, New York.

Grossman, J. H., Barnett, G. O., McGuire, M. T. and Swedlow, D. B. (1971). Evaluation of computer-acquired patient histories. *JAMA* **215**(8): 1286–1291.

Guo, D., Lincoln, M. J., Haug, P. J., Turner, C. W. and Warner, H. R. (1991). Exploring a new best information algorithm for Iliad. *Proc Annu Symp Comput Appl Med Care*: 624–628.

Guyatt, G. H., Sackett, D. L. and Cook, D. J. (1993). Users' guides to the medical literature. II. How to use an article about therapy or prevention. Are the results of the study valid? Evidence-Based Medicine Working Group. *JAMA* **270**(21): 2598–2601.

Hackbarth, G. and Milgate, K. (2005). Using quality incentives to drive physician adoption of health information technology. Medicare should help physicians overcome the barriers to adoption but should not simply pay the bill. *Health Aff (Millwood)* **24**(5): 1147–1149.

Hanley, J. A. (1989). Receiver operating characteristic (ROC) methodology: the state of the art. *Crit Rev Diagn Imaging* **29**(3): 307–335.

Haug, P. J., Gardner, R. M., Tate, K. E., Evans, R. S., East, T. D., Kuperman, G., et al. (1994). Decision support in medicine: examples from the HELP system. *Comput Biomed Res* **27**(5): 396–418.

Haux, R. (2006). Health information systems—past, present, future. *Int J Med Inform* **75**(3–4): 268–281.

Hersh, W. R. and Greenes, R. A. (1990). SAPHIRE—an information retrieval system featuring concept matching, automatic indexing, probabilistic retrieval, and hierarchical relationships. *Comput Biomed Res* **23**(5): 410–425.

Hersh, W. R. and Hickam, D. H. (1993). A comparison of two methods for indexing and retrieval from a full-text medical database. *Med Decis Making* **13**(3): 220–226.

Herxheimer, A. (1993). The Cochrane Collaboration: making the results of controlled trials properly accessible. *Postgrad Med J* **69**(817): 867–868.

Hickam, D. H., Shortliffe, E. H., Bischoff, M. B., Scott, A. C. and Jacobs, C. D. (1985). The treatment advice of a computer-based cancer chemotherapy protocol advisor. *Ann Intern Med* **103**(6 (Pt 1)): 928–936.

Hill, T. and Lewicki, P. (2006). *Statistics: Methods and Applications.* Tulsa, OK, Statsoft, Inc.

Hongsermeier, T. (1997). Managing resources. Directing patients to appropriate specialist care. *Med Group Manage J* **44**(4): 28–30, 32, 75.

Hripcsak, G. (1991). Arden Syntax for Medical Logic Modules. *MD Comput* **8**(2): 76, 78.

Hripcsak, G., Ludemann, P., Pryor, T. A., Wigertz, O. B. and Clayton, P. D. (1994). Rationale for the Arden Syntax. *Comput Biomed Res* **27**(4): 291–324.

Huang, J., Fox, J., Gordon, C. and Jackson-Smale, A. (1993). Symbolic decision support in medical care. *Artif Intell Med* **5**(5): 415–430.

Huertas-Portocarrero, D., Ruiz, P. P. and Marmol, J. P. (1988). Concurrent clinical review: using microcomputer-based DRG-software. *Health Policy* **9**(2): 211–217.

IOM. (2001). *Crossing the quality chasm: a new health system for the 21st century. Institute of Medicine.* Washington D.C., National Academy Press.

ISO. (1986). Information processing—Text and office systems—Standard Generalized Markup Language (SGML). *International Standards Organization*, from http://www.iso.org/iso/en/CatalogueDetailPage.CatalogueDetail?CSNUMBER=16387&ICS1=35&ICS2=240&ICS3=30

Jelliffe, R. W., Schumitzky, A., Bayard, D., Milman, M., Van Guilder, M., Wang, X., et al. (1998). Model-based, goal-oriented, individualised drug therapy. Linkage of population modelling, new ‘multiple model’ dosage design, bayesian feedback and individualised target goals. *Clin Pharmacokinet* **34**(1): 57–77.

Jenders, R. A., Huang, H., Hripcsak, G. and Clayton, P. D. (1998). Evolution of a knowledge base for a clinical decision support system encoded in the Arden Syntax. *Proc AMIA Symp*: 558–562.

Johnson, P. D., Tu, S., Booth, N., Sugden, B. and Purves, I. N. (2000). Using scenarios in chronic disease management guidelines for primary care. *Proc AMIA Symp*: 389–393.

Joubert, M., Peretti, A. L., Gouvernet, J. and Fieschi, M. (2005). Refinement of an automatic method for indexing medical literature—a preliminary study. *Stud Health Technol Inform* **116**: 683–688.

Kahn, C. E., Jr. (1997). A generalized language for platform-independent structured reporting. *Methods Inf Med* **36**(3): 163–171.

Kahneman, D. and Tversky, A. (1982). Judgment of and by representativeness. *Judgment under Uncertainty: Heuristics and Biases*. Kahneman, D., Slovic, P. and Tversky, A. (eds.). New York, Cambridge University Press.

Kassirer, J. P. (1976). The principles of clinical decision making: an introduction to decision analysis. *Yale J Biol Med* **49**(2): 149–164.

Kerr, E. A., Mittman, B. S., Hays, R. D., Zemencuk, J. K., Pitts, J. and Brook, R. H. (2000). Associations between primary care physician satisfaction and self-reported aspects of utilization management. *Health Serv Res* **35**(1 Pt 2): 333–349.

Kessler, D. P. and McClellan, M. B. (2002). How liability law affects medical productivity. *J Health Econ* **21**(6): 931–955.

Kimmel, Z., Greenes, R. A. and Liederman, E. (2005). Personal health records. *J Med Pract Manage* **21**(3): 147–152.

Kingsland, L. C., 3rd, Lindberg, D. A. and Sharp, G. C. (1983). AI/RHEUM. A consultant system for rheumatology. *J Med Syst* 7(3): 221–227.

Kirby, J. B., Taliaferro, G. and Zuvekas, S. H. (2006). Explaining racial and ethnic disparities in health care. *Med Care* **44**(5 Suppl): I64–172.

Kohn, L., Corrigan, J., Donaldson, M. e. and Committee on Quality of Health Care in America, I. O. M. (1999). *To Err Is Human: Building a Safer Health System*. Washington, D.C., National Academic Press.

Komaroff, A. L., Black, W. L., Flatley, M., Knopp, R. H., Reiffen, B. and Sherman, H. (1974). Protocols for physician assistants. Management of diabetes and hypertension. *N Engl J Med* **290**(6): 307–312.

Kotzin, S. (2005). Journal Selection for Medline. *World Library and Information Congress: 71th IFLA General Conference and Council*, Oslo, Norway.

Kuipers, B. and Kassirer, J. (1984). Causal reasoning in medicine: analysis of a protocol. *Cognitive Science* **8**: 363–385.

Kulikowski, C. A. and Weiss, S. M. (1982). Representation of expert knowledge for consultation: the CASNET and EXPERT projects. *Artificial Intelligence in Medicine*. Szolovits, P. Boulder, Westview Press: 21–56.

Kuperman, G. J., Teich, J. M., Bates, D. W., Hiltz, F. L., Hurley, J. M., Lee, R. Y., et al. (1996). Detecting alerts, notifying the physician, and offering action items: a comprehensive alerting system. *Proc AMIA Annu Fall Symp*: 704–708.

Larsen, G. Y., Parker, H. B., Cash, J., O'Connell, M. and Grant, M. C. (2005). Standard drug concentrations and smart-pump technology reduce continuous-medication-infusion errors in pediatric patients. *Pediatrics* **116**(1): e21–25.

Ledley, R. (1965). *Use of Computers in Biology and Medicine*. New York, McGraw-Hill.

Ledley, R. S. and Lusted, L. B. (1959). Reasoning foundations of medical diagnosis; symbolic logic, probability, and value theory aid our understanding of how physicians reason. *Science* **130**(3366): 9–21.

Lee, T. H., Juarez, G., Cook, E. F., Weisberg, M. C., Rouan, G. W., Brand, D. A., et al. (1991). Ruling out acute myocardial infarction. A prospective multicenter validation of a 12-hour strategy for patients at low risk. *N Engl J Med* **324**(18): 1239–1246.

Lindberg, D. A., Humphreys, B. L. and McCray, A. T. (1993). The Unified Medical Language System. *Methods Inf Med* **32**(4): 281–291.

Lobach, D. F., Gadd, C. S. and Hales, J. W. (1997). Structuring clinical practice guidelines in a relational database model for decision support on the Internet. *Proc AMIA Annu Fall Symp*: 158–162.

Luger, G. (2005). *Artificial Intelligence: Structures and Strategies for Complex Problem Solving*. Addison-Wesley.

Margolis, C. Z. (1983). Uses of clinical algorithms. *JAMA* **249**(5): 627–632.

McCray, A. T. (2000). Better access to information about clinical trials. *Ann Intern Med* **133**(8): 609–614.

McCray, A. T. and Nelson, S. J. (1995). The representation of meaning in the UMLS. *Methods Inf Med* **34**(1–2): 193–201.

McCulloch, W. and Pitts, W. (1943). A logical calculus of the ideas imminent in nervous activity. *Bulletin of Mathematical Biophysics* **5**: 115–133.

McDonald, C. J. (1976). Protocol-based computer reminders, the quality of care and the non-perfectability of man. *N Engl J Med* **295**(24): 1351–1355.

McNeil, B. J. and Hanley, J. A. (1984). Statistical approaches to the analysis of receiver operating characteristic (ROC) curves. *Med Decis Making* **4**(2): 137–150.

Metz, C. E. (1978). Basic principles of ROC analysis. *Semin Nucl Med* **8**(4): 283–298.

Middleton, B. (2005). Achieving U.S. Health information technology adoption: the need for a third hand. Government intervention, judiciously and gently applied, can give the extra assistance needed to boost HIT adoption nationwide. *Health Aff (Millwood)* **24**(5): 1269–1272.

Middleton, B., Hammond, W. E., Brennan, P. F. and Cooper, G. F. (2005). Accelerating U.S. EHR adoption: how to get there from here: recommendations based on the 2004 ACMI retreat. *J Am Med Inform Assoc* **12**(1): 13–19.

Miller, P. L. (1983). Critiquing anesthetic management: the "ATTENDING" computer system. *Anesthesiology* **58**(4): 362–369.

Miller, P. L. and Black, H. R. (1984). Medical plan-analysis by computer: critiquing the pharmacologic management of essential hypertension. *Comput Biomed Res* **17**(1): 38–54.

Miller, R., Masarie, F. E. and Myers, J. D. (1986). Quick medical reference (QMR) for diagnostic assistance. *MD Comput* **3**(5): 34–48.

Miller, R. A., Pople, H. E., Jr. and Myers, J. D. (1982). Internist-1, an experimental computer-based diagnostic consultant for general internal medicine. *N Engl J Med* **307**(8): 468–476.

Minsky, M. (1975). A Framework for Representing Knowledge. *The Psychology of Computer Vision*. Winston, P. New York, McGraw-Hill: 211–277.

Minsky, M. and Papert, S. (1969). Perceptrons. Report. Cambridge, MA, MIT Press.

Miyamoto, J. M. and Eraker, S. A. (1985). Parameter estimates for a QALY utility model. *Med Decis Making* **5**(2): 191–213.

Musen, M. A. (1992). Dimensions of knowledge sharing and reuse. *Comput Biomed Res* **25**(5): 435–467.

NHS. (2006). National Programme for IT in the NHS. *Connecting for Health, National HealthService*, from http://www.connectingforhealth.nhs.uk/.

Noy, N. F., Crubezy, M., Fergerson, R. W., Knublauch, H., Tu, S. W., Vendetti, J., et al. (2003). Protege-2000: an open-source ontology-development and knowledge-acquisition environment. *AMIA Annu Symp Proc*: 953.

Ogunyemi, O., Chlebowski, R., Matloff, E., Schnabel, F., Orr, R. and Col, N. (2004). Creating Bayesian network models for breast cancer risk prediction. *Cancer Risk Prediction Models: A Workshop on Development, Evaluation, and Application*, Washington, DC.

Ohno-Machado, L., Gennari, J. H., Murphy, S. N., Jain, N. L., Tu, S. W., Oliver, D. E., et al. (1998). The guideline interchange format: a model for representing guidelines. *J Am Med Inform Assoc* **5**(4): 357–372.

Overhage, J. M., Evans, L. and Marchibroda, J. (2005). Communities' readiness for health information exchange: the National Landscape in 2004. *J Am Med Inform Assoc* **12**(2): 107–112.

Oxman, A. D., Sackett, D. L. and Guyatt, G. H. (1993). Users' guides to the medical literature. I. How to get started. The Evidence-Based Medicine Working Group. *JAMA* **270**(17): 2093–2095.

Patil, R. S. (1987). Causal reasoning in computer programs for medical diagnosis. *Comput Methods Programs Biomed* **25**(2): 117–123.

Pauker, S. G. (1976). Coronary artery surgery: the use of decision analysis. *Ann Intern Med* **85**(1): 8–18.

Pauker, S. G., Gorry, G. A., Kassirer, J. P. and Schwartz, W. B. (1976). Towards the simulation of clinical cognition. Taking a present illness by computer. *Am J Med* **60**(7): 981–996.

Pauker, S. G. and Kassirer, J. P. (1978). Clinical application of decision analysis: a detailed illustration. *Semin Nucl Med* **8**(4): 324–335.

Pauker, S. G. and Kassirer, J. P. (1980). The threshold approach to clinical decision making. *N Engl J Med* **302**(20): 1109–1117.

Pauker, S. G. and Kassirer, J. P. (1987). Decision analysis. *N Engl J Med* **316**(5): 250–258.
Pauker, S. G. and McNeil, B. J. (1981). Impact of patient preferences on the selection of therapy. *J Chronic Dis* **34**(2–3): 77–86.
Pearl, J. (1988). *Probabilistic Reasoning in Intelligent Systems: Networks of Plausible Inference.* San Francisco, Morgan Kaufmann.
Peleg, M., Tu, S., Bury, J., Ciccarese, P., Fox, J., Greenes, R. A., et al. (2003). Comparing computer-interpretable guideline models: a case-study approach. *J Am Med Inform Assoc* **10**(1): 52–68.
Plante, D. A., Kassirer, J. P., Zarin, D. A. and Pauker, S. G. (1986). Clinical decision consultation service. *Am J Med* **80**(6): 1169–1176.
Poon, A. D. and Fagan, L. M. (1994). PEN-Ivory: the design and evaluation of a pen-based computer system for structured data entry. *Proc Annu Symp Comput Appl Med Care*: 447–451.
Raiffa, H. (1997). *Decision Analysis: Introductory Readings on Choices Under Uncertainty.* New York, McGraw Hill.
Ramnarayan, P., Tomlinson, A., Rao, A., Coren, M., Winrow, A. and Britto, J. (2003). ISABEL: a web-based differential diagnostic aid for paediatrics: results from an initial performance evaluation. *Arch Dis Child* **88**(5): 408–413.
Ramoni, M. and Riva, A. (1997). Basic Science in Medical Reasoning: An Artificial Intelligence Approach. *Adv Health Sci Educ Theory Pract* **2**(2): 131–140.
Rappoport, J. and Jacobs, P. (2004). *The Economics of Health and Medical Care.* Sudbury, MA, Jones and Bartlett Publishers.
Ray, G. T., Collin, F., Lieu, T., Fireman, B., Colby, C. J., Quesenberry, C. P., et al. (2000). The cost of health conditions in a health maintenance organization. *Med Care Res Rev* **57**(1): 92–109.
Rosati, R. A., McNeer, J. F., Starmer, C. F., Mittler, B. S., Morris, J. J., Jr. and Wallace, A. G. (1975). A new information system for medical practice. *Arch Intern Med* **135**(8): 1017–1024.
Rosenblatt, F. (1958). The perceptron: A probabilistic model for information storage and organization in the brain. *Psychological Review* **65**(6): 386–408.
Rosenthal, M. B., Frank, R. G., Li, Z. and Epstein, A. M. (2005). Early experience with pay-for-performance: from concept to practice. *JAMA* **294**(14): 1788–1793.
Rudowski, R., Frostell, C. and Gill, H. (1989). A knowledge-based support system for mechanical ventilation of the lungs. The KUSIVAR concept and prototype. *Comput Methods Programs Biomed* **30**(1): 59–70.
Ruland, C. M. (2004). Improving patient safety through informatics tools for shared decision making and risk communication. *Int J Med Inform* **73**(7–8): 551–557.
Rutledge, G., Thomsen, G., Farr, B., Tovar, M., Sheiner, L. and Fagan, L. (1991). VentPlan: a ventilator-management advisor. *Proc Annu Symp Comput Appl Med Care*: 869–871.
Schaeffer, L. D. and McMurtry, D. E. (2005). Variation in Medical Care: Time for Action. *Health Aff (Millwood).*
Schultz, J. (1986). A history of the Promis technology: an effective human interface. *Proceedings of the ACM Conference on the history of personal workstations*, Palo Alto, CA, ACM Press.
Schwartz, W. B., Gorry, G. A., Kassirer, J. P. and Essig, A. (1973). Decision analysis and clinical judgment. *Am J Med* **55**(3): 459–472.
Seroussi, B., Bouaud, J., Chatellier, G. and Venot, A. (2004). Development of computerized guidelines for the management of chronic diseases allowing to position any patient within recommended therapeutic strategies. *Medinfo* **11**(Pt 1): 154–158.
Shahar, Y., Miksch, S. and Johnson, P. (1996). An intention-based language for representing clinical guidelines. *Proc AMIA Annu Fall Symp*: 592–596.
Sherman, H. and Komaroff, A. (1976). Ambulatory care protocols as management tools. *Health Care Manage Rev* **1**(3): 47–52.
Shiffman, R. N. (1994). Towards effective implementation of a pediatric asthma guideline: integration of decision support and clinical workflow support. *Proc Annu Symp Comput Appl Med Care*: 797–801.
Shiffman, R. N. and Greenes, R. A. (1991). Use of augmented decision tables to convert probabilistic data into clinical algorithms for the diagnosis of appendicitis. *Proc Annu Symp Comput Appl Med Care*: 686–690.
Shiffman, R. N. and Greenes, R. A. (1992). Rule set reduction using augmented decision table and semantic subsumption techniques: application to cholesterol guidelines. *Proc Annu Symp Comput Appl Med Care*: 339–343.

Shiffman, R. N., Karras, B. T., Agrawal, A., Chen, R., Marenco, L. and Nath, S. (2000). GEM: a proposal for a more comprehensive guideline document model using XML. *J Am Med Inform Assoc* 7(5): 488–498.

Shiffman, R. N., Leape, L. L. and Greenes, R. A. (1993). Translation of appropriateness criteria into practice guidelines: application of decision table techniques to the RAND criteria for coronary artery bypass graft. *Proc Annu Symp Comput Appl Med Care*: 248–252.

Shortliffe, E. (1976). *Computer-based Medical Consultations: MYCIN*. New York, Elsevier.

Shortliffe, E. H., Davis, R., Axline, S. G., Buchanan, B. G., Green, C. C. and Cohen, S. N. (1975). Computer-based consultations in clinical therapeutics: explanation and rule acquisition capabilities of the MYCIN system. *Comput Biomed Res* 8(4): 303–320.

Slack, W. V., Hicks, G. P., Reed, C. E. and Van Cura, L. J. (1966). A computer-based medical-history system. *N Engl J Med* **274**(4): 194–198.

Smedley, B. D., Stith, A. Y. and Nelson, A. R., eds. (2003). *Unequal Treatment: Confronting Racial and Ethnic Disparities in Health Care. Institute of Medicine*. Washington, DC, National Academies Press.

Smith, A. (1987). Qualms about QALYs. *Lancet* **1**(8542): 1134–1136.

Smith, C. (2005). An evolution of experts: MEDLINE in the library school. *J Med Libr Assoc* **93**(1): 53–60.

Starmer, C. F., Rosati, R. A. and McNeer, J. F. (1974). A comparison of frequency distributions for use in a model for selecting treatment in coronary artery disease. *Comput Biomed Res* 7(3): 278–293.

Studdert, D. M., Mello, M. M., Sage, W. M., DesRoches, C. M., Peugh, J., Zapert, K., et al. (2005). Defensive medicine among high-risk specialist physicians in a volatile malpractice environment. *JAMA* **293**(21): 2609–2617.

Sturm, R. (2002). Effect of managed care and financing on practice constraints and career satisfaction in primary care. *J Am Board Fam Pract* **15**(5): 367–377.

Swartout, W. R. (1977). A Digitalis therapy advisor with explanations. Report. Cambridge, MA, MIT Lab. for Computer Science: Technical Report TR-176.

Swedlow, D. B., Barnett, G. O., Grossman, J. H. and Souder, D. E. (1972). A simple programming system ("driver") for the creation and execution of an automated medical history. *Comput Biomed Res* 5(1): 90–98.

Swets, J. A. (1979). ROC analysis applied to the evaluation of medical imaging techniques. *Invest Radiol* **14**(2): 109–121.

Tan, J. K., McCormick, E. and Sheps, S. B. (1993). Utilization care plans and effective patient data management. *Hosp Health Serv Adm* **38**(1): 81–99.

Tang, P. C., Ash, J. S., Bates, D. W., Overhage, J. M. and Sands, D. Z. (2006). Personal health records: definitions, benefits, and strategies for overcoming barriers to adoption. *J Am Med Inform Assoc* **13**(2): 121–126.

Tu, S. W. and Musen, M. A. (2001). Modeling data and knowledge in the EON guideline architecture. *Medinfo* **10**(Pt 1): 280–284.

Tu, S. W., Musen, M. A., Shankar, R., Campbell, J., Hrabak, K., McClay, J., et al. (2004). Modeling guidelines for integration into clinical workflow. *Medinfo* **11**(Pt 1): 174–178.

Uckun, S. (1994). Intelligent systems in patient monitoring and therapy management. A survey of research projects. *Int J Clin Monit Comput* **11**(4): 241–253.

van Melle, W., Shortliffe, E. and Buchanan, B. (1984). EMYCIN: A Knowledge engineer's Tool for Constructing Rule-Based Expert Systems. *Rule-Based Expert Systems: The MYCIN Experiments of the Stanford Heuristic Programming Project*. Buchanan, B. and Shortliffe, E. Menlo Park, CA, Addison-Wesley.

van Mulligen, E. M., Stam, H. and van Ginneken, A. M. (1998). Clinical data entry. *Proc AMIA Symp*: 81–85.

Vidyasagar, D. (2006). Global notes: the 10/90 gap disparities in global health research. *J Perinatol* **26**(1): 55–56.

W3C. (1997). W3C Issues XML1.0 as a Proposed Recommendation. Press Release, World Wide Web Consortium, from http://www.w3.org/Press/XML-PR.

W3C. (2004). Semantic Web. World Wide Web Consortium, from http://www.w3.org/2001/sw/.

W3C. (2006a). Extensible Markup Language (XML). World Wide Web Consortium, from http://www.w3.org/XML/.

W3C. (2006b). Web Ontology Language (OWL). World Wide Web Consortium, from http://www.w3.org/2004/OWL/.

Walton, R., Dovey, S., Harvey, E. and Freemantle, N. (1999). Computer support for determining drug dose: systematic review and meta-analysis. *BMJ* **318**(7189): 984–990.

Walton, R. T., Harvey, E., Dovey, S. and Freemantle, N. (2001). Computerised advice on drug dosage to improve prescribing practice. *Cochrane Database Syst Rev*. Report 1. CD002894.

Warner, H. R., Jr. (1989). Iliad: moving medical decision-making into new frontiers. *Methods Inf Med* **28**(4): 370–372.

Wasson, J. H., Sox, H. C., Neff, R. K. and Goldman, L. (1985). Clinical prediction rules. Applications and methodological standards. *N Engl J Med* **313**(13): 793–799.

Weed, L. L. (1968). Medical records that guide and teach. *N Engl J Med* **278**(11): 593–600.

Weinstein, M. C. (1980). Cost-effectiveness analysis for clinical procedures in oncology. *Bull Cancer* **67**(5): 491–500.

Weinstein, M. C. (1988). A QALY is a QALY—or is it? *J Health Econ* **7**(3): 289–290.

Weinstein, M. C. (1989). Methodologic issues in policy modeling for cardiovascular disease. *J Am Coll Cardiol* **14**(3 Suppl A): 38A–43A.

Weinstein, M. C. (2001). Should physicians be gatekeepers of medical resources? *J Med Ethics* **27**(4): 268–274.

Werbos, P. (1974). Beyond regression: New tools for prediction and analysis in the behavioural science. *Committee on Appl. Math*. Report. Cambridge, MA, Harvard Univ.

Yang, Y. (1996). An evaluation of statistical approaches to MEDLINE indexing. *Proc AMIA Annu Fall Symp*: 358–362.

Zadeh, L. (1965). Fuzzy Sets. *Information and Control* **8**: 338–353.

3 FEATURES OF COMPUTER-BASED CLINICAL DECISION SUPPORT

ROBERT A. GREENES

The purpose of this chapter is to dig deeper into the nature of computer-based clinical decision support, in terms of the ways in which it is or potentially could be used, its design, and its interaction with host environments. I believe that understanding these aspects of CDS is important as a foundation for serious efforts to increase its dissemination and adoption. As pointed out in previous chapters, and as will be explored in more detail in Section II, much of the success with CDS has been with on-off implementations that have been difficult to maintain over time even within their own institutions, more problematic when extended for use throughout a health care enterprise, and only rarely replicated elsewhere despite demonstrated effectiveness. If we are to tackle this issue and break down barriers to dissemination and adoption, we need to know what we are working with, which aspects are most troublesome and how they can be improved, and which components or interfaces can be standardized or made easier to deploy. We also need to understand the human factors and process and workflow implications of CDS use, so that we can determine optimal approaches to invocation and user interface design.

An underlying thesis is that CDS has a *conceptual architecture* comprised of a number of design elements or components. Many kinds of CDS are designed without the architecture being explicit, but I will try to demonstrate that all the design elements nonetheless are present, even if implicit. A second thesis is that CDS does not function in isolation, but rather that it *operates in* the context of some sort of *application environment*. However, in the discussion to follow, I will consider the chunk of software that provides CDS as a "module" of software that interacts with the application environment, recognizing full well that often CDS implementations are not modular at all, and that the dividing line between what constitutes CDS and what constitutes the application that invokes it is not clearly defined, so that the two cannot be cleanly separated. Even so, we can think of the functions that are performed as being done by either the application environment or by the CDS software, so that responsibilities for these functions can be assigned to one or the other.

These two conceptual idealizations—1) an underlying component architecture for CDS, and 2) modularity of CDS in that its tasks and responsibilities can be separated from those of the application environment with which it interacts—are helpful to better understand how to design CDS in a portable,

Clinical Decision Support: The Road Ahead

reusable, maintainable fashion. I believe that, in adopting this approach, it will also be possible to design more robust versions of existing CDS that have the ability to interoperate in other platforms, be adapted to differing application settings, be maintained and updated more readily, and thus be more widely disseminated and used.

So this chapter is about design principles. But as you will quickly conclude, this topic is a work in progress. There are many complexities, nuances, and unresolved questions yet to be answered by researchers and developers in this field. My goal is to call attention to these design principles and challenges as a framework for further work, in the belief that this will help to accelerate our progress in dissemination and adoption.

3.1 CDS AND THE HUMAN

A central characteristic of CDS is that it is intended to interact with and give advice to a human being. Our focus is primarily on clinical decision support to health care providers, generally physicians and nurses, but also sometimes pharmacists, technologists, and other personnel. Many of the same principles apply to patient-centered decision support, although our focus is not on that in this book, with its added complexities of health care literacy, language, and mental models. Nonetheless, sometime CDS will involve processes for shared decision-making between a health care provider and a patient. And, of course, the purpose is always the improvement of health and health care for the individual patient.

The ability to provide advice to a physician is in many ways a *disruptive innovation*, to borrow a phrase from the business world (Christensen and Raynor 2003), in terms of traditional perceptions by physicians of their roles and responsibilities, and the practices and relationships that derive from those perceptions. Over the years I have collected cartoons clipped out of magazines and journals, portraying the use of computers in health care. Almost all of these relate to some sort of role of computers or information technology in making decisions. This focus no doubt exemplifies the way most people first think about computers if asked to consider their potential role in health care.

In one cartoon, a patient is consulting a computer and the computer is advising, “Take two interferon tablets and call me in the morning.” In another, several surgeons around an operating table are discussing an operation they are about to perform. One is saying to the other, “Since it’s been reported that 24% of surgery is unnecessary, let’s only do 76% of the procedure.” In yet another, a patient is entering a question into a computer and the response is “Not tonight—I have a headache.” Another cartoon with an operating room venue shows a surgeon with two hearts, one in each hand, looking across the table at his colleague and saying, “Okay, the old one is in my left hand and the donor’s is in my right, correct?” A final cartoon shows a doctor examining a patient. The patient has an arrow in his back, unseen by the doctor. The doctor says, “I’m pretty sure it’s psychosomatic, but let’s run some tests to be sure.”

The humor in these cartoons has to do with the exaggeration of two opposing views of the relationship between the computer and the human, whereas a suitable relationship likely lies somewhere between these extremes.

Computer as omniscient sage

On the one hand, an extreme view is portrayed in which the computer's expertise is taken for granted, and its pronouncements are *accepted quite literally*; the computer is essentially *running the show*. The popular image of computers in medicine is that they are devices that are capable of storing vast amounts of information, performing lightning-fast computations, and making accurate decisions. The cartoons are funny because they start from that assumption and carry it to an absurd extreme. The computer may be seen as patronizing and even arrogant.

Computer as out-of-touch meddler

The opposite extreme is the traditional view of resistance by physicians to computers and to automated guidelines as being too simplistic and *representing "cookbook medicine"*—with the typical warning that computers are insensitive and incapable of recognizing the nuances of patient care, the role of physician judgment, and the prerogative of the physician as primary decision maker.

A more symbiotic view

In actuality, the use of computers in health care has taken a rather conservative, circumspect, and circuitous route to participating in clinical decision making. That route is clearly between the two extremes of the human ceding control to the computer versus the human not being willing to use the computer for decision support at all. Later in this chapter, we discuss a number of dimensions along which the nature of the human-computer interaction in CDS can be considered, including locus of control and degree of assertiveness. The extent to which the interaction is skewed in one direction or another along either of these or other dimensions will depend on the application and purpose, but it will rarely be entirely in one direction on all dimensions. As will be exemplified in a number of case studies and historical reviews in Section II, when CDS has been successfully deployed, the computers usually have been used primarily in an advisory or educational role, in providing input to the practitioner or patient, who ultimately is responsible for making all decisions. This is why, when we discuss CDS in this book, it will largely be with that perspective—in other words, the emphasis is on decision support, *not* on decision-making.

I mentioned earlier that CDS is a sort of disruptive technology. One manifestation of this is that there has been a change in patient attitudes toward their doctors, both in recognizing the limitations of knowledge and judgment of physicians as a group, and in increased tendency to question a physician's decisions and desire to participate in the decision-making process. Further, whereas in the past, physicians were reluctant to consult an information source such as a textbook or a computer in front of a patient for fear that it be regarded by the patient as a sign of weakness or lack of knowledge, there is growing evidence that the ability to look up the latest information is regarded by both patients and doctors as necessary and desirable (Ogden, Fuks et al. 2002; Weaver 2003).

Only in limited circumstances in clinical medicine might one consider using the computer directly in a closed-loop fashion to collect data, analyze the data, make decisions, and take actions without human intervention. Probably the most notable exception is the implanted electrical cardiac pacemaker, which has algorithms for determining when to stimulate the heart automatically in response to heart rate or rhythm abnormalities. Patient ventilators can do some automatic manipulations such as use of feedback control to adjust cycling thresholds to maintain a desired pressure level, adjustments to keep PEEP/CPAP pressure at specified levels to compensate for gas leaks, or modification of ventilator flow delivery rate to adapt to changes in patient inspiratory effort. However, any such use requires considerable caution and documentation of efficacy, and is done only after an arduous process resulting in regulatory approval, from the Center for Devices and Radiological Health of the Food and Drug Administration (FDA) in the United States (Hackett and Gutman 2005), or from comparable agencies in other nations (Altenstetter 2003).

But in most situations, the human remains in the middle of the decision-making loop. The guiding settings on a ventilator are still determined by a human, after viewing data obtained from the device and other information such as blood gas test results. Insulin infusion pump settings are still adjusted by humans, after reviewing laboratory and clinical data, even though the device could presumably have an algorithm for automatically responding to the most recent blood glucose laboratory result in the EHR. For clinical decisions that are not integrated with embedded or connected devices, recommendations are not implemented without the express approval or action by the human.

3.1.1 Limitations of the Technology

Given our present state of the art, computers can usually be expected to provide only relatively unsophisticated decision support, and computer models of human decision-making remain limited. This relates primarily to two factors. First, many kinds of data and nuances regarding patient findings are either not captured by a computer or encoded in a form that the computer can interpret, but which an experienced clinician not only has access to but can more effectively use. Second, some of the nuances of the decision process may not even have been encoded in the model used by the computer, but which a physician routinely considers. Because of the limitations of both the data available to the computer and the model used, it is thus important for the physician to check the reasonableness and appropriateness of a CDS recommendation before acting on it.

A further difficulty is that some of the applications of CDS that have been pursued are quite complex, for example, determining a differential diagnosis, deciding on an optimal work-up strategy, and doing treatment planning. Thus it is not surprising that most of the success to date has been in the form of simpler kinds of CDS, where the modeling of the decision problem and capture of the nuances of data are much less challenging. Examples of such kinds of CDS include the use of single-decision rules in targeted settings, such as in CPOE for providing checks of medication doses against recommended ranges, or verifying the absence of allergies or recognized drug interactions; or in checks of results of newly arrived laboratory results against normal ranges in order to alert physicians of abnormalities. Though these are considerably easier to implement, even they have subtleties and nuances that must be considered for successful implementation. For one thing, the simpler the rules, the less they take into account a variety of mitigating factors that affect the clinical significance of the potential recommendation. But increasing the patient-specificity and sophistication of the rules requires more data, and of course access to the EHR. Also, to avoid redundant alerts and "alert fatigue" (see Chapter 18), the logic should take into account past history of the condition and also whether a similar alert was generated within a specified "alert-fatigue avoidance" time window. Accounting for such factors makes the rules and the maintenance of them much more complicated, as a result of which the rules are no longer simple.

Last, replication of some of the approaches shown to be successful in early adopter settings often has been problematic for a number of other reasons beyond the decision model and the availability of the data. Factors have been both technical, cultural, and organizational in nature, which we will discuss in greater detail next, and in subsequent chapters, particularly in Sections IV and V of this book. For example, as discussed briefly in Chapter 1 and at the beginning of this chapter, when we consider the use of computers in interaction with human beings, an important issue that needs to be carefully addressed is how that interaction is regarded by the human user in terms of decision-making control, responsibility, and judgmental prerogatives. Other factors include the manner of interaction of the program with users in terms of impact on their ease of performance of operational tasks, time required, and effect on workflow procedures and processes, particularly as they relate to clinical IT services.

As we pointed out in Chapter 1, another reason for the gap in adoption has to do with underestimation of the complexity of the tasks involved in replicating an innovation such as CDS on a widespread basis, given the need for it to be well integrated with clinical information systems, actual workflow, and business and health care practice patterns in each site; and given the need for it to be readily updated and adapted to changing requirements. Challenges in deploying CDS at each site are to introduce it in a way that is acceptable to the individuals who will be required to use it, being sensitive to the culture, work style, time constraints, self-image, and other cultural and social factors of these individuals and the organizations in which they participate.

3.1.2 Considerations Regarding Human-Computer Interaction

Computers have long been regarded ambivalently as both a boon and a possible threat with respect to their interaction with humans, particularly for decision making. The field of *artificial intelligence* was specifically created almost 50 years ago with the aim of exploring the nature of intelligence, including acquisition and representation of knowledge and the ability to do reasoning and problem solving (Feigenbaum and Feldman 1963). This has involved both research studies and demonstrations centered on how to make intelligent computers as autonomous entities. We see a number of applications of this kind of pursuit in the form of advances in robotics, chess playing and other strategic game-playing programs, speech recognition, and automatic language translation.

The problem of trying to build intelligent autonomous computers is a fascinating one, but this has less direct applicability in health care than the use of computers in partnership with humans. Many of the same methodologies are used as in the pursuit of autonomous intelligent computers, but an additional focus is on the nature of the interaction between the computer and the human (Johnson 1994). Our concern here, as we have noted, is the role of the computer as a decision-*support* tool rather than as a decision-*making* entity.

What is the best way in which an "intelligent" computer should interact with a human to provide CDS? There are several possible modes of interaction: First, a computer can be *in charge* of the interaction, delivering recommendations or decisions that are expected to be carried out. This mode could be used, for example, in CPOE, when an attempt is made to order a medication with a dosage that is outside of therapeutic limits for safety. The computer should be able to either actively stop such orders from being processed or passively avoid them by not providing the means to enter (or select) such doses in the first place. In another sense, however, the choice of ordering the medication is still made by the physician, and it is only when the entered or attempted order needs to be overridden by computer because the dose is outside of therapeutic range that the computer exerts control of the interaction.

A slightly less assertive version of computer control is a mode in which a human decision maker can override the computer by providing a justification for an action to which the computer has raised a warning. An example of this mode of use in order entry systems is when a procedure is ordered by the user for an indication that is not recognized by the computer as being among those generally accepted, but which is nonetheless permissible (Harpole, Khorasani et al. 1997).

Relaxing the constraints still further, consider the mode in which a computer presents a data entry form or dialog box to be completed by the user. The entries in the form are checked for validity, e.g., to ensure that they are within range of values expected for the requested items or that they match a controlled list of possible entries; or the computer may require that entries be chosen from a drop-down list. Thus the kinds of allowed input are controlled by the computer, but within that scope, the user may make any desired choices. Following entry of data, subsequent displays or forms can be made available to the user, and the sequence and nature of the interaction guided by the computer in response to user entries. An example is a predefined order set for medications and procedures, in which a physician might be interacting with the computer to order most of the procedures but also to customize some of the options.

If we consider shifting the focus of decision-making further in the direction of user control, several modes of interaction are possible. In one mode, the user performs various actions, and the computer analyzes them in the background, displaying a warning when the action is considered to be dangerous or inappropriate. Among the earliest experiments aimed at refining this approach was a series of studies carried out by Miller and colleagues (Miller 1983; Miller and Black 1984) on "critiquing" systems. As discussed in Chapter 2, the primary applications investigated were in management of anesthesia and hypertension. In the critiquing mode, the computer made comments and recommendations for modifying notes and orders already created by the physician, which the physician could accept or reject. A more contemporary application is the typical circumstance in CPOE in which a physician is able to select choices for medications and doses directly, but the computer identifies some of the chosen orders or doses as contraindicated in this patient because of interactions or allergies or inappropriate dose (Kuperman, Teich et al. 2001), and advises the user of alternative actions that may be more appropriate.

A somewhat more passive mode of interaction is one in which the computer gathers statistics about the performance of users over some period of time and provides feedback to the users about how they compare with their peers. This can be regarded as more an educational type of intervention than as CDS unless it is provided in a highly patient-specific context. It has been shown to have mixed success, and appears to be most effective when coupled with a concerted educational initiative and buy-in by the physicians of this kind of decision support (Mugford, Banfield et al. 1991; Bindels, Hasman et al. 2003; Bodenheimer 2003; Greenhalgh, Long et al. 2005).

One of the primary targets for concerns about CDS invading a physician's autonomy is the use of clinical practice guidelines, especially those that are computer-based. Although clinical practice guidelines abound in magazines and journals, CD-ROMs, and Web sites, they are rarely used in clinical practice except as education or reference resources. In the care of a specific patient, the experience is often that the guideline does not capture the nuances of the patient, and/or does not embody what the physician believes to be best practices in his or her institution or in the current setting or to correspond with his or her own experience. There is some justification to these complaints. Clinical practice guidelines, however comprehensive they may be, usually cannot specify the details of every possible combination of circumstances that might arise in practice. If they could do so, nonetheless their rendering in a print or display medium for easy comprehension and use would be a significant challenge. Even for standardized guidelines such as those for hypertension management (NHLBI 2004), a flowchart rendering of the various alternative pathways for management would take up many pages.

Another problem with clinical practice guidelines is that many of the characteristics of patients may be outside the scope of the guideline, for example, other concurrent diseases or medications that can alter the nature of the current condition; or the presence of findings that are not available to the computer-based guideline, for example, those that relate to nonverbal subjective assessments that can best be made face-to-face and are difficult to articulate in words. Finally, even the most well-researched, evidence-based guideline may not have achieved 100 percent consensus among experts, and alternative modes of care may exist that would be equally appropriate. Thus, guidelines can be best used in a mode in which they provide suggestions or advice when requested, but do not force compliance.

Due to limitations such as those just cited, the idea of clinical guidelines tends to conjure up some of the worst associations with the term "cookbook medicine" (Liang 1992; Harding 1994; Costantini, Papp et al. 1999) that we mentioned at the outset of this chapter. At their basis, the objections relate to the view that medicine is too complex to be reduced to a set of algorithms or rules, and that it could never be codified to an extent similar, for example, to that which enables an autopilot to fly an airplane. But, in fact, no one is advocating the autopilot as a model for computer-based decision support in health care. Autopilot operation is successful in airplanes largely because the procedures and operations of normal flight are highly predictable, based on data that can be objectively gathered. As a result, rules for decision-making can be fully specified and implemented. Autopilot systems also can be made "aware" of settings in which their use is not appropriate (e.g., takeoff and landing) and circumstances in which something happens that is outside of their realm of decision-making, such as the occurrence of a combination of parameters for which there is no defined rule. In such situations, either automatically or through pilot initiative, they have a mode in which the pilot can take over control or override their operation. In the autopilot setting, the computer is more in control than the human most of the time, but the initiative can switch to the human. There are situations in health care that can approach this, for example, the aforementioned closed loop systems of implanted cardiac pacemaker devices, or other applications that potentially could become semiautomated, such as intravenous infusion systems for medication administration or patient ventilator management systems for adjusting O_2 levels or cycling thresholds of the ventilator.

In general, though, this view stems from misconceptions regarding the ways in which clinical practice guidelines can be useful in CDS. Computer-based guidelines can be made more patient-specific in several ways. First, they can be freed of the constraints of paper or screen display of static algorithms, by supporting identification of a variety of possible entry points and the eligibility/applicability criteria for these entry points, and by supporting more flexible means for browsing and navigation of pathways and access to explanatory materials (Abendroth and Greenes 1989). Second, guidelines can be made arbitrarily more complex and nuanced with respect to patient findings (subject to limitations on available evidence, author expertise, and author fortitude in delineating all of these circumstances), since the need to render the guideline in static form is no longer an issue. Third, the guideline can be decomposed into parts that can be deployed in various application settings, such as those parts best used during CPOE, those that would be most appropriate as alerts or reminders, or those that should be considered during the patient assessment and progress-note-generation process (Essaihi, Michel et al. 2003; Wang, Peleg et al. 2003). Fourth, these parts can be specified precisely so that they can operate on entered or stored EHR data and produce their recommendations automatically (Tu, Musen et al. 2004). The issues involved in automating clinical guidelines are discussed further in Chapter 13.

We will return to the general issue of interaction with the user when we discuss the process of integrating CDS into application environments later in this chapter.

3.2 DESIGN AND STRUCTURE OF CDS

Many opportunities exist for performing CDS. Two recent reviews have developed taxonomies of features (Sim and Berlin 2003) and modes of use (Kawamoto, Houlihan et al. 2005), respectively, for clinical decision support. The Kawamoto study (Kawamoto, Houlihan et al. 2005) and a follow-up to the Sim study (Berlin, Sorani et al. 2006), in particular, are noteworthy in identifying those forms of CDS that have been evaluated in clinical trials. Such schemes, as they are further refined, can be expected to be helpful in continually evaluating instances of CDS in terms of their focus and the settings and modes of their deployment to determine which are most effective.

In this section we propose our own schema for considering CDS, in terms of (1) its purpose, and (2) the architecture and component design elements required for providing it. Design elements include the decision model, knowledge content, data requirements, result specifications, and application environment factors affecting deployment and use. Since one of our main objectives is to understand what factors have held back widespread adoption of CDS and to identify ways of increasing it, we will analyze aspects of this schema to help us answer these questions. The questions to be addressed encompass the kinds of standards and infrastructure that may be needed, the kinds of business and organizational strategies that can be useful, and the kinds of approaches that can be used to encourage wider adoption.

3.2.1 Purpose

We now turn to a consideration of the many possible goals or purposes for which CDS may be intended. Purpose is somewhat orthogonal to the classification of methodologies we described in Chapter 2. Thus, as we consider the various purposes for CDS (see Table 3-1), we identify the key methodologies that have been used for their implementation. Since references to those methodologies are cited in Chapter 2, we won't repeat them here, but we do cite other aspects of methodology or examples of use not fully discussed in that chapter.

Answering questions

The simplest goal for CDS is to provide context-specific access to relevant information for a human user at the time of problem-solving or decision-making. Hyperlinks to specific resources at specific points in the interaction with a clinical IT system provide one such way to do this. An example would

TABLE 3-1 Principal purposes for CDS, and the key methodologies used.

Purpose	Key methodologies
Answering questions	Direct hyperlinks from context-specific settings, context-specific information retrieval, use of agents and information brokers, infobuttons as instance of the latter, or ultimately, a "personal guidance system"
Making decisions	Gathering data, analyzing the data, and providing recommendations for assessments or actions
• Diagnosis	Bayes theorem, algorithmic computation, heuristic reasoning, statistical data mining/pattern recognition methods
• Test selection	Decision analysis, logical rules/appropriateness criteria, and logistic models and belief networks for risk prediction (e.g., for screening decisions)
• Choice of treatment	For choosing among alternatives, decision analysis, and logical rules/appropriateness criteria. For dose modifications for age or factors such as renal function, algorithmic computation. For dosimetry or dose distribution, algorithmic computation based on geometric and pharmacokinetic models, with use of heuristics and statistical methods for optimization
• Prognosis	Logistic regression, Markov modeling, survival analysis models, and quality of life assessment scoring methods
Optimizing process flow and workflow	Multistep algorithms, guidelines, and protocols, coordination of participants by workflow modeling, scheduling, and communication methods
Monitoring actions	Use of ECA rules, with background detection of events, in real-time or asynchronously, logical evaluation of conditions, and issuing of messages. Events can be a user selection or data entry, a result arrival, or the passage of time
Focusing attention	Presentation of items in data entry or reporting applications. May be done by use of sequences to encourage intended behaviors, by a process flow model such as an underlying guideline, and/or by visual groupings based on shared attributes such as purpose, medical subdomain, or application context

be a link to laboratory tests and their normal ranges, or to a list of medications in a hospital's formulary.

More sophisticated approaches involve using intermediate search tools, such as information agents or "bots" to go out to diverse sources and report back (like Web crawlers or spiders), and "information brokers," which can map queries to the formats required as input by external knowledge bases and then map the responses to a form recognized by the requesting source. The goal is to find resources relevant to a particular context, including patient-specific parameters. For example, in a lab test result review context, the display of a lab test result might be accompanied by an infobutton (Cimino, Li et al. 2002) that, when selected, dynamically retrieves available resources about the test, such as normal ranges, textbook materials regarding the use and interpretation of the test, information about the diseases in which it is abnormal, and a list of MEDLINE references on the clinical use of the test.

To carry this to a potential extreme, we can imagine an even greater degree of context sensitivity and integration into practice embodied by an approach to information retrieval that I call a Personal Guidance System (PGS). A PGS would seek to anticipate user information needs in a way analogous to that done by a navigator in a Global Positioning System (GPS) device in an automobile, which continually tracks current location on a map. Most GPS navigator systems can display points of interest that are in the neighborhood of the user's current position or focus of interest, and can provide more specific information about any of those, should the user wish to see it. In a PGS, the system would know some details about the user, his or her role, preferences, and other characteristics, and could continuously monitor the applications being executed. If the PGS were also aware of a current patient care context, including a problem list and ideally, the current problem being focused on, the PGS could dynamically generate and continuously update a list of information resources relevant to these user/setting/patient-specific attributes. These could be available instantaneously via a menu choice, keyboard shortcut or right mouse click, infobutton selection or perhaps a sidebar, and high priority resources could be brought up for viewing automatically.

Making decisions—about diagnosis, test selection, choice of treatment, and prognosis

This purpose, in contrast to that of finding information just discussed, is for help in analyzing information. This can be for a variety of types of decisions, including making diagnoses, selecting tests, planning therapy, and estimating prognosis.

We have noted that differential diagnosis has been among the uses of CDS that have most captivated interest from the earliest days of computer use in health care. An excellent book edited by Berner (1998) focuses on diagnostic decision-making, and reviews the many approaches that have been pursued. The basic goal of differential diagnosis is to deduce, from a set of findings, the diagnosis that best explains them. This is clearly an important task, not only in order to be able to select proper treatment but also to be able to estimate prognosis and to give advice to the patient.

It should be recognized, however, that diagnosis is not usually a single event, but rather a *process* of continually refining knowledge about the patient by gathering data, performing tests, and reevaluating data, until sufficient confirmation is reached in order to take therapeutic action. Some approaches, such as decision analysis, appropriateness criteria, and clinical guidelines, have focused on structuring this process, rather than on the endpoint of diagnosis. Indeed, the decision table approach for representing a guideline (Shiffman and Greenes 1991), as illustrated in Table 2-3 in Chapter 2, doesn't even choose a diagnosis, but determines next actions based solely on combinations of findings. Issues in the view of diagnosis as a process relate to the selection of appropriate tests based on cost, risk, inconvenience, and other factors versus the potential for information gain from the tests. When one considers the fact that the institution of treatment is also a diagnostic test, in terms of information about how the patient responds to it, it can be seen that the whole patient care process continually involves diagnosis in the form of ongoing reassessment. Prognosis estimation can also be regarded as a type of diagnostic assessment, in that it characterizes the patient's current state of health as one with a particular expected survival rate and quality of life.

For the purposes of exposition, we can divide the topic of making decision into methods for hypothesis formation or refinement, both for diagnosis and prognosis estimation, and those aimed at performing an action (i.e., test selection, or choosing or detailing a treatment regimen), as depicted in Figure 3-1.

1. **Diagnosis.** The process of diagnosis can be subdivided further into detection and classification. Although some screening recommendations remain controversial, the numbers of tests in use for screening purposes is sure to increase as a result of progress in understanding of the genetic basis of disease and development of biomarkers. Screening tests in common use include, among many others, testing for phenylketonuria (PKU) in newborns, mammography in older women or those with certain risk factors, colonoscopy in average-risk patients over age 50, and prostate-specific antigen (PSA) testing in older males. Generally, screening tests have been applied primarily to alert the user to the detection of the presence of disease, rather than to make detailed specific diagnoses. The typical approach in screening is to set a liberal operating point (decision threshold) on the ROC curve for considering the test to be positive, on the basis of the view that it is preferable to err

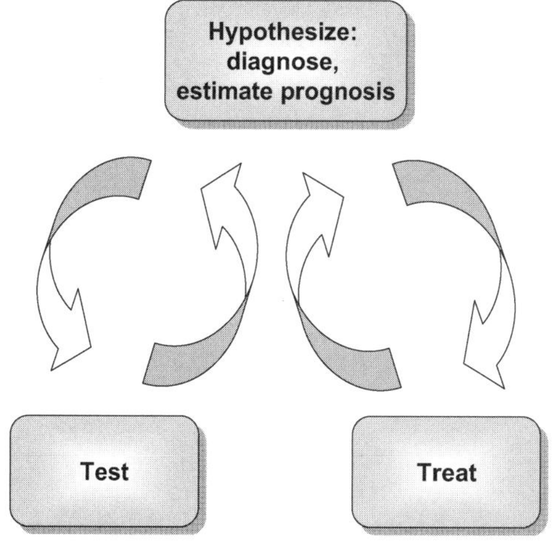

FIGURE 3-1 Medical diagnosis as an iterative process of forming hypotheses and gathering data to confirm or refute the hypotheses. Prognosis estimation is a form of hypothesizing. Treatment is a form of test, in that it also provides data that can help to confirm or refute hypotheses.

in the direction of more false positives than to fail to detect cases with disease (false negatives). Although some assessment tools for supporting CDS in this realm (such as computer-aided detection and diagnosis (CAD) image processing methods in digital mammography (Jiang, Nishikawa et al. 1999; Giger 2004)) try to classify the findings in terms of specific diagnosis, the ability in screening tests to make diagnoses is usually quite limited, and further testing typically is required. Methods used for both detection and classification (diagnosis) include Bayesian probability revision, algorithmic approaches (e.g., for electrolyte imbalance), heuristic reasoning and weighting of findings (e.g., DXplain (Barnett, Cimino et al. 1987)), and statistical data mining/pattern recognition methods such as logistic regression, classification and regression trees, and artificial neural networks.

2. **Test selection.** This set of clinical decision-making problems relates to whether and when to do screening, what test to use for screening or for diagnosis, and determination of need for and selection of follow-up testing, including referral to consultants/specialists. Primary approaches are clinical decision analysis and logic rules/appropriateness criteria. The decision to obtain screening tests or follow-up tests may be modified on the basis of CDS tools, such as those for estimating a patient's risk of breast cancer based on risk prediction models (Rockhill, Spiegelman et al. 2001; Freedman, Seminara et al. 2005), and policy recommendations/guidelines such as those of the American Cancer Society for early detection of cancer (Smith, Cokkinides et al. 2004). Methods that underlie the use of guidelines and risk assessment tools for deciding on whether to do screening or timing thereof, which we described in Chapter 2, include the use of Bayesian networks for risk assessment modeling, and ROC analysis for determining performance characteristics of tests.
3. **Treatment decisions.** Needs for CDS include picking the most appropriate therapy, and determining dose or dosage administration regimen. Picking therapy involves tradeoffs of cost, risk, benefits, and patient preferences, and thus can be modeled and supported by decision analysis. Optimal strategies may be codified as logic rules and appropriateness criteria, or in the form of clinical practice guidelines. For dose determination, CDS can be used to constrain choices, make it easier to select preferred choices through displayed groupings such as order sets, or verify dosage requests as being within acceptable ranges. In some circumstances, CDS can be used to calculate dose modifications, for example, based on body surface area in pediatrics, or as a function of renal status or age, and tend to be algorithmic. Dose administration determination may involve more elaborate considerations, as in the case of an insulin sliding scale, or calculation of portals and beam configurations for radiation therapy. These tend to be algorithmic, but may use heuristics and pattern recognition and statistics for optimization.
4. **Prognosis estimation.** The CDS question here is to predict the likelihood of good outcomes, morbidity of various types, and mortality. Thus prognostic evaluation of consequences should be an important consideration before treatment selection. Database prediction, as in systems like ARAMIS (Bruce and Fries 2005), is an ideal approach

when the data are available and sufficiently structured. Methods of analysis include logistic regression, classification and regression trees, Bayesian network modeling, and artificial neural networks. Methods for modeling of future chance processes such as Markov modeling and for assessing quality of life, such as the calculation of quality-adjusted life years (Richardson and Manca 2004) or severity of illness, such as the Apache III score (Kim, Kwon et al. 2005), are essential underpinnings.

Optimizing process flow and workflow

We have described multistep algorithms, guidelines, and protocols in Chapter 2 as a more complex form of CDS. They arise in and have use in a variety of settings, because medical care often requires sequences of tasks, with intermediate decision points and pathways. The intent is to help guide the user in the proper sequence of decisions and actions, to be sure all appropriate alternatives are considered, and to avoid proceeding along inappropriate pathways.

An example is the progression from less expensive and simpler tests to more expensive and invasive procedures in the evaluation of heart disease or breast cancer. Another is the initiation of single-drug therapy for hypertension, with adjustment, substitution, or addition/deletion of medications based on response, side effects, and complications. These are best modeled by clinical practice guidelines, flowcharts, protocols, or flow sheets.

Sometimes multiple tasks, such as orders for tests and procedures, must be done concurrently, but the next step must await at least some of the results from those tests and procedures before proceeding; and the entire process may involve multiple participants (both human and information–system-based). In these settings, the coordination and communication among participants are important, and a goal is thus to improve workflow, and to maximize speed or efficiency. In this circumstance, augmenting guidelines with workflow management capabilities is desirable (Ciccarese, Caffi et al. 2005).

In other settings, data may be changing quite rapidly; for example, in an emergency or intensive care unit setting, the statuses of patients need to be able to be assessed at a glance in order to identify who needs near-term attention versus those that are less critical. CDS in the form of dashboards and alarms/alerts that portray these changing statuses can be helpful in such settings.

Last, in clinical trials and in some procedures such as renal dialysis, strict adherence to the steps of a protocol are essential, and a CDS tool can help to ensure that this occurs.

Monitoring actions—guarding against errors, providing warnings, alerts, reminders, or feedback about performance

While decision support may provide information when sought, or by overtly asking for it (e.g., by invoking a differential diagnosis tool, a dose calculator, or a clinical guideline), another form of decision support works in the background without overt action of the user, and only interacts with the user when there is a reason to do so. This can occur either in the background of real-time interactive applications between a user and the computer, such as in CPOE, or can be asynchronous and decoupled from user actions (e.g., notification about an abnormal test result by paging or e-mail).

The computer essentially functions as a guardian or silent partner, monitoring the clinical context and what the user is doing, and either interrupting or contacting the user when situations arise that necessitate it. For example, if an order is determined to be hazardous (e.g., a medication interacts dangerously with one that the patient is already receiving, or to which he or she is allergic), a warning can be displayed. If a radiological procedure is being ordered for an indication that is not determined to be appropriate, e.g., ultrasound for question of appendicitis when a CT would be better, or if a medication is too expensive, is not on an approved formulary, or a generic medication would be an appropriate substitution, a recommendation for the alternative can be displayed. If a critical abnormal laboratory result is obtained, an alert can be triggered that notifies the physician so that appropriate action can be taken. The passage of time may cause reminders to be generated, as for periodic mammograms in a woman over 50, or for flu shots during the winter season for an elderly patient. If a patient has been in an ICU longer than an expected number of days for his diagnosis, the physician can be notified.

In all of these cases, the background alerts, reminders, or feedback may be considered useful or thought to be inappropriate or unjustified. A challenge in providing this kind of decision support is to minimize the situations in which they are inappropriate, lest the false alarms become annoying or in the worst case, are simply ignored because of their frequency. The conditions for generating messages must be properly defined to be as helpful and pertinent as possible, and the way in which alerts are provided must allow overriding of them when justified.

Focusing attention

Another form of CDS is quite indirect and subtle. This is the encouragement of best practices through use of techniques to organize and present information and options in such a way that they serve to remind or facilitate good choices. This may be done either by providing a framework for describing sequences of interactions between the computer and user, such as dialogues that are controlled by an underlying guideline or flowchart, or by associative groupings of documentation elements on display screens or printed forms and documents.

With respect to associative grouping of elements, in particular, if it is done well, it can not only focus attention, but offer benefits of improved efficiency, because groups of items can be either selected as a unit when acceptable, or at least brought together for consideration and action decisions rather than each element needing to be sought for by a user and selected or entered individually. Order sets are an example, in that the grouping of orders for a particular indication, such as admission to the cardiac ICU, are pre-established for ease of use by a physician, and to be sure that the physician does not forget to consider including medications for control of pain and anxiety, and for anticoagulation, vital sign and ECG monitoring, diet, cardiac enzyme testing, and other typically considered tasks.

Besides order sets, another application of this approach to CDS is the design of data entry forms for structured capture of information. Examples might be a form for recording a neonatal visit, an anesthesiology preoperative note, or a specialist referral for cardiac surgery evaluation. The form can include items that are automatically filled in, where possible, from stored data. It can include suggested items that are predetermined to be important, and thus serve as both a handy checklist for recording them and as a reminder to be sure to do so.

Yet another example is the generation of reports from structured elements, for example, the printout of a prescription order, or the production of a postoperative note or discharge summary. The design of the report is intended to be easy to automatically generate from stored elements, consistent in appearance, independent of the user who produced it, and well formatted and organized. Such predictability (as well as legibility) facilitates readability and usefulness.

3.2.2 Design of CDS: Components and Interactions

We now consider the structural aspects of CDS. We do so by identifying a set of functional components of CDS and its invoking environment, and how these components interact. As I noted at the beginning of this chapter, this discussion is somewhat artificial, in that it presents an idealized model of how CDS should be created. What I mean by "idealized" is that, if the design of a CDS capability clearly identifies each of these components, separates them cleanly, and addresses the design of each component in a standardized way, the goals of widespread dissemination and use of CDS can be greatly facilitated. We focus on the idealized model while recognizing that much of CDS is not implemented that way. I maintain that all the components we discuss next that are needed for CDS are present in one form or another in any implementation, but they are not always separable, in terms of the actual software code that implements them. Nonetheless, we can consider these components and the functions they perform individually, at least from a conceptual point of view.

Another idealization I adopt is to refer to a unit of software that provides CDS as a CDS *module*. The heart of a CDS module is a method of transforming input parameters to a patient-specific output. To be modular, the CDS software should be cleanly separable from surrounding or invoking software code, communicating with it via a well-defined interface. Recognizing the many possible implementation methods for CDS that may not be modular at all, we nonetheless use this term to be able to direct our focus to the portion of software directly concerned with the provision of CDS functionality, and to the nature of the interactions by which it relates to the invoking environment.

To provide CDS, several tasks must be performed, as shown in Figure 3-2:

- The CDS module is *initiated* or invoked by some process in the application environment.
- The module *obtains data* through an interface with the application environment, where the data are *entered* by a user, *retrieved* from the EHR, or *provided directly* by the invoking entity. The latter might include context-specific information about the application, user, setting, and function being performed.
- The module *applies knowledge* (e.g., facts in the form of rules, algorithms, or semantic relations), either local to the module, or retrieved from a knowledge base.
- A process is then executed that *transforms* the input parameters and knowledge according to the specification of some sort of *decision model* to generate a patient-specific output. The decision model is usually embodied in an algorithm or computational procedure of some sort.
- The module then *produces a result* that must be communicated to the application environment. That result is usually a recommendation for action.

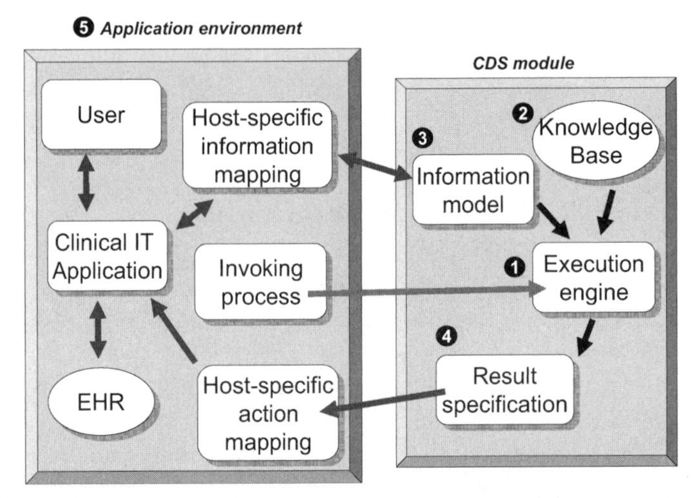

FIGURE 3-2 A conceptual model of CDS design components and their interactions with the host application environment.

To carry out these processes, the design of a CDS module conceptually has four design elements, or components, and operates in conjunction with an application environment, which is thus considered to be the fifth component (see Fig. 3-2). The application environment determines how and when the CDS module gets invoked, how it obtains data and communicates its results, how it interfaces with host software and hardware, and how it interacts with its users. The application environment can be so varied that the specification for this component is only defined with respect to CDS in terms of the nature of the CDS module's interactions with it.

Decision model/execution engine

All kinds of CDS have some kind of execution paradigm, that is, a method of organizing or processing input information, to produce some kind of output, or result. The sequence in which data are requested and the algorithm or method for processing data depend on an underlying model of the decision problem. For example, an alert or reminder may be designed to be triggered by an event, such as a mouse click, the arrival of a lab result, or the passage of time, to obtain specific data items. It then evaluates a Boolean logical condition expression about the data, to determine the truth value for the expression. If the Boolean condition evaluates to "true," the alert or reminder may then cause an e-mail, page, or displayed message to be generated, in order to notify an appropriate physician. The *decision model* in this example is *Boolean expression evaluation.*

A differential diagnosis program may collect data and evaluate the diagnostic possibilities using Bayes theorem: the Bayes theorem algorithm is the underlying decision model. A dose therapy calculation tool might use a formula that needs such parameters as body surface area, renal status, or age to make recommendations for modifications of medication dose for children, those with kidney failure or the elderly; the decision model is a computational formula.

The decision models that are used in CDS rely principally on the methods discussed in Chapter 2. These include:

1. Information retrieval; that is, the model by which data and knowledge are used to select pertinent items to retrieve

2. Logical expression evaluation
3. Probabilistic and data-driven classification/prediction
4. Heuristic methods and expert systems
5. Calculations, algorithms, and multistep processes
6. Associative grouping of elements; that is, the model determining what these associations are and under what conditions they are activated

Conceptually, we can consider that the decision model, to the extent that it involves computational processes, is embodied in an *execution engine*. The execution engine is the part of the CDS software that evaluates data to produce output. As we have noted, actual implementations may not cleanly separate this code from other parts of an application, or even from other parts of the CDS module itself, but advantages are to be realized if that can be done. Principally, this separation allows the execution engine to be refined and enhanced as improvements in the way it should operate become understood. Also, this provides flexibility and portability, in that the execution engine can be recoded and reimplemented in different platforms independently of other CDS parts, and can even be embodied in external services (see Chapter 23).

Knowledge content resources

Sometimes, as in an application for recommending electrolyte replacements in acid-base disorders, the calculations and sequences of actions are embedded in software code. However, as we noted earlier, it is often helpful to implement the general methodology of a particular decision model as an execution engine that can be used to apply the method to all knowledge of that type. For example, if the electrolyte replacement algorithm can be represented in a flowchart modeling formalism, then a guideline execution engine can run it. To be most flexible, ideally, the knowledge—the formulas or equations, the logic of production rules, the flowcharts, and so on—should exist external to the "engine" that accepts inputs of that type, processes the knowledge, and produces a result. Separating the knowledge from the engine, when it is possible to do so, provides numerous advantages:

- It enables the engine to operate on a variety of similar kinds of knowledge.
- As a consequence, the knowledge resources can be managed independently. For example, they can be authored and edited through use of a knowledge editor tool.
- With appropriate editor functionality, an editor tool can display the knowledge in a form that is readable by a human subject expert rather than requiring the skills of a software engineer or other technical support person.
- If the knowledge is made transparent in this fashion, maintenance, review, and update are easier to do.
- If a standard format is used for encoding the knowledge, or for import and export of it from external repositories, the knowledge can be shared and disseminated.
- If the decision model evolves, e.g., in terms of the ability to use more refined or detailed knowledge, the knowledge base can be updated separately to incorporate those knowledge elements.

Knowledge content resources can be structured or unstructured, depending on the purpose and the computational requirements of the decision model.

For example, if the purpose is simply to retrieve and display information in human-readable form, the only structure required might be the use of index terms or keywords, to facilitate retrieval by a search engine, although with no structure at all, a text-based search such as by Google® is still possible.

If the knowledge is a logical expression for a production rule, then the expression must obey the syntactic conventions needed to evaluate the expression in whatever language or formalism is used. In MYCIN, one of the earliest rule-based systems in medicine, production rules had the format IF *condition* THEN *action*, where *condition* was a Boolean logical expression with "certainty factors" associated with the terms. The execution engine could evaluate the conditional expression and had an algebra for combining the certainty factors to produce an updated certainty factor associated with the assertion in the *action* part of the rule (Shortliffe 1976), and could control the sequence of execution of rules through a goal-driven, backward chaining heuristic. In alerts and reminders in Arden Syntax (Hripcsak 1991), an *evoke* section defines triggering event(s), a *data* section specifies the data elements used, a *logic* section defines the procedure to evaluate the data elements in a Pascal-like syntax, and an *action* section defines the task to be carried out if the logic section evaluates to *true*. An Arden Syntax interpreter or compiler could then serve as an execution engine to process a knowledge base of Arden Syntax rules, evaluating a rule when triggered by appropriate evoking conditions. GELLO (Sordo, Boxwala et al. 2004), a new HL7-endorsed ANSI standard expression language supporting HL7's version 3 Reference Information Model (RIM) (HL7 2006), defines an object-oriented syntax for specifying logical conditions. Knowledge bases encoded in GELLO expressions could be evaluated by a GELLO execution engine. The knowledge base for a Bayesian diagnosis tool would be the prior probability distribution for the diseases to be considered, and the conditional probabilities of findings for each of the diseases (Warner, Toronto et al. 1964; Lodwick 1965; deDombal 1975). A guideline interpretation engine that is designed to support traversal of a guideline and interactive acquisition of data and evaluation of conditions to determine next steps could operate on guidelines encoded in a knowledge base according to a formalism the interpretation engine understands. Examples of this are the guideline engines supporting representation formats known as PROforma, GLIF, EON, Asbru, and SAGE, as discussed in Chapter 2 and reviewed further in Chapter 13.

Information model

CDS requires a precise specification of the kinds of information the computation model will utilize, which we refer to as its information model. The knowledge content resources typically contain statements, facts, conditional expressions, or other relations that refer to or operate on patient-specific data. If we formally specify the information model, this provides flexibility in that the same CDS resource can be used in more than one kind of setting; for example, interactively with a user as well as in background mode, retrieving data from the EHR, and in more than one platform and system environment. The specification must include not only the *format* of the data that the CDS module receives and uses but the *taxonomy* or coding scheme for its labels and also for any of its coded/categorical values. For example, if one is seeking to run a rule about medication interactions, it is important to know that they are encoded in RxNorm (NLM 2005), or that a diagnosis is encoded in SNOMED-CT (SNOMED 2006), or that a lab test is encoded in LOINC (Regenstrief 2005). The specification should also involve *grounding* the data elements in terms of precise attributes like units, method of obtaining them, time frame, etc. This is discussed further in Chapter 15, in terms of documentation elements or *archetypes*.

Note that beyond defining these requirements, the adaptations for obtaining them are not the province of the CDS module but of the application environment. If the data are to be obtained by interaction with a user, the host may also need to include an external display name for a data entry field. If the input that is allowed needs to be validated (e.g., checked for limits of a numeric range, length of a text string, the presence of valid characters, or conformance with items on a predefined pick list or dictionary), then those criteria for validation (and the content of the pick list or dictionary) need to be adapted from the information model or added by the application environment.

If the data are to be retrieved from a stored repository, then either the information model used in the host application environment should be the same, or a process for mapping the data elements from it needs to be established. Some of the data elements may need to be transformed, if differences in the definitions of those elements in the host EHR and in the CDS module's information model require it. For portability of CDS, the use of a standardized information model such as the HL7 v.3 RIM should be used, for example, to evaluate a GELLO logic expression. In general, it is unlikely that real operational clinical information systems will use a standard reference information model such as the HL7 RIM directly in its implementation. Thus for interoperability between systems, or to use externally developed CDS, a mapping would need to be developed for each implemented vendor-specific system between the HL7 RIM and the vendor system information model.

Arden Syntax uses a different approach, in which each Arden Syntax rule's data section allows customization to indicate how the various data elements should be retrieved from a particular host information system. This information is encoded within curly braces, indicating that it is not part of the general rule but host-system-specific. The disadvantage of the Arden Syntax approach is that each rule must be customized individually for each host system. Also, even if multiple rules all refer to similar data items, each rule must include the customization statements for each of the data items.

Result specification

Operation of the CDS execution engine is carried out with the goal of producing some output, whether in the form of retrieved resources, a calculated result, a recommended action, or a data entry screen or formatted report. Since the result is dynamically determined through execution, there needs to be an explicit process for determining how that output gets produced.

The result specification could be regarded as part of an expanded view of the information model, but we consider it separately because of its distinct role in the CDS process. For example, the result specification on evaluating an Arden Syntax rule to *true* might be to perform a particular action such as to notify the attending physician. A calculation of dose of a medication based on adjustments for renal function or age might produce a result in terms of a modified recommended dose. A Bayesian differential diagnosis program's result would be the set of diagnoses with their posterior probabilities. Traversal of a clinical practice guideline algorithm based on evaluation of entered or retrieved patient-specific data values would produce a result at each step, indicating the optimal next step.

Thus conveying the result to an application environment involves mapping of the result to the performance of actions or production of outputs that the application expects to carry out. This will largely be the responsibility of the application environment, as discussed in the next section. What is needed in the CDS module, and which largely does not yet exist except in some early work in the execution of clinical guidelines (Essaihi, Michel et al. 2003; Tu, Musen et al. 2004), is a taxonomy of the kinds of results that can be produced from decision support, to facilitate such host mappings.

Modularity of design is one of the reasons we consider the result specification in CDS separately from the mode of interaction with the user or applications in the application environment. Just as with the specification of the information model for data used in the decision model, separation of the result specification enables a decision support capability to be adapted to several possible modes of interaction in any of a variety of applications on different platforms. For example, the result could be processed in real-time in interactive applications, in the background in alert/reminder settings, or in batch mode in the production of reports or summaries. The result may need to be e-mailed to a physician, displayed on a screen, transmitted to another application, or stored in the host EHR.

Application environment

As we have seen, many of the features of CDS operation are not embodied in the CDS module itself but in the application environment that invokes it. The application environment determines how the CDS module communicates with a user, such as an interactive dialogue, or obtains data from the EHR, or how it conveys its results. The application environment can also pass to the CDS module certain context data such as those describing the application setting, the user, and the purpose.

The degree of integration of CDS with applications is one of the most critical ones for determining success of CDS, yet too tight an integration limits the ability to achieve portability and reuse of CDS modules. We will begin the consideration of the nature of the interaction of a CDS module with the application environment by revisiting a simple example we used in Chapter 2 to illustrate the range of options to consider and the complexities involved—even for a simple form of CDS. The example relates to the set of rules regarding the handling of abnormal and critical laboratory test result values; in other words, results that exceed predefined limits requiring flagging or, for critical results, urgent attention. The knowledge regarding such abnormal values can be found in the literature, and the simplest form of decision support is the ability to retrieve references to such abnormal values. This could be in the form of bibliographic citations or Web sites displaying laboratory values and their accepted normal ranges and critical values. Ideally, those latter sites should also cite references to the literature about how to interpret them.

The least integrated way to make this information available would be to enable the user to access the Web and to do a search for it, using his or her own search terms. Slightly more integrated access would be a resources page, which would have a set of predefined links to useful reference information that could be accessed from the clinical IT system. To be of greater value, it would be useful to have access to this information at the point at which a physician is reviewing laboratory results for a patient. The difference between looking up lab results values to determine whether they are abnormal and having a direct link to a particular citation giving that information in the context in which it is needed—that is, when the lab result is being reviewed—is that in the latter case the information needed is preselected and automatically available to the user.

A more useful way to provide this information, which is done in most clinical systems, is to automatically flag the abnormal lab result on the report or display screen that is reviewed by the provider, by a symbol indicating that it is outside of normal ranges. This could then be combined with a link to the available citations to give further information. The flag indicating abnormality could be introduced by the clinical laboratory information subsystem or by the laboratory results report display application, using a formal logical condition expression that is evaluated by the computer to determine the presence of abnormality. Thus it could occur at any of three points, either at the time the result is produced in the laboratory subsystem, when it is entered into the EHR, or when it is displayed.

Another way to deliver an abnormal result finding is by generating an alert message that is sent to the provider, perhaps by e-mail, text page, or fax. This requires integration into a clinical information system in such a way that information about the particular patient and the appropriate provider are able to be identified automatically. This would also allow more elaborate decision rules to determine whether the result is new or a repeat of an already abnormal value about which the physician has previously been notified, or if there are coexisting conditions that might explain the result (e.g., renal failure).

To be maximally useful, knowledge in the form of a decision rule such as that used for responding to abnormal laboratory test results would exist in a rules knowledge base and be triggered by a variety of different possible event scenarios—for example, the entry of an abnormal lab result into the patient's clinical record, a medication order interaction check with respect to existing lab results, or the flagging of abnormal laboratory results on review by a physician. With the knowledge in a knowledge base, different events such as result entry, order placement, or display of results could trigger evaluation of the rules indexed according to the various parameters of interest, to determine which, if any, might apply, and then to carry out actions that are appropriate based on the triggering application and depending on the result of the evaluation. For example, in an alerting application, evaluation of the rule to true would result in the generation of a warning message to the provider by e-mail, page, or fax. In an interactive CPOE application, evaluation to true might generate a recommendation to decrease the dose of a medication being prescribed. In the result display application, evaluation of truth would result in the flag symbol indicating abnormal result being appended to the value.

The first usage described, that of displaying a citation, simply requires that the knowledge about abnormal lab results be available in text form in some defined location (e.g., in a bibliographic database). It requires no computability, just the ability to retrieve it. The second usage, access via a direct link from the result information display, can be done by manually identifying the appropriate citation to be displayed whenever abnormal results occur. To be more useful, however, retrieval based on the context, for example, could work as follows: By recognizing that the context is a laboratory results display application, the CDS tool could determine that information pertaining to abnormal laboratory results would be useful, and a specialized retrieval program could be designed to select the kind of information to be retrieved from a general retrieval search engine by passing context-specific parameters related to clinical laboratory abnormal results. This constitutes a kind of "information broker" function, and is exemplified by infobuttons described in Chapter 16.

The third usage—that of flagging abnormal results—requires the presence of a formal computable expression that can be evaluated by the computer. This would be of a logical format such as "if lab test result *y* exceeds threshold *a* then return *true.*" For this to work, the value of the lab test result of interest must be assigned to the parameter *y* and the upper normal range for the lab test result must be assigned to the parameter *a*. The software application must also know that if the result *True* is returned from the evaluation, then a flag value such as * or # should be appended to the display of the laboratory result. Thus this application usage requires a formal expression, an evaluation engine, and a simple set of data parameters that can be passed to the evaluation engine and returned from it.

The fourth usage is the execution of a generalized rules interpreter in the context of an event-driven architecture, in which particular rules are evaluated as a result of triggering events, and the actions performed depending on the result of the evaluation are a function of the application that generated the trigger. This approach not only requires a knowledge base, an indexing scheme for accessing the rules in the knowledge base, and an execution engine, but it requires also a means of integration with various possible invoking applications, as well as, for example, in the case of notifications to physicians, the ability to invoke other applications. This usage has maximum flexibility and power, because the same piece of knowledge—in this case, any logical expression regarding what constitutes an abnormal laboratory test result value—can be used in a variety of different contexts. Thus the knowledge rule itself only needs to be developed once, and can be maintained or updated if necessary in one place, and if properly set up, all the applications that utilize it can be identified, should there be a reason to change the rule in the knowledge base.

We can consider a variety of other kinds of clinical decision support usages that range from passive to active and from loose to tight integration with applications, and do a similar decomposition of the necessary elements. Work needs to be done to define the extent to which application-specific behavior can be further abstracted into a taxonomy of result types, as discussed earlier, so that more of the functionality can be moved into CDS modules rather than requiring custom interfacing and handling of results in the application environment.

One of the challenges in providing CDS is that of determining the most effective way to interact with humans, so that the advice is as patient-specific and timely as possible. At the same time, it must be acceptable to the human user, by not requiring a lot of extra work, being disruptive to workflow, or being redundant. Also, given the role of decision *support* rather than decision *making*, advice must be given in a fashion that recognizes human decision-making prerogatives and avoids being inflexible or insistent when it is not necessary to do so.

Figure 3-3 lists a set of axes that can serve as a guide for thinking about the various dimensions involved in providing CDS and interacting with users, which determine the extent and manner of integration of CDS with the clinical IT application environment. We consider each of these briefly.

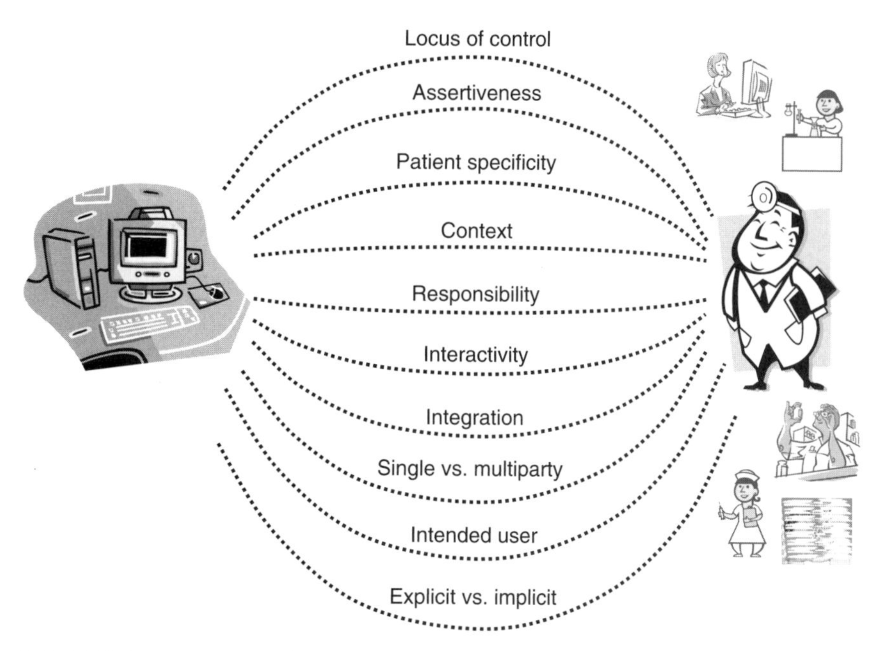

FIGURE 3-3 Dimensions of computer-user interaction in CDS.

Locus of control

A CDS instance can be *initiated* by a user when the need for help is recognized; for example, in seeking to find an answer to a question, or to obtain assistance in assessing a diagnostic or therapeutic decision. Or it can be initiated by the computer, usually in processes aimed at monitoring user actions to guard against errors or detecting suboptimal practices; for example, in detecting an inappropriate order, or a time interval at which a mammogram should be ordered or HbA1c level should be checked. The *locus of control* thus either resides with the user or the computer in these examples. There are also intermediate situations, in which CDS resources are made available and have the means to automatically be context-aware, as in infobuttons or patient-specific guidelines, but it is up to the user to initiate their use.

Degree of assertiveness

Decision support capabilities can be provided to a user with varying degrees of insistence or "assertiveness." This only applies, of course, to settings in which the computer is the locus of control and initiates the CDS instance. The most passive way in which decision support can be offered would be to simply present a discussion of a topic that can be read by the user. One step up from this would be to present specific recommendations and advice, although again in a form that is simply to be read. A slightly more insistent form of decision support would be the requirement that the information provided be acknowledged. A further increase in assertiveness might involve a list of choices of action; for example possible medication regimens, allowing "other" as an additional option or an override of the recommended dose of a medication if a justification is provided. Still more assertive would be forced choice among alternatives without the possibility of override. The most active form of decision support would be a closed-loop process in which actions occur automatically in response to inputs, although it could be inspected or monitored by a user. Implanted cardiac pacemakers have this mode of operation, but for reasons we have noted previously, there are very few other instances of such closed-loop processes in routine use.

Patient specificity

Comprehensive knowledge base resources such as the medical literature in PubMed, a guideline repository, or a collection of possible alerts and reminders, are important to the ability to provide robust decision support. But a challenge in delivering effective decision support is to be able to select resources that are relevant for a given patient and to determine when and where, or even if, it should be used in a particular setting.

To accomplish this, the CDS module needs to be able to obtain information about the clinical problems or findings, care setting (e.g., office visit, phone call, or hospitalization) and other patient-specific parameters, as specified by the logic of the rules under consideration. This involves the ability to access the EHR or obtain data from the user or from the invoking application.

Context

Beside patient characteristics and setting of the clinical encounter, context affects the nature of decision support in several ways. The principal contextual factor is the kind of application and the function being performed when CDS is invoked. For example, if the context is an interactive CPOE program, decision support would likely need to be very responsive so as not to perceptibly delay or impede the real-time interactivity. If the context is a background process aimed at collecting data on and evaluating conformity with care pathways in various units of a hospital (e.g., the coronary care unit or the postoperative orthopedic floor), deviations from expected targets on the pathway can be communicated to providers at the beginning of the next day. Event-driven panic alerts indicating critical abnormal laboratory values would need to be communicated to providers by any available means as quickly as possible and require acknowledgment. Preventive screening or immunization reminders to a patient might have a response window of a week or even a month.

The same knowledge can be used in different contexts and practice settings. To return to an example discussed earlier, consider a rule that performs an action if a lab result exceeds a threshold: this might be used in an alert that is automatically triggered when a result is produced by the laboratory, or the rule might be used during the generation of a lab test result review display screen, for flagging abnormal results. The multiplicity of uses and contexts for decision support knowledge is one of the rationales for the need to construct knowledge bases containing collections of decision support content. The content in a knowledge base can be indexed according to various axes, such as those relating to context, setting, medical problem, and CDS purpose, that enable it to be retrieved and used when needed. Another consequence of this capability is that it would be necessary for only one instance of a particular item of knowledge (e.g., a decision rule, to exist, making it easier to maintain the knowledge, review it, update it, and propagate changes to all the applications that use it).

Interactivity

Somewhat related to the classification of decision support in terms of degree of assertiveness, patient specificity, and context is the degree of interactivity. Knowledge resources may be in the form of static human-readable information, such as text or a table, retrieved in response to specific search. Such a presentation could be viewed and examined but require no specific action by the user. An interactive mode of delivery of this same information would be one for which some entry or acknowledgment is required by the user.

A more interactive form of CDS might be a computational tool that produces a similar kind of text or tabular report, but which provides the ability to manipulate parameters that are entered into it, as, for example, a computation tool for drug dose calculation based on body surface area, or a tool for performing sensitivity analysis of a decision analysis model to determine how stable its decision is as a function of change in the estimate of prior probability of disease or the riskiness of a treatment. Often, interactivity occurs in CDS in the form of requests for data to be entered or selection of options by the user.

Still other aspects relating to interactivity are illustrated by the various ways in which recommendations can be communicated to the user and the options available to the user for responding to or overriding recommendations. Some decision support messages might be provided in the form of noncritical alerts or reminders, generated in the background and only seen when a user next logs into the information system and requiring no further action on the user's part. Alternatively, for more important alerts or reminders, they might be sent by e-mail or page.

Degree of integration with clinical IT applications

A CDS capability can be integrated into the clinical information system to greater or lesser degree. CPOE is an example of an application with a need for a high degree of integration of CDS. A drug interaction checking tool would be most useful during CPOE if it were integrated with an EHR system that maintained a list of current medications for the patient. Given an EHR, a CDS tool could also include automatic checks against allergies or other patient-specific contraindications.

Single vs. multiparty focus

Some applications of decision support, such as computer-based clinical practice guidelines, have the potential for optimizing the care process by suggesting appropriate next steps. In a busy practice environment or inpatient setting, automation of guidelines could also help to optimize workflow by coordinating scheduling and use of resources and activities of participants through communication and synchronization functions (e.g., don't do task B until task A is completed), and monitoring the times, delays, and statuses of expected events. Note that the parties involved may be human or computer-based (e.g., a scheduling system or messaging system).

Notification of alerts is another example of an application that may have a multiparty aspect. Typically, if a critical lab result needs to be acted on by someone, there is a set of processes defined for notifying a patient's primary physician about such an alert (e.g., some sequence of page, telephone message, e-mail, or fax, with requirement for acknowledgment), and defined sequence for notification of other providers if that person does not respond within a specified period of time.

Intended user

Decision support for various purposes may be designed for different kinds of intended users; for example, direct support of physician decision-making; aids to nurses, pharmacists, laboratory or radiology technologists, emergency medical technicians (EMTs) or paramedics; reports of utilization of resources, errors or costs to managers; or information resources and decision aids for patients and the general public. The kind of knowledge involved, the decision-making approach used, and the mode of operation may vary considerably depending on user and purpose.

Explicit vs. indirect support

Calculation tools, guidelines, alerts, and reminders all are designed to give specific advice or recommendations. But the other kind of decision support we have included is one in which the organization, grouping, or sequencing of information presentation is intended to foster optimal decision-making in a more subtle manner, simply by focusing attention, serving to remind the user of items that might otherwise be forgotten, encouraging systematic consideration and possibly influencing prioritization. We mention again some examples of this mode of decision support, namely the use of structured data entry forms, order sets, templates for reports and summaries, dashboards, flow sheets, and protocols.

3.3 OTHER CONSIDERATIONS

We have touched on a number of settings and contexts in which decision support could be used, but which will not be discussed in detail in this book. One of the primary other uses is in the realm of education and training. Not only can decision support knowledge bases be useful as educational reference tools, but the decision support can be used directly in a dynamic way in case-based problem-solving exercises—simulations of clinical problems requiring intervention by a user and feedback about the appropriateness of the actions taken. Methods analogous to CDS may be used to generate a range of variation of clinical parameters in a simulation, with the inclusion of a random component as well. CDS-like capabilities can be used in a critiquing mode, in which actions are performed by the user first and then evaluated by the CDS-like resource for conformance with the underlying decision model (e.g., a guideline). Or decision support may be used in what is known as an "intelligent tutoring system" mode of operation, to probe student responses or actions in terms of their similarity to prototypical problems in such situations, and to tease out the underlying misconceptions.

We will not delve further into image and signal processing, pattern recognition, and feature extraction, as these are largely embedded in niche applications, since our focus is on more generic CDS capabilities and the issues of deploying them in a health care enterprise. We will also not deal in more detail with patient/consumer-oriented decision support or public health surveillance systems. Both of these topics are important and robust areas of activity, and they interact with and depend to some extent on clinical IT systems and EHRs. But again, given our focus on providing clinical decision support, we will limit our focus to the role of CDS in the operational clinical IT context.

Much of the development of CDS to date has been somewhat of an art form, with creative individuals identifying innovative and useful ways of providing it and showing effectiveness. Because of a lack of well-defined principles, the discovery process often has had to be replicated by others, sometimes with painful and disappointing results. The collective body of experience in the literature is nonetheless quite large.

Although there is new impetus to moving ahead, the lessons of the past need to be recognized if we are not to be destined to repeat the mistakes that have limited progress over the past 45 years. A goal of this book is to begin to move toward a formal understanding of the requirements for CDS, based on the lessons and experiences of the past, clarifying an understanding of the requirements for infrastructure, standards, and business/organizational strategies that will lead to success.

The task of providing and maintaining robust CDS capabilities is a long and complex undertaking. It is important not to oversimplify it, or to rush to deploy CDS without adequate preparation, lest unsatisfactory results occur, bad press be generated, and an era of discouragement take hold. We seek to increase awareness and understanding of what the effort requires and to begin a systematic approach to tackling the problems that have vexed the field and held it back over these many years.

REFERENCES

Abendroth, T. W. and Greenes, R. A. (1989). Computer presentation of clinical algorithms. *MD Comput* **6**(5): 295–299.

Altenstetter, C. (2003). EU and member state medical devices regulation. *Int J Technol Assess Health Care* **19**(1): 228–248.

Barnett, G. O., Cimino, J. J., Hupp, J. A. and Hoffer, E. P. (1987). DXplain. An evolving diagnostic decision-support system. *JAMA* **258**(1): 67–74.

Berlin, A., Sorani, M. and Sim, I. (2006). A taxonomic description of computer-based clinical decision support systems. *J Biomed Inform* (Epub ahead of print).

Berner, E. S., ed. (1998). *Clinical Decision Support Systems: Theory and Practice. Health Informatics*. Springer.

Bindels, R., Hasman, A., Kester, A. D., Talmon, J. L., De Clercq, P. A. and Winkens, R. A. (2003). The efficacy of an automated feedback system for general practitioners. *Inform Prim Care* **11**(2): 69–74.

Bodenheimer, T. (2003). Interventions to improve chronic illness care: evaluating their effectiveness. *Dis Manag* **6**(2): 63–71.

Bruce, B. and Fries, J. F. (2005). The Arthritis, Rheumatism and Aging Medical Information System (ARAMIS): still young at 30 years. *Clin Exp Rheumatol* **23**(5 Suppl 39): S163–167.

Christensen, C. M. and Raynor, M. E. (2003). *The Innovator's Solution*. Cambridge, MA, Harvard Business School Press.

Ciccarese, P., Caffi, E., Quaglini, S. and Stefanelli, M. (2005). Architectures and tools for innovative Health Information Systems: the Guide Project. *Int J Med Inform* **74**(7–8): 553–562.

Cimino, J. J., Li, J., Bakken, S. and Patel, V. L. (2002). Theoretical, empirical and practical approaches to resolving the unmet information needs of clinical information system users. *Proc AMIA Symp*: 170–174.

Costantini, O., Papp, K. K., Como, J., Aucott, J., Carlson, M. D. and Aron, D. C. (1999). Attitudes of faculty, housestaff, and medical students toward clinical practice guidelines. *Acad Med* **74**(10): 1138–1143.

deDombal, F. T. (1975). Computer-aided diagnosis and decision-making in the acute abdomen. *J R Coll Physicians Lond* **9**(3): 211–218.

Essaihi, A., Michel, G. and Shiffman, R. N. (2003). Comprehensive categorization of guideline recommendations: creating an action palette for implementers. *AMIA Annu Symp Proc*: 220–224.

Feigenbaum, E. A. and Feldman, J., eds. (1963). *Computers and Thought*. New York, McGraw Hill.

Freedman, A. N., Seminara, D., Gail, M. H., Hartge, P., Colditz, G. A., Ballard-Barbash, R., et al. (2005). Cancer risk prediction models: a workshop on development, evaluation, and application. *J Natl Cancer Inst* **97**(10): 715–723.

Giger, M. L. (2004). Computerized analysis of images in the detection and diagnosis of breast cancer. *Semin Ultrasound CT MR* **25**(5): 411–418.

Greenhalgh, J., Long, A. F. and Flynn, R. (2005). The use of patient reported outcome measures in routine clinical practice: lack of impact or lack of theory? *Soc Sci Med* **60**(4): 833–843.

Hackett, J. L. and Gutman, S. I. (2005). Introduction to the Food and Drug Administration (FDA) regulatory process. *J Proteome Res* **4**(4): 1110–1113.

Harding, J. (1994). Practice guidelines. Cookbook medicine. *Physician Exec* **20**(8): 3–6.

Harpole, L. H., Khorasani, R., Fiskio, J., Kuperman, G. J. and Bates, D. W. (1997). Automated evidence-based critiquing of orders for abdominal radiographs: impact on utilization and appropriateness. *J Am Med Inform Assoc* **4**(6): 511–521.

HL7. (2006). The Reference Information Model (RIM). Health Level Seven, Inc., from http://www.hl7.org/Library/data-model/RIM/modelpage_mem.htm.

Hripcsak, G. (1991). Arden Syntax for Medical Logic Modules. *MD Comput* **8**(2): 76, 78.

Jiang, Y., Nishikawa, R. M., Schmidt, R. A., Metz, C. E., Giger, M. L. and Doi, K. (1999). Improving breast cancer diagnosis with computer-aided diagnosis. *Acad Radiol* **6**(1): 22–33.

Johnson, H. (1994). Relationship between user models in HCI and AI. *Computers and Digital Techniques, IEE Proceedings* **141**(2): 99–103.

Kawamoto, K., Houlihan, C. A., Balas, E. A. and Lobach, D. F. (2005). Improving clinical practice using clinical decision support systems: a systematic review of trials to identify features critical to success. *BMJ* **330**(7494): 765.

Kim, E. K., Kwon, Y. D. and Hwang, J. H. (2005). [Comparing the performance of three severity scoring systems for ICU patients: APACHE III, SAPS II, MPM II]. *J Prev Med Pub Health* **38**(3): 276–282.

Kuperman, G. J., Teich, J. M., Gandhi, T. K. and Bates, D. W. (2001). Patient safety and computerized medication ordering at Brigham and Women's Hospital. *Jt Comm J Qual Improv* **27**(10): 509–521.

Liang, M. H. (1992). From America: cookbook medicine or food for thought: practice guidelines development in the USA. *Ann Rheum Dis* **51**(11): 1257–1258.

Lodwick, G. S. (1965). A probabilistic approach to the diagnosis of bone tumors. *Radiol Clin North Am* **3**(3): 487–497.

Miller, P. L. (1983). Critiquing anesthetic management: the "ATTENDING" computer system. *Anesthesiology* **58**(4): 362–369.

Miller, P. L. and Black, H. R. (1984). Medical plan-analysis by computer: critiquing the pharmacologic management of essential hypertension. *Comput Biomed Res* **17**(1): 38–54.

Mugford, M., Banfield, P. and O'Hanlon, M. (1991). Effects of feedback of information on clinical practice: a review. *BMJ* **303**(6799): 398–402.

NHLBI. (2004). The Seventh Report of the Joint National Committee on Prevention, Detection, Evaluation, and Treatment of High Blood Pressure (JNC 7). *National High Blood Pressure Education Program*; National Heart Lung and Blood Institute, National Institutes of Health, U.S. Department of Health and Human Services, from http://www.nhlbi.nih.gov/guidelines/hypertension/.

NLM. (2005). RxNorm. *Unified Medical Language System*; National Library of Medicine, from http://www.nlm.nih.gov/research/umls/rxnorm/index.html.

Ogden, J., Fuks, K., Gardner, M., Johnson, S., McLean, M., Martin, P., et al. (2002). Doctors expressions of uncertainty and patient confidence. *Patient Educ Couns* **48**(2): 171–176.

Regenstrief. (2005). Logical Observation Identifiers Names and Codes (LOINC®). Regenstrief Institute, from http://www.regenstrief.org/loinc/.

Richardson, G. and Manca, A. (2004). Calculation of quality adjusted life years in the published literature: a review of methodology and transparency. *Health Econ* **13**(12): 1203–1210.

Rockhill, B., Spiegelman, D., Byrne, C., Hunter, D. J. and Colditz, G. A. (2001). Validation of the Gail et al. model of breast cancer risk prediction and implications for chemoprevention. *J Natl Cancer Inst* **93**(5): 358–366.

Shiffman, R. N. and Greenes, R. A. (1991). Use of augmented decision tables to convert probabilistic data into clinical algorithms for the diagnosis of appendicitis. *Proc Annu Symp Comput Appl Med Care*: 686–690.

Shortliffe, E. (1976). *Computer-based Medical Consultations: MYCIN*. New York, Elsevier.

Sim, I. and Berlin, A. (2003). A framework for classifying decision support systems. *AMIA Annu Symp Proc*: 599–603.

Smith, R. A., Cokkinides, V. and Eyre, H. J. (2004). American Cancer Society guidelines for the early detection of cancer, 2004. *CA Cancer J Clin* **54**(1): 41–52.

SNOMED. (2006). SNOMED CT. *SNOMED International*; College of American Pathologists, from http://www.snomed.org/snomedct/.

Sordo, M., Boxwala, A. A., Ogunyemi, O. and Greenes, R. A. (2004). Description and status update on GELLO: a proposed standardized object-oriented expression language for clinical decision support. *Medinfo* **11**(Pt 1): 164–168.

Tu, S. W., Musen, M. A., Shankar, R., Campbell, J., Hrabak, K., McClay, J., et al. (2004). Modeling guidelines for integration into clinical workflow. *Medinfo* **11**(Pt 1): 174–178.

Wang, D., Peleg, M., Bu, D., Cantor, M., Landesberg, G., Lunenfeld, E., et al. (2003). GESDOR—a generic execution model for sharing of computer-interpretable clinical practice guidelines. *AMIA Annu Symp Proc*: 694–698.

Warner, H. R., Toronto, A. F. and Veasy, L. G. (1964). Experience with Baye's Theorem for Computer Diagnosis of Congenital Heart Disease. *Ann N Y Acad Sci* **115**: 558–567.

Weaver, R. R. (2003). Informatics tools and medical communication: patient perspectives of "knowledge coupling" in primary care. *Health Commun* **15**(1): 59–78.

II

CASE STUDIES AND CURRENT STATUS

4 REGENSTRIEF MEDICAL INFORMATICS: EXPERIENCES WITH CLINICAL DECISION SUPPORT SYSTEMS

PAUL BIONDICH, BURKE MAMLIN, WILLIAM TIERNEY, MARC OVERHAGE, and CLEMENT McDONALD

4.1 INTRODUCTION

The discipline of medical informatics endeavors to improve the process and outcomes of health care by enabling efficient access to information. Care providers can then use this information, either in the form of medical knowledge or as patient data collected during clinical practice to make decisions and comply with appropriate standards of care. The Regenstrief Institute began work on clinical information systems in 1972, when Dr. Clement McDonald and colleagues conceptualized and began construction of a computerized patient management system for outpatient diabetes care (McDonald et al. 1999). Motivating the design team was an early realization that human beings as care providers have finite capacities as information gatherers and processors, and are subject to oversights and distractions. Therefore, the system was developed to meet three primary goals. First, it was built to eliminate the problems inherent in paper records by making clinical data available to authorized users "just-in-time" (Chueh and Barnett 1997) as medical decisions are made. Second, it was designed to aid in the recognition of diagnoses and adoption of pertinent care practices by assisting clinicians during their record keeping activities. Finally, the system was designed to aggregate and analyze clinical information to be used in health care support systems, such as those for public health, health services research, and quality improvement.

The first installation of the Regenstrief Medical Record System (RMRS) at Wishard Memorial Hospital occurred in 1974, on a Digital Equipment Corporation PDP-11/44 computer with four user access lines (see Figure 4-1). Over the next few years, the use of this system expanded outside of the diabetic clinic into a few of the hospital's many general medicine clinics. During this initial development, it became clear that in order to get to the "interesting" part of the effort—clinical decision support—that significant

Clinical Decision Support: The Road Ahead

FIGURE 4-1 The first Regenstrief mainframe, a PDP-11/44. Doug Martin, an early system developer, is shown working with the system.

effort must first focus on an underappreciated complexity of health information system development: capturing structured, standardized data. Even today, this focus provides the RMRS infrastructure with the quality data substrate necessary to automate processes of care. From early in its history, the Regenstrief system has included mechanisms for tailoring rules around these data, which generate reminders and alerts to care providers (McDonald et al. 1992). What follows therefore is a history of the development and growth of the RMRS into a region-wide source of clinical data, the Indiana Network for Patient Care (INPC), and of research on the decision support interventions made possible by this infrastructure. Additionally, lessons learned throughout the more than 30 years of experience in both building and maintaining this system are detailed, alongside some reflections that may be useful for future system builders.

4.2 HISTORY

4.2.1 Early System Development and Paper-based Reminders

The early infrastructure built for the system in the diabetic and outpatient medicine clinics allowed care providers to record predefined portions of a patient visit including laboratory results, visit diagnoses, medications prescribed, and vital signs on structured paper encounter forms. These data were manually entered into a fairly general but "hard-coded" data structure. Clinical informatics researchers then developed a rule set to identify patients who were eligible for particular clinical actions, from which reminders corresponding to each rule were generated. For example, if a patient was taking aminophylline and his or her serum aminophylline level had not been measured within a designated time period, the computer generated a reminder to order this test. The reminders for each patient were delivered as a paper report that the clinic staff attached to the front of the patient's chart, along with a computer-generated flow sheet and encounter form. The flow sheet displayed the patient's active drug profile alongside other clinical data, and provided space for writing notes and orders.

In 1976, McDonald reported on the first randomized, controlled study of this intervention in the *Annals of Internal Medicine* (McDonald 1976). During the eight-month study, the reminder engine ran nightly to examine the records of patients with visits the following day. Providers were randomized to receive either a reminder report (see Figure 4-2) and supporting documentation (the clinical flow sheet and encounter form) or the supporting documentation alone. Patients inherited the study status of their respective physicians. For this initial study, nearly 300 rules had been developed, fitting into two primary categories: 1) recommendations for corollary orders based on specific medications prescribed and 2) reminders to change therapy in response to a test result abnormality.

The computer generated reminders for both intervention and control patients, but delivered them only to the physicians in the intervention group. The primary study measure was the rate at which physicians responded to the computer-suggested actions for the eligible patients. During the study period, the system registered events that required clinician action in 601 visits (301 visits by 119 study patients and 300 visits by 107 control patients). Sixty-three clinicians responded to 36 percent of events with reminders and 11 percent without ($p < 0.001$). When only the most clinically significant events were considered (e.g., because patient has a diastolic blood pressure > 110 mmHg, consider a stronger anti-hypertensive regimen), the clinicians given the

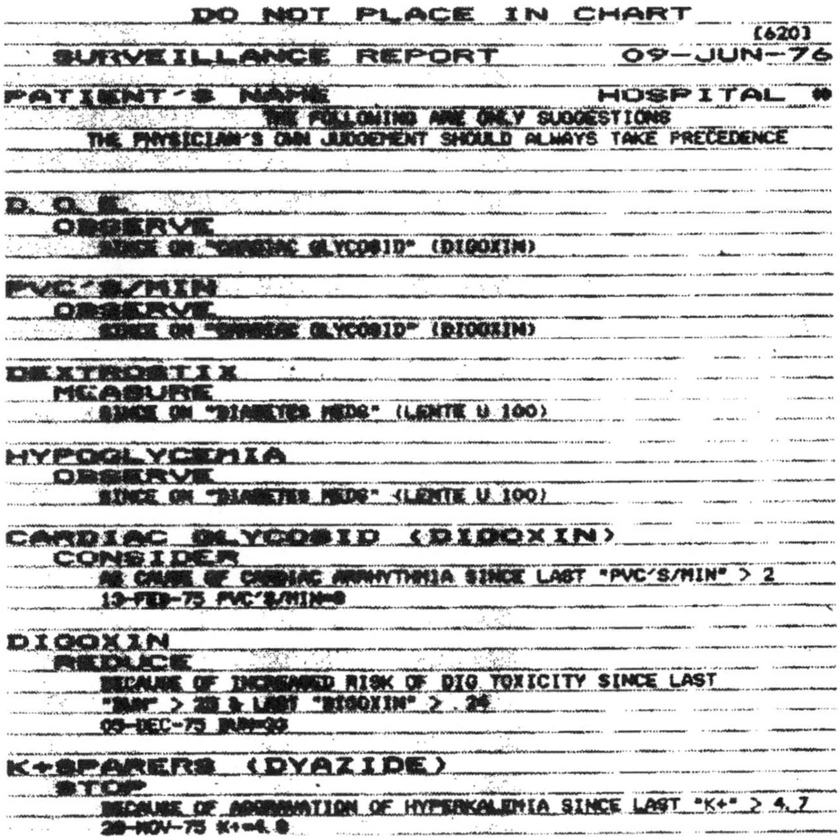

```
            DO NOT PLACE IN CHART
                                                  [620]
   SURVEILLANCE REPORT                    09-JUN-76
PATIENT'S NAME                          HOSPITAL #
               THE FOLLOWING ARE ONLY SUGGESTIONS
      THE PHYSICIAN'S OWN JUDGEMENT SHOULD ALWAYS TAKE PRECEDENCE

D. O. E.
   OBSERVE
      SINCE ON "CARDIAC GLYCOSID" (DIGOXIN)

PVC'S/MIN
   OBSERVE
      SINCE ON "CARDIAC GLYCOSID" (DIGOXIN)

DEXTROSTIX
   MEASURE
      SINCE ON "DIABETES MEDS" (LENTE U 100)

HYPOGLYCEMIA
   OBSERVE
      SINCE ON "DIABETES MEDS" (LENTE U 100)

CARDIAC GLYCOSID (DIGOXIN)
   CONSIDER
      AS CAUSE OF CARDIAC ARRHYTHMIA SINCE LAST "PVC'S/MIN" > 2
      13-FEB-75 PVC'S/MIN=[illegible]

DIGOXIN
   REDUCE
      BECAUSE OF INCREASED RISK OF DIG TOXICITY SINCE LAST
      "BUN" > [illegible] & LAST "DIGOXIN" > .2[illegible]
      05-DEC-75 BUN=[illegible]

K+SPARERS (DYAZIDE)
   STOP
      BECAUSE OF AGGRAVATION OF HYPERKALEMIA SINCE LAST "K+" > 4.7
      [illegible]-NOV-75 K+=4.[illegible]
```

FIGURE 4-2 An example of a reminder report. Processes that ran overnight culled through medical record data to generate these printed reports used during subsequent patient encounters.

reminder report had a tenfold increased response rate (47% vs. 4%). This study showed that computer-generated reminders could positively influence clinical processes, and was the first randomized, controlled study of computer-based decision support to show statistically significant effects.

The *New England Journal of Medicine* shortly thereafter published results of a larger study by McDonald that utilized a cross-over design, in which physicians served as their own controls (i.e., physicians were the intervention subjects in one phase of the study and controls in another). In this study, researchers followed nine physicians practicing in a general medicine clinic for one half-day per week for 17 weeks (McDonald 1976). During the intervention phase, clinicians were asked to put their initials on the reminder reports to indicate that they had actually seen the reminders. The 390 reminder rules used within this study fell into three main categories: 1) suggestions to observe physical findings or make symptom inquiries, 2) reminders to order a diagnostic study, and 3) reminders to change or initiate a specific care regimen. The reminders were printed on the encounter form that clinicians used for writing notes and orders, as well as on the one-page reminder sheet.

Physicians acted on the computer-suggested actions in 51 percent of 327 intervention events and 22 percent of 385 control events ($p < 0.001$). The rate at which they responded was not higher for physicians whose control periods followed their study periods; thus the computer-generated suggestions had no "training effect." In other words, the computer reminders didn't teach the physicians something, but rather activated their pre-existing knowledge and intentions. This finding is what the paper alludes to as the "non-perfectability of man." As McDonald wrote,

> Implicit in currently available remedies for medical errors is the belief that man is perfectable and his errors can be eliminated by training or coercion. However, man is not perfectable. There are limits to man's capabilities as an information processor that assure the occurrence of random errors in his activities. In studies using flight simulators, Drinkwater showed that sensory overload consistently caused pilot errors, often with "fatal" consequences. This study has obvious implications for the performance of physicians under the peak informational loads of busy practice settings. When keeping watch for random and infrequent events under experimental conditions, man predictably overlooks some target events. The physician in his watch for pathologic events is no exception (McDonald 1976).

Interestingly, during the study, one medicine resident insisted, "what was needed was education of bad physicians about what they did not know, not reminders to good physicians about what they knew." Because each physician served as his or her own control, researchers could observe the effect of the reminders on each individual physician. As it turned out, this particular physician's response to reminders was larger than that of any other study participant. He took the suggested action in 75 percent of the cases when he was reminded, but only in 25 percent of the cases when not.

A subsequent study in 1980, which used 410 computerized protocols and studied 31 physicians over 17 weeks, corroborated the findings of the two studies just described (McDonald et al. 1980). In this study, the Institute offered an improved and more sophisticated set of reminder rules along with additional access to relevant medical literature. The reminder reports given to care providers cited these references as justifications for reminders and invited the providers to obtain original copies of articles from a pharmacist stationed in the clinic during the study period. Physicians in this study had a twofold increased response rate to events in the intervention group, consistent with previous findings. However, the reminders evoked little physician interest in the supporting literature. In fact, not a single cited article was requested from the pharmacist during the study. According to informal discussions following the study, the physicians did not ask for copies of the referenced articles for two reasons. One, because of the time pressures related to other care duties on the wards and two, because they largely agreed with, and knew the evidence that justified, the reminder (i.e., they did not need convincing).

4.2.2 RMRS' Maturation into a Hospital-wide Medical Record System

The successes of these initial studies drove the continued development and acceptance of computer systems in Wishard Hospital. However, much of the initial development work of the RMRS quickly taught the team that programming into fixed data structures was a poorly scalable model. Collaborations with computer scientists from Purdue University helped to restructure the initial system into a relational data model that allowed one to describe medical concepts and their definitions in separate tables alongside the clinical data collected (McDonald et al. 1988). The team also continued to deal with the difficulties inherent in manual data capture, so processes and applications were developed to collect data at their sources (pharmacy, laboratory, etc.). By the late 1970s, the RMRS had grown into a full electronic medical record system for the hospital. Additionally, the initial work in building reminder rules led to compilation of a formal computer language called CARE (McDonald et al. 1984). These two developments significantly increased both the scope and potential impact of the reminder-based decision support tools and led to a second wave of studies based on this new development work.

In 1984, the *Annals of Internal Medicine* published the results of the research group's largest reminder study (McDonald et al. 1984). This two-year, randomized, controlled trial included 1,490 different reminder rules, 130 different providers (including house officers, nurse practitioners, and faculty members), approximately 14,000 patients, of whom 90 percent were eligible for one or more of the reminder actions, and more than 50,000 visits. The study design incorporated a more rigorous randomization by practice (called "teams" in the Wishard environment) to minimize the possibility of contamination between intervention and control care providers. The team also used more sophisticated analytic techniques (Zeger and Liang 1986) to take into account possible clustering effects of patients within providers and/or providers within study teams.

The RMRS employed the same general strategy of delivering reminders on paper reports used in all previous studies. During this study, the computer generated 140,000 reminders for both intervention and control visits, and an average of six different clinical actions per patient. The reminders increased the physicians' response to actionable events from 29 percent in control practices to 49 percent in the intervention practices. Of note, the reminders had the greatest effect on potential preventive care interventions (e.g., influenza vaccination, pneumonia vaccination, occult blood testing), for which reminders demonstrated up to a fourfold increase in clinician response rates.

In all the studies discussed thus far, intervention groups responded more often to clinical events aided by reminder systems than to those that were not. However, even with these positive outcomes, physicians receiving reminders failed to respond to a relatively high percentage (40–50%) of potentially harmful events. There are many circumstances that contributed to this phenomenon. In some cases, physicians did not see the reminder reports. In McDonald's 1976 "nonperfectability" study, researchers asked physicians to indicate all reminder reports they had seen or read by their initials. The results showed that approximately 15 percent failed to complete this step. In other cases, a nonresponse is the correct response. Finally, despite best efforts, the computer did not always have all the relevant information about the suggested action. After auditing a sample of the charts in a 1984 study, the development team found gaps in the electronic patient record that would have invalidated the reminders in 5 percent (pneumonia vaccine) to 50 percent (cervical Papanicolaou smear testing) of cases, depending upon the action suggested.

Some of these realizations led to refinements in the way reminders were administered. In one study, Litzelman assessed the effects of forcing physicians to respond to selected reminders for cancer screening tests (Litzelman et al. 1993). In a randomized, controlled trial, intervention physicians either had to order the suggested test (cervical Papanicolaou smear, mammogram, or fecal occult blood test) or explain why they did not do so. This study included 145 physicians and 5,407 patients. Intervention physicians followed 46 percent of the reminders compared with 38 percent for control physicians ($p < 0.05$). Intervention physicians' reasons for not adhering to the guidelines included the physician being too busy or the patient being too sick that day (23%), the reminder being inappropriate (23%, mostly due to missing data on prior hysterectomies), or the patient's refusal to take the test (10%) (Litzelman and Tierney 1996).

Additional work was conducted in a McDonald study that provided a similar style of reminders for patients eligible for influenza immunizations (McDonald et al. 1992). In this study, large effects on care processes were demonstrated, and with a secondary analysis, the team additionally detailed reductions in morbidity measured through statistically significant decreased hospitalizations, decreased emergency room visits, and less blood gas tests ordered within the intervention cohort. This effect was limited to patients who were eligible for influenza vaccines in the years when a large excess in pneumonia mortality occurred. Nonetheless, it is unrealistic to expect a clear demonstration of improved outcomes in every process intervention, because individual small scale process interventions often do not have the huge sample sizes that would be needed to show outcome effects. The HIP Mammography study included nearly 60,000 eligible women and followed them for over seven years (Shapiro 1997). The previously described Litzelman reminder study, by contrast, included less than 3,000 eligible women who were followed for two years.

Other research in the Institute assessed whether it was necessary to deliver reminders in real time. In other words, could similar results to the previously described studies be replicated by providing monthly reports to physicians instead of reminders? Researchers compared these two interventions in a 2 × 2 factorial randomized, controlled trial (Tierney et al. 1986). This seven-month study included 6,045 patients and targeted 13 preventive care rules, both involving testing (e.g., mammograms) and treatment (e.g., oral calcium carbonate for osteoporosis prophylaxis). Each rule could be delivered as a monthly feedback report or as a printed reminder in the clinic (immediate feedback). Each feedback report contained a list of patients that were noted by the RMRS to be eligible for preventive care, but had not received it. Suggestions on the feedback reports were followed by a series of specific actions the physician could take including scheduling of an earlier return visit, ordering extra diagnostic work, and triggering the computer to send a reminder at next visit.

The study demonstrated that both reminders and feedback reports increased adherence to the preventive care suggestions, although reminders were significantly more efficacious. Notably, the combination of reminders and feedback reports was no better than feedback reports alone. This can be explained by an evaluation of clinician responses to the feedback report, 80 percent of whom requested that reminders be printed at the next visit.

4.2.3 Early Development of Computerized Physician Order Entry

Most of the initial development work of the RMRS was built around the assumption that care providers would be unable or unwilling to incorporate a computer workstation into their clinical routines. Thus, interfaces were primarily designed to be built around paper. During the 1980s, as computer workstations became more commoditized, Regenstrief began development of computer applications to give care providers more direct control of a patient's medical record. Many of these utilities and tools took the form of a menu-based system known as the Medical Gopher (McDonald and Tierney 1986). The Gopher contained physician order entry (CPOE), note writing, and data query abilities from its outset, and was designed to interact directly with the repository that had developed over the past 10 years (see Figures 4-3 and 4-4). Although CPOE can exist apart from decision support, developers integrated it into the Medical Gopher from its inception and leveraged the new direct physician interaction opportunities to facilitate future studies. Regenstrief's first deployment of the Gopher occurred in the outpatient medicine clinics, where researchers performed studies on three different interventions during order entry designed to affect subsequent laboratory test ordering behaviors: display of previous test results, display of test cost, and indication of the likelihood of a positive result when tests were being ordered.

The first study included 111 internal medicine physicians and ran for 16 weeks (Tierney et al. 1987). Scheduled patients were randomized to intervention or control groups, and previous test results were displayed when any of a selected subset of the most common test panels (e.g., electrolyte panel) was ordered on intervention patients. When the computer displayed previous test results for commonly ordered tests, the physicians ordered 16.8 percent fewer of such tests and generated 13 percent lower test charges.

Displaying the charges for individual outpatient tests, along with the total charge for all tests ordered during the outpatient visit, also reduced test ordering. Researchers studied the orders of 121 physicians in an outpatient medicine clinic over 59 weeks (Tierney et al. 1990). Half of the physicians were randomized to the intervention group and the other half to the control group. For 14 weeks before the intervention began, the test ordering rates and related charges were the same across both groups. During the 26-week intervention, when the computer displayed the charge of each new test ordered and the cumulative charges for all tests ordered during that session, physicians ordered 14 percent fewer tests per patient visit and generated charges that were 13 percent lower ($6.68 per visit). The effect was greater for scheduled visits than nonscheduled visits and fell back to near baseline in the six months

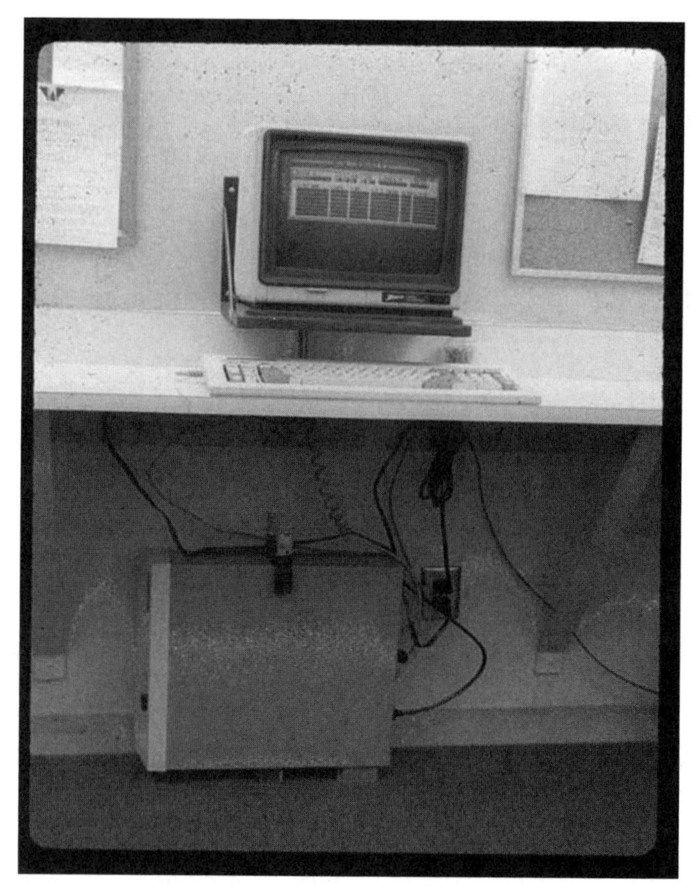

FIGURE 4-3 An early Medical Gopher station in Wishard Memorial Hospital. This is an example of the first of many iterations of this computerized physician order entry system.

following the intervention. No change in adverse outcomes (hospitalizations, ER visits, and clinic visits) occurred in the intervention group.

In the third study, Tierney developed logistic regression equations, based on data in the RMRS about medication use, previous test values, and demographics to predict the likelihood that a test result would be abnormal (Tierney et al. 1988). This was done for eight different laboratory tests. When the computer displayed the predicted probabilities that a test would be abnormal—most of which were much lower than the physicians expected—physicians ordered significantly fewer tests, resulting in a 9 percent reduction in charges.

During the late 1980s, the Regenstrief informatics team extended the use of the Medical Gopher from the outpatient clinics into the inpatient wards of Wishard Memorial Hospital, which allowed for the first randomized study of CPOE compared to traditional, paper order entry on the inpatient medicine service. In this study, the order entry system provided problem-specific menus, order-specific templates, a display of the patient's charge for each item, and warnings for allergies, drug–drug, or drug–diagnosis interactions (Tierney et al. 1993). The menus and templates were designed to encourage cost-effective ordering and discourage expensive treatments. There were no active reminders included in this study. Six inpatient medicine ward services were randomized to intervention or control groups.

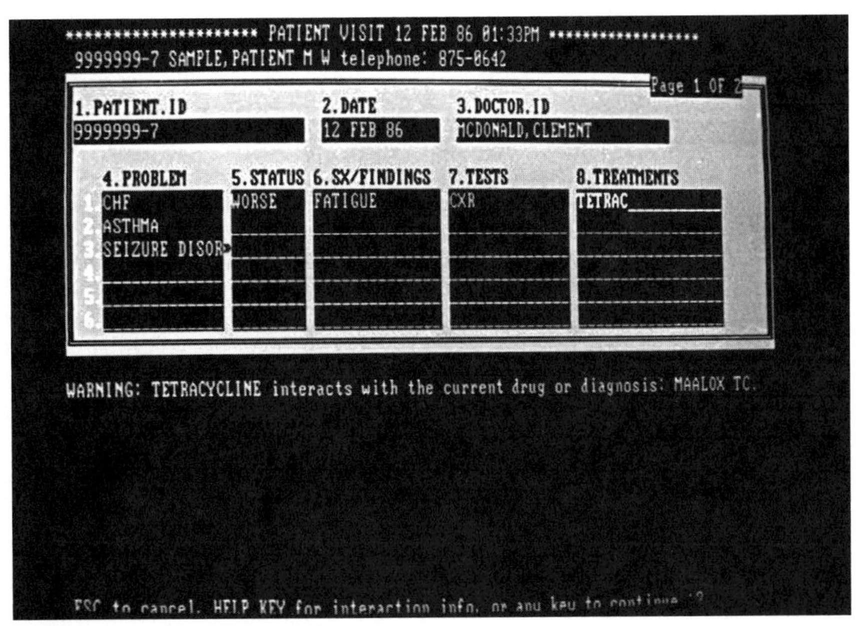

FIGURE 4-4 Screen shot from the first version of the Medical Gopher. It provided an immediate forum to generate reminders in real time for providers as they ordered therapies for patients.

Over 19 months, 68 teams, each comprised of a faculty member, residents, and medical students, were randomly assigned to these services and cared for 5,219 patients. Researchers accounted for the fact that house staff returned for additional rotations by placing them onto a service with the same status—intervention or control—as their initial rotation. Physicians and students on ward services randomized to the intervention arm used the PC-based order entry system for all orders, whereas control services continued to use a traditional paper-based, order-writing chart and had no access to the Gopher system. The patients seen on intervention services were not significantly different from those on control services. However, there was a demonstrated savings of 12.7 percent (nearly $900) per admission when the CPOE system was used. Further, hospital stays for intervention admissions were 10.5 percent (0.89 days) shorter than controls ($p = 0.11$). No differences in hospital readmissions, ER visits, or clinic visits were seen at one and three months following discharge. When the cost savings were extrapolated to all of the hospital's medical services, the predicted savings were $3 million per year.

4.2.4 Continued Development of the Medical Gopher

Once the Medical Gopher order entry system was used regularly in both inpatient and outpatient settings, the system evolved into a platform for real-time, complex decision support. Developers created a new, rule-writing language for the Gopher order-entry system called G-CARE (short for Gopher-CARE), which is the basis for the CPOE system still being used. Care providers can respond to reminders generated from G-CARE in real time as part of the order entry process, while also considering data already stored in the RMRS. These rules can be activated at many steps during an order, including at the start of a session (before the physician enters any data), immediately after entry of a medical problem, at the time an order is selected, when the order is completed, and at the completion of a session. These alternatives are further described elsewhere (Overhage et al. 1995). G-CARE also can be used to provide prior test results, suggest orders for baseline testing or follow-up monitoring, or block contraindicated orders and suggest alternatives. For example, the computer might suggest a nuclear medicine renal study instead of an intravenous pyelogram (IVP) in a patient with renal insufficiency.

Given previous successes with preventive care reminders in the outpatient setting, the research team expected similar results in a study of preventive care reminders on inpatient care. However, the initial evaluation proved to be a negative study (Overhage et al. 1996). These negative findings were attributed to two factors: 1) providers often think of preventive care (e.g., immunizations) as a distinctly outpatient practice, and 2) the delivery method for reminders was too gentle—a message advised the physician that reminders were available and required him or her to press a special key to see more information.

Changes were, therefore, made to reminders that integrated potential preventive care orders into a list that a provider could accept or reject but not ignore. On a subsequent inpatient study that focused this variation on four preventive measures (pneumococcal vaccination, influenza vaccination, aspirin for vascular disease risk, and subcutaneous heparin for DVT prophylaxis), real-time reminders to inpatient physicians ultimately had the large effect originally expected (Dexter et al. 2001). In this study, providers were randomized into intervention and control groups, and intervention physicians received reminders as orderable pop-ups. Over half of the 6,371 patients admitted to a general medicine service during an 18-month period were eligible for one or more of the four preventive measures. Their ordering rates (intervention vs. control) were 35.8 percent vs. 0.8 percent for pneumococcal vaccination, 51.4 percent vs. 1.0 percent for influenza vaccination, 32.2 percent vs. 18.9 percent for prophylactic heparin, and 36.4 percent vs. 27.6 percent for prophylactic aspirin ($p < 0.001$ in all cases).

Real-time, computer-generated reminders within the Gopher have also demonstrated a capability to decrease errors of omission. In a 1997 study, Overhage developed a set of corollary orders for 87 selected test and treatment requests (Overhage et al. 1997). For example, an order for a heparin drip would produce a corresponding order for an APTT measurement to follow heparin's effect on the clotting cascade. These were studied formally over a six-month trial, where clinicians were randomized to receive corollary orders in direct patient care processes. The physicians who had been given these reminders ordered them for 46.3 percent of their eligible patients versus a 21.9 percent ordering rate in comparable circumstances within the control group ($p < 0.0001$). Pharmacists intervened for errors considered to be life threatening, severe, or significant 33 percent less often for intervention physicians than for control physicians. No significant change in length of stay or hospital charges was detected.

As the medical institution became more familiar with the versatility of reminders, their new uses grew exponentially, and some particular styles were formally evaluated. For example, Dexter et al. studied reminders for advanced directive discussions between clinicians and their patients (Dexter et al. 1998). If the system had no record of an advanced directive, and a given patient met prespecified criteria (such as an age of 75 or older, or serious chronic conditions such as end-stage renal disease), the order entry system was programmed to remind clinicians of the need to obtain different types of advanced directives for the patient. Two main types facilitated by the Gopher are instruction directives (how clinicians should care for the patient in certain circumstances) and proxy directives (who should make decisions for the patient or who receives power of attorney). In a 2 × 2 factorial design, the study randomized physicians to one of four groups: no reminders, both types of reminders, or one of each type of reminder. The results demonstrated that both types of reminders increased the rate at which physicians discussed advanced directives with eligible patients, with the greatest effect—a sixfold increase—observed for physicians who received both types of reminders over the one-year study period (4% for controls compared to 24% for these physicians).

Reminders were also written to suggest less expensive medications and therapies when a physician chose the more expensive medicine in a given therapeutic class. For example, when physicians ordered one of the patented and expensive anti-depressants, their attention was directed to fluoxetine (Prozac), which at the time was roughly one-fifth the cost but equally efficacious (see Figure 4-5). In the case of fluidized beds, providers are reminded

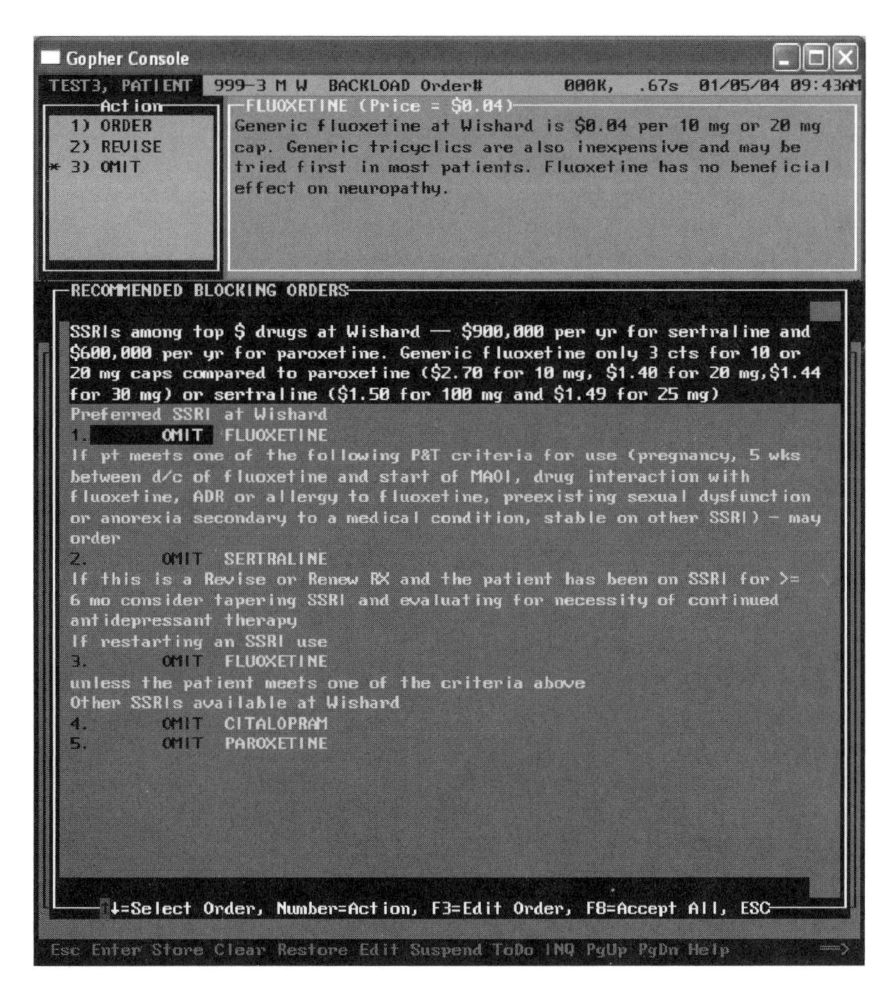

FIGURE 4-5 An example of suggested orders in the Medical Gopher based upon cost and efficacy data.

about static air mattresses, which some studies have shown to be as good but less expensive. The Institute has also written rules and logic that help the hospital physicians adhere to Medicare regulations about short-stay patients. For example, the system reminds physicians to correctly classify their patients as short-stay patients at admission, and then shortly before the 36-hour limit, reminds the responsible physician(s) to write full admissions orders if necessary.

4.2.5 Operationalizing Clinical Guidelines within the Gopher

As guidelines were introduced into the order entry system using G-CARE, Regenstrief investigators documented some valuable insights (Tierney et al. 1995, 1996; McDonald et al. 1996). Tierney described some of the problems encountered when developing computerizing guidelines for heart failure. For example, information in published guidelines can be rather inadequate for defining a computer-executable reminder rule. Published guidelines often include vague terminology, omit branch points, rely on data not available in the electronic records, and fail to address comorbidities or concurrent therapy.

Despite these limitations, developers were able to program detailed guidelines for managing ischemic heart disease, hypertension, and reactive airway disease into the outpatient Gopher workstations. In a two-year randomized, controlled trial, 2,123 patients were prospectively enrolled, with 700 having each of these conditions. Half of the physicians were randomized to receive suggested orders for management of these chronic conditions, half did not. All physicians used Gopher to write all orders. A retrospective review of 10 percent of the charts of patients included in this study showed that the suggested care was indeed indicated. However, receiving suggested orders had no effect on adherence to any of the evidence-based suggestions, clinical outcomes (e.g., hospitalizations or emergency department visits for heart or lung disease, blood pressure control), or health-related quality of life for patients with heart disease (Tierney et al. 2003), lung disease (Tierney et al. 1997), or hypertension (Murray et al. 2004). These same physicians, who adhered to preventive care reminders, ignored most reminders about chronic disease management. Querying them on their responses to guidelines for managing chronic conditions elicited mixed sentiments. Although they found the guidelines to be good sources of information, they also felt they were intrusive and mostly aimed at controlling costs. The management of chronic disease was also felt to be more complicated and subject to more special cases and alternatives than can be easily incorporated into computer reminder rules.

4.2.6 Growth of the Indianapolis Network for Patient Care

While the Regenstrief system continued to scale and address the needs of the Indiana University health care campus, health care within Indianapolis evolved in many important ways. For example, more hospitals entered into the community, creating more choices for providers and patients alike. Hospitals also began to evolve affiliated outpatient clinic networks. Laboratory and other diagnostic studies were increasingly performed outside of primary hospital settings, as commercial diagnostic laboratories became commonplace.

In the meantime, the public health system continued to provide more preventive care services and treatments to better care for the underserved. These many realities made it increasingly less likely that any one medical record system could accurately represent the care of an individual patient. In fact, it was quickly apparent that to provide better care, and provide the next generation of substrate for decision support systems, that the Institute must devote significant resources to the development of federated, regional clinical data repositories.

The Indiana Network for Patient Care (INPC) evolved from dedicated work that began in earnest in 1994, but had its foundations in the standards development work, which started in the early 1980s (Biondich and Grannis 2004; McDonald et al. 2005). Regenstrief's early work on standards led to more organized efforts such as the HL7 messaging standard and LOINC, which to this day serve as foundations for the INPC. The network and its federated data repository are built on the transformation, standardization, and aggregation of pre-existing electronic data using these standards. This network currently serves the Indianapolis metropolitan statistic area but is actively evolving to include hospitals throughout the entire state of Indiana.

Some early research has begun to show the value of this process. A recent study by Overhage et al. highlighted both the feasibility and cost savings for Indianapolis area emergency rooms (Overhage et al. 2002). Additionally, investigators are actively building technologies on top of the INPC that enable reminders and alerts to reach large numbers of care providers through clinical result messaging services and adaptive turnaround documents (Biondich et al. 2003). Future work is intended to formally evaluate interventions such as "enhanced laboratory reports," which bundle corollary reminders on top of typical printed laboratory results; and a quality improvement program, which prints information about a given patient's compliance with national quality measures alongside other result information.

4.3 CONCLUSION

Regenstrief scientists have created a 40-year legacy of experience with decision support systems. Among these experiences are many lessons that are likely helpful to the larger informatics community. Probably most importantly, the Institute has demonstrated the value of a concurrent focus on standardized, structured data acquisition alongside the development of decision support systems. Having access to these data allowed researchers to perform dozens of reminder studies, many of which serve as foundations for other work in the medical informatics community. The research team has demonstrated that computerized reminders can change clinical behaviors (McDonald 1976), reduce errors (McDonald 1976; Overhage et al. 1997), and improve adherence to practice guidelines (McDonald 1976). These changes have a strong and persistent effect on patient care (McDonald et al. 1984). They also can promote preventive medicine in both the outpatient (McDonald et al. 1984) and inpatient (Dexter et al. 2001) settings, and have a greater effect than delayed feedback for enhancing preventive care (Tierney et al. 1986). However, they do not necessarily provoke providers to review the associated literature, and they must be crafted in ways that encourage active participation by end-users (Tierney et al. 1986, 1995).

Investigators also have documented many lessons about the specific content of reminders and how they affect behavior. For example, reminders that include either prior test results (Tierney et al. 1987) or predictions of abnormal results (Tierney et al. 1988) can reduce unnecessary testing. Additionally, displaying the charges for diagnostic tests significantly reduces the number and cost of tests ordered, especially for patients with scheduled visits (Tierney et al. 1990). This effect does not persist if charges are no longer displayed. Finally, requiring physicians to respond to computer-generated reminders improves their compliance with preventive care protocols (Litzelman et al. 1993); however, promoting preventive care through computerized reminders presents further challenges in the inpatient setting (Overhage et al. 1996).

There are also a series of lessons culled from the experience inherent in not only developing these systems from scratch, but additionally in serving as both the implementation and direct support team.

For example, it is important to start with the assumption that "the user is always right," because data repositories often lack fine details, and reminder rules cannot anticipate every situation. As a corollary, users should be able to override nearly every decision. When reminders for mammograms first were created, researchers found that users were dismissing them. Why? Because the computer system was unaware of a prior mastectomy, a dying patient, or a recently obtained mammogram from another institution. Simply appending these conditions as selectable options after the reminder both acknowledged the system's limitations and regained the users' confidence.

Workflow has proven to be one of the most critical aspects of delivering excellent, efficient patient care within the Regenstrief environment. Decision support often introduces new steps (whether it is a new piece of paper to be reviewed or an alert within CPOE that must be assessed). Implementers of decision support must be cognizant of the impact on workflow. They should avoid punishing the user with additional obstacles when simple rewording or changing a default value will do. The same result often can be achieved in either a user-friendly or not-so-friendly manner. For example, if providers are disregarding decision support that suggests a more effective test or treatment, developers should first consider where, when, and how the message is being delivered (e.g., could it be conveyed more concisely or at a more appropriate position in workflow?) before introducing extra steps (e.g., forcing the user to acknowledge the message with an extra key press). It is very important that the user not be overwhelmed. Researchers at the Institute have found that too many reminders or too many choices are worse than none, and that the best reminders are ones that are short and focused.

Finally, developers should never underestimate the power of user feedback. In fact, they should actively seek it out. Early in the development of the Medical Gopher, the development team introduced the weekly pizza meeting for gaining user feedback. In these meetings, pizza was traded for house staff and student feedback on the system. This feedback was critical in both forming a user-friendly system and addressing system problems early. Even though the team has found that listening is often 90 percent of the solution, the ability to respond rapidly with improvements or fixes proves to cover the last 10 percent.

REFERENCES

Biondich, P. G., Anand, V., Downs, S. M., McDonald, C. J. (2003). Using adaptive turnaround documents to electronically acquire structured data in clinical settings. *AMIA Annu Symp Proc/AMIA Symp* 86–90.

Biondich, P. G., Grannis, S. J. (2004). The Indiana network for patient care: An integrated clinical information system informed by over thirty years of experience. *J Public Health Manag Pract* S81–86.

Chueh, H., Barnett, G. (1997). "Just-in-time" clinical information. *Acad Med* **72**(6): 512–517.

Dexter, P. R., Wolinsky, F. D., Gramelspacher, G. P., Zhou, X. H., Eckert, G. J., Waisburd, M. et al. (1998). Effectiveness of computer-generated reminders for increasing discussions about advance directives and completion of advance directive forms. A randomized, controlled trial. *Ann Intern Med* **128**(2): 102–110.

Dexter, P. R., Perkins, S., Overhage, J. M., Maharry, K., Kohler, R. B., McDonald, C. J. (2001). A computerized reminder system to increase the use of preventive care for hospitalized patients. *N Engl J Med* **345**(13): 965–970.

Litzelman, D. K., Dittus, R. S., Miller, M. E., Tierney, W. M. (1993). Requiring physicians to respond to computerized reminders improves their compliance with preventive care protocols. *J Gen Intern Med* **8**(6): 311–317.

Litzelman, D. K., Tierney, W. M. (1996). Physicians' reasons for failing to comply with computerized preventive care guidelines. *J Gen Intern Med* **11**(8): 497–499.

McDonald, C. J. (1976). Protocol-based computer reminders, the quality of care and the non-perfectability of man. *N Engl J Med* **295**(24): 1351–1355.

McDonald, C. J., Wilson, G. A., McCabe, G. P., Jr. (1980). Physician response to computer reminders. *JAMA* **244**(14): 1579–1581.

McDonald, C. J., Blevins, L., Glazener, T. (1984). CARE: A real world medical knowledge base. COMPCON '84: the 28th Institute of Electrical and Electronic Engineers (IEEE) Computer Society international conference; 1984 February 27–March 1, 1984; San Francisco, CA.

McDonald, C. J., Hui, S. L., Smith, D. M., Tierney, W. M., Cohen, S. J., Weinberger, M. et al. (1984). Reminders to physicians from an introspective computer medical record. A two-year randomized trial. *Ann Intern Med* **100**(1): 130–138.

McDonald, C. J., Tierney, W. M. (1986). The Medical Gopher—A microcomputer system to help find, organize and decide about patient data. *West J Med* **145**(6): 823–829.

McDonald, C. J., Blevins, L., Tierney, W. M., Martin, D. K. (1988). The Regenstrief medical records. *MD Comput* **5**(5): 34–47.

McDonald, C. J., Hui, S. L., Tierney, W. M. (1992). Effects of computer reminders for influenza vaccination on morbidity during influenza epidemics. *MD Comput* **9**(5): 304–312.

McDonald, C. J., Tierney, W. M., Overhage, J. M., Martin, D. K., Wilson, G. A. (1992). The Regenstrief Medical Record System: 20 years of experience in hospitals, clinics, and neighborhood health centers. *MD Comput* **9**(4): 206–217.

McDonald, C. J., Overhage, J. M., Tierney, W. M., Abernathy, G. R., Dexter, P. R. (1996). The promise of computerized feedback systems for diabetes care. *Ann Intern Med* **124**(1 Pt 2): 170–174.

McDonald, C. J., Overhage, J. M., Tierney, W. M., Dexter, P. R., Martin, D. K., Suico, J. G. et al. (1999). The Regenstrief Medical Record System: A quarter century experience. *Int J Med Inform* **54**(3): 225–253.

McDonald, C. J., Overhage, J. M., Barnes, M., Schadow, G., Blevins, L., Dexter, P. R. et al. (2005). The Indiana network for patient care: a working local health information infrastructure. An example of a working infrastructure collaboration that links data from five health systems and hundreds of millions of entries. *Health Aff (Millwood)* **24**(5): 1214–1220.

Murray, M. D., Harris, L. E., Overhage, J. M., Zhou, X-H., Eckert, G. J., Smith, F. E. et al. (2004). Failure of computerized treatment suggestions to improve health outcomes of outpatients with uncomplicated hypertension: Results of a randomized controlled trial. *Pharmacotherapy* **24**(3): 324–337.

Overhage, J. M., Mamlin, B., Warvel, J., Tierney, W., McDonald, C. J. (1995). A tool for provider interaction during patient care: G-CARE. *Proc Annu Symp Comput Appl Med Care* 178–182.

Overhage, J. M., Tierney, W. M., McDonald, C. J. (1996). Computer reminders to implement preventive care guidelines for hospitalized patients. *Arch Intern Med* **156**(14): 1551–1556.

Overhage, J. M., Tierney, W. M., Zhou, X. H., McDonald, C. J. (1997). A randomized trial of "corollary orders" to prevent errors of omission. *J Am Med Inform Assoc* **4**(5): 364–375.

Overhage, J. M., Dexter, P. R., Perkins, S. M., Cordell, W. H., McGoff, J., McGrath, R. et al. (2002). A randomized, controlled trial of clinical information shared from another institution. *Ann Emerg Med* **39**(1): 14–23.

Shapiro, S. (1997). Periodic screening for breast cancer: The HIP Randomized Controlled Trial. Health Insurance Plan. *J Nat Cancer Inst* (22): 27–30.

Tierney, W. M., Hui, S. L., McDonald, C. J. (1986). Delayed feedback of physician performance versus immediate reminders to perform preventive care. Effects on physician compliance. *Med Care* **24**(8): 659–666.

Tierney, W. M., McDonald, C. J., Martin, D. K., Rogers, M. P. (1987). Computerized display of past test results. Effect on outpatient testing. *Ann Intern Med* **107**(4): 569–574.

Tierney, W. M., McDonald, C. J., Hui, S. L., Martin, D. K. (1988). Computer predictions of abnormal test results. Effects on outpatient testing. *JAMA* **259**(8): 1194–1198.

Tierney, W. M., Miller, M. E., McDonald, C. J. (1990). The effect on test ordering of informing physicians of the charges for outpatient diagnostic tests [comment]. *N Engl J Med* **322**(21): 1499–1504.

Tierney, W. M., Miller, M. E., Overhage, J. M., McDonald, C. J. (1993). Physician inpatient order writing on microcomputer workstations. Effects on resource utilization. *JAMA* **269**(3): 379–383.

Tierney, W. M., Overhage, J. M., Takesue, B. Y., Harris, L. E., Murray, M. D., Vargo, D. L. et al. (1995). Computerizing guidelines to improve care and patient outcomes: The example of heart failure. *J Am Med Inform Assoc* **2**(5): 316–322.

Tierney, W. M., Overhage, J. M., McDonald, C. J. (1996). Computerizing guidelines: Factors for success. *Proc AMIA Ann Fall Symp*: 459–462.

Tierney, W. M., Murray, M. D., Gaskins, D. L., Zhou, X. H. (1997). Using computer-based medical records to predict mortality risk for inner-city patients with reactive airways disease. *J Am Med Inform Assoc* **4**(4): 313–321.

Tierney, W. M., Overhage, J. M., Murray, M. D., Harris, L. E., Zhou, X-H., Eckert, G. J. et al. (2003). Effects of computerized guidelines for managing heart disease in primary care [see comment]. *J Gen Intern Med* **18**(12): 967–976.

Zeger, S. L., Liang, K. Y. (1986). Longitudinal data analysis for discrete and continuous outcomes. *Biometrics* **42**(1): 121–130.

5 PATIENTS, DOCTORS, AND INFORMATION TECHNOLOGY: CLINICAL DECISION SUPPORT AT BRIGHAM AND WOMEN'S HOSPITAL AND PARTNERS HEALTHCARE

DAVID W. BATES and HELEN G. LO

Brigham and Women's Hospital (BWH) has been a pioneer in the development of clinical information systems and implementation of clinical decision support within the United States. In the mid-1980s, the hospital leadership elected to develop its own clinical information system, largely believing that an in-house production would provide better overall functionality and, especially, clinical decision support than would vendor-developed applications available at the time. Throughout, delivery of high-quality, safe, cost-effective care has been at the core of the BWH mission and the BWH leadership. In particular, Richard Nesson, MD, the BWH CEO at the time the decision was made, believed that clinical information systems represented a key tool in achieving that vision.

In this chapter, the evolution of clinical decision support at the BWH is discussed, mostly with respect to two application suites: the inpatient Computerized Physician Order Entry (CPOE) and outpatient Longitudinal Medical Record (LMR). Over the years, a series of studies have been performed to assess the impact of clinical decision support on a wide array of parameters, including safety, quality, costs, satisfaction, and provider time. In this chapter, the results of these studies are reviewed. Finally, the generalizable lessons learned across the studies are discussed, as are the next set of frontiers.

5.1 HISTORY

The BWH clinical information systems originally were derived from those developed by Warner Slack, MD, and Howard Bleich, MD, at what was then the Beth Israel Hospital. Nesson convinced Slack and Bleich to build the initial version of the Brigham Integrated Computer System, or BICS. BICS was developed as an entirely MUMPS-based system. Subsequently, Bleich and Slack's responsibility for BWH systems ended, and John Glaser, PhD, was brought in as chief information officer. At that time, BICS was perceived as very physician-friendly, and it contained a high proportion of all clinical results, such as those from laboratory, radiology, and cardiology tests and procedures. With Nesson's blessing, Glaser decided to implement Computerized Physician Order Entry (CPOE), which was first brought up in 1993. Jonathan Teich, MD, was the primary architect of the application, and he designed it to be fast, adaptable to clinicians' workflow, and enabled with real-time clinical decision support. The entire design and implementation team was very clinically oriented, and Cynthia Spurr, RN, in particular, played an important role, especially with respect to implementation.

Subsequently, Teich led the development of an outpatient electronic medical record application called Miniamb, which was implemented in 1989. The first versions included a problem list with uncoded problems, a medication list, allergy list, visits, and notes, which were generally entered as free text that had been dictated and uploaded. Initially, almost no decision support was included. Subsequently, in 1997, the Longitudinal Medical Record (LMR) was introduced. Initially, it ran on a Visual Basic platform, integrated with the MUMPS database, but the application was eventually migrated to use a Web front end. Now, all development is done on the Web version, which is being widely deployed across the Partners network, which includes several thousand physicians (see Figure 5-1).

In 1996, Partners Health System was formed as an integrated delivery system that included both BWH and the Massachusetts General Hospital (MGH). CPOE was introduced at MGH in 1996. Very recently, a set of initiatives called the Signature Initiatives has been developed under the direction of Dr. James Mongan, the current CEO of Partners. Mongan recognized that for Partners to demonstrate leadership and excel in the areas of quality, safety, and

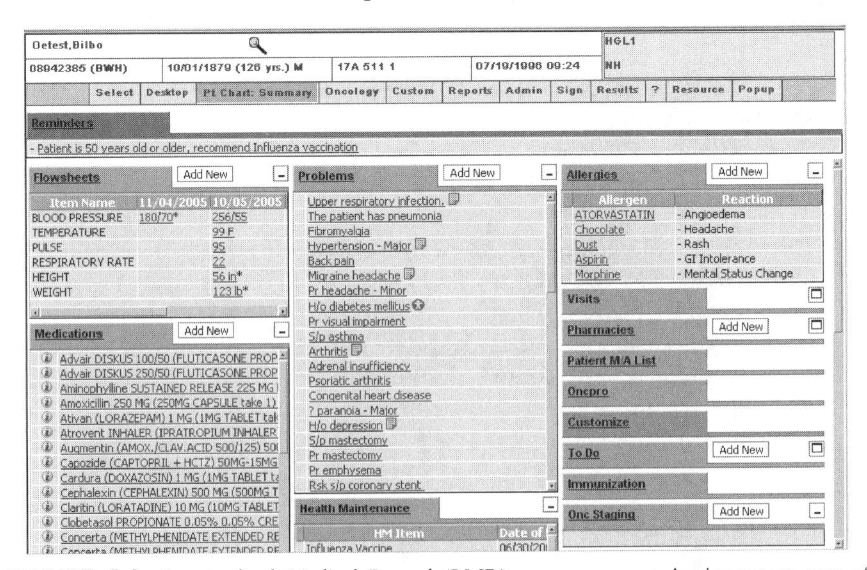

FIGURE 5-1 Longitudinal Medical Record (LMR) summary page. An important part of clinical decision support is simply making available the key information that the clinician needs. Here, the clinician can see at a glance that the only reminder active for this visit is that the patient should receive a flu shot.

Status FY06	BWH	MGH	DFCI	NWH	FH	SH/Charter	UH/Charter	SRH	SKRH	PHC	PCHI	McLean*
Inp CPOE	PHS	PHS	N/A	Medtch	Medtch	Siemens	Siemens	Medtch	Siemens	N/A	N/A	Medtch
Amb CPOE - MEDs	LMR	LMR, OnCall	LMR	LMR, GE		LMR	LMR	LMR	TBD	N/A	LMR, GE	TBD
ED CPOE	PHS	PHS	N/A	Medtch/TBD	Medtch	TBD	TBD	N/A	N/A	N/A	N/A	N/A
OncAmb eMAR	LMR	TBD	LMR	TBD	LMR	TBD	TBD	N/A	N/A	N/A	N/A	N/A
Inp MAR -nonbar	N/A	N/A	N/A	Medtch	Medtch	Siemens	Siemens	Medtch	Siemens	N/A	N/A	Medtch
Inp eMAR -bar code	PHS	TBD	N/A	Medtch	Medtch	Siemens	Siemens	Medtch	Siemens	N/A	N/A	Medtch
Inp Progress Notes	TBD	TBD	N/A	TBD	TBD	TBD	TBD	TBD	TBD	N/A	N/A	Medtch
Inp Pt Assessment	TBD	TBD	N/A	Medtch	Medtch	Siemens	Siemens	Medtch	Siemens	N/A	N/A	Medtch
Inp Nursing Notes	TBD	TBD	N/A	Medtch	Medtch	Siemens	Siemens	Medtch	Siemens	N/A	N/A	Medtch
Amb Nursing Notes	LMR	LMR, OnCall	LMR	LMR, GE		LMR	LMR	LMR	LMR	(PTCT) Medtch	LMR, GE	TBD
Amb Visit Notes	LMR	LMR, OnCall	LMR	LMR, GE		LMR	LMR	LMR	LMR	N/A	LMR, GE	TBD
Consult Notes	TBD	TBD	TBD	TBD	TBD	TBD	TBD	TBD	TBD	N/A	N/A	TBD
Op Notes, D/C Sum	BICS, CDR	PCIS, CDR	N/A	Medtch, CDR	Medtch, CDR	Siemens, RV	Siemens, RV	N/A	N/A	N/A	N/A	N/A
ED visit notes	LMR	MGH EDDS	N/A	PICIS	Medtch	Vitalworks	Vitalworks	N/A	N/A	N/A	N/A	N/A
ED tracking	BWH	MGH EDDS	N/A	PICIS	Medtch	Vitalworks	Vitalworks	N/A	N/A	N/A	N/A	N/A
Results Viewing and Repository	BICS, RV, CDR	PCIS, RV	BICS, RV	Medtch, RV	Medtch, RV, BICS	Siemens, RV	Siemens,RV	PCIS, RV, Medtch	Siemens, RV	Medtch, RV	RV/CDR	Medtch
Pt Computing	PG	PG	PG							N/A	PG	

KEY: PG = Patient Gateway
RV = Results Viewer
EDDS = Emergency Dept Documentation System
CDR = Clinical Data Repository

* contract signed with Meditech Sept '05

FIGURE 5-2 Implementation of clinical systems at Partners Healthcare in 2005. This matrix illustrates the large number of systems in place, and the heterogeneity of systems that are in the Partners network, despite conscious efforts to avoid heterogeneity. This illustrates why efforts to implement standard clinical decision support across a network like Partners is so challenging.

efficiency, information systems represent a critical cornerstone. Although there are islands of progress, there is tremendous variability among the various Partners institutions in this area (see Figure 5-2). Accordingly, the first of six Partners initiatives in this effort focuses on implementing inpatient CPOE at all Partners hospitals, including the smaller community hospitals, and encouraging all Partners physicians to begin using an electronic health record.

5.2 CLINICAL DECISION SUPPORT AND INPATIENT CPOE AT BWH

The initial versions of CPOE, implemented in 1993, included relatively little in the way of clinical decision support compared to today. The main areas of focus related to referential knowledge, anticipated needs, alerts, reminders, order sets, guidelines, and feedback. An array of reference materials was initially accessible from CD-ROMs. In addition, a bank of calculation tools was compiled to aid clinicians with frequently performed calculations, such as the Cockcroft-Gault equation, for estimated creatinine clearance, and the arterial-alveolar oxygen (A-a) gradient. Since then, the breadth of reference information has been greatly increased, with corresponding increase in usage. The reference library is now available on the network. All these tools were passive, however, and were consulted only when the clinician elected to look at them. A more productive approach has been to integrate a subset of the resources into routine clinician workflow, resulting in essentially universal utilization of selected tools.

Another key effort made in the early versions of BICS Order Entry (OE) was to anticipate clinician needs. For example, a physician ordering digoxin will want to know the patient's renal function, serum potassium, and last digoxin level, if any. BICS OE pulled these values and made them available when the provider was writing the order. In addition, the application suggested starting doses for medications. These were determined by reviewing a large database of the recent instances when the medication had been given and by selecting the mode (the dose that appeared most frequently), unless an individual case had some compelling reason to do otherwise.

Only a very small set of alerts were delivered initially. Specifically, the system checked for allergies to a few of the most important drug classes, such as penicillins and sulfa drugs, and for approximately 10 of the most important drug–drug interactions. Subsequently, the sophistication of alerts has grown substantially. Some of the first additions were comprehensive rule sets for checking drug allergies and drug–drug interactions, although these were not added until one to two years after the initial implementation. The overall approach has been to gradually layer on additional decision support, to maximize the likelihood of clinician acceptance.

The first versions of CPOE also included few reminders, although a large number have been added over time. One example is "corollary orders" (Overhage et al. 1997), of which there are now many in the computer system—for instance, presenting the provider with the opportunity to prescribe heparin in the context of bed rest, which has increased compliance with recommended thrombotic prophylaxis (Teich et al. 2000).

A considerable number of order sets were included from the beginning, although not all order sets were available in the first versions. Time-motion studies showed that submitting groups of orders was five times faster than writing them individually, which gave the development team the impetus to increase the number and build tools surrounding order sets, and to ensure that additional order sets were included.

5.2.1 Medication-related Decision Support

One of the key reasons for implementing CPOE was that it might improve medication safety. The first phase of the Adverse Drug Event Prevention Study was observational, and its main goal was to describe the epidemiology of preventable adverse drug events (ADEs) and potential adverse drug events in hospitalized patients (Bates et al. 1995). A key finding was that 60 percent of serious medication errors (those that harmed someone or had the potential to do so) occurred at the prescription or transcription stages and were possibly preventable using order entry. Furthermore, many of the serious prescribing errors appeared to be related to access to knowledge (Leape et al. 1995), even though the key piece of information was usually somewhere in the computer. This emphasized the need to anticipate the knowledge needs of providers.

Phase 2 of the ADE Prevention Study tested two interventions, CPOE and a team-based intervention, targeted at the administration and dispensing stages of the process. The team-based intervention had no effect. CPOE, however, reduced the serious medication error rate by 55 percent, despite minimal decision support consisting mainly of default dosages and limited drug–allergy and drug–drug interaction checking (Bates et al. 1998). Subsequently, much more decision support has been added, so this clearly represents a conservative estimate of its effect. A comprehensive drug–allergy module was added, followed by comprehensive drug–drug interaction checking and more minor changes. The cumulative impact of these serial changes was studied in another report that examined all medication errors as the primary

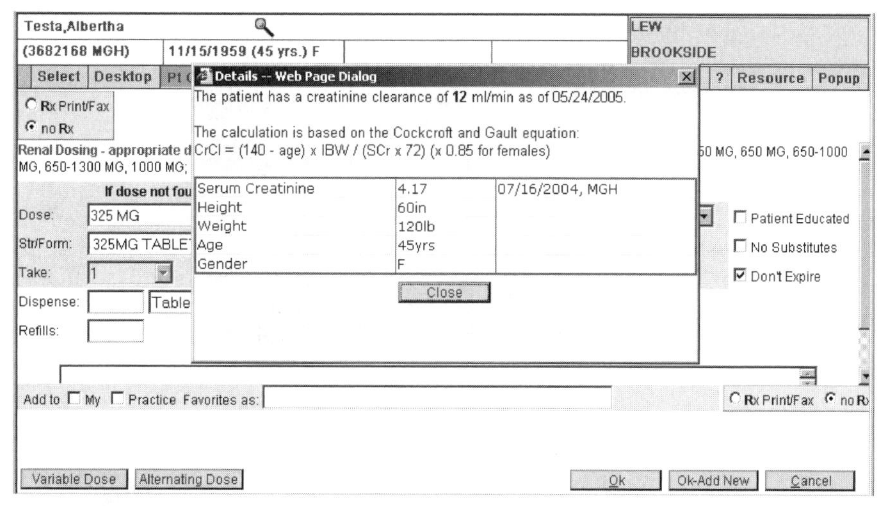

FIGURE 5-3 Renal dosing. The clinician is ordering acetaminophen, for which the clinician should adjust the dose based on the patient's level of renal function. If the clinician chooses to see how the dosing suggestion was reached, he or she can click on a button and see the Cockcroft-Gault calculation.

outcome (Bates et al. 1999). In an interrupted time series design, with samples at one-year intervals, the overall medication error rate fell 83 percent simply with computerization of prescribing. It rose slightly in the next year for unclear reasons while more comprehensive allergy checking was added, then dropped in the final year with the two main changes being the introduction of comprehensive drug–drug interaction checking and elimination of a problem with the ordering of potassium.

Many more types of decision support have been added subsequently. Two of the most important additions have involved special medication dosing for patients who have renal compromise or are part of the geriatric population (Chertow et al. 2001; Peterson et al. 2005). The renal dosing application, called Nephros, utilizes existing data for the patient's age, gender, and most recent creatinine, coupled with weight information, entered by the clinician. For drugs that need to be renally dosed, the computer performs a Cockcroft-Gault calculation and suggests a drug dose appropriate for the patient's level of renal function (see Figure 5-3). In a controlled trial, the appropriate dose was selected 67 percent of the time in the intervention group compared to 54 percent in the control group. In addition, dosing frequency was more often appropriate—59 percent in the intervention group, compared to 35 percent in the control group (Chertow et al. 2001). Moreover, patients stayed in the hospital a half day less. This intervention was likely successful in part, because it was transparent to the physician—that is to say, when a physician orders a medication, the computer automatically suggests the appropriate dosage.

For geriatric dosing, the major problem involved physicians selecting initial dosages that were too high for geriatric patients (Peterson et al. 2005). Thus, the investigators had an expert panel develop initial dosing recommendations for psychoactive medications, for which the application then suggested an appropriate default starting dosage (see Figure 5-4). Key results were that patients more often got the recommended dosage (29% vs. 19%), had a lower rate of tenfold overdose (2.8% vs. 5%), and were less likely to fall (2.8 vs. 6.4 falls per 1,000 patient-days). There was no difference in the frequency of mental status change or

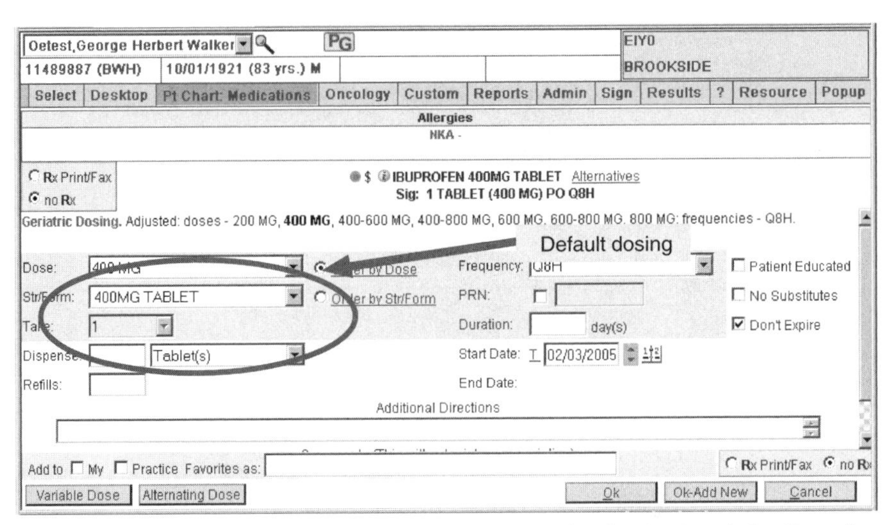

FIGURE 5-4 Drug–age or Gerios dosing support. Here, the clinician is ordering ibuprofen, and the default initial dosing frequency has been set to 400 mg.

in length of stay. Although the dosing improvements were highly significant statistically, clearly there is a great deal of room for improvement.

Another approach taken has been to implement drug-specific guidelines for a number of medications. For example, because of baseline overuse of vancomycin, the Centers for Disease Control's guidelines for prescribing the drug were adapted into electronic format (Shojania et al. 1998). After implementation of the guideline, the number of vancomycin-days per provider decreased. Notably, this effect was seen more through shortening courses of vancomycin than decreasing the number of times vancomycin was started. Guidelines have been implemented for many other medications, including human growth hormone and a number of expensive antibiotics. These interventions often have been highly effective, in part because CPOE requires the ordering physicians to clearly identify themselves, making it easy to provide individual follow-up, if there appears to be any "gaming" activity.

5.2.2 Laboratory Interventions

A number of interventions have been implemented in an attempt to improve the appropriateness of use of the clinical laboratory, including the display of charges for tests, reminders if redundant tests are ordered, and suggestions for ordering specific tests such as antiepileptic drug levels (Bates et al. 1997, 1998; Chen et al. 2003). In a randomized controlled trial evaluating the impact of charge display, the charge for each test and the cumulative charges for a session were shown using a "cash register" function. Physicians liked seeing the charges, but no statistically significant impact on ordering practices was observed. The intervention group, however, showed a beneficial trend of 4.5 percent fewer tests performed, with this figure being greater for tests that are ordered infrequently, which tend to be more expensive. The annual estimated cost reduction benefit in the intervention group was $1.7 million, so the institution elected to continue to display charges.

Regarding redundant tests, a study was performed demonstrating that 28 percent of 12 target tests were ordered earlier than a test-specific predefined minimum interval (Bates et al. 1998), thus showing substantial unnecessary utilization, which amounted to an estimated $930,000 per year in charges. Alerts for these tests then were implemented and studied in a randomized controlled trial (Bates et al. 1998). Even though the tests considered to be redundant by the interval criteria were performed only 24 percent of the time in the intervention group versus 51 percent of the time in the control group, the savings realized were a mere $35,000 versus a prior projection of $436,000. The reasons for the difference were multifactorial. Only 44 percent of the redundant tests performed were ordered by computer (many were being sent to the laboratory without an order). In addition, 31 percent of the reminders were overridden. Also half of the orders were not screened for technical reasons—specifically the software team had elected to exempt order sets from the screening without realizing the full impact this would have. There are at least two take-away messages from this experience: First, it is important to design "closed-loop" systems to ensure inclusion of all utilities. Second, it is important to include in the screening those orders that are in order sets, which may be especially prone to redundancy.

In another evaluation, antiepileptic drug level testing was targeted. Prior work demonstrated that only 26 to 29 percent (depending on the drug) of antiepileptic drug level tests performed in the hospital appeared to be appropriate (Schoenenberger et al. 1995). Subsequently, an intervention in which guidelines for drug level testing were displayed at the time of ordering resulted in a 19.5 percent decrease in the use of these levels, despite a 19.3 percent increase in overall test volume during the study period (Chen et al. 2003); there was also a major decrease in the proportion of tests that appeared to be inappropriate.

5.2.3 Radiology Interventions

In the inpatient setting, nearly all radiographs have been ordered electronically since the initial implementation of CPOE. When providers order each a radiograph, they are asked to enter coded historical findings and the clinical question (i.e., what they would like assessed or ruled out). Simply computerizing the process has improved the likelihood that the radiologists will receive useful information. In one trial of decision support for abdominal radiographs (Harpole et al. 1997), providers were unlikely to cancel orders, even if the examinations were virtually certain to provide no useful information, unless an alternative was offered. However, when the option of choosing alternative views or studies was offered, suggestions were much more likely to be accepted, although they were still accepted only about half the time. Displaying the charges for radiographs had no impact on the overall level of utilization (Bates et al. 1997).

5.2.4 Signout

One study demonstrated that inpatients being cross-covered by another physician had nearly a fivefold increase in risk of suffering an adverse event (O'Neil et al. 1993). This led to the development of *Signout*, an application that allows house officers to sign out their patients electronically. Information including the medication list, key recent laboratory tests, and the code status are abstracted by the information system, and the provider is asked to enter additional data including a problem list and a description of the hospital course. This information can then be exchanged when providers "hand off" their patients. An evaluation of the impact of this application demonstrated that after implementing it, the additional risk associated with being cross-covered was eliminated (Petersen et al. 1998).

5.2.5 Assessment of Satisfaction with CPOE

Formal study of the impact of CPOE on provider satisfaction (Lee et al. 1996) also has been done. On medicine, all the ordering physicians were house officers, whereas on surgery and in obstetrics and gynecology, attendings also wrote orders. Whether satisfaction was associated with attending status was not measured, although there was no correlation between satisfaction and provider age. Physicians and nurses were quite satisfied with the application overall, including the imbedded decision support, although among physicians, internists were more satisfied than surgeons. Satisfaction was highly correlated with the perceived impact of CPOE on productivity, ease of use, and speed, and was less strongly associated with features directed at improving the quality of care. This suggests that decision support must be fast to be tolerated and confirms that users may not perceive the need to improve quality even if it is present.

5.2.6 Impact of CPOE on Provider Time

In a formal time-motion study, interns spent 9 percent of their time ordering after implementation of CPOE versus 2.1 percent before, although CPOE saved them an additional 2 percent of time because they spent less time looking for charts, and could write orders from remote locations, so that the net difference was 5 percent of their total time (Shu et al. 2001). However, this is counterbalanced by decreased time required by other personnel such as nursing and pharmacy. Overall, the impact on time of any decision-support related intervention must be considered carefully.

5.3 DECISION SUPPORT DELIVERED USING THE OUTPATIENT ELECTRONIC HEALTH RECORD

5.3.1 Medication-related Decision Support

An array of medication-related decision support functions also have been implemented in the outpatient setting, starting with relatively simple suggestions such as drug–allergy and drug–drug interaction checking, and then adding a variety of more sophisticated decision support, including renal dosing, drug–pregnancy checks, drug–age checking, and drug–disease checking (see Figure 5-5). Most results of the studies on impact of these interventions have yet to be published. However, one published report focused on improving the acceptance of medication-related alerts (Shah et al. 2006). Alerts were divided into those that were clinically important enough to make them interruptive, with the remainder classified as noninterruptive or informational. In that study, over a six-month evaluation period, there were 18,115 drug alerts, among which 71 percent were noninterruptive and 29 percent were interruptive. Of the interruptive alerts, 67 percent were accepted, which compares very favorably to some other reports, which have found acceptance levels of only 10 to 30 percent (Payne et al. 2002; Weingart et al. 2003). Some of the keys to this

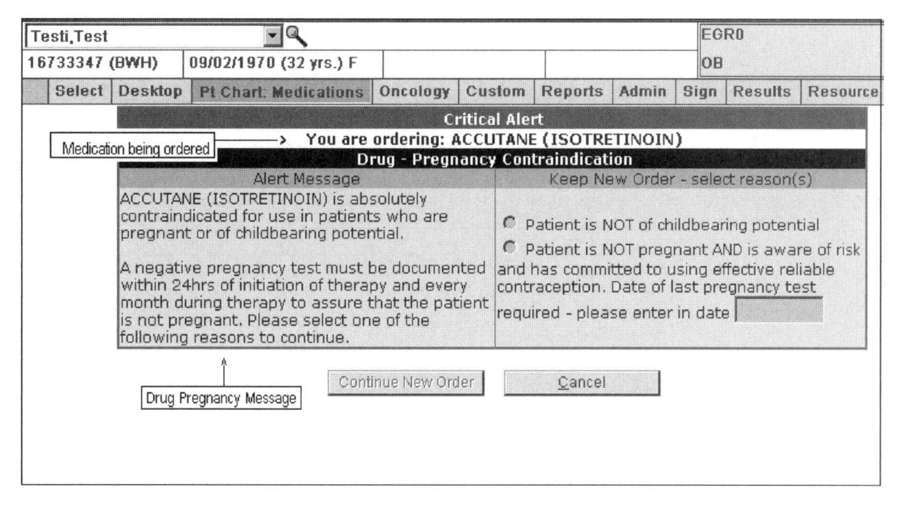

FIGURE 5-5 Drug–pregnancy alert. A number of drugs should never be used if a woman is pregnant, and many more are relatively contraindicated. For Accutane, the restrictions are especially strict; a negative pregnancy test is required if the woman is of child-bearing age. From the informatics perspective, the most challenging part of delivering the alerts appropriately was determining whether a woman is pregnant.

success were being highly selective in which alerts to display, iterating to identify alerts with high override rates, and using the interruptive approach only for truly important alerts (see Figure 5-6).

Another evaluation focused on drug–allergy alerts, and found that 80 percent of the drug–allergy alerts were overidden (Hsieh et al. 2004). However, only 10 percent of alerts were triggered by an exact match between the drug prescribed and the allergy listed. On close evaluation, all the overrides appeared clinically justifiable. To address this, a group of recommendations was developed to fine-tune the specificity of warnings, thereby increasing the utility of the allergy alerting system (see Figure 5-7).

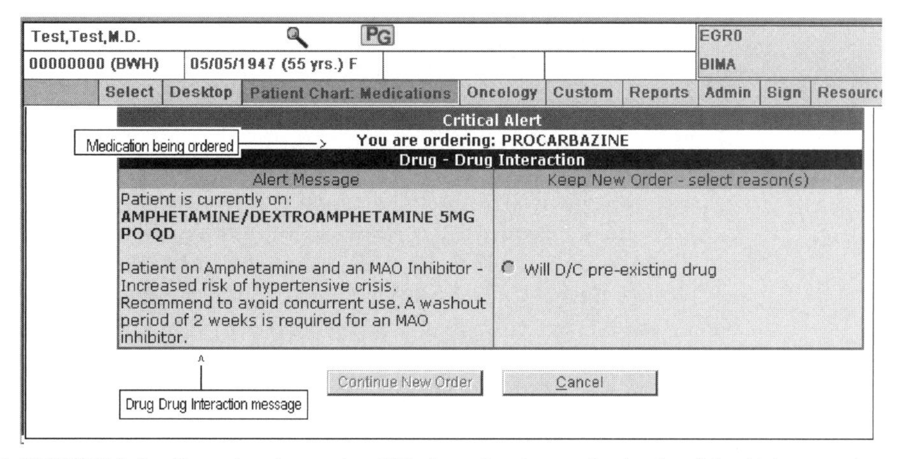

FIGURE 5-6 Drug–drug interaction. This drug–drug interaction is a Level 1, which means that the clinician is not allowed to bypass it. The clinician's only option is to discontinue one of the drugs. There are very few such drug–drug interactions.

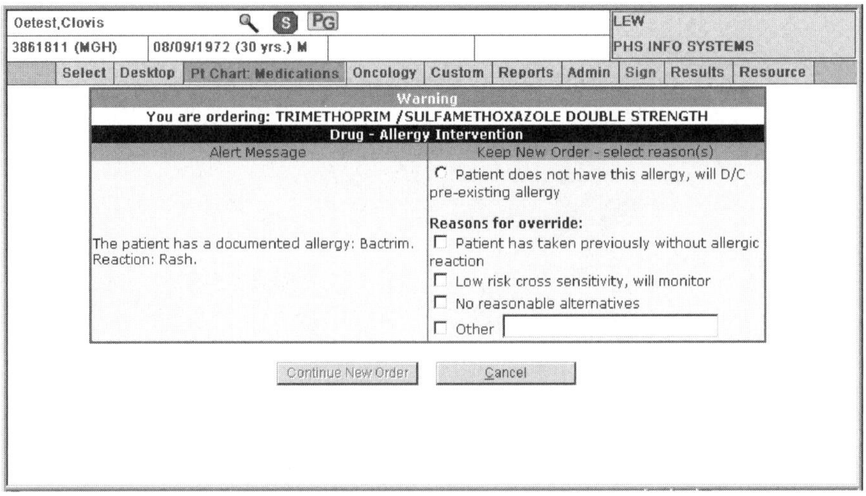

FIGURE 5-7 Drug–allergy alert. Here, the patient has a prior rash to sulfa, and trimethoprim/sulfamethoxazole has been ordered. This warning is a Level 2, which means in these systems that the alert is interruptive, and that the clinician must provide a reason for overriding, but is allowed to do so.

5.3.2 Laboratory-related Decision Support

Less laboratory-related decision support for ordering has been done in the outpatient setting, since BWH still has not implemented computerized laboratory ordering for ambulatory patients. However, a number of reminders to monitor for specific medications such as nonsteroidals and other medications have been implemented, and these are undergoing evaluation.

Another key issue is follow-up of abnormal results, which is often suboptimal (Poon et al. 2004). A number of studies have suggested that about a third of abnormal test results, even for tests such as Pap smears and mammograms, do not receive appropriate follow-up (Poon et al. 2004). To help address this issue, Dr. Eric Poon led the development of a tool called Results Manager, which makes it easier for physicians to handle test results, by aggregating, organizing, and prioritizing them, and then making it easy for providers to generate letters (Poon et al. 2003). Use of the application has grown rapidly, as it has been very popular with clinicians (see Figure 5-8). An analysis currently is being performed to determine whether it decreases time to follow-up for abnormal tests.

Another issue concerns tests ordered during a hospitalization for which reports come back after the patient has left the hospital, as a result of which they may not be evaluated by a physician. Roy et al. (2005) studied the incidence and consequences of potentially actionable test results returning after discharge, and found that it was a frequent issue with substantial potential for harm. Consequently, a trial is now being performed to assess whether the Results Manager described earlier improves the likelihood that results get appropriate follow-up.

5.3.3 Radiology Decision Support

Electronic ordering and mapping of all key historical factors and indications for radiographs are in place, so that providers request studies using controlled

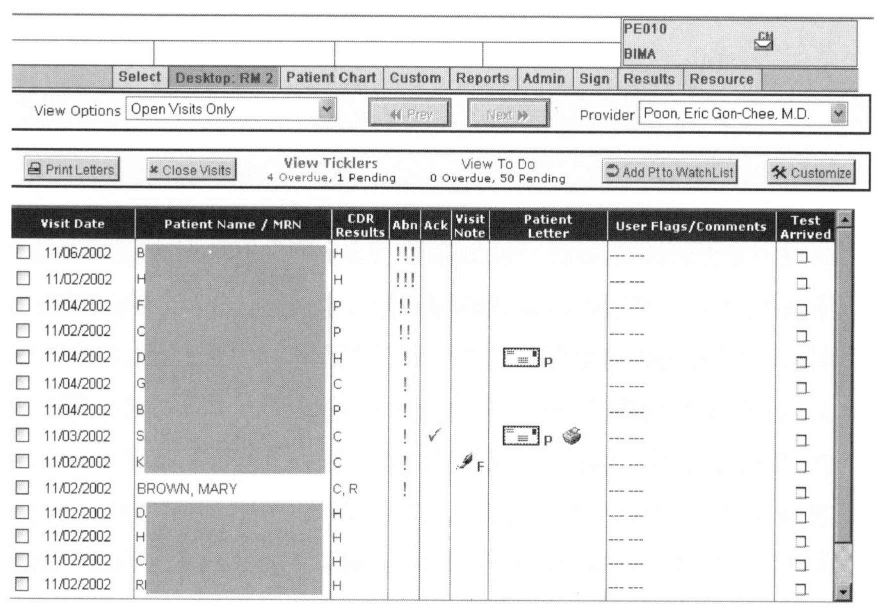

FIGURE 5-8 Results Manager. This screen shows what a physician sees when he or she is reviewing a patient's test results; the results are prioritized according to how abnormal they are, and the clinician can generate a letter about them with only a few clicks.

vocabularies. Indications for procedure requests can be filtered for appropriateness, and checking for redundancy is also done. One next step will be to expand monitoring to include a variety of conditions, deliver feedback, and evaluate the impact. In one such study, an evaluation was done of patients who underwent abdominal imaging for abnormal liver function tests, and found that of all modalities evaluated, CT scan had the highest yield, and that unexpected new findings that appeared to be clinically important were found in a higher proportion of patients than anticipated (Rothschild et al. 2001).

5.3.4 Impact on Provider Time

A recent formal time-motion study by Pizziferri demonstrated that electronic health records (EHR) neither significantly decrease nor increase clinic time for primary care physicians (Pizziferri et al. 2005). The results contradict a major barrier to EHR adoption, which is the perception that converting to EHR is slower for clinicians than using the paper-based status quo. Furthermore, a majority of providers studied believed that EHRs would increase quality of care, access, and communication. Additional studies, however, are needed to examine an EHR's impact on a provider's nonclinic activities, such as post-visit documentation. Combined with the study findings, efforts to identify physicians for whom integrating an EHR may be more difficult will be important in aiding the transition from paper-based records.

5.4 OVERARCHING STUDIES

In addition to the many individual studies described earlier, the cost-effectiveness of computerized physician order entry (CPOE) and the electronic health record have been assessed, which required collecting results from all the individual studies.

A detailed report of the cost-effectiveness of CPOE has been submitted for publication. A much less sophisticated analysis suggested that CPOE cost approximately $1.4 million to implement, with $900,000 for software and $500,000 for hardware. However, the benefits in terms of charges—mostly in terms of drug savings, ADE prevention, and more appropriate use of the clinical laboratory—came to $5 to $10 million annually.

In addition, the cost-effectiveness of implementing the outpatient electronic health record has been evaluated (Wang et al. 2003). Wang estimated the net financial benefit and costs for a primary care provider over a five-year period. The estimated net benefit was $86,400 per provider. Benefits accrued mostly from savings related to fewer drug expenditures, more appropriate utilization of radiology tests, better capture of charges, and decrease in the rate of billing errors. The results were most sensitive to the proportion of patients whose care was capitated. These data suggest that high-yield areas to focus on when selecting decision support for an electronic health record include drug-cost suggestions and radiology and laboratory recommendations, especially in settings where more appropriate usage affects physician reimbursement.

5.5 OVERARCHING LESSONS

Over the years, there have been a number of successes with clinical decision support at BWH, but there have also been probably just as many, if not more, failures. In particular, efforts to support documentation for diabetes and congestive heart failure failed, because the tools were sufficiently complex that clinicians would not use them. A number of reminders and guidelines have had little or no impact, especially when clinicians were not fully convinced that the message being delivered was correct. For instance, a suggestion not to use intravenous ketorolac was routinely ignored, because physicians believed it was more effective than the alternatives suggested, despite lack of supporting evidence.

A summary of the lessons learned over the years has been published (Bates et al. 2003) as "ten commandments for clinical decision support" (see Table 5-1).

1. "Speed is everything." A routine goal is subsecond screen flips, since providers will not tolerate much longer than that, and minimizing the number of screens used is also important.
2. "Anticipate needs and deliver in real time." One example is showing the potassium lab values when a drug that lowers potassium is prescribed.
3. "Fit into the user's workflow." If a suggestion seems to come out of left field or at a time when the user is focused on another issue, it is much less likely to be heeded.
4. "Little things can make a big difference." In the prototypical example, the decision regarding how a default is set can have an enormous impact on the frequency that a provider will choose a specific action. Generally, it is good informatics practice to set the default to an action that is most likely to be correct.
5. "Physicians resist stopping." Here, the point is that, when you suggest that a physician not take an action, but fail to provide an alternative,

TABLE 5-1 Ten commandments for effective clinical decision support.

1. Speed is everything	7. Simple interventions work best
2. Anticipate needs and deliver in real time	8. Asking for information is OK—but be sure you really need it
3. Fit into the user's workflow	9. Monitor impact, get feedback, and respond
4. Little things can make a big difference	10. Knowledge-based systems must be managed and maintained
5. Physicians resist stopping	
6. Changing direction is fine	

the initial action is likely to be continued even if it is virtually certain to have little or no yield.

6. The corollary to 5 is that "changing directions is fine." In contrast to the previous suggestion, when one does suggest a superior clinical alternative, physicians are fairly willing to accept the recommendation.
7. "Simple interventions work best." Here, the point is that the level of success has been highest for straightforward guidelines and much less for more complex guidelines, nearly all of which have required substantial adaptation before they could be computerized.
8. "Ask for additional information only when you really need it." Implementation of many guidelines or pieces of clinical decision support has required some information, such as the weight for renal dosing, which was not already available. Although clinicians eventually supplied the weight in most instances, even getting this small piece of clinical information routinely required an effort, which seemed completely disproportionate. Getting multiple pieces of data would undoubtedly prove even harder.
9. "Monitor impact, get feedback, and respond." For most pieces of clinical decision support implemented, at least some additional changes are required. Failure to make multiple incremental changes can result in lack of benefit, and even promote errors, as found by Koppel et al. (2005).
10. "Manage and maintain your knowledge-based systems." This is related to the preceding tenet, but it is useful to routinely track how often each piece of decision support is triggered, and try to ensure that there is an "owner" for each rule, and that each will get periodic follow-up to make sure it still applies.

5.6 FUTURE DIRECTIONS

Many additional challenges remain. In particular, it is still unclear how to deliver clinical decision support for complex conditions, such as chronic diseases in the outpatient setting, when physicians are recording notes. Randomized controlled trials in this area are currently being conducted using "smart forms"—documentation tools that incorporate decision support—with three of the initial conditions being acute respiratory infections, coronary heart disease, and diabetes. Clearly, a challenge is that patients often have multiple chronic conditions that interact and require more complex algorithms for data processing.

Another key challenge is achieving much higher levels of performance for specific measures. Often, while there are clear-cut advantages of using a specific piece of decision support, substantial room for improvement remains, for example with renal dosing. Possibilities include using more feedback to providers and using "bundles"—an approach developed by the Institute for Healthcare Improvement, which targets multiple, simultaneous processes to be carried out in a highly reliable way.

There are also many novel opportunities to deliver decision support in the inpatient setting, especially as coded documentation becomes available. Partners is currently in the process of implementing a full electronic health record, which will open many new windows. For example, rapid and easy access will make it possible to virtually assess a patient's stability based on vital signs information and to more accurately assess a patient's mental status using nursing notes.

Despite all the decision support that has been incorporated over the years at Brigham and Women's Hospital, there continues to be a long list of changes to make and additional pieces of decision support that would be desirable to provide. It is unlikely that this task will be complete any time in the foreseeable future. Overall, clinical decision support has the potential to revolutionize clinical care in the coming years, but many lessons remain to be learned about what best to deliver and how to deliver it.

REFERENCES

Bates, D. W., Cullen, D. J., Laird, N., Petersen, L. A., Small, S. D., Servi, D. et al. (1995). Incidence of adverse drug events and potential adverse drug events. Implications for prevention. ADE Prevention Study Group. *JAMA* **274**(1): 29–34.

Bates, D. W., Kuperman, G. J., Jha, A., Teich, J. M., Orav, E. J., Ma'luf, N. et al. (1997). Does the computerized display of charges affect inpatient ancillary test utilization? *Arch Intern Med* **157**(21): 2501–2508.

Bates, D. W., Boyle, D. L., Rittenberg, E., Kuperman, G. J., Ma'luf, N., Menkin, V. et al. (1998). What proportion of common diagnostic tests appear redundant? *Am J Med* **104**(4): 361–368.

Bates, D. W., Leape, L. L., Cullen, D. J., Laird, N., Petersen, L. A., Teich, J. M. et al. (1998). Effect of computerized physician order entry and a team intervention on prevention of serious medication errors. *JAMA* **280**(15): 1311–1316.

Bates, D. W., Teich, J. M., Lee, J., Seger, D., Kuperman, G. J., Ma'luf, N. et al. (1999). The impact of computerized physician order entry on medication error prevention. *J Am Med Inform Assoc* **6**(4): 313–321.

Bates, D. W., Kuperman, G. J., Wang, S., Gandhi, T., Kittler, A., Volk, L. et al. (2003). Ten commandments for effective clinical decision support: Making the practice of evidence-based medicine a reality. *J Am Med Inform Assoc* **10**(6): 523–530.

Chen, P., Tanasijevic, M. J., Schoenenberger, R. A., Fiskio, J., Kuperman, G. J., Bates, D. W. (2003). A computer-based intervention for improving the appropriateness of antiepileptic drug level monitoring. *Am J Clin Pathol* **119**(3): 432–438.

Chertow, G. M., Lee, J., Kuperman, G. J., Burdick, E., Horsky, J., Seger, D. L. et al. (2001). Guided medication dosing for inpatients with renal insufficiency. *JAMA* **286**(22): 2839–2844.

Harpole, L. H., Khorasani, R., Fiskio, J., Kuperman, G. J., Bates, D. W. (1997). Automated evidence-based critiquing of orders for abdominal radiographs: Impact on utilization and appropriateness. *J Am Med Inform Assoc* **4**(6): 511–521.

Hsieh, T. C., Kuperman, G. J., Jaggi, T., Hojnowski-Diaz, P., Fiskio, J., Williams, D. H. et al. (2004). Characteristics and consequences of drug allergy alert overrides in a computerized physician order entry system. *J Am Med Inform Assoc* **11**(6): 482–491.

Koppel, R., Metlay, J. P., Cohen, A., Abaluck, B., Localio, A. R., Kimmel, S. E. et al. (2005). Role of computerized physician order entry systems in facilitating medication errors. *JAMA* **293**(10): 1197–1203.

Leape, L. L., Bates, D. W., Cullen, D. J., Cooper, J., Demonaco, H. J., Gallivan, T. et al. (1995). Systems analysis of adverse drug events. ADE Prevention Study Group. *JAMA* **274**(1): 35–43.

Lee, F., Teich, J. M., Spurr, C. D., Bates, D. W. (1996). Implementation of physician order entry: User satisfaction and self-reported usage patterns. *J Am Med Inform Assoc* **3**(1): 42–55.

O'Neil, A. C., Petersen, L. A., Cook, E. F., Bates, D. W., Lee, T. H., Brennan, T. A. (1993). Physician reporting compared with medical-record review to identify adverse medical events. *Ann Intern Med* **119**(5): 370–376.

Overhage, J. M., Tierney, W. M., Zhou, X. H., McDonald, C. J. (1997). A randomized trial of "corollary orders" to prevent errors of omission. *J Am Med Inform Assoc* **4**: 364–375.

Payne, T. H., Nichol, W. P., Hoey, P., Savarino, J. (2002). Characteristics and override rates of order checks in a practitioner order entry system. *Proc AMIA Symp* 602–606.

Petersen, L. A., Orav, E. J., Teich, J. M., O'Neil, A. C., Brennan, T. A. (1998). Using a computerized sign-out program to improve continuity of inpatient care and prevent adverse events. *Jt Comm J Qual Improv* **24**(2): 77–87.

Peterson, J. F., Kuperman, G. J., Shek, C., Patel, M., Avorn, J., Bates, D. W. (2005). Guided prescription of psychotropic medications for geriatric inpatients. *Arch Intern Med* **165**(7): 802–807.

Pizziferri, L., Kittler, A. F., Volk, L. A., Honour, M. M., Gupta, S., Wang, S. et al. (2005). Primary care physician time utilization before and after implementation of an electronic health record: a time-motion study. *J Biomed Inform* **38**(3): 176–188.

Poon, E. G., Wang, S. J., Gandhi, T. K., Bates, D. W., Kuperman, G. J. (2003). Design and implementation of a comprehensive outpatient Results Manager. *J Biomed Inform* **36**(1–2): 80–91.

Poon, E. G., Haas, J. S., Louise, P. A., Gandhi, T. K., Burdick, E., Bates, D. W. et al. (2004). Communication factors in the follow-up of abnormal mammograms. *J Gen Intern Med* **19**(4): 316–323.

Rothschild, J. M., Khorasani, R., Silverman, S. G., Hanson, R. W., Fiskio, J. M., Bates, D. W. (2001). Abdominal cross-sectional imaging for inpatients with abnormal liver function test results: yield and usefulness. *Arch Intern Med* **161**(4): 583–588.

Roy, C. L., Poon, E. G., Karson, A. S., Ladak-Merchant, Z., Johnson, R. E., Maviglia, S. M. et al. (2005). Patient safety concerns arising from test results that return after hospital discharge. *Ann Intern Med* **143**(2): 121–128.

Schoenenberger, R. A., Tanasijevic, M. J., Jha, A., Bates, D. W. (1995). Appropriateness of antiepileptic drug level monitoring. *JAMA* **274**(20): 1622–1626.

Shah, N. R., Seger, A. C., Seger, D. L., Fiskio, J. M., Kuperman, G. J., Blumenfeld, B. et al. (2006). Improving acceptance of computerized prescribing alerts in ambulatory care. *J Am Med Inform Assoc* **13**(1): 5–11.

Shojania, K. G., Yokoe, D., Platt, R., Fiskio, J., Ma'luf, N., Bates, D. W. (1998). Reducing vancomycin use utilizing a computer guideline: Results of a randomized controlled trial. *J Am Med Inform Assoc* **5**(6): 554–562.

Shu, K., Boyle, D., Spurr, C., Horsky, J., Heiman, H., O'Connor, P. et al. (2001). Comparison of time spent writing orders on paper with computerized physician order entry. *Medinfo* **10**(Pt 2): 1207–1211.

Teich, J. M., Merchia, P. R., Schmiz, J. L., Kuperman, G. J., Spurr, C. D., Bates, D. W. (2000). Effects of computerized physician order entry on prescribing practices. *Arch Intern Med* **160**(18): 2741–2747.

Wang, S. J., Middleton, B., Prosser, L. A., Bardon, C. G., Spurr, C. D., Carchidi, P. J. et al. (2003). A cost-benefit analysis of electronic medical records in primary care. *Am J Med* **114**(5): 397–403.

Weingart, S. N., Toth, M., Sands, D. Z., Aronson, M. D., Davis, R. B., Phillips, R. S. (2003). Physicians' decisions to override computerized drug alerts in primary care. *Arch Intern Med* **163**(21): 2625–2631.

6 CASE STUDIES IN CLINICAL DECISION SUPPORT: LDS HOSPITAL EXPERIENCE

R. SCOTT EVANS

LDS Hospital in Salt Lake City, Utah, has been developing decision support applications on the HELP (Health Evaluation through Logical Processing) System for over 30 years. The HELP System was designed to be an electronic health record with decision support and research capabilities. Numerous applications have been developed that use different levels and methods of decision support to improve the patient care process and the quality of patient care. Many of these applications have been scientifically evaluated to measure their impact on patient care. This chapter will use a number of these applications as examples to demonstrate different methods of design, development, implementation, and evaluation along with what has worked, not worked, and why. The HELP System architecture and capabilities that are needed to develop and run the applications also is described and highlighted within the examples.

6.1 INTRODUCTION

Medical decision-making requires the clinician to apply accumulated knowledge to a specific amount of patient information to produce a result that may be a diagnosis, prognosis, course of therapy, or the selection of further tests. Too often, the decisions are based on limited knowledge, the information is incomplete or imperfect, and the decisions must be made during a limited period of time. The improvement in health care quality and safety expected by the public will depend in part on appropriate use of computerized applications (Institute of Medicine 2001).

For over 30 years, computerized applications have been developed to aid the clinician in the medical decision-making process. These applications have been collectively called by a number of different terms. One of these terms is *clinical decision support*, which has been defined as "any computer program designed to help health professionals make clinical decisions, deal with medical data about patients or with the knowledge of medicine necessary to interpret such data" (Shortliffe et al. 1987). An important word in this

Clinical Decision Support: The Road Ahead

definition is "help." These tools are designed to help or support health professionals, not replace them. These decision support tools can be classified into three different categories: 1) tools for information management, 2) tools for focusing attention, and 3) tools for patient-specific consultation. Improvement in health care quality and safety also would include computer applications to identify and reduce the rate of errors, inappropriate or inefficient actions, and adverse events (Bates et al. 2003).

A variety of computer applications that can be defined as decision support tools have been developed at LDS Hospital in Salt Lake City, Utah. This chapter describes a number of those applications along with evaluations with respect to the impact on the quality and safety of patient care. The applications discussed in this chapter are not all-inclusive of the applications developed and used at LDS Hospital, but are chosen to illustrate the kinds of functionality provided, and technology approach used. Also, we concentrate especially on those for which impact on health care quality and safety has been evaluated. It should be noted that the term "computer application" as used in this chapter might be comprised of more than one computer program and/or different types of computer functionalities and data storage.

6.1.1 Key Features for Clinical Decision Support Tools

The computer applications described in this chapter are dependent on the architecture, features, and capabilities built into the HELP (Health Evaluation through Logical Processing) hospital information system developed at LDS Hospital (Warner et al. 1972; Pryor et al. 1983; Pryor 1990; Kuperman et al. 1991; Haug et al. 1994; Gardner et al. 1999). These capabilities are not unique to the HELP System. Other systems that provide these same capabilities must have an analogous architecture with similar features that provide these capabilities. Unlike most computer systems used in hospitals during the 1970s, HELP was designed from the outset as a clinical system, specifically with the aims of being able to provide decision support and to be used as a research tool, rather than just to provide administrative and financial functionality (Warner 1966). The design of the HELP system was influenced by experience using early computer applications developed in cardiology and intensive care (Warner et al. 1961; Warner et al. 1968; Pryor et al. 1969; Gardner et al. 1982). Figure 6-1 displays the essential components of the HELP System.

A key feature of the system is the electronic health record (EHR). Most patient information is stored in the EHR. Some of the information stored in the EHR comes from applications that are part of the HELP System, and some comes from other applications that are interfaced to the HELP System. A number of medical devices also are interfaced to the HELP System, and patient vital signs, medication pump, and ventilator information is stored automatically in the EHR (Gardner et al. 1991).

The main point here is that the clinical information is stored in a common database. The transcribed dictations from x-rays, history and physical exams, and other reports, and admission diagnoses are stored as free-text, whereas most of the EHR data are stored in a coded format. Coded data are needed for the decision support process. Applications have been developed on the HELP System to code some of the patient information contained in free-text documents (Haug et al. 1997). Each coded element in the database and the free-text data are stored with an event time. Thus, although the code for the drug

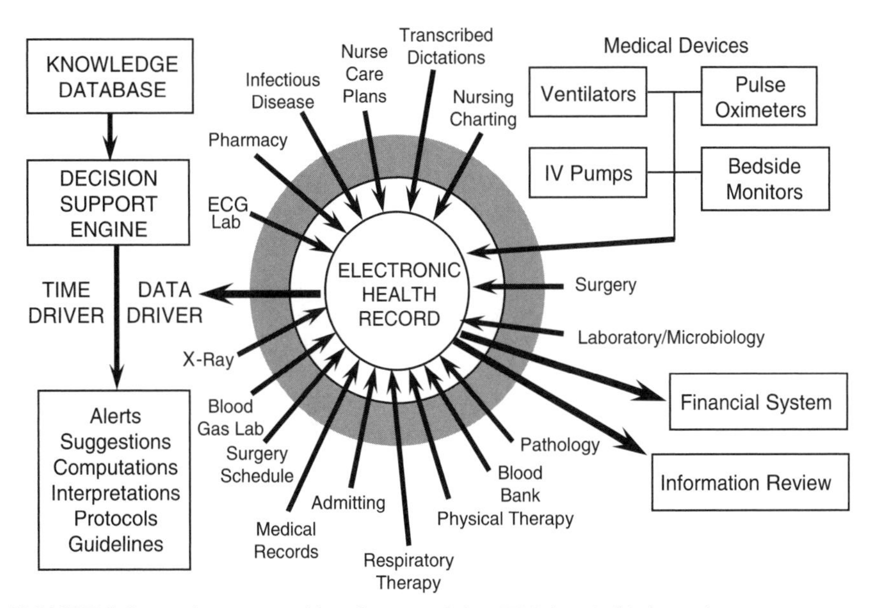

FIGURE 6-1 Architecture and key features of the HELP hospital information system.

cefazolin is always the same when it is stored as a data element in the database, the event time for each data element would be different when the drug is ordered and each time the drug is administered. Although the HELP System was designed as a clinical system, the data elements that are "billable," such as laboratory tests, also contain billing codes that are sent to the financial system. This provides a method to automatically capture much of the patient billing information.

The first benefit from having an EHR is that all the patient information is not just contained in a single paper chart, but can be accessed, formatted, and displayed wherever user authorization allows.

A second key feature of the HELP System is a knowledge base that contains thousands of medical logic modules (MLMs). The MLMs contain medical logic that has been developed by medical experts from different knowledge domains. Some of the MLMs contain simple rules to identify patients with elevated potassium levels based on laboratory results; others may contain complex logic and require patient information from a number of data sources in the EHR. Each of the MLMs contains two main parts. The first part identifies which data elements in the EHR are needed for the logic. Each data element retrieved from the EHR can be restricted to be from within a time interval specified in the MLM. The second part contains the computer logic used to analyze the data elements. Not all medical logic on the HELP System is contained in MLMs. There are a number of applications that contain different levels of programmed logic in the computer code. The decision to use separate MLMs vs. integrated logic depends on the specified purpose of the application and must balance the need for enhanced response time and ease of knowledge management.

The third key feature of the HELP System is the ability to data- and time-drive the knowledge base. The shaded circle around the EHR in Figure 6-1 depicts the data-driver. All data that are stored in the EHR pass though the data-driver, and each data element is screened. The first task the data-driver

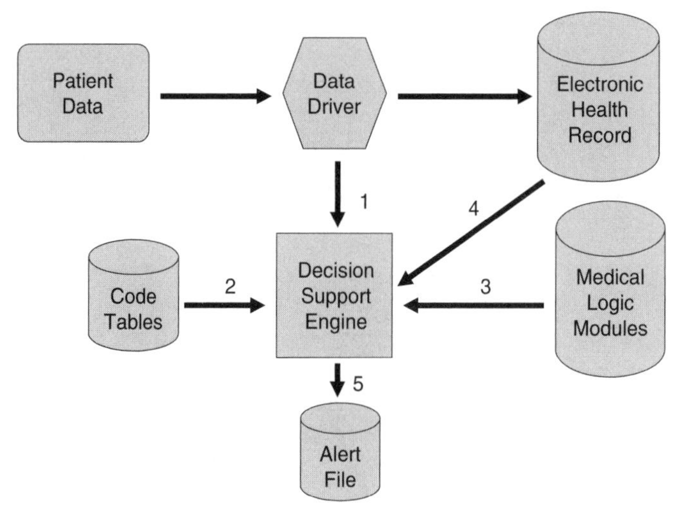

FIGURE 6-2 Main functional processes of the data-driver on the HELP hospital information system.

does is to make a copy of the data string being stored into the EHR and to send that copy to the decision support engine (see Figure 6-2). The decision support engine parses the data string into separate data elements. Each data element is then compared with the code tables to see if there are any MLMs that should be activated based on that data element. If so, each MLM is loaded into the decision support engine. When the decision support engine runs the MLM, it retrieves additional needed information from the EHR as specified in the MLM. If the logic in the MLM generates an alert, another record is built and stored in the alert file. This enables the system to continuously monitor patients. The time-driver is simply a program on the system that checks a table each minute to see if there are MLMs or other applications that should be run. Examples of applications that are dependent on the data- and time-driver will be discussed in this chapter.

Another key feature of the HELP System is that the data in the EHR are never deleted. As patients are discharged from the hospital and all billing is completed, the patient record is moved to an archival EHR. All clinical patient data from the HELP System since 1983 are stored in the current or archival EHRs. The archival storage of the EHR provides data that are often analyzed and used to develop the medical logic contained in the MLMs and has been essential for numerous retrospective research studies.

6.2 TOOLS FOR INFORMATION MANAGEMENT

Computer applications that fall under this category are programs that manage the entry, storage, retrieval, and reporting of patient information. Applications on the hospital radiology, respiratory therapy, and laboratory systems are examples. These applications usually contain different amounts of clinical decision support. The listing of high-low and critical high-low ranges along with the reporting of laboratory test results aids clinicians in medical decision-making that goes beyond just reporting the test results (Clayton et al. 1987). The integration of different data types and organization of specific displays and panels can be very helpful to improve the clinicians' use of the information and thus help improve the quality and safety of health care. Hospital pharmacy systems provide patient-allergy, drug–drug, drug–laboratory, and drug–food alerts when drugs are entered in the electronic medication administration records. Overall, these systems definitely improve the quality and safety of patient care by making information more accessible and easier to read.

6.2.1 Pharmacy System

The pharmacy application developed on the HELP System accesses patient data from the EHR to generate alerts of potential adverse drug events; drug–drug, drug–allergy, drug–laboratory, drug–disease, drug–dose, drug–diet, and drug–interval (Hulse et al. 1976). The application also generates prescription labels and patient drug profiles that are used for unit dose dispensing. The alerts are displayed to the pharmacists as they enter the hand-written physician orders into the application. The pharmacists then inform the physicians or nursing staff of the potential problems. An evaluation of the pharmacy application showed that 5 percent of patients and 0.8 percent of drug orders generated alerts, and that physicians changed patient therapy for 77 percent of the alerts.

Physician acceptance was enthusiastic and pharmacists were more efficient and accurate in their monitoring of prescriptions as a result of the application. The system was found to cost $0.35 per patient day and was included in the pharmacy charge. The problem with this approach is that patients may receive the drugs before the pharmacists enter the orders into the pharmacy application, receive the alerts, and contact the appropriate medical staff. An approach to this problem is that physicians should not handwrite their orders but enter them directly using provider order entry (POE) applications. That way the physicians would immediately get the alerts and change the order before the drug is administered.

6.2.2 Blood Gas Reports

The computerization of laboratory instruments during the 1960s met the need to provide more medical information and in less time. However, this led to medical staff being presented with large amounts of patient data with the expectation that they could thus make medical decisions sooner. Often this increase in information, although making the medical decisions more accurate, required the medical staff to take more time to gather and assimilate the pertinent information. This situation resulted in computers being used to provide or assist in the interpretation of laboratory test results.

The early development of the HELP System paralleled a number of advances in arterial blood gas analysis, and the initial use of computers to provide interpretation of blood gas results (Bleich 1969; Goldberg et al. 1973). The blood acid-base map developed by Goldberg et al. was modified based on the altitude of Salt Lake City, Utah, and used to develop decision support logic. This resulted in the reporting of blood gas results on the HELP System to automatically include the interpretations without any direct physician interaction (Gardner et al. 1975). The accuracy of the computer interpretations was compared with the interpretations of four pulmonary and three nephrology experts. The results ranked the computer interpretations second among the experts. A physician survey found that 80 percent of the blood gas interpretations were helpful and 28 percent changed patient care. This study also demonstrated an important issue that often emerged during the development of many of the computerized applications discussed in this chapter. Before we can develop computer logic, we need to fully understand the previously used manual process and often need to standardize the numerous different processes that are used to achieve the same results. Thus, in this case, the standard interpretation of blood gas data facilitated the development of the computer logic to interpret the blood gas results.

6.2.3 Emergency Department Infection Report

An example of a computer program that manages patient information and that was developed specifically to improve the safety of health care is the emergency department infection report used at LDS Hospital. Thousands of patients visit emergency departments every day and based on their specific clinical manifestations have specimens collected and sent for microbiology examination. Often the laboratory tests are ordered only for precautionary purposes and the patients are sent home. Every emergency department can relate stories about patients who were sent home and subsequently had laboratory test results that contained important information that was overlooked or not followed up. The emergency department infection report is a simple printout that contains all the microbiology and other infection-related test results for all the emergency department patients during the past 10 days. During each shift, a member of the emergency department staff is assigned to run the program and examine the report for any new infection information. When important information is found, the patients are contacted and given specific instructions based on the test results. Use of this information management tool results in the emergency department at LDS Hospital contacting an average of two patients each day and informing them that they need an antibiotic or that their previously prescribed antibiotic needs to be changed (Gibbons 2005). This application demonstrates that the value of decision support applications is not determined by the sophistication or complexity of the program(s) or database. Over the past 20 years, the emergency department at LDS Hospital has changed thousands of therapeutic decisions based on the information contained in the emergency department infection reports. Improvement in physician decision-making directly results in improved health care.

6.2.4 Nurse Bedside Charting

Without the entry of medical data, tools for information management and decision support could not function. A major question is how, where, and when the information should be entered. In some situations, direct data access is provided by interfaces to medical devices, laboratory instruments, or other computer systems. The data provided from interfaces to other computer systems often requires initial data entry by medical staff. An important message of this chapter is that accurate and timely computer decision support is dependent on the data available to the decision logic in the knowledge base, hence the importance of the information contained in the EHR.

An important source of medical information is that which is acquired and documented by the nursing staff. Generally this information is available only in the paper chart. In the 1980s, an electronic nurse-charting program was developed on the HELP System (Pryor 1989). The program also contained some decision logic that would alert the nurse when patient information that was entered was out of range or inappropriate. Moreover, this application initiated important debates concerning where and when the nurse documentation should take place. LDS Hospital installed bedside computers in every room and nurse management stressed the need to document patient care at the bedside as it was given. This decision was based on the initial evaluation of the nurse-charting program that showed that nurses preferred to use the bedside computers and showed an increase in the acquisition of nursing data in real-time and improved patient care (Halford et al. 1989).

LDS Hospital is the largest of 20 hospitals owned by Intermountain HealthCare (IHC) and as the nurse-charting program began to be installed at other IHC hospitals, the question regarding the additional cost of the bedside computers resulted in a moratorium on further installation of the computers until justification could be demonstrated. This resulted from a second study at another IHC hospital that showed that the nurses were documenting most of their care at the nurse station rather than at the bedside (Hinson et al. 1994). As expected, nurse documentation done at the nurse station was generally found to take place at the end of the shift, probably less accurately, and thus was believed to have a reduced value for decision support. An additional study on the value of using the nurse-charting program at the bedside subsequently demonstrated the value of the bedside computers, and the moratorium was lifted (Wilson 1994). This nurse-charting example was included in this chapter for a number of reasons. It is a good example of computerized data entry of important patient information, but it also demonstrates that these applications can generate issues that can impede their widespread adoption and impact on patient care. There are often differences in their acceptance and use from one facility to another, which can raise organizational, technical, and business challenges that need to be met. Implementation of these applications is almost always met with the need to make major process changes and overcome a number of social and political issues. A recent study on the importance of where and when nurse charting should be done demonstrates that these applications are never finished, that they require continual monitoring and enhancements, and that user education is an ongoing process (Nelson et al. 2005). This latest study showed that nursing education about medical error avoidance and performance feedback helped to increase the charting of nursing information as soon as it was done from 59 percent to 73 percent and charting at the bedside also increased from 40 percent to 63 percent.

6.2.5 Respiratory Therapy Charting

Although the HELP System was developed at LDS Hospital, many of the developers were also faculty members of the University of Utah-based Department of Medical Informatics. Thus, some of the clinically used applications initially were designed, developed, implemented, and evaluated as graduate student projects. The respiratory care charting system is an example. The respiratory therapists found the new system so functional that it completely replaced the old manual methods for charting and reviewing respiratory care information. The fact that the system also provided automatic billing, management functionality, alerts, and reports was an additional incentive to use the system and a reason for management and administration acceptance. An evaluation of the system showed that the computer charting was more complete and extensive than manual charts, and that productivity increased by 18 percent (Andrews et al. 1985). Properly designed computer charting programs act like constant reminders of what information needs to be entered. The evaluation also showed that the old manual four-step charge capture process could be completely replaced by the computerized documentation. The respiratory care application has since been updated, improved, and installed at other IHC hospitals, and the respiratory care departments continue to do all their documentation and reporting through the computerized system.

The respiratory charting program also was used in one of the first evaluations of bedside documentation (Andrews et al. 1988). The program was ported to a few Radio Shack TRS-80 portable computers, and six therapists were selected to test the application. The respiratory therapists could document at the bedside and then transfer the data to the EHR via phone lines. A comparison of the portable computers to the ward computer terminals showed no difference in productivity or promptness of documentation. Due to the need to carry the portable computers around and the extra time to connect to phone jacks, the therapists preferred using the ward computer terminals. This preference was also partially due to the fact that during that time the ward computer terminals were usually available. As nurse charting and other applications were implemented and use of the ward terminals increased, the respiratory therapists quickly adapted to use the new bedside computers.

6.3 TOOLS FOR FOCUSING ATTENTION

6.3.1 Infectious Disease Monitor

The infectious disease monitor is a good example to illustrate the benefits of using the data-driver and the time-driver together. This application identifies patients who have conditions that infection control practitioners and infectious disease physicians want to be aware of: 1) patients with hospital-acquired infections, 2) patients with reportable diseases, 3) patients with antibiotic-resistant pathogens, and 4) patients with infections in sterile body sites (Evans et al. 1985). The code for a microbiology test directs the decision support engine to load MLMs that contain logic to identify pathogens based on the specimen and/or body site. Most pathogens found in sterile body sites (blood, cerebral spinal fluid, pericardial fluid, etc.) generate alerts. Other specimen and body site locations such as sternum, knee, and hip wounds are also high interest infections. Some MLMs contain logic to determine which infections need to be reported to state and federal health departments whereas others contain the Center for Disease and Control (CDC) criteria for the identification of hospital-acquired or nosocomial infections. Thus, as patient information from microbiology culture results, urinalyses, and chest X-rays are stored in the EHR, the data-driver provides 24 hour and hospital-wide surveillance.

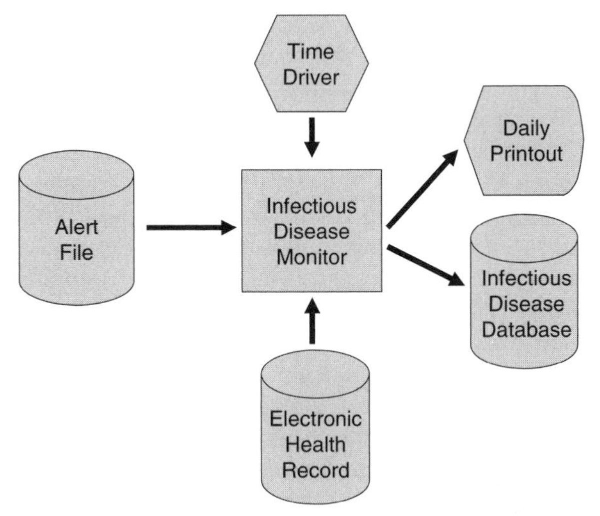

FIGURE 6-3 Example of using the time-driver to notify infection control of patients with possible hospital-acquired infections and provide needed information.

Since most microbiology results are updated in the morning, the time driver activates the infectious disease monitor at 12:30 P.M. each day and the program retrieves all the infectious disease data-driven alerts that have been stored in the alert file during the past 24 hours (see Figure 6-3). Depending on the type of alert, the infectious disease monitor program then accesses the EHR and appends other pertinent patient information needed by infection control practitioners or infectious disease physicians.

The program builds a listing of reports that is sent to the printer in the infectious disease or quality assurance offices. Each report contained in the listing displays the condition(s) that generated the alert at the top of the report. Since microbiology cultures and antibiotic susceptibilities take days or weeks to complete, many of the alerts are first generated based on Gram stain and other preliminary findings and results. Figure 6-4 shows the other pertinent patient information that typically is included with each report. The hospital-acquired

LDSH Infectious Disease Monitor for Aug 25, 1999

```
***** PATIENT WITH POSSIBLE HOSPITAL-ACQUIRED WOUND *****
***** WOUND AND PREV. DSCH WITHIN 30 DAYS *****
***** ANTIBIOTIC ALERT *****

PAT: 11111111 Jones, David          76     M          E801         MR#: 000000
ADMITTED: 08/21/1999.15:55                         ADMIT DIAG: Pulmonary embolism
PREV. ADMIT: 08/06/1999          PREV. DSCH: 08/13/1999
DOC: 9999 Smith, Ralph              SERVICE: Cardiovascular
SURGERY: 08/21/1999.18:35        Debreadment         Contaminated  SURGEON: 00000
CURRENT ANTIBIOTICS: 08/22/1999.11:10 Fluconazole (Diflucan)        IV   200   q24hrs
CULTURE RESULTS                  -FINAL REPORT-                   ROUTINE CULTURE
SOURCE: Wound, Chest                                    COLLECTED: 08/21/1999.21:54
STAIN: 4+ WBCs, 3+ Gram positive cocci, 1+ Gram negative bacilli
RESULT: Streptococcus milleri, 4+
SUSCEPTIBLE: Ampicillin, Cefaxolin, Ciprofloxacin, Clindamycin, Erythromycin,
             Levofloxacin, Penicillin, Trovafloxacin, Vancomycin
!!!!! VERIFIED !!!!!
```

FIGURE 6-4 Example of the daily printout from the infectious disease monitor from LDS Hospital.

infections are also stored in the infectious disease database, where the information is used by other programs to prepare monthly, quarterly, and ad hoc infection control reports, patient-infection summaries, and epidemiology graphs with control limits. The text "!!!!! VERIFIED !!!!!" included at the bottom of the report notifies the infection control practitioner that the culture is now finalized by the microbiology lab and the computer logic has determined the infection to be hospital-acquired and will be stored in the infectious disease database. The infection control practitioners can use a program to delete the computer verified infection if they do not agree with the computer logic.

An evaluation of the infectious disease monitor found that the computer surveillance identified more hospital-acquired infections than the manual surveillance by infection control practitioners and required less time (Evans et al. 1986). This study showed an additional value of evaluating decision support applications. The evaluation identified infections that were missed by the computer surveillance, and subsequent investigation demonstrated that changes and additions to the computer logic would enable the application to identify most of the missed infections. This application is now being used as the primary surveillance tool by 10 hospitals at IHC.

6.3.2 Therapeutic Antibiotic Monitor

The therapeutic antibiotic monitor is another example to illustrate the functionality of the data- and time-drivers. The purpose of this computer application is to identify patients who may not be receiving appropriate therapeutic antibiotics based on the results from microbiology culture and susceptibility results. While fulfilling a different purpose, the application process is similar to the infectious disease monitor. This MLM is activated by the data-driver based on the presence of antibiotic susceptibility codes. The logic first determines from the microbiology results whether the patient should be treated with an antibiotic based on the identification of a pathogen and the specimen type and/or body site. If the logic identifies the need for an antibiotic, a list of appropriate antibiotics is created based on the antibiotic susceptibility results. The patient's current antibiotics are retrieved from the EHR and compared with the list of appropriate antibiotics. If the patient is not receiving an antibiotic that will cover the identified pathogen(s), an alert is sent to the alert file. Each morning, the time-driver activates a program that sends a printed list of the identified patients to the pharmacy. Pharmacists follow up on the identified patients and notify the physicians, if they agree that the patient is still not receiving appropriate antibiotics.

During a 12-month study, the therapeutic antibiotic monitor identified 696 instances in which patients appeared not to be receiving appropriate antibiotics (Pestotnik et al. 1990). Of those, 420 were judged to be true, and the physicians either changed or started antimicrobial therapy 30 percent of the time when pharmacists contacted them. When contacted by the pharmacists, the physicians stated that they were not aware of the culture and susceptibility results 49 percent of the time. This application was included in this chapter to illustrate that decision support that requires patient data from just two sources, in this case microbiology and pharmacy, can help improve the quality and safety of patient care. Although physicians do not always change antimicrobial therapy based on the alerts, they have always been appreciative of the information when contacted by pharmacists. For hospitals that are just getting into or are planning to implement computer decision support, this is an example of a fairly simple application that can impact patient care and get a positive response from the users. This is important because first applications that are too sophisticated and don't work can decrease the chance of success for future applications.

6.3.3 Adverse Drug Event Monitor

The experience and knowledge gained from developing and using the infectious disease monitor and the therapeutic antibiotic monitor led to an increased interest in adverse drug events (ADEs). The therapeutic antibiotic monitor selected the less expensive antibiotic from a list of clinically equal antibiotics. In 1988, we began to question if a drug that cost a few dollars less per day to use but caused a number of ADEs was really less expensive than another drug that caused only a few ADEs. We decided to add logic about the frequency of ADEs to the selection of the least expensive antibiotic. However, we soon found out that no one at LDS Hospital had any idea what the actual ADE rate was, nor which drugs caused the ADEs. An average of nine to ten ADEs were reported each year. Based on the literature, 2 to 10 percent of hospitalized patients experienced ADEs. If computerized surveillance could help identify more hospital-acquired infections, why not ADEs also? Thus, MLMs were developed that monitored: 1) laboratory test results that could be indicative of a possible ADE, 2) elevated serum drug levels, 3) the ordering of drugs that are commonly used to treat ADEs, and 4) physiologic data that could signal possible ADEs (Evans et al. 1991). Logic was included to help reduce the number of false positive alerts. For example, a doubling in serum creatinine would not generate an alert if the patient's diagnosis included renal failure or the patient was scheduled for a kidney transplant. Pharmacists received printouts containing alerts of patients with possible ADEs and verified the ADEs based on information in the paper chart, the EHR, and clinical judgment.

Use of the ADE monitor increased the annual number of verified ADES from 9 to 10 to over 500 (Classen et al. 1991). In addition, the analyses of the subsequent ADE database allowed us to do root-cause analyses on each of the ADEs and identify potential methods to reduce the numbers of certain types of ADEs (Evans et al. 1994) and identify risk factors for ADEs (Evans et al. 2005a). We were able to verify that the use of drug-allergy alerts did lead to the reduction in the number of type B, allergic or idiosyncratic, ADEs. We also found that early physician notification of mild and moderate ADEs by pharmacists helped to decrease the number of severe ADEs.

The adverse drug event monitor is a good example of how the knowledge and information gained from the implementation and use of one decision support application can lead to the development of another. Our interest in improving the logic to identify less expensive antibiotics resulted in a completely new and exciting application that has had a large impact on patient care. The development of the ADE monitor application also led to the development of the ADE verification program that helps pharmacists determine and report whether the patients' clinical manifestations are caused by ADEs. The rules for identifying ADEs have been updated numerous times over the past 15 years to identify ADEs that were previously missed. Between the end of 1999 and early 2000, a number of new rules were added to see if this application could be used to identify adverse medical device events (AMDEs). Like ADEs, we found that no one had a good handle on the surveillance or rates for AMDEs (Samore et al. 2004). We found that although the computerized surveillance yielded higher rates of AMDEs than traditional voluntary reporting, many device-related problems were not identified due to sparse or nonexistent event documentation in the EHR and the absence of routine or organized surveillance for these events. We found that when problems with bedside medical devices were identified, the nurses usually would replace them without any further documentation. When medical devices caused adverse events, they were occasionally reported through an electronic incident report, but except for line occlusions, no relevant documentation was entered into the nurse-charting program.

6.3.4 Lab Alerts

Laboratory information systems usually include references for critical low and high ranges and attach an L or H to the value in the display. When the laboratory test results are reviewed by medical personnel, the flagged values allow them to recognize patient situations that need immediate actions. However, in this approach, medical personnel can remain unaware of the abnormalities until the results are reviewed. Therefore, laboratories are supposed to call and notify medical personnel of critical laboratory test results as soon as possible. It is then the responsibility of the person receiving the call to see that appropriate action is taken.

In 1985, the data driver in the HELP System was used to identify patients with certain potentially life-threatening conditions based on their laboratory test values (hypokalemia, falling potassium, hyperkalemia, etc. (Tate et al. 1990)). The computerized laboratory alerting system (CLAS) displayed an alert message to the user the next time the laboratory test results were reviewed for the patients. Because of this built-in time delay, an enhanced version was tried in which yellow lights were attached to each computer terminal and would flash as soon as an alert was generated for a patient on that nursing division. However, the flashing lights were considered obtrusive by nursing staff and subsequently removed. An evaluation of the system without the flashing lights showed that the medical staff responded to the critical laboratory values sooner with the computer alerts and that the system significantly reduced the time patients were in potential life-threatening situations. The value of the system was further supported by a later study that showed that only 9.5 percent of critical laboratory values were telephoned to the nursing floor by the laboratory and 15 percent of audited patient charts contained no documentation indicating clinician awareness of critical laboratory values. CLAS was then updated to use patient-specific criteria and to automatically page the nurses taking care of patients on whom alerts existed. With this approach 51 percent of alerts were received within 12 minutes, 92 percent of the alerts were considered important, and 67 percent of the time the nurses were previously unaware of the critical laboratory test values (Tate et al. 1995).

6.3.5 Antibiotic Duration Monitor

During the early 1980s, infectious disease physicians considered the prolonged use of prophylactic antibiotics for surgical operations to be a major cause for the increase in antibiotic-resistant pathogens. This was compounded by the fact that prophylactic antibiotics usually were continued until the patient was discharged from the hospital. To address these issues, a computer application is activated each morning by the time-driver and checks the EHR of each hospitalized patient to determine if the patient had a surgical operation and is receiving antibiotics that were initiated within 24 hours of the latest operation. For identified patients, the EHR is then examined for any indication that antibiotics are needed; for example, a subsequent scheduled surgery within 24 hours, the previous surgery was classified as contaminated or dirty, identified pathogens from microbiology examinations, pending microbiology cultures, Gram stain showing bacteria or numerous white blood cells, fever, chest x-ray dictations suggesting the need for antibiotics, bacteria detected by urinalysis, admission diagnosis of infection, or the patient being in isolation. Like microbiology and other laboratory information, x-ray dictations are interfaced automatically with the HELP System so they can be used for this purpose (Haug et al. 1992).

Patients that are receiving prophylactic antibiotics longer than 48 hours after the operation and do not have any evidence in the EHR of infection are added to a printed list sent to the pharmacy each day. Pharmacists follow up on the identified patients to verify a need for antibiotics. If physicians did not specify a stop time on the order and the pharmacists do not identify a need for the antibiotics, the pharmacists place a stop order in the chart. For the six months before the antibiotic duration application was installed, surgery patients for whom antibiotics were ordered received an average of 19 doses compared to 13 during the first six months the application was used (Evans et al. 1990). The time required by a single pharmacist was 45 minutes a day. The reduction in antibiotic doses during six months was determined to yield a cost savings of over $44,000. Continued use of the application along with pharmacist education at surgical department meetings eventually reduced the average number of doses to 5.3.

6.3.6 Preoperative Antibiotic Monitor

During the evaluation of the antibiotic duration application, it was noticed that most of the surgical prophylactic antibiotics were not given within two hours before the start of surgery. Numerous studies since the 1960s demonstrated that prophylactic antibiotics should be given within the two-hour time window to allow maximum antibiotic concentrations in tissue and blood during the procedure. Many antibiotics were started earlier than two hours before and most were started during or hours after the surgical procedure. A computer application was developed to generate reminders of the importance of starting prophylactic antibiotics within two hours before the start of the operation and placed in selected patients' charts (Larsen et al. 1989). The application is activated by the time-driver at numerous times of the day and identifies the patients that need antibiotics by accessing the surgery schedule and identifying the procedures that needed preoperative antibiotics. Lists of the patients that should receive preoperative antibiotics are printed out in the holding rooms and admitting area beginning at 11:00 A.M. the day before elective surgeries. A need for a preoperative antibiotic is also noted for the selected patients on the electronic surgery schedule.

Before the computer-generated preoperative antibiotic reminders, 40 percent of the surgical patients who should have received preoperative antibiotics had the antibiotics started within two hours before surgical incision compared to 58 percent during the next six months. This rate increased to 96 percent within the next year. Analysis of hospital-acquired infections during the pre- and postintervention periods showed that the postoperative wound rate was significantly lower with the computer-generated reminders of preoperative antibiotic use (Classen et al. 1992). This application is another example of how the development of one application can provide valuable information that leads to the development of other applications that directly impact the quality of health care. Moreover, it is often the evaluations of applications that identify processes that do not follow specified protocols and provide additional opportunities for improvement in health care. Thus, evaluations not only demonstrate the benefits of the applications, but can also identify inappropriate processes and practices in health care.

6.3.7 High-Risk Alerts for Hospital-Acquired Infections

The infectious disease monitor described previously identifies patients with hospital-acquired infections based on specific information that is entered into the EHR. In 1989, we wondered if we could use statistical methods to identify patients at high risk of developing an infection in the hospital before the infection onset. A study database was created with 3,151 patients with hospital-acquired infections and 3,152 control patients. Stepwise logistic regression was used to develop a predictive model for high-risk patients based on 10 of 18 putative risk factors tested. A computer program was activated each day to use an equation based on the model to monitor all hospitalized patients and create a computer printout of the high-risk patients. During the first six months of 1990, 78 percent of hospitalized infections occurred in high-risk patients and 63 percent were predicted before the documented onset of the infection (Evans et al. 1992). A subsequent study, not published, used the high-risk program to monitor all hospitalized patients each day, and random patients were included on the daily printout. Infection control practitioners then placed stickers on the doors of the high-risk patient rooms. The stickers identified the patients as high risk for infection and contained general methods to prevent infection. During the six-month study period, the infection control practitioners received numerous questions concerning the stickers and a few physicians wondered why only one of two very similar patients had the sticker on the door. The evaluation of the randomized process showed that there was not a significant difference in hospital-acquired infection rates between study and control patients.

This study demonstrated two important lesions for evaluation studies on the impact of decision support programs. First, decision support applications are designed to help clinicians and thus, by nature, are almost always educational. Randomizing by patient gives the physician information on some patients that often is used to treat other patients. That's a good thing. Randomizing by physicians is difficult, since most measurable outcomes are patient-specific, and usually multiple physicians participate in the treatment of the same hospitalized patients. Second, identifying a patient as being at high risk of an infection was too general. The suggestions of decision support need to be very specific. Based on our experience with preoperative antibiotics, we should have known better and developed a program to give patient-specific suggestions to prevent a particular type of infection.

6.3.8 Drug-Dose Monitor

An example of an application that is comprised of just one computer program and activated by the time-driver is the drug-dose monitor. This program calculates the renal function of every hospitalized patient each day and determines whether the patient is receiving a drug dosage that is too high. The program uses logic that contains the recommended 24-hour dosages for specific drugs based on the patient's creatinine clearance and underlying disease. Patients who are identified as receiving excessive dosages along with a computer-suggested dosage are included on a list that is sent each morning to the pharmacy. Pharmacists follow up on the alerts and notify the physicians when they agree with the alerts. Initially, the program only monitored anti-infective agents, but a number of other drugs were added that need to be monitored due to their frequent association with ADEs (e.g., morphine, meperidine, Lovenox, Toradol, procainamide, etc.) or excessive costs (H_2-blockers).

The program initially was evaluated in terms of its impact on the top five antibiotics that caused ADEs at LDS Hospital (vancomycin, cefazolin, gentamicin, imipenem, and cefuroxime). During a 12-month study, pharmacists contacted the physicians of each of the patients for whom they were alerted, and discussed the possible excessive dosages (Evans et al. 1999). Compared to the preintervention period, the number of days of excessive dosage was reduced from an average of 4.7 to 2.9, the number of drug doses was reduced from 13.4 to 10.9, average antibiotic costs decreased from $128 to $98, and ADEs caused by the five antibiotics were significantly reduced.

6.3.9 Enhanced Notification of Ventilator-related Events

The unintentional disconnection of patients who are dependent on mechanical ventilators is a definite quality and safety issue faced by all hospitals. This problem and our earlier interest and experience with adverse medical device events led us to develop an enhanced alerting system to notify medical personnel whenever patients become disconnected for longer than 10 seconds (Evans et al. 2005b). New updates and programs added to our medical device interfaces allowed us to determine when patients become disconnected and take control of all the computers in the same unit as the disconnected patient and flash an alert on the screen identifying the disconnection and the patient room. The alert also sends a "submarine dive-horn" audio alert to the computers outside of the patient rooms. This program is now running in four ICUs at LDSH and we plan to add other hospitals. With this system, every critical ventilator event is identified and the duration is logged. Initial evaluations show that patients are now disconnected only for an average of 20 seconds. Acceptance by medical personnel has been very high, and patient safety was improved through early intervention that avoids prolonged hypoxia. In addition, the system has facilitated root-cause analyses and new safety strategies to prevent some disconnections.

This application is another example demonstrating the value of collaboration between the clinical team and medical informatics. Respiratory Care management saw the need to solve this problem and initially approached the medical informatics group for help. It needs to be stated that much of the success of the decision support programs at LDS Hospital stems from the support and innovativeness of the medical personnel. The environment, culture, and mutual trust created by the early computer applications developed at LDS Hospital greatly facilitate the development, acceptance, and even the expectation of new applications to improve health care and patient safety.

6.4 TOOLS FOR PATIENT-SPECIFIC CONSULTATION

6.4.1 Blood Ordering

As stressed in this chapter, the ideas or incentives for developing the computerized decision support applications at LDS Hospital were generated through the combined efforts of clinical and medical informatics staff trying to improve the quality and safety of patient care. One of the first applications developed at LDS Hospital to aid in patient-specific consultation during the ordering process was the result of an unsatisfactory review in 1986 by the Joint Commission of Accreditation of Healthcare Organizations (JCAHO). This review identified the lack of a number of blood usage measures including ordering practices, and use of clinically valid criteria. The underlying problem was that written blood orders did not contain the reasons the blood was being ordered. Pharmacy had tried to solve this problem for a number of years through use of a special blood-order form that included a place to write the reason for the order. However, the reasons were seldom included.

In an effort to solve this problem, a blood ordering application was developed where medical staff would electronically order blood products and be required to enter the reason before the order could be completed (Gardner et al. 1990). The blood products and reasons could be selected from drop-down lists and the program would validate the reason based on patient laboratory, surgery, and other clinical information contained in the EHR. For example, if the selected reason was "anemia" and the patient's hemoglobin was greater than 12 g/dl, the hematocrit was greater than 35 percent, and the patient's age was greater than 35 years, the program would notify the user that the reason did not meet approved criteria. Another reason could be selected or the user could override the criteria check. Overrides were flagged and followed up by quality management and monthly reports were sent to department medical directors. Initially, user acceptance was good, and physicians ordered 45 percent of the blood products through the program and 7 percent were from physician standing orders. Nurses were authorized to order blood products based on verbal orders (14%) and phone orders (8%). However, nurses also ordered blood products for the remaining 26 percent based on physician written orders. Although there was strong support and a mandate by administration and most of the clinical leadership for physicians to use the program, a number of physicians continued to handwrite orders for blood products. Over time, given the fact that almost all other orders had to be handwritten, physician use of the blood products ordering program decreased to the point where most of the blood orders are entered in the computer by nurses and they use their judgment as to why the blood product was needed.

6.4.2 Ventilator Protocols

In an effort to improve the treatment of acute respiratory distress syndrome (ARDS), critical care physicians at LDS Hospital assisted in the development of complex paper flowcharts to be used as treatment protocols. A computer application was developed on the HELP System to see if pertinent patient information could be accessed from the EHR and to recommend patient-specific ventilator adjustments based on the logic contained in the protocols (Sittig et al. 1990). Respiratory therapists and physicians could run the computer protocols from the bedside computers and receive the ventilator recommendations. The user could accept or reject all or each of the specific recommendations. The recommendations were developed to be clear, concise, and specific (see Figure 6-5).

The application was initially tested on eight ARDS patients. As a result, computer logic and data entry errors were identified and fixed, and improvements were made in the timeliness of data access. An evaluation of the computer protocols for 72 additional patients showed that recommendations were followed 92.3 percent of the time compared to 63.9 percent for the initial eight patients and the accuracy of the recommendations improved to 92.8 percent compared to 71.5 percent (Henderson et al. 1991). The protocols were continuously improved based on assessment of their clinical use. The evaluation of 111 patients showed that the computer protocols were used for over 35,000 hours and controlled the decision-making for ventilator adjustments 95 percent of the time. The survival rate of the ARDS patients on the computer protocols was 67 percent compared to 33 percent without protocol use (East et al. 1999). In addition, the ventilator protocols have been used to show that computerized decision support tools can be exported to and function at other hospitals.

A multicenter randomized trial including nine other hospitals, six not affiliated with IHC, showed that although ARDS patients on the computer protocols did not experience a significant reduction in mortality or ICU length of stay, there was a significant reduction in morbidity with respect to multiorgan dysfunction score and severity of overdistension lung injury. Moreover, although some of the hospitals, based on relative incompleteness of their EHRs, had to enter some patient information when the protocols were run, the computer protocols used the same logic and presented the recommendations in the same manner.

6.4.3 Anti-infective Agent Assistance

In 1989, an infection database was set up on the HELP System to include all positive microbiology cultures with antibiotic susceptibilities for the latest five-year period. The time driver is used to automatically update the database each month and build antibiotic antibiograms (Evans et al. 1993). Stepwise

Increase PEEP by 2 cm H_2O from 8 to 10 cm H_2O
Maintain FiO_2 = 70%
Maintain VT set = 500 ml
Increase Rate set by 2 from 28 to 30 breaths/min
Adjust Peak Flow to maintain I:E near 1:2.3
Draw an ABG 20 min after vent change

FIGURE 6-5 Example of patient-specific ventilator adjustment recommendations from the computerized ventilator protocols. (PEEP = positive end-expiratory pressure, FiO_2 = fraction of inspired oxygen, VT = tidal volume, I:E = inspiratory/expiratory ratio, ABG = arterial blood gas.)

logistic regression models were used to identify patient variables contained in the infection database that can help predict which pathogens a patient may have before microbiology culture results are available. The program also contains therapeutic rules developed by infectious disease specialists to help in the selection of probable pathogens and appropriate treatment. A computer program is available to allow medical staff to select a patient and the suspected type of infection (urinary tract, respiratory, blood, etc.) and receive a display that predicts pathogens along with a list of the most likely empiric antibiotic regimens to cover the pathogens. The program then selects an appropriate antibiotic regimen for the patient based on probability of clinical success, patient allergies, toxicity, and cost. An evaluation of the program found the empiric antibiotics selected by the computer were appropriate 94 percent of the time whereas physician-ordered antibiotics were appropriate 77 percent of the time. A follow-up study showed that physicians ordered appropriate antibiotics significantly more often when they used the empiric antibiotic program (Evans et al. 1994).

Based on the physician use and approval of the computerized antibiograms, empiric antibiotic suggestions, and the therapeutic antibiotic monitor, an anti-infective agent management program was developed (Evans et al. 1995). This program is used as an information tool to help physicians in the selection of appropriate anti-infective agents. A single screen was designed to display all patient information that physicians should be aware of for the selection process (see Figure 6-6). Although the screen may look cluttered at first, it was designed by physicians, and new users quickly learn the layout of the information. There are three parts to the display: 1) pertinent patient information and calculations used by infectious disease specialists to determine the need for and selection of appropriate anti-infective agents, 2) suggested anti-infective agents along with the dosage, route, and interval based on imbedded computerized logic, and 3) options to quickly access

IHC Antibiotic Assistant & Order Program

```
000000000 Doe, Jane Q     E606     67yr     F          Dx:ABD SEPSIS
» Max 24 hr WBC=21.0↓ (21.3)        Admit:07/27/98.14:55        Max 24hr    Temp=38.7↑ (38.2)
Patient's Diff shows a left shift, max 24hr bands = 22 ↑ (11)
» RENAL FUNCTION: Decreased, CrCl = 50, Max 24hr Cr= 1.0↓ (1.1)        IBWeight: 58kg
» ANTIBIOTIC ALLERGIES: Ampicillin,
» CURRENT ANTIBIOTICS:
1. 07/29/98       5DAYS   TROVAFLOXACIN (TROVAN),            VIAL 300.Q 24 hrs
2. 08/01/98       2DAYS   AMPHOTERICIN B (FUNGIZONE),        VIAL 35.             Q 24 hrs
  Total amphotericin given = 70mg    K= 3.6mg/dl  08/03/98    MAG= 2.5mg/dl   08/03/98 » »
» IDENTIFIED PATHOGENS                      SITE                      COLLECTED
p Gram negative Bacilli                        Peritoneal Fluid               07/27/98.17:12
  Yeast                                        Peritoneal Fluid               07/27/98.17:12
  Torulopsis glabrata                          Peritoneal Fluid               07/27/98.17:12
» THERAPEUTIC SUGGESTION        DOSAGE ROUTE  INTERVAL
          Imipenem            500mg          IV           *q12h      (infuse over 1hr)
          Amphotericin B               35mg             IV           q24h      (infuse over 2-4hrs)
***** Antiinfective suggestions should not replace clinical judgment *****
*Adjusted based on patient's renal function.
P=Preliminary status: Susceptibilities based on antibiogram or same pathogen w/susceptibilities
<1>Micro   <2>OrganismSuscept,   <3>Drug Info,   <4>ExplainLogic,   <5>Empiric Abx,
<6>Abx Hx   <7>ID Rnds,   <8>Lab/Abx Levels,   <9>Xray,   <10>Data Input Screen,
<Esc>EXIT,   <F1>Help,   <0>UserInput,   <.>OutpatientModels, <+orF12>Change Patient
↑↓,         ORDER:<*>Suggested Abx,   <Enter>Other Abx,   </>D/C Abx,   <->Modify Abx,
```

FIGURE 6-6 Example of the information screen presented by the anti-infective agent management program.

detailed patient information such as the antibiograms (OrganismSuscept) and empiric antibiotic predictions.

The ability to suggest anti-infective agents was first requested by intensive medicine physicians who pointed out that the therapeutic antibiotic monitor suggested antibiotics based on a single culture result. The patients for whom they provided care often had many pathogens from multiple sites, needed anti-infective agents due to aspiration, contaminated surgical procedures, or based on the admission diagnosis, or for other reasons. They wanted the computer logic to help assimilate all anti-infective agent requirements into one regimen.

In July 1994, the anti-infective agent management program was installed in the 12-bed shock/trauma ICU at LDS Hospital and for 12 months all anti-infective orders had to be entered through the program. The program allowed the physicians to electronically send the orders for their selected anti-infectives directly to the clerks' computers and printers. If physicians did not agree with the computer-suggested anti-infective agents, they had to select the reason before the order could be completed. The evaluation of this study showed that physician use of the program reduced the number of times patients received inappropriate anti-infective agents, reduced the number of excessive anti-infective dosages, reduced the number of adverse drug events caused by anti-infective agents, reduced the number of times patients received anti-infective agents to which they had documented sensitivities, and reduced the cost of anti-infective agents (Evans et al. 1998).

Based on the positive results of the evaluation, nine other IHC hospitals that have the HELP System requested to have the anti-infective program installed. The program was installed as-is in eight of the hospitals. The other hospital was Primary Children's Medical Center, and most of the computer logic to suggest anti-infective agents was not appropriate for pediatric and neonatal patients. Pediatric infectious disease and intensive care specialists evaluated each of the rules used by the program, including the need to treat, anti-infective agent selection, and appropriate dosages. Around 75 percent of the rules were changed, and many new ones were added. In most cases, the screen displays were kept similar to the adult version. The new logic could determine whether the patient was a neonate, pediatric, or adult patient.

The pediatric version was installed in the pediatric intensive care unit (PICU) and all anti-infective agents had to be ordered through the program. Use of the program during a six-month period was compared to the previous six months. The study was initiated at the end of January so seasonal infections would be similar in the intervention and preintervention populations (Mullett et al. 2001). The evaluation of the impact of the program showed that during the intervention period there was: 1) a 58 percent decrease in physician requests for pharmacy help in dosage selection, 2) a 59 percent decrease in pharmacy interventions due to erroneous dosage selection, 3) a 28 percent decrease in the number of days patients received excessive dosages, 4) a 36 percent decrease in the number of days patients received subtherapeutic dosages, 5) an 11.5 percent decrease in the number of anti-infective orders, and 6) a 9 percent decrease in the cost of anti-infectives. No impact on the rate of ADEs was found. All but one of the ADEs due to anti-infective agents during both study periods were allergic reactions caused by first-time use of the agents. There was one ADE in each period due to use of an agent to which there was a noted allergy in the patient record. During the intervention period, root-cause analysis showed that the allergic antibiotic order had been handwritten in the operating room, and the computer program was not used.

The computer logic contained in the anti-infective agent management program is extensive and many of the rules are dependent on the use of new agents and current knowledge. The program is used about 90 times a day just at LDS Hospital alone, and clinician feedback has always been encouraged.

This program illustrates several important points. Use of computer applications that contain decision support logic need to be constantly evaluated and have the logic changed and updated as needed. Routine clinical use of an application provides an opportunity for constant evaluation by the users. Methods and processes need to be in place to provide instant user input of any questions or problems. However, user feedback may not be enough. Routine or occasional reevaluations of the applications should be made. During 2001, three non-IHC infectious disease physicians, two primary and one arbitrator, who did not participate in the development of the logic and rules used by the anti-infective agent program, compared the computer-suggested and the physician-ordered anti-infective agents to established infectious disease guidelines (ID-supported). Each patient in the shock/trauma ICU at LDS Hospital was evaluated each afternoon Monday through Friday for four months. The logic in the computer program was updated based on the evaluation and feedback from the infectious disease physicians. The same comparison of computer-suggested and physician-ordered anti-infective agents was then repeated for another four months. During the first phase of the study, 70 percent of the computer-suggested agents were ID-supported compared to 72 percent of the physician orders. In the second phase, 84 percent of the computer suggestions were ID-supported compared to 70 percent physician's orders. There was a 33 percent concordance between the computer-suggested and physician orders for which 98 percent were ID-supported (Tettlebach et al. 2002).

In 2003, IHC organized a software oversight committee (SOC). One of the first programs examined was the anti-infective agent program. Some members of the committee wondered if physicians blindly followed the computer-suggested anti-infectives without using their own clinical judgment. All the previous evaluations of the program were made in ICUs where physician training in infectious diseases was felt to be higher than in other divisions of the hospital. The committee selected the orthopedic division of LDS Hospital, and an infectious disease physician examined each case during a six-month period where the anti-infective agent program suggested anti-infectives. The infectious disease physician determined that the computer suggestions were not blindly followed.

An anti-infective management committee comprised of members from infectious diseases, pharmacy, intensive care, and medical informatics was formed in February of 2004. This committee meets monthly to review any reported issues and direct changes or additions that need to be made to the computer logic. Based on the monthly committee meetings, the computer logic has been changed every month. The updates have varied from simple changes in the suggested dosage of an antibiotic to new logic to incorporate natural language processing from dictated echocardiograms to detect vegetations on heart valves.

6.4.4 Patient Isolation Program

The patient isolation program developed at LDS Hospital is included in this chapter because is falls into a category that provides disease-specific consultation rather than patient-specific consultation. Patients with certain infectious diseases need to be placed in isolation in an effort to prevent the infectious pathogen(s) from spreading to other patients. Different types of isolation are used based on the specific disease. In 1984, infection control practitioners at LDS Hospital were concerned about the problem that nurses were placing patients in isolation who didn't need it, ordering the wrong type of isolation, and not placing patients in isolation who needed to be.

An ordering program was developed where nurses could use a number of different menus to determine whether their patients needed to be placed in isolation. Nurses can choose from alphabetical lists of diseases or specific pathogens, lists by pathogen type (bacterial, viral, parasitic, etc.), infection site (cutaneous, gastrointestinal, CNS, etc.), or from a list of most common infections. Then, based on the selected disease, the program determines whether the patient needs to be in isolation and what type. The nurse can then order the isolation and receive a printout containing instructions for the appropriate use of gowns, masks, gloves, and other necessary supplies. The program then electronically sends the order to equipment management where the disease-specific supplies and cart are prepared and the bill is sent to the financial system.

A study of the impact of the isolation program found that more patients were appropriately placed in isolation and with the correct type of isolation when the program was used (Jacobson et al. 1986). This program uses a large amount of decision logic to determine the need for isolation and the correct type but uses only patient identification information from the EHR. In fact, if the patient's disease does not require isolation, no other data from the EHR is used.

6.5 CONCLUSION

Many of the applications reported in this chapter have been installed successfully in 10 other IHC hospitals that have the HELP System. In most cases, the installation process was greatly facilitated when driven by local clinical champion(s) and/or administration support. In some cases, a local difference in the process of care required some modification in the computer logic. In some clinical domains, IHC has established "clinical programs" such as Women's and Newborns, Cardiovascular, Intensive Medicine, and so on, where enterprise-wide practice standards and board goals facilitate the use of common computer logic. The use of Arden syntax (see Chapter 12) has enabled medical logic modules to be transported to hospitals outside of IHC in a number of cases (Pryor et al. 1993).

Although the HELP System has proven to be an excellent and dependable platform to develop decision support applications, it is built on old technology. IHC has been developing a new Web-based HELP^2 System that was slated to eventually replace HELP (Huff et al. 1994; Clayton et al. 2003; Haug et al. 2003). A number of HELP applications and functionalities have been migrated to HELP^2 and the new system has some new capabilities and benefits that were not found on the HELP System. These new capabilities are due to the fact that on HELP^2, all IHC hospitals, clinics, instacares, and physician offices share a common EHR. On the HELP System, each hospital's EHR was contained on a separate database and did not include most outpatient information.

The new Web-based and fully integrated system has allowed a pediatric intensivist to use the Internet and access a child's laboratory, medication, and ECG information at a hospital 305 miles away and make a life-saving diagnosis and therapeutic change. The new anti-infective management program being developed on HELP^2 is more accurate, with access to microbiology, chest x-ray and other patient information obtained at one IHC facility before the patient is transferred to another IHC hospital. These new benefits are possible because IHC decided to assign and use a unique patient number enterprise-wide. Although the future database design, architecture, and access tools are being reevaluated at IHC, the functionality of the HELP System and benefits of the integrated HELP^2 System provide a stable roadmap that should be followed by any new system.

Medical decision support is analogous to some of the computerized tools on aircrafts that provide information and alerts. We all agree, and there are numerous instances that show, that we need a human pilot in charge of flying the aircraft (even when autopilot mode is used). However, human pilots cannot look out the window to tell how fast or how high they are flying or the exact direction. The computerized tools provide this information. Numerous computer alerts are built into the system to notify the human pilot of specific events or problems. This same type of information and alerting functionality are needed by medical personnel as they provide medical care each day. It is hoped that this chapter provides enough information on the LDS Hospital experience to show the current capabilities of medical decision support and help enlighten future applications that others will develop to improve the quality and safety of patient care.

There are a number of take-home messages that come out of the LDS Hospital experience with medical decision support over that past 30 years that are summarized in Table 6-1.

TABLE 6-1 Take-home messages provided from the LDS Hospital experience with medical decision support.

- The timing of data entry is critical. Patient information needs to be entered into the EHR as soon as possible, including interfaces, medical devices, and manual data entry.
- Successful decision support applications are developed by a team consisting of clinical domain experts providing the why and what needs to be done and the medical informaticists providing the how.
- Decision support should be integrated with the daily work processes of the medical staff and occur at the appropriate point of patient care. Patient alerts should be sent directly to the most appropriate people as soon as possible.
- Decision support applications need to be tested for safety before they are made available for general use. One bad experience can create barriers or restrictions for any future applications.
- Often large patient care improvement projects need to be broken down into smaller more manageable processes.
- The medical logic and rules need to be evidence based and match local processes of patient care.
- The logic and rules need to be periodically reviewed and updated as patient care and technology change.
- The applications must be easy to use and training should not be so difficult that patient safety could be compromised.
- Evaluation of medical decision support applications is often the hardest part.
- The applications need to be cost effective and reasonable to implement and maintain in order to gain administration support as well as clinical support.
- Physician support of order entry is easier to get if all orders, laboratory, medication, radiology, and so on, can be made at the same time using the same application.

Computers are being assimilated in many aspects of our everyday lives and are excellent tools to provide medical decision support. The automobile (transportation) and telephone (communication) were probably the two technological advances developed during the 1900s that had the largest impact on improving patient care overall—the computer will be shown to have a similar position during the next hundred years. However, the computer is just the tool and in itself has not improved patient care. Medical decision support has improved the quality and safety of health care by providing medical staff with the information they need, when they need it.

REFERENCES

Andrews, R. D., Gardner, R. M., Metcalf, S. M., Simmons, D. (1985). Computer charting: An evaluation of a respiratory care computer system. *Respir Care* **30**: 695–707.

Andrews, R. D., Gardner, R. M. (1988). Portable computers used for respiratory care charting. *Int J Clin Monit Comput* **5**: 45–52.

Bates, D. W., Evans, R. S., Murff, H., Stetson, P. D., Hripcsak, G. (2003). Detecting adverse events using information technology. *JAMA* **10**: 115–128.

Bleich, H. L. (1969). Computer evaluation of acid-base disorders. *J Clin Invest* **48**: 1689–1696.

Classen, D. C., Pestotnik, S. L., Evans, R. S., Burke, J. P. (1991). Computerized surveillance of adverse drug events in hospital patients. *JAMA* **266**: 2847–2851.

Classen, D. C., Evans, R. S., Pestotnik, S. L., Burke, J. P. (1992). The timing of prophylactic administration of antibiotics and the risk of surgical-wound infection. *N Engl J Med* **326**: 281–286.

Classen, D. C., Pestotnik, S. L., Evans, R. S., Lloyd, J. F., Burke, J. P. (1997). Adverse drug events in hospitalized patients: Excess length of stay, extra costs and attributable mortality. *JAMA* **277**: 301–306.

Clayton, P. D., Evans, R. S., Pryor, T. A., Gardner, R. M., Haug, P. J., Wigertz, O. B., Warner, H. R. (1987). Bringing HELP to the clinical laboratory—Use of an expert system to provide automatic interpretation of laboratory data. *Ann Clin Biochem* **24**: 5–11.

Clayton, P. D., Narus, S. P., Huff, S. M., Pryor, T. A., Haug, P. J., Larkin, T. et al. (2003). Building a comprehensive clinical information system from components. The approach at Intermountain Health Care. *Methods Inf Med* **42**: 1–7.

Evans, R. S., Gardner, R. M., Bush, A. R., Burke, J. P., Jacobson, J. A., Larsen, R. A. et al. (1985). Development of a Computerized Infectious Disease Monitor (CIDM). *Comp Biomed Res* **18**: 103–113.

Evans, R. S., Larsen, R. A., Burke, J. P., Gardner, R. M., Meier, F. A., Jacobson, J. A. et al (1986). Computer surveillance of hospital-acquired infections and antibiotic use. *JAMA* **256**: 1007–1011.

Evans, R. S., Pestotnik, S. L., Burke, J. P., Gardner, R. M., Larsen, R. A., Classen, D.C. (1990). Reducing the duration of prophylactic antibiotic use through computer monitoring of surgical patients. *DICP, Ann Pharmacother* **24**: 351–354.

Evans, R. S., Pestotnik, S. L., Classen, D. C., Bass, S. B., Menlove, R. L., Burke, J. P. (1991). Development of a computerized adverse drug event monitor. *Proceedings from the Fifteenth Annual Symposium on Computer Applications in Medical Care*: 23–27.

Evans, R. S. (1991). The HELP system: A review of clinical applications in infectious diseases and antibiotic use. *MD Comput* **8**: 282–288.

Evans, R. S., Burke, J. P., Classen, D. C., Gardner, R. M., Menlove, R. L., Goodrich, K. M. et al. (1992). Computerized identification of hospital-acquired infections and high-risk patients. *Am J Infect Control* **20**: 4–10.

Evans, R. S., Classen, D. C., Stevens, L. E., Pestotnik, S. L., Gardner, R. M., Lloyd, J. F., Burke, J. P. (1993). Using a hospital information system to assess the effects of adverse drug events. *Proceedings from the Seventeenth Annual Symposium on Computer Applications in Medical Care*: 161–165.

Evans, R. S., Pestotnik, S. P., Classen, D. C., Horn, S. D., Bass, S. B., Burke, J. P. (1994). Preventing adverse drug events in hospitalized patients. *Ann Pharmacother* **28**: 523–527.

Evans, R. S., Pestotnik, S. P., Classen, D. C., Lundsgaarde, H. P., Burke, J. P. (1994). Improving empiric antibiotic selection using computer decision support. *Arch Intern Med* **154**: 878–884.

Evans, R. S., Pestotnik, S. L., Classen, D. C., Clemmer, T. P., Weaver, L. K., Orme, J. F. et al. (1998). A computer-assisted management program for antibiotics and other antiinfective agents. *N Engl J Med* **338**: 232–238.

Evans, R. S., Pestotnik, S. L., Classen, D. C., Leavitt, B. R., Burke, J. P. (1999). Evaluation of an antibiotic drug-dose monitor. *Ann Pharmacother* **33**: 1026–1031.

Evans, R. S., Lloyd, J. F., Stoddard, G. J., Nebeker, J. R., Samore, M. H. (2005a). Risk factors for adverse drug events: A 10 year analysis. *Ann Pharmacother* **39**: 1161–1168.

Evans, R. S., Johnson, K. V., Flint, V. B., Kinder, A. T., Lyon, C. R., Hawley, W. L. et al. (2005b). Enhanced notification of critical ventilator events. *J Am Med Inform Assoc* **12**(6).

Gardner, R. M., Cannon, G. H., Morris, A. H., Olsen, K. R., Price, W. G. (1975). Computerized blood gas interpretation and reporting system. *Computer* **1**: 39–45.

Gardner, R. M., West, B. J., Pryor, T. A., Larsen, K. G., Warner, H. R., Clemmer, T. P., Orme, J. F. Jr. (1982). Computer-based ICU data acquisition as an aid to clinical decision-making. *Crit Care Med* **10**: 823–830.

Gardner, R. M., Golubjatnikov, O. K., Laub, R. M., Jacobson, J. A., Evans, R. S. (1990). Computer critiqued blood ordering using the HELP system. *Comp Biomed Res* **23**: 514–528.

Gardner, R. M., Hawley, W. L., East, T. D., Oniki, T. A., Young, H. F. (1991). Real time data acquisition: Recommendation for the medical information bus (MIB). *Int J Clin Monit Comput* **8**: 251–258.

Gardner, R. M., Pryor, A. T., Warner, H. R. (1999). The HELP hospital information system: Update 1998. *Int J Med Inform* **54**: 69–82.

Gibbons, M. B. (2005). Manager, Emergency Department, LDS Hospital, Salt Lake City. (Personal Communication).

Goldberg, M., Green, S. B., Moss, M. C., Marbach, C. B., Garfinkel, D. (1973). Computer-based instruction and diagnosis of acid-base disorders. *JAMA* **223**: 269–275.

Halford, B., Burkes, M., Pryor, T. A. (1989). Measuring the impact of bedside terminals. *Nursing Management* **20**: 41–45.

Haug, P. J., Pryor, T. A., Frederick, P. R. (1992). Integrating radiology and hospital information systems: the advantage of shared data. *Proc Annu Symp Comput Appl Med Care* 187–191.

Haug, P. J., Gardner, R. M., Tate, K. E., Evans, R. S., East, T. D., Kuperman, G. et al. (1994). Decision support in medicine: Examples from the HELP system. *Comput Biomed Res* **27**: 396–418.

Haug, P. J., Christensen, L., Gundersen, M., Clemons, B., Koehler, S., Bauer, K. (1997). A natural language parsing system for encoding admitting diagnosis. *Proc AMIA Annu Fall Symp,* 814–818.

Haug, P. J., Rocha, B. H., Evans, R. S. (2003). Decision support in medicine: Lesions from the HELP system. *Int J Med Inform* **69**: 273–284.

Hinson, D. K., Huether, S. E., Blaufuss, J. A., Neiswanger, M., Tinker, A., Meyer, K. J., Jensen, R. (1994). Measuring the impact of a clinical nursing information system on one nursing unit. *Proceedings of the Seventeenth Annual Symposium on Computer Applications in Medical Care* 203–210.

Henderson, S., Crapo, R. O., Wallace, C. J., East, T. D., Morris, A. H., Gardner, R. M. (1991). Performance of computerized protocols for the management of arterial oxygenation in an intensive care unit. *Int J Monit Comput* **8**: 271–280.

Huff, S. M., Haug, P. J., Stevens, L. E., Dupont, R. C., Pryor, T. A. (1994). HELP the next generation: A new client-server architecture. *Proc Annu Symp Comput Appl Med Care* 271–275.

Hulse, R. K., Clark, S. J., Jackson, C., Warner, H. R., Gardner, R. M. (1976). Computerized medication monitoring system. *Am J Hosp Pharm* **33**: 1061–1064.

Institute of Medicine. (2001). Crossing the quality chasm: A new health system for the 21st century. Washington, DC: National Academy Press.

Jacobson, J. T., Johnson, D. S., Ross, C. A., Conti, M. T., Evans, R. S., Burke, J. P. (1986). Adapting disease-specific isolation guidelines to a hospital information system. *J. Infect Control* 7: 411–418.

Kuperman, G. J., Gardner, R. M., Pryor, T. A. (1991). HELP: A dynamic hospital information system. New York: Springer-Verlag.

Larsen, R. A., Evans, R. S., Burke, J. P., Pestotnik, S. L., Gardner, R. M., Classen, D. C. (1989). Improved perioperative antibiotic use and reduced surgical wound infections through use of computer decision analysis. *Infect Control Hosp Epidemiol* **10**: 316.

Mullett, C. J., Evans, R. S., Christenson, J. C., Dean, J. M. (2001). The development and impact of a pediatric antiinfective decision support tool. *Pediatrics* **108**: 1–7. www://pediatrics.org/cgi/content/full/108/4/e75

Nelson, N. C., Evans, R. S., Samore, M. H., Gardner, R. M. (2005). Detection and prevention of medication errors using real-time nurse charting. *J Am Med Inform Assoc* **12**: 390–397.

Pestotnik, S. L., Evans, R. S., Burke, J. P., Gardner, R. M., Classen, D.C. (1990). Therapeutic antibiotic monitoring: Surveillance using a computerized expert system. *Am J Med* **88**: 43–48.

Pryor, T. A., Russell, R., Budkin, A., Price, W. G. (1969). Electrocardiographic interpretation by computer. *Comput Biomed Res* **2**: 537–548.

Pryor, T. A., Gardner, R. M., Clayton, P. D., Warner, H. R. (1983). The HELP system. *J Med Systems* **7**: 87–102.

Pryor, T. A. (1989). Computerized nurse charting. *Int J Clin Monit Comput* **6**: 173–179.

Pryor, T. A. (1990). Development of decision support systems. *Int J Clin Monit Comput* **7**: 137–146.

Pryor, T. A., Hripcsak, G. (1993). Sharing MLMs: An experiment between Columbia-Presbyterian and LDS Hospital. *Proc Annu Symp Comput Appl Med Care* 399–403.

Samore, M., Evans, R. S., Lassen, A., Gould, P., Lloyd, J. F., Gardner, R. M. et al. (2004). Surveillance of medical device-related hazards and adverse events in hospitalized patients. *JAMA* **191**: 325–333.

Shortliffe, E. H. (1987). Computer programs to support clinical decision making. *JAMA* **258**: 61–66.

Sitig, D. F., Gardner, R. M., Morris, A. H., Wallace, C. J. (1990). Clinical evaluation of computer-based respiratory care algorithms. *Int J Clin Comput* **7**: 177–185.

Tate, K. E., Gardner, R. M., Weaver, L. K. (1990). A computerized laboratory alerting system. *MD Comput* **7**: 296–301.

Tate, K. E., Gardner, R. M., Sherting, K. (1995). Nurses, pagers, and patient-specific criteria: three keys to improved critical value reporting. *Proc Annu Symp Comput Appl Med Care* 164–168.

Tettlebach, W. H., Ergonul, M. O., Samore, M., Ruben, M., Evans, R. S. (2002). Evaluation of antiinfective orders supported by computer assistance. Abstract presentation, ICAAC. San Diego, CA.

Warner, H. R., Toronto, A. F., Veasey, L. G., Stephenson, R. (1961). A mathematical approach to medical diagnosis. Application to congenital heart disease. *JAMA* **177**: 177–183.

Warner, H. R. (1966). The role of computers in medical research. *JAMA* **13**: 944–949.

Warner, H. R. (1968). Computer-based monitoring of cardiovascular functions in postoperative patients. *Circulation* **37**: 1168–1174.

Warner, H. R., Olmsted, C. M., Rutherford, B. D. (1972). HELP—A program for medical decision-making. *Comp Biomed Res* **5**: 65–74.

7 PENETRATION AND AVAILABILITY OF CLINICAL DECISION SUPPORT IN COMMERCIAL SYSTEMS

KENT GALE and JASON HESS

7.1 INTRODUCTION

Commercially available software for clinical decision support (CDS) did not appear until the early 1970s. Through the 1970s it is estimated that 10 to 15 organizations attempted to build their own clinical systems with CDS as a key component. The intellectual property from many of these early pioneers is now part of several commercial systems today. KLAS Enterprises has validated 10 commercially available, clinician-oriented software solutions that have CDS capability as manifested in complex clinical alerts.

KLAS Enterprises is a research and analysis firm specializing in independently monitoring and reporting health care information technology (HIT) vendor performance and the performance of health care professional services vendors/firms/organizations (PSFs). Information is collected by KLAS in an unbiased and independent manner. KLAS performs in-depth, confidential interviews with health care executives, department directors, and managers, complementing a previously completed questionnaire, to gather valuable insight into specific strengths, weaknesses and future expectations for products and/or services provided to the industry. From these interviews representing 4,500 hospitals and 2,500 clinics, KLAS maintains a live database of vendor performance information, rating over 500 products and services from more than 300 vendors. The information is refreshed with new performance evaluations and interviews daily. Using this methodology, KLAS has interviewed virtually every live site where clinicians interact with commercially available clinical software capable of utilizing rules from CDS logic embedded in the software systems.

KLAS has specifically targeted the use of CDS (alerts, prompts, guidance, etc.) in commercially available software for acute care organizations in four different studies since 2001, and in three of the four studies for both acute and ambulatory environments associated with acute care organizations since 2002. The research data from these studies, ongoing interviews with provider

Clinical Decision Support: The Road Ahead

organizations, consultants, and vendors and access to historical data serve as the basis for this chapter.

For the purposes of this discussion, commercially available software for CDS will cover only vendors and products that are designed and used generally (by clinicians and mainly physicians) and not focused on a narrow niche (respiratory therapy, medication administration, oncology, etc.). This chapter will not discuss CDS that occurs specifically within a typical pharmacy information system where checks for interactions such as drug–drug, drug–allergy, and so forth represent a commodity functionality, available in virtually every pharmacy information system marketed today. These pharmacy systems use First Data Bank, Multum, and other content vendors to provide much of the database for CDS.

7.2 CDS PENETRATION AND AVAILABILITY IN COMMERCIAL SYSTEMS

In the early 1970s, Homer Warner, MD, PhD, developed a clinical decision support (CDS) program that led to one of the first offerings by a software vendor for commercial use. The program was developed at LDS Hospital in Salt Lake City, and piloted at St. Luke's Medical Center in Phoenix, Arizona. The program began in the cardiac catheterization laboratory and served to provide physicians with immediate alerts based on rules. The rules interpretation became the basis for the later development of the HELP (Health Evaluation through Logic Processing) system. The Laboratory Information Systems (LIS) built to feed lab results to the HELP system ultimately became a successful LIS and was sold to hundreds of hospitals around the country, beginning in 1972, by the software vendor, Medlab Computer Services, Inc. (MCSI).

With funds from LIS sales, MCSI developed the first commercially available version of HELP. The first three years yielded no commercially viable sites using the CDS component provided by HELP other than LDS Hospital and the pilot site, St. Luke's Medical Center. Control Data Corporation (CDC) acquired MCSI in 1975 and put the majority of CDC's effort into the LIS, with the HELP system component waiting for market demand. Subsequently, 3M Health Information Systems acquired the rights to HELP from CDC. 3M was successful in contracting with three to four organizations, early on, to install HELP.

One of the first integrated delivery networks (IDNs) in Kentucky contracted for HELP to replace their Technicon Data Systems (TDS) solution operating in several hospitals. That effort was not successful, and the IDN installation reverted back to the long-standing TDS system. Several hospitals were able to put the HELP database, health data dictionary, and the Enterprise Master Person Index in place, but ultimately reported they were minimally successful in implementing alerts using the rule base from LDS Hospital. 3M elected to move the software to a more open architecture (ORACLE) and named the product Care Innovation. 3M signed 16 contracts to deliver Care Innovation to hospital organizations across the country. 3M sold a large contract to deliver computer-based CDS to the Department of Defense for CHCS II (Composite Healthcare System II). Shortly thereafter, 3M discontinued commercial enhancements to Care Innovation, and since that time many of the clients moved to other commercially available products and replaced or are replacing Care Innovation.

TABLE 7-1 Care Innovation hospitals and current status as of July 2005.

Facility	Status
Rex Healthcare, Raleigh, NC	Still in use
Rush University Hospital, Chicago, IL	Still in use
University of Colorado Hospital, Denver, CO	Still in use
University Hospital, Augusta, GA	Discontinued
Meriter Hospital, Madison, WI	Discontinued
Mercy Hospital, San Diego, CA	Discontinued
Central DuPage Hospital, Central DuPage, Illinois	Discontinued
Deaconess Billings Hospital/Clinic	Discontinued
Poudre Valley Hospital	Discontinued
Driscoll Childrens, Corpus Christi, TX	Discontinued
Community Health Care, Wasau, WI	Discontinued
Mt. Carmel Health, Columbus, OH	Discontinued
ProMedica Health System, Toledo, OH	Discontinued
Tucson Medical Center, Tucson, AZ	Discontinued

Virtually all of the Care Innovation customers had aggressive plans to move to CDS, stimulated by the real-time alerting that was successfully being used at LDS Hospital. Several sites implemented deeper and more complex clinical alerting, and in those cases, report using it for a small number of alerts (five to ten at a hospital) unlike the 3,000+ reported to be in use at LDS Hospital. The addition of new medical decision rules and alerts within the Care Innovation product has been slow, reportedly, from reports by the remaining operational clients, due to the expectation that Care Innovation will be discontinued as a product. Sample hospitals that utilized Care Innovation and their status are noted in Table 7-1.

In February 2005, Intermountain Healthcare (IHC, the parent of LDS Hospital and development site for HELP and one of the nation's largest integrated health care systems) and GE Healthcare, a division of General Electric Company, announced a $100 million, 10-year collaboration to enhance the patient care process in hospitals and clinics and accelerate the adoption of electronic health records among health systems in the United States. The announcement stated that the organizations' first joint project would be aimed at preventing adverse drug events and increasing patient safety. According to GE and IHC, a new joint clinical research center, expected to create more than 100 in-state jobs, will provide a central location for researchers to combine IHC's clinical data with GE's clinical IT programs.

In addition, GE is providing its Centricity IT technologies across institutions within IHC's network, which serve more than two million patients. These installations will enable the widespread use of new electronic pharmaceutical profile software throughout the IHC network, which is made up of 21 hospitals and 92 clinics (GE 2005).

GE is the third major vendor (after CDC and 3M) to seek to convert intellectual property from IHC's LDS Hospital into a commercial off-the-shelf product.

Cerner launched Discern Expert (Cerner's CDS system) with their Cerner Classic product in the mid-1990s and was successful in testing its use at Good Samaritan Regional Medical Center in Phoenix, Arizona, now part of Banner Health. Based upon a JAMA report (October 21, 1998), a pilot project to reduce ADEs used Cerner's commercially available software, including

Discern Expert, to carry out the study. The application of CDS (alerting) was retrospective, with printed alerts to be reviewed by the pharmacist or radiology technician and then acted upon. As measured by KLAS with our CPOE studies since 2001, none of the Cerner Classic sites employed concurrent use of CDS for physicians at the time of physician interaction, like ordering. Cerner Millennium sites are doing some concurrent alerting as discussed later in this chapter. Other Cerner Classic sites were confirmed to be doing alerting for certain administrative decision-making tasks. Some types of checking for duplicate orders or procedure redundancy were in place, but more clinically oriented decision support was not included as part of the Cerner Classic product (as reported by the Classic customers) (Raschke et al. 1998).

In the mid-1990s Sunquest Information Systems (mainly an LIS vendor) developed a CDS solution that was to operate independently of any other system, with the purpose of monitoring all relevant clinical IT transactions in the provider organization. The project was headed up by Homer Warner, Jr., Ph.D., the son of Homer Warner, Sr., the developer of the HELP system. Sunquest called the product Clinical Event Monitor, and it reportedly was purchased by as many as seven provider organizations by early 2000. Misys purchased Sunquest and took over the rights to CEM and renamed it Misys Insight. As reported by Misys, "Misys Insight runs on a low cost Windows 2000 server connected to a LAN or WAN and collects HL7 messages from ancillary systems. The data is analyzed by an event processor, which compares clinical data with knowledge rules to determine if the data is clinically significant. The system then transmits the patient data to the appropriate clinician's pager, Web browser or any e-mail addressable device." In a 2001–2002 KLAS study, there were no provider organizations found by KLAS actively using CEM. In terms of provider-sponsored CDS solutions, some acute care provider organizations have had limited success in small pilots during the 1980s and 1990s with their core clinical IT vendor. However, none of these products resulted in widespread use of CDS. Things are changing and the recent studies clearly show an upward trend.

7.3 CLINICAL ALERTING IN 2002

During 2001 to 2002, KLAS conducted an initial study to determine which commercially available systems were providing "real" CDS solutions and what barriers existed in preventing full utilization of CDS software. KLAS interviewed 72 unique health care professionals at hospitals in the United States including sites that were rumored to utilize CDS software (clinical alerting). KLAS left the ultimate definition of CDS or "alerting" up to the interviewee. Based upon the definitions by the respondents, KLAS grouped the types of CDS into three categories:

- **Reminders**. Simple rules resulting in reminders such as those relating to performance of immunizations. For example, when a patient arrives at the physician's office for a check-up, a nurse may be alerted, based on the patient's age and immunization status, that the patient needs an immunization. The nurse administers the immunization and updates the medical record.
- **Simple Alerts**. Alerts based on rules that are simple to set up and ubiquitous, such as drug–drug and drug–allergy conflicts. For example,

a nurse, upon entering an order for a certain medication, is alerted of a patient's allergy to the selected medication.
- **Complex Rules.** Rules that check multiple parameters from different clinical and administrative systems to generate a reminder or alert. An example would be a situation in which an important deviation in blood gas results, based on analysis of other charted data and an expected result protocol, generates an alert to the clinician and/or respiratory therapist suggesting an immediate action step. Another example might be when a physician entering a new medication order is immediately notified that a recently completed procedure will counteract the initial administration of the medication (KLAS 2001).

Despite reports of hundreds of clinical alerting software solutions being available for IDNs and hospitals in 2001, KLAS was unable to find evidence of commercially available complete alerting systems in live use. KLAS did find that virtually every hospital or IDN had at least one software application that provided a minimal level of CDS, the most common of which were pharmacy information systems where drug–drug interaction alerting was reported and in the laboratory where simple duplicate order checking and/or abnormal results checking occurred (KLAS 2001).

Through the course of the interviews KLAS found the following with regard to the availability of software and vendor experience necessary to provide more thorough CDS capability:

- Every major clinical data repository (CDR) or health care information system (HIS) vendor had at least talked about providing comprehensive CDS to customers, and the vendors with the largest market share were talking about it the most.
- Providers reported certain requirements for their comfort to fully implement CDS software including:
  - A CDR or a virtual CDR
  - All of the clinical feeder applications (laboratory, pharmacy, radiology, nursing documentation, etc.) in place
  - Coded data from a lexicon or data dictionary
  - Comprehensive order entry and result reporting
  - Physician-friendly workstations
  - Track record referenced to a library of alerts accepted by the medical community
- Every vendor, as reported by providers, was missing at least one component from the preceding list.

In this study KLAS noted that virtually every person interviewed understood the concept of CDS capabilities, but most reported that the only real alerting taking place was in ancillary applications such as laboratory or pharmacy. Providers look forward to the day when clinical alerting is routinely performed in the core clinical system used by physicians and nurses (all clinicians) supporting the care delivery process (KLAS 2001).

Of the health care professionals interviewed:

- One-third reported some alerting and excitement in doing so.
- Most alerts, though designed and developed, were not active and have not yet benefited the organization.

- The alerts most heavily in use compared administrative and financial data, and in 51 percent of cases, were designed to inform the business side of the institution.
- The most common alert was the 72-hour rule (reimbursement significance).
- Clinical alerts comprised 49 percent of the alerts.
- In total, 64 types of examples were given, nearly half of which were administrative or financial (72-hour rule, duplicate orders, etc.).

The bottom line in 2001 into early 2002 was that few acute care hospitals had any CDS performed by the software and virtually no complex computer-aided decision-making (KLAS 2001). Further evidence of this is noted in Tables 7-2 and 7-3.

- Six institutions reported having 50 or more alerts in place.
- Three institutions reported 100+ alerts in place.
- A 3M HELP site reported over 3,000 alerts, a majority of them highly clinical, and a Cerner Classic site reported over 900 alerts. Neither of these sites utilized products that were currently marketed as a vendors "product" for future clinical alerting, and both were anxious to move to the vendor's commercially available product once available.
- An EPIC site reported over 100 alerts and EpicCare was reported to be commercially available as reported by the institution.
- Several sites reported use of custom alerting software systems actively in place with plans to be replaced by commercially available products from their current vendors in the future.
- The remaining sites were excited about growing their base of alerts (4 to 35) in the immediate future.
- Several SMS (Siemens) and Meditech sites were entertaining new alerting software purchases from their vendors (KLAS 2001).

7.4 OBSTACLES TO CDS IN 2002

In 2002, two common reasons cited by most IDNs and hospitals for not having complex CDS implemented were the lack of alerting viability in the software products they were using, and the extreme difficulties physicians had in electronic ordering and in interacting with the electronic record, which is where CDS alerts would have the most immediate impact (KLAS 2001).

Fully 75 percent of those sites interviewed planned to have alerts solidly in place in the next three years (measurement again at the time of this writing, in 2006, shows that it is still in the future). Based upon their implementations and track record, 3M, Eclipsys, Cerner, and IDX seemed poised at the time to deliver (KLAS 2001).

- 3M with the experience of HELP at IHC, Billings Deaconess, and San Diego Mercy
- Eclipsys with Brigham and Woman's Hospital in Boston
- Cerner with Oklahoma Baptist Healthcare and Phoenix Samaritan
- IDX with Nebraska Health and Mayo Clinic

TABLE 7-2 Details on commercially available Clinical Decision Support systems in early 2002.

Vendor	Product	Comment
Cerner	Classic	Initial rules able to transcend Cerner's own ancillaries and either were built into the software or used a tool called CCL (Cerner Command Language), subsequently replaced by Discern Expert (rules).
Cerner	Millennium	Unsure of data dictionary plan. Confidence in Cerner's previous learning experience and apparent ease in setting up rules/alerting. No Millennium sites with complex alerting. User friendly with strong CDR.
Eclipsys	Sunrise (SCM)	Brigham and Women's development to be strong foundation. Database of alerts from HealthVision not commercially transportable. No integrated Rx feeding decision software. No real successful site. User-friendly workstation. Reported standard set of alerts available but not in use yet.
Epic	EpicCare Inpatient	Ambulatory focus successful with proven medical decision support. Move to inpatient unproven. No coded data apparent yet. Strong ambulatory EHR targeted for inpatient.
IDX	LastWord	One site with some success in a custom environment. Intense effort to build decisions. Intense physician involvement and solid CDR.
McKesson HBOC	Pathways HNS	Expected alerting to be in place shortly. No sites really live with decision software. Components not solidly in place to support medical decisions. Other McKesson products behind Pathways HNS (Star, HealthQuest, Series, Precision) in plans to do CDS.
Meditech	Magic	Software for end-user management of decision system not in place. Use of data dictionary not apparent to clients. Lack of physician involvement in building alerts and decision points. Integration a potential future benefit.
SMS (Siemens)	Invision LCR	CDR capability validated. Plan to actually have user-built alerts not validated by clients. No track record of decision-making software. No other Siemens products were described as having CDS capabilities (MedSeries4 and Unity).
QuadraMed	Affinity	Planned for future but not live.
3M	Care Innovation	No physician ordering in place. Medical decisions after the fact at pilot site. New database was not fully designed and useful yet with coded data. Building new bridges to foreign feeder systems. Complex nature of building decisions. Confident in long-term experience.

TABLE 7-3 Installation status of commercially available CDS software in early 2002.

Vendor	Sites interviewed	Hospitals with clinical rules and alerts (Few with complex CDS)
Cerner Millennium	9	5
Eclipsys Sunrise Clinical Manager	8	5
Epic (mainly ambulatory)	2	1
IDX LastWord	5	3
McKesson HBOC Pathways HNS and Care Manager	4	1
McKesson HBOC Star	6	1
MedScape Logician (ambulatory)	6	2
Meditech Magic	8	2
SMS Invision	9	0
SMS MedSeries4	2	0
QuadraMed Affinity	5	0
3M Care Innovation	6	1

7.5 WHAT CHANGED SINCE EARLY 2002?

7.5.1 A Preview of Findings from CPOE Inquiries 2002–2005

KLAS measured CDS software as part of the process of looking at direct interaction between the caregiver (mainly the physician) and the patient. One of the areas where rules, knowledge-based actions, and alerting comes heavily into play is during computerized physician/provider order entry (CPOE).

After assessing the depth of CDS software during the study during 2001 to 2002, KLAS began an in-depth effort during 2002 to 2003 to look primarily at CPOE activity in the United States. This was the first of three studies KLAS conducted looking at CPOE in years 2003, 2004, and 2005. Although the primary goal of this new effort was to understand CPOE use/depth, KLAS was also able to further validate the usage of CDS software because of the close linkage between CPOE and CDS (KLAS 2001).

7.5.2 Changes from 2002 to 2003

In February of 2003, KLAS completed the first inquiry into CPOE, or the CPOE Digest. KLAS asked additional questions about CDS alerting at the time of ordering. KLAS found that the ability for physicians to be notified of simple alerts from decision logic at the time of medication ordering (drug–drug, drug–allergy) was widely available as indicated in the alerting study from the previous year. KLAS further asked about complex CDS software (multiple parameters from multiple domains), and similar to findings in the previously quoted 2002 study, the usage continued to be in its infancy with little growth in complex CDS.

Table 7-4 breaks out the number of live inpatient CPOE sites as of February 2003. It also shows the number of those sites doing CDS with the clinical information system (CIS) software as well as the total number of individual complex alerts reported across all sites (KLAS 2003).

TABLE 7-4 Live inpatient CPOE sites as of February 2003.

Vendor and product	Number of inpatient CPOE sites interviewed	Number of inpatient CPOE sites interviewed doing complex CDS	Total number of active complex alerts across all inpatient CPOE sites interviewed
Cerner Millennium	5	2	60
Eclipsys SCM	11	9	200
Eclipsys TDS	10	2	59
Epic EpicCare Inpatient	1	1	2
IDX LastWord	5	3	60
Meditech C/S	3	0	0
Meditech Magic	3	1	5
Per-Se Patient1	6	3	508
Siemens Invision	8	4	364
Other*	4	1	10

* Other category includes CliniComp and GE.

Although the 2003 CPOE Digest differed from the earlier quoted 2002 study in that it focused more on *complex* CDS versus simple alerting, the chart from the 2003 CPOE Report in Table 7-4 isolates the following:

- Of the five Cerner sites doing clinical alerting in the 2002 study, none reported any real complex alerting. In 2003, five different organizations were interviewed, with two of those five reporting limited complex CDS.
- Eclipsys grew from five to nine sites doing complex CDS in 2003.
- Epic had one inpatient site go live on CPOE in 2002; this site reported one to two complex rules in place.
- GE was a new entrant in 2002, having one site live with CPOE reporting limited complex CDS.
- Of the three IDX LastWord sites doing some alerting in 2002, all three reported doing limited complex CDS in 2003 (KLAS 2003).
- SMS Invision (Siemens) went from zero sites doing alerting during 2001 to 2002, to four organizations doing limited complex CDS in 2003.

Figure 7-1 shows the number of complex alerts by inpatient organization and vendor/product in the 2003 report. Eclipsys SCM was the leader with the most alerts, followed by Eclipsys TDS, SMS (Siemens) Invision, and Per-Se Patient1. GE and Epic reported one site each doing complex CDS, and CliniComp, although live on CPOE, had no complex CDS taking place. Organizations such as Intermountain HealthCare in Salt Lake City and Partners HealthCare in Boston were not using commercially available software at this time and were not included in these statistics (KLAS 2003).

KLAS further queried providers in the 2002 CPOE Digest with regard to technology being used for concurrent alerting. As noted in the chart in Figure 7-2, providers overwhelmingly indicated that *On-Screen at Time of Order* was the most common technology used for CDS software (KLAS 2003).

Some alerting also took place with pagers, cell phones, via e-mail, and so on, but overwhelmingly on screen alerting was the leader.

In summary, from 2002 to 2003, KLAS was able to validate some growth in complex CDS. However, the use was still very limited, with only 25 out of

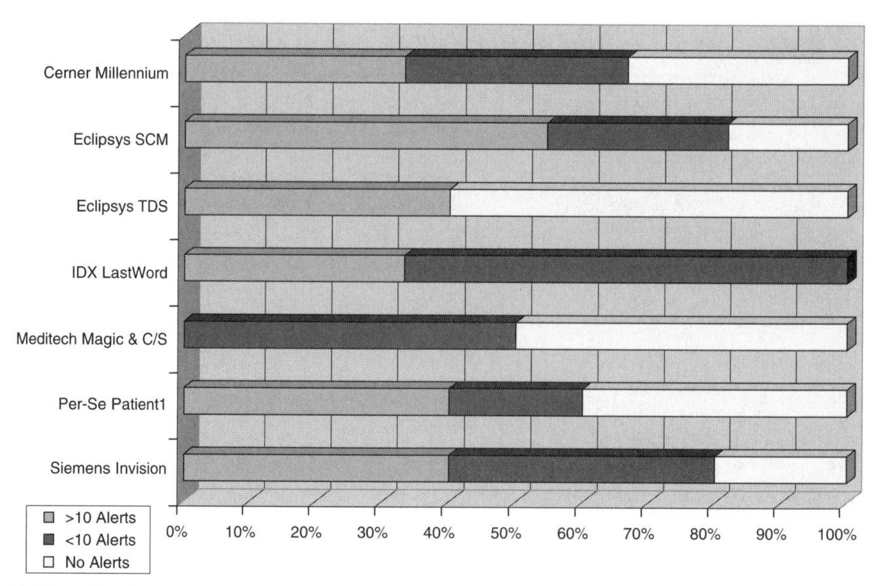

FIGURE 7-1 Inpatient organizations doing alerting.

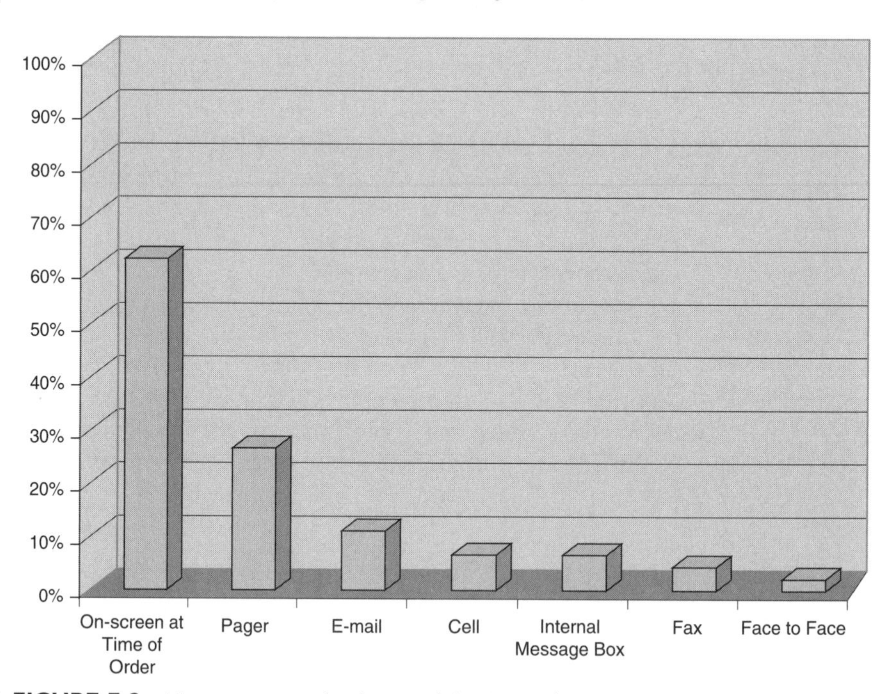

FIGURE 7-2 Most common technology used for CDS software (2003).

125 live CPOE sites validated as doing any complex decision-making with their CIS software. In addition, the ability for physicians to enter all medication orders and be notified of simple alerts from decision logic at the time of the medication order was widely available (though sparingly in use at less than 2 percent of acute care hospitals across the United States); however, the physician and the pharmacist were not always using the same medication ordering and alerting system in the process. Also making the environment difficult was the fact that nearly half (48 percent) of all medication orders were re-entered by the pharmacy (KLAS 2003).

TABLE 7-5 Live inpatient CPOE sites as of 2004.

Vendor and product	Number of inpatient CPOE sites interviewed	Number of inpatient CPOE sites interviewed doing complex CDS	Total number of active complex alerts across all inpatient CPOE sites interviewed
Cerner Millennium	13	10	664
Eclipsys SCM	17	13	385
Eclipsys TDS	15	7	224
Epic EpicCare Inpatient	3	0	0
IDX LastWord	7	7	153
Meditech C/S	7	5	124
Meditech Magic	7	3	33
Misys Patient1	7	6	584
Siemens Invision	18	8	488
Other*	8	1	30

*Other category includes CliniComp, GE, IDX Carecast, McKesson Horizon Expert Orders, and Siemens Soarian.

7.5.3 Changes from 2003 to 2004

In 2003, KLAS conducted the second inquiry into CPOE usage and depth of CDS software, or the 2004 CPOE Digest (see Table 7-5). KLAS was able to validate with few exceptions increases in the number of CPOE sites doing complex CDS, and an increase in the total number of complex medical decision alerts taking place (KLAS 2004).

Complex reporting saw many changes in the 2004 CPOE Report, compared to the 2003 report. Of note was the following:

- Cerner Millennium increased from two to ten sites doing complex CDS, and showed a huge jump in the number of complex alerts taking place at its sites (from 60 to 664 alerts).
- CliniComp had one site begin using clinical rules for decision-making.
- Eclipsys gained more sites doing complex CDS/alerting.
- GE remained constant with one live site doing CPOE, but had a major shift with the site previously reporting ten complex alerts used, now reporting no complex CDS alerts being used.
- Epic gained two additional CPOE sites; however, none of the three reported use of software with complex CDS as the effort to engage complex alerting was on hold.
- IDX LastWord added four more sites using complex rules (three to seven). IDX Carecast (replacement for LastWord) reported one site live on CPOE, but no complex CDS taking place.
- McKesson Horizon Expert Orders reported one site live on CPOE, but no use of CDS software.
- Meditech Client Server version realized three new sites doing complex CDS with Meditech software, and Meditech Magic version gained two. Both client sets were making limited use of such decision-making software.

- Misys Patient1 (formerly Per-Se) continued to have a large number of total complex alerts at Misys sites (584), and also gained three sites doing complex rules.
- Siemens Invision (formerly SMS), gained ten sites doing CPOE and four more sites doing complex CDS with software (KLAS 2004).

7.5.4 Changes from 2004 to 2005

In 2005, KLAS published the third in-depth CPOE Digest (see Table 7-6). This report was the culmination of three years of comparative data added to the 2005 data, and for which KLAS spoke with virtually every inpatient CPOE site across the United States. From this research KLAS was able to note some valuable trends. The expected increase in physicians doing CPOE was confirmed, as the numbers initially started at 45,000 physicians doing CPOE in 2003, grew to 69,000 in 2004, and then made a sizeable jump to more than 113,000 in 2005. Also increasing year-to-year was the percentage of hospitals in the United States doing CPOE with a commercially available product: 1 percent of hospitals in 2003, 2 percent in 2004, and a jump to 4 percent in 2005 (KLAS 2005).

With a consistent increase in hospitals doing CPOE over three years—2003 (125 hospitals), 2004 (159 hospitals), 2005 (233 hospitals)—as far as sheer numbers go, KLAS found no consistent pattern for growth of complex CDS by vendor, with only a few exceptions (see Figures 7-4 through 7-6).

From the standpoint of percentage increase in CPOE sites compared to complex CDS, the numbers are somewhat different. Figure 7-3 compares the 2004 to 2005 percentage growth of CPOE in hospitals compared to the total number of complex CDS alerts. Whereas Figures 7-4 and 7-5 indicate that

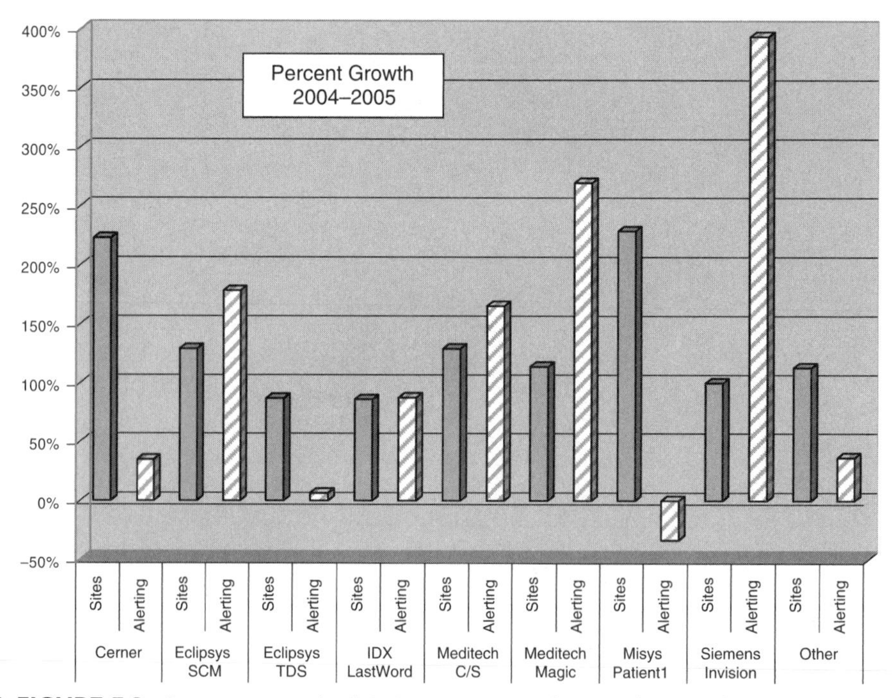

FIGURE 7-3 Percentage growth of CPOE sites compared to complex CDS from 2004–2005.

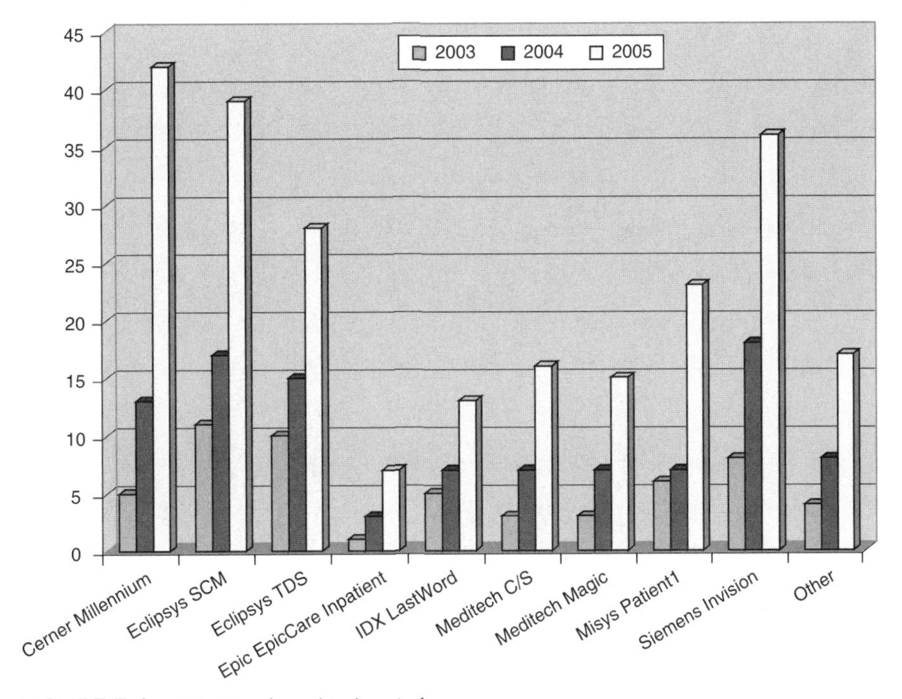

FIGURE 7-4 CPOE confirmed in hospitals.
*Other category includes CliniComp, GE, IDX Carecast, McKesson Horizon Expert Orders, and Siemens Soarian.

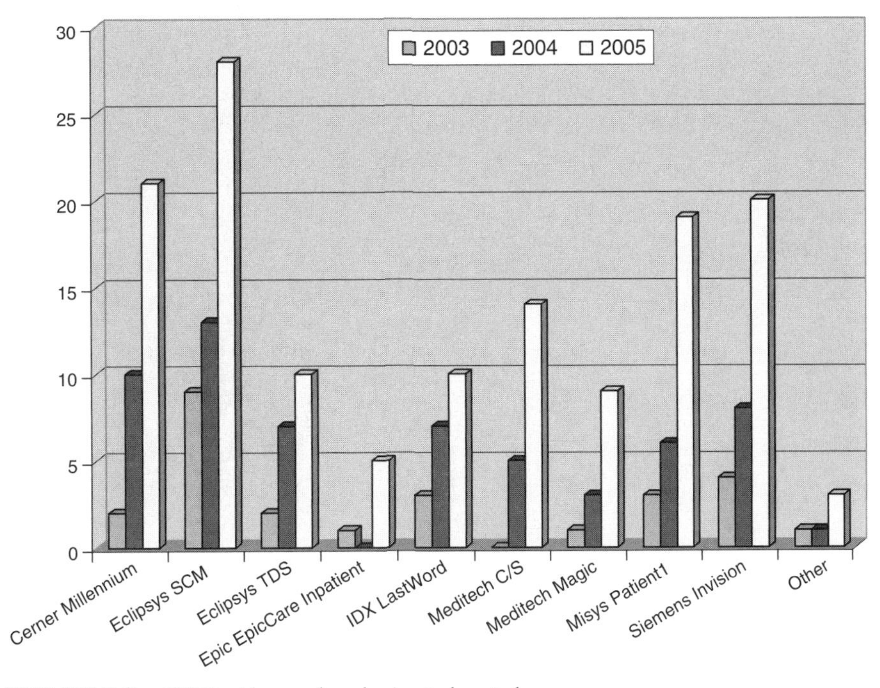

FIGURE 7-5 CPOE with complex alerting in hospitals.
*Other category includes CliniComp, GE, IDX Carecast, McKesson and Siemens Soarian.

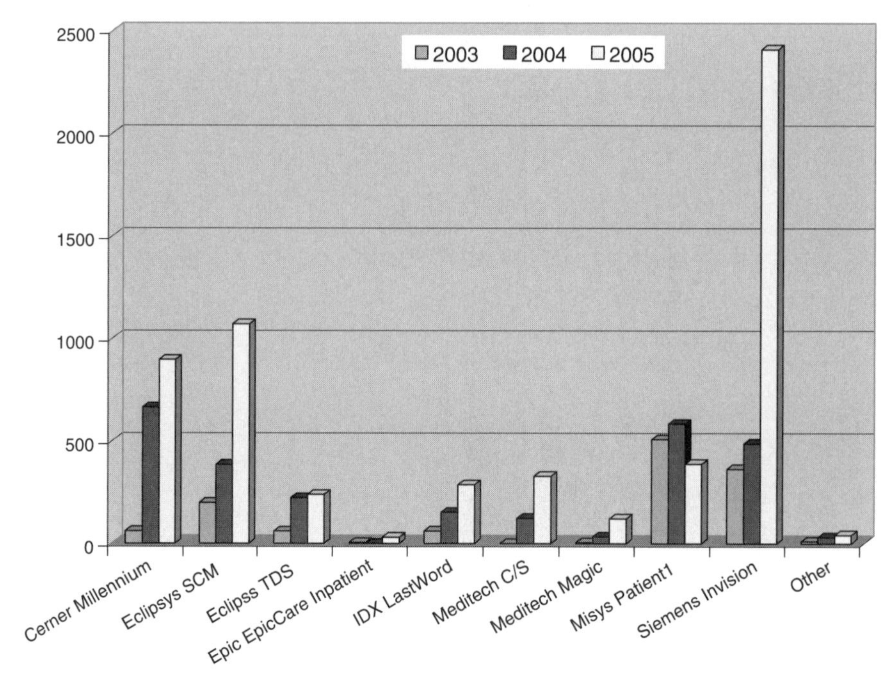

FIGURE 7-6 Total number of alerts across all acute care sites with CPOE.
*Other category includes CliniComp, GE, IDX Carecast, McKesson and Siemens Soarian.

CPOE and clinical alerting are both growing at varied rates, Figure 7-3 points out the erratic nature of that growth. For example, in 2004 to 2005, Siemens Invision experienced nearly 100 percent growth in CPOE going live in hospitals, but experienced nearly 400 percent growth in total number of complex CDS alerts taking place. Conversely, Cerner Millennium experienced nearly 220 percent growth in CPOE sites in 2004 to 2005, but saw only an increase in total CDS alerts of 35 percent (KLAS 2005).

Also of note is Misys Patient1, whose CPOE sites grew 229 percent, but which actually saw a 33 percent drop in complex CDS alerts during that same

TABLE 7-6 Live inpatient CPOE sites as of 2005.

Vendor and product	Number of inpatient CPOE sites interviewed	Number of inpatient CPOE sites interviewed doing complex CDS	Total number of active complex alerts across all inpatient CPOE sites interviewed
Cerner Millennium	42	21	897
Eclipsys SCM	39	28	1071
Eclipsys TDS	28	10	239
Epic EpicCare Inpatient	7	5	30
IDX LastWord	13	10	287
Meditech C/S	16	14	329
Meditech Magic	15	9	122
Misys Patient1	23	19	389
Siemens Invision	36	20	2410
Other	17	3	41

*Other category includes CliniComp, GE, IDX Carecast, McKesson, and Siemens Soarian.

time period. This points to the fact that while it is one thing to go live with CPOE, it is entirely another to do so with complex CDS alerting. Commentary from the 2005 CPOE Report speaks to alerting challenges, alert fatigue and alerts being turned off, as may have been the case with Misys Patient1 total alerts in 2004 to 2005. Alerts are ignored if they occur too frequently. Some users have established low and high level alerts. The low level alerts are seen but can be ignored versus high level alerts that "require action steps." The challenge appears to be in the development of more precise criteria that eliminate the firing of both false positives and false negatives (KLAS 2005).

With the results from the 2005 CPOE Report, KLAS found the following:

- Cerner Millennium hospitals with complex CDS grew from 10 to 21 in 2005; more than 100 percent. The increase in the number of defined and active complex alerts did not match the CPOE growth, and was about 35 percent.
- Eclipsys SCM more than doubled the number of sites doing complex CDS and the number of complex alerts more than doubled in that time frame.
- Epic EpicCare Inpatient, IDX LastWord, and McKesson Horizon Expert Orders showed slight growth in complex CDS.
- Meditech Magic, Meditech Client Server, and Misys Patient1 experienced at least a doubling of hospital clients using their clinical decision software. The actual growth in number of complex alerts either did not match the trend in clients or declined as in the case of Misys.
- GE Centricity, IDX Carecast, and Siemens Soarian live CPOE hospitals reported no alerting.
- Siemens had the highest increase in hospitals that were using the INVISION software for CDS with a much larger than expected growth in number of active complex alerts. Soarian is the planned replacement of Siemens INVISION as reported by Siemens clients, and the CDS progress is expected to move to that product set (KLAS 2005).

General findings from 2005 CPOE Digest can be summarized as follows:

- As voiced by actual provider organizations, the use of Complex CDS software from commercial vendors is challenging because (KLAS 2005):
  - A normative set of "decisions" acceptable to the health care and legal community is not yet available. Building the decisions from scratch is overwhelming when considering the testing and validation necessary prior to buy-in by physicians.
  - Acceptable response to an alert is still undefined, as is acknowledgement of an alert from a medical decision.
  - Ineffective rules and decisions based upon general medical practice, creating false alerts, are thorny issues for specialists. A reduction of the number of alerts from automated decision support appeared in a number of hospital installations during 2004.
  - Effective CDS is impeded when orders and discrete results are not mapped accurately. Issues of one-to-many and many-to-one surface, leaving committees that design the rules challenged to move forward effectively. These issues may be tied to database design and the use of a lexicon or health data dictionary.

- Complex alerting is taking place and increasing. KLAS was able to validate somewhat mixed results regarding use of complex CDS. There are more complex active alerts (in the magnitude of thousands) taking place.
- Even with the increase in CPOE sites reported, there is an increase in the number of sites reporting no alerting. This explains the decrease in the percentage of alerts from year to year (KLAS 2005).

7.6 HOW FAR ALONG ARE CDS VENDORS IN 2005?

7.6.1 Miscellaneous Commentary from Live Users

We have a rules and alerts oversight committee of physicians, nurses, and pharmacists from across the system. They monitor alerts, so we look at them periodically to see how many of them are being overridden, if they are written the way they should be, and if they are giving the type of benefit that we hope to get from them. We do this on a regular basis. The demand for rules in the outpatient clinic is not nearly as great as it is on the inpatient side (KLAS 2005).

Physician satisfaction is kind of a mixed bag. They understand the importance of it so they are happy about the evidence-based order sets and the lack of interpretation of handwriting. From a safety standpoint, they are happy with the system. We are starting to experience a little alert fatigue at one site. We are revisiting some of the alerts that are in the system (KLAS 2005).

We have had to abandon doing any rules or alerts beyond the most simplistic ones. It takes a lot of concurrence to get anything out and live and we do not have the time to work on it right now. If we had something off the shelf, it would be a different story, like what they use for the standard drug interactions in the pharmacy system (KLAS 2005).

We have seen very positive benefits as a result of implementing this version. We have much better decision support capabilities and better safety. Now, when users want to override an alert, we can require them to enter a reason for the override or not allow them to bypass the alert at all (KLAS 2005).

With our application we are seeing evidence that the rules and alerts that are firing are having an impact on ordering practices. We can run reports that show us how many times an alert about a redundant test fires and whether the physician cancels or modifies the order as a result. We can see this not only for redundant tests, but for allergies and for dosage. The system is catching those errors and the physicians are not just blowing right by them but are actually paying attention to them (KLAS 2005).

7.7 SELF-REPORTED VENDOR DATA AS OF FEBRUARY 2005

7.7.1 Cerner Self-Reported

A knowledge-based system is effective only if medical and organizational knowledge is constantly updated and accessible at the point of decision. Cerner Millennium supports care delivery by analyzing activities and data across the organization and then proposes directions or alerts to providers about issues that occur at any point in the care process. Cerner CPOE infuses clinical knowledge throughout the care process, guiding clinicians to the latest evidence at the point of care via order sets, plans of care, alerts and notifications, and documentation. Additionally, clinical performance improvement reports allow benchmarking adherence to key quality standards, and knowledge management processes streamline the creation, review, and maintenance of content within Cerner Millennium. Knowledge is captured and codified against a standard vocabulary and stored in a single data repository. Without such infrastructure, data and content cannot be successfully deployed, maintained, and analyzed.

It is imperative to realize the database management effort involved in an interfaced system. When each component of the medication ordering process is located in a separate system, each one requires maintenance of clinical reference information, allergy files, formularies, decision support (rules engine), interfaces from lab and radiology, and data from point-of-care administration. When considering all of the transactions and synchronization that have to occur with an interfaced system, safety becomes a critical issue. With information duplication and updates made in separate systems, there is a large potential for error.

Our suggestion for alerts and decision support is that one of the systems must be the master system storing the largest amount of patient data and should then be leveraged as the decision support and alert master for clinicians. Due to the limitation of interfaces, this master system may not contain the same level of detail as the feeder systems.

Cerner's Executable Knowledge solutions, such as *Executable Knowledge for Regulatory Standards* package, include executable order sets, clinical rules and alerts, and documentation and reports that deliver the best in evidence-based medicine by rigorously and scientifically evaluating findings published by major regulatory organizations (such as Joint Commission on Accreditation of Healthcare Organizations [JCAHO], Centers for Medicare and Medicaid Services [CMS]) and in the body of medical journals. These solutions provide citations and links to the clinical evidence that is the source for the rule, alert, order set, or documentation.

Complex alerting includes the ability to go beyond simple notification, presenting interactive alerts, alerts occurring in the background, and providing clinicians with actionable information at the point of care. Aside from the event-driven decision support capabilities such as drug--drug interactions or duplicate order alerting, Cerner can also support a more complex, proactive alerting process. The ability to do this is directly related to a single data platform that supports patient care across multiple environments and encounters, and an ability to use that architectural model fully to leverage existing and new data elements, alerting providers outside of the context of a clinical event. For instance, based on a patient's current plan of care, there may be opportunities to alert a provider of a change in the patient's status that may require an analysis or adjustment of the current care plan. Through a decision support engine working from the comprehensive data elements for that patient, sophisticated rules and alerts can be designed to create the ultimate care environment for the patient and the clinician. Data documented as a part of the normal care process can trigger events or orders to enhance patient safety. For example:

- The nurse documents an elevated temperature during her routine assessment.
- The elevated temperature triggers an alert.
- Cerner's rules engine queries the database for previous blood culture orders in the last 24 hours.
- When none are found, the system activates an order for blood cultures, and the collection task populates the appropriate resources task list, prompting for collection without the need to place any phone calls or interrupt the physician or nurse's workflow process (KLAS 2005).

7.7.2 Eclipsys Self-Reported

The Sunrise Advanced Clinical Management solutions alerting and CDS is fully integrated with our Knowledge-Based Orders™—our order entry module. We provide both synchronous (real-time) alerts and asynchronous (background) alerts, fully integrated with patient orders. If a disparate system is used to enter orders into Sunrise Advanced Clinical Management solution, our CDS rules will fire in Sunrise upon entry of the orders into the Sunrise database.

The Sunrise Advanced Clinical Management solutions' CPOE is fully integrated from our foundation to the end-user interface. Orders are entered directly into Sunrise, and our CDS alerts fire on a real-time basis. All necessary components are included in our Sunrise clinical solution. The only third-party software required is used for reporting functionality (i.e., Crystal Reports).

Sunrise Advanced Clinical Management solutions have two types of alerts—synchronous and asynchronous alerts. Synchronous, real-time alerts are those alerts that appear at the time of order, and are rules-driven. Sunrise has numerous clinical and/or financial rules that can be modified by an organization and placed into production. Sunrise's multidimensional alerts can make any item in the database an evoking object to fire an alert. Therefore, orders, notes, lab results, allergies, and so forth can make the rules fire.

When a clinician places an order, the system checks all the rules. If the order the clinician is about to place violates a rule, it will trigger an alert. For example, if the clinician were about to order a medication that is too low of a dose for a patient, a rule would be violated and an alert would appear. Sunrise brings information to physicians and clinicians while they are in the midst of their clinical decision-making process—before the order is submitted or before the order is filed in the Health Data Repository. This enables the clinician to have actionable information in a timely fashion, while still in the clinical thought process and while still engaged with an individual patient.

When the alerts are activated, the clinician has the ability to override the alert. Alerts will not stop the clinician from actually placing the order, but the clinician has an opportunity to write a comment as to why he or she is overriding the alert. This provides an audit trail for all clinicians. Additionally, an upcoming enhancement will provide advanced alert functionality where a facility can require certain alerts not to be ignored. This will allow for hard stops on alerts—that is, alerts that physicians cannot override.

Some of the types of synchronous and asynchronous alerts provided with Eclipsys' CDS and Knowledge-Based Orders are:

- Duplicate checking
- Drug–drug, food–drug interaction warnings and drug–allergy checking
- Venipuncture (schedule-based) alerts
- Abnormal lab values: Absolutes and trends
- Clinical algorithms
- Diagnostic indications
- Drug–lab interactions and warnings
- Alternative suggestions or reminders
- Consequent orders
- Clinical protocols and guidelines
- Infection control procedures
- Alert escalation and automatic paging
- Standard calculations (BSA) (KLAS 2005)

7.7.3 Epic Self-Reported

Epic considers alerting and decision support to be an essential component of CPOE, and real-time alerts at the point of order entry (drug interactions, duplicate therapy, prompts for preferred alternate orders, min/max dose alerts) are available regardless of ancillary systems used. Interfaced results are associated with the originating order record and filed with the patient's electronic record. Decision support alerts can trigger based on discrete external data (e.g., lab results) interfaced into Epic's underlying data repository.

The system executes decision support rules based on information in the repository regardless of the source of that information. For example, a potassium value received from an interfaced system that is significantly lower than the previous value can trigger an alert if the patient is on digoxin. EpicCare can also use interfaced insurance data to warn providers when they order outpatient prescriptions outside a patient's payer/plan formulary.

Clients use third-party medications and interaction rules databases such as Medi-Span or NDDF Plus with Epic's CPOE system, and import standard code sets such as CPT® or HCPCS from third-party databases. All required components and data are facilitated by Epic.

EpicCare seamlessly incorporates decision support into the workflow, managing the wealth of data available in EpicCare to present the clinician with key information at appropriate points. Clinicians can take action directly from an alert prompt. For example, when EpicCare notifies users that preferred alternatives exist for a particular order, users can immediately replace the original order with an alternate. The data-driven alerts described later can provide links to appropriate order sets, making it simple to select treatments relevant to the condition(s) that triggered the alert. Epic's CPOE also incorporates knowledge management, providing access to specific internal and external references at the point of ordering.

EpicCare supports rules-based alerts (minimum/maximum single and daily inpatient dose checks based on conditions such as age, for example) and allows for active, data-driven alerts triggered at the point of care by complex combinations or comparisons of criteria over time. A flexible GUI rules editor gives organizations the ability to base these criteria on best practice protocols and include any element in the repository, including clinical, demographic, and administrative data. In addition to intelligent alerts at the time of ordering, the system can include complex data-driven alerts in result reports.

Epic's experience has shown that intrusive alerting contributes to "alert fatigue," so we allow the use of filters to specify who sees an alert. For example, EpicCare can filter IV incompatibility alerts for ordering physicians but display them for pharmacists, who can take appropriate action. Epic also supports both passive and active forms of most alerts. For example, a passive allergy alert can advise clinicians that a patient is allergic to a selected medication while it is still open in the order entry screen, but at the point of signing that order, an active interaction check occurs. The same central decision support system used for order alerting can proactively guide appropriate paths of care based on documented patient data such as admitting or encounter diagnosis, chronic conditions, chief complaint, and preventive care schedules.

Clients can readily configure alerts in EpicCare, and most rules are defined through GUI screens. EpicCare also supports the use of custom programming routines for complex decision support (automatically adjusting a medication dosage based on critical lab values, for example).

Examples of alerts available in Epic's CPOE system include:

- BestPractice Alerts driven by a wide range of clinical data: Result trends, data recorded in configurable flow sheet rows, orders, diagnoses, current medications, histories data, patient age, and more
- Identification of order sets relevant to patients' diagnoses
- Medication interactions: Drug–drug, drug–allergy, drug–food, drug–alcohol, IV incompatibility
- Duplicate therapy warnings
- Prompts for preferred alternatives to selected orders
- Min/max warnings for inpatient single and daily doses, based on patient criteria such as age and weight
- Lifetime dose alerts (KLAS 2005)

7.7.4 GE Self-Reported

At GE Healthcare, we do offer integrated pharmacy, laboratory, and radiology systems, all from a single vendor. As such we are able to provide an award-winning CPOE system that is capable of managing all orderables within a hospital: bedside care, ancillary care, labs, radiology, medications, case management, consultants, and so on. Further, we are able to leverage our advanced decision support, beyond the traditional third-party drug databases that most vendors offer, across all these patient orders. We also have an open architecture and are standards-based, meaning that customers can leverage existing investments in legacy systems and still benefit from our natural-language-based rules engine and notification engine.

We successfully manage a variety of complex alerting on a variety of orders and workflow, including protocols (insulin and prednisone sliding scales), multiday and skip-day protocols, multistep oncology protocols, infusions, and so on. One example of our complex alerting might include an oncology protocol where one of the 20 steps is a hydration step linked to a single onco-therapeutic. When the oncology medication is discontinued, the provider will be alerted to the hydration step as well. This would be critically important when the patient cannot afford to be fluid overloaded (KLAS 2005).

7.7.5 IDX Self-Reported

Carecast CDS is performed through:

1. Built-in active logic, which utilizes tables configured by the customer and/or First DataBank:
   - Allergy checking, driven by data and an algorithm supplied by First DataBank.
   - Redundant order checking, table-driven and supplied by the customer as entries in the order master table.
   - Drug–drug interaction through detection rules based on tables and rules supplied by First DataBank; the explanation of the conflict are available to the user online.
   - Maximum dose checking for neonate, newborn, pediatric, adult, and geriatric patient populations. Doses are checked per drug and per patient type based either on weight (kg), body surface area (m^2), or absolute dose; rules are table-driven based on maximum dose data supplied by the user.
   - Integrated critical pathways are table-driven based on tables built by the customer.
2. A powerful rules engine, Blaze Advisor, can provide active logic triggered at many different points in Carecast, either synchronous with a user action or asynchronous (in the background, e.g., by the filing of a new laboratory result). Using Blaze's sophisticated graphical user interface, the customer can write rules applying complex logic to both transactional and database data, with the outcome modifying the database and/or affecting the screen flow and data presented to the user.

   The Carecast Rules engine allows user-defined rules to extend and automate the appropriate response for many different types of patient care and decision-making conditions. By automating rules-driven processes, health care organizations can improve efficiency and reduce costs while significantly reducing errors in managing patients, processing orders, performing care tasks, and monitoring patient information. The Carecast Rules Management System can be used to develop clinical rules and alerts for many processes in health care that rely on the consistent application of instructions.
3. Passive decision support is provided by Carecast's extensive capabilities for predefined orders with contextually appropriate default values, either individually or in order sets, and also by Carecast's innovative "information windows" that display pertinent data (such as allergy warnings or recent laboratory results) at the point of ordering and charting.

In addition, eMedicine's peer-reviewed Clinical Knowledge Base content is available through Carecast. With articles on more than 6,500 diseases and conditions, eMedicine is one of the most comprehensive and current sources of peer-reviewed clinical reference materials on the Web. Context-sensitive access to the eMedicine Clinical Knowledge Base within Carecast enables clinicians to immediately access relevant medical reference information at the point of care.

Clinical Lexicon® is a clinically balanced entrance terminology for use in implementing coded problem lists, medical record interfaces for clinician data entry, or clinical ICD coding tools. The lexicon provides the data to support clinical term selection in a manner to be implemented within the vendor system, and provides validated mapping to SNOMED RT, ICD-9-CM, and problem maintenance features. The lexicon is supported longitudinally with updates issued twice annually.

IDX offers another flexible decision support product that is integrated with Carecast. It consists of an inference engine that can be configured to execute a wide variety of site-specific knowledge bases or rules. Carecast provides many application hooks and alerting outcomes so that user interaction with the system can be evaluated by customer-defined rules (KLAS 2005).

7.7.6 McKesson Self-Reported

Complex alerting is more of a concept than a definition, and is one that will evolve over time as more clinical and informatics research is applied to an ever-expanding utilization of information technology in delivering care. As do others, we currently support active and passive alerts, which typically are triggered by a single value, a defined trend, or the comparison of a relatively few number of clinical variables. These trigger alerts to a user who is logged on to a system (active) or generate alerts and notifications based on conditions being systematically monitored while a user is off-line (passive).

We believe a significant role for "complex alerting" will be to know when not to provide an unnecessary alert based upon then-current circumstances. The term, complex alerting, may best apply when multiple, dynamic variables are considered in the context of probable clinician response. Many of our clinical advisories approach the concept of complex alerting by taking into consideration the patient's problem set, clinical indicators of the patient's condition, and institutional best practices. We then provide an advisory that anticipates the clinician's response to the data with other relevant information at just the time it is needed for clinical decision-making. Although this is a huge step forward for most care settings, it pales to what will be possible as more and more evidence-based protocols and research-driven outcome probabilities are proven and made available for broad utilization. Future complex alerts will combine clinical factors, business rules, provider profiles, evidence base protocols, and resource capacity planning to allow rapid modeling of multiple scenarios that anticipate not only the clinicians' response but the health system's capacity and configuration to attain the optimum medical, social, and financial outcomes.

Horizon Expert Orders is a prepackaged CPOE solution engineered around the best practices of Vanderbilt University Medical Center and the McKesson Clinical Leadership Panel (McKesson Horizon Expert Orders customers comprised of community and integrated delivery network hospitals). The prepackaged medical content, rules, and protocols may be tailored to the clinical practices adopted by acute care facility purchasing Horizon Expert Orders (KLAS 2005).

7.7.7 Meditech Self-Reported

Meditech does have a complete CPOE solution for physicians throughout the health care continuum. Caregivers are able to order medications and diagnostics as well as materials from a single point of entry. Decision support tools such as rules-based logic, associated data from other applications such as nursing and lab, interaction checking from the pharmacy formulary, and access to evidence-based medicine are all provided in real-time to the caregiver. This includes inpatient and outpatient orders as well as the ability to roll medications that the patient is currently taking into the inpatient or ambulatory setting. This process is completely automated and integrated with all of the Meditech applications including pharmacy, laboratory, and radiology. However, should a customer decide not to utilize Meditech's laboratory, radiology, and pharmacy modules in their CPOE solution the responses are the following.

Rules-based logic can be used to prompt and alert physicians and other caregivers to new information including, but not limited to, significant changes in patient condition and abnormal results. Alerts can generate pages, faxes, e-mail, and warnings within the software for the physician and/or caregiver. The physicians can also tailor the notifications they receive to meet their needs through the preferences feature on their desktop. A log of alert activity is maintained (KLAS 2005).

7.7.8 Misys Self-Reported

Providing clinicians with CDS is one of the key criteria for the Misys third, fourth, and fifth generation CPRs (clinical patient records). CDS includes tools to assist organizations in reducing medical errors, improving patient safety, decreasing unwanted practice variation, and achieving efficiency gains. Misys CPR provides for real-time alerts and comprehensive CDS.

Misys alerts and warnings are an inherent feature of data entry and review processes, including patient registration, visit creation and edit, patient transport, order entry, result entry, and patient data review. Embedded and configurable tools available to support process automation and alert checking include therapeutic conflict checking, drug and IV incompatibility checking, allergy checking, dose checking, duplicate order checking, alert record, critical value processing delta checking, absurdity checking, scheduling conflict checking, and display of current data.

The Criteria Evaluation Engine (CEE) is a real-time CDS tool that allows Misys CPR customers to determine the conditions they want to monitor and what actions should be taken (e.g., notification) when the conditions become true. CEE has a wide range of application and action options to perform when a condition is met with hundreds of data links to all significant data in the patient record.

Some examples of how Misys CPR customers are using the CEE include:

- **Infection Control and Microbiology**: CEE produces lists of resistant organisms (methicillin-resistant *S. aureus*), reportable organisms and diseases (*E. coli*, *Salmonella*, *Shigella*, hepatitis, HIV, tuberculosis, etc.)
- **Medical Records**: CEE automatically orders procedures when LOS, physician service, and other visit conditions are met for required documentation such as discharge summary, history, and physical, and so on

- **Surgery and Anatomic Pathology**: CEE reports selected malignancy cases for Tumor Board and retrospective review
- **Clinicians**: CEE provides interactive patient condition, order validity, or cost alerts during order entry

In addition, CEE can be used for monitoring conditions and conditional alerting. Misys is continually working to extend CPR alerting, surveillance, and CDS capabilities.

7.7.9 Siemens Self-Reported

In the actual writing of orders, the physician has access to all the needed information, but also is assisted by several levels of CDS. Clinical checks such as allergy interactions, drug–drug interactions, min/max dose checking, and weight-based dose calculations are built into Invision with the embedded National Drug Data File from First Data Bank. In addition, Invision works closely with Siemens Rules Engine, which is delivered as part of Invision, to provide more complex alerts and reminders. The Rules Engine uses Arden Syntax to build and run rules. The Rules can be triggered in-line within an ordering process, or asynchronously to monitor changes in the patient condition that would warrant a change in the patient's orders.

Being able to assess data from multiple sources and provide clinicians with alerts to potential problems before an adverse patient event occurs is critical. An example of such alerting is for drug-induced nephrotoxicity. By monitoring serum creatinine levels, the rule can warn about potential nephrotoxicity for over 400 drugs. The alert can be customized to adjust for different creatinine trigger levels and age-specific trends. The Siemens Rules Engine can run rules at the time of ordering to alert physicians of contraindications, or make suggestions of differing courses of action. The rules can run calculation and assess data from multiple sources and return responses that can be evaluated by specific patient parameters such as age, diagnoses, gender, and so on (KLAS 2005).

7.8 CONCLUSION

The status of commercially available software solutions for CDS continues to change. The status as of August 2005 is as follows:

Cerner Corporation continues to implement and bring live the largest number of new CPOE sites, where the provider organizations intend to leverage Cerner's clinical decision tools. In that regard, Cerner has Multum as a component aimed at the medication side of the clinical data. Zynx, a content company and an early Cerner offering, is now divested from Cerner.

The U.S. Department of Defense is fully engaged in taking commercial off-the-shelf software (COTS) like 3M Care Innovation and adding it to its comprehensive health care system (CHCS II) for the 150+ military hospitals. The plan for the 3M components was to get CDS automated at all of the military hospitals. Some delays have been encountered but the goal remains the same.

Eclipsys Corporation has many new clients aiming to leverage the knowledge-based decision capabilities of Sunrise Clinical Manager with some of the new. Net components available with version 3.5 and later.

Epic clients talk of a core rules engine that makes it easy to trigger clinical decisions. Few clients have used it in the acute care environment though 2005 will be the real measurement with the number of live inpatient sites almost doubling during that time.

GE is moving ahead with IHC. Based upon information reported to KLAS as of August 2005, GE and IHC have combined resources to rewrite the CDR, health data dictionary, and rules connections for a fully enhanced POE solution to come to the market in late 2007. The goal is for IHC to have the latest technology including a reliable and responsive set of clinician-based applications that support the robust CDS rules that are currently in place at IHC. The first components to go live will be the CDR and a comprehensive eMar for patient safety at one of IHC's smaller hospitals close to IHC headquarters. The applications will be built in Java or a similar language and will be based on Eclipse's architecture or potentially. Net, as an option. The most important piece of the plan is for IHC to be able to transport the thousands of rules and decision points running at LDS Hospital onto the new technology with an advanced end-user interface, which enhances the overall effectiveness of the system.

IDX Corporation has doubled the number of clients on Carecast. With the maturing of Carecast, clients are reporting that automation of CDS will be a top priority once the core pieces are solidly in place. Recently, however, GE Healthcare acquired IDX, so the future evolution of that system is unclear at the time of this writing.

McKesson Corporation is seeing 12 to 20 hospitals testing and implementing Horizon Expert Orders and Horizon Expert Documentation. These products are aimed at automating the decision processes. McKesson HED and HEO clients report that 2005 will be a benchmark year to see if the development is on target and bearing the expected fruit.

Meditech continues to work with the existing product and clients to provide rules and alerts. Use so far has been limited.

Misys offers Insight for those who want clinical decision logic as a bolt-on piece and the Misys CDR for acute care serves about 20 hospitals, several of them making limited use of medical decision-making software.

Siemens continues to enhance and support INVISION clients while Soarian is maturing. The long-term solution is Soarian and early sites are not doing any clinical decision automation yet.

The overall conclusion from these analyses and reports is that, though adoption of CDS remains a challenge, growth is taking place and software vendors and provider organizations are making strides in successful implementation and utilization.

REFERENCES

GE Press Release. (2005). GE News and Events. http://www.gehealthcare.com/company/pressroom/releases/pr_release_10225.htm

Raschke, R. A., Gollihare, B., Wunderlich, T. A., Guidry, J. R., Leibowitz, A. I., Peirce, J. C., et al. (1998). A computer alert system to prevent injury from adverse drug events. *J Am Med Assoc* **280**(15): 1317–1320.

KLAS Enterprises, LLC. (2001). Clinical information systems: Meeting the clinical needs of the enterprise by delivering robust systems for 2001. KLAS Enterprises, LLC, Draper.

KLAS Enterprises, LLC. (2003). CPOE Digest (Computerized Physician Order Entry). KLAS Enterprises, LLC, Orem.

KLAS Enterprises, LLC. (2004). CPOE Digest (Computerized Physician Order Entry). KLAS Enterprises, LLC, Orem.

KLAS Enterprises, LLC. (2005). CPOE Digest (Computerized Physician Order Entry). KLAS Enterprises, LLC, Orem.

VENDOR WEB SITE REFERENCES

Cerner: www.cerner.com
Eclipsys: www.eclipsys.com
Epic: www.epicsys.com
GE: www.ge.com
IDX: www.idx.com
McKesson: www.mckesson.com
Meditech: www.meditech.com
Misys: www.misyshealthcare.com
Siemens: www.siemens.com

8 LESSONS LEARNED

ROBERT A. GREENES

In this chapter we reflect on aspects of the experiences with CDS reported in the preceding chapters of this section. We are interested in identifying factors that were instrumental to success and those that have held back replication of demonstrated successes. It seems apparent that most of the innovation in this realm has come from academic sites such as those described in Chapters 4 through 6, rather than from commercial systems. Some movement of academically validated approaches into commercial systems has indeed occurred, as described in Chapter 7, and the offerings of commercial systems are gaining momentum. However, as portrayed in that chapter, commercial vendors have tended to be cautious and deliberate in introducing CDS functionality into their markets, to be sure that their customers are prepared for and want it. So we are likely to continue to look to academic sites to innovate in this realm.

A factor that may have a role in the generally slow adoption of practices shown to be effective is the often-quoted "17-year rule" of Balas and Boren (2000) (see also Chapter 3). They showed that it takes an average of 17 years for medical discoveries to make their way into routine practice, and that at the end of that period only 14 percent of research knowledge actually has been adopted. It is likely that a similar inertial process is at play with respect to the diffusion and adoption of CDS, but there may be a number of additional technical, organizational, and sociocultural factors in this sphere of activity, in which use of information technology is actually quite a disruptive innovation (Christensen and Raynor 2003) (see also Chapter 3) that requires pervasive changes in the way people do their jobs and even think about their jobs. Also, dissemination of CDS must overcome differences in system platforms, design, and functionality, and adoption must rely not only on incorporation of the methodology but acceptance or adaptation of the knowledge content of CDS; that is, involving at least two of the three life cycle processes we introduced in Chapter 1. Thus the human adaptations and the infrastructure requirements are considerably more complicated than for adopting circumscribed or niche applications that are not so disruptive to the nature of practice—even costly "big-ticket" items. Examples of the latter are CT and MRI scanning, which had to overcome significant hurdles in demonstrating their efficacy, and faced many regulatory and economic obstacles and barriers to wide diffusion, yet went from no use to ubiquitous use in less than a decade (Hillman and Schwartz 1985; Durick and Phillips 1988; Oh, Imanaka et al. 2005).

Clinical Decision Support: The Road Ahead

Recognizing these generic issues, we seek nonetheless to identify specific characteristics of the sites of innovation and the approaches taken that may have particularly impeded diffusion. This can be helpful if, having identified such characteristics, special efforts to address them can be devised.

8.1 ACADEMIC PROTOTYPES

In looking at the experiences of individual academic sites, it is quite clear that one of the impediments to widespread adoption has been the difficulty in replicating elsewhere the receptive environment and salutary conditions that made the application a success. Many of these conditions make the academic sites in which the initial work was done quite nonrepresentative with respect to either other peer academic sites or the vastly larger number of nonacademic sites that are potential targets for adoption.

8.1.1 Nature of the Project

In the typical scenario an application was built in the form of a pilot or prototype. It was implemented by a team led by an academic researcher with the primary goal of evaluating effectiveness of a particular kind of CDS intervention. In some circumstances, the prototype even may have not been implemented by the IT organization of the medical center, although with its cooperation and support. Attention typically was not focused on making the implementation "industrial strength," with the planning, design, architecture, and necessary tools and resources for ease of long-term maintenance and update.

This was demonstrated dramatically in a 2002 "knowledge inventory" study, in which the use of CDS in various subsystems and applications in operation at Brigham and Women's Hospital was analyzed (Boxwala, Denekamp et al. 2004) Clinical knowledge content, as we generally think of it, usually has been formulated by analysis of databases, review of the literature, establishment of opinion or consensus of experts, or indirectly, by study of the performance of experts, as reviewed in Section III. Knowledge derived by any of these approaches, however, rarely is expressed initially in a form that can be directly processed by a computer to make inferences, and thus is not initially able to be provided in the form of CDS. It must be refined and formalized in a way that is unambiguous and executable; that is, it must have the components of CDS described in Chapter 3: a computable decision model, using specified data elements and producing well-defined outputs. In the CDS applications that have been implemented and evaluated in academic centers, the project leader was often a physician who believed that a particular aspect of practice could be improved or even the local subject-matter expert on that aspect of practice, and who either developed a set of decision rules or led a team that did so, based on his or her experience. This typically was done in an ad hoc fashion. As many such applications and experiments occurred, a collection of such CDS knowledge began to be introduced, similarly ad hoc, or at best in a fashion unique to particular suites of applications.

8.1.2 Consequences for Operation

Not only were industrial strength, robustness, and long-term support usually not primary goals of the investigators in the typical project, but it can be fairly argued that attention to the issues involved in pursuing them would have substantially increased the cost and time required. If anything, inclusion of such capabilities was usually an afterthought, once the application was shown to be successful, and the institution decided to adopt it and keep it running. After an application was shown to be effective, the institution usually did create some mechanism to address long-term support. This was, nonetheless, typically ad hoc, partly out of necessity, given the limited institution-specific resources and many other competing priorities, and the one-off design of the application, which did not allow for leveraging of software development and integration tools that might have been used in production systems.

Eventually, these needs have come to be recognized. Examples are the substantial efforts that have been made in recent years in building knowledge management infrastructure, tools, and resources, by Partners Healthcare and Intermountain Healthcare, as described in Chapters 21 and 22, respectively. But the recognition of need for them came about only after a long period of homegrown development of prototypes, operational adoption of successful projects, and a growing need to maintain and support these applications in increasingly complex and diverse health care enterprises. The 2002 knowledge inventory study cited earlier (Boxwala, Denekamp et al. 2004), for example, demonstrated the presence of tens of thousands of knowledge content items, multiple ad hoc ways of encoding and representing the knowledge, several homegrown editors, and many instances of redundancy and inconsistency. No systematic curation of this content was done, and it was difficult, when a new project was initiated, for the experts even to assess what was already in place and how the proposed CDS would relate to those other elements.

The preceding points should not be regarded as criticisms of the process for establishing the effectiveness of the CDS approaches that were pursued. After all, attempting to change clinical practice through computer systems is a big undertaking, and it is important to know whether the approach one is taking is going to work before making substantial investment and commitment of effort. Also, unlike the case with a laboratory investigation, it is not usually possible to construct an artificial or isolated experiment to determine effectiveness. The approach needs to be implemented in a live operational setting, it must be put into use long enough for novelty and learning effects to settle down, the users must believe that the system is intended for real operation and thus make a commitment to its use and cooperate with it fully, and actual impacts on process and outcome need to be assessed. It is hard to imagine another way to do such an evaluation than the way these studies were done.

Another factor is that such experimentation often took place in the framework of clinical IT systems that were themselves homegrown, in academic medical centers such as those reviewed in this section. Without the expectation of future replications of the implementation that a commercial product would have, over which to amortize costs, such medical center-specific implementations typically have lacked a robust set of software development and integration tools and resources. Of course, the evolution of these environments into large and diverse academic medical centers and integrated delivery systems over the years has changed this situation considerably, so that such software development tools and resources are also increasingly needed and deployed in these environments, and, as is the case for knowledge management infrastructure noted earlier, these investments are now being made for the systems as a whole. Nonetheless, the goal of replication and adaptation elsewhere was not an initial driver in the design and development of such systems.

Another related factor is that, since these systems were homegrown, there was generally a fairly open attitude toward the prospect of creating and testing extensions to the existing system when reasonable proposals for doing so surfaced. Thus the barriers to designing and conducting an experiment with a prototype intervention to be studied in live routine operation were not high. This was not only salutary for being able to do these experiments, but a corollary of this openness and flexibility was the scant attention given to the long-term requirements should the prototypes become permanent functions.

Superimposed on the preceding is the general evolution of the state of the art of hardware, software, and communication technologies that has occurred over the past 20 to 30 years. Designs of both production applications and the prototype add-ons that became operational in academic centers such as we have reviewed in this section evolved within older system platforms, generally. Partly because of legacy investments, academic clinical systems have only slowly adapted to and taken advantage of newer design approaches like the use of middleware, thin clients, and Web services. Although we can now use these capabilities to facilitate experimentations in user interface and interaction with CDS modules, while insulating the underlying architecture of systems, such was not the case at the time of much of the reported development and evaluation of CDS interventions in the homegrown system environments of academic centers.

8.1.3 Local Differences

A major factor resulting from homegrown development of systems and applications was platform dependence. Not only was this reflected in the way the CDS was encoded, in terms of its dependence on particular computer languages, database structures, user interfaces, operating systems, application tools and interfaces, and modes of interaction with users, it also was reflected in the very kinds of applications available, into which the CDS could be incorporated. An example is revealed from the discussions of medication interaction checks in the Brigham and Women's Hospital system (see Chapter 5) and those at LDS hospital (see Chapter 6). Given that the Brigham was one of the early implementers of CPOE, this application and direct interaction with physicians became a target for incorporation of various forms of CDS and their evaluation. At LDS, where CPOE came later, some of the drug interaction CDS capabilities initially were targeted at the pharmacists as the primary users instead. Other differences in system functionality, workflow practices, and conventions also made differences in the choice of approach to implementation that was taken.

8.1.4 Maintenance and Update

Typically, the primary champion for an academic project moved on to exploring another idea or even to leave for another institution, at which point the future maintenance and support of the application fell to software engineers. Further, the applications often were implemented in the form of direct programming-language level coding, as a result of which there was no easy way to make the embedded logic transparent or easily inspectable by a subject matter expert, further complicating the maintenance. Unless meticulous efforts were made to keep human-readable separate logs of all places where knowledge was incorporated in applications, and all modifications and updates of the implementation, human-readable and computer-encoded representations often became badly "out of sync" with one another.

In many of the academic projects described in this section of the book, a life cycle process for management of CDS capabilities was not formally articulated at least initially. However, review of the work at these sites indicates that its de facto nature was often something like that depicted in Figure 8-1, which characterizes the way rules-based knowledge was managed for many years at Brigham, as documented in the 2002 knowledge inventory study (Boxwala, Denekamp et al. 2004). Maintenance of CDS in a clinical environment is difficult because of the need to keep the knowledge content on which it is based transparent, so that it is understood and accepted by its users—health care professionals and the public. Also, the formulation of knowledge must be capable of being readily modified and updated as biomedical science and technology evolve, and recommendations for optimal practice change.

To the extent that CDS knowledge is embedded in a computational algorithm, or in the format of a display screen, and is tightly integrated with a host IT system, these modifications require considerable technical effort to accomplish, and are not transparent to subject experts or clinical users. If different IT systems are incompatible with one another, the integration with IT systems needs to be redone for every system in which the CDS capabilities are to be made available, and modified in each such system as knowledge is

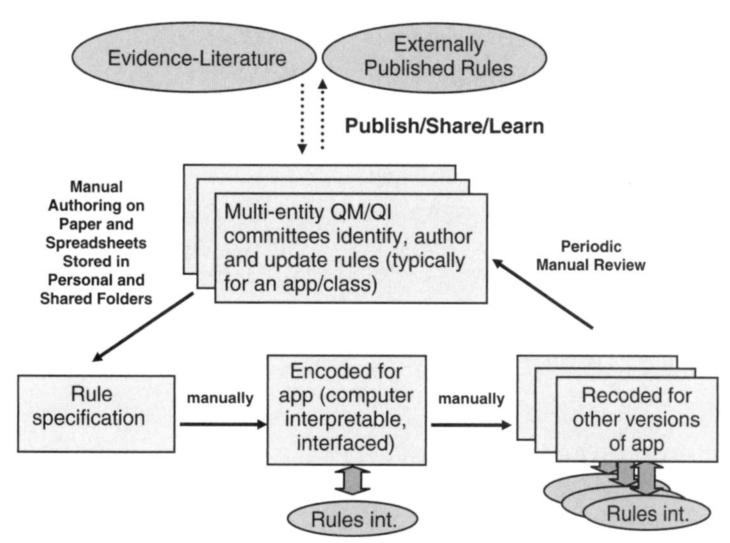

FIGURE 8-1 De facto life cycle process for rule-based clinical decision support at Partners HealthCare before formal knowledge management process was adopted. Note the disconnect between the knowledge generation process by committees or experts and the encoding in applications, and the manual processes of translation into computer format and recoding for different applications or versions. The dotted-line processes of interchange with external knowledge sources were ideals that did not effectively occur.

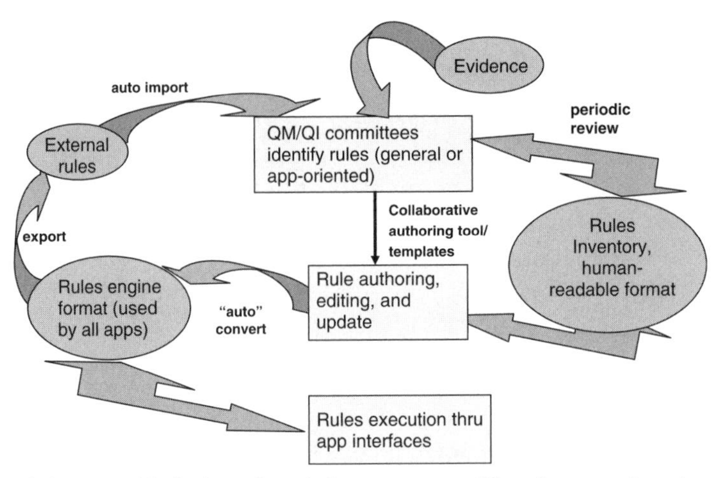

FIGURE 8-2 A model of a future knowledge management life cycle process for rules authoring and update, driving current efforts at formalization of the process at Partners HealthCare. Note the automatic or semi-automatic conversion and re-use that are intended.

updated. Also, not only does knowledge content change, but the decision models that are used may themselves evolve, and thus the representation of the knowledge to be compatible with those evolved models may also change.

These difficulties limit the ability to disseminate and reuse knowledge in different environments or even in different applications within the same environment. An optimal approach to CDS implementation, therefore, seeks to:

- Develop standardized representations of the formal decision model, data elements, and result specifications
- Specify the knowledge in a form that the model can process, but separate from the model itself
- Keep the decision model's execution "engine" separate from the host IT environments in which it will operate

An idealized target life cycle for rule-based knowledge that was adopted as a model underlying the investment in knowledge management by Partners Information Systems (Partners HealthCare System, Inc., is the parent of the Brigham) in 2003, as shown in Figure 8-2, is quite different from the de facto process that has been in place; it replaces many manual processes that can get badly out of sync, or just not get done, with ones that are more tightly integrated and automated. (The process of implementing that model of a life cycle process for knowledge management, not only for rules-based knowledge but for CDS knowledge generally, is described further in Chapter 21.)

8.2 STANDARDS AND SHARING OF INTEROPERABLE CONTENT AND TOOLS

In most of CDS applications, there was little or no use of standard dictionaries, common reference information models, knowledge models, external rules engines, and message or service interfaces, which might have enhanced the ability of a successful local CDS implementation to be adopted or reimplemented elsewhere. This critique is, of course, somewhat unfair. Even vendor implementations such as those discussed in Chapter 7 generally have not been standards-based; although vendors often used standards, it was generally primarily for internal portability and support within the company's products, or for specific interfaces between systems such as for integrating a third-party laboratory system or for communicating with a billing system.

Further, as elaborated in Section IV, many of the standards needed for truly portable CDS are still very much in flux. The Arden Syntax, which was developed for the specification of Medical Logic Modules (see Chapter 12), and adopted by HL7 and ASTM, is the longest-standing CDS-specific standard. Yet the sharing of knowledge in Arden Syntax across platforms and vendors is virtually nonexistent. Attempts to foster this have been made, but the sharing is usually limited to occurring within the user groups and implementers of a particular vendor's products.

Several efforts of the Clinical Decision Support Technical Committee (CDS TC) of HL7 underway at the time of this writing are potentially steps in the right direction. The GELLO expression language (Sordo, Boxwala et al. 2004) has been adopted as an HL7 and ANSI standard, and is being explored for use in applications such as the encoding of prior authorization rules for medication prescribing (Sordo, Dunlop et al. 2006). Proposals for standardizing the specification of order sets and infobutton managers are under consideration in the CDS TC. A specification for a Web services interface for execution of CDS is being developed as a joint effort of the CDS TC and a Service Oriented Architecture Special Interest Group of HL7. Assuming these efforts will come to fruition, they may stimulate the availability of commercial and noncommercial offerings of knowledge bases of content and the incorporation of their use in vendor-based and noncommercial clinical information system platforms.

Currently a number of knowledge content resources are provided by vendors or sometimes in the public domain; for example, vocabularies/taxonomies, clinical statement specifications (i.e., *archetypes* of documentation elements, built on standard vocabularies, as described in Chapter 15), and drug interaction tables. However, these are usually available through proprietary or custom interfaces, or they are commodities provided in simple tabular form that can be accessed through relational databases, but requiring uploading and management by the host environment. Very few if any external knowledge resources are now available that can be automatically invoked by a clinical information system to provide CDS. It is too soon to know to what extent the adoption of standards such as those currently in process will stimulate the marketplace for such resources, and enable existing IT systems to utilize them directly.

8.3 USERS

The fact that many of the successful forms of CDS were first introduced in teaching hospitals such as the Brigham or LDS has other important implications. House officers can be motivated to put in the extra effort for CPOE or to respond to alerts and reminders more easily than attending physicians can, and probably have a higher tolerance for the learning of new approaches and the disruption of traditional practice patterns, and the impact on their time. How easy the experiences in a teaching hospital can be translated to a community hospital setting or an office practice has rarely been studied. Chapter 18 addresses some of these issues based on experience with adoption of CPOE.

8.4 OTHER CONSIDERATIONS

Even with new impetus for adoption, it is important to avoid embracing CDS as a magic bullet, a cure-all that in one fell swoop can address the complexities, inefficiencies, and safety and quality problems of health care. If a naïve view is adopted, a danger is that the tasks and challenges involved in introducing and managing CDS will be drastically underestimated and oversimplified. As we noted in Chapter 2, in the past four years, several news items and published reports have cited undesirable experiences resulting from installation of CPOE (e.g., see Shabot 2004; Han, Carcillo et al. 2005; Koppel, Metlay et al. 2005), which caused considerable reaction and discussion among the academic informatics, hospital IT management, and IT vendor communities. Such experiences highlight some of the pitfalls of introducing computer systems in health care. The difficulties experienced in these instances are in part attributable to inadequate preparation, training, or time allowed for adaptation that such systems require, and which take considerable effort. But they also may reflect a mismatch between the experiences in one setting and the ability to replicate them in a setting with quite different motivations, target users, workflow processes, and expectations.

CDS shares many of the same challenges for implementation as CPOE, in that it directly impacts on practitioner workflow, interactions with other people as well as with the computer, and style of practice. Much can go wrong if the effects on these dimensions are not considered. Also, deployment itself is insufficient, as maintenance, update, and continuing education must also be supported. It is thus necessary to realize that deployment of CDS capabilities is a multilayered, multistep process that requires considerable underlying functionality. It requires major investment, and understanding of the need for infrastructure, training, and cooperative effort by many parties.

Regarding the decision to adopt a technology and then attempting to go about it, we have mentioned the Balas and Boren study (2000), documenting the slow pace and severe winnowing process that occurs as technologies progress from research results to adoption in practice. A factor related to this is the problem that unsuccessful efforts are often not published. As a result, ineffective approaches tend to get replicated, wasting effort and resources, and the knowledge underlying best practices for CDS deployment only slowly gets enhanced.

There is no consensus on the best ways to integrate decision support into the flow of health care practice, in terms of interactions with patients, nurses, and physicians. We have mentioned the differences in capabilities among systems and applications as one factor that influences integration decisions, but another may also be that there are alternative approaches that simply may work better in one setting than in another.

Other concerns relate to the possibility of misapplication or inappropriateness of knowledge for a specific patient, concerns about "cookbook medicine," and the odious image of "big brother"—style monitoring, accountability, and oversight. See also Chapter 18 on organizational and cultural factors.

Finally, because of the difficulties in assembling, validating, documenting, and deploying decision support knowledge, the critical mass and investment required for mounting appropriate infrastructure, training of personnel, and adapting the processes to their own systems are large. If each institution must face these challenges on their own, the effort and costs required are likely to severely limit progress. However, only limited standards and tools for addressing knowledge management challenges exist, as we have noted, and more importantly, there is no framework for sharing and reuse that is generally endorsed, subscribed to, and supported by stakeholders. Thus it is not surprising that progress is piecemeal, and in fact, without addressing this, it is likely to remain that way. We return to this challenge in Section VII, as we consider the road ahead.

REFERENCES

Balas E. A. and Boren, S. A. (2000). Managing clinical knowledge for health care improvement. *Yearbook of Medical Informatics*. Van Bemmel, J. and McCray, A. T. (eds.) Stuttgart, Schattauer Verlagsgesellschaft mbH: 65–70.

Boxwala, A. A., Denekamp, Y., Greenes, R. A., Kuperman, G. J. and Middleton, B. L. (2004). Survey and evaluation of knowledge bases for clinical decision support. *Medinfo*, San Francisco, CA, IOS Press.

Christensen, C. M. and Raynor, M. E. (2003). *The Innovator's Solution*. Cambridge, MA, Harvard Business School Press.

Durick, D. A. and Phillips, M. L. (1988). Diffusion of an innovation: adoption of MRI. *Radiol Technol* **59**(3): 239–241.

Han, Y. Y., Carcillo, J. A., Venkataraman, S. T., Clark, R. S., Watson, R. S., Nguyen, T. C., et al. (2005). Unexpected increased mortality after implementation of a commercially sold computerized physician order entry system. *Pediatrics* **116**(6): 1506–1512.

Hillman, A. L. and Schwartz, J. S. (1985). The adoption and diffusion of CT and MRI in the United States. A comparative analysis. *Med Care* **23**(11): 1283–1294.

Koppel, R., Metlay, J. P., Cohen, A., Abaluck, B., Localio, A. R., Kimmel, S. E., et al. (2005). Role of computerized physician order entry systems in facilitating medication errors. *JAMA* **293**(10): 1197–1203.

Oh, E. H., Imanaka, Y. and Evans, E. (2005). Determinants of the diffusion of computed tomography and magnetic resonance imaging. *Int J Technol Assess Health Care* **21**(1): 73–80.

Shabot, M. M. (2004). Ten commandments for implementing clinical information systems. *Proc (Bayl Univ Med Cent)* **17**(3): 265–269.

Sordo, M., Boxwala, A. A., Ogunyemi, O. and Greenes, R. A. (2004). Description and status update on GELLO: a proposed standardized object-oriented expression language for clinical decision support. *Medinfo* **11**(Pt 1): 164–168.

Sordo, M., Dunlop, R., Martin, R. D., McKinnon, B. M., Schueth, A. J. and Greenes, R. A. (2006). GELLO and ePrescribing: Exploring the Use of a Standard for Prior Authorization in Electronic Prescribing. *DSG Technical Report, DSG_TR_2006_002*. Report. Boston, Decision Systems Group, Brigham and Women's Hospital.

GENERATION AND FORMULATION OF KNOWLEDGE

9 HUMAN-INTENSIVE TECHNIQUES

EDWARD H. SHORTLIFFE and VIMLA L. PATEL

9.1 INTRODUCTION

When we consider what makes human beings excellent at medical decision-making, we generally acknowledge that there are two key determinants: how much the experts know, and how well they apply what they know when devising solutions to problems that may arise. Thus, as we consider the creation of optimal decision support systems, we must similarly consider both the knowledge that they embody and the processes they adopt when applying that knowledge. A system can be "dumb" if the knowledge it needs is lacking or faulty, and it can demonstrate "poor judgment" if it reaches inappropriate conclusions despite a wealth of necessary factual knowledge. We recognize that it means little if we cram huge amounts of knowledge into a system but the program subsequently cannot use it wisely or appropriately.

In this chapter we focus on the acquisition of knowledge so that it can be encoded for use in decision support systems. As we have suggested, that means that we need to understand both the factual knowledge that is required to solve the relevant problems and the judgmental knowledge that characterizes a decision maker who gets to the heart of a problem effectively, discards irrelevant information, and demonstrates an ability to be creative rather than to solve problems by rote formula every time they arise.

Chapters 10 and 11 discuss analytical methods for identifying new or relevant knowledge from databases or the literature, such as data mining techniques and meta-analysis. Here, rather, we focus on the acquisition of knowledge by interacting with human beings—analyzing their behaviors, inferring their beliefs and knowledge, asking them to explain their thought processes and actions, and applying formal or informal methods for extracting from those behaviors and explanations the factual and judgmental knowledge that they appear to be applying. Such interactions can be undertaken by human beings interacting with experts (often called *knowledge engineering*) or by computer programs that experts can use to convey what they know for capture in a computer-based representation (often called *interactive transfer of expertise*).

There are a number of reasons why we want to capture expert knowledge (Crandall, Klein, and Hoffman 2006). These include:

- **Knowledge preservation.** We want to capture "wisdom," which develops with expertise. Such knowledge usually is not documented in any formal way, and we lose it once the expert retires or otherwise leaves the job.

- **Knowledge sharing.** Captured expert knowledge, meaningfully represented, can be reused in training programs, where trainees can train to develop expert strategies and functional efficiency. Such knowledge also can be shared among those who need to use it for a wide variety of decision-making tasks.
- **Knowledge to form the basis for decision aids.** New technology can be created based on the expert knowledge to help practitioners make better decisions. The technology, properly implemented, must embody the concepts, principles, and procedures of the work domain.
- **Knowledge that reveals underlying skills.** As the use of expert knowledge is explicated, it also reveals underlying strategies and skills.

Although the computer-based representation of knowledge is covered in Section IV of this volume, it is difficult to discuss the acquisition of knowledge without considering the representational issues that motivate and guide the acquisition process. Furthermore, the entire effort to capture and utilize knowledge in computer programs is predicated on the recognition that knowledge has a central role to play in providing tailored guidance through decision support systems. For example, cognitive psychologists have now recognized the centrality of domain-specific knowledge in the skilled solving of complex problems (Glaser and Chi 1988; Patel and Groen 1991b). Researchers in artificial intelligence have also realized that "knowledge is power" and that general representations and search strategies, once a primary focus in that field, are limited in their ability to create intelligent behavior in machines (Feigenbaum, Buchanan, and Lederberg 1971). Knowledge-dependent computer applications, such as *expert systems* that use expert knowledge to perform complex problem-solving and decision-making tasks, are intended for use when the real experts are scarce, expensive, inconsistent, or simply unavailable on a routine basis. This characterization begs the questions "What is an expert?" and "How do we distinguish the knowledge and abilities of experts from those who are novices, or less expert, in a field?" Although we can easily agree that experts are those who have special skills or knowledge derived from extensive experience in their domain of expertise, their ability to achieve accurate and reliable performance also shows flexibility and adaptiveness in their environment that is difficult to explain by factual knowledge alone. We recognize that experts know "how," not just "what," and any attempt to capture knowledge for computer representation and use must recognize that these two general classes of knowledge are equally important.

Knowledge acquisition (KA) may be defined as the process of identifying and eliciting knowledge from existing sources—from domain experts, from documents, or inferred from large datasets—and subsequently encoding that knowledge so that it can be verified and validated. The biomedical informatics literature reports often the design of knowledge-based systems and the evaluation of the performance of those systems (Bell, Pattison-Gordon, and Greenes 1994; Evans, Cimino et al. 1994; Patel, Allen et al. 1998; Achour, Dojat et al. 2001; van der Maas, ter Hofstede, and ten Hoopen 2001; Peleg, Boxwala et al. 2004). However, reproducible methods to acquire such knowledge, and to assure its accuracy, are typically discussed separately, even though they are intimately related to the design and construction of decision support programs. A knowledge base used in a clinical decision support system might contain knowledge structures that represent potential findings and diagnoses and the relationships among them (*conceptual* or *factual* knowledge), a knowledge structure representing guidelines or algorithms used to operate on this knowledge structure (*procedural* knowledge), and possibly also a knowledge structure with application logic used to apply these guidelines and algorithms to the underlying conceptual structure (*strategic* knowledge). All these types of knowledge must be combined to achieve a functioning decision support facility, and knowledge acquisition must address each type of knowledge during the elicitation process.

The techniques and theories that enable knowledge acquisition can be viewed within the context of the process illustrated in Figure 9-1. The process begins with the acquisition of knowledge from human experts (knowledge acquisition, or KA), followed by the representation of that knowledge (KR) in a computationally tractable form that supports knowledge-based agents or applications (Hoffman, Shadbolt et al. 1995). Many people would then include the verification and validation of the output of those knowledge-based agents or applications as part of the KA process, since they provide feedback regarding the quality of the contents of the underlying knowledge structures.

There is a variant of Figure 9-1 in which the knowledge is acquired not from a single expert collaborator, but rather, from a group of experts, perhaps through a consensus-development process or by studying several experts and merging what one has learned into a single knowledge base. The field of cognitive science offers several methods for understanding the reasoning processes, mental models, and knowledge used by experts when they solve problems, as well as for dealing with team decision-making and consensus development. We shall present some of those notions in the subsection that follows. There are also formal methods by which experts work together, supported by the literature and formal research studies, to reach consensus in formulating knowledge (e.g., the process of evidence-based guideline development (Peleg, Gutnik et al. 2006)). The acquisition and representation of consensus guidelines is further discussed in Chapter 13.

Finally, there has been substantial work to develop computer programs that acquire knowledge directly from experts (see Figure 9-2). Termed

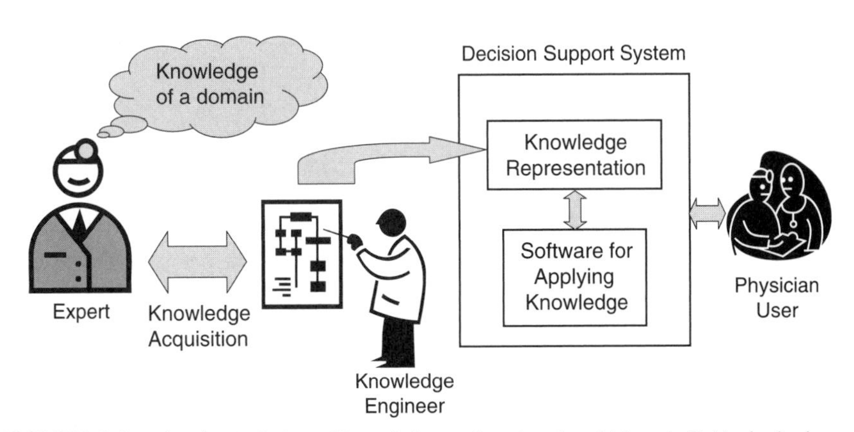

FIGURE 9-1 The classical view of knowledge engineering, in which an individual who knows the technical details of a system's representational conventions also has the skills of interviewing and observation necessary to work closely with an expert (or a group of experts) in order to obtain the needed knowledge and to convert it to a computationally useful form.

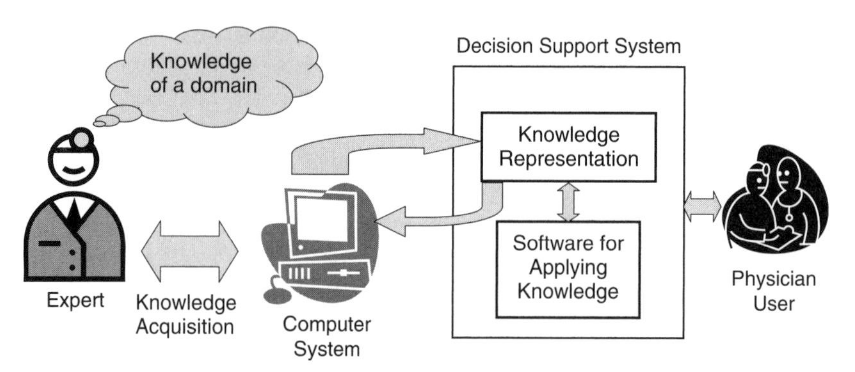

FIGURE 9-2 The interactive transfer of expertise using a computer program for knowledge acquisition. Note that such programs will generally both create new knowledge *and* use pre-existing knowledge to guide the knowledge acquisition process.

knowledge acquisition systems or *knowledge authoring systems*, these programs fill the role of knowledge engineer, providing human beings with a computational environment for assessing what knowledge is missing from a system and transferring their knowledge so that it can be encoded for that system's use. Such programs are often tightly coupled with the decision support system itself, allowing the system's decision-making abilities to be assessed and debugged as part of the knowledge acquisition/enhancement process. They always rely on access to the pre-existing knowledge in the system, as is indicated by the arrows going in both directions between the computer and the knowledge base in the figure.

9.2 THEORETICAL BASIS FOR KNOWLEDGE ACQUISITION

We now focus on the frequently cited theoretical basis that underlies the numerous methods and techniques that exist to elicit domain knowledge from sources such as relevant experts. The currently accepted psychological basis for KA depends on defining and acknowledging the concept of expertise. Two major goals of expertise research have been to understand what distinguishes outstanding individuals in a domain from less outstanding individuals and to characterize the development of expertise. This approach originated with the pioneering research of deGroot (1965) in the domain of chess, from which it extended to investigations of expertise in a range of content domains, including physics (Larkin, McDermott et al. 1980; Chi, Feltovich, and Glaser 1981), music (Sloboda 1991), sports (Allard and Starkes 1991), and medicine (Patel and Groen 1991b). This research has shown that, on average, the achievement of expert levels of performance in any domain requires about ten years of full-time experience. An "expert" is someone who has achieved a high level of proficiency, as indicated by various measures, such as international "Elo" ratings in chess (named for the system's creator, Árpád Élö, a Hungarian-born American physics professor), world rankings in various athletic endeavors, and certification by a sanctioned licensing body, as in medical subspecialties.

In medicine, the expert–novice paradigm has contributed to our understanding of the nature of medical expertise and skilled clinical performance. Expert physicians have extensive general knowledge of medicine (acquired through medical school and residency training) and deep, detailed knowledge of their relatively narrow areas of specialization (acquired from both training and clinical experience). Every experienced physician has acquired common wisdom and medical knowledge as well as certain mastery in the application of medical skills; this constitutes generic expertise. Investigators have suggested the following classification of levels of expertise (Patel and Groen 1991b):

- A *beginner* is a person who has only everyday, lay knowledge of a domain; an example is a typical patient.
- A *novice* is someone who has begun to acquire the prerequisite knowledge assumed in the domain, such as a medical student; novices have a basic familiarity with the core concepts, the language, and to a lesser extent, the culture of medicine.
- An *intermediate* is above the beginner level but below the subexpert level and is typically a senior medical student or a junior resident.
- A *subexpert* (e.g., a specialist solving a clinical problem outside his or her domain of expertise) possesses generic knowledge and experience that exceeds that of an intermediate but lacks specialized knowledge of the medical subdomain in question.
- An *expert* (e.g., a cardiologist or an experienced intensive care nurse) has specialized knowledge of the subdomain in addition to broad generic knowledge.

The development of expertise has been shown to follow a somewhat counterintuitive trajectory. It is often assumed that the novice becomes an expert by a steady, gradual accumulation of knowledge and fine-tuning of skills. That is, as a person becomes more familiar with a domain, his or her level of performance (e.g., accuracy and quality) gradually increases. It turns out, however, that one generally can document a degradation in performance as a subject moves from novice to expert. This has been referred to as the *intermediate effect* (Patel and Groen 1991a). It has been repeatedly demonstrated that superior expert performance is mediated by highly structured and richly interconnected domain-specific knowledge. Experts' knowledge is hierarchical and densely interconnected, which allows new pieces of information to become well integrated. Given that a novice's knowledge base is sparse and an expert's knowledge base is intricately interconnected, an intermediate may have many of the pieces of knowledge in place but lack the extensive connectedness of an expert, leading to the intermediate effect just mentioned. For example, expert cardiologists are routinely called upon to integrate clinical findings at various levels of aggregation, from biochemical abnormalities evidenced in blood tests to perturbations at the system level to clinical manifestations as expressed in the patient's complaints. After the performance degradation phase due to the intermediate effect, practitioners develop the missing connections among concepts in their knowledge base and, as they gain experience in the execution of a task, their performance becomes increasingly smooth, efficient, and automatic.

A great deal of experts' knowledge is finely tuned and highly automated, enabling them to execute a set of procedures in an efficient, yet highly adaptive manner, which is sensitive to shifting contexts. They can readily filter out irrelevant information. Novices, as opposed to intermediates, do not conduct irrelevant searches, simply because they lack knowledge rich enough to generate such searches. Studies demonstrate that expert performance is not a result of generally superior memory skills, but it is a function of a well-organized knowledge base adapted to recognizing familiar configurations of stimuli. The nature of experts' organized knowledge can also account for their superior perceptions of patterns. This is demonstrated compellingly in studies of expert radiologists, where they can be shown to look at the x-ray image at a glance, to develop an immediate impression, and then to search the image for findings that fail to fit or otherwise modify the initial impression. For more details on the nature of expertise, refer to several of the key papers in the field (Chi, Glaser, and Farr 1988; Ericsson and Smith 1991; Ericsson 1996; Feltovich, Ford, and Hoffman 1997).

One of the things that domain experts know about is the procedures they use in their practice. They learn many "heuristics" or rules of thumb (Chapman and Elstein 2000). These compiled, top-level procedures can lead experts to skip steps when they describe the processes by which they carry out their task. Some such heuristics are shared with other experts, but others are ones they have created on their own (Patel, Arocha, and Kaufman 1994). In addition, experts have metacognitive awareness of their own strategies and how they manage their resources (Glaser 1996). Metacognition refers to the collection of cognitive process and functions that individuals use when thinking about their own cognition (about the way that they think).

Thus, when such experts work with knowledge engineers or KA programs, their goal is to transfer their existing knowledge to the computer so that it is able to replicate human expert performance in the task for which they have specialized expertise. Given the complexity of the types of knowledge and perceptual issues that characterize human expertise, it is challenging to capture such knowledge and to encode it for computer use so that expert performance by a decision support system can be achieved. What, then, are the approaches that have allowed nonexperts to analyze, understand, and encode the ways that individual experts make decisions?

9.3 COGNITIVE TASK ANALYSIS

The general approach that cognitive scientists use in analyzing the basis for human performance is known as *cognitive task analysis* (CTA). Its purpose is to capture the way the mind works—to capture *cognition.* CTA should describe the basis for skilled performance that is being studied. The methods in this field are varied, and a detailed exposition is beyond the scope of this book, but in using CTA, cognitive scientists try to capture what people are thinking about, what they are paying attention to, the strategies they are using in making decisions, what they are trying to accomplish, what information they discard, and what they know about the way a process works (Crandall, Klein, and Hoffman 2006). The three key aspects of CTA are: 1) knowledge elicitation, 2) data analysis, and 3) "knowledge representation," where in this case the representation of knowledge conforms to formal criteria and methods that may not be inherently computational, even though they might provide insight in constructing a computer system's knowledge base in the same domain. Cognitive scientists will utilize one of a variety of knowledge representation schemes to describe and capture what they have learned and to compare the expertise and reasoning processes of individuals (for example, novices versus experts when presented with the identical problems). In the following sections, we briefly describe each of these three key aspects of CTA.

9.3.1 Knowledge Elicitation Methods

Conducting KA studies is often complex and resource-intensive. As a result, it is important to select the appropriate KA methods and tools at the outset of such projects in order to ensure that the end product is amenable to the planned application domain. One of the key issues to consider when planning a KA study is the source of the knowledge to be elicited. The use of domain experts is probably the most common and simultaneously problematic source of knowledge (Scott, Clayton, and Gibson 1991). The use of domain experts presupposes the selection of individuals with sufficient domain knowledge, interest in participating in the KA process, and minimal bias—a combination of attributes not always easily attained.

Further complicating the use of domain experts is the frequent need to collect knowledge from multiple experts. Groups of experts often are needed to mitigate the problems associated with using single experts, which may include individual biases or the limitations of a single expert's knowledge or line of reasoning in the given domain (Liou 1990)—all of which may lead to knowledge elicitation with incomplete or potentially noneffective contents. However, though the use of multiple experts has the potential benefit of utilizing group synergies to generate consensus knowledge that is greater than the sum of the contributing individual knowledge (Boy 1997; Morgan and Martz 2004), it is also not without its potential pitfalls, most notably the difficulties surrounding the merging of multiple-experts' knowledge (Morgan and Martz 2004) and the potential for the resulting knowledge to represent a single expert's opinion or input, rather than a true group consensus (Liou 1990). Despite these potential concerns, the benefits of using multiple experts in KA generally outweigh the disadvantages.

Straightforward interview techniques often are used because they require a minimum level of resources, can be performed in a relatively short time frame, and can yield a significant amount of qualitative knowledge. The disadvantages of interview techniques include a frequent lack of quantitative data, which are needed for the input into the next step in the process. Furthermore, the results often can be biased due to the framing or presentation of questions or the selection of topics that are of interest only to researchers (Hawkins 1983; Wood and Roth 1990; Boy 1997; Morgan and Martz 2004). But, perhaps most importantly, interviews simply lead to introspective opinions of the collaborating experts, and the knowledge elicited may *not* correspond to what they actually do when solving problems in the domain. For this reason, most knowledge engineers and psychologists who perform knowledge elicitation would prefer to observe the experts as they carry out tasks, either in simulated or "real world" environments. In order to gain insight into their mental processes, the experts may be asked to talk aloud about what they are doing and thinking *while they are performing the task*. In the world of cognitive science, such responses generated during problem solving are known as *think-aloud protocols* (Ericsson and Simon 1993).

In contrast, ethnographic evaluations of expert performance are observational studies conducted in context, with a minimum of knowledge engineer's or psychologist's involvement in the workflow or situation under consideration. Such studies also implicitly evaluate the knowledge used by those experts and have been have been used in a variety of domains, ranging from air traffic control systems to complex health care delivery applications

(Hughes, King et al. 1995; Liszka, Stubblefield, and Kleban 2003; Laxmisan, Malhotra et al. 2005; Cohen, Blatter et al. 2006). One of the primary benefits of contemporary ethnographic research methods is that they are specifically tailored to minimize potential observational or researcher-induced biases (e.g., the Hawthorne effect), while maximizing the role of collecting information in context, providing situational-specific knowledge. The resulting qualitative data generated by observational studies are often characterized as being "rich" or "concrete" (Iqbal, Gatward, and James 2005). The advantages of observational techniques are similar to interviews in that they require a minimum of resources, and further, provide for the capture of generally unbiased and contextual information. The disadvantages of observational techniques are again similar to interviews, in that they are time-intensive and do not easily yield large amounts of quantitative data. When quantitative data are generated from the observational studies, it is often a time and resource-intensive task to code generated transcripts to extract data. Furthermore, in the absence of think-aloud protocols, it is left to the researchers to infer thought processes and knowledge structures from the behaviors that they have observed.

9.3.1.1 Group Techniques

A number of group techniques for expert KA have been reported, including brainstorming (Osborn 1953), nominal group studies (Delbecq, Van de Ven, and Gustafon 1986; Jones and Hunter 1999), presentation discovery (Payne and Starren 2005), Delphi studies (Adelman 1989), consensus decision-making (McGraw and Seale 1988), and computer-aided group sessions (Adams, Toomey, and Churchill 1999). All of these techniques focus on the elicitation of consensus-based knowledge. It has been argued that such consensus-based knowledge is superior to the knowledge that may be gained from a single expert, since the group techniques used to generate such knowledge may reduce individual biases, increase the potential for the incorporation of multiple lines of reasoning, and account for potentially incomplete domain knowledge on the part of individuals (McGraw and Seale 1988). However, conducting such group technique KA studies can be difficult; it may be difficult to recruit appropriate experts to participate or to schedule mutually agreeable times and locations for such groups to meet. Furthermore, a forceful or coercive minority of experts or single experts might exert disproportionate influence over the contents of the resulting knowledge collection (Liou 1990).

9.3.1.2 Biases in Logical and Probabilistic Reasoning

In clinical medicine, much of what experts report during knowledge elicitation is inherently uncertain. Although physicians, including experts, have been shown to be poor at the formal estimation of probabilities associated with relationships (Leaper, Horrocks et al. 1972; Berwick, Fineberg, and Weinstein 1981), they will frequently use terms that show that they are managing uncertainty in their approach to problems (e.g., "suggests," "supports," "goes against," "often," "evokes the possibility"). Despite the challenges, many knowledge engineers and psychologists have sought to obtain true probabilities from experts as part of their knowledge elicitation activities. In addition to poor estimation of probabilities by human beings, bias in their probabilistic reasoning has also been well documented (Lichtenstein and Fischoff 1980; Kahneman and Tversky 1982; Tversky and Kahneman 1983), and types of bias have been categorized (Fraser, Smith, and Smith 1992). These bias types include tendencies (a) to allow undue influence of cognitive availability (recency) of information, mistaking this characteristic for frequency, (b) to anchor judgments on initial estimates, (c) to assess the likelihood of an event based on familiarity or stereotypic rather than objective frequency, and (d) to overestimate the frequency of rare events.

Following the demonstrations of Tversky and Kahneman, some researchers speculated that various biases might also be manifest in experts (Fischhoff 1989), and they suggested that knowledge engineers should avoid the use of probabilistic or statistical judgments in knowledge elicitation altogether (Hink and Woods 1987). The work on probabilistic reasoning bias became a red flag, because the notion of uncertainty is crucial in many expert systems (Fox 1986; Kuipers, Moskowitz, and Kassirer 1988; Zadeh and Kacprzuk 1992). For example, in diagnostic domains one may need to formulate such rules as: "If the patient has spots, then the patient has measles with certainty X" (see, for example, the *certainty factor* uncertainty model used in the MYCIN expert system (Shortliffe and Buchanan 1975)). If experts provide biased probability estimates, there could be substantial problems for those building expert systems containing rules that are triggered when particular probability values are in effect for specific variables.

In many applications, statistical judgment and the sorts of judgments involved in decision analysis are contrived in that they can take experts away from their usual way of thinking about problems. However, some investigators have argued that people have little trouble in giving probabilities, and that decision analysis can be used in knowledge elicitation (Fischhoff 1989), where the focus is on improving judgment by making decision processes and judgment criteria explicit. Some researchers have expressed doubt that the biases in probabilistic reasoning that have been observed in laboratory research occur with the same frequency and magnitude in any real-world problem solving situations (Beyth-Marom and Arkes 1983; Christensen-Szalanski and Beach 1984).

Bias in logical reasoning also has been observed in the laboratory, where the problems include:

- A tendency to assign undue weight to the first evidence obtained
- Over-reliance on variables that have taken on extreme values
- The tendency to seek evidence that confirms the current hypothesis
- The tendency to reason about only one or two hypotheses at a time
- The tendency to be overconfident
- The desire to maintain consistency even if that means devaluing or ignoring important information
- Belief in illusory correlations
- The tendency to be overly conservative
- Basing conclusions on hindsight (Johnson-Laird 1983; Evans 1989; Fischhoff 1989; Fraser, Smith, and Smith 1992)

In their studies of medical decision-making, Schwartz and Griffin (1986) cited over 20 relevant papers supposedly demonstrating that experts rely on heuristics. However, they argued that experts do not seem to be prone to biases to such an extent that the concern should have practical import in knowledge elicitation work.

9.3.2 Data Analysis Methods

9.3.2.1 Protocol and Discourse Analysis

The techniques of protocol and discourse analysis are very closely related, and concern themselves with the elicitation of knowledge from individuals while they are engaged in problem-solving or reasoning tasks (i.e., *think-aloud* studies, as mentioned earlier). Such analyses may be performed in order to determine the conceptual entities and relationships between those entities used by individuals while they reason about a problem domain. The basic premises of these techniques are derived from the domains of psychology and cognitive science (Kintsch and Greeno 1985; Groen and Patel 1988; Patel, Arocha, and Kaufman 2001). In this approach, not only are a job's task activities charted, but also problem solvers are instructed to explain what they are doing and thinking while they are performing the task. The think-aloud procedure generates a response protocol, which is a recording of the deliberations that is subsequently transcribed and analyzed for propositional content and semantic content. The process of verbalization typically does not significantly affect the normal course of cognitive processes (Ericsson and Simon 1993), and it can yield information about the reasoning sequences and goal structures in experts' problem solving (Patel and Groen 1991b; Patel and Ramoni 1997).

The think-aloud problem solving/protocol analysis technique has been used extensively in cognitive research on medical expertise (Johnson, Duran et al. 1981; Kuipers and Kassirer 1984; Patel and Groen 1986; Kuipers, Moskowitz, and Kassirer 1988). For example, Kuipers and Kassirer (1984) found that in a routine case, experts tended to produce very sparse protocols that did not provide much basis for characterizing reasoning patterns. The authors suggested that expert knowledge is so compiled that it is difficult to articulate intermediate steps. This led to using clinical probes to elicit constrained information within the think-aloud paradigm (Groen and Patel 1988). Patel, Arocha, and Kaufman (1994) showed that experts interpret clinical data from the first few segments of the patient problem in terms of high-level hypotheses, which later they evaluate. This serves to partition the problem into manageable units, thus reducing the load on working memory. In contrast, experts out of their domain of expertise (subexperts) generate hypotheses mostly at lower levels, and they keep generating new hypotheses instead of evaluating them.

During such protocol analysis studies, the recorded explanations by subjects are codified for analysis at varying levels of granularity (Feltovich, Spiro, and Coulson 1989; Patel and Groen 1991b; Polson, Lewis et al. 1992). Discourse analysis is the process by which an individual's intended meaning within a body of text or some other form of narrative discourse is analyzed into discrete units of thought (propositions) and the subsequent analysis of the contexts in which those units appear (propositional relations in semantic structures), as well as the quantification and description of the relationships existing between those same units (Davidson 1977; Alvarez 2002). The advantage of this approach to conceptual knowledge acquisition is that it situates the overall elicitation process within the broader distributed socio-cognitive context in which individuals perform real-world reasoning and problem solving (Patel, Arocha, and Kaufman 2001; Patel, Kaufman, and Arocha 2002).

9.3.2.2 Concept Analysis

In recent years, some CTA researchers have adopted a technique called concept mapping as a method of both eliciting and representing knowledge (Novak 1990; Crandall, Klein and Hoffman 2006). The modern idea of concept map can be interpreted as a "user-friendly" expression of meaning in a text. Concept maps have been used in many studies of the psychology of expertise, and this work has shown that these maps can support the formation of consensus among experts (Gordon, Schmierer and Jill 1993). Concept maps constructed by domain experts clarify what they wish to express, and they eventually show high levels of agreement (Gordon 1992). In concept mapping knowledge elicitation, the researchers help the domain practitioners build representation on their domain knowledge, merging the activities of knowledge elicitation and representation. This technique also is proven to be useful as a tool for creating knowledge-based performance support systems (Cañas, Coffey et al. 1997; Dorsey, Campbell et al. 1999). Concept maps are labeled node-link structures, like semantic networks described elsewhere in the chapter, but are less formal than the networks based on formal propositional representations.

9.3.2.3 Verification and Validation of Knowledge Acquisition

As mentioned earlier, the process of verification and validation of knowledge is ideally and most effectively applied throughout the entire knowledge engineering spectrum. Therefore, it is important to understand the types of verification and validation metrics and techniques available for use within the specific context of KA. *Verification* is the evaluation of a knowledge-based system to ensure that it satisfies the end-user or domain-specific requirements used to define the design of that system (logical consistency, general notions of completeness, avoidance of redundancy, and the like). For example, Suwa and colleagues developed techniques for analyzing a knowledge base of rules, derived from a knowledge engineering process, to demonstrate that the rules were both complete and consistent (Suwa, Scott, and Shortliffe 1982). *Validation* is the evaluation of a knowledge-based system to ensure that it satisfies the end-user or domain-specific requirements to be realized upon implementation and refinement of that system. An example of a verification measurement for a knowledge-based system would be the concordance between the system's reasoning concerning a given a set of "real world" input data in comparison to the reasoning that would be used by a domain expert assessing the same input data within the same real world context. The MYCIN system (Shortliffe 1976) pioneered these kinds of knowledge base validation experiments (Yu, Buchanan et al. 1979; Yu, Fagan et al. 1979).

To summarize the distinction, verification is the evaluation of whether a knowledge-based system meets the perceived requirements of the end users or application domain, and validation is the evaluation of whether that system meets the realized (e.g., real-world) requirements of the end users or application domain. However, in both instances, similar evaluation metrics may be used. A number of critical verification and validation criteria exist, such as multiple-source or expert agreement, degree of interrelatedness of the knowledge, and consistency of the generated knowledge.

9.3.2.4 Heuristic Methods

The most commonly used approach to evaluating knowledge is the use of heuristic evaluation criteria (Neilsen 1994). The advantage of this approach is the obvious simplicity of the evaluation method (e.g., knowledge engineers or experts may manually review the knowledge generated and determine if its contents are consistent with the heuristics being used for the purpose of evaluation). However, methods for doing this are limited in their tractability when applied to large knowledge sets, since they are difficult, if not impossible, to automate. Furthermore, they make comparison of knowledge "quality" across multiple sets infeasible, because of the qualitative nature of the evaluation results being generated.

9.3.3 Representational Methods

Cognitive task analysis also speaks to representation of interpreted data, rather than just the collection of primary data. CTA techniques generally provide abstract frameworks that assume particular types of knowledge structures as well as underlying reasoning processes.

In representation of verbal data, investigators have made use of two kinds of representational formalisms: propositional representations and semantic networks. Intuitively, a proposition is an idea underlying the surface structure of a text. The notion's usefulness arises from the recognition that a given piece of discourse may have many related ideas embedded within it. A propositional representation provides a means of representing these ideas, and the relationships between them, in an explicit fashion. In addition, it provides a way of classifying and labeling these ideas. Systems of propositional analysis (Kintsch 1974; Frederiksen 1975) are essentially languages that provide a uniform notation and classification for propositional representations. In all these approaches, as in case grammars, a proposition is denoted as a relation (predicate) over a set of arguments (concepts). Sowa's system of conceptual graphs provides another example of a language of this type (Sowa 1984). Although there are notational differences in the formalisms, the underlying assumption is that propositions correspond to the basic units of the representation of discourse and form manageable units of knowledge representation.

The primary challenge is to represent the structure of verbal or written data arising from observations and interviews as well as from think-aloud protocols. The first stage of analysis involves generating a propositional representation of the acquired text. This is then transformed into a semantic network representation. The network consists of propositions that describe attribute characteristics, which form the nodes of the network, and propositions that describe relational information, which form the links.

The primary relations of interest in these networks are binary dependency relations, specifically, *causal, conditional*, and Boolean connectives (*and*, *inclusive or*, and *exclusive or* relations). In addition, algebraic relations (e.g., *greater than*), identifying relations, and categorical relations (i.e., category membership, part-whole relations) can be expressed. One can also distinguish between the *source* of a process and the *result* of a process. Uncertainty in relations can be represented by modal qualifiers (e.g., *can*), and truth values can be indicated when they deviate from the default value (truth with certainty).

A semantic network is a directed graph formed by nodes and labeled connecting paths. Nodes may represent either clinical findings or hypotheses, whereas the paths represent directed connections between such nodes. These networks also provide a relatively precise means for characterizing the directionality of reasoning.

Figure 9-3 shows a semantic structure generated using discourse analysis to understand the implied and explicit knowledge contained in a specific text taken from a think-aloud protocol. The example is based on a diagnostic explanation offered by a psychiatrist when presented with a case from cardiology (Patel, Groen, and Arocha 1990). The case is not within the subject's domain of specialization, and the diagnosis of a *shock state* is inaccurate. Because of this, the representation is lacking in coherence and contains one possible inconsistency; that is, a patient cannot have both high and low blood pressure at the same time. Furthermore, the underlying mechanism that explains the signs and symptoms in this patient is attributed to toxicity of drugs that the patient has injected in an effort to respond to external stress. This is an inaccurate description of the patient's problem.

The diagram consists of nodes linked by arrows. The arrows have labels indicating the relationship between nodes. The two most important are CAU:, which means that the source node causes the target (e.g., *upsurge in blood pressure* causes *flame-shaped hemorrhage*), and COND:, which means that the source node is an indicator of the target (e.g., *tachycardia* indicates *shock state*). The arrows labeled CAU: represent causal relations, and those labeled COND: represent conditional relations. A difference between the two is in the strength of implication: COND: expresses a directional conditionality, $P1 \rightarrow P2$, which implies if proposition $P1$ is true then $P2$ is true. CAU: $P1 \rightarrow P2$, is a stronger relation indicating that one variable, $P2$ is a functional result of another, $P1$.

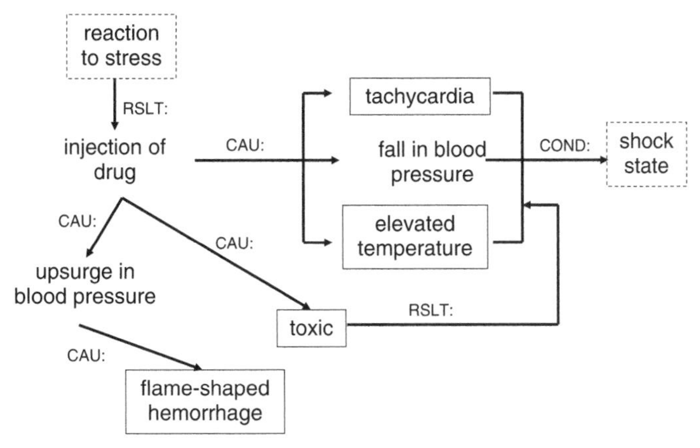

FIGURE 9-3 Semantic analysis of a clinical text. In the diagram, solid rectangles indicate cues from the text, broken lines indicate diagnostic hypotheses, and arrows indicate directionality of relations. COND: = *conditional* relation, CAU: = *causal* relation, RSLT: = *resultive* relation. In this case, the text is taken from an explanation protocol provided by a psychiatrist who had been challenged by a case from the field of cardiology: "*The patient has been reacting to stress likely by his injecting a drug (or drugs), which has resulted in tachycardia, a fall in blood pressure, and elevated temperature. These findings are due to the toxic reaction caused by the injected drugs. He is in or near shock. The flame-shaped hemorrhage may represent a sequel of an upsurge in blood pressure possibly as a result of his injection of drugs.*"

9.4 COMPUTER-BASED KNOWLEDGE ACQUISITION

The knowledge contained in any large-scale decision support system is so extensive and complex that it has become unreasonable to consider managing such knowledge bases manually. As a result, specialized environments have been constructed that allow trained individuals to enter new knowledge, and maintain or "curate" what is already there. Such systems often require structural knowledge of a domain over which the inferential knowledge is overlaid. Today, that structural knowledge, which defines the concepts in a domain and some aspects of the hierarchical relationships among them, is known as an *ontology* of that domain. Knowledge engineers typically begin with the creation of a basic ontology for a field and then build inferential structures and relationships that allow a knowledge system to draw conclusions and generate advice. These knowledge representation issues are discussed in detail in Section IV.

We mention this topic here because there is a continuum in the development of computer systems for knowledge acquisition between those that are used for entering knowledge acquired through another means and those that actually interact with experts to extract, encode, and maintain that knowledge. Among those in the former category is the well-known Protégé system, which supports the creation of ontologies and the encoding of related complex knowledge in a domain (Musen 1992; Tu, Eriksson et al. 1995). But it would be rare to identify clinical experts who would be able to sit down with Protégé and "teach" it what they know about their domains of expertise. Protégé is for programmers and knowledge engineers to use after they have identified the knowledge that needs to be encoded.

The notion of obtaining knowledge directly from experts using an interactive dialog had its roots in the field of artificial intelligence in the early 1970s. For example, Carbonell pioneered the notion of computer-based mixed initiative dialogs, focusing on educational uses but recognizing that an ability to interview and interact with a knowledgeable user had broad implications for computational extraction of knowledge, as well as its conveyance in the educational setting (Carbonell 1970). Heavily influenced by Carbonell's work, Shortliffe later experimented with a mixed-initiative dialog system that would allow physicians to teach the MYCIN program new knowledge (rules) in the domain of infectious disease therapy (Shortliffe 1976). The basic notion was that an expert ought to be able to challenge MYCIN with a new patient case, use its explanation facilities to determine what faulty or missing knowledge explained any errors in performance (see Chapter 18 in Buchanan and Shortliffe (1984)), and then to enter new or corrected rules for MYCIN to incorporate into its knowledge base. Consider, for example, Figure 9-4, in which an expert attempts to enter a new rule using natural language methods. In this example, the program has incorrectly interpreted the expert's statement regarding the gastrointestinal (GI) track. The physician has therefore indicated that the program's understanding of the rule is not yet correct. The completion of the session, leading to correct interpretation of the rule, is shown in Figure 9-5.

MYCIN's early foray into knowledge acquisition was later extensively expanded and enhanced by Davis (1979). It was he who coined the phrase "interactive transfer of expertise" to describe the notion of an expert interacting with an intuitive, natural interface implemented in a computer program to "teach" the machine about his or her knowledge of a domain. Davis's system, which generalized the methods beyond the MYCIN domain, was known as

** **NR** *(the expert physician indicates that he would like to enter a New Rule)*
The new rule will be called RULE200

IF: 1** **ORGANISM IS A GRAM NEGATIVE ROD**
and 2** **IT IS ANAEROBIC**
and 3** **IT ENTERED THE BLOOD FROM THE GI TRACK**
and 4** **<CR>** *(the physician enters a carriage return to indicate completion)*
THEN 1** **IT IS PROBABLY A BACTEROIDES**
On a scale of 1 to 10, how much certainty would you affix to this conclusion?
** **9**
And 2** **<CR>**

This is my understanding of your rule:
RULE200

IF: 1) The site of the culture is blood, and
2) The site of the culture is GI, and
3) The stain of the organism is gram neg, and
4) The morphology of the organism is rod, and
5) The aerobicity of the organism is anaerobic
THEN: There is strongly suggestive evidence (.9) that the identity of the organism is bacteroides
Okay? (YES or NO)
** **NO**

FIGURE 9-4 A mixed-initiative knowledge-acquisition dialog between MYCIN and an infectious disease expert. The physician's entries are indicated by bolded text in capital letters, with MYCIN's responses in mixed upper- and lowercase. Comments are italicized.

Teiresias. The program's power came in part from the creation of a notion known as *rule models*, metalevel descriptions of the knowledge categories and relationships that characterized the domain in question and guided the interactive knowledge acquisition process. A full dialog with Teiresias would

Change PREMISE? (YES or NO)
** **YES**
Want to see the current version? (YES or NO)
** **NO**
Change (C), Delete (D), or Add (A) a component:
** **C**
Which component? (enter number):
** **2**
Enter the new component:
** **PORTAL OF ENTRY IS THE GI TRACK**
More changes to this PREMISE? (YES or NO)
** **NO**
Change rule's conclusion? (YES or NO)
** **NO**

This is my understanding of your rule:
RULE200

FIGURE 9-5 In this continuation of the session from Figure 9-4, the expert indicates what corrections need to be made and is able to verify that the revised rule is what was intended.

be too detailed for this volume, but several examples can be found in Chapter 9 of Buchanan and Shortliffe (1984) and in Davis's article, which shows how the approach could be used in a totally different domain (Davis 1979).

Although the performance of these early programs was promising, the complexity of their creation, maintenance, and use made it difficult to get experts to work with them directly. They much preferred to work with knowledge engineers and psychologists who used the knowledge elicitation techniques we have previously described. Thus, in the 1980s, there was a gradual move toward creating powerful knowledge authoring and editing tools that could be used by knowledge engineers *after* they had elicited the pertinent knowledge from human experts. Graphical user interfaces, unavailable in the 1970s when MYCIN and Teiresias were created, encouraged the adaptation of visual programming concepts for use in knowledge base construction and maintenance. One of the earliest efforts was Musen's creation of OPAL, a graphical authoring environment for entering and maintaining cancer chemotherapy research protocols that were used by ONCOCIN to guide oncologists in the treatment of cancer patients (Shortliffe, Scott et al. 1981; Musen, Combs et al. 1988). OPAL later was generalized to be used for knowledge entry and editing in any domain, and this led to the creation of Protégé, which is today heavily used for knowledge base (especially ontology) construction and maintenance (Gennari, Musen et al. 2003).

Today, although experiments continue, it is a rare knowledge elicitation tool that is designed and successfully implemented for use directly by physicians or other clinical experts. We instead see continued emphasis on the specialized skills of individuals who know the computational systems but who also have the interpersonal skills, and ability to learn what is often a new domain, in order to work closely with experts, and groups of individuals, in order to elicit the knowledge that is needed for medical decision support.

9.5 CONCLUSION

In the modern world, knowledge management has become a major focus of activity in diverse businesses, including health care. Because of the effort required to develop and validate such knowledge, there is growing recognition of the need to share knowledge components when they are developed and optimally to involve experts in providing, assessing, and maintaining knowledge that is needed. Although we are creating large institutional, local, regional, and national databases, only some of the knowledge that we require to inform practice and policy can be derived solely by analyzing those data or the literature (see Chapters 10 and 11). Many areas of clinical endeavor still depend heavily on the kind of judgmental knowledge and experience that is difficult to acquire from anyone other than those who have the wisdom and efficiency that comes with experience and lifelong learning. Thus, despite the formal analytical methods that are appropriately being used to make sure that we learn as much as we can from our accumulated experience stored in pooled databases and in the literature, knowledge elicitation from experts, and groups of experts, will continue to be a crucial component of knowledge creation and management for medical decision support. The early promise of computer-based transfer of expertise to knowledge systems has not been borne out, although significant research opportunities and potential continue to exist, and may be facilitated by our increasing knowledge of human problem-solving methods and by enabling improvements in technology. For now, however, it is the direct interaction among experts, and between experts and knowledge engineers, that will serve a crucial role in assuring the development of high quality and accepted knowledge bases that in turn enable the development and effective use of decision support systems.

REFERENCES

Achour, S. L., Dojat, M. et al. (2001). A UMLS-based knowledge acquisition tool for rule-based clinical decision support system development. *J Am Med Inform Assoc* **8**(4): 351–360.

Adams, L., Toomey, L., and Churchill, E. (1999). Distributed research teams: Meeting asynchronously in virtual space. *Journal of Computer Mediated Communication* **4**(4).

Adelman, L. (1989). Management issues in knowledge elicitation. *IEEE Transactions on Systems, Man, and Cybernetics* **19**: 448–461.

Allard, F. and Starkes, J. L. (1991). Motor skill experts in sports, dance and other domains. In *Toward a General Theory of Expertise: Prospects and Limits*, A. Ericsson and J. Smith, eds., 126–152. New York: Cambridge University Press.

Alvarez, R. (2002). *Discourse Analysis of Requirements and Knowledge Elicitation Interviews*, IEEE Computer Society.

Bell, D. S., Pattison-Gordon, E., and Greenes, R. A. (1994). Experiments in concept modeling for radiographic image reports. *J Am Med Inform Assoc* **1**(3): 249–262.

Berwick, D., Fineberg, H., and Weinstein, M. (1981). When doctors meet numbers. *American Journal of Medicine* **71**: 991–996.

Beyth-Marom, R. and Arkes, H. R. (1983). Being accurate but not necessarily Bayesian: Comments on Christensen-Szalanski and Beach. *Organizational Behavior and Human Decision Processes* **31**: 255–257.

Boy, G. A. (1997). The group elicitation method for participatory design and usability testing. *Interactions* **4**(2): 27–33.

Buchanan, B. G. and Shortliffe, E. H., eds. (1984). *Rule-Based Expert Systems: The MYCIN Experiments of the Stanford Heuristic Programming Project*. Reading, MA: Addison-Wesley.

Cañas, A. J., Coffey, J. et al. (1997). EI-Tech: A performance support system with embedded training for electronics technicians. *Proceedings of the 11th Florida Artificial Intelligence Research Symposium*, Sanibel Island, FL.

Carbonell, J. R. (1970). AI in CAI: An articial intelligence approach to computer-assisted instruction. *IEEE Transactions on Man-Machine Systems* MMS **11**: 190–202.

Chapman, G. B. and Elstein, A. S. (2000). Cognitive processes and biases in medical decision making. In *Decision Making in Health Care: Theory, Psychology, and Applications*, G. B. Chapman and F. A. Sonnenberg, eds., 183–210. New York: Cambridge University Press.

Chi, M. T. H., Feltovich, P. J., and Glaser, R. (1981). Categorization and representation of physics problems by experts and novices. *Cognitive Science* **5**: 121–152.

Chi, M. T. H., Glaser, R., and Farr, M. J. (1988). *The nature of expertise*. Hillsdale, NJ: Lawrence Erlbaum.

Christensen-Szalanski, J. J. and Beach, L. R. (1984). The citation bias: Fad and fashion in the judgment and decision making literature. *American Psychologist* **39**: 75–78.

Cohen, T., Blatter, B., et al. (2006). Distributed cognition in the psychiatric emergency department: A cognitive blueprint of a collaboration in context. *Artificial Intelligence in Medicine* **37**(2): 73–83.

Crandall, B., Klein, G., and Hoffman, R. R. (2006). *Working minds: A practitioner's guide to cognitive task analysis*. Cambridge, MA: MIT Press.

Davidson, J. E. (1977). *Topics in Discourse Analysis*. University of British Columbia.

Davis, R. (1979). Interactive transfer of expertise: Acquisition of new inference rules. *Artif Intell* **12**: 121–158.

deGroot, A. D. (1965). *Thought and Choice in Chess*. The Hague, The Netherlands: Mouton.

Delbecq, A. L., Van de Ven, A. H., and Gustafon, D.H. (1986). *Group Techniques for Program Planning: A Guide to Nominal Group and Delphi Processes*. Middleton, Wisconsin: Green Briar Press.

Dorsey, D. W., Campbell, G. E. et al. (1999). Assessing knowledge structures: Relations with experience and post-training performance. *Human Performance* **12**: 33–57.

Ericsson, A. (1996). *The Road to Excellence: The Acquisition of Expert Performance in the Arts and Sciences, Sports and Games.* Hillsdale, NJ: Lawrence Erlbaum Publishers.
Ericsson, K. A. and Simon, H. A. (1993). *Protocol Analysis: Verbal Reports as Data.* Cambridge, MA: MIT Press.
Ericsson, Y. A. and Smith, J. (1991). *Toward a general theory of expertise: Prospects and limits.* New York: Cambridge University Press.
Evans, D. A., Cimino, J. J. et al. (1994). Toward a medical-concept representation language. The Canon Group. *J Am Med Inform Assoc* **1**(3): 207–217.
Evans, J. (1989). *Bias in Human Reasoning: Causes and Consequences.* Hillsdale, NJ: Erlbaum.
Feigenbaum, E. A., Buchanan, B. G., and Lederberg, J. (1971). On generality and problem solving: A case study using the DENDRAL program. In *Machine Intelligence*, B. Meltzer and D. Michie, eds., 165–190. Edinburgh: Edinburgh University Press.
Feltovich, P. J., Ford, K. M., and Hoffman, R. R. (1997). *Expertise in Context.* Menlo Park, CA: AAAI/MIT Press.
Feltovich, P. J., Spiro, R. J., and Coulson, R. L. (1989). The nature of conceptual understanding in biomedicine: The deep structure of complex ideas and the development of misconceptions. In *Cognitive science in medicine*, D. Evans and V. Patel, eds. Cambridge, MA: MIT Press.
Fischhoff, B. (1989). Eliciting knowledge for analytical representation. *IEEE Transactions on Systems, Man, and Cybernetics* **19**: 448–461.
Fox, J. (1986). Knowledge, decision making, and uncertainty. In *Artificial Intelligence and Statistics*, W. A. Gale, ed., 57–76. Reading, MA: Addison-Wesley.
Fraser, J. M., Smith, P. J., and Smith, J. W. (1992). A catalog of errors. *Int J Man-Mach Stud* **37**: 265–307.
Frederiksen, C. H. (1975). Representing logical and semantic structure of knowledge acquired from discourse. *Cognitive Psychology* **7**: 371–458.
Gennari, J. H., Musen, M. A. et al. (2003). The evolution of Protégé: An environment for knowledge-based systems development. *Int J Hum-Comp Stud* **58**(1): 89–123.
Glaser, R. (1996). Changing the agency for learning: Acquiring expert performance. In *The Road to Excellence: The Acquisition of Expert Performance in the Arts and Sciences, Sports and Games*, K. A. Ericsson, ed. Hillsdale, NJ: Lawrence Erlbaum.
Glaser, R. and Chi, M. (1988). Overview. In *The nature of expertise*, M. Chi, R. Glaser, and M. Farr, eds., xv–xxviii. Hillsdale, NJ: Erlbaum.
Gordon, S. E. (1992). Implications of cognitive theory for knowledge acquisition. In *The Psychology of Expertise: Cognitive Research and Empirical AI*, R. R. Hoffman, ed., 99–120. New York: Springer Verlag.
Gordon, S. E., Schmierer, K. A., and Jill, R. T. (1993). Conceptual graph analysis: Knowledge acquisition for instructional system design. *Journal of Human Factors* **35**(3): 459–481.
Groen, G. J. and Patel, V. L. (1988). The relationship between comprehension and reasoning in medical expertise. In *The Nature of Expertise*, M. Chi, R. Glaser, and M. Farr, eds., 287–310. Hillsdale, NJ: Lawrence Erlbaum.
Hawkins, D. (1983). An analysis of expert thinking. *Int J Man-Mach Stud* **18**(1): 1–47.
Hink, R. F. and Woods, D. L. (1987). How humans process uncertain knowledge. *The AI Magazine* **8**: 41–53.
Hoffman, R. R., Shadbolt, N. et al. (1995). Eliciting knowledge from experts: A methodological analysis. *Organizational Behavior and Human Decision Processes* **62**: 129–158.
Hughes, J., King, V. et al. (1995). The role of ethnography in interactive systems design. *Interactions* **2**(2): 56–65.
Iqbal, R., Gatward, R., and James, A. (2005). A general approach to ethnographic analysis for systems design. *Proceedings of the 23rd Annual International Conference on Design of Communication: Documenting & Designing for Pervasive Information*, 34–40, Coventry, United Kingdom: ACM Press.
Johnson-Laird, P. N. (1983). *Mental Models: Towards a Cognitive Science of Language, Inference, and Consciousness.* Cambridge, MA: Harvard University Press.
Johnson, P. E., Duran, A. S. et al. (1981). Expertise and error in diagnostic reasoning. *Cognitive Science* **5**: 235–283.
Jones, J. and Hunter, D. (1999). Using the Delphi and nominal group technique in health services research. In *Qualitative Research in Health Care*, N. Mays and C. Pope, eds. London: BMJ Books.
Kahneman, D. and Tversky, A. (1982). On the study of statistical intuitions. *Cognition* **11**: 123–141.

Kintsch, W. (1974). *The Representation of Meaning in Memory*. Hillsdale, NJ: Lawrence Erlbaum.

Kintsch, W. and Greeno, J. G. (1985). Understanding and solving word arithmetic problems. *Psychological Review* **92**(1): 109–129.

Kuipers, B., Moskowitz, A. J., and Kassirer, J. P. (1988). Critical decisions under uncertainty. *Cognitive Science* **12**: 177–210.

Kuipers, B. J. and Kassirer, J. P. (1984). Causal reasoning in medicine: Analysis of a protocol. *Cognitive Science* **8**: 363–385.

Larkin, J. H., McDermott, J. et al. (1980). Expert and novice performances in solving physics problems. *Science* **208**: 1335–1342.

Laxmisan, A., Malhotra, S. et al. (2005). Decisions about critical events in device-related scenarios as a function of expertise. *Journal of Biomedical Informatics* **38**(3): 200–212.

Leaper, D., Horrocks, J. et al. (1972). Computer-assisted diagnosis of abdominal pain using estimates provided by clinicians. *British Medical Journal* **4**: 350–354.

Lichtenstein, S. and Fischoff, B. (1980). Training for calibration. *Organizational Behavior and Human Performance* **26**: 49–171.

Liou, Y. I. (1990). Knowledge acquisition: Issues, techniques, and methodology. *Proceedings of the 1990 ACM SIGBDP Conference on Trends and Directions in Expert Systems*, 212–236. Orlando, Florida: ACM Press.

Liszka, A., Stubblefield, W., and Kleban, S. (2003). GMS: Preserving multiple expert voices in scientific knowledge management. *Proceedings of the 2003 Conference on Designing for User Experiences*, 1–4. San Francisco, California: ACM Press.

McGraw, K. L. and Seale, M. R. (1988). Knowledge elicitation with multiple experts: Considerations and techniques. *Artificial Intelligence Review* **2**: 31–44.

Morgan, M. and Martz, W. (2004). Group consensus: Do we know it when we see it? *Proceedings of the Proceedings of the 37th Annual Hawaii International Conference on System Sciences (HICSS '04)—Track 1, Volume 1*, IEEE Computer Society.

Musen, M. A. (1992). Dimensions of knowledge sharing and reuse. *Comput Biomed Res* **25**(5): 435–467.

Musen, M. A., Combs, C. M. et al. (1988). OPAL: Toward the computer-aided design of oncology advice systems. In *Selected Topics in Medical Artificial Intelligence*, P. L. Miller, ed., 167–180. New York: Springer-Verlag.

Neilsen, J. (1994). *Usability Engineering*. Boston: Academic Press Professional.

Novak, J. D. (1990). Concept maps and vee diagrams: Two metacognitive tools for mathematics and science education. *Instructional Science* **19**: 29–52.

Osborn, A. (1953). *Applied Imagination: Principles and Procedures of Creative Thinking*. New York: Scribner.

Patel, V. L., Allen, V. G. et al. (1998). Representing clinical guidelines in GLIF: Individual and collaborative expertise. *J Am Med Inform Assoc* **5**(5): 467–483.

Patel, V. L., Arocha, J. F., and Kaufman, D. R. (1994). Diagnostic reasoning and expertise. *Psychology of Learning and Motivation: Advances in Research and Theory* **31**: 137–252.

Patel, V. L., Arocha, J. F., and Kaufman, D. R. (2001). A primer on aspects of cognition for medical informatics. *J Am Med Inform Assoc* **8**(4): 324–343.

Patel, V. L. and Groen, G. J. (1986). Knowledge-based solution strategies in medical reasoning. *Cognitive Science* **10**: 91–116.

Patel, V. L. and Groen, G. J. (1991a). Developmental accounts of the transition from medical student to doctor: Some problems and suggestions. *Medical Education* **25**(6): 527–535.

Patel, V. L. and Groen, G. J. (1991b). The general and specific nature of medical expertise: A critical look. In *Toward a General Theory of Expertise: Prospects and Limits*, A. Ericsson and J. Smith, eds., 93–125. New York: Cambridge University Press.

Patel, V. L., Groen, G. J., and Arocha, J. F. (1990). Medical expertise as a function of task difficulty. *Memory and Cognition* **18**(4): 394–406.

Patel, V. L., Kaufman, D. R., and Arocha, J. F. (2002). Emerging paradigms of cognition in medical decision-making. *Journal of Biomedical Informatics* **35**(1): 52–75.

Patel, V. L. and Ramoni, M. (1997). Cognitive models of directional inference in expert medical reasoning. In *Human & Machine Cognition*, K. Ford, P. Feltovich, and R. Hoffman, eds., 67–99. Hillsdale, NJ: Lawrence Erlbaum Associates.

Payne, P. R. and Starren, J. B. (2005). Quantifying visual similarity in clinical iconic graphics. *J Am Med Inform Assoc* **12**(3): 338–345.

Peleg, M., Boxwala, A. A. et al. (2004). The InterMed approach to sharable computer-interpretable guidelines: A review. *J Am Med Inform Assoc* **11**(1): 1–10.

Peleg, M., Gutnik, L. A. et al. (2006). Interpreting procedures from descriptive guidelines. *J Biomed Inform* **39**(2): 184–195.

Polson, P. G., Lewis, C. et al. (1992). Cognitive walkthroughs: A method for theory-based evaluation of user interfaces. *Int J Man-Mach Stud* **36**(5): 741–773.

Schwartz, S. and Griffin, T. (1986). *Medical Thinking: The Psychology of Medical Judgment and Decision Making*. New York: Springer Verlag.

Scott, A. C., Clayton, J. E., and Gibson, E. L. (1991). *A Practical Guide to Knowledge Acquisition*. Reading, MA: Addison-Wesley.

Shortliffe, E. H. (1976). *Computer-Based Medical Consultations: MYCIN*. New York: Elsevier/North Holland.

Shortliffe, E. H. and Buchanan, B. G. (1975). A model of inexact reasoning in medicine. *Math. Biosci.* **23**: 351–379.

Shortliffe, E. H., Scott, A. C. et al. (1981). An expert system for oncology protocol management. *7th International Joint Conference on Artificial Intelligence*, 876–881. Vancouver, BC.

Sloboda, J. (1991). Musical expertise. In *Toward a General Theory of Expertise: Prospects and Limits*, K. A. Ericsson and J. Smith, eds., 153–171. New York: Cambridge University Press.

Sowa, J. F. (1984). *Conceptual Structures*. Reading, MA: Addison Wesley.

Suwa, M., Scott, A. C., and Shortliffe, E. H. (1982). An approach to verifying completeness and consistency in a rule-based expert system. *AI Magazine* 3(4): 24–27.

Tu, S., Eriksson, H. et al. (1995). Ontology-based configuration of problem-solving methods and generation of knowledge-acquisition tools: Application of PROTEGE-II to protocol-based decision support. *Artif Intell Med* 7(3): 257–289.

Tversky, A. and Kahneman, D. (1983). Extensional versus intuitive reasoning: The conjunction fallacy in probability judgment. *Psychological Review* **4**: 293–315.

van der Maas, A., ter Hofstede, A., and ten Hoopen, A. (2001). Requirements for medical modeling languages. *J Am Med Inform Assoc* **8**(2): 146–162.

Wood, W. C., and Roth, R. M. (1990). A workshop approach to acquiring knowledge from single and multiple experts. *SIGBDP '90: Proceedings of the 1990 ACM SIGBDP Conference on Trends and Directions in Expert Systems*, 275–300. Orlando, Florida: ACM Press.

Yu, V., Buchanan, B. G. et al. (1979). Evaluating the performance of a computer-based consultant. *Comput Prog Biomed* **9**: 95–102.

Yu, V., Fagan, L. M. et al. (1979). Antimicrobial selection by a computer: A blinded evaluation by infectious disease experts. *JAMA* **242**: 1279–1282.

Zadeh, L. A. and Kacprzuk, J. (1992). *Fuzzy Logic for the Management of Uncertainty*. New York, Wiley.

10 GENERATION OF KNOWLEDGE FOR CLINICAL DECISION SUPPORT: STATISTICAL AND MACHINE LEARNING TECHNIQUES

MICHAEL E. MATHENY and LUCILA OHNO-MACHADO

10.1 INTRODUCTION

Clinical decision support (CDS) systems must rely on knowledge that originates from a variety of sources. However, selecting the sources and integrating this knowledge into a functional system are not trivial tasks. In early CDS systems, knowledge acquired directly from medical experts was encoded in the form of rules that were triggered and chained according to an embedded or an external inference engine (Shortliffe 1976). As discussed in Chapters 1 and 9, some examples include early systems from the 1970s and early 1980s, such as MYCIN (Shortliffe et al. 1975), Internist/QMR (Miller et al. 1982, 1986), and more recent ones, such as DXplain (Barnett et al. 1987). MYCIN was composed of expert-derived rules with associated certainty factors, whereas QMR and DXplain were based on a set of physician-based assessments of (a) the strength with which clinical findings evoke a certain diagnosis, (b) prevalence of diseases, and (c) related indices. As noted by Heckerman and colleagues, the formal mathematical definitions of these indices in terms of probabilities have not been fully elucidated (Heckerman and Miller 1986).

Even when newer knowledge representation strategies such as Bayesian networks (Pearl 1988) were proposed by some researchers in the late 1980s (Beinlich et al. 1989; Heckerman 1990), definition of the graph structure and probabilities involved in the model were usually still assessed by experts. For example, Swhe and colleagues (1991) "translated" the QMR representation into Bayesian networks, and showed that the same diagnostic quality could be achieved with a representation that made explicit important modeling assumptions. However, the popularity of Bayesian networks in the medical community did not grow as expected, and this type of model still is used primarily in the medical domain for research purposes, with very few exceptions.

Reasons for this limitation may include the need for severe model simplifications in order to make these models practical for clinical use. These simplifications in turn may reduce the main advantages of using Bayesian networks, which is their explicit knowledge representation combining a sound probabilistic modeling of dependencies with a visually appealing display. Algorithms for learning Bayesian networks from data have evolved in the past two decades but also are used primarily in research applications (Cooper and Herskovits 1992; Buntine 1996; Moore and Lee 1998).

Most clinical decision support systems in current use do not learn from data and still rely on the rule-based paradigm. There are at least two factors that contribute to this predominant reliance on expert assessments for the construction of CDS systems:

- Data are simply not available or not structured enough to allow knowledge to be "learned" from them.
- Techniques to discover patterns from data are not well disseminated or not well evaluated in the biomedical community.

A potential third factor may be that systems derived from human knowledge in which nonprobabilistic rules are defined by experts may be more clearly understandable by clinicians. For example, if an expert can articulate all the rules that were used to make a diagnosis and how they were chained, then a system based on these rules can potentially explain its reasoning in a way that clinicians would be more familiar with (Clancey 1983). Whether understanding and agreement by clinicians is necessary for the underlying logic in CDS systems remains a controversial issue. Decision support applications currently in use in clinical environments largely rely on rules for their "logic" (see Section II of this book), but this should not necessarily mean that other approaches are not as good or perhaps even better. For domains in which structured data are abundant and the decisions are made at times in which a snapshot of these data could help identify specific patterns, pattern recognition algorithms from the fields of statistical and machine learning can be of great value.

There have been extensive new developments in statistical and machine learning research in the past few decades. These advances have coincided with improvements in data quality and quantity from the implementation of large repositories of structured electronic data, some of which are based on domain-specific data element standards (Pollock et al. 1998; Cannon et al. 2001; Wattigney et al. 2003). Increased availability of data has allowed further development of several models that can detect patterns in biomedical data and generalize well to previously unseen cases. Clinical decision support systems that utilize statistical or machine learning techniques are now available in virtually every medical specialty (Goldman et al. 1982; Knaus et al. 1985; Baxt 1991; O'Leary et al. 1998; Grundy et al. 1999; Shaw et al. 2002). Just as in the foregoing discussion relating rule-based systems and more sophisticated knowledge representation paradigms, simple understandable models (e.g., linear and logistic regression, score systems) have far outweighed in number and utilization the more sophisticated machine learning models (e.g., support vector machines, neural networks, and recursive partitioning algorithms), many of which remain limited to research applications.

In this chapter, we will review the methodologies of the most commonly used diagnostic and prognostic models for deriving knowledge from data in the medical domain, and discuss specific strengths and weaknesses of alternative modeling methods. Popular examples of some modeling methods will be discussed. Since our focus is on models that have been utilized in practice, the discussion will concentrate on logistic regression models. We conclude with a discussion on current directions for the field.

Note the absence of sections dedicated to topics that have received wide coverage in the computer science literature, but that in fact have limited representation in biomedical informatics applications and are beyond the scope of an introductory chapter. For example, although rule-induction algorithms and kernel-based classifiers such as support vector machines (Boser et al. 1992) often have been utilized in research applications, few actual applications are used in medical practice and therefore we elected not to cover these models in this chapter. Refer to statistical and machine learning textbooks for a review of these topics (Duda et al. 2001; Hastie et al. 2001).

Another omission is the discussion of optimization techniques such as genetic algorithms and evolutionary computing (Koza 1992), and formalism extensions such as fuzzy logic (Zadeh 1994) and rough sets (Pawlak 1982). Elements of these techniques can be used in conjunction with classifiers discussed here in a number of different ways, but they do not constitute classifiers themselves. Furthermore, there are no examples of practical use of these techniques in clinical decision support.

10.2 LEARNING FROM DATA

Statistical and machine learning pattern recognition algorithms have been in existence for several decades. These algorithms recognize regularities in data and construct a model that can be utilized in new cases. Interest in these types of methods has increased in the past two decades, with the addition of new algorithms such as neural networks and support vector machines (Vapnik 1995). A myriad of publications in the scientific and lay literature can now be found under the rubric of "data mining." Data mining techniques are pattern recognition techniques intended to find correlations and relationships in the plethora of data. The term is intriguing, but also somewhat misleading. Most pattern recognition or predictive models used in clinical domains are confirmatory rather than exploratory in nature. The distinction between unsupervised and supervised learning models is directly related to this issue.

Unsupervised learning models are not based on predefined classifications, and are used frequently for exploratory data analyses in domains in which knowledge is sparse. For example, high-throughput micro-array data are often subject to unsupervised learning modeling so that "clusters" of variables or objects can be revealed without guidance from the users or the existing literature. The objective is to unveil hidden patterns in the data that were not previously anticipated, and label these patterns *a posteriori*. This is in sharp contrast with supervised learning models, in which the objective is to determine how to best classify objects with predefined labels representing classes of interest (e.g., malignant versus benign cases) using the data at hand. As expected, unsupervised learning models have not been applied in clinical decision support systems and have a limited role in this area. All models in current use for clinical decision support have been based either on expert knowledge or supervised learning models. The latter is the subject of this chapter.

In order to understand how a model can be derived from data, it is useful to construct an artificial example. Suppose a researcher did not know the range of normal values for a new diagnostic test, but she did have a large data set indicating, for a set of patients, the value of the test and the actual diagnosis for each patient. Also suppose that there are missing and noisy data in the data set. The task is to determine the range of normal values for the test, so that when anyone examines the value for a new patient, it would be possible to declare, with a certain level of confidence, whether the result pointed to an abnormality or not. Although one might not need a sophisticated model to answer this simple question, it would be necessary to review all labeled data to determine optimal thresholds to label a result as "normal" or "abnormal."

This analysis can extend to several tests and clinical findings, and multiple possible diagnoses, in which case the task is to find optimal combinations of values that are most frequently associated with particular diagnoses, since a single test or clinical finding in isolation may not suffice. Researchers would have to examine several thousands of records containing dozens of attributes for each patient to determine which combinations of variable values seemed most likely to be associated with each diagnosis. Given time and memory limitations, it might be difficult to build this type of classifier. For this type of problem, utilizing multivariate techniques that "learn" from data can be very helpful.

10.3 OVERVIEW OF LOGISTIC REGRESSION

The first step toward the construction of a predictive model is the selection of which variables are going to be considered from a data set containing large numbers of cases. The number of cases needs to exceed the number of variables; a well-known heuristic is that the number of variables utilized in a model should not exceed one-tenth of the number of cases. The type of modeling technique also needs to be selected. Logistic regression is by far the most popular method for constructing predictive models in medicine (Lemeshow and Le Gall 1994). This type of classification model usually deals with binary outcomes such as diagnosis of a certain disease or condition (e.g., myocardial infarction), or prognosis within a certain period of time (e.g., death while in hospital). Using a large number of training cases, it is possible to estimate the parameters of a logistic regression model with a certain level of confidence and estimate the future performance of the model in previously unseen cases. The level of confidence will depend on the number and quality of cases (e.g., presence of outliers, noise), as well as how well the model fits the training data.

The logistic function links i predictors, or independent variables, each denoted by x_i and collectively represented by the vector $\mathbf{x}$, to the dependent variable being predicted, represented by Y using the logistic function as in the equation:

$$Y = \frac{1}{1 + e^{-(\beta \mathbf{x} + c)}}$$

This function tries to model a step function with two possible values for Y, and it is therefore used to classify binary outcomes. Figure 10-1 illustrates a logistic regression model and also a class of models known as artificial neural

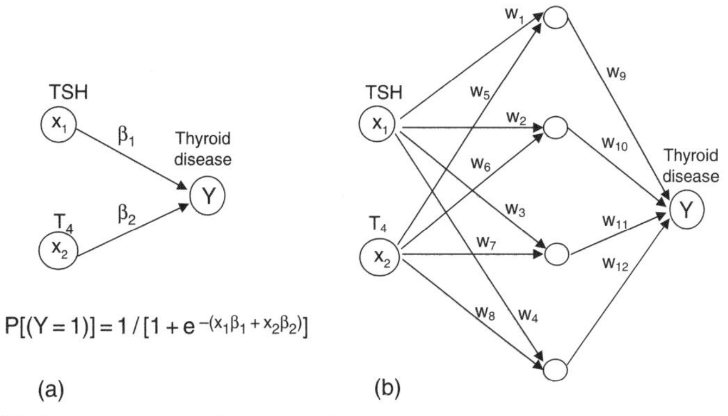

FIGURE 10-1 (a) Example of a simple bivariate logistic regression model (no intercept is included for simplicity). (b) Example of an artificial neural network constructed for the same purpose.

networks, which we will describe in subsection 10.4. In most models, Y is a binary variable representing patient status as having a certain disease or condition ($Y = 1$) or not ($Y = 0$), or prognostic class, and the vector $\mathbf{x}$ represents the clinical, laboratory, and demographic predictors (e.g., x_1 may represent *age*, x_2 may represent *TSH*, and so on). The vector $\boldsymbol{\beta}$ represents the coefficients that are estimated for each predictor and c is a constant. The parameters of the logistic function usually are obtained by maximum likelihood estimation using iterative algorithms (Hosmer and Lemeshow 1989). The coefficients correspond directly to the log of the odds ratio associated with each variable. The parameter c calibrates the model for the baseline rate of the outcome of interest. These features make the model somewhat easy to interpret, as the sign and magnitude of the coefficients (when standardized) may provide direct indication of how much each particular predictor is associated with an increased risk of a certain outcome (e.g., large positive coefficients will usually increase the probability of $Y = 1$ for variables such as those representing most laboratory assays).

In certain datasets, predictors may need to be combined in interaction terms or transformed so that a good fit to the data can be obtained. Consider the example in Figure 10-2: a laboratory test value that is considered normal if within a certain range (e.g., *TSH* within $0.4-6\,\mu$U/ml), and abnormal otherwise. Even in this simple univariate problem of classifying the values into normal and abnormal, a logistic regression model in which variables are not transformed will not be able to correctly classify all cases, even in the absence of noise. The reason is simple: the logistic regression function is monotonic and would necessarily classify a portion of the abnormal cases (either the low or high values) as being normal. However, a simple transformation of the variable such as by means of the quadratic shown in the figure might allow the logistic regression model to correctly classify all cases.

Figure 10-3 illustrates a bivariate problem in which values for two laboratory tests have to be within a certain range for the patient to be considered healthy. In this example, both free T_4 and *TSH* need to be within normal limits for the classification "euthyroid" to be made. It is easy to see that no single line would separate the shaded area from the rest, which means that no linear model can produce correct classifications for all cases. Variable transformations or interaction terms are necessary.

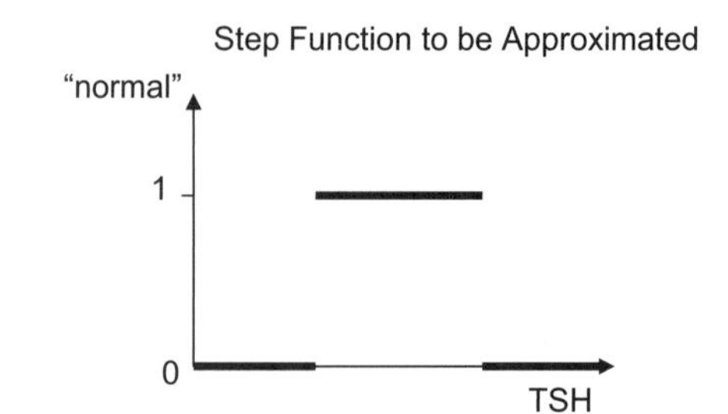

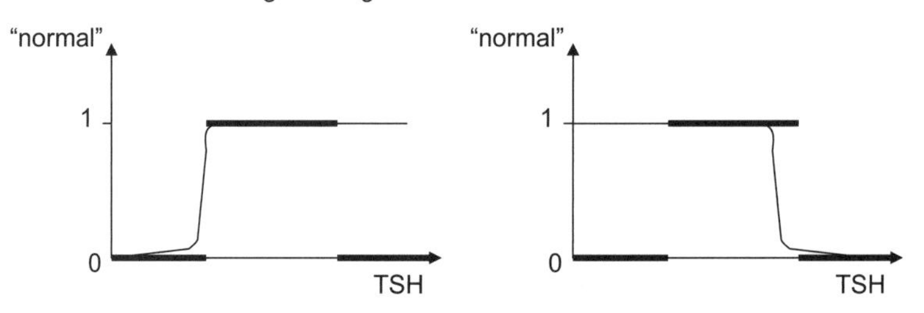

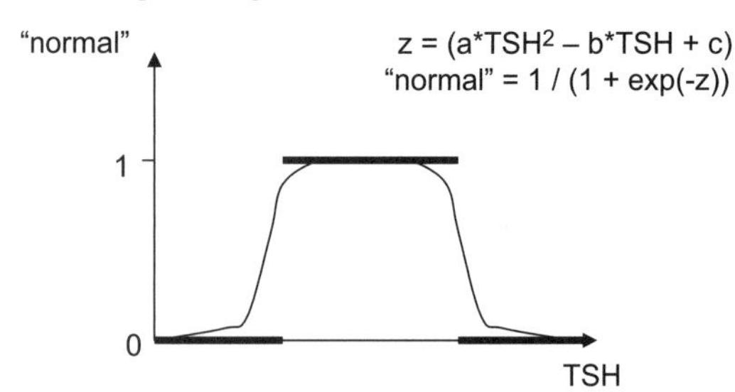

FIGURE 10-2 A step function (bold) indicating "normal" laboratory values within a certain range. The step function is overlaid with logistic functions for illustration purposes. Without variable transformation, the logistic regression function will always miss one of the extremes of values, misclassifying values within that range as not "normal." A simple quadratic transformation can make logistic regression work for this example.

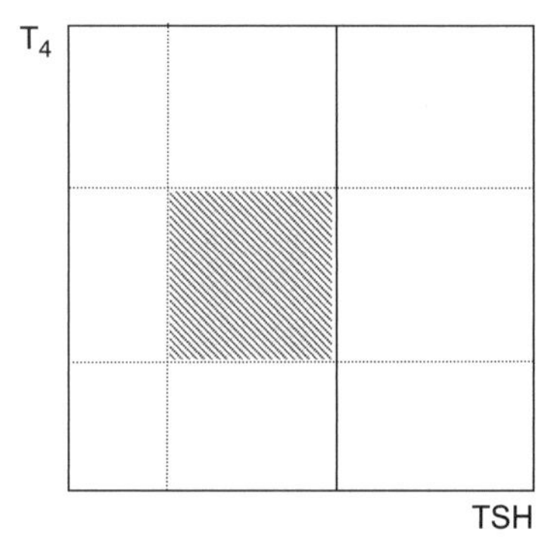

FIGURE 10-3 Simplified bivariate example. For a case to be considered "euthyroid" (shaded area), values for both tests have to be within a certain range. Without variable transformation, logistic regression will not work for all cases because the problem is not linearly separable.

Problems that are not solvable by linear or semilinear models such as logistic regression without variable transformations or addition of interaction terms are known as nonlinearly separable problems. Although logistic regression models can be used in nonlinearly separable problems, predetermining which transformations or interactions are necessary is a laborious ad-hoc process that is computationally intractable if the number of variables is not very small. Furthermore, the interpretation of a model that uses transformed variables or interaction terms is difficult. Therefore, most models that are used in practice do not make use of interaction or transformed terms.

Techniques originated in the computer science community have addressed nonlinearly separable problems in different ways. Next, we review some of these techniques.

10.4 OVERVIEW OF SOME MACHINE LEARNING MODELS

Artificial intelligence techniques such as those commonly referred to as *machine learning* techniques have been explored to address some potential limitations of standard modeling techniques. Among these techniques, classification trees and artificial neural networks have been the most popular in the medical domain.

10.4.1 Classification Trees

Classification trees recursively and univariately partition cases into two subgroups (Breiman et al. 1984). At each branch in an upside-down tree, as illustrated in Figure 10-4, the attribute-value pair that best partitions the cases

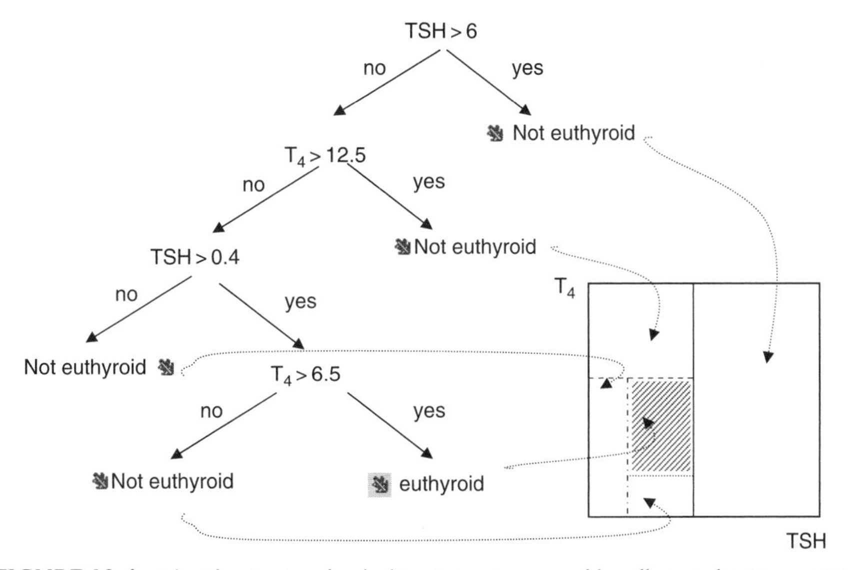

FIGURE 10-4 Classification tree for the bivariate outcome problem illustrated in Figure 10-3. Cases are recursively partitioned according to the attribute-value pair that best divides the cases into "euthyroid" or not. The resulting partitions easily can be visualized in this simplified two-dimensional problem.

into the categories of interest (e.g., "euthyroid" or not) is chosen. A simple step function assigns "yes" or "no" to the criterion in question (e.g., $TSH > 6 = \text{yes}$). This is repeated until the partitions that represent the "leaves" of the tree have only cases from a single category. Figure 10-4 illustrates the simplified bivariate example from Figure 10-3. The first attribute-value pair to be chosen is (TSH, 6 μU/ml). Cases in the right branch/leaf ($TSH > 6$) are not euthyroid. Cases in the left branch ($TSH <= 6$) may be euthyroid or not. The next attribute is T_4 at 12.5. Cases in the right branch/leaf ($T_4 > 12.5$) are not euthyroid. Cases in the left branch may be euthyroid. The pair (TSH, 0.4 μU/ml) is then chosen, and cases are classified into "Not euthyroid" if $TSH <= 0.4$. Otherwise (T_4, 6.5) is chosen and those cases with $T_4 > 6.5$ are classified as "euthyroid."

Note that classification trees can solve nonlinearly separable problems, since the number of branches is not limited. However, given their limitation of using only univariate cuts at each branching point, there may be too many branches for the tree to be easy to interpret. Pruning algorithms have been developed to address this issue (Gelfand et al. 1991). The Goldman tree (shown in Figure 10-5) for deciding whether a patient with chest pain should be admitted to the emergency department is the prime example of the application of a classification tree (Goldman et al. 1982). This study identified nine important clinical factors that enabled the system to correctly categorize all (60) patients with myocardial infarction (MI) in the sample (482). Sensitivity was an absolute priority in this model, and a portion (71) of patients without MI was categorized as false positives. The clinical factors were age, duration of pain, chest pain +/− radiation, presence of diaphoresis, history of angina (and severity of pain) or prior MI, local pressure causes reproduction of pain, EKG ST-segment changes, Q waves, or T-wave changes not known to be old.

10.4.2 Artificial Neural Networks

The use of artificial neural networks (ANNs) has been reported in several medical domains, particularly in critical care (Fraser and Turney 1990; Tu and Guerriere 1993; Dybowski et al. 1996; Kayaalp et al. 2000; Frize et al. 2001). ANNs are highly flexible models composed of several processing units. Each of these units processes incoming information and may propagate information forward if warranted by their activation function. The most common activation function is the logistic, which has been presented earlier in subsection 10.3. The logistic function tries to model a step function that has been widely used to represent the electrical conduction in real neurons, which only propagate electric impulses if a certain threshold value is achieved. Although it is possible to build ANNs without utilizing an intermediate "hidden" layer of neurons, the flexibility of ANNs comes from the inclusion of more than one nonlinear "hidden" node in this layer. In fact, the limitation of perceptrons, which were precursors to ANNs and were subject of much interest in the 1950s and 1960s, was noted by several authors (Minsky and Papert 1969). The same authors noted that multilayered perceptrons did not suffer from this limitation, but at that time there were no algorithms for estimating weights of multilayered perceptrons.

The field was stagnant until Rumelhart et al. (1986) published the back-propagation algorithm in the mid-1980s (developed originally by Werbos 1974). In the following two decades, a plethora of successful applications

FIGURE 10-5 Computer-Derived Decision Tree for the Classification of Patients with Acute Chest Pain. Reproduced (with permission) from Goldman and colleagues (Goldman et al. 1982). "Each of the 14 letters (A through N) identifies a terminal branch of the tree." In the Goldman study, seven terminal branches (C, D, H, I, K, M, and N) contained all the patients with acute myocardial infarction, along with a portion of the patients with other diagnoses. [Publisher to obtain permission to reproduce classification tree]

were reported in and out of the medical literature, but many of these research models did not translate into real clinical applications. Some, however, have been evaluated in real applications, such as automated analysis of Pap smears (Baxt 1991; O'Leary et al. 1998). There is no theoretical advantage of using ANNs over logistic regression in binary classification problems unless the ANNs have a hidden layer of nonlinear neurons. Hence, we will limit our discussion to this type of ANNs.

Figure 10-1 illustrated the similarities and differences between binary logistic regression and commonly used ANNs with a single output unit. ANNs and logistic regression models have several differences:

- The activation function of the output unit in ANNs need not be a (sigmoid logistic).
- ANNs have intermediate processing units, often called hidden units or hidden nodes.
- ANNs can have multiple output units, so different classification problems can be modeled with a single network (although one could argue that polytomous logistic regression also allows for multiple outcomes to be modeled).

The hidden units in ANNs operate between the inputs and the outputs to process information to be sent to the output unit. Figure 10-6 illustrates how an ANN might solve the nonlinearly separable problem of classifying cases

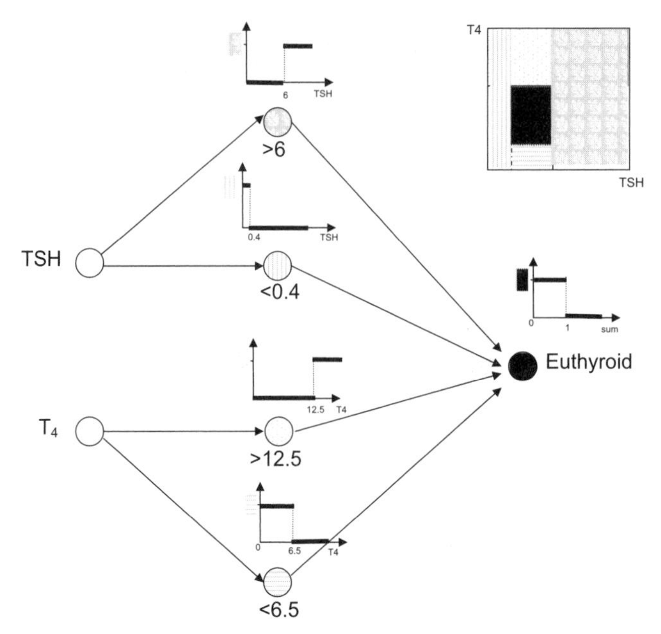

FIGURE 10-6 Artificial neural network with a hidden layer of nodes. For didactic purposes, activation functions in this example correspond to step functions that define partitions similar to the ones in the classification tree. Corresponding sigmoid (logistic) functions would be used in practice. As opposed to the classification tree example, the partitions here are overlapping. The outputs of the step functions are multiplied by their respective weights and combined as inputs to the output unit. The output unit has a step function that determines whether a case is "euthyroid" or not.

into "euthyroid" or not based on values of two laboratory tests, as illustrated in Figure 10-3. In this example, the activation functions of the intermediate layer of neurons correspond to the branching points that define the partitions of the classification tree presented in subsection 10.4.1, but such correspondence will often not be the case. Furthermore, in the example we used step functions in the hidden layer. Nonlinearity in the hidden layer is a necessary feature for ANNs to offer an advantage over logistic regression. We use step functions in our example for illustration purposes only, but remind you that they are seldom used in ANNs, as opposed to the sigmoid (logistic) function.

The outputs of these hidden-layer functions are multiplied by the weights that lead into the output node, and summed to serve as input to the output node, which classifies cases into "euthyroid" or not. We do not represent every possible weight between the input layer and the hidden layer to allow better visualization in the picture, but you can consider that the nondisplayed connections are associated with null weights.

10.5 PREDICTION MODELS IN MEDICINE

Most of the medical classification and prediction models currently in use are research tools with limited utilization in clinical care. Even though early CDS systems addressed mostly diagnostic aspects of clinical care, many of the most popular statistical and machine learning classification models in medicine used today address *prognostic* aspects of care.

In this section, we will discuss some applications of the modeling techniques described earlier, although each clinical example will not include all methods. In almost all cases, logistic regression modeling techniques are the most commonly reported and used in clinical practice for a variety of prognostication and classification purposes. The other techniques presented generally have been compared to the standard of logistic regression, and have rarely outperformed LR. The lack of widespread use of a number of these models can be attributed to the lack of general knowledge of the methods and increased complexity of the techniques. In many cases, the amount of data available does not allow the construction of complex models with many parameters, as there is a tendency of these models to overfit the data and not generalize well to new cases. As data become more abundant, this limitation is likely to play a smaller role.

The most common indices of model performance are discrimination and calibration. Discrimination assesses how well the models can potentially discriminate positive and negative cases in general. Models that estimate higher probabilities of outcome "1" for cases that actually had that outcome have high discrimination, which usually is measured as the area under the ROC curve (Hanley and McNeil 1982). However, even models with high areas under the ROC curve may estimate probabilities that are very far from reality, or "uncalibrated." Calibration assesses how close the model's estimated probability is to the "true" underlying probability of the outcome of interest. Calibration in logistic regression models usually is assessed by the Hosmer-Lemeshow goodness-of-fit test (Lemeshow and Hosmer 1982).

10.5.1 Prognosis of ICU Mortality

The Acute Physiology and Chronic Health Evaluation series of models (APACHE-II (Knaus et al. 1985) and APACHE-III (Knaus et al. 1991)) constitute some of the most widely used logistic regression-based predictive models. These tools are used in intensive care units (ICUs) to predict in-hospital mortality based on a variety of physiologically based variables. The initial version of APACHE (Knaus et al. 1981) was notable as the first clinical predictive model to exclusively use objective physiologic parameters to predict outcome, and was an expert-based scoring system that used these parameters to estimate the risk of outcome.

Both APACHE-II and APACHE-III remain in use today for research, quality control, and clinical applications. APACHE-II was published in 1985 using a much larger development data set (5,815 admissions from 13 hospitals) than APACHE, and improved upon the expert-based scoring system with the inclusion of a logistic regression model using a patient's expert-based physiology score, emergency status, and adjustments for certain diagnostic categories. The model showed good discrimination on different independent evaluation sets (Jacobs et al. 1987; Giangiuliani et al. 1989; Chisakuta and Alexander 1990; Teskey et al. 1991; Turner et al. 1991; Wong et al. 1995), but its calibration was found to be highly variable.

The latest version, APACHE-III, was published in 1991 in response to criticisms regarding case-mix and generalizability of APACHE-II. The system was developed from a database of 17,440 patients across 40 ICUs in the United States. APACHE-III is a commercial product, and was not made as easily available to the medical community at large as APACHE-II, but external evaluations conducted were similar to APACHE-II, indicating good discrimination and highly variable calibration (Bastos et al. 1996; Carneiro et al. 1997; Rivera-Fernandez et al. 1998; von Bierbrauer et al. 1998; Zimmerman et al. 1998; Pappachan et al. 1999; Cook 2000; Ihnsook et al. 2003).

These models remain useful in research, but limitations in calibration and across disparate patient populations have restricted use in some clinical situations (particularly with respect to application to the individual patient). Other prognostic systems for the adult ICU, more common in Europe, are the Simplified Acute Physiologic Score SAPS-II (Le Gall et al. 1993), and the Mortality Prediction Model MPM-II (Lemeshow et al. 1993). These models have been extensively compared all over the world in disparate patient populations. Several reviews and comparisons among these models have been published to date (Castella et al. 1991, 1995; Rowan et al. 1994; Del Bufalo et al. 1995; Wilairatana et al. 1995; Moreno et al. 1998; Nouira et al. 1998; Tan 1998; Patel and Grant 1999; Vassar et al. 1999; Capuzzo et al. 2000; Katsaragakis et al. 2000; Livingston et al. 2000; Markgraf et al. 2000; Beck et al. 2003; Ohno-Machado et al. 2006).

Multiple studies have compared logistic regression to artificial neural networks in this domain. Clermont and colleagues (2001) found that with a development data set of sufficient size (1,200), locally developed logistic regression and artificial neural networks performed equivalently in terms of both calibration (adequate) and discrimination (AUCs ranging from 0.80 to 0.84). However, both models experienced performance degradations as the development sample size decreased. Another smaller study with a development set of 168 by Dybowski and colleagues (1996) showed superior discrimination of the ANN compared to LR (0.863 vs. 0.753 AUCs, respectively).

Two additional studies have compared the APACHE-II LR model to ANNs. Nimgaonkar and colleagues (2004) found that after developing an ANN on 1962 patients in an Indian ICU with the 22 APACHE-II variables, the ANN had superior discrimination to APACHE-II (0.87 and 0.77 AUCs, respectively). Wong and colleagues (1999) performed a similar comparison with a development data set of 2,932 patients in the United Kingdom, and found that the two methods had equivalent discrimination (0.82 and 0.83 ACC for ANN and APACHE, respectively).

Comparisons of calibration also were done in both APACHE comparison studies, but are problematic because the LR model was developed on external patient populations disparate from the locally derived United Kingdom and Indian populations utilized for the ANN models. Comparisons of discrimination do not suffer from this problem in the same way.

10.5.2 Cardiovascular Disease Risk

Another category of extremely well-known prediction tools in medicine provides estimates of the risk of developing future heart disease in patients. U.S. medical practice almost exclusively uses the most recent 10-year heart disease risk model (Wilson et al. 1998) developed from patients in the eleventh examination of the original Framingham cohort (Anderson et al. 1991b) or the initial examination of the Framingham Offspring Study (Kannel et al. 1979). This is one of the most famous patient cohorts, and has been followed in the community of Framingham, Massachusetts, since the early 1950s. All the well-known models developed in this domain are based on logistic regression methods, and these have been developed in many nations. Most of the cardiovascular risk models were developed from the Framingham patient cohort (Kannel et al. 1976; Anderson et al. 1991a; Haq et al. 1995; Anonymous 1996, 1998; Wilson et al. 1998; Wallis et al. 2000), although some used other sources (Stevens et al. 2001; Assmann et al. 2002; Conroy et al. 2003).

The widespread use of these models is related to a number of key factors that influence the utility and generalizability of the prediction model. First, the modeled outcome is of paramount importance, as heart disease is the number one cause of mortality in the United States, accounting for 28.5 percent of all deaths (2004). Effective treatments exist for many of the outcome predictors, such as hypertension, hyperlipidemia, and smoking. Second, the patient population used in model development was in many ways representative of the American population. The Framingham cohort was an excellent source of data because the longitudinal nature of the cohort allowed reliable discrimination of patients at higher risk but who had not yet presented any sign or symptom of heart disease. One of the primary limitations of the cohort was the lack of racial diversity.

External validation of these models has shown good discrimination and moderate calibration, with some limitations when applied to populations with significantly different demographics and specific comorbidities (such as diabetes) (Lenz and Muhlhauser 2004; Song and Brown 2004; Stephens et al. 2004; Guzder et al. 2005). Recalibration strategies have been used to remediate this problem.

These models have also clearly delineated the relative magnitude of various risk factors associated with heart disease, and have been used by a number of medical associates to establish guidelines of care (Grundy et al. 1998). In addition, the models are distributed as simple equations that can be quickly scanned by clinicians and patients, or embedded in calculators or computer-based software (Hingorani and Vallance 1999).

10.5.3 Prognosis in Interventional Cardiology

Another widely studied area of risk prediction and stratification has been for the outcomes of death and significant morbidity in interventional cardiology. Risk modeling in this domain has been particularly popular for a number of reasons. First, the treatments (balloon angioplasty or coronary artery stenting) are therapies directed at preventing myocardial infarction in patients who have developed significant heart disease. All the remarks on the importance of this disease process in the prior section apply; these therapies can become necessary when prevention strategies fail.

Perhaps even more importantly, since that treatment is relatively intensive and done within a medical center, detailed data collection can provide high quality source data. In addition, a number of the adverse outcomes associated with the treatment (or lack of treatment) are realized quickly. This is important because, in general, a model's performance is inversely related to the distance in time of the prediction from the occurrence of the outcome of interest. These factors have allowed the resulting models for this domain to attain high levels of discrimination. In recent years, this has been further facilitated by the establishment of a national standard for the collection and storage of interventional cardiology data (Cannon et al. 2001).

Development of logistic regression prediction models for post-procedural mortality following angioplasty has followed a path similar to other medical domains. In general, the development populations were initially small and originated from a single center, which resulted in low generalizability (Hannan et al. 1992; Resnic et al. 2001). These were followed by regional, multi-institutional models (Ellis et al. 1997; Hannan et al. 1997; O'Connor et al. 1999; Moscucci et al. 2001). Finally, the American College of Cardiology aggregated data from centers across the United States to generate a mortality risk prediction model (Shaw et al. 2002, 2003).

These models have been externally validated on a number of independent data sets. In general, discrimination has remained excellent across disparate patient populations and over more than a decade of changing clinical care. However, as noted in many of the other models, calibration remains a problem and seems to be directly related to both the size of the development data, and how far in the past they were collected (Moscucci et al. 1999; Holmes et al. 2000, 2003; Rihal et al. 2000; Singh et al. 2003; Matheny et al. 2005). Resnic and colleagues compared the classification performance of logistic regression and ANNs in a single center study and concluded that the differences were not significant (Resnic et al. 2001).

10.5.4 Pneumonia Severity of Illness Index

Finally, another logistic regression risk model example that has had a significant impact in the emergency department for both workflow (documentation requirements) and treatment is the Pneumonia Severity Index developed from the Pneumonia Patient Outcomes Research Team (PORT) (Fine et al. 1997).

The team developed a prediction rule for the risk of death within 30 days for adult patients with community-acquired pneumonia. This disease is diagnosed in approximately 4 million adults each year in the United States, and over 600,000 of these are hospitalized (Garibaldi 1985). The aggregate cost of hospitalization for this disease was estimated at 4 billion dollars per year (Dans et al. 1984; La Force 1985). The results of the PORT study suggested that, if the risk model had been used to treat patients based on the risk categories suggested, 26 to 31 percent of patients who were hospitalized for care could have been treated safely as outpatients, and an additional 13 to 19 percent could have been hospitalized only for brief observation (Fine et al. 1997).

The key factors that led to the widespread use of this risk prediction tool were a combination of coinciding interest in evidence-based medical practice and cost containment as well as the high quality of the risk prediction tool. The model was validated on over 50,000 patients in 275 U.S. and Canadian hospitals in the PORT study. Prior pneumonia risk prediction tools had suffered from small development population sizes (Daley et al. 1988; Keeler et al. 1990; Kurashi et al. 1992; Fine et al. 1995) and limited external validation (Marrie et al. 1989; Kurashi et al. 1992; Fine et al. 1995).

The model has been widely used, and incorporated in both paper (Dean et al. 2000) and electronic (Aronsky et al. 2001) decision support tools for use in determining hospital admission from an emergency department. A number of subsequent multicenter randomized prospective studies have supported the use of the PSI as an appropriate admission tool (Atlas et al. 1998; Marrie et al. 2000). It was incorporated into the American Thoracic Society's (ATS) Community-Acquired Pneumonia guidelines (Niederman et al. 2001), although the society emphasized limitations of the model in populations that were not well represented in the development data set (such as outpatient clinic patients), echoing findings from a few studies (Marras et al. 2000). In addition, there are a number of factors that physicians must take into account such as the presence of coexisting conditions, patients' preferences, and inadequate home support (Halm et al. 2000).

Cooper and colleagues (Cooper et al. 2005) reported that several types of classifiers can achieve similar performance in this domain.

10.6 CONCLUSION

The utilization of statistical and machine learning techniques to discover knowledge from existing clinical data is a relatively recent development in biomedical informatics. The techniques for constructing and evaluating classification and prediction models are still evolving, and there are few theoretical justifications to prefer one learning technique over another. Some models, notably those constructed using logistic regression techniques, have been popularized in the medical domain, especially for research. These models span a limited number of specialties, and are for the most part concerned with prognostication. To our knowledge, there have been no formal large-scale studies documenting the utilization of these models by nonacademic clinicians at large. Even though some models are widely available on the Web, there is currently no information on how many times they actually have been used in the provision of care. Several questions still remain:

- What types of data repositories can reasonably be used for medical pattern discoveries? Can data collected during clinical care be used to build decision support models? If so, what types of learning methods are adequate for sparse and noisy data?
- When can models originated from data of a single population be generalized to other populations? How can researchers assess the generalizability of such models?
- How can knowledge acquired from experts be integrated with knowledge discovered from real data?

None of these questions has been fully answered by the medical informatics community, but research in this area is encouraging. The popularity of some data-derived classification and prediction models, and their endorsement by health care institutions (an online model for assessing the risk of breast cancer is available at the NCI Web site, for example), indicate that there is increasing interest in their use as diagnostic or prognostic tools. The availability of such models on the Web also contributes to their utilization by the public at large.

It is important that clinicians utilize classification and prediction models. However, the integration of any computer system in the process of care is challenging. The electronic medical record is still not a reality in most settings in which medicine is practiced. The effective utilization of CDS systems depends on their seamless integration in a computer environment that is effectively used by practicing clinicians, and hence it is premature to expect that predictive models will be largely utilized until this barrier is removed. The absence of a suitable computer environment is the first obstacle, but other issues also need further consideration. In order to provide counseling at the individual level, predictive models have to improve enough so that the uncertainty and imprecision of the estimates are acceptable from a clinical perspective. The poor calibration of estimates is often caused by limited representation of the population to which models will be applied at the model construction phase. Yet, collecting proper data from the institutions in which models are expected to be applied is not a trivial task. Until these types of deficiencies are properly acknowledged and fixed, and studies show that the predictive models perform at least at the same level as humans, the utilization of predictive models for individual care may remain limited. However, given the rapid pace of technological advances in biomedicine and the increasing acceptance of computers by health care providers, it is expected that better models will continue to be developed that may soon be incorporated as additional tools in the provision of individualized care.

REFERENCES

(2004). 202 deaths and death rates for the 15 leading causes of death in 5-year age groups. *CDC/NCHS, National Vital Statistics System.*

Anderson, K. M., Odell, P. M., Wilson, P. W., and Kannel, W. B. (1991a). Cardiovascular disease risk profiles. *American Heart Journal* **121**: 293–298.

Anderson, K. M., Wilson, P. W., Odell, P. M., and Kannel, W. B. (1991b). An updated coronary risk profile. A statement for health professionals. *Circulation* **83**: 356–362.

Anonymous. (1996). 1996 National Heart Foundation clinical guidelines for the assessment and management of dyslipidaemia. Dyslipidaemia Advisory Group on behalf of the scientific committee of the National Heart Foundation of New Zealand. *New Zealand Medical Journal* **109**: 224–231.

Anonymous. (1998). Joint British recommendations on prevention of coronary heart disease in clinical practice. British Cardiac Society, British Hyperlipidaemia Association, British Hypertension Society, endorsed by the British Diabetic Association. *Heart (British Cardiac Society)* **80** (Suppl 2): S1–29.

Aronsky, D., Chan, K. J., and Haug, P. J. (2001). Evaluation of a computerized diagnostic decision support system for patients with pneumonia: Study design considerations. *Journal of the American Medical Informatics Association* **8**: 473–485.

Assmann, G., Cullen, P., and Schulte, H. (2002). Simple scoring scheme for calculating the risk of acute coronary events based on the 10-year follow-up of the prospective cardiovascular Munster (PROCAM) study. *Circulation* **105**: 310–315.

Atlas, S. J., Benzer, T. I., Borowsky, L. H., Chang, Y., Burnham, D. C., Metlay, J. P. et al. (1998). Safely increasing the proportion of patients with community-acquired pneumonia treated as outpatients: An interventional trial. *Archives of Internal Medicine* **158**: 1350–1356.

Barnett, G. O., Cimino, J. J., Hupp, J. A., and Hoffer, E. P. (1987). DXplain. An evolving diagnostic decision-support system. *JAMA* **258**: 67–74.

Bastos, P. G., Sun, X., Wagner, D. P., Knaus, W. A., and Zimmerman, J. E. (1996). Application of the APACHE III prognostic system in Brazilian intensive care units: A prospective multicenter study. *Intensive Care Medicine* **22**: 564–570.

Baxt, W. G. (1991). Use of an artificial neural network for the diagnosis of myocardial infarction. *Ann Intern Med* **115**: 845–848.

Beck, D. H., Smith, G. B., Pappachan, J. V., and Millar, B. (2003). External validation of the SAPS II, APACHE II and APACHE III prognostic models in South England: A multicentre study. *Intensive Care Medicine* **29**: 249–256.

Beinlich, I. A., Suermondt, H. J., Chavez, R. M., and Cooper, G. F. (1989). The ALARM monitoring system: A case study with two probabilistic inference techniques for belief networks. *Proceedings of the Second European Conference on Artificial Intelligence in Medicine* 247–256.

Boser, B., Guyon, I., and Vapnik, V. (1992). A Training Algorithm for Optimal Margin Classifiers. *Proceedings on the fifth annual workshop on Computational Learning Theory.*

Breiman, L., Friedman, J., Olshen, R., and Stone, C. (1984). *Classification and regression trees.* Wadsworth and Brooks.

Buntine, W. L. (1996). A guide to the literature on learning probabilistic networks from data. *IEEE Trans Knowl Data Eng* **8**: 195–210.

Cannon, C. P., Battler, A., Brindis, R. G., Cox, J. L., Ellis, S. G., Every, N. R. et al. (2001). American College of Cardiology key data elements and definitions for measuring the clinical management and outcomes of patients with acute coronary syndromes. A report of the American College of Cardiology Task Force on Clinical Data Standards (Acute Coronary Syndromes Writing Committee). *Journal of the American College of Cardiology* **38**: 2114–2130.

Capuzzo, M., Valpondi, V., Sgarbi, A., Bortolazzi, S., Pavoni, V., Gilli, G. et al. (2000). Validation of severity scoring systems SAPS II and APACHE II in a single-center population. *Intensive Care Medicine* **26**: 1779–1785.

Carneiro, A. V., Leitao, M. P., Lopes, M. G., and De Padua, F. (1997). [Risk stratification and prognosis in critical surgical patients using the Acute Physiology, Age and Chronic Health III System (APACHE III)]. *Acta Medica Portuguesa* **10**: 751–760.

Castella, X., Artigas, A., Bion, J., and Kari, A. (1995). A comparison of severity of illness scoring systems for intensive care unit patients: results of a multicenter, multinational study. The European/North American Severity Study Group. *Critical Care Medicine* **23**: 1327–1335.

Castella, X., Gilabert, J., Torner, F., and Torres, C. (1991). Mortality prediction models in intensive care: Acute physiology and chronic health evaluation II and mortality prediction model compared. *Critical Care Medicine* **19**: 191–197.

Chisakuta, A. M. and Alexander, J. P. (1990). Audit in intensive care. The APACHE II classification of severity of disease. *Ulster Medical Journal* **59**: 161–167.

Clancey, W. J. (1983). The epistemology of a rule-based expert system: A framework for explanation. *Artificial Intelligence* **20**: 215–251.

Clermont, G., Angus, D. C., DiRusso, S. M., Griffin, M., and Linde-Zwirble, W. T. (2001). Predicting hospital mortality for patients in the intensive care unit: A comparison of artificial neural networks with logistic regression models. *Critical Care Medicine* **29**: 291–296.

Conroy, R. M., Pyorala, K., Fitzgerald, A. P., Sans, S., Menotti, A., De Backer, G. et al. (2003). Estimation of ten-year risk of fatal cardiovascular disease in Europe: The SCORE project. *European Heart Journal* **24**: 987–1003.

Cook, D. A. (2000). Performance of APACHE III models in an Australian ICU. *Chest* **118**: 1732–1738.

Cooper, G. and Herskovits, E. (1992). A Bayesian method for the induction of probabilistic networks from data. *Machine Learning* **9**: 309–347.

Cooper, G. F., Abraham, V., Aliferis, C. F., Aronis, J. M., Buchanan, B. G., Caruana, R. et al. (2005). Predicting dire outcomes of patients with community acquired pneumonia. *J Biomed Inform* **38**: 347–366.

Daley, J., Jencks, S., Draper, D., Lenhart, G., Thomas, N., and Walker, J. (1988). Predicting hospital-associated mortality for Medicare patients. A method for patients with stroke, pneumonia, acute myocardial infarction, and congestive heart failure. *JAMA* **260**: 3617–3624.

Dans, P. E., Charache, P., Fahey, M., and Otter, S. E. (1984). Management of pneumonia in the prospective payment era. A need for more clinician and support service interaction. *Archives of Internal Medicine* **144**: 1392–1397.

Dean, N. C., Suchyta, M. R., Bateman, K. A., Aronsky, D., and Hadlock, C. J. (2000). Implementation of admission decision support for community-acquired pneumonia. *Chest* **117**: 1368–1377.

Del Bufalo, C., Morelli, A., Bassein, L., Fasano, L., Quarta, C. C., Pacilli, A. M., and Gunella, G. (1995). Severity scores in respiratory intensive care: APACHE II predicted mortality better than SAPS II. *Respiratory Care* **40**: 1042–1047.

Duda, R., Hart, P., and Stork, D. (2001). *Pattern Classification*. Wiley Interscience.

Dybowski, R., Weller, P., Chang, R., and Gant, V. (1996). Prediction of outcome in critically ill patients using artificial neural network synthesized by genetic algorithm. *Lancet* **347**: 1146–1150.

Ellis, S. G., Weintraub, W., Holmes, D., Shaw, R., Block, P. C., and King, S. B., 3rd. (1997). Relation of operator volume and experience to procedural outcome of percutaneous coronary revascularization at hospitals with high interventional volumes. *Circulation* **95**: 2479–2484.

Fine, M. J., Auble, T. E., Yealy, D. M., Hanusa, B. H., Weissfeld, L. A., Singer, D. E. et al. (1997). A prediction rule to identify low-risk patients with community-acquired pneumonia. *New England Journal of Medicine* **336**: 243–250.

Fine, M. J., Hanusa, B. H., Lave, J. R., Singer, D. E., Stone, R. A., Weissfeld, L. A. et al. (1995). Comparison of a disease-specific and a generic severity of illness measure for patients with community-acquired pneumonia. *Journal of General Internal Medicine* **10**: 359–368.

Fraser, R. B. and Turney, S. Z. (1990). An expert system for the nutritional management of the critically ill. *Computer Methods & Programs in Biomedicine* **33**: 175–180.

Frize, M., Ennett, C. M., Stevenson, M., and Trigg, H. C. (2001). Clinical decision support systems for intensive care units: using artificial neural networks. *Medical Engineering & Physics* **23**: 217–225.

Garibaldi, R. A. (1985). Epidemiology of community-acquired respiratory tract infections in adults. Incidence, etiology, and impact. *American Journal of Medicine* **78**: 32–37.

Gelfand, S. B., Ravishankar, C. S., and Delp, E. J. (1991). An iterative growing and pruning algorithm for classification tree design. *IEEE Transactions on Pattern Analysis and Machine Intelligence* **13**: 163–174.

Giangiuliani, G., Mancini, A., and Gui, D. (1989). Validation of a severity of illness score (APACHE II) in a surgical intensive care unit. *Intensive Care Medicine* **15**: 519–522.

Goldman, L., Weinberg, M., Weisberg, M., Olshen, R., Cook, E. F., Sargent, R. K. et al. (1982). A computer-derived protocol to aid in the diagnosis of emergency room patients with acute chest pain. *NEJM* **307**: 588–596.

Grundy, S. M., Balady, G. J., Criqui, M. H., Fletcher, G., Greenland, P., Hiratzka, L. F. et al. (1998). Primary prevention of coronary heart disease: Guidance from Framingham: A statement for healthcare professionals from the AHA task force on risk reduction. American Heart Association. *Circulation* **97**: 1876–1887.

Grundy, S. M., Pasternak, R., Greenland, P., Smith, S., Jr., and Fuster, V. (1999). Assessment of cardiovascular risk by use of multiple-risk-factor assessment equations: A statement for healthcare professionals from the American Heart Association and the American College of Cardiology. *Circulation* **100**: 1481–1492.

Guzder, R. N., Gatling, W., Mullee, M. A., Mehta, R. L., and Byrne, C. D. (2005). Prognostic value of the Framingham cardiovascular risk equation and the UKPDS risk engine for coronary heart disease in newly diagnosed Type 2 diabetes: Results from a United Kingdom study. *Diabetic Medicine* **22**: 554–562.

Halm, E. A., Atlas, S. J., Borowsky, L. H., Benzer, T. I., Metlay, J. P., Chang, Y. C., and Singer, D. E. (2000). Understanding physician adherence with a pneumonia practice guideline: Effects of patient, system, and physician factors. *Archives of Internal Medicine* **160**: 98–104.

Hanley, J. A. and McNeil, B. J. (1982). The meaning and use of the area under a receiver operating characteristic (ROC) curve. *Radiology* **143**: 29–36.

Hannan, E. L., Arani, D. T., Johnson, L. W., Kemp, H. G., Jr., and Lukacik, G. (1992). Percutaneous transluminal coronary angioplasty in New York State. Risk factors and outcomes. *JAMA* **268**: 3092–3097.

Hannan, E. L., Racz, M., Ryan, T. J., McCallister, B. D., Johnson, L. W., Arani, D. T. et al. (1997). Coronary angioplasty volume-outcome relationships for hospitals and cardiologists. *JAMA* **277**: 892–898.

Haq, I. U., Jackson, P. R., Yeo, W. W., and Ramsay, L. E. (1995). Sheffield risk and treatment table for cholesterol lowering for primary prevention of coronary heart disease. *Lancet* **346**: 1467–1471.

Hastie, T., Tibshirani, R., and Friedman, J. (2001). *The Elements of Statistical Learning*. New York: Springer.

Heckerman, D. (1990). A tractable inference algorithm for diagnosing multiple diseases. *Proceedings of Fifth Conference on Uncertainty in Artificial Intelligence* 163–171.

Heckerman, D. and Miller, R. A. (1986). Towards a better understanding of the INTERNIST-1 knowledge base. *MEDINFO* 86.

Hingorani, A. D. and Vallance, P. (1999). A simple computer program for guiding management of cardiovascular risk factors and prescribing. *BMJ* **318**: 101–105.

Holmes, D. R., Jr., Berger, P. B., Garratt, K. N., Mathew, V., Bell, M. R., Barsness, G. W. et al. (2000). Application of the New York State PTCA mortality model in patients undergoing stent implantation. *Circulation* **102**: 517–522.

Holmes, D. R., Selzer, F., Johnston, J. M., Kelsey, S. F., Holubkov, R., Cohen, H. A. et al. (2003). Modeling and risk prediction in the current era of interventional cardiology: A report from the National Heart, Lung, and Blood Institute Dynamic Registry. *Circulation* **107**: 1871–1876.

Hosmer, D. W. and Lemeshow, S. (1989). Applied Logistic Regression.

Ihnsook, J., Myunghee, K., and Jungsoon, K. (2003). Predictive accuracy of severity scoring system: a prospective cohort study using APACHE III in a Korean intensive care unit. *International Journal of Nursing Studies* **40**: 219–226.

Jacobs, S., Chang, R. W., and Lee, B. (1987). One year's experience with the APACHE II severity of disease classification system in a general intensive care unit. *Anaesthesia* **42**: 738–744.

Kannel, W. B., Feinleib, M., McNamara, P. M., Garrison, R. J., and Castelli, W. P. (1979). An investigation of coronary heart disease in families. The Framingham offspring study. *American Journal of Epidemiology* **110**: 281–290.

Kannel, W. B., McGee, D., and Gordon, T. (1976). A general cardiovascular risk profile: The Framingham Study. *American Journal of Cardiology* **38**: 46–51.

Katsaragakis, S., Papadimitropoulos, K., Antonakis, P., Strergiopoulos, S., Konstadoulakis, M. M., and Androulakis, G. (2000). Comparison of acute physiology and chronic health evaluation II (APACHE II) and simplified acute physiology score II (SAPS II) scoring systems in a single Greek intensive care unit. *Critical Care Medicine* **28**: 426–432.

Kayaalp, M., Cooper, G. F., and Clermont, G. (2000). Predicting ICU mortality: A comparison of stationary and nonstationary temporal models. *Proceedings/AMIA Annual Symposium* 418–422.

Keeler, E. B., Kahn, K. L., Draper, D., Sherwood, M. J., Rubenstein, L. V., Reinisch, E. J. et al. (1990). Changes in sickness at admission following the introduction of the prospective payment system. *JAMA* **264**: 1962–1968.

Knaus, W. A., Draper, E. A., Wagner, D. P., and Zimmerman, J. E. (1985). APACHE II: A severity of disease classification system. *Critical Care Medicine* **13**: 818–829.

Knaus, W. A., Wagner, D. P., Draper, E. A., Zimmerman, J. E., Bergner, M., Bastos, P. G. et al. (1991). The APACHE III prognostic system. Risk prediction of hospital mortality for critically ill hospitalized adults [see comment]. *Chest* **100**: 1619–1636.

Knaus, W. A., Zimmerman, J. E., Wagner, D. P., Draper, E. A., and Lawrence, D. E. (1981). APACHE-acute physiology and chronic health evaluation: A physiologically based classification system. *Critical Care Medicine* **9**: 591–597.

Koza, J. R. (1992). *Genetic Programming: On the Programming of Computers by Means of Natural Selection.* Cambridge, MA: MIT Press.

Kurashi, N. Y., al-Hamdan, A., Ibrahim, E. M., al-Idrissi, H. Y., and al-Bayari, T. H. (1992). Community acquired acute bacterial and atypical pneumonia in Saudi Arabia. *Thorax* **47**: 115–118.

La Force, F. M. (1985). Community-acquired lower respiratory tract infections. Prevention and cost-control strategies. *American Journal of Medicine* **78**: 52–57.

Le Gall, J. R., Lemeshow, S., and Saulnier, F. (1993). A new Simplified Acute Physiology Score (SAPS II) based on a European/North American multicenter study [erratum appears in *JAMA* 1994 May 4; **271**(17): 1321]. *JAMA* **270**: 2957–2963.

Lemeshow, S. and Hosmer, D. W., Jr. (1982). A review of goodness of fit statistics for use in the development of logistic regression models. *American Journal of Epidemiology* **115**: 92–106.

Lemeshow, S. and Le Gall, J. R. (1994). Modeling the severity of illness of ICU patients. A systems update. *JAMA* **272**: 1049–1055.

Lemeshow, S., Teres, D., Klar, J., Avrunin, J. S., Gehlbach, S. H., and Rapoport, J. (1993). Mortality Probability Models (MPM II) based on an international cohort of intensive care unit patients. *JAMA* **270**: 2478–2486.

Lenz, M. and Muhlhauser, I. (2004). Cardiovascular risk assessment for informed decision making. Validity of prediction tools. *Medizinische Klinik* **99**: 651–661.

Livingston, B. M., MacKirdy, F. N., Howie, J. C., Jones, R., and Norrie, J. D. (2000). Assessment of the performance of five intensive care scoring models within a large Scottish database. *Critical Care Medicine* **28**: 1820–1827.

Markgraf, R., Deutschinoff, G., Pientka, L., and Scholten, T. (2000). Comparison of acute physiology and chronic health evaluations II and III and simplified acute physiology score II: A prospective cohort study evaluating these methods to predict outcome in a German interdisciplinary intensive care unit [see comment]. *Critical Care Medicine* **28**: 26–33.

Marras, T. K., Gutierrez, C., and Chan, C. K. (2000). Applying a prediction rule to identify low-risk patients with community-acquired pneumonia. *Chest* **118**: 1339–1343.

Marrie, T. J., Durant, H., and Yates, L. (1989). Community-acquired pneumonia requiring hospitalization: 5-year prospective study. *Reviews of Infectious Diseases* **11**: 586–599.

Marrie, T. J., Lau, C. Y., Wheeler, S. L., Wong, C. J., Vandervoort, M. K., and Feagan, B. G. (2000). A controlled trial of a critical pathway for treatment of community-acquired pneumonia. CAPITAL Study Investigators. Community-acquired pneumonia intervention trial assessing levofloxacin [see comment]. *JAMA* **283**: 749–755.

Matheny, M. E., Ohno-Machado, L., and Resnic, F. S. (2005). Discrimination and calibration of mortality risk prediction models in interventional cardiology. *J Biomed Inform* **38**: 367–375.

Miller, R., Masarie, F. E., and Myers, J. D. (1986). Quick medical reference (QMR) for diagnostic assistance. *MD Computing* **3**: 34–48.

Miller, R. A., Pople, H. E., Jr., and Myers, J. D. (1982). Internist-1, an experimental computer-based diagnostic consultant for general internal medicine. *New England Journal of Medicine* **307**: 468–476.

Minsky, M. L., and Papert, S. (1969). *Perceptrons*. Cambridge, MA: MIT Press.

Moore, A. and Lee, M. S. (1998). Cached sufficient statistics for efficient machine learning with large datasets. *JAIR* **8**: 67–91.

Moreno, R., Apolone, G., and Miranda, D. R. (1998). Evaluation of the uniformity of fit of general outcome prediction models. *Intensive Care Medicine* **24**: 40–47.

Moscucci, M., Kline-Rogers, E., Share, D., O'Donnell, M., Maxwell-Eward, A., Meengs, W. L. et al. (2001). Simple bedside additive tool for prediction of in-hospital mortality after percutaneous coronary interventions. *Circulation* **104**: 263–268.

Moscucci, M., O'Connor, G. T., Ellis, S. G., Malenka, D. J., Sievers, J., Bates, E. R. et al. (1999). Validation of risk adjustment models for in-hospital percutaneous transluminal coronary angioplasty mortality on an independent data set. *Journal of the American College of Cardiology* **34**: 692–697.

Niederman, M. S., Mandell, L. A., Anzueto, A., Bass, J. B., Broughton, W. A., Campbell, G. D. et al. (2001). Guidelines for the management of adults with community-acquired pneumonia. Diagnosis, assessment of severity, antimicrobial therapy, and prevention. *American Journal of Respiratory & Critical Care Medicine* **163**: 1730–1754.

Nimgaonkar, A., Karnad, D. R., Sudarshan, S., Ohno-Machado, L., and Kohane, I. (2004). Prediction of mortality in an Indian intensive care unit. Comparison between APACHE II and artificial neural networks. *Intensive Care Medicine* **30**: 248–253.

Nouira, S., Belghith, M., Elatrous, S., Jaafoura, M., Ellouzi, M., Boujdaria, R. et al. (1998). Predictive value of severity scoring systems: comparison of four models in Tunisian adult intensive care units. *Critical Care Medicine* **26**: 852–859.

O'Connor, G. T., Malenka, D. J., Quinton, H., Robb, J. F., Kellett, M. A., Jr., Shubrooks, S. et al. (1999). Multivariate prediction of in-hospital mortality after percutaneous coronary interventions in 1994–1996. Northern New England Cardiovascular Disease Study Group. *Journal of the American College of Cardiology* **34**: 681–691.

O'Leary, T. J., Tellado, M., Buckner, S. B., Ali, I. S., Stevens, A., and Ollayos, C. W. (1998). PAPNET-assisted rescreening of cervical smears: Cost and accuracy compared with a 100% manual rescreening strategy. *JAMA* **279**: 235–237.

Ohno-Machado, L., Resnic, F. S., and Matheny, M. E. (2006). Prognosis in Critical Care. In *Annual Review of Biomedical Engineering*, M. L. Yarmush and K. R. Diller, eds., Vol. 8. Nonprofit Publisher of the Annual Review of TM Series, Palo Alto, CA.

Pappachan, J. V., Millar, B., Bennett, E. D., and Smith, G. B. (1999). Comparison of outcome from intensive care admission after adjustment for case mix by the APACHE III prognostic system. *Chest* **115**: 802–810.

Patel, P. A. and Grant, B. J. (1999). Application of mortality prediction systems to individual intensive care units. *Intensive Care Medicine* **25**: 977–982.

Pawlak, Z. (1982). Rough sets. *Int J Inf Comput Sci* **11**: 341–356.

Pearl, J. (1988). *Probabilistic Reasoning in Intelligent Systems*. San Mateo, CA: Morgan-Kaufmann.

Pollock, D. A., Adams, D. L., Bernardo, L. M., Bradley, V., Brandt, M. D., Davis, T. E. et al. (1998). Data elements for emergency department systems, release 1.0 (DEEDS): A summary report. DEEDS Writing Committee. *Journal of Emergency Nursing* **24**: 35–44.

Resnic, F. S., Ohno-Machado, L., Selwyn, A., Simon, D. I., and Popma, J. J. (2001). Simplified risk score models accurately predict the risk of major in-hospital complications following percutaneous coronary intervention. *American Journal of Cardiology* **88**: 5–9.

Rihal, C. S., Grill, D. E., Bell, M. R., Berger, P. B., Garratt, K. N., and Holmes, D. R., Jr. (2000). Prediction of death after percutaneous coronary interventional procedures. *American Heart Journal* **139**: 1032–1038.

Rivera-Fernandez, R., Vazquez-Mata, G., Bravo, M., Aguayo-Hoyos, E., Zimmerman, J., Wagner, D., and Knaus, W. (1998). The Apache III prognostic system: Customized mortality predictions for Spanish ICU patients. *Intensive Care Medicine* **24**: 574–581.

Rowan, K. M., Kerr, J. H., Major, E., McPherson, K., Short, A., and Vessey, M. P. (1994). Intensive Care Society's Acute Physiology and Chronic Health Evaluation (APACHE II) study in Britain and Ireland: A prospective, multicenter, cohort study comparing two methods for predicting outcome for adult intensive care patients. *Critical Care Medicine* **22**: 1392–1401.

Rumelhart, D. E., Hinton, G. E., and Williams, R. J. (1986). Learning representations by back-propagating errors. *Nature* **323**: 533–536.

Shaw, R. E., Anderson, H. V., Brindis, R. G., Krone, R. J., Klein, L. W., McKay, C. R. et al. (2002). Development of a risk adjustment mortality model using the American College of Cardiology-National Cardiovascular Data Registry (ACC-NCDR) experience: 1998–2000. *Journal of the American College of Cardiology* **39**: 1104–1112.

Shaw, R. E., Anderson, H. V., Brindis, R. G., Krone, R. J., Klein, L. W., McKay, C. R. et al. (2003). Updated risk adjustment mortality model using the complete 1.1 dataset from the American College of Cardiology National Cardiovascular Data Registry (ACC-NCDR). *Journal of Invasive Cardiology* **15**: 578–580.

Shortliffe, E. H. (1976). *Computer-Based Medical Consultations, MYCIN*. New York: Elsevier.

Shortliffe, E. H., Davis, R., Axline, S. G., Buchanan, B. G., Green, C. C., and Cohen, S. N. (1975). Computer-based consultations in clinical therapeutics: Explanation and rule acquisition capabilities of the MYCIN system. *Computers & Biomedical Research* **8**: 303–320.

Singh, M., Rihal, C. S., Selzer, F., Kip, K. E., Detre, K., and Holmes, D. R. (2003). Validation of Mayo Clinic risk adjustment model for in-hospital complications after percutaneous coronary interventions, using the National Heart, Lung, and Blood Institute dynamic registry [see comment]. *Journal of the American College of Cardiology* **42**: 1722–1728.

Song, S. H., and Brown, P. M. (2004). Coronary heart disease risk assessment in diabetes mellitus: Comparison of UKPDS risk engine with Framingham risk assessment function and its clinical implications. *Diabetic Medicine* **21**: 238–245.

Stephens, J. W., Ambler, G., Vallance, P., Betteridge, D. J., Humphries, S. E., and Hurel, S. J. (2004). Cardiovascular risk and diabetes. Are the methods of risk prediction satisfactory? *European Journal of Cardiovascular Prevention & Rehabilitation* **11**: 521–528.

Stevens, R. J., Kothari, V., Adler, A. I., Stratton, I. M., and United Kingdom Prospective Diabetes Study, G. (2001). The UKPDS risk engine: A model for the risk of coronary heart disease in Type II diabetes (UKPDS 56). *Clinical Science* **101**: 671–679.

Swhe, M., Middleton, B., Heckerman, D., Henrion, M., Horvitz, E., and Lehmann, H. (1991). Probabilistic diagnosis using a reformulation of the INTERNIST-1/QMR knowledge base I: The probabilistic model and inference algorithms. *Methods of Information in Medicine* **30**: 241–255.

Tan, I. K. (1998). APACHE II and SAPS II are poorly calibrated in a Hong Kong intensive care unit. *Annals of the Academy of Medicine, Singapore* **27**: 318–322.

Teskey, R. J., Calvin, J. E., and McPhail, I. (1991). Disease severity in the coronary care unit. *Chest* **100**: 1637–1642.

Tu, J. V. and Guerriere, M. R. (1993). Use of a neural network as a predictive instrument for length of stay in the intensive care unit following cardiac surgery. *Computers & Biomedical Research* **26**: 220–229.

Turner, J. S., Mudaliar, Y. M., Chang, R. W., and Morgan, C. J. (1991). Acute physiology and chronic health evaluation (APACHE II) scoring in a cardiothoracic intensive care unit. *Critical Care Medicine* **19**: 1266–1269.

Vapnik, V. N. (1995). *The Nature of Statistical Learning Theory*. New York: Springer-Verlag.

Vassar, M. J., Lewis, F. R., Jr., Chambers, J. A., Mullins, R. J., O'Brien, P. E., Weigelt, J. A. et al. (1999). Prediction of outcome in intensive care unit trauma patients: a multicenter study of Acute Physiology and Chronic Health Evaluation (APACHE), Trauma and Injury Severity Score (TRISS), and a 24-hour intensive care unit (ICU) point system. *Journal of Trauma-Injury Infection & Critical Care* **47**: 324–329.

von Bierbrauer, A., Riedel, S., Cassel, W., and von Wichert, P. (1998). [Validation of the acute physiology and chronic health evaluation (APACHE) III scoring system and comparison with APACHE II in German intensive care units]. *Anaesthesist* **47**: 30–38.

Wallis, E. J., Ramsay, L. E., Ul Haq, I., Ghahramani, P., Jackson, P. R., Rowland-Yeo, K., and Yeo, W. W. (2000). Coronary and cardiovascular risk estimation for primary prevention: Validation of a new Sheffield table in the 1995 Scottish health survey population. *BMJ* **320**: 671–676.

Wattigney, W. A., Croft, J. B., Mensah, G. A., Alberts, M. J., Shephard, T. J., Gorelick, P. B. et al. (2003). Establishing data elements for the Paul Coverdell National Acute Stroke Registry: Part 1: Proceedings of an expert panel. *Stroke* **34**: 151–156.

Werbos, P. (1974). Beyond regression: New tools for prediction and analysis in the behavioural science. Cambridge, MA: Harvard University.

Wilairatana, P., Noan, N. S., Chinprasatsak, S., Prodeengam, K., Kityaporn, D., and Looareesuwan, S. (1995). Scoring systems for predicting outcomes of critically ill patients in northeastern Thailand. *Southeast Asian Journal of Tropical Medicine & Public Health* **26**: 66–72.

Wilson, P. W., D'Agostino, R. B., Levy, D., Belanger, A. M., Silbershatz, H., and Kannel, W. B. (1998). Prediction of coronary heart disease using risk factor categories. *Circulation* **97**: 1837–1847.

Wong, D. T., Crofts, S. L., Gomez, M., McGuire, G. P., and Byrick, R. J. (1995). Evaluation of predictive ability of APACHE II system and hospital outcome in Canadian intensive care unit patients. *Critical Care Medicine* **23**: 1177–1183.

Wong, L. S. and Young, J. D. (1999). A comparison of ICU mortality prediction using the APACHE II scoring system and artificial neural networks. *Anaesthesia* **54**: 1048–1054.

Zadeh, L. A. (1994). Fuzzy logic, neural networks, and soft computing. *Communications of the ACM* **37**: 77–84.

Zimmerman, J. E., Wagner, D. P., Draper, E. A., Wright, L., Alzola, C., and Knaus, W. A. (1998). Evaluation of acute physiology and chronic health evaluation III predictions of hospital mortality in an independent database [see comment]. *Critical Care Medicine* **26**: 1317–1326.

11 EVIDENCE-BASED MEDICINE AND META-ANALYSIS: GETTING MORE OUT OF THE LITERATURE

JOSEPH LAU

We are in the era of evidence-based medicine (EBM). Recent medical literature is replete with articles bearing the terms "evidence-based," "systematic review," or "meta-analysis" in their titles. The term EBM was first coined in 1992 (Evidence Based Medicine Working Group 1992). A Google® search of the Internet using the term "evidence-based medicine" yielded over three million hits in August 2005. Although probably only a small fraction of this number truly represents unique Web sites that deal with the methodologies and applications of EBM, it still represents a very large number of sites and signifies a rapidly growing interest in this area. Many organizations in the United States—including government agencies such as the Agency for Healthcare Research and Quality (AHRQ), the Centers for Disease Control and Prevention, the Center for Medicare and Medicaid Services and the National Institutes of Health, professional medical societies, and health care payers and managed care companies—have embraced the EBM approach of evaluating evidence to inform practices and policies. EBM also has become a global activity; numerous participants from many countries are involved worldwide.

EBM formalizes the principles and methods of reviewing and synthesizing evidence that have been developing over several decades. An often-used definition of EBM states, "Evidence-based health care is the conscientious use of current best evidence in making decisions about the care of individual patients or the delivery of health services. Current best evidence is up-to-date information from relevant, valid research about the effects of different forms of health care, the potential for harm from exposure to particular agents, the accuracy of diagnostic tests, and the predictive power of prognostic factors" (Sackett et al. 1996).

This chapter focuses on systematic reviews and meta-analyses, which are the fundamental tools of EBM. The methodologies of carrying out these approaches are reviewed so that you can better appreciate how to interpret and utilize their results. Also discussed are some sites where these products can be accessed to assist medical or health policy decision-making, as well as their limitations.

Clinical Decision Support: The Road Ahead

11.1 SYSTEMATIC REVIEWS AND META-ANALYSES

Systematic reviews and meta-analyses grew out of the need to synthesize the large volume of biomedical literature as it was expanding rapidly during the second half of the last century (Cook et al. 1997). Clinicians are faced with an ever-increasing array of diagnostic tests and medical interventions. The basic premise of the *systematic review* is that comprehensive, rigorous, and unbiased review and synthesis of up-to-date evidence provides the most reliable information to inform health practice. A *meta-analysis* is a systematic review that uses statistical methods to combine results across several studies to address specific questions. It can be used to provide more precise overall estimates of effects and to explore heterogeneity across studies to understand discrepancies. In the rest of this chapter, all discussions on systematic review are also relevant to meta-analysis. Many hundreds of systematic reviews and meta-analyses are published each year. MEDLINE currently indexes over 13,000 items under the medical subject heading of "meta-analysis." These articles include clinical topics and methodological topics.

Systematic reviews and meta-analyses have contributed important insights to interpreting clinical trial results and are making an important impact on clinical practices, health policies, and biomedical research. Discussed here are lessons learned from a cumulative meta-analysis of thrombolytic therapy for acute myocardial infarction (Lau et al. 1992). Thirty-three randomized controlled trials published between 1959 and 1990 involving almost 37,000 patients were included in a meta-analysis of the use of intravenous streptokinase to reduce overall mortality of acute myocardial infarction (see Figure 11-1). All except six of these trials reported nonsignificant results. Thus it should not be too surprising that it was not until 1988, with the publication of two large trials in 1986 and 1988, each with over 10,000 patients demonstrating the unequivocal efficacy of this treatment, did the FDA approve the use of this treatment for acute myocardial infarction; expert recommendations for routine use followed. A meta-analysis combining the 33 earlier studies, however, found a highly statistically significant result of approximately 20 percent reduction of overall mortality, and thus could have resulted in earlier adoption of streptokinase therapy, had it been done prior to the two later definitive studies.

A cumulative meta-analysis, a method of updating and displaying the results of a meta-analysis with the inclusion of each new study, readily demonstrates that statistically significant evidence of efficacy of intravenous streptokinase in reducing overall mortality of AMI was reached as early as 1973, after combining the first eight trials involving about 2,400 patients. Additional studies, including the two largest trials, did not substantially alter the estimate of the magnitude of treatment effect. The later studies mostly narrowed the confidence interval of the estimate. Several other treatments of acute myocardial infarction that were analyzed using cumulative meta-analysis displayed similar findings (Antman et al. 1992). Thus, routinely updated meta-analyses could provide the earliest indication of the benefit or harm of an intervention, as well as minimizing the ethical issue of conducting additional trials in areas where there is already sufficient evidence (for or against a procedure), and avoiding the expenditure of unnecessary resources.

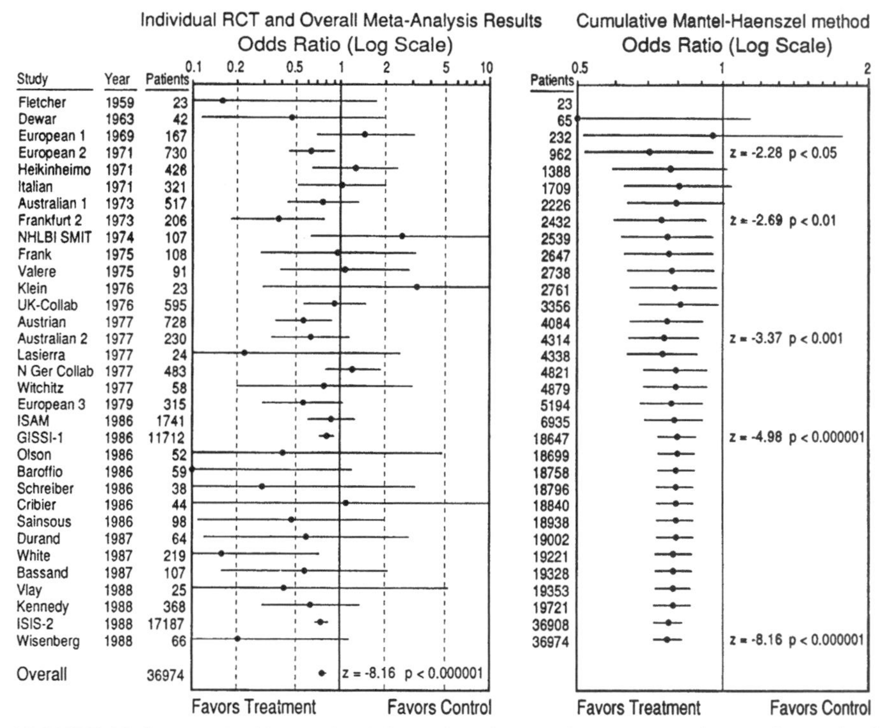

FIGURE 11-1 Standard forest plot (left panel) and a cumulative meta-analysis (right panel) of intravenous streptokinase therapy for acute myocardial infarction; see text for explanation. (Reproduced with permission from the *New England Journal of Medicine*) (Lau et al. 1992).

11.2 METHODOLOGIES OF SYSTEMATIC REVIEW AND META-ANALYSIS

Traditional narrative review articles on a medical topic typically cover a broad range of issues such as etiology, pathology, methods of diagnosis, range of treatments available, and prognosis. Systematic reviews differ from traditional narrative review articles in that the former seeks to answer one or several well-focused research question(s) rather than attempting to provide a broad overview of a topic (Mulrow 1987). Systematic reviews have covered health care interventions, evaluated diagnostic tests, and assessed the association of factors with clinical conditions or outcomes. A systematic review follows a well-defined protocol to identify, appraise, and synthesize the available evidence to minimize bias and to arrive at reliable conclusions. These methods follow precise steps, and each step has its own challenges. Although there are different methodological issues in reviewing studies of intervention for treatment efficacy, diagnostic tests for accuracy, and observational studies for associations, the overall principles of systematic review are similar. The basic principles and methods of systematic review and meta-analysis of intervention studies will be the focus of this chapter.

11.3 DEVELOPING A SYSTEMATIC REVIEW PROTOCOL

Systematic reviews and meta-analyses generally are retrospective analyses of published data. A prospectively formulated protocol for such reviews and meta-analyses that is carefully followed will minimize bias. A protocol should clearly describe the specific research question(s), literature search strategy, selection criteria, approach to critical appraisal of the studies, methods of statistical analyses, and interpretation of the results. Conducting systematic reviews and meta-analyses require specific methodological knowledge. Collaboration between clinical and methodological experts will enhance the validity and usefulness of the review.

11.4 FORMULATING THE RESEARCH QUESTION

Formulating the research question is the most critical step in any systematic review or meta-analysis. The question should be clinically important and framed in a way that it could potentially be directly answered with available studies. For systematic reviews of interventions, the commonly referred to PICO (Population, Intervention, Comparator, Outcome) approach has been found to be very useful to define research questions (Counsell 1997). The research question thus formulated will guide every phase of the review process, from searching the literature to interpreting the results. A typical literature search may start off with hundreds or thousands of citations, most of which will be irrelevant. The PICO review criteria serve as a sieve through which only the studies most likely to be relevant will be retrieved and analyzed. For example, the question "What drugs should be used to treat hypertension?" is not directly answerable from the EBM perspective. Instead, a well-formulated systematic review question would ask, "How does nifedipine compare with hydrochlorothiazide in patients with moderately elevated diastolic blood pressure (100–110 mmHg) in clinical trials evaluating long-term (one-year or more) mortality and morbidity?" By constructing multiple questions each containing various combinations of the PICO elements and then performing systematic reviews to seek answers to each of these questions, one then can begin to develop a fuller appreciation of the evidence available to answer the initial broad and unfocused question. Evidence may often not be available on a specific question; these areas are identified as future research needs.

Formulating a research question is an iterative process and involves a compromise to create a question that is answerable by available evidence. A question that is very narrowly focused may be directly applicable to specific individuals, but the evidence to address this question may be very difficult to come by. In contrast, whereas many studies would likely be available to address a broadly formulated question, the applicability of such results to a specific population will be more uncertain.

A now common use of meta-analyses is to provide data for decision analyses (Jordan and Lau 2003). A decision tree provides a structure that defines needs for specific systematic reviews. Meta-analyses can provide more precise estimates of the probability and utility data in the decision model. A decision analysis model often requires many pieces of data and hence many meta-analyses might be needed. The challenge of finding high quality evidence for conducting meta-analyses for this need is thus magnified.

Developing an analytic framework (sometimes called evidence model or causal pathway) can be very useful in some settings to help formulate research questions (see Figure 11-2). It is particularly useful when multiple interrelated outcomes and factors must be considered to arrive at a recommendation. This

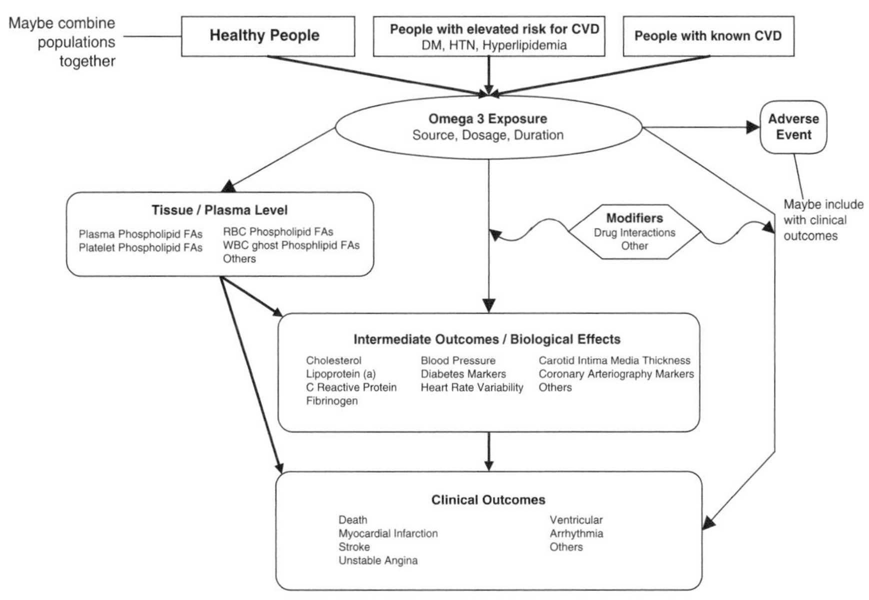

FIGURE 11-2 Analytic framework concerning the effect of omega-3 fatty acid exposure (as supplement or from food sources) on cardiovascular disease. Population of interest in top rectangles; exposure in oval; outcomes in rounded rectangles; effect modifiers in hexagon. Thick connecting lines indicate associations and effects reviewed in this report. Lists in smaller font indicate the specific factors reviewed. (Reprinted from AHRQ Evidence Report: Wang, C., Chung, M., Litchtenstein, A., Balk, E., Kupelnick, B., Devine, D., Lawrence, A., Lau, J. Effects of omega-3 fatty acids on cardiovascular disease. Evidence Report/Technology Assessment No. 94 (Prepared by Tufts-New England Medical Center Evidence-based Practice Center under Contract No. 290-02-0022). AHRQ Publication No. 04-E009-2. Rockville, MD: Agency for Healthcare Research and Quality. March 2004.)

approach is often taken by Evidence-based Practice Centers (EPCs) sponsored by the AHRQ, and routinely used in the U.S. Preventive Services Task Force in their process of reviewing evidence (see subsection 12) (Harris et al. 2001). Even when data are unavailable, an evidence model may still be useful to guide the thinking process.

11.5 LITERATURE SEARCH

A systematic review should be comprehensive and all relevant literature should be reviewed. However, the literature search and selection of articles are often constrained by available time and resources. The literature search strategy and selection of inclusion language for the review should be guided by careful forethought and understanding the nature of the evidence. The literature search generally begins with a search of the MEDLINE database because it is free, readily available electronically, and it indexes over 4,000 biomedical journals that are most likely to be useful in systematic reviews of clinical and health policy topics. Many authors also search the Cochrane Central Register of Controlled Trials, part of the Cochrane Library, which has indexed over 400,000 controlled trials (see subsection 12). This register is a product of many volunteers worldwide hand-searching journals to identify clinical trials not indexed originally in MEDLINE. This database can be searched using standard search terms and Boolean operators and it is updated quarterly. For most mainstream medical topics, the incremental yield of searching additional databases such as EMBASE appears to be minor, and its exclusion seldom affects the overall conclusion (Sampson et al. 2003).

The importance of including non-English language articles in a systematic review is also likely to be topic-dependent. In general, the exclusion of non-English language articles in a meta-analysis has not been found to cause major differences in results (Moher et al. 2000). For certain topics, unrestricted inclusion of articles of all languages may actually result in a biased assessment of the overall effect, if in certain countries there is a tendency to publish only positive results (Vickers et al. 1998).

11.6 DATA EXTRACTION

Data needed for analyses in systematic reviews and meta-analyses must be extracted from the original studies. This is a tedious and time-consuming effort. Considerable skill and experience are also needed to ensure the reliability of the data extraction process. Quite often, two individuals extract data independently, and discrepancies are reconciled, to improve the reliability. Data needed for systematic reviews and meta-analyses include information pertaining to the PICO parameters discussed earlier, as well as to methodological issues pertaining to the design and conduct of a study. This information should be made available to the readers so that they can draw conclusions about the methodological quality and applicability of a study and the reliability of the results. Typically a data collection form is developed for each systematic review. Although there is information common to all studies, each systematic review will also need to collect data unique to the topic.

Data extraction is often a challenging exercise. Data reported across studies often are not standardized, and important information is often missing. The same information may be reported inconsistently within a study, leading to uncertainties about the correct answer. The need for subjective judgment in the collection of data potentially contributes bias to the systematic review process.

11.7 ASSESSING THE QUALITY OF STUDIES

Conclusions drawn from systematic reviews should be based on good quality studies. It is also important to convey to the readers of the systematic review the conclusions drawn by the reviewers regarding the quality of the evidence. However, the evaluation of study quality is not straightforward. One definition of study quality is, "The confidence that the trial design, conduct, and analysis has minimized or avoided biases in its treatment comparison (Moher et al. 1995)." Various methods have been proposed to assess the methodological quality of a study, which include checklists of specific study elements, and numerical quality scores based on some schemes of weighting of elements believed to contribute to quality (Moher et al. 1995). It is often assumed that poor quality studies report exaggerated effect size. Several empirical assessments in limited topics reported that the lack of concealment of random allocation and blinding has led to exaggerated effects (Schulz et al. 1994). However, the development of a list of quality factors is hampered by the fact that there is no true reference standard to determine quality. In addition, weights assigned to quality-related factors in coming up with a numerical score are arbitrary. Quality scores based on various study design and conduct features have been found to yield inconsistent results (Juni et al. 1999; Balk et al. 2002). Another problem is that the assessment of study quality is based on author-supplied information in the article. It has been found in many instances that the absence of information in the paper does not necessarily mean that a specific feature of the trial was not performed (Hill et al. 2002). Because of word count limitations in journals, descriptions of study methods are often minimized in favor of discussions. Guidelines, such as the CONSORT statement, have been published to improve the conduct and reporting of future clinical trials (Moher et al. 2001). Hopefully, these efforts will lead to improved quality of future trials, and of the documentation thereof. In the meantime, assessing the quality of a report will remain a challenge.

11.8 COMBINING DATA IN A META-ANALYSIS

Meta-analysis is a systematic review in which the reviewers have decided that sufficient data are available, from studies meeting inclusion criteria, to address a specific question, and that it is reasonable to combine them to provide an overall answer. Many textbooks and tutorial articles have been written about these methods (Cooper and Hedges 1994; Normand 1999; Egger et al. 2001). The most common form of meta-analysis aims to determine an overall weighted average of the effect size, by combining data using a fixed-effect or a random-effects model. Both continuous data (e.g., blood pressure measurements recorded as mmHg) and dichotomous data (e.g., dead or alive expressed as odds ratio, risk ratio, or risk difference) can be used in a meta-analysis (Lau et al. 1997). A fixed-effect model assumes that all studies are estimating a single true value. Variations around the true value are due only to the sampling error affected by the number of events and the size of the study. A fixed-effect model meta-analysis weighs studies by the inverse of the within-study variance. Thus, large studies and studies with more events tend to receive the most weights in a fixed-effect model meta-analysis (i.e., they are most influential). The random-effects method incorporates both the fixed-effect weight and the between-study weight due to heterogeneity of results across studies. The random-effects model distributes the weight more evenly (i.e., small studies receive more weight) across studies when there is heterogeneity across studies. The random effects model tends to give more conservative results (i.e., wider confidence interval) when there is heterogeneity. The choice of the statistical model used to combine studies should be based on *a priori* understanding of the studies, but the random effects model is the method generally recommended by most experts as the default analysis.

11.9 EXPLORING HETEROGENEITY WITH SUBGROUP AND META-REGRESSION ANALYSES

No two studies are identical. Differences in results across several studies in a meta-analysis are to be expected, and efforts should be made to understand the reasons for these discrepancies. The common method of combining data to provide a point estimate, using either a fixed-effect model or a random-effects model, fails when significant heterogeneity of treatment effect among trials is present. As an alternative to either ignoring differences, as in the fixed effects model, or incorporating the differences in the form of pooling with a random effect model, subgroup analysis and meta-regression could be used to explore heterogeneity across studies (Lau et al. 1998).

Subgroups within individual studies may be too small to yield significant results. By combining similarly defined subgroups across several studies, a meta-analysis may reveal consistent trends and statistical significance when combined. Age and sex are natural covariates for subgroup analyses in which data are frequently available. However, subgroup results may not be consistently reported across studies. Therefore, meta-analyses of subgroup data should be viewed with caution, because their summaries may be based on selectively reported significant subgroup results.

Meta-regression is a technique for performing a regression analysis to assess the relationship between the treatment effects (e.g., risk ratio) and the study characteristics of interest (e.g., dosage, severity of illness, or duration of treatment) or factors concerning the execution of the study (e.g., proper blinding) (Schmid 1999; Thompson and Higgins 2002). The method provides a means to explore sources of heterogeneity and therefore is amenable to explaining discrepancies that may be found across clinical trials. Meta-regression methods are even more important for the interpretation of observational data, due to the inability to control for potential confounders that either were not, or were inadequately, measured. Without the benefit of individual patient data, these meta-regression models must rely on the summary results of published studies. These summary results describe only between-study, not between-patient, variation in the risk factors and are therefore most useful for a study of characteristics that differ across studies (e.g., drug dosage).

A recent meta-analysis of vitamin E supplementation illustrates the usefulness of meta-regression (Miller et al. 2005). Nineteen studies evaluated the effects of various dosages of supplemental vitamin E, ranging from 16 IU to 2000 IU. None of the individual studies that qualified for this meta-analysis had explored the relationship of dosage with overall mortality. A standard meta-analysis found no overall effect using either a fixed-effect or random-effects model. However, a meta-regression found a clear relationship of increased mortality with increasing dosage of vitamin E supplementation. The threshold at which increased mortality was observed was far lower than the safe upper limit level recommended by authoritative bodies.

There are limitations to meta-regression. Key risk factors that vary across patients and that can be measured only as aggregate values, such as age and gender, are difficult to address adequately by meta-regression. One reason for this is that aggregated values tend to exhibit little between-study variation, thus providing minimal information across the potential range of the factor. Use of aggregated values may also introduce ecological bias when they are used to estimate effects for individuals of factors that vary within study by patient. A final, major difficulty with meta-regression is the lack of consistently reported covariates in clinical trials.

11.10 ISSUES IN CONDUCTING META-ANALYSES

Many issues are encountered in the conduct of a meta-analysis. The reviewer must consider the trade-offs between thoroughness and feasibility in formulating research questions and in performing the literature review. One also needs to decide how best to handle heterogeneity of clinical study designs and quality issues, as well as the choice of outcome metric and the method of synthesis.

11.10.1 Publication Bias

A meta-analysis should assess the potential for various factors that may affect its validity. Unpublished studies with negative results threaten the validity of a meta-analysis: this is known as publication bias. Various methods have been proposed to detect or to adjust for unpublished studies (Thornton and Lee 2000). The inverted funnel plot is the most popular method used to detect publication bias. This idea is based on the premise that an unbiased collection of studies of various sizes should all be scattered symmetrically around a common effect. Small studies inherently have greater variability of results than large studies and will display a wider scattering. In a typical funnel plot the effect size of the study is plotted against the weight (study size or inverse of the variance) of a study. The greater variability of the small studies will appear as the wide end of the inverted funnel, whereas larger studies with lesser variability represent the narrow neck of the funnel. If small studies with negative results are not published, a funnel plot of available studies will show an asymmetric funnel suggesting publication bias. Because of its intuitive appeal, this method has been in popular use since its introduction about 20 years ago.

The validity of this method to detect publication bias has been challenged. Funnel plots of the same data in a meta-analysis using different effect-size scales (e.g., risk difference vs. odds ratio) and different methods of determining the study weight (e.g., variance vs. number of subjects) have been demonstrated to result in opposite appearance of the shape of the funnel, and therefore different interpretations (Tang and Liu 2000). Also, most meta-analyses have fewer than 10 studies, which may not be a sufficient number of studies to form a valid interpretation of the funnel plot. Furthermore, the funnel plot is not based on statistical principles; it is a subjective visual interpretation of a scatter plot that could be inconsistently evaluated by different interpreters. It has also been demonstrated that readers cannot differentiate between plots generated by computers simulating true publication bias and heterogeneity across studies (Terrin et al. 2005).

Asymmetric funnel plots should be interpreted more appropriately as heterogeneous data, for which publication bias is only one possible explanation. Other causes of heterogeneity could be study quality, differences in patient population, or differences in intervention. Other methods to detect or to adjust for publication bias have been proposed that are based on statistical principles (Thornton and Lee 2000). However, all these methods are based on certain assumptions that are difficult to verify (Terrin et al. 2003). The only foolproof method to reduce the risk of publication bias is the mandatory registration of all human clinical trials prior to their conduct.

11.10.2 Large Trial vs. Meta-Analysis of Small Trials

Large clinical trials often are considered as the definitive last word in clinical evidence. Several empirical evaluations have been performed to compare the results of large trials with meta-analyses of small trials as a means to assess the validity of meta-analyses (Villar et al. 1995; Cappelleri et al. 1996; LeLorier et al. 1997). These evaluations differ in the selection of comparison studies, their methods of analysis, and definition of large (based on power calculations or arbitrarily defining large as 1,000 patients or more), and these factors may contribute to some discrepancies in these evaluations (Ioannidis et al. 1998). Overall, these evaluations found that disagreements between large trials and the corresponding meta-analysis of small trials occur in about 10 to 30 percent of the comparisons. The high rate of discrepancy raises the question of validity of meta-analyses (LeLorier et al. 1997). However, another study that compared the rate of discrepancies between large trials, defined as at least 1,000 patients, within the same meta-analysis, reported that disagreements among large trials were just as common as disagreement between results of large trials and meta-analyses of small trials (Furukawa et al. 2000). These observations point to the fact that heterogeneity across clinical trials addressing the same problem is very common, regardless of study size.

The interpretation of the results of a meta-analysis should be made with respect to all factors that might affect the results. All studies, large or small, should be considered together in summarizing evidence. Figure 11-3 depicts the factors that contribute to the observed results of an individual clinical trial. Figure 11-4 depicts the factors that must be considered in the assessment of results from a meta-analysis.

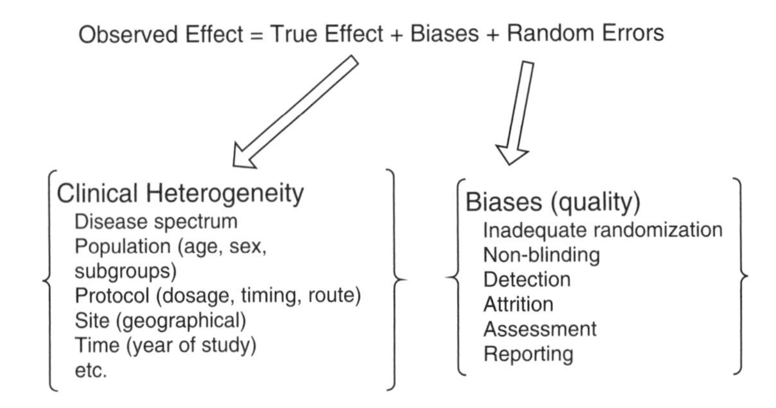

$$TE_{obs} = \beta_0 + \beta_1 x_1 + \beta_2 x_2 + \cdots + \beta_i x_i + \beta_j x_j + \cdots + \varepsilon$$

FIGURE 11-3 Factors that contribute to the observed treatment effects in a randomized controlled trial.

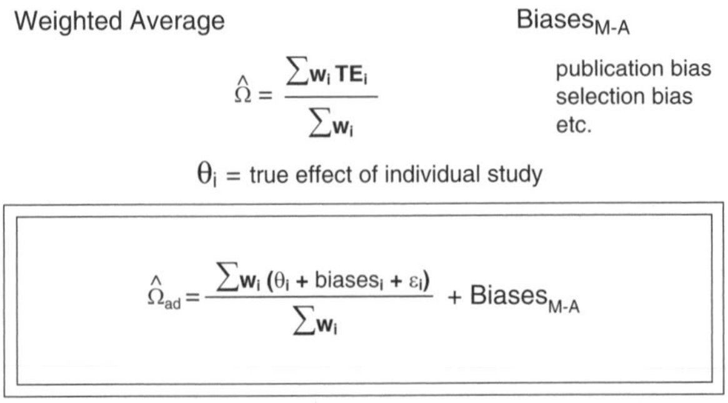

Weighted Average

$$\hat{\Omega} = \frac{\sum w_i TE_i}{\sum w_i}$$

Biases$_{M\text{-}A}$

publication bias
selection bias
etc.

θ_i = true effect of individual study

$$\hat{\Omega}_{ad} = \frac{\sum w_i (\theta_i + \text{biases}_i + \varepsilon_i)}{\sum w_i} + \text{Biases}_{M\text{-}A}$$

FIGURE 11-4 Estimated treatment effect in a meta-analysis.

11.10.3 Updating Systematic Reviews and Meta-Analyses

The need to routinely update meta-analyses has been amply demonstrated (Antman et al. 1992). Most systematic reviews and meta-analyses are published in peer-reviewed printed journals. With the publication of new research findings, the conclusions of these reviews may change and their usefulness may become obsolete. Printed articles do not readily lend themselves to updating already published results by the same author. Subsequently, in areas of rapidly developing interest and many publications, one might see several meta-analyses published by different groups on the same topic appearing in different journals over a short period of time.

Even after a treatment has been found to be efficacious, new trials may still need to be conducted to evaluate different uses of the interventions or in different populations or in different settings. Routine updates of completed reviews should be conducted to keep the information current. Even with the advent of electronic publication, there is neither incentive nor mechanism to encourage authors to keep the published meta-analyses current. The Cochrane Library is the only entity that has a built-in mechanism to routinely update published systematic reviews.

11.11 USES OF SYSTEMATIC REVIEWS AND META-ANALYSES

EBM has disseminated throughout health care. Initially used to assess evidence in clinical medicine, it has now been applied to surgery, nursing, mental health, public health, dentistry, veterinary medicine, and genetics. Even the judicial system is interested in learning about its methods (Ioannidis and Lau 2004). Systematic reviews and meta-analyses have been used to assess the evidence of interventions, diagnostic tests, risk factor associations, and prognosis. Meta-analyses have covered many topics in virtually every clinical discipline. Types of studies that have been synthesized in these reviews included randomized controlled trials, cohort studies, case-control studies, as well as case reports.

The largest number of published systematic reviews and meta-analyses is in the area of randomized controlled trials of interventions. Meta-analyses of interventions have been published in all major clinical areas and specialties of medicine. Because of the availability of a large number of trials, cardiovascular diseases have received the most attention.

Compared with meta-analyses of interventions, which number in the thousands, there are only several hundred meta-analyses of diagnostic tests. Studies evaluating the use of diagnostic tests can generally be grouped into six categories: 1) technical feasibility, 2) diagnostic accuracy, 3) impact on diagnostic thinking, 4) impact on therapeutic decision-making, 5) impact on clinical outcome, and 6) societal impact (Tatsioni et al. 2005). There are far more publications on evaluating various aspects of technical feasibility of diagnostic tests than all other categories combined, but these studies have limited direct clinical relevance. Because of the immediate clinical usefulness and the availability of studies, most meta-analyses of diagnostic tests have focused on diagnostic performance (i.e., sensitivity and specificity). Detailed discussion on issues related to systematic reviews of diagnostic tests is beyond the scope of this chapter; we refer you to articles for examples of these assessments and articles that discuss their methodologies (Irwig et al. 1994; Vamvakas 1998).

There are many meta-analyses of associations of factors with health outcomes. These meta-analyses typically assess the associations reported in observational studies such as cohort or case-control studies. Examples of these include associations of second-hand cigarette smoke with cancers, or associations of nutrition intake (fish or fish oil or antioxidants) with cancer or cardiovascular diseases. Because observational studies cannot fully account for confounders, meta-analyses of these studies must be interpreted with greater care than randomized trials (Ioannidis et al. 2001).

11.12 ACCESSING SYSTEMATIC REVIEWS AND META-ANALYSES AND RELATED PRODUCTS

Systematic reviews and meta-analyses, like other journal articles, are increasingly available online soon after they are published. Meta-analyses published in journals indexed in MEDLINE or other major electronic databases (e.g., EMBASE) can readily be identified by using the term "meta-analysis." Meta-analysis has been a recognized medical subject heading in MEDLINE since the early 1990s, but the term "systematic review" is not. Most systematic reviews include this phrase in their titles or abstract, so a text-word search will identify them. Using the term meta-analysis will find almost all of the clinical applications of meta-analysis, as well as most of the methodological articles concerning this topic. However, some articles that report having quantitatively combined results of several studies, without carrying out the more rigorous step of systematic review, sometimes also have been incorrectly indexed as meta-analysis in MEDLINE. The number of articles thus classified appears to be small; nonetheless, readers should be aware of this problem.

The Cochrane Library is the product of the Cochrane Collaboration, which is an international voluntary organization with the aim of identifying, synthesizing, and disseminating the information about the effects of health care intervention. The Library currently represents the single most comprehensive source of high quality systematic reviews that are routinely updated. It has over 2,200 completed systematic reviews and 1,500 more protocols in various stages of preparation (www.Cochrane.org). Cochrane systematic reviews are indexed in MEDLINE. The abstract of the review is available but the full text can be accessed only through a subscription via the Internet or on CD-ROM.

A major EBM initiative in the United States was undertaken by the AHRQ, which created the Evidence-based Practice Center Program in 1997 to produce evidence reports and technology assessments. The EPCs develop evidence reports and technology assessments based on rigorous, comprehensive reviews of relevant scientific literature, emphasizing explicit and detailed documentation of methods, rationale, and assumptions. These reports are intended to be used for informing and developing coverage decisions, quality measures, educational materials and tools, guidelines, and research agendas. More than 120 of these reports covering a wide array of topics have been completed and indexed in MEDLINE. These reports are freely available to the public at the AHRQ Web site (http://www.ahrq.gov/clinic/epc).

AHRQ also manages the National Guideline Clearinghouse (NGC) (www.guidelines.gov), which is a Web site that has a database of evidence-based clinical practice guidelines. This database currently contains over 1,700 guidelines from various organizations around the world. Guidelines must meet certain methodological criteria to be included. The guidelines database is updated weekly and it provides structured and standardized summaries of each guideline. The Web site also has a facility to allow a user to make parallel comparison of two or more guidelines, as well as links to the full text of the guidelines when they are available.

EBM is a global activity. Many individuals, organizations, and government agencies around the world are participating in this activity. Numerous Web sites provide tutorials, repository of completed systematic reviews or technology assessments, as well as links to other evidence-based Web sites.

11.13 CONCLUSION

In a little over a decade, EBM has captured the attention of the medical community as an invaluable approach to inform health care practices and policies. The number of publications and Web sites with information on EBM has rapidly increased. Although many methodological issues remain, there are no longer debates on whether systematic reviews are useful or whether meta-analysis is a valid statistical method to combine evidence. However, their limitations must be recognized.

The practice of EBM needs data. It is sometimes discouraging to carry out a systematic review and find that there are few or no studies of sufficient quality upon which to draw conclusions and make recommendations. From the perspective of a user looking for systematic reviews to guide patient management, many clinical questions for routine patient care have yet to be analyzed, and the lack of evidence may also be disappointing.

Sometimes the best available evidence for a research question is in the form of nonrandomized observational studies or those reporting surrogate outcomes. Multiple sources of evidence, including observational studies, are sometimes needed to supplement systematic reviews of randomized controlled trials to further inform health care decisions. Using information from observational studies to inform health practices has its benefits and drawbacks. Systematic reviews and meta-analyses that have insufficient evidence to answer specific questions could nonetheless be useful as they identify areas of research gaps, and this information could be used to propose future research agenda.

The practice of EBM needs skills. Systematic reviews and meta-analyses of randomized controlled trials often are based on studies with restricted patient inclusion criteria. The applicability of their results to the general population may be uncertain. Interpreting results from systematic reviews for care of individual patients as well as understanding the jargons of systematic review and meta-analysis needs training. The quality of systematic reviews also varies across journals, even though they may have been peer reviewed. Guidelines have been proposed to improve the reporting of systematic reviews of randomized controlled trials (QUORUM) (Moher et al. 1999), observational studies (MOOSE) (Stroup et al. 2000), and diagnostic test evaluations (STARD) (Bassuyt et al. 2003). Cochrane reviews, through the standardization of the methodologies and infrastructure support, have been found to be of higher quality than those published in journals (Jadad et al. 1998).

The practice of EBM should be based on evidence. Methodologies used in meta-analyses often are adapted from other areas without further evaluation, specifically for the setting of meta-analysis. Methodological decisions also often are made on assumptions that are not based on evidence (e.g., including all languages in a systematic review may not necessarily be desirable). Sometimes methods are proposed and widely used without a formal evaluation (e.g., funnel plot to detect publication bias). The large number of systematic reviews and meta-analyses published over the past 20 years has fostered the development of better methods of synthesis and the appreciation of methodological issues. Empirical studies have been performed to elucidate how best to synthesize and interpret evidence.

The practice of EBM needs to be more efficient. Systematic reviews save users from countless hours of having to conduct their own research. The availability of online EBM products such as Cochrane reviews and AHRQ evidence reports make immediate access feasible. However, users may still need to spend hours sifting through these publications to digest the information. Efforts are being made to streamline the information from systematic reviews and meta-analyses so that the key messages can be used in real-time by practicing clinicians for patient care purposes. Currently, the practice of EBM at the point of care is haphazard, mostly the efforts of individuals. System-wide implementation of resources to assist users to identify relevant evidence is needed to increase the impact of EBM in real world settings.

The success of EBM also has attracted some to use the term "evidence-based" loosely and perhaps inappropriately for various purposes ranging from continuing medical education (CME) to clinical practice guidelines. In the case of CME, publications and workshops that are sponsored by pharmaceutical companies primarily for the purpose of promoting their products may label their activities as "evidence-based," when only several selected randomized controlled trials in favor of their products are discussed. Some clinical practice guidelines make claims of being evidence-based, but the methodologies of the review of evidence and their connection to the recommendations and the bibliography used to support their recommendations are neither explicit nor transparent. Randomized controlled trials are invaluable in the armamentarium of EBM; however, arbitrarily selecting a few studies to support a particular viewpoint does not constitute evidence-based method. The result of a single study is seldom able to fully inform general clinical practice.

Evaluating and summarizing clinical evidence and using the analyses for patient care or health policy decisions are complex activities. A great deal of research has been done over the last 20 years to improve the methods and the understanding of the issues. The quality of the primary research is also being improved because of it. The number of systematic reviews will continue to increase and EBM has now permeated all disciplines of health care and has become a household word (The Year in Ideas 2001). There are important limitations, and the quality of EBM products is only as good as the primary research that they summarized. Much work remains to be done to realize the promises of EBM.

REFERENCES

Antman, E. M., Lau, J., Kupelnick, B., Mosteller, F., Chalmers, T. C. (1992). A comparison of results of meta-analyses of randomized control trials and recommendations of clinical experts. Treatments for myocardial infarction. *JAMA* **268**: 240–248.

Balk, E. M., Bonis, P., Moskowitz, H., Schmid, C. H., Ioannidis, J. P. A., Wang, C., Lau, J. (2002). Correlation of quality measures with estimates of treatment in meta-analyses of randomized trials. *JAMA* **287**: 2973–2982.

Bossuyt, P. M., Reitsma, J. B., Bruns, D. E., Gatsonis, C. A., Glasziou, P. P., Irwig, L. M. et al. (2003). Standards for reporting of diagnostic accuracy. Towards complete and accurate reporting of studies of diagnostic accuracy: the STARD initiative. *BMJ* **326**: 41–44.

Cappelleri, J. C., Ioannidis, J. P. A., deFerranti, S. D., Schmid, C. H., Aubert, M., Chalmers, T. C., Lau, J. (1996). Large trials versus meta-analyses of smaller trials: How do their results compare? *JAMA* **276**: 1332–1338.

Cook, D. J., Mulrow, C. D., Haynes, R. B. (1997). Systematic reviews: Synthesis of best evidence for clinical decisions. *Ann Intern Med* **126**: 376–380.

Cooper, H., Hedges, L. V. (1994). *The handbook of research synthesis.* New York: Russell Sage Foundation.

Counsell, C. (1997). Formulating questions and locating primary studies for inclusion in systematic reviews. *Ann Intern Med* **127**: 380–387.

Egger, M., Smith, G. D., Altman, D. G., eds. (2001). Systematic Reviews in Health Care: Meta-analysis in context. London: BMJ Publishing Group.

Evidence-Based Medicine Working Group. (1992). Evidence-based medicine: A new approach to teaching the practice of medicine. *JAMA* **268**: 2420–2425.

Furukawa, T. A., Streiner, D. L., Hori, S. (2000). Discrepancies among megatrials. *J Clin Epidemiol* **53**: 1193–1199.

Harris, R. P., Helfand, M., Woolf, S. H., Lohr, K. N., Mulrow, C. D., Teusch, S. M., Atkins, D. (2001). Current methods of the U. S. Preventive Services Task Force: A review of the process. *Am J Prev Med* **20**: 21S–35S.

Hill, C. L., LaValley, M. P., Felson, D. T. (2002). Discrepancy between published report and actual conduct of randomized clinical trials. *J Clin Epidemiol* **55**: 783–786.

Ioannidis, J. P. A., Cappelleri, J. C., Lau, J. (1998). Issues in comparisons between meta-analyses and large trials. *JAMA* **279**: 1089–1093.

Ioannidis, J. P. A., Haidich, A. B., Pappa, M., Pantazis, N., Kokori, S. I., Tektonidou, M. G. et al. (2001). Comparisons between randomized and non-randomized evidence. *JAMA* **286**: 821–830.

Ioannidis, J. P. A., Lau, J. (2004). Systematic review of medical evidence. *Brooklyn Law J* **12**: 509–535.

Irwig, L., Tosteson, A. N. A., Gatsonis, C., Lau, J., Colditz, G., Chalmers, T. C., Mosteller, F. (1994). Guidelines for meta-analyses evaluating diagnosis tests. *Ann Intern Med* **120**: 667–676.

Jadad, A. R., Cook, D. J., Jones, A., Klassen, T. P., Tugwell, P., Moher, M., Moher, D. (1998). Methodology and reports of systematic reviews and meta-analyses: A comparison of Cochrane reviews with articles published in paper-based journals. *JAMA* **280**: 278–280.

Jordan, H. S., Lau, J. (2003). Linking pharmacoeconomic analyses to results of systematic review and meta-analysis. *Expert Rev Pharmacoeconomics Outcomes Res* **3**: 89–96.

Juni, P., Witschi, A., Bloch, R., Egger, M. (1999). The hazards of scoring the quality of clinical trials for meta-analysis. *JAMA* **282**: 1054–1060.

Lau, J., Antman, E. M., Jimenez-Silva, J., Kupelnick, B., Mosteller, F., Chalmers, T. C. (1992). Clinical implications of cumulative meta-analyses of randomized control trials: Acute myocardial infarction as an example. *N Engl J Med* **327**: 248–254.

Lau, J., Ioannidis, J. P. A., Schmid, C. H. (1997). Quantitative synthesis in systematic reviews. *Ann Intern Med* **127**: 820–826.

Lau, J., Ioannidis, J. P. A., Schmid, C. H. (1998). Summing up evidence: One answer is not always enough. *Lancet* **351**: 123–127.

LeLorier, J., Gregoire, G., Benhaddad, A., Lapierre, J., Derderian, F. (1997). Discrepancies between meta-analyses and subsequent large randomized, controlled trials. *N Engl J Med* **337**: 536–542.

Miller, E. R. 3rd, Pastor-Barriuso, R., Dalal, D., Riemersma, R. A., Appel, L. J., Guallar, E. (2005). Meta-analysis: High-dosage vitamin E supplementation may increase all-cause mortality. *Ann Intern Med* **142**: 37–46.

Moher, D., Jadad, A. R., Nichol, G., Penman, M., Tugwell, P., Walsh, S. (1995). Assessing the quality of randomized controlled trials: An annotated bibliography of scales and checklists. *Control Clin Trials* **16**(1): 62–73.

Moher, D., Cook, D. J., Eastwood, S., Olkin, I., Rennie, D., Stroup, D. F., for the QUOROM group. (1999). Improving the quality of reports of meta-analyses of randomized controlled trials: The QUOROM statement. *Lancet* **354**: 1896–1900.

Moher, D., Pham, B., Klassen, T. P., Schulz, K. F., Berlin, J. A., Jadad, A. R., Liberati, A. (2000). What contributions do languages other than English make on the results of meta-analyses? *J Clin Epidemiol* **53**: 964–972.

Moher, D., Schulz, K. F., Altman, D. G. (2001). CONSORT. The CONSORT statement: Revised recommendations for improving the quality of reports of parallel group randomized trials. *BMC Medical Research Methodology* **1**: 2.

Mulrow, C. D. (1987). The medical review article: state of the science. *Ann Intern Med* **106**: 485–488.

Normand, S. T. (1999). Meta-analysis: Formulating, evaluating, combining, and reporting. *Stat Med* **18**: 321–359.

Sackett, D. L., Rosenberg, W. M., Gray, J. A., Haynes, R. B., Richardson, W. S. (1996). Evidence based medicine: What it is and what it isn't. *BMJ* **312**: 71–72.

Sampson, M., Barrowman, N. J., Moher, D., Klassen, T. P., Pham, B., Platt, R. et al. (2003). Should meta-analysts search Embase in addition to Medline? *J Clin Epidemiol* **56**: 943–955.

Schmid, C. H. (1999). Exploring heterogeneity in randomized trials via meta-analysis. *Drug Information J* **33**: 211–224.

Schulz, K. F., Chalmers, I., Grimes, D. A., Altman, D. G. (1994). Assessing the quality of randomization from reports of controlled trials published in obstetrics and gynecology journals. *JAMA* **272**: 125–128.

Stroup, D. F., Berlin, J. A., Morton, S. C., Olkin, I., Williamson, G. D., Rennie, D. et al. (2000). Meta-analysis of observational studies in epidemiology: A proposal for reporting. Meta-analysis Of Observational Studies in Epidemiology (MOOSE) group. *JAMA* **283**: 2008–2012.

Tang, J. L., Liu, J. L. Y. (2000). Misleading funnel plot for detection of bias in meta-analysis. *J Clin Epidemiol* **53**: 477–484.

Tatsioni, A., Zarin, D. A., Aronson, N., Samson, D. J., Flamm, C. R., Schmid, C., Lau, J. (2005). Challenges in systematic reviews of diagnostic technologies. *Ann Intern Med* **142**: 1048–1055.

Terrin, N., Schmid, C. H., Lau, J., Olkin, I. (2003). Adjusting for publication bias in the presence of heterogeneity. *Stat Med* **22**: 2113–2126.

Terrin, N., Schmid, C. H., Lau, J. (2005). Researchers cannot visually identify publication bias from funnel plots in typical-size systematic reviews. *J Clin Epidemiol* **58**: 894–901.

Thompson, S. G., Higgins, J. P. Y. (2002). How should meta-regression analyses be undertaken and interpreted? *Stat Med* **21**: 1559–1573.

Thornton, A., Lee, P. (2000). Publication bias in meta-analysis: Its causes and consequences. *J Clin Epidemiol* **53**: 207–216.

The Year in Ideas. *New York Times Magazine*, Section 6, December 9, 2001.

Vamvakas, E. C. (1998). Meta-analyses of studies of the diagnostic accuracy of laboratory tests. A review of the concepts and methods. *Arch Pathol Lab Med* **122**: 675–686.

Vickers A., Goyal N., Harland R., Rees R. (1998). Do certain countries produce only positive results? A systematic review of controlled trials. *Control Clin Trials* **19**: 159–166.

Villar, J., Carroli, G., Belizan, J. M. (1995). Predictive ability of meta-analysis of randomized controlled trials. *Lancet* **345**: 772–776.

IV

REPRESENTING THE KNOWLEDGE: STANDARDIZATION EFFORTS

12 DECISION RULES AND EXPRESSIONS

ROBERT A. JENDERS

12.1 INTRODUCTION

Deterministic reasoning is a key type of decision-making process in which a decision maker applies branching logic and deduction against the information of a particular situation in order to arrive at a plan of action. A decision rule is a representation of knowledge in a particular domain that encapsulates the flow of logic employed in deterministic reasoning to make a decision. Decision rules, then, represent a form of algorithm, typically represented as discriminating questions or logical IF-THEN statements that may be followed to reach some conclusion. They map the circumstances of a particular situation, such as the case of an ill patient for whom a diagnosis must be chosen, to a particular choice, whether that be a diagnosis, a treatment plan, or an inferred observation that, in turn, may lead to another decision.

In a computer-based clinical decision support (CDS) system, decision rules often are represented in one of two formats: procedures and production rules. Like a subroutine in a programming language, a procedure is a collection of references to data together with logical statements that manipulate them and execute, largely serially, using control structures to direct the flow of decision-making through the procedure. In a system based on production rules, each unit of knowledge is a single IF-THEN logical statement, and an inference engine, evaluating the available data and statements, chooses which statement to execute next.

Although these formalisms have been applied to address a wide range of problems, lack of specificity for the medical domain and lack of standardization have impaired both use and sharing of knowledge bases encoded using them. Recognition of these impediments led to development of a standard approach that combines these formalisms, represented by the Arden Syntax. Perceived limitations with this standard and the need to encode a growing body of computable clinical practice guidelines has led to the examination of other approaches, including the use of a standard expression language in the context of a guideline formalism.

This chapter examines the use of decision rules as a knowledge representation formalism for CDS. The details of such a formalism are explored, including inference mechanisms that are employed in order make decisions

Clinical Decision Support: The Road Ahead

using knowledge encoded in this fashion. Further work to adapt these formalisms as standards is reviewed, with an emphasis on the Arden Syntax and a common expression language. Advantages and disadvantages of these approaches are explored.

12.2 PROCEDURAL KNOWLEDGE

Some of the earliest work (Miller 1994) in implementing decision rules for CDS used procedures written in conventional programming languages. Two key features characterize this representation. First, clinical knowledge and inferencing or control knowledge are mixed in the same representation. This means that instructions to the computer about how to use the clinical knowledge, such as which statement to execute next, is mixed with logical statements about the clinical domain, such as a laboratory test threshold that must be exceeded in order for the diagnosis of a particular disease state to be made.

Second, the flow of control is made explicit. A procedure typically is a series of statements that are executed serially—in the order that they appear in the unit of knowledge. Control statements, such as GO TO and iterations, interrupt the serial execution but still specify explicitly the next statement to be executed, although that may be dependent on data available only at the time of execution. Control knowledge includes not only specification of the flow of execution but also how communication with users occurs (e.g., synchronously via a computer terminal), conditions under which the procedure will execute (e.g., when called from an electronic medical record), and methods for displaying output (e.g., sending a fax to a clinician).

Decision rules characterized by an explicit flow of control in accord with a series of branching questions or logical statements sometimes are represented graphically as decision trees (see Figure 12-1). In a typical decision

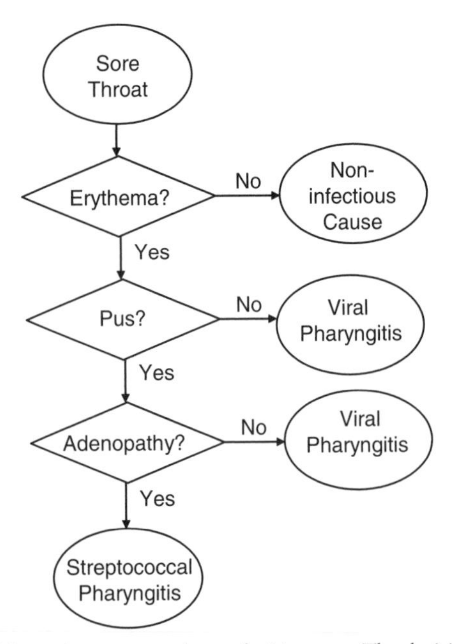

FIGURE 12-1 Decision rule represented as a decision tree. The decision rule helps determine the diagnosis in a case of a patient with a sore throat based on physical examination findings.

tree, each node in the tree may ask a different yes/no question, and the appropriate branch of the tree is followed depending on the response to the question. Ultimately, a conclusion of the decision rule is reached when the traversal encounters a terminal or leaf node of the tree that offers no further refining questions.

This approach offers many advantages. Nearly any programming language that supports subroutines, functions, or procedures may be used to encode the clinical knowledge in executable format. In turn, this means that commonly available programming tools for these languages, such as compilers or debuggers, may be used. If a programming language used is one that is supported on many different types of computers, development and maintenance of the knowledge may be done on multiple platforms without the need to acquire specialized software. Further, because flow control is explicit, the knowledge engineer can tightly control the order of execution of statements, improving the predictability of the results of executing the software and thus improving its accuracy. Moreover, conventional programming languages, such as C++ or Java, typically offer libraries of preprogrammed functions to perform common tasks, such as retrieving data from databases, thus facilitating the interface between the decision rule and data repositories.

However, while advantageous in many respects, the procedural approach to knowledge representation also has significant flaws. Key among these is the mixture of control and clinical knowledge. This makes it difficult to acquire and to maintain the knowledge, because the author must be familiar not only with the clinical domain but also with the syntax and control features of the programming language. Moreover, subsequent edits of the clinical knowledge may inadvertently alter the control structures embedded in the same statements, thus adversely affecting the execution and possibly the accuracy of the CDS. Also, any changes to the clinical knowledge may require recompilation of the software for the decision rule—an expensive and time-consuming process, magnified if the updated decision rule then must be distributed to many different places.

To avoid these challenges, many CDS systems employ an architecture that separates control knowledge from clinical knowledge. This allows the clinical knowledge to be maintained separately from the control knowledge, thus allowing the clinical domain expert or knowledge engineer to focus on just the expert decision rules without having to be concerned about control structures or the need to recompile the entire CDS system each time a new decision rule is introduced or an old one updated. Moreover, the clinical knowledge may be represented in a format more understandable to clinical experts than a typical programming language, thus facilitating validation of executable clinical knowledge. An early form of knowledge representation that fulfills these advantages is the production rule system.

12.3 KNOWLEDGE AS PRODUCTION RULES

Production rules were first studied in the 1940s, when they were developed as axioms that could be used to rewrite strings as part of the specification of a formal grammar. Because each such rule specified a new string that could be produced based on an extant string compliant with the grammar, these axioms were known as production rules (Jackson 1990). Applied to solving problems, a production rule maps from the characteristics of a situation to the behavior that should be performed or the conclusion that should be reached in that situation. Consequently, they are sometimes called condition-action rules. The conventional format for a production rule is the IF-THEN statement:

IF <condition> THEN <action>

where *<condition>* represents a logical statement that, if true, leads to the *<action>* being undertaken. The condition part is sometimes known as the left-hand side (LHS) of the statement, and the action is known as the right-hand side (RHS). The condition may be a simple, single comparison involving data available to the CDS system, or it may be an arbitrarily complex statement in Boolean logic, using association, conjunction, disjunction, and negation related to data (see Figure 12-2). Typical clinical conditions might be:

potassium > 5.5
(potassium > 5.5) and (creatinine < 2.0)
(diagnosis = 'acute renal failure') or ((potassium > 5.5) and (creatinine < 2.0))

The action or RHS of a production rule may be an instruction to generate a message, usually a recommendation that some action be performed by a person, or a conclusion, typically represented by an assignment statement, that contributes another fact or data element available to the CDS system. Typical actions include

write 'Consider reducing the dose of the drug'
diagnosis:= 'acute renal failure'
creatinine_clearance:= 54

In effect, the RHS is a modification to be performed to the data available to the CDS system if the rule were to execute or "fire." The effect of such an action may be to negate a previously established data element or conclusion. In addition, like the LHS of a rule, the RHS may be arbitrarily complex and consist of several actions. A knowledge base represented using production rules would consist of a collection of these condition–action statements. The data elements against which the knowledge base would apply may consist of data about a patient, possibly retrieved by the CDS system from a clinical data repository.

IF NOT (erythema AND pus AND adenopathy) THEN
CONCLUDE "non-infectious cause"

IF erythema AND NOT (pus AND adenopathy) THEN
CONCLUDE "viral pharyngitis"

IF erythema AND pus AND NOT adenopathy THEN
CONCLUDE "viral pharyngitis"

IF erythema AND pus AND adenopathy THEN
CONCLUDE "streptococcal pharyngitis"

FIGURE 12-2 Decision rule represented as production rules. This collection of production rules represents the same knowledge as the decision tree in Figure 12-1. Each rule associates a Boolean condition that evaluates to true or false with an action (in this case, a diagnosis). The terms "erythema," "pus," and "adenopathy" are Boolean variables that evaluate to true or false based on data available to the CDSS.

The operation of a production rule CDS system consists of repeated cycles of match, select and execute, applying the knowledge base against data available to the CDS system in order to reach a desired conclusion, such as establishing a diagnosis or recommending a treatment. In the first step, matching, the LHS of the rules is compared to data available to the CDS system to see which ones could be executed. Because often more than one rule may be eligible for execution, the result of the match may be a conflict set: a collection of rules that are all true and eligible for execution at the same time. Because a production rule system, like the central processing unit of a computer, typically can execute only one instruction at a time, the second step in the process—selection—then occurs. Sometimes called conflict resolution when applied to a conflict set, the selection process identifies which rule will be executed next. Finally, one or more rules are executed, with the result specified by the RHS being carried out. In the case where the RHS specifies a conclusion or an assignment, this new fact or data element, in addition to whatever other new data may have been acquired by the CDS system from sources external to it, may render the LHS of additional rules true (or change those presently true to false), and the cycle begins anew.

Rules may be applied against data in one of two basic ways or inferencing mechanisms. In forward chaining, the inference engine of the CDS system attempts to match data elements against the LHS of rules, executing the actions of those rules that match until some goal state—for example, establishment of a diagnosis—is reached. In backward chaining, the inference engine initially finds rules that conclude whatever goal state the system is attempting to satisfy, and then it tries to ascertain which LHS of these, if any, can be satisfied by data. Forward chaining typically is employed when there is a large amount of data relative to the possible conclusions to be drawn from those data or if the CDS system is triggered or driven by the arrival of new data. By contrast, if the CDS system is used to critique a selection such as a treatment or a diagnosis made by a clinician, then backward chaining might be used.

The hallmark feature of a production-rule CDS system that distinguishes it from one that uses procedural knowledge is that each IF-THEN rule is independent of every other one and can be executed without regard to the execution state of any other rule. Thus, the order of execution of rules cannot be guaranteed. Some systems do include features, such as meta-rules or priority scores, to try to force a certain order of execution, particularly during conflict resolution, but even in these situations the order of execution cannot be predetermined completely.

One variation on the condition-action rule formalism is the event-condition-action (ECA) rule, used to represent expert knowledge in databases and in World Wide Web programming (Papamarkos 2003). In this variation, an event is defined that specifies when the conditions should be evaluated, and if the conditions are true at that point the action is undertaken.

Just as a decision tree is a graphical representation of procedural knowledge, a decision table may be used to summarize the knowledge in a production-rule knowledge base. A decision table is a graphical structure in which each column is headed by a data element deemed important in making decisions in a particular domain, along with a column for the action (see Figure 12-3). Each tuple in the table is equivalent to a single IF-THEN rule. It associates the values of one or more of the variables (LHS), not all of which need be represented in

Erythema?	Pus?	Adenopathy?	Diagnosis
no	no	no	non-infectious cause
yes	no	no	viral pharyngitis
yes	yes	no	viral pharyngitis
yes	yes	yes	streptococcal pharyngitis

FIGURE 12-3 Decision rule represented as a decision table. This decision table represents the same knowledge as the decision tree in Figure 12-1. Each tuple of the table represents an association between specific values for clinically important variables (conditions) and the diagnosis (action) that can be inferred from those findings.

any given tuple, with an action (RHS). Indeed, this technique sometimes has been used to identify duplicate rules in a production rule database by allowing easy detection of those that have the same conditions and actions. This allows compression of the resulting knowledge base, which facilitates maintenance of the knowledge base. It also can be used to identify missing values for some of the variables in the LHS, as an aid to ensuring completeness of the production rule set, as well as conflicts, in terms of an identical LHS but different actions.

The key advantage of a production-rule CDS system over a procedural representation is that the representation of knowledge is independent of the control knowledge needed to operate the CDS system and manage the inferencing process. Because of this, production rule knowledge bases can be acquired, maintained, and shared without having to alter or recompile the inference engine or the CDS system itself. Also, the rules are represented in a way that resembles natural language (using only IF-THEN logical statements). Although IF-THEN statements may be similar to those in a programming language, other statement types used in a programming language are not included in the rules. This makes it easier for the clinical domain expert to manipulate and understand the knowledge directly than would be the case with procedural knowledge, in which logic and control statements are intertwined. This feature supports relatively easy acquisition of expert knowledge as production rules, because they are encoded in a format familiar to most people. Indeed, this resemblance to natural language provides another advantage of production-rule CDS systems: easy provision of explanation of reasoning to the user. The CDS system can collect all the rules that fire and display them in their order of execution, which allows the recipient of system advice to see how the system's conclusion was reached. Finally, the modularity of production rules allows them to be manipulated individually, without needing to edit a large amount of procedural code in the process.

However, the independence of production rules also is a disadvantage. Because of this and the sometimes-unpredictable way that the rules may interact under various combinations of input data, the output of a production-rule CDS system may be difficult to predict. Indeed, changes to a single rule may have difficult-to-predict interactions with other rules, leading to unexpected changes in CDS system behavior. This challenge is magnified when the knowledge base grows beyond a hundred or so rules, as would be required for a CDS system addressing any meaningful set of clinical problems. A large number of rules makes it difficult for a knowledge engineer to locate related rules, so that the effect of any changes in the knowledge base can be understood. However, many rule-based CDS systems offer special tools for managing the knowledge base, which allow searching for related rules or provide simulations to predict the response to knowledge base changes under various conditions. A further disadvantage of the production-rule approach is that it uses declarative logic. In its conventional form as described here, production rules do not incorporate probabilities. On the other hand, much of medical decision-making involves probabilistic reasoning to a certain extent. To compensate for this defect, some production-rule systems have incorporated measures of probability, such as certainty factors, as part of the rule format. In this way, not only does each rule identify some consequent, but also it assigns a probability to that consequent. The inference engine then must take into account these factors and their propagation, as rules are chained together in order to make a probabilistic recommendation to the clinician.

A seminal system that demonstrated the use of decision rules implemented as production rules was MYCIN (Shortliffe 1976). MYCIN was a computer-based consultation system developed in the mid-1970s that gave advice about diagnosis and treatment of infectious diseases. MYCIN used primarily backward chaining to reach conclusions. It also introduced certainty factors in order to incorporate probabilistic reasoning into an otherwise deterministic, decision-rule system (Carter 1998). Other systems that incorporated similar decision-rule technology as part of clinical information systems included the HELP system (Haug 1994) at LDS Hospital in Salt Lake City and the Regenstrief Medical Record System at Indiana University (McDonald 1999).

As additional institutions began to implement such technology, it became clear to researchers that a considerable amount of redundant effort was being mounted to encode the same or similar decision rules in formats that differed at least slightly from place to place. Moreover, decision rules encoded at one institution could not be used readily at another, thus inhibiting sharing of knowledge and increasing the cost of knowledge engineering. This underscored the need for a standard representation for encoding decision rules. In addition, considering that both the procedural and the production-rule approach each offered advantages when implementing decision rules, it seemed that a hybrid of these approaches might be ideal. These lines of thought eventually culminated in efforts to compose standards for knowledge representation, and an early product of such efforts was the Arden Syntax.

12.4 THE HYBRID APPROACH: ARDEN SYNTAX

In 1989, a consensus conference was held, bringing together workers in academia, industry, and government with the goal of creating a standard for representing clinical logic in a shareable format. The eventual product of this effort, published as a standard in 1991 under the auspices of the American Society for Testing and Materials (ASTM), was the Arden Syntax for Medical Logic Systems (Pryor 1993). Arden Syntax was moved under the auspices of another standards development organization, Health Level Seven (HL7) in 1998, where it has subsequently evolved, culminating in the release of version 2.5 of the standard in 2005. Simultaneously, the American National Standards Institute (ANSI) certified Arden Syntax as a standard.

The unit of representation in the Arden Syntax is the medical logic module (MLM) (Hripcsak 1994). Each MLM contains sufficient logic and references to data to make a single clinical decision. Each MLM is a procedure, in which the logical statements execute serially. However, each MLM also functions independently like a production rule, with a separate trigger that, when satisfied by data, causes the inference engine to execute it and produce some action. Thus, this approach is considered a hybrid of the procedural and the production-rule forms of knowledge representation (see Figure 12-4).

```
maintenance:
     title:        Screen for positive troponin I;;
     filename:     troponin;;
     version:      1.40;;
     institution:  World-Famous Medical Center;;
     author:       Robert A. Jenders, MD, MS (jenders@ucla.edu);;
     specialist:   ;;
     date:         2005-08-15;;
     validation:   research;;

library:
     purpose:      Screen for evidence of recent myocardial infarction;;
     explanation:  Triggered by storage of troponin result.  Sends message
                      if result exceeds threshold;;
     keywords:     troponin; myocardial infarction;;
     citations:    ;;

knowledge:
     type:         data-driven;;
     data:
        troponin_storage := event {storage of troponin};

        /* get test result */
        tp := read last {select result from test_table where
              test_code = 'TROPONIN-I'};

        threshold := 1.5;

        /* email for research log */
        email_dest := destination {'email', 'name'= "jenders@ucla.edu"};
     ;;

     evoke:  troponin_storage;;

     logic:

       if (tp is not number) then conclude false;
       endif;

       if tp > threshold then conclude true;
       else conclude false;
       endif;

     ;;

     action:
       write "Patient may have suffered a myocardial infarction."  ||
         "Troponin I = " || tp || " at " time of troponin
       at email_dest;

     ;;
    urgency:    50;;
end:
```

FIGURE 12-4 Sample Arden Syntax MLM. A medical logic module consists of slots organized into three categories: maintenance, library, and knowledge. Site-specific mappings—in this example, an event definition, a query string, and a destination definition from a fictional organization—are enclosed by curly braces. Comments are delimited by /**/.

A medical logic module is a text file consisting of English-language-like statements. Each MLM is organized into three labeled sections, called categories. Each category, in turn, has one or more attribute-value pairs known as slots that express in statement form clinical knowledge about the domain in question or knowledge about the MLM itself. The first category is the maintenance category. The MLM author uses this category to document the software engineering aspects of the MLM—who wrote it, when and where it was written, the version of Arden Syntax used, which version of this MLM this is, and so on. In order to facilitate this, the maintenance category contains these slots: title, mlmname, Arden Syntax version, version, institution, author, specialist, date, and validation. The statements in these slots are unstructured text.

The second category is the library category. The MLM author uses this category to describe the medical knowledge that underlies the logic of the MLM. In particular, the category is used to describe in narrative format the rationale behind the logic of the MLM, to explain how the MLM functions and to identify references to the biomedical literature and to other knowledge sources pertinent to the logic of the MLM. In order to facilitate this, the library category contains the following slots: *purpose*, *explanation*, *keywords*, *citations*, and *links*. With the exception of the latter two slots, the other slots of this category have unstructured values. By contrast, the Arden Syntax documents a structure for citations and links. The library category allows the reader to discern at a glance the function of the MLM, without having to review its executable statements.

The third category, which contains the branching logic actually executed by the CDSS, is the knowledge category. The slots in this category are *type* (with a fixed value of "data-driven"), *data*, *priority*, *evoke*, *logic*, *action*, and *urgency*. The values of these slots are structured in order to facilitate execution by the computer. The *data slot* is used to represent all the data elements needed in the logic slot in order to render a medical decision. Typically these data elements are assigned to variables as the result of queries executed against a clinical database. In most cases, variables are very simple objects with two attributes (a value and a primary time) and no methods. The most recent version of the Arden Syntax provides a mechanism for building objects with multiple attributes. Because the developers of the Arden Syntax recognized that agreement on a common clinical database schema, query language, and vocabulary would require many years, if it ever would occur at all, they introduced a construct known as the curly braces for the characters used to enclose it ({ }). This provides a mechanism for the author to include institution-specific database mappings and links in an otherwise standard syntax. When an MLM is transferred from one institution to another, the statements in the curly braces may have to be adjusted to reflect the database mappings of the new host institution. Site-specific mappings enclosed in curly braces also are used to define events that are included as triggers in the evoke slot as well as delineation of destinations for messages from the CDS system.

Contributing additional structure to the knowledge category, the *evoke slot* identifies conditions, typically defined as constraints on data values, that state when the MLM should be executed—functioning, in effect, as the LHS of a production rule. MLMs also may be called directly from other MLMs, through the action slot (described in the next paragraph), thus executing as a type of subroutine. The *logic slot* contains the IF-THEN statements and calculations that represent the deterministic reasoning over the available data. A number of operators are available to manipulate data in the logic slot. Among the more important ones are those that allow temporal reasoning. Arden Syntax offers a number of powerful operators for extraction of temporal information from data and reasoning over these times. This is especially important in medical reasoning, and these operators help facilitate the representation of clinical logic in this computable format.

The *action slot* specifies, like the RHS of a production rule, what is supposed to occur if the result of processing the logic slot returns a true value. Typically this involves sending a message to a clinician or writing a value to the database that can be used to trigger or process another MLM, thus facilitating forward chaining. The *priority slot* contains a numeric score that can be used in conflict resolution to order the execution of MLMs that may be triggered at the same time. Finally, the *urgency slot* contains a numeric score that represents the clinical importance of the alert or reminder being encoded; the value in this slot can be used, for example, to decide what of several routes can be used to communicate an alert to a clinician, with faster routes (beepers) being associated with higher values for urgency.

Arden Syntax has been adopted as the knowledge representation formalism for several large vendors of CDS software. It is used mainly in transaction-oriented clinical information systems, in which storage of discrete data elements, such as test results and visit information, represent individual events that can trigger execution of MLMs. In such systems, the Arden Syntax has been used to implement relatively simple alerts and reminders. Though capable of doing so, it has not been used, by and large, to represent the declarative knowledge of the typical clinical practice guideline. It has been adopted at a number of medical centers in the United States, principally customers of those vendors that have introduced it into their CDS systems, but it has seen some limited use outside the United States, primarily in Europe.

Arden Syntax offers a number of advantages as a hybrid formalism for knowledge representation in CDS systems. Its key advantage is that it is a standard. This facilitates sharing of computable knowledge by reducing the amount of revision that must be performed on each MLM in order for it to execute properly at an institution other than the one at which it was composed originally. It also facilitates development of tools for acquiring, debugging, and maintaining knowledge encoded in this format. The hybrid nature of the Arden Syntax offers the best features of both the procedural and the production-rule formalisms: the control of flow available in a procedural representation with the modularity and separation of inferencing control from knowledge available in a production-rule representation.

Nevertheless, Arden Syntax has some disadvantages, too, some of which are not specific to it but pertain to any formalism that might be used for knowledge sharing. One important challenge is the lack of standardized database mappings: the curly braces problem. Although the curly braces highlight those parts of the MLM that require attention as part of the process of knowledge transfer, thus helping to ensure that these mappings are addressed, the absence of standard mappings still requires time-intensive and potentially error-prone manual revision. Although work is under way at the time of publication to standardize these mappings, they still remain a challenge. This work includes using the HL7 Reference Information Model (RIM) as the standard data model for composing queries. However, the use of the object-oriented

RIM as the foundation for all HL7 standards highlights another potential deficiency of the Arden Syntax: its relatively simple object model. In this regard, the Arden Syntax is not consistent with other RIM-based standards in the HL7 family of standards. Another disadvantage of Arden Syntax is its procedural code. Clinicians are more accustomed to viewing clinical knowledge in a declarative format, such as a decision tree or a narrative clinical practice guideline. The lack of familiarity with procedural code can make validation of the knowledge a challenge.

Indeed, although Arden Syntax has been used to encode clinical guidelines,a consensus has developed that it is best used for relative simple alerts and reminders. By contrast, under this consensus, workers believe that a declarative formalism that captures the specific features of guidelines, such as eligibility criteria (instead of transaction-based triggers) is more appropriate than procedural logic to express a clinical guideline in computable form. Although this has led to the creation of a number of different formalisms (Peleg 2003), such as the Guideline Elements Model (GEM—a standard of ASTM for marking up narrative guideline content in a structured fashion) and the Guideline Interchange Format (GLIF), no widespread agreement has yet been reached regarding a standard formalism (see Chapter 13). Accordingly, workers in HL7 and other organizations have created standard components of an overall formalism as a decomposition of the problem. Approved as a distinct HL7 and ANSI standard in 2005, the Guideline Expression Language–Object-oriented (GELLO), represents a standard formalism intended to address these challenges.

12.5 EXPRESSION LANGUAGES

The purpose of an expression language is to allow the knowledge engineer to build up statements that query data, logically manipulate them, provide for reasoning over them, and facilitate calculations and other formulae involving them in a variety of applications. GELLO (Sordo 2004) was designed specifically to do this for the case of representing the logic in clinical guidelines, although it need not be restricted to this particular representation. GELLO is based on the Object Constraint Language (OCL), itself a standard of the Object Management Group (OMG), and can be used with any object-oriented data model. As a result, GELLO can be used with the standard HL7 RIM to extract data from clinical repositories and to manipulate those data, thus facilitating closer integration with other HL7 standards that use the RIM and taking advantage of the rich object model and relationships that this approach offers. An example of GELLO, including object references, queries, calculations, and logical manipulations of data, is seen in Figure 12-5.

GELLO was developed initially to represent the procedural component of the declarative guideline formalism GLIF. However, its generic nature allows it to be used in a number of applications. Some have suggested that it can replace the knowledge category of the Arden Syntax MLM. Other applications that use GELLO are being explored. These include representation of medication prior authorization rules and delineation of rule-based mapping between the standard terminologies ICD-9 and SNOMED CT.

A key advantage of GELLO is its use of OCL, which allows leverage of tools that manipulate OCL to be used to represent knowledge in GELLO.

```
let lastTroponin: Observation = Observation→select(code=
    ("SNOMED-CT", "102683006")).sortedBy(effectiveTime.high).last()

let threshold : PhysicalQuantity =
    Factory.PhysicalQuantity( "1.5, ng/dl")

let threshold_for_osteodystrophy : int = 70

let myocardial_infarction :Boolean =   if lastCreatinine <> null and
    lastCreatine.value.greaterThan(threshold)
  then
    true
  else
    false
  Endif

if myocardial_infarction then
    whatever action or message
else
    whatever action or message
endif
```

FIGURE 12-5 Example of GELLO encoding a simple guideline. This guideline represents the same knowledge contained in the Arden Syntax MLM in Figure 12-4. Because GELLO was created to extract data from clinical repositories, to manipulate those data and to reason over them, it does not have an explicit syntax for sending messages to clinicians. The GELLO code would be embedded in complete guideline representation or other application for use by the CDSS.

Further, because it is object-oriented, GELLO facilitates manipulation of data that are represented and manipulated more conveniently as objects with heterogeneous attributes than as other data structures such as matrices (for example, the different vital signs obtained during a patient visit). GELLO addresses the curly braces challenge by facilitating the use of standard vocabularies and data models, thus enhancing the possibility of knowledge transfer. Although GELLO is more complex than the Arden Syntax logic and data slots, tools can hide this complexity in a way that allows knowledge authors to create computable knowledge in a straightforward fashion. Accordingly, GELLO is a useful contribution in the effort to create a standard formalism for representing clinical guidelines as a common instance of deterministic reasoning represented in a decision rule.

12.6 FUTURE WORK

With increasing emphasis on patient safety and prevention of medical errors, coupled with increasing use of electronic health records, demand for computer-based CDS will grow. Responding to this demand will require leveraging the considerable investment in the creation of clinical practice guidelines to adapt them for use in CDS systems. A parallel trend is the emphasis on interoperability of clinical information systems. These trends will prompt convergence on a standard for representing decision rules in general and clinical practice guidelines in particular in a computable format, one component of which will be a standard expression language that can be executed in many different CDS systems with a minimum of adaptation. This will facilitate knowledge sharing by reducing the cost of knowledge engineering, which in turn will foster compliance with clinical practice guidelines and other evidence-based medicine, leading to an improvement in patient safety and clinical outcomes.

12.7 CONCLUSION

A decision rule is a representation of deterministic reasoning in which branching logic is used in combination with data to reach conclusions regarding diagnosis, treatment, and other important clinical goals. One way to represent a decision rule in a computable format is through the sequential execution and explicit flow of control of a procedure. Another approach that separates the clinical knowledge from the inferencing mechanism and other control processes is the production rule, in which the knowledge of a domain is represented by a collection of modular IF-THEN expressions. Efforts to incorporate the advantages of these two approaches as well as create a standard formalism that would facilitate knowledge sharing led to the development of the HL7 standard Arden Syntax. In this formalism, knowledge is represented as modular procedures known as medical logic modules, which also can be triggered independently like a production rule. However, challenges with this approach, including nonstandard data mappings and a possibly inadequate data model, coupled with the need to implement the declarative knowledge of clinical practice guidelines, have led to pursuit of other approaches. Expression languages, such as the HL7 standard GELLO, have been developed in order to extract data from clinical repositories and manipulate those data, using standard data models and vocabularies, and thus address the challenge of creating an overall formalism to represent computable clinical practice guidelines.

REFERENCES

Carter, J. H. (1998). Design and implementation issues. *Clinical Decision Support Systems: Theory and Practice*, 169–197. New York: Springer.

Haug, P. J., Gardner, R. M., Tate, K. E. et al. (1994). Decision support in medicine: Examples from the HELP system. *Comput Biomed Res* **27**(5): 396–418.

Hripcsak, G. (1994). Writing Arden Syntax medical logic modules. *Comput Biol Med* **24**(5): 331–363.

Jackson, P. (1990). Production systems. *Introduction to Expert Systems*, 135–151. Wokingham, England: Addison-Wesley.

Health Level Seven (HL7). HL7 is an international standards development organization certified by the American National Standards Institute (ANSI). HL7 focuses on standards related to health care computing, including CDSS standards such as the Arden Syntax, GELLO and the Reference Information Model. Copies of these standards may be obtained at http://www.hl7.org.

Miller, R. A. (1994). Medical diagnostic decision support systems—Past, present and future: A threaded bibliography and brief commentary. *J Am Med Inform Assoc* **1**: 8–27.

McDonald, C. J., Overhage, J. M., Tierney, W. M. et al. (1999). The Regenstrief Medical Record System: A quarter century experience. *Int J Med Inform* **54**(3): 225–253.

Papamarkos, G., Poulovassilis, A., Wood, P. T. (2003). Event-Condition-Action rule languages for the Semantic Web. In Proc Workshop on Semantic Web and Databases, 29th Annual International Conference on Very Large Data Bases (VLDB'03). San Francisco: Morgan Kaufmann.

Peleg, M., Tu, S., Bury, J. et al. (2003). Comparing computer-interpretable guideline models: A case-study approach. *J Am Med Inform Assoc* **10**(1): 52–68.

Pryor, T. A., Hripcsak, G. (1993). The Arden Syntax for medical logic modules. *Int J Clin Monit Comput* **10**: 215–224.

Shortliffe, E. H. (1976). Computer-Based Medical Consultations: MYCIN (Artificial Intelligence Series). New York: Elsevier.

Sordo, M., Boxwala, A. A., Ogunyemi, O., Greenes, R. A. (2004). Description and status update on GELLO: A proposed standardized object-oriented expression language for clinical decision support. *Medinfo* **11**(Pt 1): 164–168.

13 GUIDELINES AND WORKFLOW MODELS

MOR PELEG

Clinical guidelines aim to improve quality of care, decrease unjustified practice variations, and save costs. In order for guidelines to affect clinicians' behavior, they should provide patient-specific decision support during patient encounters. Specifying guidelines in computer-interpretable guideline (CIG) formalisms that could provide automatic inference based on patient data may achieve this goal. The knowledge contained in guidelines is difficult to formalize due to the fact that despite efforts made to improve the quality of narrative guidelines, evidence-based recommendations are often incomplete and vague, and do not constitute a full care process. Several methodologies have been developed to support the transition from narrative guidelines into CIG implementations. They include methodologies for markingup narrative guideline elements in order to assess a guideline's quality and completeness and map it to CIG formalisms. Many CIG formalisms exist, differing in their goals, computation model, the elements used to structure guideline knowledge, and the degree to which they support workflow integration. Specifying a narrative guideline as a CIG is a difficult task, yet the resulting application cannot be shared easily by different institutions and software systems. Therefore, sharing encoded knowledge is a challenging goal. The specification of standard methods to support such sharing is a major focus in the field. The road to achieving widespread use of guideline-based decision-support systems is long and difficult. This chapter reviews the current state-of-the-art in guideline-based decision support research and considers likely future directions that can be taken to reach the ultimate goal.

13.1 INTRODUCTION: CLINICAL GUIDELINES AND ALGORITHMS

Traditionally, medicine has been viewed as an art, and medical practice has been based to a large extent on individual clinical experiences and in keeping with the accepted practices and opinions of experts and opinion leaders. In the last two decades, we have been witnessing a movement toward evidence-based medicine (see Chapter 11), which seeks to base medical practice on evidence-based studies (such as clinical trials), employ outcome measures, and perform clinical audits. The influential report of the Institute of Medicine (IOM),

To Err Is Human (Kohn, Corrigan et al. 1999), bolsters this movement by setting an agenda for reducing medical errors and improving patient safety through the design of a safer health system. The report recommends making greater use of evidence-based approaches to health care and incorporation of information technology. The 2001 IOM report, *Crossing the Quality Chasm: A New Health System for the 21st Century* (Institute of Medicine 2001), suggests ways to make scientific evidence more useful and accessible to clinicians and patients, such as authoring and dissemination of clinical practice guidelines. Although clinical guidelines have been used in health care since at least the early 1970s, the emphasis in the current health care agenda on a safer, evidence-based medical practice has brought about a resurgence of interest in them.

Clinical guidelines are systematically developed statements to assist practitioner and patient decision-making about appropriate health care for specific clinical circumstances (Field and Lohr 1990). Aims of clinical guidelines are to eliminate errors, reduce unjustified practice variation and wasteful commitment of resources, and encourage best practices and accountability in medicine (Timmermans and Berg 2003). Clinical guidelines typically are created by medical experts or panels convened by specialty organizations, who review the relevant evidence-based studies and, using a consensus-based process, compile a set of recommendations. Their focus may be on screening, diagnosis, management, treatment, or referral of patients with specific clinical conditions. The recommendations typically are written as narrative text and tables, which point back to background material and evidence, ranking the strength of clinical validity, and the strength by which recommendations should be followed according to the guideline authors.

Although the recommendations aim to be based on evidence, they are often not constructed in a way that reflects the flow of actual patient encounters, and thus are sometimes difficult to apply. In order to solve this problem, clinical guidelines are sometimes portrayed as algorithms (flowcharts) to more directly specify for providers the recommended steps of data gathering, decision-making, and actions (i.e., process flow) during patient encounters. The algorithms are based on the guidelines, but where evidence is not available, the gaps are filled in based on expert opinion.

A cognitive study has shown that different knowledge engineers/algorithm authors create dissimilar clinical algorithms using the same clinical guideline as a starting point, depending in part on their degree of prior experience and knowledge of the domain (Patel, Allen et al. 1998). In that study, the authors found that physicians who created algorithms tended to add organization and detail that were based on their knowledge, and which was not explicitly contained in the narrative guideline, whereas computer scientists tended to produce more consistent algorithms, but which reflected more literal interpretations of the narrative text. The algorithms of highest quality were created by teams involving clinicians and computer scientists. Variation in structure and detail also was seen between algorithms created by different computer scientists, or different clinicians. The particular computerized authoring tools used to create the algorithms also had an effect on the algorithms produced.

Studies have shown that guidelines have the most effect on clinician behavior if they are made available during patient encounters, and if they deliver patient-specific advice at key points of decision-making during those encounters (Shea, DuMouchel et al. 1996; Overhage, Tierney et al. 1997). This can be achieved by representing guideline knowledge in a formalism that enables computer-based execution and supports automatic inference. Such formalisms are known as Computer-Interpretable Guideline (CIG) modeling methodologies, or CIG formalisms.

13.2 THE KNOWLEDGE CONTAINED IN CLINICAL GUIDELINES

Unlike clinical trial protocols, which constrain clinical practice to clearly defined steps, narrative guideline documents contain a recommendation set that suggests options for optimal care. Because the nature of clinical guidelines is to suggest rather than impose a strict procedure for care, they often are written in a relaxed language that emphasizes the fact that the judgment of the clinician should determine the care process. However, the relaxed language used in narrative guidelines is not formal enough for computer processing, and the knowledge presented in a narrative guideline is thus often unclear, vague, incomplete, ambiguous, and even contradictory, which creates a problem in interpreting the guideline in order to computerize it.

13.2.1 The Quality of Narrative Guidelines

Many approaches have been developed to improve the quality of narrative guidelines. Some approaches concentrate on the methodological quality of guideline development, that is, the nature of evidence and methodologies for aggregating research results of different studies that differ in patient population and settings, rather than on structuring the representation of guideline knowledge. For example, the 1992 Institute of Medicine's report on the development of clinical guidelines (Field and Lohr 1992) suggests eight attributes for assessing guideline quality. Four attributes relate to guideline content: validity, reliability and reproducibility, clinical applicability, and clinical flexibility. The other attributes relate to the process of guideline development or representation: clarity, multidisciplinary process, scheduled review, and documentation. A variety of guideline assessment tools have been published. The two most prominent tools are the Appraisal of Guidelines Research and Evaluation (AGREE) instrument (http://www.agreecollaboration.org/) and Shaneyfelt's appraisal tool (Shaneyfelt, Mayo-Smith et al. 1999). The Guide-Line Implementability Appraisal (GLIA) (Shiffman, Dixon et al. 2005) complements the guideline quality appraisal instruments and addresses potential difficulties in implementation. These tools evaluate guidelines according to desirable attributes that can be mapped to the IOM attributes (Field and Lohr 1992).

The Australian Health Information Council (AHIC) suggests criteria that should be confirmed to ensure that a narrative guideline is reliable and valid (http://www.ahic.org.au/evaluation/knowledge.htm). These criteria include the validity of the knowledge source, which depends on systematic review of evidence and rating of levels of evidence, and the currency of the guideline (i.e., that the guideline is up to date).

The reports of the IOM and the AHIC do not provide precise schemas for representing algorithmic guideline knowledge. Other organizations, such as the Agency for Healthcare Research and Quality (AHRQ) (Hadorn 1995) and the Society for Medical Decision Making (Society for Medical Decision

Making 1992) have published models for algorithm development, which include a precise syntax for algorithm steps and informal definitions of such steps.

13.2.2 The Types of Knowledge Contained in Narrative Guidelines

The Guidelines Elements Model (GEM) (Shiffman, Karras et al. 2000) is an XML-based knowledge model for guideline documents. GEM elements relate to a guideline's identity, developer, purpose, intended audience, method of development, target population, knowledge components, testing, and review plan. The Knowledge Components subtree in GEM includes 44 of GEM's 110 elements and, combined with the Target Population tree, supports decision making. *Knowledge* components in guideline documents include tags for marking names of terms and their definitions and are used to structure guideline recommendations as conditional recommendations (i.e., decision rules) and imperative recommendations (i.e., clinical actions). Recommendations can be sequenced using a link element to represent guidelines that unfold over time. GEM is a standard of the American Society for Testing and Materials (ASTM) and is supported by many tools, available at the GEM Web site (ycmi.med.yale.edu/GEM), which include GEM-Cutter for marking up guidelines according to GEM elements, GEM-Q and GEM-COGS, for assessing the quality of marked-up guidelines, Extractor for review of recommendations, and GEM-Arden for translation of guidelines marked up in GEM into medical logic modules.

The Clinical Practice Guideline–Reference Architecture (CPG-RA) (http://www.cpg-ra.net/) is an XML-Schema based knowledge model for structuring guidelines. It is being developed by the Sowerby Centre for Health Informatics at Newcastle, England, and by members of the Guidelines International Network (GIN). According to CPG-RA, the clinical content of a guideline is structured into sections containing a summary of the evidence (methods and reliability), with its reference sources and clinical recommendations. Each section is classified into categories describing its content: background, prevention, screening, diagnosis, and management, and can be hierarchically decomposed. The *diagnosis* category is structured as a sequence of the alternative diagnoses that should be considered, followed by appropriate clinical activities to perform, and definitions of terms. The *management* category is structured as a sequence of management actions for a clinical context of a patient, relevant management options, including issues of process flow and ordering of actions. The structure of the diagnosis and management elements is similar to two categories of computational models appearing in many formal guideline modeling languages: decision maps and activity graphs, respectively (Tu, Campbell et al. 2003), as described in the subsection 13.3, "Formal Methods for Specifying CIGs."

Like CPG-RA, the Stepper tool (Ruzicka and Svatek 2004) aims to use a markup model to facilitate guideline execution. Narrative portions of guidelines are marked with respect to the following knowledge components: procedural elements, causality, goal statements, and concept definitions. These structures are iteratively refined, by providing tree structures of subelements. For example, procedural elements are refined into a structure called *scenario*, consisting of a condition and a recommendation part, the former corresponding to a potentially complex expression over patient states and/or history. At later stages of transformation, goals and scenarios can also be aggregated. Currently, the structured knowledge components cannot be converted into operational code directly applicable to patient data. However, being a customizable document transformation tool, Stepper has been used to transform parts of a narrative guideline document into Asbru representation by defining transformation rules from the source document into narrative categories (e.g., definitions, conclusions, and recommendations), medical categories (e.g., interventions, drugs, symptoms), control structures (e.g., action sequencing, decomposition, synchronization), and Asbru elements (see the subsection, "Formal Methods for Specifying CIGs" and http://www.cs.vu.nl/~serbanr/Research/Protocure/ExperGuidFormal.pdf).

Another classification of guideline knowledge is based on the kinds of decision-support tasks that a guideline involves (Tu and Musen 2000). The tasks include making decisions, setting goals, specifying work to be performed, data abstraction/interpretation, and generating alerts or reminders. Apart from the last task, the other four tasks identify knowledge components of narrative guidelines: decisions, goals, actions, and definitions used for data abstraction and interpretation.

A different classification of guideline content was developed by Berrios and colleagues (Berrios, Cucina et al. 2002). They developed an ontology for indexing medical knowledge according to questions that the knowledge answers. The questions are formed from relationships among four basic concepts: pathology, manifestation, investigation, and therapy (e.g., how does chemotherapy compare with hormonal therapy in the setting of pregnancy (manifestation)?).

13.3 FORMAL METHODS FOR SPECIFYING CIGS

Specifying guideline knowledge formally, as CIGs, allows computer-based execution—implementation of which, as noted previously, is more likely to affect clinician's behavior than availability of narrative guidelines (Shea, DuMouchel et al. 1996; Overhage, Tierney et al. 1997). During creation of the formal representation, ambiguities are removed, and areas are identified in which evidence is missing or for which no recommendations are given. Medical organizations that are adapting the recommendations into a care-flow process may fill in the gap with their opinions or leave the decision up to the end user. Many formalisms exist for specifying CIGs, each with its own motivations and features (Shahar, Miksch et al. 1998; Tu and Musen 1999; Johnson, Tu et al. 2000; de-Clercq, Hasman et al. 2001; Sutton and Fox 2003; Boxwala, Peleg et al. 2004; Terenziani, Montani et al. 2004; Tu, Campbell et al. 2004; Ciccarese, Caffi et al. 2005). Several papers have reviewed and compared formal methods for CIG specification (Tu and Musen 2000; Wang, Peleg et al. 2002; Peleg, Tu et al. 2003; de-Clercq, Blom et al. 2004). Wang's review has focused on guideline representation primitives, process models, and their relationship to a patient's clinical status (Wang, Peleg et al. 2002). Tu and Musen's comparison (Tu and Musen 2000) focused on the computational methods of the formalisms. De Clercq's paper addresses, in addition to guideline representation issues, aspects concerning guideline acquisition, verification, and execution (de-Clercq, Blom et al. 2004). The comparison by Peleg and colleagues identified eight components that capture the structure of CIGs (Peleg, Tu et al. 2003). These dimensions fall into two broad categories—structuring guidelines as plans for decisions and actions, and linking the guideline to patient data and medical concepts.

In this section, we describe several well-known approaches for formally representing guidelines as CIGs.

13.3.1 Task-Network Models

Many of the approaches have in common a process–flow-like model termed *Task-Network Model* (TNM) (Peleg, Tu et al. 2003)—a hierarchical decomposition of guidelines into networks of component tasks that unfold over time (see Figure 13-1). The task types vary in different TNMs, yet all of them support modeling of medical actions, decisions, and nested tasks.

Following is a short review of some well-known TNMs, which highlights the distinguishing features of each methodology.

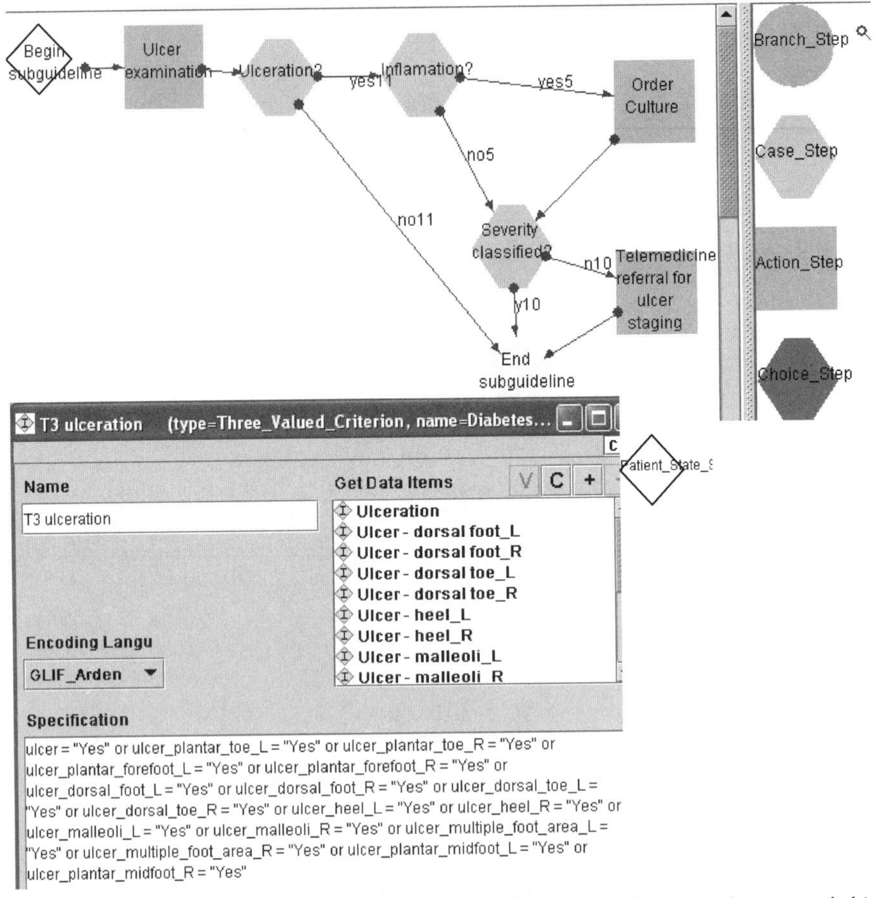

FIGURE 13-1 Part of a diabetes foot management algorithm, dealing with ulcers, encoded in GLIF3. The insert shows a formal specification of the decision criterion belonging to the case step "Ulceration?"

13.3.1.1 Asbru

Asbru (Shahar, Miksch et al. 1998) represents guidelines as skeletal plans that can be hierarchically decomposed into subplans or actions. The main emphasis is on guideline intentions, not only action prescriptions. Skeletal plans capture the essence of a procedure, but leave enough room for execution-time flexibility in the achievement of particular intentions. Intentions are specified as temporal patterns of actions and external-world states that should be maintained, achieved, or avoided, during, or at the completion of a plan. The same temporal expression language is used for representing time-oriented actions, conditions, and intentions in a uniform fashion. The temporal expression language uses time annotations consisting of a time range (i.e., range of the start time, the end time, and the duration) and a time reference (i.e., point in time, or the time at which a plan changes state). Parameter definitions can be abstracted from raw data and may depend on context (e.g., pregnancy). Figure 13-2 shows an example of an intention.

Several tools support authoring of guidelines in Asbru: Delt/A (http://ieg.ifs.tuwien.ac.at/projects/delta/) and URUZ (Shahar, Young et al. 2003), both focusing on easing the transition from narrative to formal representations via a markup stage; AsbruView (http://www.ifs.tuwien.ac.at/asgaard/asbru/tools.html), which focuses on visualization and user interface for authoring; and CareVis (http://ieg.ifs.tuwien.ac.at/projects/carevis/), which provides multiple simultaneous views to cover different aspects of a complex underlying data structure of treatment plans and patient data. The developers of Asbru are involved in the Protocure II project (http://www.protocure.org/), which addresses the important topic of quality improvement of guidelines and protocols by integrating formal methods of software engineering in the lifecycle of guidelines development and maintenance ("living guidelines").

In the Protocure II project, an original textual guideline is translated into an intermediate representation (Seyfang, Miksch et al. 2005) and then into Asbru. A semiautomatic translator converts the Asbru model into the specification format used by the Karlsruhe Interactive Verifier (KIV), which is an interactive theorem prover, (http://homepages.inf.ed.ac.uk/wadler/realworld/kiv.html). This procedure enables formal verification of Asbru-encoded guidelines.

13.3.1.2 EON, PRODIGY, and GLIF

EON (Tu and Musen 1999), **PRODIGY** (Johnson, Tu et al. 2000), and **GLIF** (Boxwala, Peleg et al. 2004) have strongly influenced each other. In addition to including the generic tasks used by all TNMs, they use scenarios—partial specification of patient states allowing classification of a patient into an appropriate state within a CIG. **EON** uses a task-based approach to define decision-support services that can be implemented using alternative techniques (Tu and Musen 2000). The decision-making task is supported by two classes of decision steps: simple if-then-else constructs and rule-in and rule-out criteria (that correspond to argumentation rules that confirm or refute a decision option; see Figure 13-3)[1] as a way of setting qualitative preferences. Goals in EON are specified in a criteria language that uses patient data and abstractions based on classification hierarchies (e.g., disease hierarchies). Actions to

[1]Argumentation rules originated in the PROforma formalism.

```
<intentions>
  <intention type="intermediate-state" verb="maintain">
    <parameter-proposition parameter-name="blood-glucose">
      <value-description type="equal">
        <qualitative-constant value="HIGH"/>
      </value-description>
      <context>
        <context-ref name="GDM-Type-II"/>
      </context>
      <time-annotation>
        <time-range>
          <starting-shift>
            <earliest>
              <numerical-constant unit="week" value="24"/>
            </earliest>
            <latest>
              <numerical-constant unit="week" value="24"/>
            </latest>
          </starting-shift>
          <finishing-shift>
            <earliest>
              <qualitative-constant value="delivery"/>
            </earliest>
            <latest>
              <qualitative-constant value="delivery"/>
            </latest>
          </finishing-shift>
        </time-range>
        <time-point>
          <qualitative-constant value="conception"/>
        </time-point>
      </time-annotation>
    </parameter-proposition>
  </intention>
</intentions>
```

FIGURE 13-2 A specification of an intention in Asbru: "In the context of GDM-Type-II, maintain blood glucose state at high level, starting week 24 and ending at delivery."

be performed are represented as Management Diagrams (also known as activity graphs)—networks of scenarios, actions, decisions, subguidelines, and branch and synchronization steps for modeling parallel paths (see Figure 13-4a). Data interpretation can be achieved using: 1) abstraction based on classification hierarchies, 2) definition of terms referring to values of patient data items, and 3) temporal abstractions. EON CIGs can be authored in Protégé-2000 (protégé.stanford.edu) and executed by an execution engine that uses a temporal data mediator to support queries involving temporal abstractions and temporal relationships. A third component provides explanation services for other components.

The Guideline Interchange Format version 3 (GLIF)3 (Boxwala, Peleg et al. 2004) stresses the importance of sharing guidelines among different institutions and software systems, building on the most useful features of other CIG models, and incorporating standards. GLIF3 represents guidelines as clinical algorithms, similarly to EON's Management Diagrams. Its model of medical knowledge is used by action and decision steps to formally refer to patient data items, clinical concepts, and clinical knowledge. Patient data items are specified by a *medical concept*, the code for which is taken from a controlled clinical vocabulary and by a *data structure*, taken from a standard

```
decision :: Additional_drug_choice ;
  caption :: 'Additional drug choice' ;
  choice_mode :: single ;
  support_mode :: symbolic ;

  candidate :: ISA_beta_blocker ;
    argument :: excluding, ( Asthma = Yes or COPD = Yes ) ;
    argument :: excluding, ( STD_Heart_block = Yes ) ;
    argument :: excluding, ( Current_Rx = Ca_blocker_non_DHP ) ;
    argument :: for, ( Current_Rx = Thiazide_diuretic ) ;
    argument :: for, ( Current_Rx = Ca_blocker_DHP_long or Current_Rx =
          Ca_blocker_DHP_short ) ;
    argument :: excluding, ( Current_Rx = Beta_blocker_ISA or Current_Rx =
          Beta_blocker_non_ISA ) ;
    argument :: against, ( Type_1_Diabetes = Yesand Proteinuria <> None ) ;
    recommendation :: Netsupport( Additional_drug_choice, ISA_beta_blocker ) >= 1 ;
    ...
end decision .
```

FIGURE 13-3 Part of the argumentation rule-set for selecting a second antihypertensive drug, in PRO*forma* syntax (full example provided in http://www.openclinical.org/docs/ext/cigs/comparison/Hypertension_model_PROforma.txt).

reference information model (RIM), such as the Observation, Medication, and Procedure classes of the Health Level 7 (HL7) RIM (Schadow, Russler et al. 2000). Clinical knowledge is expressed as relationships between medical concepts (e.g., contraindication relationships between a drug and a disease). GLIF3 has a formal language for expressing decision and eligibility criteria. This expression language originally was based on the Arden Syntax (Hripcsak, Ludemann et al. 1994) and was replaced by an object-oriented language, called GELLO (http://cslxinfmtcs.csmc.edu/hl7/arden/2004-09-ATL/v3ballot_gello_aug2004.zip), which has been recently accepted as an HL7 and ANSI standard (see Chapter 12). GLIF3 is supported by two authoring and validation tools (Peleg, Boxwala et al. 2004) and two execution engines. Figure 13-1 shows part of a GLIF3-encoded guideline.

The **PRODIGY** (Johnson, Tu et al. 2000) project has aimed at producing the simplest, most readily comprehensible model necessary to represent chronic disease management guidelines. PRODIGY-3 emphasizes a scenario-based approach, in which a guideline is organized as a collection of clinical contexts; in each context, selection among relevant clinical actions is made. This formalism is known as a Decision Map (see Figure 13-4b). This approach

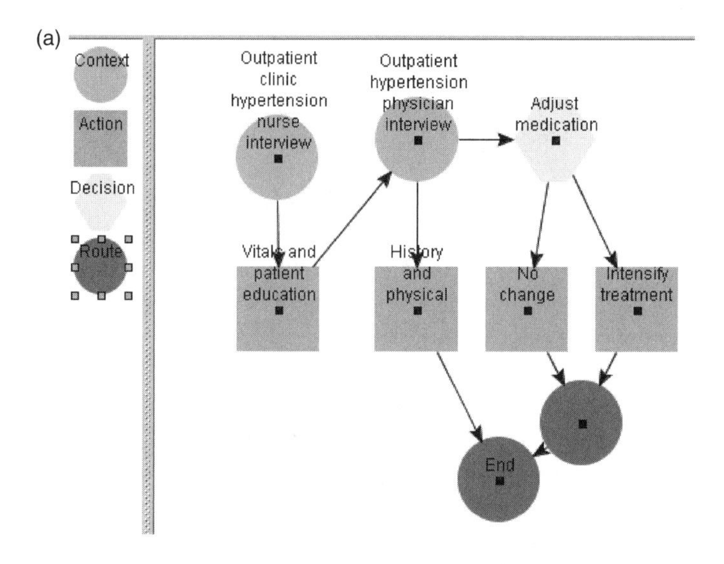

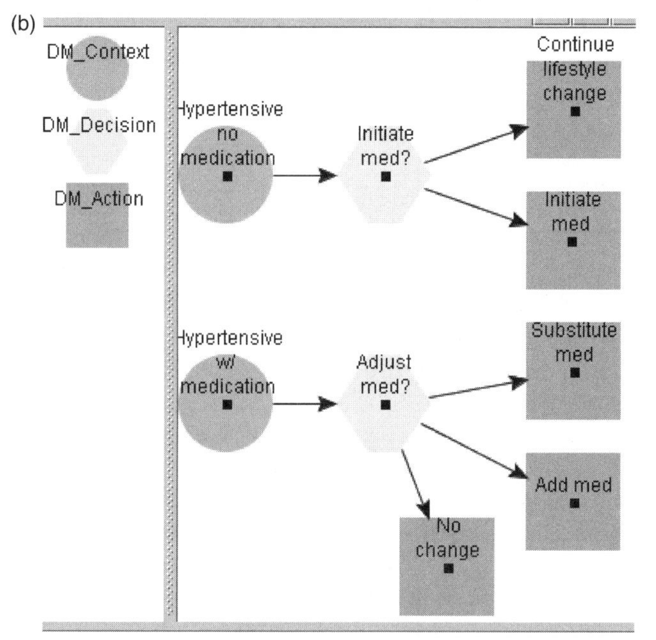

FIGURE 13.4 Part of a SAGE guideline model showing (a) an Activity Graph for outpatient immunizations for both adult and pediatric patients on a primary care encounter, and (b) a Decision Map for determining appropriate immunization at a pediatric encounter.

inspired the use of scenarios and patient-state steps in EON and GLIF3, respectively. (A state step in GLIF3 is essentially an entry point into a CIG.) The PRODIGY-3 model can represent many specialized medical actions, including referrals, creating prescriptions, scheduling, asserting making conclusions, and modifying drug treatments. Decisions in PRODIGY-3 always require confirmation. They are modeled as for-and-against rule-sets, where the rules can be structured using predefined templates, refer to classification hierarchies in order to perform term abstractions, or refer to patient data, which are viewed as instances of virtual medical record[2] (vMR) classes. Using Protégé-2000, several complex chronic disease management guidelines have been encoded in PRODIGY-3 and over 150 simple guidelines have been translated from earlier version of PRODIGY.

13.3.1.3 GUIDE/NewGuide

The **GUIDE/NewGuide** project (Ciccarese, Caffi et al. 2005) is an approach for modeling and executing guidelines in the context of organizational workflow, founded on two major paradigms: Component-Oriented Programming (COP) and Separation of Concerns (SoC). NewGuide requires considering three components: the Guideline Management System (GLMS), dealing with the representation of medical knowledge; the Workflow Management System (WfMS, http://www.wfmc.org/standards/docs/), dealing with the representation of organizational knowledge; and the electronic health records (EHRs), dealing with data. Interaction among these components is achieved via SoC, where the knowledge representation process required for each component is done separately by three roles: medical expert, organizational expert, and formalization expert, hiding specific details as much as possible to allow minimization of the knowledge gaps among these roles. Communication is message-based, according to specific contracts, and uses ontologies and terminologies (ICD9-CM, SNOMED, LOINC).

Each concern has its own representation models and languages. The specific health care organization's EHR is interfaced through a vMR model. GMLS components are organized in a distributed architecture: an editor to formalize guidelines via a flowchart-like approach, a repository to store and publish them, an enactment system to implement guideline instances in a multiuser environment, and a reporting system able to completely trace any individual physician guideline-based decision process. In this way, it is possible to detect noncompliances, which users can justify. Different organizations can get guidelines from the repository, adapt and introduce them in clinical practice. At this level, the communication with an external system, such as a WfMS, is managed by the message manager, which delegates requests and responses to the Web user interface or to a SOAP interface. A distinguishing feature of NewGuide is the possibility for the user to access external decision support systems, such as decision trees or influence diagrams, in case of nondeterministic decisions.

13.3.1.4 SAGE

The **SAGE** (Standards-Based Shareable Active Guideline Environment) aims to create an infrastructure that will allow execution of standards-based clinical practice guidelines across heterogeneous clinical information systems (Tu, Campbell et al. 2004). SAGE uses a deployment-driven methodology to formalize the guideline knowledge required to provide clinical decision support. It involves identification of usage scenarios of guideline-based care in clinical workflow and encoding them and their appropriate guideline recommendations.

[2]A vMR is a view of a patient medical record that is simplified for clinical decision support purposes. The vMR supports a structured data model for representing information related to individual patients, domains for values of attributes in the data model, and queries through which guideline CDS can test the states of the patient.

Data items used by decision expressions are identified in the narrative guidelines and instantiated as detailed data models that correspond to constraints on classes of a vMR that are ultimately based on the HL7 RIM. Clinical concepts are specified in terms of codes and concept definitions that are based on standard terminologies. In the SAGE guideline model, *contexts*—characterized by a triggering event (e.g., patient checking in), patient characteristics, organizational setting (e.g., primary care outpatient clinic), organizational roles specifying who should respond to the event (e.g., nurse), and needed resources—coordinate the activation of guideline-based decision support.

Procedural guideline logic is represented as an activity graph (Tu, Campbell et al. 2003) (see Figure 13-4a), which specifies how guideline-based clinical decision support (CDS) should behave for a given scenario, and is assembled from steps that are based on the Workflow Management Coalition's process model. Recommendations that do not need to be organized and executed as part of a process are represented as decision maps (see Figure13-4b), as in the PRODIGY-3 formalism. The SAGE project has demonstrated the use of an execution engine that interprets encoded clinical guideline content and executes that content via functions of a target clinical information system (Ram, Berg et al. 2003).

13.3.1.5 PRO*forma*

PRO*forma* (Fox, Johns et al. 1996) advocates the support of safe guideline-based decision support and patient management by combining logic programming and object-oriented modeling, and its syntax and semantics are formally defined. One aim of the PRO*forma* project is to explore the expressiveness of a deliberately minimal set of modeling constructs. PRO*forma* supports four tasks: actions, compound plans, decisions, and inquiries of patient data from a user. All tasks share attributes describing goals, control flow, preconditions, and postconditions. An underlying premise is that the simple task ontology should make it easier to demonstrate soundness and to teach the language to encoders.

PRO*forma*'s decisions are represented as argumentation rule-sets, where different candidate options are associated with arguments—conditions, which if true provide different degrees of support for that option: for, confirming, against, and excluding (see Figure 13-3). This approach was later adopted by EON, PRODIGY, GLIF3, and SAGE. A number of software components (e.g., Tallis, Arezzo at http://www.openclinical.org/gmmsummaries.html) have been written to create, visualize, and enact PRO*forma* guidelines (Sutton and Fox 2003).

13.3.1.6 GLARE

GLARE (Terenziani, Montani et al. 2004) is an approach for modeling and executing clinical guidelines, which emphasizes management of temporal knowledge. The GLARE TNM has the following kinds of nodes: 1) atomic actions, which can be work actions, query actions, decisions, and conclusions, and 2) composite actions, which are composed of actions that can be assembled in sequence, in parallel, iterated, or done in branching paths. In addition, temporal constraints can be defined between component actions. GLARE is supported by an authoring and validation tool, and by an execution engine, which supports temporal reasoning and a hypothetical reasoning facility that makes it possible to compare different paths in the guideline, by simulating what could happen if a certain choice was made.

13.3.2 Other CIG Modeling Methods

13.3.2.1 Arden Syntax

The Arden Syntax (Hripcsak, Ludemann et al. 1994) is a standard of HL7 and ASTM suitable for representing individual decision rules in self-contained units called Medical Logic Modules (MLMs), which usually are implemented as event-driven single-step alerts or reminders (see Chapter 12). MLMs are triggered by events (e.g., admission of a patient, or storage of certain medical data in an EHR) or called by another MLM. Once triggered, MLMs evaluate logical decision criteria (e.g., *potassium* < *3.5*), and, if the criteria hold, they perform an action, such as sending an alert to a clinician. The mappings between the institution-specific terms and the MLM's variables are only partly defined by the syntax. Arden Syntax was not meant to be used for encoding complex guidelines that involve multiple decisions or process flow sequences, although one example of a guideline encoded as a set of interacting MLMs has been created (Peleg, Boxwala et al. 2000). However, there is no support to aid human understanding of the way MLMs interact with one another. In addition, only if-then-else representation of decision rules is possible. The Arden Syntax is supported by several vendor applications and has been used to implement MLMs in many institutions (Peleg, Boxwala et al. 2000).

13.3.2.2 GASTON

GASTON (de-Clercq, Hasman et al. 2001) represents CIGs using primitives and ontologies to represent the medical domain (e.g., entities such as drugs, diseases, and treatments, and relationships among them) and problem-solving methods (PSMs). The primitive classes are based on version 2.0 of GLIF: action, decision, branch, and synchronization steps. PSMs contain a high-level description that details a strategy for solving a problem. An example is the selection PSM that reports conflicts (e.g., drug interactions) resulting from a user's choice of an action (drug prescription) or a decision. A guideline is associated with a task it has to solve. The task can be specified as a set of primitives or as an appropriate PSM. For the former, the guideline's structure is specified in terms of primitives and subguidelines, where decision criteria refer to concepts defined in the domain ontology. When a guideline is to be executed by a PSM, its control structure doesn't need to be specified. GASTON is supported by an authoring tool and an execution engine.

13.3.2.3 OncoDoc

OncoDoc (Seroussi, Bouaud et al. 2005) is a method for representing guidelines halfway between formal knowledge representation and textual reading. Guideline knowledge is represented formally as decision trees. However, instead of automatically executing the decision tree, the user browses it as hypertext and flexibly interprets both patient data and guideline content, thus controlling the interpretation of the guideline knowledge in the specific context of a patient situation. The decision tree is built from clinical parameters that are identified in the guideline narrative and given labels chosen from standard classifications. All theoretically possible clinical situations are represented.

This approach has been applied first to breast cancer therapeutic management. Handling of chronic diseases, such as hypertension, involves considering the patient's therapeutic history (e.g., inadequate response to past treatment, or adverse effects, to select relevant patient-specific therapy among the recommendations). Therefore, OncoDoc was extended (Seroussi, Bouaud et al. 2005) by introducing a level of therapeutic strategies, structured along lines of therapy and levels of therapeutic intention. For each theoretical clinical situation, a range of pharmacological treatments is recommended. Matching patient's therapeutic history elements along with recommended therapies allows OncoDoc to rule out nontolerated or nonefficient past treatments to finally select the best ones.

13.4 FROM NARRATIVE TO FORMAL REPRESENTATIONS OF GUIDELINES

As discussed previously, narrative guidelines are written in a form that makes it extremely difficult to convert them automatically into their formal representation as CIGs. Several approaches have been developed in order to facilitate the transition from narrative to formal representations. These approaches mark up narrative text to indicate that they belong to certain structural components of guidelines, according to markup ontologies. Structuring the narrative document is a step toward creating a computable implementation, and can be used to link a formal representation to the narrative text. In addition, the markup process often can identify ambiguity that needs to be resolved, as well as areas where evidence is lacking and recommendations are not provided.

A variety of markup models exist; each model has different aims and a different view of the structure of guideline content. CPG-RA, discussed in subsection 13.2.2, aids in defining structural components, but has not yet demonstrated transition from the markup into a formal representation that can provide automatic decision support.

Georg and coauthors (Georg, Seroussi et al. 2005) developed an approach for automatically generating decision rules from GEM-encoded guidelines. This approach can be used for assessing the consistency of textual guidelines at the time of writing, as well as providing assistance for guideline implementation.

Other markup methods try to close the gap between narrative guidelines and their implementation by providing modelers with tools for marking up narrative as components of a computable representation (Shahar, Young et al. 2003; Votruba, Miksch et al. 2004). The aim is that authoring of guidelines will be driven and supported by an implementation and a formal model. Nonetheless, studies have not been done yet to show that these markup methods ease the process of transforming narrative into formal representations.

According to the Digital electronic Guideline Library framework (DeGeL) approach, a CIG is developed via a process in which conventional narrative guidelines gradually are transformed from traditional narrative form to fully formal representations (Shahar, Young et al. 2003). Currently, guidelines may be formalized using the Asbru or GEM guideline modeling languages. The URUZ tool was developed to assist modelers in marking up text according to the Asbru guideline modeling language, discussed in the previous section. It enables a user to decompose the actions embodied in the guideline into atomic actions and other subguidelines, and to define the control structure relating them (e.g., sequential, parallel, repeated application). Clinician users can create a semiformal representation in semiformal-Asbru, where temporal patterns, the building blocks of a guideline in Asbru, are expressed with combinations of text and time-annotations instead of Asbru's complicated formal expressions (Shahar, Young et al. 2003). Another tool is used to search for vocabulary terms in controlled vocabularies (ICD-9-CM for diagnosis codes, CPT-4 for procedure codes, and LOINC-3 for observations and laboratory tests) and embed them in the guideline document (Shahar, Young et al. 2003). Knowledge engineers can then transform the semiformal representation into formal Asbru.

Another tool that facilitates the translation of narrative guidelines into a formal representation of CIGs is the Document Exploration and Linking Tool (Delt/A), which is the further development of the Guideline Markup Tool (GMT) (Votruba, Miksch et al. 2004; also see http://ieg.ifs.tuwien.ac.at/projects/delta/). Delt/A supports the transformation process of clinical guidelines from their original textual form (HTML) through an intermediate and a semiformal representation (XML) to a formal representation (and vice versa), providing two main features: links and macros. Links are used to show and connect related parts in HTML and XML markup. Macros combine multiple XML elements together with their attributes and can be used for simple construction of new XML documents. These macros are typical patterns of clinical guideline components (e.g., two mutually exclusive plans), which ease the implementation process. These structural patterns differ from the clinical patterns of GLIF3 macros. A Macro Step in GLIF3 is a special class that has attributes that define the information required to instantiate a set of underlying GLIF steps. Those underlying steps represent a pattern that appears in clinical guidelines (Peleg, Boxwala et al. 2000). In this way, macro steps provide a means to specify declaratively a procedural pattern, using a single construct that is realized by a set of GLIF3 steps. For example, a macro step for risk assessment is mapped into a pattern of underlying GLIF3 steps: an action step for collecting patient data, an action step for computing the risk of the patient based on collected data, and a decision step that is linked to alternative action steps, corresponding to a decision among recommendations appropriate for the various risk levels.

13.5 INTEGRATION OF GUIDELINES WITH WORKFLOW

Guideline modeling languages can be stratified according to the level of their support for integration with the organization's workflow and information systems. The basic CIG languages support modeling of guideline knowledge but, except for definitions of variables used in the encoding, they do not support data modeling intended to facilitate interfacing the guideline model with an EHR. The Arden syntax, Asbru, PRO*forma*, and the model developed by Seroussi belong to this category. These guideline modeling methodologies are very useful for implementing guidelines that require manual data entry or conducting a dialog of questions and answers. An example is supporting electronic prescription of drugs by developing a set of criteria that checks whether to authorize high-cost drug prescription.

Formalisms belonging to the second level of integration support include a patient information model. For example, EON, SAGE, and PRODIGY use a vMR model for representing patient data that is used by the guidelines, and tools supporting the execution of these methodologies use mappings defined from the guideline's patient data EHR schemas to retrieve relevant data. GLIF3 uses a model based on the Medication, Observation, and Procedure classes of the HL-7 RIM to represent patient data referenced by the guideline. GLARE has a database ontology for modeling patient data.

The third level of integration support considers the workflow of activities that are taking place in the setting of the implementing institution, and fits the guideline model within that workflow. This approach considers available resources, organizational roles that perform activities (e.g., a clinician ordering a prescription), care setting, and timing constraints. It helps in identifying the best way to implement a task in a given setting, taking into consideration the health care information system and environment factors. NewGuide and SAGE are CIG formalisms that support this level of integration.

The NewGuide model is based on the Workflow model of the Workflow Management Coalition, and includes an organizational model and a care process (guideline). The organizational model represents entities such as organization, organizational unit, resources, agents that execute activities and consume resources, organizational roles, goals and subgoals that roles try to achieve for the benefit of the organization, as well as relationships among those entities. Each organization that implements a guideline creates instances of the organization model entities. A NewGuide guideline can be translated automatically into a Petri Net formalism—a bipartite graph of places (i.e., conditions) and transitions that can be simulated to study possible behaviors of a modeled guideline. By translating a guideline into a Petri Net, the effects of implementing a guideline at that particular facility can be simulated and measures of system performance, costs, resource overload, and bottlenecks can be obtained before the system is put into use and can be used to achieve optimal resource allocation. By using the Oracle Workflow tool, for example, the guideline can drive resource allocation and task management in a clinical setting.

SAGE takes a different approach to considering integration into workflow. The approach is based on identifying opportunities for decision support within the health care process. These opportunities are modeled as contexts, as discussed previously, which consider the triggering event, care setting, organizational roles involved, and available EHR functions. The availability of EHR functions implies the possible modes of implementation. For example, if the EHR has an order entry function and we want to perform a drug ordering task, then we might want to implement this task by linking to the EHR's order-entry function. But if order-entry is not supported by the EHR system, then we may choose to implement the task by sending an alert to the physician's inbox.

Shiffman and coauthors suggest that the gap that exists between marked-up guidelines and their implementations as workflow-integrated decision-support systems could be closed by mapping the marked-up guideline into generic action types (Action Palette) associated with services that could fulfill them (Shiffman, Michel et al. 2004). At the end of the markup stage, the guideline is tagged as precise logical statements that can be translated into computable statements. The first stage in linking these statements into clinical workflow involves identifying 1) the data sources needed to assess the logical statements, and 2) insertion points for the generic actions within the workflow (e.g., patient registration, history recording, physical examination and laboratory testing, assessment of findings, and plan formulation). Next, the implementer should select an appropriate action type and an associated service type to implement it. Available action types include gathering information (testing and monitoring), interpreting information, performing a task (prescription, therapeutic procedure, education, documentation, advocating a policy, and preparation for performing a task), and arranging for or organizing additional care (referring). The associated service types offer design patterns for facilitating clinical care. The output of the action-mapping process is a requirement specification that should be operationalized by information systems personnel.

The Decision-Support Opportunity aims to identify opportunities for decision support during clinical workflow (Osheroff, Pifer et al. 2005). CDS support could be provided in a system-facilitated manner, where the user initiates a request for assistance from the system, or in a system-initiated form. For example, an opportunity for decision support can be a need to refresh a clinician's memory on diagnostic or management essentials before, during, or after a visit. In a system-facilitated approach, the CDS service could be in the form of a context-specific link to a database, whereas in a system-initiated approach, the advice could be delivered in the form of alerts regarding omission or commission errors. Another CDS opportunity is for documenting the patient's condition that would enable safe drug prescription based on patient-condition-specific factors. In a system-facilitated approach, the CDS service could be in the form of documentation templates and order sets, whereas in a system-initiated approach the advice could be delivered by checking drug–drug interactions and allergies.

13.6 METHODS FOR SHARING OF CIG CONTENT

A number of studies of computer applications that can deliver patient-specific clinical knowledge at the point of care during patient encounters have shown positive impacts on clinicians' behavior. Since developing such resources requires much effort, we would like to be able to share them, enabling the community of clinical guideline developers, publishers, and users to work collaboratively, leveraging prior work. However, with the variety of CIG formalisms, sharing CIG components that conform to different formalisms is not a straightforward task. In this section, we discuss approaches to sharing of CIG content.

13.6.1 Interchanging among CIG Formalisms

Creating a translation among different representation formalisms is an appealing idea, since it enables different CIG formalisms to coexist and evolve concurrently. The InterMed project, which was responsible for the development of GLIF, started out in 1996 with the aim of creating an interchange format among several guideline formalisms, but it soon became clear that practical interchange among the formalisms could not be achieved due to differences in functionality supported by the formalisms (Peleg, Boxwala et al. 2004).

13.6.2 Adopting a Single Formalism as a Standard

If the CIG community were to adopt a single CIG formalism as a standard, and tools for authoring, validation, execution, and maintenance of CIGs in this formalism were to be developed, then sharing encoded CIGs among CIG users would not be problematic. For such a common platform to be accepted and widely used, a broad spectrum of participants must have a stake in it, and contribute to its further growth and development. At a guideline workshop that was hosted by InterMed in March 2000, participants explored the issues involved in progressing toward a shareable standardized representation of clinical guidelines (Peleg, Boxwala et al. 2004). Later that year, InterMed helped to establish the HL7 Clinical Guidelines Special Interest Group (CGSIG), under a reorganized Clinical Decision Support Technical Committee (CDSTC). It soon became apparent that agreeing on a fully comprehensive CIG formalism that would be accepted by the entire CGSIG was not achievable, due to differences in opinion among CGSIG members. Therefore, the goal became that of developing and standardizing *components* of CIG models on which consensus could be established among members of the CIG community.

13.6.3 Standardizing CIG Components and Fitting Them Together

Members of the CIG community have all been in favor of achieving sharing of CIG content. Standardizing CIG components would enable sharing of significant parts of encoded guidelines across different CIG modeling methods. The selection of CIG components by the CGSIG was influenced by the results of a study that compared CIG formalisms in terms of components that capture the structure of CIGs (Peleg, Tu et al. 2003).

Three components were selected to be the initial foci for standardization: 1) an object-oriented guideline expression language, 2) a patient data model based on a vMR that would be derived from the HL7 RIM, and would specifically enable reference to the subset of EHR data needed for guideline-based decision support, and 3) guideline control flow. In addition, work is in progress concerning standardizing the documentation attributes of CIGs, relying on the work of the GEM and CPG-RA groups.

A standard expression language could be used for specifying and sharing decision logic and eligibility criteria, calculations, patient state definitions, conditions, and system actions. The specification of GELLO (http://cslxinfmtcs.csmc.edu/hl7/arden/2004-09-ATL/v3ballot_gello_aug2004.zip), as discussed in Chapter 12, arose from this need, and has been adopted as an HL7 and ANSI standard since 2005. GELLO, based on the Object Constraint Language (OCL) (http://www-306.ibm.com/software/rational/uml/resources/documentation.html) of the Object Management Group (OMG), is an object-oriented expression language that is vendor-independent, side-effect-free, and extensible.

An object-oriented virtual vMR would ease the process of mapping guideline patient data items to EHRs, allowing decision criteria, eligibility criteria, and patient states to be defined in guideline models by reference to the vMR rather than to specific EHRs. Members of the CGSIG are developing the vMR, based on experiences with the patient information models of PRODIGY, EON, SAGE, and the HL7 RIM.

All the TNMs, except for PRODIGY, organize guidelines as plans that unfold over time, by linking plan components in sequence, in parallel, and in iterative and cyclic structures, thus defining control-flow. In addition, all the models support nesting of plans, as well as expression of temporal constraints on plan components. It therefore seemed reasonable to standardize CIG control flow. Several CGSIG members (Peleg, Tu et al. 2004) have been evaluating the Workflow Management Coalition's (WfMC) Workflow model (http://www.wfmc.org/standards/docs/) as a common control-flow model. This workflow model supports the control flow of TNMs,[3] is a standard of the WfMC, and has well-defined formal foundations derived from Petri Nets that enable formal verification and simulation. By specializing activities into guideline-specific tasks, such as inquiries and decisions, the different guideline models should be able to map to this formalism. Yet much more work is required in order to achieve a standard representation of control flow and define its mapping to the existing TNMs.

In order to experiment in fitting the standard components together, members of the CGSIG are working on tasks involving several components; for example, writing a query in GELLO for collecting data from an EHR, based on the vMR model.

13.6.4 Sharing Guidelines at the Execution Level

Guidelines encoded in different formalisms can be shared at the execution level (Wang, Peleg et al. 2003). The underlying model for this approach includes a set of generalized guideline execution tasks that were extracted from existing guideline representation models. The generalized tasks are modeled in an ontology that represents 1) primary tasks, such as *data collection*, *clinical intervention*, *medical decision-making*, *patient state verification*, *branching*, *synchronization*, and *subguideline invocation*, which constitute the nodes of a TNM; and 2) the auxiliary tasks, such as *criterion evaluation*, *event registration*, and *event invocation*, which are used to support the execution of the primary tasks. Mappings between specific guideline representation models and the ontology of the common guideline execution tasks were defined and an execution engine was developed that can execute guidelines belonging to different formalisms, according to this approach (Wang, Peleg et al. 2003).

13.6.5 Assembling CIGs from Executable Components

The OpenClinical Group suggested a model for publishing CIGs on the Web and executing them over the Web to deliver patient-specific management advice (Fox, Bury et al. 2001). In this model, executable guidelines are published as Web-accessible services, called *publets*. Building on this work, a framework for creating, representing, and indexing a Medical Knowledge Repository of resources that could be used as components for developing executable guidelines was suggested (Peleg, Steele et al. 2005). The repository could include, among others, resources such as medical calculators, drug databases, tools for authoring and executing CIGs, standard expression languages, and standard patient data models. This framework leverages ideas

[3]In fact, the Workflow model of the WfMC is the basis for NewGuide's CIG formalism.

from the CORBA architecture, the Web Services Architecture, and the semantic Web Services Framework.

13.6.6 Libraries of CIGs

Sharing of narrative guidelines is facilitated by electronic libraries of evidence-based narrative guidelines (The National Guideline Clearinghouse, http://www.guidelines.gov, and the International Guideline Library http://www.g-i-n.net/). To ease sharing of CIGs, they too should be stored in libraries that could be browsed and searched. The DeGel (Shahar, Young et al. 2003) and Open Clinical (http://www.openclinical.org) groups are setting up such libraries.

13.7 CONCLUSION

Implementing clinical guidelines as decision-support systems that provide patient-specific recommendations during clinical encounters increases the chances of affecting clinicians' behavior and achieving the benefits of guidelines. The road to achieving widespread use of such CDS systems is long and difficult. Along this road, we have known successes and failures, and there are many future directions that can be taken to reach this goal.

13.7.1 Successes

A number of CIGs have been represented using CIG modeling languages and implemented. Some of these systems also have been evaluated and shown to be effective and beneficial.

Eleven CDS applications have been implemented using PRO*forma* technology (http://www.openclinical.org/gmmsummaries.html). Quantitative trials have been carried out for seven of these systems. All seven have shown major positive effects on a variety of measures of quality and/or outcomes of care. The seven systems include CAPSULE (assisting general practitioners in prescribing for common conditions), LISA (advising on dose adjustment in treatment of children with acute lymphoblastic leukemia), Retrogram (advice on the use of antiretroviral therapy for HIV+ patients), a treatment planner for patients with type 2 diabetes and hypertension, RAGs (helping GPs take the family history, assess risk, and explain risk factors to patients), CADMIUM (combining conventional image processing with automated interpretation of images and diagnosis), and initial assessment of women referred to specialist breast clinics. CREDO is a current clinical trial of decision-making and workflow management in the care of women at risk for or with a proven diagnosis of breast cancer.

Four NewGuide guidelines were implemented for different hardware, users, and health care settings (Ciccarese, Caffi et al. 2005). They include pressure ulcer prevention, acute ischemic stroke treatment, post-stroke rehabilitation, and heart failure management. The evaluation of the stroke management guideline showed that health outcomes and costs are related to guideline compliance (http://www.openclinical.org/gmmsummaries.html).

ATHENA is an EON-based implementation of the JNC-VI hypertension guideline that has been deployed successfully as a CDS system at clinics in three medical centers of the Department of Veterans Affairs (http://www.openclinical.org/gmmsummaries.html). Analysis of the data collected during a clinical trial to test the impact of the ATHENA system is under way. Preliminary results indicate that, during the 15-month clinical trial, clinicians interacted with the advisory screen for 63 percent of patients eligible for guideline-based CDS. Use of the system remained high throughout the 15 months.

An immunization guideline and a diabetes guideline have been encoded in the SAGE formalism and guideline scenarios were executed using the SAGE engine and the vMR/Clinical Information System (CIS) services of Carecast™ (Ram, Berg et al. 2003). Demonstration of the integration of a SAGE guideline CDS system into clinical information systems of Mayo Clinic and University of Nebraska Medical Center is under way (Ram, Berg et al. 2003).

Two guidelines, management of diabetes-related foot disorders and post-coronary artery bypass surgery patient care planning, have been implemented in GLIF3, linked with an EHR, and executed using the GLEE engine in an educational setting (http://www.openclinical.org/gmmsummaries.html).

Three complex chronic disease management guidelines and over 150 simple guidelines were implemented in PRODIGY-3. Two vendors have integrated identical PRODIGY components into their clinical information systems for general practitioners (Peleg, Tu et al., 2003).

A diabetes (Marcos, Roomans et al. 2002) guideline and a jaundice guideline have been implemented in the Asbru formalism (http://www.openclinical.org/gmmsummaries.html).

Several systems have been implemented using GASTON technology. The CritICIS system is a real-time reminder system used in critical care environments such as intensive care units in the Netherlands (de-Clercq, Hasman et al. 2001). A validation study showed that 88 percent of all issued reminders were classified as correct. The GRIF (de-Clercq, Hasman et al. 2001) system was developed to support test ordering by general practitioners. It too has undergone evaluation, showing that the amount of information and the level of detail in which the practitioner describes the patients' medical status are crucial for the reminder system to react correctly. M-PADS (de-Clercq, Hasman et al. 2001) is a system developed for selecting the most appropriate psychoactive drug in order to treat psychiatric patients. Additionally, CDS systems were developed that provide advice through the Internet, integrated in a Web-based consumer health record system. Using this technology, two guidelines were implemented: treatment of diabetes and treatment of hypertension.

A cardiac rehabilitation decision support system that uses the GASTON framework has been implemented and tested during a six-week pilot study in four cardiac rehabilitation centers in the Netherlands (http://www.openclinical.org/prj_cardss.html). After the pilot study, several new functions were added to the system. A randomized trial has started in 38 Dutch hospitals to assess the effect of the CDS on guideline adherence.

The implementation of OncoDoc for breast cancer management has been evaluated in a study that compared physicians' compliance and patient accrual in clinical trials before the use of OncoDoc, and after its routine use (Seroussi and Bouaud 2003). Compliance increased from 61.5 percent to 85 percent, and patient accrual increased by 50 percent.

A second line of success is the agreement within the community of CIG developers that sharing CIG content is important and that work should be done toward such a goal, while diversity of guideline formalisms that are evolving should be encouraged. The agreement was evident in a series of guideline-specific meetings (Boston 2000 http://www.glif.org/workshop/workshop.htm; Leipzig 2000 http://www.onto-med.de/en/events/EWGL-P2000/; London 2001 http://www.openclinical.org/gmmworkshop2001.html; Prague 2004 http://www.openclinical.org/cgp2004.html) and in a case-based guideline comparison paper in which six CIG formalisms were compared (Peleg, Tu et al. 2003). People vary in their approaches to achieving CIG content sharing. HL7 CGSIG members are focusing on standardization efforts, whereas other efforts try to avoid premature standardization, and instead are focusing on setting up libraries of encoded CIGs and CIG components, as discussed earlier.

In the standardization arena, one success story has been the standardization of the GELLO guideline expression language that can be used for formally defining decision and eligibility criteria, as well as patient states.

13.7.2 Limitations and Challenges

The process of creating a standard guideline model, while receiving wide support in HL7, has many limitations. The field of computerized guidelines may be too immature for starting the standardization while requirements and goals are still changing. Members of the community have disagreed about developing a full guideline model and instead are focusing on standardizing components of a guideline model. But disagreement still exists about which components should be included in such a standard, and it is not yet clear that a complete CIG standard could be assembled from the component standards in the making. The CIG standard should be easy to use and should be supported by authoring, markup, and execution tools for it to be widely adopted. Standards would be more effective if consensus has been achieved and if users and industry are involved in the standard development process. Up to now, the standardization process has been driven by the developers of guideline modeling methodologies instead of other stakeholders, such as users, payers, and vendors. There is no process for identifying urgent user needs.

Another area that has not been adequately studied is how to integrate guidelines into clinical information systems. Preliminary work has been done by the NewGuide and SAGE groups, as well as the Action Palette and the Decision-Support Opportunity Map.

Furthermore, CDS systems that use CIGs should not be seen as a silver bullet for the dissemination and implementation of clinical guidelines. Most of the literature that demonstrates the efficacy of computerized CDS discusses reminder-based systems. A few recent evaluation studies show that evidence-based CDS systems may not necessarily have clinical impact (Eccles and Grimshaw 2004; Tierney, Overhage et al. 2005). Much more study is needed to understand the factors that make such an implementation successful.

13.7.3 Future Research

Future research should concentrate on the areas that constitute the major limitations of widespread successful implementations of CIGs:

- Filling in the gap between authoring and implementation by providing authoring support that is driven by an implementation and a formal model. This includes communication and visualization methods and tools to communicate the CIG to the domain experts during the authoring and execution phase.
- Creating a model for integrating a CIG with an actual CIS.
- Involving stakeholders in standardization of a CIG formalism or components of such a formalism and having them set priorities for developing a guideline standard.
- Further development in the direction of sharing executable CIG components and assembling CIGs from them.
- Evaluating the cost and impact of deployed CDS systems for guideline-based care, and understanding the barriers and facilitators in the acceptance of such systems.

13.8 RECOMMENDED RESOURCES

Field, M. J., Lohr, K. N. (1992). Guidelines for Clinical Practice: From development to use. Washington DC: Institute of Medicine, National Academy Press.

This book was written by an expert committee appointed by the Institute of Medicine, which examined clinical guidelines, focusing on their development and implementation. The book discusses the strengths and limitations and how they can be used more effectively to benefit health care.

The National Guideline Clearinghouse; www.guidelines.gov.

A public resource for evidence-based clinical practice guidelines, initiated by the Agency for Healthcare Research and Quality (AHRQ), U.S. Department of Health and Human Services.

Peleg, M., Tu, S. W., Bury, J., Ciccarese, P., Fox, J., Greenes, R. A. et al. (2003). Comparing computer-interpretable guideline models: A case-study approach. *J Am Med Inform Assoc* **10**(1): 52–68.

The comparison of Peleg and colleagues identified eight components that capture the structure of CIGs. These dimensions fall into two broad categories: structuring guidelines as plans of decisions and actions, and linking the guideline to patient data and medical concepts.

de-Clercq, P. A., Blom, J. A., Korsten, H. H. M., Hasman, A. (2004). Approaches for creating computer-interpretable guidelines that facilitate decision support. *Artif Intell Med* **31**: 1–27.

De Clercq's paper includes a review of well-known CIG formalisms, addressing, in addition to guideline representation issues, aspects concerning guideline acquisition, verification, and execution.

The OpenClinical Web site http://www.openclinical.org/.

OpenClinical is an international organization that has been created to promote awareness and use of decision support, clinical workflow, and other advanced knowledge management technologies for patient care and clinical research. The Web site provides a substantial and growing set of resources for technologists, clinicians, healthcare providers, and suppliers who wish to find out more about this field (from the OpenClinical Web site).

Guidelines International Network (G.I.N.); Web site http://www.g-i-n.net/.

G.I.N. is a major international initiative that seeks to improve the quality of health care by promoting systematic development of clinical practice guidelines and their applications into practice. The Web site offers guideline resources such as a guideline library, development tools and resources, training material of guidelines, and patient resources (from the G.I.N. Web site).

ACKNOWLEDGMENTS

I would like to thank Samson Tu, Richard Shiffman, Dean Sittig, Vojtech Svatek, Silvia Miksch, Brigitte Seroussi, Jacques Bouaud, Silvana Quaglini, John Fox, and Richard Thomson for their very helpful comments and suggestions.

REFERENCES

Agrawal, A. and Shiffman, R. N. (2001). Evaluation of guideline quality using GEM-Q. *Medinfo* 1097–1101.

Berrios, D. C., Cucina, R. J., and Fagan, L. M. (2002). Methods for semi-automated indexing for high precision information retrieval. *J Am Med Inform Assoc* **9**(6): 637–652.

Boxwala, A. A., Peleg, M., Tu, S., Ogunyemi, O., Zeng, Q., Wang, D. et al. (2004). GLIF3: A representation format for sharable computer-interpretable clinical practice guidelines. *J Biomed Inform* **37**(3): 147–161.

Ciccarese P., Caffi, E., Quaglini, S., and Stefanelli, M. (2004). An innovative Health Information System: The Guide Project. *International Journal of Medical Informatics*, Special Issue on Medinfo 2004, in press.

de-Clercq, P. A., Blom, J. A., Korsten, H. H. M., and Hasman, A. (2004). Approaches for creating computer-interpretable guidelines that facilitate decision support. *Artif Intel Med* **31**: 1–27.

de-Clercq, P. A., Hasman, A., Blom, J. A., and Korsten, H. H. (2001). Design and implementation of a framework to support the development of clinical guidelines. *Int J Med Inform* **64**(2–3): 285–318.

Eccles, M. P. and Grimshaw, J. M. (2004). Selecting, presenting and delivering clinical guidelines: Are there any "magic bullets"? *Med J Aust* **180**(6 Suppl): S52–S54.

Field, M. J. and Lohr, K. N. (1990). Guidelines for clinical practice: Directions for a new program. Washington, DC: Institute of Medicine, National Academy Press.

Field, M. J. and Lohr, K. N. (1992). Guidelines for clinical practice: From development to use. Washington, DC: Institute of Medicine, National Academy Press.

Fox, J., Bury, J., Humber, M., Rahmanzadeh, A., and Thomson, R. (2001). Publets: Clinical judgement on the web. *Proc AMIA Symp* 179–183.

Fox, J., Johns, N., Rahmanzadeh, A., and Thomson, R. (1996). PROforma: A method and language for specifying clinical guidelines and protocols. Amsterdam: Medical Informatics Europe.

Georg, G., Seroussi, B., and Bouaud, J. (2005). Extending the GEM model to support knowledge extraction from textual guidelines. *Int J Med Inform* **74**(2–4): 79–87.

Hadorn, D. C. (1995). Use of algorithms in clinical guideline development. In *Clinical Practice Guideline Development: Methodology Perspectives*. McCormick, K. A., Siegel, R. A., eds., AHCPR Pub. No. 95-0009, 93-104. Rockville, MD: Agency for Health Care Policy and Research.

Hripcsak, G., Ludemann, P., Pryor, T. A., Wigertz, O. B., and Clayton, P. D. (1994). Rationale for the Arden Syntax. *Comput Biomed Res* **27**(4): 291–324.

Institute of Medicine (2001). Crossing the Quality Chasm: A New Health System for the 21st Century. Washington, DC: National Academy Press.

Johnson, P. D., Tu, S. W., Booth, N., Sugden, B., and Purves, I. N. (2000). Using scenarios in chronic disease management guidelines for primary care. *Proc AMIA Symp* 389–393.

Kohn, L. T., Corrigan, J. M., and Donaldson, M. S. (1999). To Err Is Human: Building a Safer Health System. Committee on Quality of Health Care in America. Washington, DC: Institute of Medicine, National Academy Press.

Osheroff, J. A., Pifer, E. A., Teich, J. M., Sittig, D. F., and Jenders, R. A. (2005). Improving Outcomes: A Practical Guide to Clinical Decision Support Implementation. Chicago, IL: HIMSS Press.

Overhage, J. M., Tierney, W. M., Zhou, X. H., and McDonald, C. J. (1997). A Randomized Trial of "Corollary Orders to Prevent Errors of Omission." *J Am Med Inform Assoc* **4**(5): 364–375.

Patel, V. L., Allen, V. G., Arocha, J. F., and Shortliffe, E. H. (1998). Representing a clinical guideline in GLIF: Individual and collaborative expertise. *J Am Med Inform Assoc* **5**(5): 467–483.

Peleg, M., Boxwala, A. A., Bernstam, E., Tu, S., Greenes, R. A., and Shortliffe, E. H. (2000). Sharable representation of clinical guidelines in GLIF: Relationship to the Arden Syntax. *J Biomed Inform* **34**(3): 170–181.

Peleg, M., Boxwala, A. A., Tu, S., Zeng, Q., Ogunyemi, O., Wang, D. et al. (2004). The InterMed approach to sharable computer-interpretable guidelines: A review. *J Am Med Inform Assoc* **11**(1): 1–10.

Peleg, M., Steele, R., Thomson, R., Patkar, V., Rose, T., and Fox, J. (2005). Open-source publishing of medical knowledge for creation of computer-interpretable guidelines. *Artif Intel Med Europe 2005*, in press.

Peleg, M., Tu, S., Mahindroo, A., and Altman, R. (2004). Modeling and analyzing biomedical processes using workflow/petri net models and tools. *MedInfo* 74–78.

Peleg, M., Tu, S. W., Bury, J., Ciccarese, P., Fox, J., Greenes, R. A. et al. (2003). Comparing computer-interpretable guideline models: A case-study approach. *J Am Med Inform Assoc* **10**(1): 52–68.

Ram, P., Berg, D., Tu, S. W., Mansfield, J. G., Ye, Q., and Abarbanel, R. (2003). Executing clinical practice guidelines using the SAGE execution engine. Stanford University, Report number SMI-2003-0971.

Ruzicka, M. and Svatek, V. (2004). Mark-up based analysis of narrative guidelines with the Stepper tool. *Stud Health Technol Inform* **101**: 132–136.

Schadow, G., Russler, D. C., Mead, C. N., and McDonald, C. J. (2000). Integrating medical information and knowledge in the HL7 RIM. *Proc AMIA Symp* 764–768.

Seroussi, B. and Bouaud, J. (2003). Using OncoDoc as a computer-based eligibility screening system to improve accrual onto breast cancer clinical trials. *Artif Intell Med* **29**(1–2): 153–167.

Seroussi, B., Bouaud, J., and Chatellier, G. (2005). Guideline-based modeling of therapeutic strategies in the special case of chronic diseases. *Int J Med Inform* **74**(2–4): 89–99.

Seyfang, A., Miksch, S., Conde, C. P., Wittenberg, J., Marcos, M., and Rosenbrand, K. (2005). A many-headed bridge between informal and formal guideline representations. *Proc. 10th Conf Art Intell Med (AIME)*, in press.

Shahar, Y., Miksch, S., and Johnson, P. (1998). The Asgaard project: A task-specific framework for the application and critiquing of time-oriented clinical guidelines. *Artif Intell Med* **14**(1–2): 29–51.

Shahar, Y., Young, O., Shalom, E., Mayaffit, A., Hessing, R. M., and Galperin, M. (2003). DeGeL: A hybrid, multiple-ontology framework for specification and retrieval of clinical guidelines. *Proc Artif Intell Med Europe* 122–131.

Shaneyfelt, T. M., Mayo-Smith, M. F., and Rothwangl, J. (1999). Are guidelines following guidelines? The methodological quality of clinical practice guidelines in the peer-reviewed medical literature. *JAMA* **281**(20): 1900–1905.

Shea, S., DuMouchel, W., and Bahamonde, L. (1996). A meta-analysis of 16 randomized controlled trials to evaluate computer-based clinical reminder systems for preventative care in the ambulatory setting. *J Am Med Inform Assoc* **3**(6): 399–409.

Shiffman, R. N., Dixon, J., Brandt, C., Essaihi, A., Hsiao, A., Michel, G., O'Connel, R. (In press). The GuideLine Implementability Appraisal (GLIA): Development of an instrument to identify obstacles to guideline implementation. *BMC Med Inform Decis Making.*

Shiffman, R. N., Karras, B. T., Agrawal, A., Chen, R., Marenco, L., and Nath, S. (2000). GEM: A proposal for a more comprehensive guideline document model using XML. *J Am Med Inform Assoc* 7(5): 488–498.

Shiffman, R. N., Michel, G., Essaihi, A., and Thornquist, E. (2004). Bridging the guideline implementation gap: A systematic, document-centered approach to guideline implementation. *J Am Med Inform Assoc* **11**(5): 418–426.

Society for Medical Decision Making (1992). Proposal for clinical algorithm standards. *Med Decis Making* **12**: 149–154.

Sutton, D. R. and Fox, J. (2003). The syntax and semantics of the PROforma guideline modeling language. *J Am Med Inform Assoc* **10**(5): 433–443.

Terenziani, P., Montani, S., Bottrighi, A., Torchio, M., Molino, G., and Correndo, G. (2004). The GLARE approach to clinical guidelines: main features. *Stud Health Technol Inform* **101**: 162–166.

Tierney, W. M., Overhage, J. M., Murray, M. D., Harris, L. E., Zhou, X. H., Eckert, G. J. et al. (2005). Can computer-generated evidence-based care suggestions enhance evidence-based management of asthma and chronic obstructive pulmonary disease? A randomized, controlled trial. *Health Serv Res* **42**(2): 477–498.

Timmermans, S. and Berg, M. (2003). From autonomy to accountability? Clinical practice guidelines and professionalism. In *The Gold Standard: The Challenge of Evidence-Based Medicine and Standardization in Health Care*. Philadelphia, PA: Temple University Press.

Tu, S., Campbell, J., and Musen, M. A. (2004). The SAGE guideline modeling: Motivation and methodology. *Stud Health Technol Inform* **101**: 167–171.

Tu, S. and Musen, M. (2000). Representation formalisms and computational methods for modeling guideline-based patient care. *First European Workshop on Computer-based Support for Clinical Guidelines and Protocols* 125–142. Leipzig, Germany.

Tu, S. W., Campbell, J., and Musen, M. A. (2003). The structure of guideline recommendations: A synthesis. *Proc AMIA Symp* 679–683.

Tu, S. W. and Musen, M. A. (1999). A flexible approach to guideline modeling. *AMIA Symp* 420–424.

Tu, S. W. and Musen, M. A. (2000). From guideline modeling to guideline execution: Defining guideline-based decision-support services. *Proc AMIA Symp* 863–867.

Votruba, P., Miksch, S., Seyfang, A., and Kosara, R. (2004). Tracing the formalization steps of textual guidelines. *Stud Health Technol Inform* **101**: 172–176.

Wang, D., Peleg, M., Bu, D., Cantor, M., Landesberg, G., Lunenfeld, E. et al. (2003). GESDOR—A generic execution model for sharing of computer-interpretable clinical practice guidelines. *Proc AMIA Symp* 694–698.

Wang, D., Peleg, M., Tu, S. W., Boxwala, A. A., Greenes, R. A., Patel, V. L., and Shortliffe, E. H. (2002). Representation primitives, process models and patient data in computer-interpretable clinical practice guidelines: A literature review of guideline representation models. *Intl J Med Inform* **68**(1–3): 59–70.

14 ONTOLOGIES, VOCABULARIES, AND DATA MODELS

STANLEY M. HUFF

14.1 INTRODUCTION

Previous publications have described correct principles for developing and maintaining terminologies (Humphreys and Lindberg 1989; Cimino 1994, 1998). Other articles describe the details of how particular clinical terminologies like SNOMED CT, the UMLS Metathesaurus, LOINC, and RxNorm have been developed and the conceptual content, relationships, and capabilities of those systems (Lindberg et al. 1993; Forrey et al. 1996; Huff et al. 1998; Humphreys et al. 1998; Stearns et al. 2001; Nelson et al. 2002; Wang et al. 2002; McDonald et al. 2003). The purpose of this chapter is to describe vocabulary and terminology issues and challenges related specifically to successful implementation of clinical decision support (CDS) systems. We will discuss:

- Why standard coded data are essential for accurate and reliable execution of decision logic
- How to unambiguously reference data in the electronic medical record (EMR) from CDS expressions
- Alternatives for pre- and postcoordinated representations of data
- Representation of patient data as name-value pairs
- The relationship between terms and information/data models that provide the context of use
- Terminology in the life cycle of CDS programs
- The next steps that are needed in standardizing models and terminology for use in CDS modules

14.2 THE NEED FOR CODED DATA

One of the first issues that presents itself when considering terminology issues related to CDS applications is the need for coded data. Why do data need to be encoded? The primary answer is that encoded data are required in order to have accurate and reproducible execution of decision logic. Unstructured free text is too ambiguous and imprecise to support valid reasoning by computers. An example may help illustrate the point. A CDS program was developed at

Clinical Decision Support: The Road Ahead

LDS Hospital that watched for people who were on an intravenous antibiotic that could be switched to an equivalent oral antibiotic. The program looked for indications that a person on I.V. antibiotics was capable of oral intake. At LDS Hospital, oral intake could be deduced from the existence of an oral diet order or the existence of coded orders for oral medications. There was a desire to export this decision logic to other hospitals. However, in exporting the program it was found that many institutions did not have the coded data to support the protocol, so an attempt was made to interpret free text orders to determine the existence of oral intake, but processing the free text for real time decision-making was found to be impractical. There were just too many ways to represent the fact that the patient was on oral intake. Some of the variations found were *oral, take orally, by mouth, per os, via nasogastric tube, po, p.o., PO, swallow*, with all variations, abbreviations, and misspellings of these representations. It was impossible to anticipate the multiple ways that oral intake could be represented. Successful implementation of the logic depended on getting coded data as the initial input.

Coding of data has advantages other than just the execution of the logic. For example, coded data can be used to support different report formats for particular users; when appropriate, nurses can display the data differently from respiratory therapists. It also makes the maintenance of decision logic easier since the logic references a code rather than referencing all the words that might be used to represent the needed concept. Use of codes makes it possible to translate more easily to different languages. Storing codes may save storage space, depending on the implementation, although this is not an important consideration in most EMR systems today.

14.3 REFERENCING DATA IN DECISION LOGIC

CDS systems are highly dependent on clinical information to function. To be useful, the clinical information must contain sufficient detail and must be structured in a way that the CDS system understands. In proprietary decision support applications that are tightly bound to a particular clinical application, the representation of clinical information used in decision support may be the same as the representation used in the associated clinical application. However, when the decision support system is based on a more portable standard, such as the Arden Syntax (Hripcsak 1991; Health Level Seven 1999) or GELLO (Sordo et al. 2003; Health Level Seven 2005a), the clinical information must be transformed from one format to another.

The Arden Syntax does not specify the format of a reference to data in an associated clinical application. Rather, such references are implementation-specific and demarcated by curly braces within the code (see Chapter 12). Since these references to external data vary from one implementation to the next, the issue of dealing with references to external data has been dubbed "the curly braces problem." The most significant consequence of the curly braces problem is that CDS modules are not readily portable between different systems implementing the Arden Syntax. Implementers of the logic must fill in the curly braces with statements that will reference data from their local patient data store.

The curly braces problem is not unique to the Arden Syntax. All decision support systems that strive to maintain portability across disparate clinical applications must address the issue of referencing external clinical information. In the worst case, a data reference in a given logic module would need to be mapped to a particular data element in every system where the logic was deployed (a many-to-many map from logic modules to local data elements). In the best case, the logic module is mapped to a shared common data model like the HL7 RIM (a many-to-one map from logic modules to the RIM), and each system that deploys the logic needs to map to the HL7 RIM, but a many-to-many map for each data reference is avoided.

Several computable syntaxes have been proposed for representing the form and structure of clinical information, including GALEN Representation and Integration Language (GRAIL) (Rector and Nowlan 1993; Rector et al. 1997), Abstract Syntax Notation 1 (ASN.1) (ISO/IEC 8824-1 1990; ISO/IEC 8825-1 1990; ISO/IEC 8825-2 1996), Archetype Definition Language (ADL) (Beale and Heard 2006), Web Ontology Language (OWL) (W3C 2004), Clinical Element models (Coyle et al. 2003), DICOM templates (NEMA PS 3 Supplement 23 1997), General Purpose Information Components (GPICs) models from CEN (CEN TC 251 2003), description logics (Wikipedia 2006), and the Health Level Seven Reference Information Model (HL7 RIM) (Health Level Seven 2005c). Although we will not attempt to provide a comprehensive comparison of these different syntaxes, it is useful to give a couple of examples of models to clarify their purpose and use. Using ASN.1 as the formal representation, a simple medication order can be represented as follows:

```
MedicationOrder ::= SET {
  drug          Drug,
  dose          Decimal,
  route         DrugRoute,
  frequency     DrugFrequency }
```

In human language the meaning of this model would be described in the following way. *MedicationOrder* is defined as a *SET* of elements. The *SET* of elements in *MedicationOrder* consists of a *drug*, a *dose*, a *route*, and a *frequency*. The value of the *drug* element is a coded item whose value comes from the set of codes that are children of the *Drug* node in the terminology. The *dose* element in the model is a decimal number. The *route* element of the model is a coded item whose value comes from the set of codes that are children of the *DrugRoute* node in the terminology. The *frequency* element of the model is a coded item whose value comes from the set of codes that are children of the *DrugFrequency* node in the terminology.

Given this definition of the model, the model is used as a guide or template for creating instances of patient data. For example, a medication order for a particular patient could be represented as:

```
MedicationOrder {
  drug          Penicillin VK,
  dose          500 mg,
  route         Oral,
  frequency     Every 6 Hours }
```

The definition of *MedicationOrder* implies that there is a terminology that contains a *Drug* concept where *Drug* has a computable relationship to other concepts that are drugs. A graphical representation of the drug hierarchy is

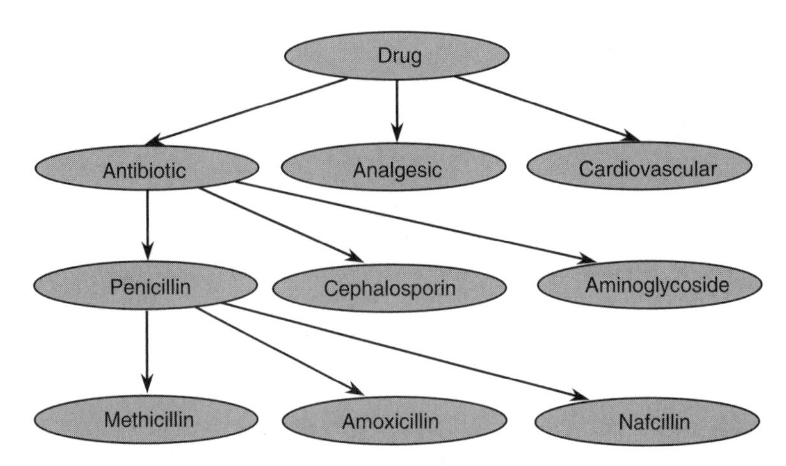

FIGURE 14-1 A graphical representation of a simple semantic network of drugs. The arcs in the diagram represent is-a relationships between the concepts in the network.

shown in Figure 14-1. Similar hierarchies would exist for *DrugRoute* and *DrugFrequency*.

The same drug information hierarchy can be represented in relational table form, which makes the information computable. For example, a table can be created with three columns: concept, relationship, and target concept as shown in Table 14-1.

The example shown in Table 14-1 is only one way in which the relationships between concepts in a terminology can be represented. A different mechanism is used in GALEN. The important point is that the coded attributes or fields of the information model (drug, route, frequency) are linked to the allowed values for that attribute in a terminology. Again, similar tables would exist to represent *DrugRoute* and *DrugFrequency*.

Besides defining value sets for elements, the representation of the relationships between concepts in a terminology can also serve a second important function. The table can be used to perform hierarchical inference. For example, if a rule needs to evaluate whether Nafcillin is-a Antibiotic, this can be determined by querying the table and finding the row that asserts that

TABLE 14-1 An example of how a simple drug hierarchy can be represented in a relational table. Each arc that was shown in Figure 14-1 is now represented by a row in this table. A program can determine hierarchical inference by querying the table and testing whether a given relationship exists between two concepts.

Concept	Relationship	Target Concept
Antibiotic	is-a	Drug
Analgesic	is-a	Drug
Cardiovascular	is-a	Drug
Penicillin	is-a	Antibiotic
Cephalosporin	is-a	Antibiotic
Aminoglycoside	is-a	Antibiotic
Pen VK	is-a	Penicillin
Amoxicillin	is-a	Penicillin
Nafcillin	is-a	Penicillin

Nafcillin is-a Penicillin, followed by a query that locates the row that asserts that Penicillin is-a Antibiotic. This example further illustrates the interdependencies that exist between the information model and the structure, content, and capabilities of the terminology system to which the model is linked. Not only does the information model link to the terminology, but the terminology model also provides the ability to do hierarchical inferencing on values in instances of data created in accordance with the model.

Another important point is that, taken together, the model and its associated terminology represent an instance of an ontology. An ontology is "a specification of a conceptualization ... a description (like a formal specification of a program) of the concepts and relationships that can exist for an agent or a community of agents" (Gruber 2006). In the preceding example, the model defines the relationship between a *MedicationOrder* and its component parts (drug, dose, route, frequency), and also defines the relationship of the coded components to the terminology. The rows in the drug hierarchy table further define relationships between the concepts that exist in the terminology.

The expectation is that models will be defined for each kind of data that can be stored on a patient and that might be referenced in decision logic. Thus, a model for a medication reaction could be defined as follows:

```
MedicationReaction ::= SET {
   drug                 Drug,
   dose                 Decimal,
   route                DrugRoute,
   manifestation        Manifestation,
   reactionTime         DateTime }
```

The human interpretation of this model is: A *MedicationReaction* is defined as a *SET* that contains a *drug, dose, route, manifestation*, and *reactionTime*. The purpose in defining the models is to overcome the curly braces problem and serve as a way of referencing clinical data in decision support modules. For example, with the two models that are defined earlier, one could create decision logic of the following form:

> *If there EXISTS a* ***MedicationOrder*** *and a* ***MedicationReaction*** *for the patient WHERE* ***MedicationOrder.drug*** *is-a* ***MedicationReaction.drug*** *THEN alert the ordering clinician.*

The model provides the overall structure of the data representation, and the standard terminologies that are referenced by the model provide the meaning or semantics. The terminologies provide computable semantics because they include definitions, relationships, synonyms, and in some cases description logic definitions that make the meaning of the terms computable. If the models and their associated standard terminologies are shared across institutions that are implementing the decision logic, then the models provide a clear and unambiguous way of referencing clinical data from the decision logic.

14.4 ISSUES OF PRE- AND POSTCOORDINATION

The problem of referencing clinical data from decision logic is not completely solved even if implementers are using a common syntax and terminologies, since it is possible to represent the same information in multiple ways while using standard terminologies and information models. For example, consider the following representations for the concept "Ibuprofen, 200 mg oral tablet."

Example 1
Medication: [RX563605, RxNorm, Ibuprofen 200 MG Oral Tablet]

Example 2
Medication: [RX503378, RxNorm, Ibuprofen]
Dose:
Value: 200
Units: [258684004, SNOMED-CT, mg]
Form: [385055001, SNOMED-CT, Tablet (a subtype of Oral dosage form)]

In this example, fields in the information model are shown as labels followed by a colon, and concepts from coded terminologies are shown as a triplet of code, terminology name, and text in square brackets. Both examples represent the same information, however, they are structured quite differently. In the first example, the substance, dose, and form are all represented by a single concept code (RX563605), whereas in the second example, several concepts (RX503378, 258684004, 38505001) are compositionally combined to represent the same concept. This example illustrates the alternative approaches of pre- and post-coordination. The ISO/CD 17115 *Health informatics—Vocabulary of terminology* (ISO/CD 17115 2003) defines *precoordinated concept representation* as "compositional concept representation within a formal system, with an equivalent single unique identifier," and *postcoordinated concept representation* is defined as "compositional concept representation using more than one concept from one or many formal systems, combined using mechanisms within or outside the formal systems." Thus, in the preceding example, representing the concept of *Ibuprofen 200 MG Oral Tablet* with the single code of RX563605 is an example of precoordination, and representing the same concept as a combination of multiple codes and values is an example of postcoordination.

There are trade-offs between precoordinated and postcoordinated representations. Precoordinated representations are often easier to use due to the fact that they behave more like complete sentences. They can simplify data entry screens by requiring fewer fields for the user to fill out and they are easy to reference in decision logic. Instead of having to specify the substance, dose, and route for every medication being entered, the user only needs to select the appropriate precoordinated term. This is very convenient when the possible number of precoordinated concepts is relatively small and manageable. In addition, precoordination avoids the problem of being able to combine pieces in a way that is incorrect or nonsensical.

However, when there are many pieces of information that can be combined in multiple ways, a precoordinated approach can lead to a combinatorial explosion of the number of concepts needed. For example, creating precoordinated concepts to record family history of diseases (e.g., "family history of breast cancer," and "family history of coronary artery disease") effectively doubles the number of disease concepts in a terminology. Taking this a step farther, creating precoordinated concepts such as "maternal family history of breast cancer," or "paternal family history of coronary artery disease" causes an even greater combinatorial explosion of concepts. As the specificity of the precoordinated concepts increases, so does the magnitude of the combinatorial explosion.

Postcoordination of concepts helps to avoid the problem of combinatory explosion. In a postcoordinated approach, rather than working with a set of preformed sentences, we have a dictionary of words from which we can generate an almost unlimited number of sentences. Although the flexibility of this approach can keep the representation of clinical information manageable, it may also allow the creation of statements that do not make sense. For example, breaking a medication and its route into separate fields may allow a user to specify a route of "topical" for a medication that may be administered only intravenously. To prevent situations like this, postcoordination models need extra metadata and rules to specify which values are allowable under various circumstances.

The choice between pre- and postcoordinated approaches must take into consideration the nature of the concepts and the way they will be used. When the number of things that can be said is relatively small and well-constrained, precoordinated concepts may be the most useful. When pieces of information can be combined in many different ways, one should consider using postcoordination.

Postcoordination relies on the appropriate use of clinical terminologies (e.g., LOINC and SNOMED CT) and clinical information models (e.g., HL7 RIM, OpenEHR Archetypes, Clinical Element models) (Coyle et al. 2003; Health Level Seven 2005c; Beale and Heard 2006). At a gross level, information models can be thought of as a series of fields with well-defined semantic relationships, and terminologies may be considered as the source of possible values for these fields. In this simplistic approach, terminologies are little more than organized sets of concepts.

In reality, terminology models and information models often overlap. For example, either of these models may provide mechanisms for negation (the ability to declare that something was NOT present or that it did NOT exist), or for qualification (the ability to qualify the meaning of a primary term by adding adjectives or adverbs describing such characteristics as severity, intensity, and quality). Though it may be desirable to consistently represent such information in either the information model or the terminology model, we must often deal with overlapping representations. The reason for this is that not all models, either terminology or information, are created equally. Although some clinical terminologies support features such as negation and qualification, others do not. Likewise, not all information models provide comparable sets of features.

When designing a decision support solution, the designers must evaluate the information and terminology models that may be appropriate for the problem at hand. Careful consideration must be given to areas where the information models and terminology models overlap in order to develop a strategy that is consistent and sufficiently expressive. To this end, the creators of information models and terminology models are increasingly specifying best practices for using various models together. For example, the TermInfo project represents a significant collaboration between HL7 and SNOMED to actively define how to best use the SNOMED CT terminology within HL7's Reference Information Model (RIM) (HL7 TermInfo 2006).

14.5 DATA REPRESENTATION USING NAME-VALUE PAIRS

As noted in the previous paragraphs, complete data representation requires the combination of an information model and a terminology. The representation of data is further complicated when the data are represented in a database or in a standard message. In these cases, the information model must be mapped to the structures and data types that are available in the physical database or message. In the case of a relational database, the information model must be mapped to the columns and associated data types (real, integer, string, Boolean) that are available in the database. Likewise, if the data are represented in an HL7 message, the information model must be mapped to constructs (the HL7 RIM attributes) and data types available in the HL7 Version 3 messaging standard.

We will step through an example to illustrate the interactions among information models, terminologies, and physical data structures. The phrase "numbness of right arm and left leg" could be represented using the following combination of codes from the SNOMED CT coding system:

Numbness (44077006)
Right (24028007)
Arm (40983000)
Left (7771000)
Leg (30021000)

By changing the order of the codes as shown next, the meaning can be changed to "numbness of left arm and right leg":

Numbness (44077006)
Left (7771000)
Arm (40983000)
Right (24028007)
Leg (30021000)

The fact that order is needed within the set of codes reaffirms the need for a reference information model in addition to just the codes from the terminology. In the HL7 Version 3 standard, a RIM class named *Observation* represents clinical observations. Among many attributes, an *Observation* has a *code*, a *value*, and a *targetSiteCode*. *TargetSiteCode* can occur more than once. An example instance of data in this form that represents "numbness of left arm and right leg" can be represented in the following way:

Observation.code: Diagnosis.primary (LOINC code 18630-4)
Observation.value: Numbness (SNOMED CT 44077006)
Observation.targetSiteCode: Arm (SNOMED CT 40983000)
  Qualifier.name: Laterality; Qualifier value: Left (7771000 SNOMED CT)
Observation.targetSiteCode: Leg (SNOMED CT 30021000)
  Qualifier.name: Laterality; Qualifier value: Right (24028007 SNOMED CT)

Using a combination of a code that names the kind of observation (the question) and a value that is the actual result of the observation is a common practice in data representation. This practice is commonly called a name-value pair approach, or an entity-attribute-value approach (Nadkarni 1997; Nadkarni et al. 1999). The name-value pair approach is used in HL7 Version 2, Version 3, and in the HL7 Clinical Document Architecture (CDA) as well as in the DICOM standard. It is a very flexible representation that allows coded concepts like "numbness" to be reused in the context of a primary diagnosis, a final diagnosis, the reason for a test, or as the name of a complication. Note that in the previous example, *Observation.targetSiteCode* is itself a name-value pair.

If we look more closely at the definitions within the HL7 RIM, we find that *Observation.code* is of **type** *ConceptDescriptor*, which is one of the coded data types that is allowed in HL7 Version 3. *Observation.value* is of **type** ANY, which means that depending on the situation *Observation.value* can be any of the HL7 Version 3 data types (real number, integer number, physical quantity, time stamp, etc.). Because primary diagnosis is a coded value, the *ConceptDescriptor* (CD) data type is also the appropriate data type for *Observation.value*. The CD data type is the most sophisticated data type in the HL7 version 3 specification for representing coded data. The parts of the CD data type that are pertinent to our discussion are as follows.

Part	Description
code	The plain code symbol defined by the code system. For example, "784.0" is the code symbol of the ICD-9 code "784.0" for headache.
codeSystem	A UID (Universal Identifier) that specifies the code system that defines the code.
codeSystemVersion	If applicable, a version descriptor defined specifically for the given code system.
Qualifier	Specifies additional codes that increase the specificity of the primary code. *Qualifier* is of type *ConceptRole*.

Note that the CD data type represents coded information as the combination of a code, code system identifier, and a code system version, and that a given code can have qualifiers that are other codes. The substructure of a qualifier includes the following parts:

Part	Description
name	Specifies the manner in which the concept role value contributes to the meaning of a code phrase.
value	The concept that modifies the primary code of a code phrase through the role relation.

Thus, an HL7 representation for "numbness of left arm and right leg," which is just as valid as the preceding representation is as follows:

Observation.code: Diagnosis.primary (LOINC code 18630-4)
Observation.value: Numbness (SNOMED CT 44077006)
Qualifier.name: Has-Location: Arm (SNOMED CT 40983000)
Qualifier.name: Laterality; Qualifier value: Left (7771000 SNOMED CT)
Qualifier.name: Has-Location: Leg (SNOMED CT 30021000)
Qualifier.name: Laterality; Qualifier value: Right (24028007 SNOMED CT)

In this representation, the information that was represented within the Observation.targetSiteCode of the RIM model is now represented simply as nested qualifiers inside of the CD qualifier structure of Observation.value.

There are two important conclusions that can be drawn from this example: 1) Even when there is an integrated information and logical model, there can be complex issues related to how the shared model maps to a particular database or message structure. This is true for HL7 messaging situations as shown earlier, but it would be equally true if the data were being stored in a relational database. In a relational database, the issue would be whether to store the qualifiers in the same table as the primary finding, or other issues related to normalization of the data into relational form; and 2) Even when the logical structure of the data is consistent, a patient data service that was retrieving patient data to support the execution of decision logic would need to be aware of the physical differences in an HL7 message or in the physical database schema. Since this is true, every effort should be made to store only one canonical representation of data in the patient database. Otherwise, there is likely to be unrecognized synonymy in the database that will lead to inaccurate results from decision logic execution.

Note that sharing of decision logic does not require that all participants use the same physical database or the same standard codes. In fact, for practical reasons it is often best to store enterprise specific codes in the patient database in order to keep errors in standard coded terminologies from being propagated into the longitudinal patient record. If this approach is taken, it is possible to introduce corrections and changes from the standard terminologies into the longitudinal patient database in a controlled fashion. The important principle is that whatever terms and models are used in the electronic patient record, they must be able to be mapped to the standard shared models and codes.

We also learn from the earlier examples that coded concepts can play at least two different roles in a name-value pair strategy: 1) as the name of the kind of observation being made, or 2) as the value of the observation being made. LOINC codes are designed specifically to be used as observation names/identifiers, but there are also a set of "observable" entities within SNOMED CT that can play the same role. However, LOINC does not currently make codes for items that can be the values of coded observations. The values for coded items are traditionally held in a terminology like SNOMED CT or Medcin, or in the case of drug names, in a drug-specific terminology from First Data Bank, Medispan, Multum, or the U.S. Metathesaurus (RxNorm).

14.6 TERMINOLOGY IN THE LIFE CYCLE OF DECISION SUPPORT PROGRAMS

Terminology plays an important role in all aspects of the life cycle of decision support programs. The usual life cycle consists of the following phases:

1. Initial authoring of the program
2. Testing of the program using test data, or using real data in a test environment
3. Revision of the logic based on testing
4. Deployment of the program into a production system
5. Monitoring of the program's behavior in the live environment
6. Repeating steps 1 through 5 as needed
7. Retirement

In situations where decision logic is shared between heterogeneous systems, the following two steps also occur:

1. Export of logic modules to external collaborators
2. Import of logic modules from external collaborators

Different tools and capabilities are needed at some of these different phases. For example, in the initial authoring phase (and in subsequent revision cycles) the author must be able to find the definitions of data that are available in the EMR to participate in the logic being authored. Previous publications have described ways in which the author can find the models (Huff et al. 2004; Parker et al. 2004). Based on how the models are expressed, one might:

- Query a model from an XML database (Knowledge repository)
- Use a graphical tool to look up and display models
- Use simple indexes on models represented in text files
- Integrate models into a program for authoring applications

Besides finding and displaying the model, there is a need to show the connection between the model and the terminology. For instance, in a medication order model, the author needs to be able see that the "ordered drug" refers to a hierarchy within the terminology that contains orderable drugs. The author should be able to branch from the model into the terminology and see the kinds of drugs that are available and the relationships that exist between drugs, such as the ingredients contained by a combination preparation, or the usual route of administration for a preparation. Linking browsing of the terminology to the context of use of a code in an information model allows an author to understand how to formulate a rule against the available data, or to recognize the need for additional detail in the information model, or additional content or relationships in the data dictionary.

In the testing and deployment phases, the system must provide run-time services for accessing relationships in the data dictionary. For example, a decision engine needs to be able to ask a terminology server whether a particular drug like Nafcillin is-a Penicillin. The ability to answer questions about relationships between concepts is an essential capability for a terminology server.

The capabilities needed in the authoring and run-time phases point to an important principle: Terminology services should be considered a modular part of the EMR and decision support infrastructure. In particular, the terminology should be accessed by a set of standardized service calls rather than by direct access to vocabulary database tables. Making vocabulary contents available via services accomplishes several goals.

1. It makes it possible to change the structure of the vocabulary database without changing the logic in the decision module (if you have created your own vocabulary database), or it allows you to change vendors with a minimal amount of change (if you are using services from one of the commercial suppliers of terminology services).
2. Software can be optimized to keep commonly used concepts and relationships in cache for improved performance.
3. It promotes sharing of standard codes, concepts, and relationships between developers of decision support programs. For example, use

of terminology services will lead to a consistent operational meaning to relationships like is-a, has-part, has-ingredient, and so on.

4. It allows consistent use of terminology capabilities like translation between code systems, management of synonyms, hierarchical inferencing, and translation to different human languages.

Standards have been created for supporting access to terminologies in these different modes. The first standard was created by the Object Management Group and was called TQS–Terminology Query Services or LQS–Lexicon Query Services (Object Management Group 1998). Recently, the Health Level Seven (HL7) organization has approved an initial standard for Common Terminology Services (CTS Standard) (Health Level Seven 2005b), and is working on a second release of the standard (CTS Release II). All these standards provide the definition of Application Programmer Interfaces (APIs) to terminology services. All modern EMR and decision support systems should be designed with terminology content and services as an integral part of the system infrastructure.

Importing decision logic poses particular challenges in the situation where the local data are not stored using the same model and terminology as is used in the shared decision module. There are two major ways that externally authored logic modules can be implemented. First, the local EMR can be designed so that it has services that can answer queries based on standard shared information models and terminologies, or second, the queries in the imported logic must be translated from the standard model to the local model and terminology. In either case, the steps involved in importing decision logic include:

1. Finding correspondence between models used in the external logic modules and models used in the local EMR.
2. Translating terms and codes used in the external model to terms and codes used in the local EMR.
3. Making changes to logic as determined by local policies, workflow, reference ranges, data availability, and so on.

As demonstrated in the earlier examples, consistent execution of decision logic implies that data in the EMR have a consistent representation. The other alternative is to have the decision logic reference all the various forms in which the data might be found. The latter approach is impractical, because the possible ways that any particular kind of data can be represented are quite varied, and they can change over time without warning. So the most practical approach is to transform incoming data from their native form into a single canonical form when they are stored in the EMR. This is accomplished by having the interface software be aware of the various expected models for a given kind of data and using a library of model mappings to convert from the inbound form of the data to the canonical EMR model of the data. This sort of data model normalization function should be a part of all HL7-like interface programs, and part of the function of EMR database services.

Experience has led to another important requirement related to maintaining decision logic that is running in a production environment. We have had the situation where a concept that is being used in decision logic changes. For example, a particular drug or lab test may become obsolete and be replaced by a new test, or there may have been an error in the creation of a given standard terminology that resulted in two codes existing for the same concept. When these situations occur, decision logic will cease to function correctly until changes are made so that the logic references the new concept. The best strategy is to approach the problem prospectively. That is, when a concept needs to be updated, authors or maintainers of decision logic that references the concept should be notified before the change is implemented in the production system. At Intermountain Healthcare (IHC), the approach to addressing this has been to create a registry of concept usage to support this strategy. The registry is a database that contains rows that show what objects (queries, rules, data entry screens, reports, etc.) use a particular concept or class of concepts. Again, services are provided so that authoring programs can update the registry as concepts are included in decision logic, or a batch process can be run routinely to find concepts in decision logic and add or modify entries in the registry.

14.7 CONTEXTUAL RESTRICTIONS WITHIN THE TERMINOLOGY

Over the years, the need to more effectively express the restrictions imposed by *context* to a variety of terminology entities has been recognized. These contextual restrictions can help determine "defaults" and manage "exceptions" that arise in the clinical environment. For example, standard terminology services normally take into account the desired language of the designation, returning the preferred designation for the concept code in the supplied language (Health Level Seven 2005b). In the UMLS Metathesaurus, the concept *Diabetes Mellitus, Insulin-Dependent* (CUI *C0011854*), can be expressed in English using the following designations (synonyms): *Diabetes Mellitus, Insulin-Dependent*; *Diabetes Mellitus, Brittle*; *Diabetes Mellitus, Juvenile-Onset*; *Diabetes Mellitus, Ketosis-Prone*; *Diabetes Mellitus, Sudden-Onset*; *Diabetes Mellitus, Type 1*; *Diabetes Mellitus, Type I*; and *IDDM*. Although all these designations arguably convey the same meaning, the appropriateness of the designations *IDDM* and *Diabetes Mellitus, Ketosis-Prone* seems restricted to clinical providers that have familiarity with diabetes, and designations like *Diabetes Mellitus, Juvenile-Onset* and *Diabetes Mellitus, Insulin-Dependent* are more widely used by nonspecialized providers. If *English* is supplied as the language context and *Endocrinologist* as the medical specialty context to a terminology server, the server would be able to return IDDM as the appropriate designation for concept *C0011854* for a user that was an English-speaking endocrinologist.

Allowing for different designations in different contexts is clearly important. If additional contextual details are taken into account, the returned designation can potentially be more appropriate to the needs of the clinical user. It is important to clarify that the appropriateness of the returned designation does not influence the decision support system reasoning, but instead how results are communicated to the person who reads and acknowledges the results of decision logic.

Although context has been applied most commonly to designations, it is also extremely useful in qualifying or restricting relationships between concepts. For example, a drug ontology might record relationships to indicate which drugs are metabolized in the kidney. The usual rule would be that drugs metabolized in the kidney should not be ordered for patients that have renal failure. However, because some drugs are only partially metabolized in the kidney, they are still safe to use in renal failure patients. Given that the knowledge that a request was initiated from an order entry application, a context-aware terminology server could exclude only the list of drugs that are metabolized to a sufficient degree as to preclude them from being ordered for a patient with renal failure.

Context dimensions can include characteristics of the patient (e.g., age, sex, clinical condition), of the clinical provider (e.g., role, discipline), and of the specific clinical setting (e.g., inpatient vs. outpatient, ICU, rural hospital), among others. Designations, relationships, and concepts are the terminology entities most frequently affected by contextual restrictions. However, contextual restrictions can potentially apply to other terminology entities, including value sets and pick-lists, or even entire code systems. The representation of contextual restrictions within the terminology conveniently isolates these nuances of meaning from information models and inference models (Rector 2001).

Most clinical systems and terminology servers currently in operation have some mechanism to handle contextual restrictions, at least for designations. Within IHC, these restrictions have been handled in many different, and frequently inconsistent, ways. It is recognized, however, that the current solutions are quite cumbersome to maintain, leading to the proliferation of overlapping domains and/or to the adoption of precoordinated descriptors that eventually become unable to represent all the desired combinations of contextual restrictions.

If a mechanism to represent and apply contextual restrictions is not available within the terminology server, clinical systems and decision support systems normally are forced to adopt a restrictive approach where all designations are previously identified as an explicit list in the source code logic, or the choices are preconfigured using application-specific dictionary tables.

More recent efforts dealing with terminologies and concept models have embraced OWL or related logic-based formalisms (Rector 2001). However, contextual constraints likely require additional reasoning methods that go beyond what logic-based formalisms can effectively represent (Rector 2004).

14.7.1 Current Proposal for Representing Contextual Restrictions

As an example of how to approach the issue of context, the current proposal at IHC for representing contextual restrictions identified the following context dimensions:

1. *Patient*: age group (e.g., 17 years old or younger, 29 days old or younger), sex (e.g., male, female), clinical condition (e.g., diabetic, pregnant, immunosuppressed), health insurance plan.
2. *User*: discipline (e.g., physician, nurse, pharmacist), specialty (e.g., cardiologist, endocrinologist), role (e.g., care manager, attending physician, discharge planner), group (e.g., hospitalists at LDS Hospital, ICU nurses at Primary Children's Medical Center).
3. *Location*: unit (e.g., Trauma ICU at LDS Hospital, Neonatal ICU at Primary Children's Medical Center), unit type (e.g., ICU, ED, OR), department (e.g., Intensive Care Medicine, Cardiology), facility (e.g., LDS Hospital, Primary Children's Medical Center), region (e.g., Urban

Central, South), enterprise (e.g., IHC, University of Utah Medical Center), clinical setting (e.g., inpatient, outpatient).

4. *Information System*: clinical system (e.g., Computerized Provider Order Entry, Emergency Department Information System), decision support module (e.g., Glucose management protocol, drug–drug interaction rules), network (e.g., Intranet, Internet).
5. *Language*: human language (e.g., English, Spanish), controlled terminology or coding system (e.g., UMLS, LOINC, ICD9-CM).

Each dimension defines a context attribute, where the attribute value is obtained from a predefined domain of valid context values. Depending on the type of terminology entity, some of the contextual dimensions must be specified. For instance, for a designation, *Language* is a required context attribute. If a given context dimension is not specified, assuming it is not required, the terminology services can infer that the particular terminology entity is not constrained by it. For instance, a given problem like appendicitis can occur in male or female patients, so if *Patient sex* is not specified as part of the context when a set of concepts are included as members of a problem pick-list, the terminology server can imply that the membership relation will be true irrespective of the patient sex. Similarly, when a given attribute is required, but not applicable to the terminology entity in question, the context value can be set to include all possible values. For example, if *Patient age group* is a required element but not applicable to a given concept-to-concept relationship, the value of this context attribute can be set to *All patient age groups*.

When multiple context dimensions are used, they should be interpreted as summative restrictions, unless their domains are not independent. For example, combining *Patient age group* and *Patient sex* will result in a context expression that will restrict on both age group and sex. However, when context dimensions are not independent (e.g., *is-a* or *part-of*), the most specific (restrictive) value should be used. For example, instead of specifying both *Unit* and *Unit type*, *Unit* should be used, since *Unit type* can be inferred from *Unit*. Notice that the context domains and their values and relationships are also specified within the terminology. Consequently, if the clinical system and the decision support system know where the patient is currently located (the actual *Unit*), the *Unit type*, *Facility*, *Region*, and *Enterprise* can be obtained by querying the hierarchical relationships that exist in the terminology server.

In terms of implementation, the standard terminology services are being extended at IHC to accept context expressions whenever applicable. Before calling a terminology service, the clinical application or the decision support system will have to assemble the appropriate context expression, using the available information from the patient, provider (user), and particular system module or application. Similarly, the terminology server will have to know when and how to enforce the restrictions implied by the received context expression. The database implementation options that have been explored include an XML-based implementation, where context expressions are stored as XML fragments and are selectively retrieved using XPaths, and a more flexible option where context expressions are represented as *tokenized* strings that are indexed and queried using high-performance textual retrieval methods.

14.8 WHAT NEEDS TO BE DONE

It is currently impossible to share decision logic developed in one institution with other institutions without extensive remapping and reworking of the logic module. This inability to share decision support modules is directly traceable to the fact that information models for clinical data are either nonexistent or nonstandard and that few institutions are using standard coded terminologies. Several steps need to be taken to get to a future state where sharing of decision support logic across institutions is feasible.

1. The medical informatics community needs to adopt a standard language for representing detailed clinical models. The modeling language must have a mechanism of linking to standard coded terminologies. Candidates for the common language include the Archetype Definition Language (ADL) proposed by the openEHR group, a language based on the HL7 RIM, a language based on the internal representations used by Protégé, a language based in OWL, or an entirely new language (Noy et al. 2003; W3C 2004; Health Level Seven 2005c; Beale and Heard 2006). These are not necessarily mutually exclusive options because it is probably possible to translate among some of these languages. A first step toward adoption of a standard modeling language would be to agree on a set of requirements for model representation and then do some comparative studies of the available languages.
2. There will need to be adoption of standard terminologies that are referenced by the detailed clinical models. Within the United States, terminologies have been approved for use by the National Committee on Vital and Health Statistics and by the Consolidated Health Informatics consortium (CHI 2006). These decisions are focusing attention on a few terminologies that are likely to be adopted within the United States and perhaps worldwide.
3. With a standard modeling language and standard terminologies in place people can begin to produce a library of detailed clinical models coupled to standard terminologies. In order for the models to be shared and approved, there will need to be a common repository where the models can be stored and accessed. The repository will need to record mappings between different models that represent the same clinical data (families of iso-semantic models). Also, the repository will need to record metadata about the models. The metadata must include (but not be limited to):
   a. Creator
   b. Creation date/time
   c. Last updated date and time
   d. Status
   e. Name of approving body
   f. First date of clinical deployment
   g. Decision modules that reference the model
4. A process for approving a single model from a family of iso-semantic models that will be used as the reference model for decision logic will need to be developed. Again, a first step in developing the process will be to define generally accepted characteristics of a good model. The

TermInfo work (HL7 TermInfo 2006), which is a collaboration between HL7 and the College of American Pathologists, is creating an implementation guide that proposes conventions for SNOMED CT use in HL7 Version 3 messages.

REFERENCES

Beale, T. and Heard, S. (2006). The Archetype Definition Language Version 2 (ADL2). The openEHR Foundation, Australia.

CEN TC 251. (2003). Health informatics—General purpose information components—Part 1: Overview; English version prEN 14822-1: 2003.

CHI. (2006). Consolidated Health Informatics. http://www.hhs.gov/healthit/chi.html.

Cimino, J. J. (1994). Controlled medical vocabulary construction: Methods from the Canon Group [editorial]. *J Am Med Inform Assoc* **1**: 296–297.

Cimino, J. J. (1998). Desiderata for controlled medical vocabularies in the twenty-first century. *Methods Inf Med* **37**: 394–403.

Coyle, J. F., Mori, A. R., and Huff, S. M. (2003). Standards for detailed clinical models as the basis for medical data exchange and decision support. *Int J Med Inform* **69**: 157–174.

Forrey, A. W., McDonald, C. J., DeMoor, G., Huff, S. M., Leavelle, D., Leland, D. et al. (1996). Logical observation identifier names and codes (LOINC) database: A public use set of codes and names for electronic reporting of clinical laboratory test results. *Clin Chem* **42**: 81–90.

Gruber, T. (2006). What is an ontology? http://www-ksl-svc.stanford.edu:5915/doc/frame-editor/what-is-an-ontology.html.

Health Level Seven. (1999). Arden Syntax for Medical Logic Systems. Health Level Seven, Ann Arbor, Michigan.

Health Level Seven. (2005a). GELLO: A Common Expression Language, ANSI/HL7 V3 GELLO, R1-2005. Health Level Seven, Ann Arbor, Michigan.

Health Level Seven. (2005b). HL7 Common Terminology Services, ANSI/HL7 V3 CTS, R1-2005. Health Level Seven, Ann Arbor, Michigan.

Health Level Seven. (2005c). HL7 Reference Information Model. Health Level Seven, Ann Arbor, Michigan.

HL7 TermInfo. (2006). HL7 TermInfo Project. http://www.hl7.org/Special/committees/terminfo/index.cfm.

Hripcsak, G. (1991). Arden syntax for medical logic modules. *MD Comput* **8**: 76, 78.

Huff, S. M., Rocha, R. A., Coyle, J. F., and Narus, S. P. (2004). Integrating detailed clinical models into application development tools. *Medinfo* **11**: 1058–1062.

Huff, S. M., Rocha, R. A., McDonald, C. J., De Moor, G. J. E., Fiers, T., Bidgood, W. D. et al. (1998). Development of the LOINC (Logical Observation Identifier Names and Codes) Vocabulary. *JAMIA* **5**: 276–292.

Humphreys, B. and Lindberg, D. A. B. (1989). Building the unified medical language system. In *Symposium on Computer Applications in Medical Care*. L. Kingsland, ed., 475–480. IEEE Computer Society Press.

Humphreys, B. L., Lindberg, D. A., Schoolman, H. M., and Barnett, G. O. (1998). The Unified Medical Language System: An informatics research collaboration. *J Am Med Inform Assoc* **5**: 1–11.

ISO/CD 17115. (2003). Health informatics—Vocabulary of terminology.

ISO/IEC 8824-1. (1990). Specification of Abstract Syntax Notation One (ASN.1). International Organization for Standardization, Geneva, Switzerland.

ISO/IEC 8825-1. (1990). Specification of Basic Encoding Rules for Abstract Syntax Notation One (ASN.1). International Organization for Standardization, Geneva, Switzerland.

ISO/IEC 8825-2. (1996). Information technology—ASN.1 encoding rules: Specification of Packed Encoding Rules (PER). International Organization for Standardization, Geneva, Switzerland.

Lindberg, D. A., Humphreys, B. L., and McCray, A. T. (1993). The Unified Medical Language System. *Methods Inf Med* **32**: 281–291.

McDonald, C. J., Huff, S. M., Suico, J. G., Hill, G., Leavelle, D., Aller, R. et al. (2003). LOINC, a universal standard for identifying laboratory observations: A 5-year update. *Clin Chem* **49**: 624–633.

Nadkarni, P. M. (1997). QAV: Querying entity-attribute-value metadata in a biomedical database. *Comput Methods Programs Biomed* **53**: 93–103.

Nadkarni, P. M., Marenco, L., Chen, R., Skoufos, E., Shepherd, G., and Miller, P. (1999). Organization of heterogeneous scientific data using the EAV/CR representation. *J Am Med Inform Assoc* **6**: 478–493.

Nelson, S. J., Brown, S. H., Erlbaum, M. S., Olson, N., Powell, T., Carlsen, B. et al. (2002). A semantic normal form for clinical drugs in the UMLS: early experiences with the VANDF. *Proc AMIA Symp* 557–561.

NEMA PS 3 Supplement 23. (1997). Digital Imaging and Communications in Medicine (DICOM), Supplement 23: Structured Reporting. The National Electrical Manufacturers Association, Rosslyn, VA.

Noy, N. F., Crubezy, M., Fergerson, R. W., Knublauch, H., Tu, S. W., Vendetti, J., and Musen, M. A. (2003). Protégé-2000: An open-source ontology-development and knowledge-acquisition environment. *AMIA Annu Symp Proc* 953.

Object Management Group. (1998). Lexicon Query Service. TC Document CORBAmed 98-03-22. Object Management Group.

Parker, C. G., Rocha, R. A., Campbell, J. R., Tu, S. W., and Huff, S. M. (2004). Detailed clinical models for sharable, executable guidelines. *Medinfo* **11**: 145–148.

Rector, A. L. (2001). The interface between information, terminology, and inference models. *Medinfo* **10**(Pt 1): 246–250.

Rector, A. L. (2004). Defaults, context, and knowledge: Alternatives for OWL-indexed knowledge bases. *Pac Symp Biocomput* 226–237.

Rector, A. L., Bechhofer, S., Goble, C. A., Horrocks, Nowlan, W. A., and Solomon, W. D. (1997). The GRAIL concept modelling language for medical terminology. *Artificial Intelligence in Medicine* **9**: 139–171.

Rector, A. L., and Nowlan, W. A. (1993). The GALEN Representation and Integration Language (GRAIL) Kernel, version 1. In *The GALEN Consortium for the EC*. University of Manchester, Manchester.

Sordo, M., Ogunyemi, O., Boxwala, A. A., and Greenes, R. A. (2003). GELLO: An object-oriented query and expression language for clinical decision support. *AMIA Annu Symp Proc* 1012.

Stearns, M. Q., Price, C., Spackman, K. A., and Wang, A. Y. (2001). SNOMED clinical terms: Overview of the development process and project status. *Proc AMIA Symp* 662–666.

W3C. (2004). OWL Web Ontology Language Semantics and Abstract Syntax. http://www.w3.org/TR/owl-semantics/.

Wang, A. Y., Sable, J. H., and Spackman, K. A. (2002). The SNOMED clinical terms development process: Refinement and analysis of content. *Proc AMIA Symp* 845–849.

Wikipedia. (2006). Description Logics. http://en.wikipedia.org/wiki/Description_logic.

15 GROUPED KNOWLEDGE ELEMENTS

MARGARITA SORDO and MATVEY B. PALCHUK

15.1 INTRODUCTION

A decision support system is a computer-based system that analyzes available data to guide people through a decision-making process. The availability of data may be considered to be the most fundamental part of CDS, because analysis and guidance depend on it. Usefulness further depends on the data being well-structured and unambiguous. Coding of data items is essential in order to understand and be able to manipulate them. Chapter 14 reviewed approaches to standardizing the terminology used for data items. Equally important is that the data type of each item be specified, including units for quantifiable data types or categorical values for dictionary/directory-based data types, and that they mean the same thing. Lack of ambiguity of these aspects of a data item is essential when data are communicated between the user and the computer. User-computer communication usually is done by means of *documentation systems*.

By *documentation*, we mean the *assembly* of information, whether it is for *input* as data entry forms or dialogue boxes, or for *printout* or *display* as narratives, tables, flow sheets, dashboards, specialty-oriented views, or other types of reports. Documentation systems specify which data items are to be included in documents and how they are to be organized, in relation to one another. Documentation systems designed to capture or present highly structured text and/or coded data can thus act as decision support systems. By prompting the user and exerting control over captured information, as well as by predetermining which data elements or requests for data are presented together and in which sequences, structured input systems provide a sort of passive decision support, reminding the user about the particular elements to be included, the format of the input, the allowable values or ranges, and which other items may be associated with it. Output documentation similarly can organize information in such a way as to facilitate comprehension, recognition of trends, or making other associations. With coded data, it is possible to automatically reason about information and derive new information, which can also be included in documents.

The purpose of documents, from the preceding point of view, is to obtain *data* from or to present data to a user. The specification of the document's

Clinical Decision Support: The Road Ahead

structure is a form of *knowledge*. Standards are being developed for such specification to encourage the collection of higher quality, more interpretable, more comprehensive data, and to encourage reuse of document specifications, or parts thereof, where appropriate. This chapter will review those efforts, in terms of their degree of maturity and harmonization, and how they relate to clinical decision support.

Regarding the relation to CDS, two aspects of documentation systems are important. One aspect is the organization or grouping of data items in documents; as we have noted, the specification of this in evolving standards is a form of knowledge, driven by an underlying rationale or purpose for such organization or grouping. The second aspect is the management of a collection of documentation specifications. This is a knowledge management (KM) task, the purpose of which is to reconcile collections of document specifications in an enterprise, encourage convergence on and reuse of specific ones that encourage best practices and conformance with enterprise goals. Since documents are often complex entities with many parts, the KM also extends to parts of such documents for which convergence on and reuse of them is desirable.

This second aspect of documentation—its knowledge management—has not received much attention in the clinical informatics and standards development communities, and is only recently being recognized as an important challenge, as KM systems begin to be introduced into health care enterprises to manage their knowledge content. Examples of approaches to curating documentation specifications, and supporting authors and editors in creating and updating them, are discussed in Chapters 21 and 22. In this chapter, we will briefly describe an approach to the KM for documentation that the authors have been exploring, which is in preliminary stages, but which needs to be considered as further progress is made in the standardization of the documentation systems themselves.

In order to consider both of these two perspectives, one aimed at producing documents, the other at the management of collections of their specifications, we will introduce two definitions:

Documentation Knowledge Element (DKE). This is a unit of a document not divided further (described in more detail later in this chapter). From the point of view of a documentation specification, a DKE can be referred to simply as a Documentation Element (DE). This latter is used in order to be consistent with the way DKEs are referred to by those developing documentation standards specifications. When we consider it from the point of view of indexing and retrieving it in a KM system, we refer to it as a Knowledge Element (KE), along with other types of knowledge elements that are managed by the KM system.

Knowledge Element Group (KEG). A KEG is a grouping of DKEs. KEGs can be nested, so they can also be groupings of KEGs, or just DKEs themselves. A KEG has attributes relating to the rationale for its position in a document, its purpose, and other aspects of it that can be used to index and retrieve it in a KM system. The documentation specification part of a KEG that governs its appearance in a document, in terms of physical layout, font, color, types of responses permitted when on an input form (e.g., check boxes or drop-down list entry), type of behavior (e.g., single choice vs. multiselect), is referred to in the documentation standards realm as a *template* (also described further later in this chapter). In the KM world, presentation details are of less relevance than the attributes governing rationale for document position and grouping. By grouping based on common features and purposes, we are able to identify or foster commonalities of parts of document specifications, and to support authors seeking to reuse existing parts. The goal is to define the DKEs and KEGs in such a way that their meaning remains unchanged regardless of the contexts where they appear. By creating documentation systems that rely on DKEs and KEGs, we can improve consistency and reduce redundancy of captured information while tapping into the potential of these underlying structures to facilitate decision support.

15.2 CLINICAL DOCUMENTATION

15.2.1 Data Capture

A multitude of information is generated during the interactions between a care provider and a patient. A wide variety of mental and physical tasks take place during those brief encounters. The clinician is interviewing a patient, making various observations, reviewing previous records, studying results of tests, reading reports of studies, performing an examination, doing procedures, and possibly even consulting external knowledge sources. All this information is then synthesized to form a mental picture of the patient state. Taking into account all the uncertainties and the unknowns, a roadmap is then created, outlining subsequent steps necessary to answer outstanding questions, refine a diagnosis, or arrive at or modify a therapeutic plan.

Raw observations as well as differential diagnoses and treatment plans derived in the course of an encounter constitute important sources of information. Capturing information at the point of care on an ongoing basis creates a historical record of wellness or disease process, as well as the assessments, plans, and actions of the provider. Such a record is an indispensable tool in patient care. Ideally, it not only provides a longitudinal perspective for any member of the care team, but also serves to encapsulate a provider's thoughts and reasoning, act as a foundation for enhancing patient safety, contribute to overall progress of medical knowledge, aid in processes of regulatory compliance, and provide a basis for the provider's compensation.

Although there are clear benefits of capturing data at the point of care, there have always been and continue to be significant challenges associated, not only with this task itself, but also with ensuring continued and maximized usefulness of such information. In the world of pen and paper, various mechanisms have been developed over time to aid in the process. All of us are familiar with dangerous-looking addressographs for stamping a card with demographic information onto a piece of paper, or complicated constructs with multiple color-coded pages with carbon paper between them, intended to be disassembled and routed to various destinations.

15.2.2 Forms

From big stamps that could be applied to blank pages to exceedingly complex structures—think income tax returns—forms have become the mainstay of information capture. Although a blank sheet of paper is an exceedingly versatile and comfortable device for capturing information, a form introduces substantial improvements, since it can specify the kind of information to be recorded and can dictate or even enforce the format of data. More interestingly, whether by design or not, forms frequently act as a guide, reminding whoever is filling it out of the kinds of data that are required. For example, a History and Physical (H&P) Exam form typically is designed primarily to standardize the format of such a note, so that the information is reliably found and is communicated in a familiar fashion. But to an overworked intern in the middle of the night, such a form may serve as a reminder to ask a certain question or perform a particular exam maneuver that would otherwise have been forgotten.

In this capacity as guides, forms acquire the characteristics of a decision support system. It is a fairly primitive system—a paper form is incapable of analyzing the entered data or performing any reasoning, and is merely a static set of prompts. It is also passive; to get the benefits of this basic decision support, one needs to pick up and use the appropriate form. Nonetheless, by virtue of presenting a set of prompts and exerting a degree of control over content, it is clear that even a static paper form is capable of aiding its user in performing whatever task the form was designed to accomplish.

Many things in the physical world are recreated *in silico* on computer screens. Borrowing from the physical world is a common approach in software design, since familiar-looking objects ease the cognitive load and make it easier for users to interact with computers. In the world of electronic health records (EHRs), paper forms are a natural framework for creating interfaces that are familiar to clinical users. Electronic equivalents of paper forms prompt users to enter specific kinds of data into text boxes, only this time a keyboard and mouse are used for data input. With electronic forms, however, form creators are able to introduce significant enhancements that take advantage of dynamic behaviors made possible through the use of computers. Electronic forms are capable of analyzing user input and interacting with other electronic sources of data and reacting in real time by adapting themselves accordingly. As such, the potential for acting as a CDS mechanism is greatly increased. Building on dynamic capabilities, the forms acquire the capability to move beyond mere passive decision support and become proactive real-time "collaborators," anticipating the needs of their users and adjusting in real time to provide the optimal path for completion of a task at hand.

In addition to acting as a decision support modality, electronic forms have another very important feature we alluded to earlier—they can specify and enforce content that is being entered or displayed. In this capacity, forms promote the capture of data in structured fashion. For example, instead of entering "John Q. Public" into a word-processing document as a string of characters, a form might prompt the user to enter first name, followed by middle initial, followed by last name into separate fields. Once the user is done and the data are saved, the computer knows unambiguously that "Public" is a last name. To carry this example further, the underlying software for our form might use a code, such as Health Level 7 (HL7) (HL7; http://www.hl7.org) version 3 Reference Information Model [11] concept C10654 to identify this particular piece of data. Now imagine that the entered data are exchanged from one system to another within an organization or even between organizations. In this case, as long as each system is able to understand that C10654 stands for last name, even if one refers to it as "surname" and another as "family name," the data element will continue to be unambiguously understood.

Forms consist of individual elements. These may be simple, such as the field for a last name in our example, or more complex, such as an element that automatically calculates the value of body mass index, based on height and weight. Elements may contain other elements; for example, a form element that captures demographic information will contain an element for capturing last name, among others. Essentially, a form is a collection of nested elements.

Another example of a dynamic behavior of a form is a conditional organization of form sequences found in certain dialog systems. A sequence between one form screen and a subsequent one is governed by decision logic based on the data that have been entered. For example, in a form in which the user is asked about presence of family history of disease relevant to the topic being discussed, a positive response could lead to invocation of a form that requests details of the family history.

15.2.3 Templates

A *template* is defined as "a pattern used to create documents." In the paper world, if a form is created by stamping a rubber stamp onto a piece of paper, that rubber stamp is the template. In the digital world, a form rendered on a screen of a computer is an instance of a template used to create it. We've established that forms are collections of nested elements. Templates, therefore, must contain sufficient information to specify the kind of elements that are being used and in what order they are to be presented. In addition, a template should have instructions on how the elements should appear, such as descriptions of fonts, colors, and user interface widgets. We will define a template as a software artifact used to define lineup of elements in a form and to specify their presentation.

Templates could be used to create a wide variety of forms. Free from the constraints of paper and going beyond simple data capture devices, our forms can look like flow sheets for presenting temporal changes in data, serve as dynamically adjusting order sets, or act as vehicles for presenting reports.

Templates found in contemporary EHR systems typically are limited to imposing a structure on mostly text-entry documents such as a visit note or H&P note. This structure does not provide a high degree of granularity and usually is limited to defining sections of a document. For H&P, these sections include chief complaint, history of present illness, past medical history, and so on. Documents created using such templates are rich in terms of capturing the nuances of clinical reasoning in the form of narrative text. Predefined structure of document sections enhances the efficiency of interaction and definitely provides a degree of basic decision support. However, these structured templates typically have low reusability potential and are often difficult to maintain to ensure that knowledge contained within them is up-to-date. Furthermore, data captured in such structured documents amounts to blocks of plain text, and is therefore not well suited for advanced decision support or interoperability. Overreliance on these coarsely structured data capture mechanisms and the benefits they provide may, in the long run, adversely impact the quality of resulting documentation.

More sophisticated templates consist of much finer elements and aim to capture specific individual pieces of data in coded fashion. Provided there is a robust infrastructure in place to support creation, use, and maintenance of such granular elements and templates, there is a high potential for reuse of individual elements across various templates and for various purposes. Coded data resulting from instantiation of these templates can be used to drive advanced decision support systems and be interoperable across different systems. The downside of using templates that demand a high degree of coded data capture is increased data entry burden on system users. In addition, if a majority of data needs to be entered in coded fashion, the users will notice that they are unable to express certain fine distinctions or nuances in the resulting document due to excessive constraints imposed during data entry.

It is clear that a fine balance must be found between coarse structure and fine coding of data so that a documentation system remains usable for those who need to interact with it, while maintaining high quality and usefulness of data and ensuring long-term reusability and interoperability.

Templates, as defined in this chapter, are static in nature, in that they represent a fixed collection of individual elements. Any dynamic behavior in a system based on such templates must come from the individual elements themselves. As forms-based electronic systems evolve, it is possible that the need for static templates will diminish. In their place, knowledge-driven systems will point to existing elements in real time and assemble them into collections ready to be presented to users and instantiated as a document, a flowsheet, an order set or a report.

15.2.4 Elements

We postulated that forms consist of nested elements and used the idea of individual elements to define templates. Let us consider elements in more detail. We will begin with concepts. A *concept* represents a discrete idea. Consider a juicy fruit that grows on trees; may be green, yellow, or red; tastes sweet when ripe; has seeds in the middle; and makes a crunchy noise when you bite into it. You probably formed a mental image of this fruit—that is a concept. Concepts are described by terms, and one of the terms for our concept is, of course, "apple." For various knowledge domains, there exist standard collections of concepts and their associated terms. SNOMED is one such standard (SNOMED; http://www.snomed.org/).

Each documentation element is a representation of a concept. Blood pressure serves as a good example of a prototypical documentation element. In SNOMED, it is found as concept 75367002. But a concept alone—basically a term like "blood pressure"—is not sufficient to create a documentation element to be used in a real software system. To record blood pressure, a system needs to capture the value and value units for systolic and diastolic pressure, the side of the body where the measurement took place, which extremity was used, the position of the patient, when the measurement was taken, what kind of equipment was used, and whether the patient's feet were touching the floor. And this is only a subset of various pieces of data that are related to blood pressure.

Each piece of data in our example is associated with a particular data type. Systolic pressure value is a number (which could be constrained to be nonnegative and not greater than 300), and patient position is a string taken from a predefined collection that would include *standing*, *sitting*, or *lying*.

Taken together, the underlying concept plus all the associated pieces of data, each conforming to its data type, form a documentation element. A blood pressure documentation element can be thought of as an object in object-oriented programming, with variables comprised of our additional pieces, such as values, position, and equipment. An *information model* must govern the internal structure of documentation elements, the mechanism for nesting elements within elements, and the relationships between elements. Additionally, data type definitions must also be standardized.

A documentation element, as the representation of a discrete idea, is bound to a concept from a standard terminology. This lays the groundwork for future interoperability by ensuring that the meaning of data captured via this element remains unambiguously understood even if the data are shared among different systems. Additionally, a terminology such as SNOMED provides a way to relate elements to each other by utilizing relationships. For example, traversing a SNOMED hierarchy, we find that "Blood Pressure" is a child of "Vital Signs." One can imagine a system that dynamically assembles all siblings of Blood Pressure into a form for entering or displaying Vital Signs. Augmenting relationships derived from standard terminologies, documentation elements could support relationships based on location in an overall documentation model (e.g., Vital Signs element appears immediately before Physical Exam), application in a particular specialty, or intention for action.

A system based on documentation elements enjoys the benefits of decoupling the data from presentation (handled by templates) and business logic. A library of elements provides a set of building blocks for assembling various documentation artifacts for creation of templates. It also aids in processes of updating and maintaining the clinical knowledge contained in documentation elements.

15.3 CURRENT APPROACHES TO CLINICAL DOCUMENTATION

There are many approaches to architecting a documentation system. System designers might choose to use some, all, or none of the constructs we just introduced. We will illustrate two different paths being investigated currently—one in the United States by HL7 (HL7; http://www.hl7.org) and another by the European OpenEHR community (http://www.openehr.org/).

15.3.1 HL7 Templates

At HL7, the efforts surrounding development of templates are shepherded by the Templates Special Interest Group (SIG) of the Modeling and Methodology Technical Committee. The initial impetus for the development of HL7 templates came from creators of specifications for Clinical Document Architecture (CDA) as a way of controlling the content of the body of a clinical document for the purposes of ensuring that individual implementers can create and work with customized documents while preserving the structure and meaning of data captured within those documents.

The HL7 CDA, currently in Release 2, is a rich and flexible standard for exchanging clinical documents among electronic health record systems (Dolin et al. 2006). The CDA separates the document into a header and a body. The header contains metadata about the document (who and what it is about, who the authors are, etc.), and is specified with a high degree of granularity in the CDA. The body, which contains the "payload" of the document, is purposefully left very generic to accommodate a wide variety of clinical content. The CDA ensures human readability of the documents by requiring a narrative version of content. Coded entries are used to make data machine-processable, and each such entry must be derived from the Reference Information Model. In order to successfully create and share documents with agreed-upon content, there needs to be a mechanism to impose constraints on the order and the types of coded entries comprising the document body. According to the CDA authors, "Templates and/or implementation guides can be used to constrain the CDA specification within a particular implementation and to provide validating rule sets that check conformance to these constraints" (Dolin et al. 2006). In addition to CDA, there are other areas within the HL7 family of standards (Orders and Observations, for example), which would benefit from an agreed-upon template specification. The work of the Templates SIG is ongoing, and there is a present need for a templates specification, although an agreed upon standard does not yet exist.

The description of templates we use in this chapter is similar but not exactly equivalent to what HL7 currently is considering. According to a working definition, an HL7 template is "a registered set of constraints on a balloted HL7 static model. HL7-balloted static models are all derived from the HL7 Reference Information Model" (RIM) (HL7 Templates SIG Project outline). The HL7 RIM is a high-level information model encompassing an entire field of health care with such constructs as "Entity" and "Act" at its core. We wont delve into its details here, but the power of this model becomes clear when you consider that "Entity" can be specialized to represent a patient, a doctor, or a hospital, and various actions such entities undertake can be derived from "Act."

A template as we defined it earlier can indeed be viewed as a set of constraints. From an entire library of documentation elements, it specifies which kinds of elements are to be used, and in what order, in the final document. Our template also defines the way elements are to be rendered—for example, to represent a multiple-select set of choices, use either an array of checkboxes or a list where more than one option can be selected at a time. The difference from the HL7 template definition is that the latter must constrain only models or model fragments derived from the RIM (see Figure 15-1). As such, whenever a new template needs to be created, its author must find model fragments within HL7 artifacts that correspond exactly to what he or she is trying to achieve, and use them as the basis for building said template.

15.3.2 OpenEHR Archetypes and Templates

The OpenEHR foundation adopted an approach pioneered by T. Beale and S. Heard of Ocean Informatics in Australia, who proposed to use the term *archetypes* in describing computable expressions of concepts (Beale and Heard 2003a), and created an Archetype Definition Language (ADL) (Beale and Heard 2003b)—a formal language used to represent and share archetypes. Currently, ADL is only one of several ways of expressing archetypes; OWL, the web ontology language, is another W3C Web Ontology Language.

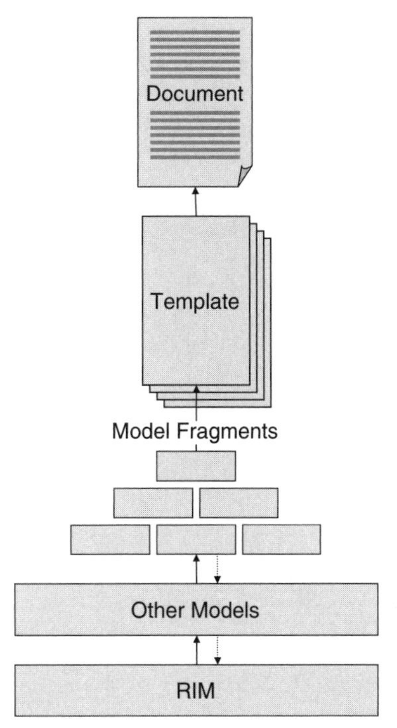

FIGURE 15-1 HL7 Templates Cascade. RIM—Reference Information Model; all other HL7 models are derived from the RIM. Typically in illustrations of this type, the RIM appears on the top of the diagram. Placing it at the bottom affords a view that highlights the similarities between HL7 modeling efforts and those of the European community. Note the downward arrows from Model Fragments to Other Models and eventually to the RIM—if HL7 were to institute a process whereby, if preexisting model fragments do not exist, required changes can be incorporated back into the RIM, it would ease the process of creating individual building blocks for HL7 templates.

As defined by Beale (2003b): "An archetype is a reusable, formal model of a domain concept expressed in the form of constraints on data whose instances conform to some class model known as a reference model. The main purposes of archetypes are (Open EHR; http://www.openeha.org):

- To allow domain experts such as clinicians to create the definitions that will define the data structuring in their information systems
- To provide run-time validation of data input via a GUI or any batch process
- To provide a basis for intelligent querying of data

It is our belief that the description of documentation elements presented here is essentially compatible with archetypes. In a documentation system, the archetypes conform to information models, are "rooted" in controlled medical terminologies and ontologies, and act as building blocks in construction of templates. For example, the BP archetype measuring systemic arterial blood pressure in Figure 15-2 is a model of what information should be captured when measuring blood pressure. The BP archetype is an aggregation of five concepts: systolic and diastolic pressure values, instrument or other protocol information, cuff and patient position. A comprehensive list of archetype examples can be found at http://oceaninformatics.biz/archetypes/. Templates, in turn, are instantiated to create any number of documentation artifacts, such as notes, flow sheets, forms, order sets, or reports (see Figure 15-4). Users interact with these artifacts and patient data are captured as a result.

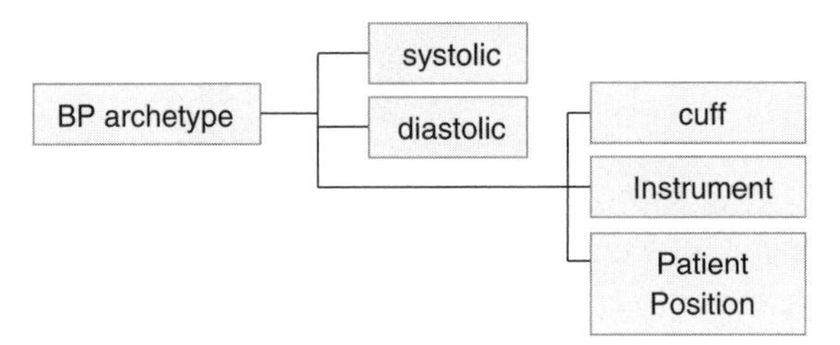

FIGURE 15-2 A simplified archetype for blood pressure. The BP archetype is an aggregation of five concepts: systolic and diastolic pressure values, instrument or other protocol information, cuff and patient position.

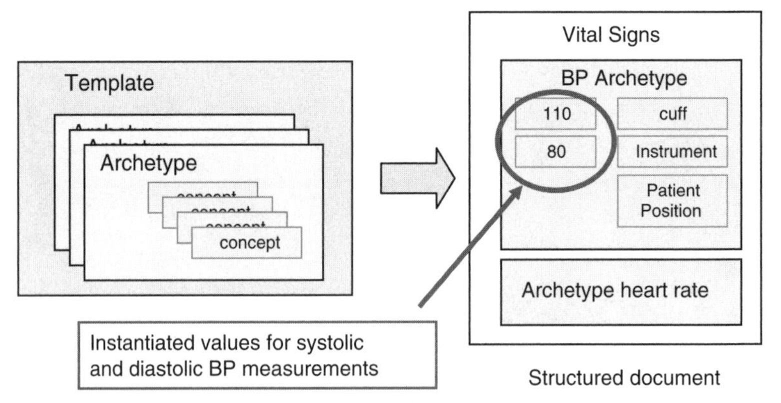

FIGURE 15-3 Concepts, archetypes, templates, and structured documents. Archetypes are aggregations of concepts to create more complex structures. Templates are aggregations of archetypes, and a structured document is the instantiation of a template.

Figure 15-3 illustrates the different levels of aggregation from concepts to structured documents:

1. Concepts
2. Archetypes (transaction-oriented data structures based on the concepts) (Beale 2002; Kernberg 2004)
3. Templates (groups of reusable archetypes plus a set of context-specific constraints) (Heard et al. 2003; Kernberg 2004)
4. Structured documents (Alschuler 2002) resulting from the instantiation of a template

The bottom-up method of building archetypes and templates afforded by this approach is beneficial. Knowledge engineers can freely utilize readily available sources of information (expert debriefing, for example) to rapidly prototype and create working documentation artifacts. Since individual elements are "bound" to controlled vocabularies and ontologies, they have a high degree of reusability and interoperability.

We made an assertion earlier that as documentation systems evolve, the need for templates might decrease due to their static nature. In analyzing the various layers in the documentation "stack" (see Figure 15-4), it is worth noting that potential for interoperability resides in archetypes. Templates are

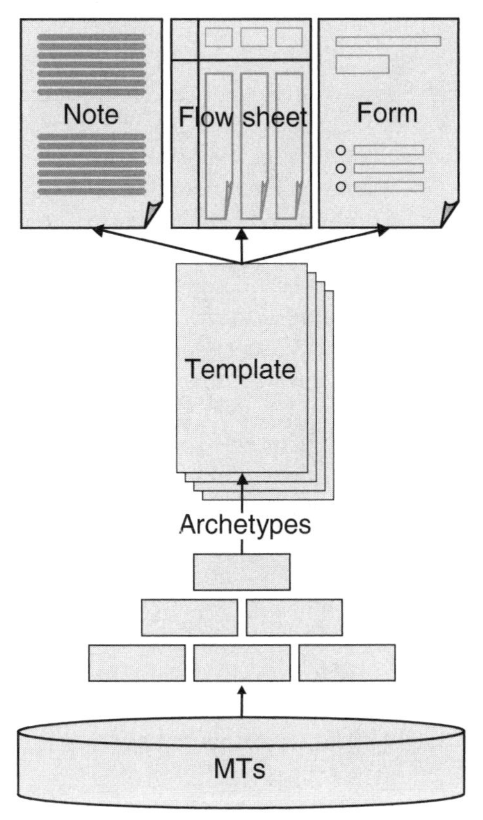

FIGURE 15-4 Archetypes and Templates Stack. Controlled medical terminologies (MTs), along with appropriate information models, ontologies, and data type definitions (not shown), provide the foundation necessary for achieving interoperability, reusability, and ease of maintenance of knowledge encoded in archetypes. Archetypes are essentially the same as documentation elements—they are the building blocks for construction of templates. Templates are instantiated to create various types of documents, such as notes, forms, or flow sheets.

needed to constrain content and provide a high degree of customization, but they are not necessary for interoperability. In other words, in order to preserve the meaning of data contained in the document when it is shared, both the sender and the receiver must "understand" the archetypes used within it, but it is not necessary for these systems to agree on or even be aware of the templates used in creating the documents.

The two approaches described here, though different, are not at all incompatible. It is hoped that as HL7 is developing the standard for Templates and OpenEHR is working on implementing Archetypes and Templates, the outcome of these endeavors can be reconciled into a single way of representing biomedical information and a method of constraining documentation artifacts, so that we can get closer to the goal of reaching ever-higher degrees of syntactic and semantic interoperability.

15.4 GROUPED KNOWLEDGE ELEMENTS (KEGs)

Forms, structured reports, protocols, and order sets are just a few examples of documents designed to collect and handle information. All these documents serve as structured repositories to capture, display, or process information in a specific manner.

Simple forms and reports are commonly used to collect and/or display information about a patient. In general, forms and reports can be customized to adapt to clinical settings and requirements. Structured reports, for example, normally are used to present the results of tests in a clear, organized format. Description of findings may be done using predefined vocabularies to express the characteristics of results. Protocols and clinical guidelines are a type of structured document that incorporates vocabularies for terminology, and also provides decision support in the form of suggestions, reminders, and links to auxiliary information. Order sets are predefined groups of orders for specific clinical settings, diagnoses, or treatments. They provide decision support to help clinicians select procedures and treatments appropriate to a clinical problem, by prepackaging them and then allowing customization within them. These prepackaged procedures provide continuous support through the ordering process, hence improving the tailoring of treatments, by using specific information about a patient's current condition; and highlighting key symptoms and findings. Similarly, medication ordering can be tailored by suggesting appropriate medications and dosages based on the medical condition of a patient; or cost reduction by suggesting effective alternative diagnostic procedures or treatments. Moreover, intentions and goals are a form of meta-knowledge reflective of the purpose of using an order set to perform a series of tasks. In summary, decision support can be incorporated into documentation systems in a variety of ways:

- Representation and interpretation of data
  a. Modeling patient information
  b. Representation of medical concepts
  c. Abstraction and interpretation of data
- Modeling processes
  a. Representation of actions
  b. Organization of plans
  c. Modeling decisions
  d. Representation of goals and intentions

Within all these types of documents, there are collections of specific data/knowledge elements or grouped knowledge elements that we referred to as KEGs in the introduction to this chapter. KEGs can be considered as clusters of elements gathered together for a particular purpose. KEGs are groups of elements that share common features: for example, a person's name, age, and gender can be grouped as "personal data"; current and past diseases can be grouped as "patient health history," and other clusters of grouped elements could contain information about lab test results; medications prescribed; referrals ordered to health care providers; educational materials provided; or plans for further care and return visits. As we have noted, KEGs can also be organized as a hierarchy of elements containing meta-level information for KM purposes, providing extra dimensions for organizing, indexing, and retrieving information. In summary, a KEG is a collection, for a particular purpose, of elements that share some kind of common association either relevant to:

- **A specific clinical setting**, for example, preoperative order sets, order sets for ICU admission for chest pain, groups of data elements on a surgical operative note, or elements on a dialysis protocol data entry form; or demographics, for example, patient social or family history.

- **A type of encounter** between physician and patient (e.g., whether for an initial consult, follow-up, review of symptoms, chronic disease management, preprocedure or postprocedure).
- **An intention** or overall purpose, which can be defined as the reason(s) for carrying out an action within a context (see Figure 15-5). The intention could be to gather knowledge/information about a patient in order to provide a diagnosis; to execute an action such as admitting/discharging a patient, to monitor a patient's status, and so on. Figure 15-6 contains a complete ontology of intentions.

At Partners Healthcare System, we have been working on a series of activities aimed at facilitating knowledge management for the enterprise. The process began with an evaluation of the KM capabilities in place at the hospitals in the Partners system (Boxwala et al. 2002), many of which had been developed ad

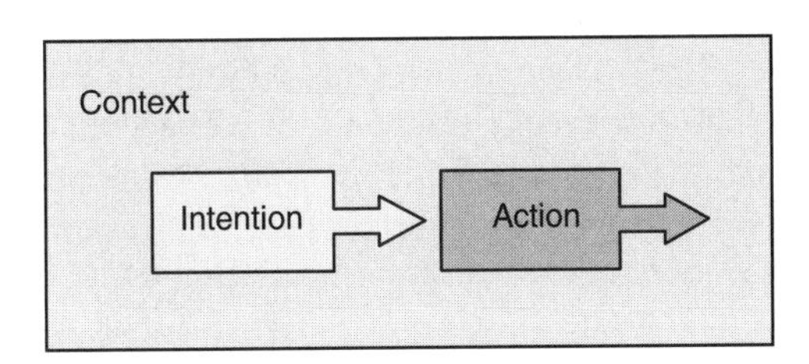

FIGURE 15-5 Context-Intention-Action triplet. An intention is defined as the reason(s) for carrying out an action within a specific context.

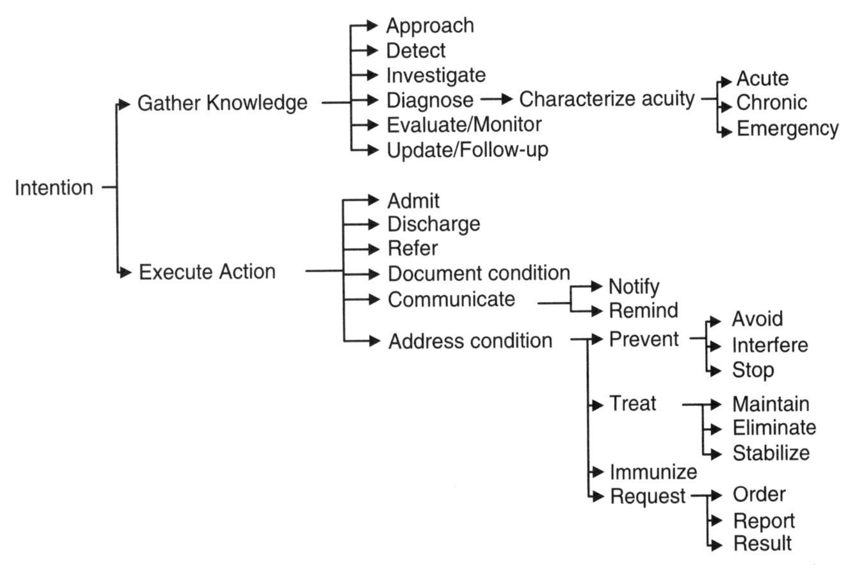

FIGURE 15-6 Ontology of intentions. An intention can be for gathering knowledge about a patient during a physician-patient encounter, or as the execution of an action such as admit a patient, or address the condition of a patient by preventing or treating a disease.

hoc to support various applications of CDS such as those described in Chapter 5. We identified six types of knowledge resources that needed to be managed:

1. Reference information models
2. Dictionaries, ontologies, and semantic structures of knowledge
3. Calculation or derivation of quantities
4. Decision rules
5. Guided sequences of actions
6. Clustering or grouping of knowledge elements

In two subsequent projects, we designed and evaluated possible approaches to the formal representation of rules knowledge (Greenes et al. 2004) and the development of knowledge management (KM) infrastructure for authoring, subject matter expert elicitation, versioning, publishing, and sharing of this knowledge (Sordo et al. 2004).

One of the least well-developed capabilities we found was the ability to identify and manage the sixth type of knowledge resource, that is, clusters or groups of knowledge elements (KEGs). The two main use cases for KEGs that we found were order sets (Sordo et al. 2006) and documentation models. For both order sets and documents, groups of elements are designed for a particular purpose, based on clinical intent or goal (indication) that allows them to be indexed and retrieved. Cataloging of KEGs by those purposes would allow the contained archetypes and the associated templates to be viewed in relation to other similar templates and archetypes for the same purposes, to facilitate review by clinical oversight committees, eliminate redundancies, and converge on best practice approaches. Grouping them also would facilitate establishing constraints for elements that are purpose-specific, e.g., diuretic doses in the setting of renal failure in a patient with CHF. This would also facilitate modularity of development and maintenance of repositories. We describe our approach here, which is being incorporated in the knowledge management infrastructure for Partners HealthCare described in Chapter 21, because we believe this offers a needed generalizable model that is not currently being addressed by standards efforts. Although this is a relatively new pursuit that needs to be further refined, we think this is a promising approach to bridging the gap between the need for a sound ontology representation of clinical meta-data, and the need for facilitating interoperability and exchange of semantically rich clinical information.

As we have discussed earlier, templates are made up of groups (possibly nested) of archetypes, ultimately specifying data elements to be retrieved or entered, and to be included in a report or form. The architecture of a document can be defined as a set of component parts (e.g., sections and subsections where specific templates would be allocated), each containing data elements and their values. Order sets, radiology reports, discharge summaries, progress notes, operative notes, and other documents have structure that can be described by groupings of components.

We address with KEGs an aspect of the management of these documents not covered by the HL7 Structured Documents Technical Committee (authors of the CDA—Clinical Document Architecture) and the Templates SIG of the Modeling and Methodology Technical Committee in HL7, which are focusing on development of standards for clinical documents (Alschuler 2002) and templates (Kernberg 2004).

The problem addressed is that knowledge content of forms, documents, and order sets can often be inventoried and compared only in the form of screen shots. At the Brigham and Women's Hospital (BWH), one of the

Partners hospitals, knowledge engineers have indicated that numerous rounds of e-mail and communication are required to help clinicians proposing new order templates to conform to the guidelines of the drug safety committee. Several potential benefits can be expected from a structured approach to knowledge management of KEGs, including:

- **Reusability**: the ability to create modular, shareable component subgroups for particular purposes (e.g., orders for routine admission, emergency anticoagulation, or CCU vital signs), to index them, and to be able to retrieve and incorporate them in new documents or order sets that are being created.
- **Increased interoperability**: KEGs maintained and managed in a common fashion, can be used by multiple applications (potentially running on different platforms) across an organization.
- **Transparency and ease of maintenance**: given a common way of representing KEGs, content contained in production systems can be kept up-to-date and additionally, enterprise-wide policies can be propagated to individual users at the point of care.
- **Encouragement of best practices**: by indexing KEGs in terms of their appropriateness indications, those that optimize care in various settings can be retrieved together, and an author/editor can view all KEGs related to that intention. Other benefits accrue from use of this approach in particular for KEGs that specify parts of documents and forms.
- **Ease of adaptation to different display platforms or form factors**: if the data format is platform-independent, e.g., through use of XML and style sheets, it can be rendered on various print devices, tablets, or PDAs without extensive reprogramming.

Work in HL7 on archetypes, templates, and compound documents have addressed parts of what is needed, although a standard has yet to emerge. The work on KEGs is intended as a complement to those activities.

In the remainder of this section we present an example of the use of KEGs as part of the ongoing KM infrastructure development process at Partners, focusing on order sets. Order sets are structured documents that organize complex health care protocols and plans into a series of units of work aimed to ease interaction, reduce errors, standardize procedures, provide decision support, and improve care practices.

The development of an order set requires close collaboration between users who request the creation/update of an order set, domain experts who provide specific knowledge to be incorporated into the order set, special committees and review panels who validate the accuracy of content, and developers who implement the order set, and after approval, bring it into the system (Sordo et al. 2004). Direct participation from all parties involved is of essence to guarantee a sound implementation process.

The structure and functionality of order sets is of vital importance in fostering consistency, shareability, and interoperability of information to support effective clinical use among institutions. After analyzing two order entry systems at Brigham and Women's Hospital and Massachusetts General Hospital—both part of Partners—we developed an order set schema that can be represented as XML semantic-Web-based RDF tags to facilitate indexing and retrieval. KEGs are the foundation for the proposed order set schema. Depicted in Figure 15-7 is a template containing nine KEGs, each KEG being

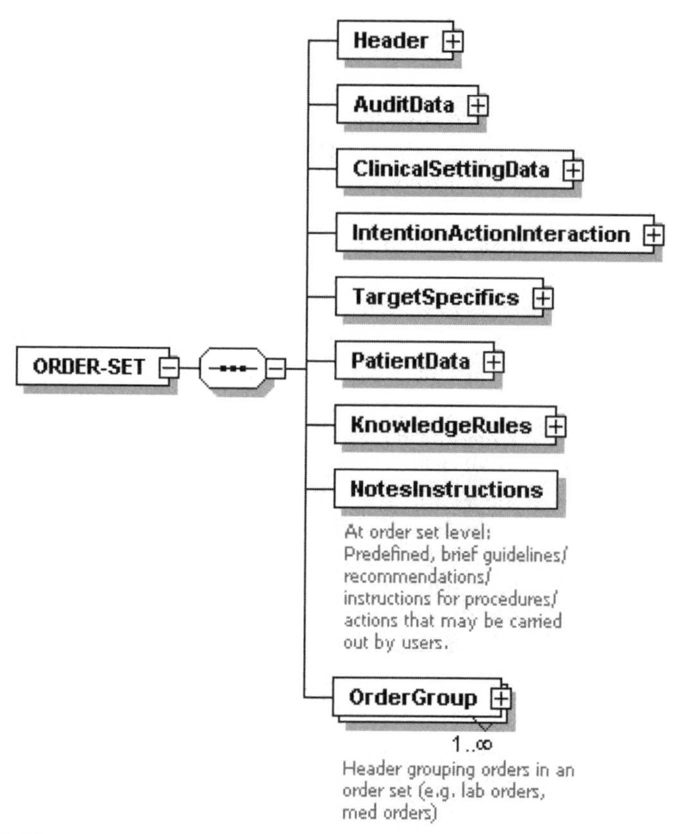

FIGURE 15-7 Order Set Schema representation showing main KEGs.

an aggregation of KEGs and complex objects containing specific data (a full description can be found at (Sorda et al. 2006)).

The modular structure of nested KEGs in the order set schema supports:

- A clear structure that separates content from format.
- Development of a knowledge repository where information can be encoded and accessed in multiple ways. A Partners Healthcare internal knowledge portal contains an inventory of encoded knowledge assets, and a meta-knowledge document library of the knowledge specifications for encoded knowledge.
- Adoption and reuse of current standards in multiple clinical applications by allowing audiences (champions, developers, users) to incorporate structured elements into production without loss of consistency of information.
- Adoption of a unified content strategy to encourage authors to collaborate, resulting in processes that are repeatable and transparent across the organization, regardless of author or department.
- Adoption of authoring environments and Web applications to ease the creation of order sets and other resources based on XML schemas to ensure content is written consistently for all environments: same content, different uses.

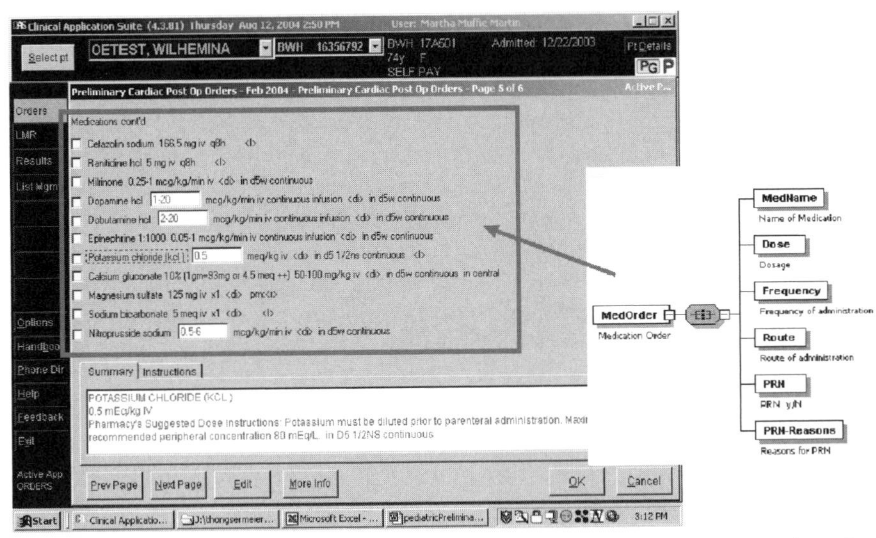

FIGURE 15-8 Schema for medication order. Each of the medication orders in the order set has an internal structure with the necessary information about the medication name, dose, frequency of administration, route, PRN or "as needed," and reasons.

KEGs provide the basis for reusability, and are critical for the construction of consistent content that can be retrieved, tracked, and assembled as part of a unified content strategy. Further, a significant advantage of KEGs is the fact that templates and documents built upon KEGs are transportable across systems and applications. The separation of content from format is at the core of the KEGs approach: same content, different uses. To illustrate this assertion, the red rectangle in Figure 15-8 shows a set of KEGs for medication orders. All these orders share a common KEG structure that contains all the relevant elements for medication orders (right-hand side of Figure 15-8). This approach facilitates the representation and integration of clinical data in multiple contexts.

15.5 CONCLUSION

This chapter has focused on the knowledge both relating to 1) documentation specifications themselves, and 2) the management and retrieval of documentation components that are incorporated in such specifications. Both are important aspects of the way in which documents can be used to provide CDS: the first, by governing how documents will be used to facilitate decision making by humans; the second, by providing a means whereby components that have particular purposes can be identified from KM repositories and incorporated in the design of the specifications for such documents.

Thinking about both aspects is important in the development of an enterprise-wide KM approach for handling information. With KEGs, information can be "chunked" into elements that can later be assembled into applications. In this building blocks approach, each element is identified by its properties, regardless of the context where it is used. It is clear that standards are needed in order to communicate knowledge in a consistent manner, so authors can create knowledge that can be grouped, reused, and integrated based on specific purposes without the need of duplication. Further, with this approach, we envision that authors would be able to create new knowledge by augmenting existing knowledge.

As exemplified by the differences between the HL7 and OpenEHR views of templates, a problem with current documentation specification standards is that they are still evolving. They have also generally lacked explicit delineation of the roles that documents or parts thereof are to play in CDS. Unlike other forms of CDS where there are identifiable components relating to knowledge content and inferencing methods, as discussed in Chapter 3, in documentation, the knowledge content is to a large extent implicit in the way the document is organized, and the inferencing method used is "association" of document elements into groups. The goal of the KEGs structure we have introduced is an effort to make this knowledge explicit, by categorizing the reasons for association.

As we have noted, the approach for an order set schema based on KEGs is part of the infrastructure being installed at Partners for KM. Evaluation of the proposed approach to date has consisted of mapping existing content in order sets into KEGs in a reverse engineering fashion to determine the suitability of the proposed schema. Next steps will involve 1) implementing a taxonomy for cataloging KEGs to improve reusability; 2) putting in place integration approaches among vendor solutions for content management; and 3) establishing an authoring environment to support authors in creating consistent, structured knowledge. The ultimate goal is a more collaborative enterprise-wide environment where experts can create, share, and maintain knowledge.

Although we presented a specific example of the use of KEGs as a means for structuring clinical information in an order set schema, KEGs can serve as building blocks for the definition of any schema, document, or template that can benefit from the extra dimensions KEGs add to data representation, indexing, and retrieval. We anticipate that the structured representation of knowledge provided by KEGs easily could be incorporated into current HL7 standards to enrich metadata and to provide contextual validation to archetypes and templates. One test of that will be to determine how the Partners order set schema can be harmonized with an order set schema standards specification currently under development by the Clinical Decision Support Technical Committee of HL7.

In its simplest form, KEGs are ontology aids for representing and handling metadata about elements and their relationships. Hence they provide the necessary mechanisms to translate business rules into system implementations, ensuring that business concepts remain aligned with deployed procedures, while expanding the ability to integrate clinical knowledge by feature (e.g., intention, clinical setting, type of user, or patient-specific characteristics).

REFERENCES

Alschuler, L. (2002). HL7's CDA, Clinical Document Architecture: An overview. Available from www.hl7.org as Document ID# 620.

Archetypes from *OpenEHR*. http://oceaninformatics.biz/archetypes/.

Beale, T. (2002). Archetypes: Constraint-based domain models for future-proof information systems. OOPSLA 2002 workshop on behavioral semantics. http://www.openehr.org/downloads/archetypes/archetypes_new.pdf.

Beale, T., Heard, S. (2003a). Archetype Definition Language (ADL). The OpenEHR Foundation. http://www.openehr.org.

Beale, T., Heard, S. (2003b). Archetype definitions and principles. OpenEHR Foundation. http://www.openehr.org.

Boxwala, A. A., Kuperman, G. J., Denekamp, Y., Scott-Wright, A., Middeleton, B. L., Greenes, R. A. (2002). Survey and evaluation of electronic knowledge use for clinical decision support at Brigham and Women's Hospital (Knowledge Inventory Report), DSG Technical Report DSG-TR-2002–07, http://dsg.bwh.harvard.edu/gello/KnowledgeInventoryReport.doc.

Dolin, R. H. et al. (2006). HL7 clinical document architecture, release 2. *JAMA* **13**: 30–39.

Greenes, R. A., Sordo, M., Zaccagnini, D., Meyer, M., Kuperman, G. J. (2004). Design of a standards-based external rules engine for decision support in a variety of application contexts: Report of a feasibility study at partners healthcare system. *Medinfo*.

Heard, S., Beale, T., Freriks, G., Rossi Mori, A., Pishev, O. (2003). Templates and archetypes: How do we know what we are talking about? Version 1.2. http://www.openehr.org/downloads/archetypes/templates_and_archetypes.pdf.

HL7. http://www.hl7.org.

HL7 Templates SIG Project Outline. (2006). informatics.mayo.edu/wiki/index.php/HL7_Templates_SIG; accessed 01/28/06.

Kernberg, M. (2004). Template and archetype architecture draft for San Antonio. Available from www.hl7.org as Document ID #1209.

Open EHR. http://www.openehr.org/.

Partners Healthcare System. http://www.partners.org/.

The Reference Information Model (RIM). http://www.hl7.org/Library/data-model/RIM/model page_mem.htm.

SNOMED. http://www.snomed.org/.

Sordo, M., Hongsermeier, T., Greenes, R. A. (2004). An enterprise-wide method for managing clinical knowledge for decision support: A partners healthcare experience. *DSG Technical Report* DSG-TR-2004-03.

Sordo, M., Hongsermeier, T., Kashyap, V., Greenes, R. A. (2006). Partners healthcare order set schema: An information model for management of clinical content. In *Computational Intelligence in Healthcare*. Germany: Springer-Verlag. (In Press).

W3C Web Ontology Language (OWL). http://www.w3.org/2004/OWL. Goldfarb, C. F., Rubinsky, Y. (1991). *The SGML Handbook*. Oxford University Press.

WC3. The Extensible Markup Language (XML) 1.0 Third Edition. http://www.w3.org/TR/REC-xml/.

16 INFOBUTTONS AND POINT OF CARE ACCESS TO KNOWLEDGE

JAMES J. CIMINO and GUILHERME DEL FIOL

Clinicians' information needs are frequent and, all too often, unresolved. Online information resources are available to address many of these needs, but they are underutilized for a variety of reasons. The information needs that arise while clinicians use clinical information systems are particularly disposed to resolution with such resources, and many researchers have investigated approaches to integrating these systems with each other. One method that is being tried by the authors, in two separate efforts, is called *infobuttons*. We describe work underway to develop infobutton managers that provide a flexible, generic method for integrating infobuttons with clinical systems. These infobutton managers have been integrated successfully with several different clinical information systems in an institution-independent manner. HL7 has an effort underway to standardize methods for communicating with infobutton managers.

16.1 INTRODUCTION

Much of this book deals with the use of computer algorithms and heuristics for providing clinical decision support (CDS). There is still a place, however, for the use of health information resources (such as the published literature) to support decision-making simply by educating the decision maker. That is, the clinician can make a better-informed decision by reading (or listening to or watching) relevant knowledge, which can then be incorporated into the clinician's cognitive processes. The educational process used to train clinicians already overwhelms them with more knowledge than they can possibly record, let alone retain and recall when needed, and staying up to date after training is even more difficult. However, if the appropriate knowledge can be invoked at the time that the clinician needs it, then it has the potential to truly support clinical decision making as "just-in-time" information (Chueh 1997).

More and more, clinicians are making their decisions while using a computer. The need to make decisions may be triggered when a clinician receives new information about a patient and that information is often computer-based. The act of making decisions is often operationalized in the form of writing an order, and order writing is increasingly a computer-based activity.

Clinical Decision Support: The Road Ahead

Thus the clinician sitting in front of a computer presents an opportunity for CDS. First, the clinician is carrying out some limited set of activities, which suggests that the types of decision support needs that arise may be similarly limited. This makes automated solutions numerable, if not always tractable. Second, the clinician's task, as well as the specific patient information involved, can help to further narrow the prediction about the kind of decision support and specific kinds of knowledge that are most likely needed. Third, the user is already in front of the computer—the perfect place to retrieve and present knowledge resources that can address the need. Fourth, the information that triggers the request for decision support can be exploited not only to identify the need but to help get the answer.

In other words, a clinician using a clinical information system may be expected to have some typical, common CDS needs that can be suggested by what the clinician is doing and seeing on the system. A clever CDS capability can anticipate the needs and attempt to automatically satisfy them. For example, consider a nurse practitioner who is reviewing a urine culture result. When he sees that the organism is *Proteus mirabilis*, he might wonder several things: "How did this patient get Proteus mirabilis in her urine?", "Is this clinically significant?", "What is the best treatment for this?", "Are diagnostic studies of the urinary tract warranted?", and so on. If the computer system presents these questions, and their answers, the nurse practitioner can learn exactly what he needs to know at the exact moment he needs to know it: just-in-time learning.

This sort of integration between clinical and knowledge systems (Cimino and Sengupta 1991) and architectures for their integration (Greenes 1991) have been envisioned for some time. Clearly, there are several challenges, not all of them technical, to realizing the preceding scenario. A number of recent developments, including better understanding of clinician information needs, more sophisticated controlled terminologies, and the advent of the World Wide Web (with all its attendant standards and resources) have facilitated the development of working solutions to the just-in-time education challenge.

This chapter describes one approach, called *infobuttons*, that addresses this challenge. We first review what has been learned of clinician information needs. We then examine a variety of projects that have integrated health knowledge resources into clinical information systems. We focus on the use of infobuttons, as one of those projects, and review the origins and evolution of the infobutton approach. We then describe how infobuttons are being implemented today in a variety of settings, and the emerging strategies for managing them. Finally, we describe the emerging standard for integrating infobuttons into clinical systems.

The *emphasis* for infobuttons, as well as for other related approaches to point-of-care access to knowledge, is on methods for *automatically selecting and retrieving* appropriate knowledge resources, *rather than* on methods for automatically interpreting them, as is the case with many other CDS capabilities.

16.2 UNDERSTANDING AND ADDRESSING CLINICIAN INFORMATION NEEDS

16.2.1 Information Needs in Clinical Practice

A well-known study from Covell et al. (1985) found that physicians in one outpatient clinical setting had two questions for every three patients, and that only 30 percent of these questions were answered during the patient visit, most commonly by another physician or other health professional. This study predated the Web or even most electronic textbooks, so one might suspect that the identification by practitioners of potential needs for resources was on the low side, since the possibilities for Google® search and other instantaneous methods of gratification were not even on the horizon yet. Nonetheless, other studies of various types of clinicians in various settings have had similar findings, especially with regard to the frequency of unanswered questions (Osheroff 1991; Ely 1992; Dee 1993, Green 2000; Wyatt 2000).

In order to better understand the types of clinical questions that commonly arise at the point-of-care, Ely et al. (2000) created a taxonomy with 64 question types. The taxonomy was then used to classify 1,396 clinical questions from primary care and family care physicians. The three most common types were "What is the drug of choice for condition X?" (11%); "What is the cause of symptom X?" (8%); and "What test is indicated in situation X?" (8%). A previous study from the same group indicated that the most common question topics among a group of 103 family doctors were about drug prescribing (19%) (Ely 1999).

Subsequently, Cimino led a set of studies to examine the specific information needs that arise while clinicians are using clinical information systems. In an observational study, nurses and physicians were asked to think aloud as they used a clinical information system to review and enter patient data. Information needs arose most often while reviewing laboratory results and medication orders (Currie 2003). Analysis of the user interactions showed that fully half of the information needs were requests for health knowledge and, of these, 40 percent of the questions were medication-related (Allen 1993). Over half of health knowledge needs (55%) were not successfully resolved.

16.2.2 Use and Impact of Online Information Resources

A number of studies have examined, usually via surveys, how clinicians use online resources to help answer their questions. These studies invariably find that, despite the availability of these resources, clinicians still seldom use them. For example, one systematic review found that clinicians search resources from 0.3 to 9 times a month (Hersh 1998). A more recent study looking at the utilization of a resources portal (Clinical Information Access Portal, CIAP) available to 55,000 clinicians in Australian public hospitals, found a rate of 0.48 search sessions per clinician per month, despite the fact that 88 percent of the users reported that the resources portal had the potential to improve patient care (Westbrook 2004).

Less is known about use of resources and evidence-based practices among nonphysician clinical disciplines, but utilization seems to be equally low or even lower. One study found that nurses use colleagues, particularly senior nurses, as a primary source of information to answer their clinical questions (Thompson 2001). However, when resources are easily available, utilization among nurses seems to improve. In a survey of users of a Web portal for accessing health resources, 74 percent of respondents who were aware of the portal reported having used it (Gosling 2004). The main factors associated with higher use were perceived support from hospital leadership and use by their peers.

In addition to survey results, Cimino (1993) studied clinical information system log files to determine the rate of resource usage. In a population of approximately 4,000 users over a six-month period, 65 percent of users accessed a resource at least once, but these users, on average, made only one access every two weeks. Most accesses were made while users were reviewing laboratory results, with a pharmacy knowledge base being the most popular resource (Cimino 1993).

Despite the apparent underutilization of online resources, online summarizations of evidence and best practices have been considered a sound strategy for meeting clinicians' information needs (McColl 1998; Rousseau 2003). In a controlled laboratory study, clinicians' (nurses' and physicians') performance on answering a set of clinical questions improved by 21 percent when they were provided with access to a set of resources (Westbrook 2005). The strongest effect was found on the group of nurses, who performed as well as the physicians when both groups had access to resources. In another study, physicians found at least partial answers to their questions in 73 percent of the cases in which they pursued an answer using resources (Magrabi 2005). In a third study, the presentation of results of literature searches to physicians was associated with changes to treatment decisions in 18 percent of the patients in the study, and 78 percent of these changes were judged by experts as representing an improvement in or maintenance of a treatment strategy (Lucas 2004).

16.2.3 Barriers to Use of Online Information Resources

A number of barriers preclude more effective use of online resources at the point-of-care. In an observational study followed by interviews with 48 generalist physicians, Ely (2005) identified and classified barriers into "obstacles preventing pursuit of answers" and "obstacles to finding answers to pursued questions." The most common obstacles in the first category were, from most to least frequently cited: 1) doubt that an answer existed; 2) ready availability of consultation leading to a referral rather than a search; 3) lack of time; 4) question not important enough; and 5) uncertainty about where to look for information. In the second category, the most common obstacles were: 1) topic not included in the selected resource; and 2) failure of the resource to anticipate ancillary information needs. In interviews, physicians highlighted the need for comprehensive resources that directly answer questions that are likely to occur in practice.

Other studies have demonstrated that the amount of time taken to find an answer seems to represent a critical barrier. Covell's original study found that lack of time was the most frequently stated barrier (Covell 1985). A study conducted by Hersh found that although medical students were able to find answers to 85 percent of a set of clinical questions using Medline, the time taken to find those answers was, on average, 30 minutes (Hersh 1996). More recently, Westbrook found that the time taken for clinicians to find an answer to clinical questions using a set of resources was six minutes on average (Westbrook 2005). Although this seems to represent an improvement over Hersh's results, Westbrook's methodology was different in that experienced clinicians were the study subjects, and a set of online evidence summaries was available for searching. Both studies recognized that the time required to find an answer constitutes a major barrier to the use of the evaluated resources, given the time constraints of real clinical practice.

Some studies have shown that clinicians' literature search skills are generally suboptimal. Osheroff and Bankowitz found that clinicians using online resources were able to get answers to only 40 percent of their questions (Osheroff 1993). A survey of 294 general practitioners in New Zealand indicated that most of them limited themselves to keyword searching and only 10 percent knew how to refine searches by using MeSH terms (Cullen 2002). On the other hand, Hersh's systematic review showed that advanced search methods were not more effective than simple text word methods. Still, the general performance of users and information retrieval systems was suboptimal, with most searches retrieving only one fourth to one half of the relevant articles on a given topic (Hersh 1998).

In summary, the lack of easy access to resources that can provide high quality and objective answers in a timely manner is a barrier to the use of online resources at the point of care. By lowering this barrier, the utilization and effectiveness of resources at answering clinical questions should improve.

16.3 LINKING CLINICAL INFORMATION SYSTEMS TO ONLINE RESOURCES

The integration of clinical information systems with online health knowledge resources has the potential to address the dual barriers of access and time constraints. A number of prototype systems have attempted to explore this approach. We describe some of this work here, but we direct you to more extensive published reviews of the topic (Cimino 1996; Stead 2000).

Because the National Library of Medicine's Medline database was one of the first and most prevalent online resources, Medline searches were typically the target of this initial work. Among the earliest such systems were Hepatopix (Powsner 1989) and Psychtopix (Powsner 1992). These systems contained sets of topic-specific Medline search strategies that could be matched to "topics of interest" encountered in reports (liver biopsy reports and psychiatric records, respectively). A user reading a report thus automatically could perform a bibliographic search relevant to information encountered in a clinical record. Although these systems were effective, they required extensive manual effort by experts to create the search strategies, which were only relevant to specific topics. Scaling these systems to be able to handle larger domains required a proportional scaling of expert effort.

Another approach to automating bibliographic searches was to allow the user to select a topic of interest from the clinical record and, using a cut-and-paste approach, transfer it to Medline for use in searching. The Term Linker (Loonsk 1991) and the Meta-1 Front End (Powsner 1991) both took this approach and each employed the Unified Medical Language System to support translation to the Medical Subject Headings.

The Chartline system (Miller 1992) took the process a step further by allowing the user to specify particular topics of interest related to the term of interest in the clinical record. The term was translated to MeSH and inserted into a predefined Medline search strategy, in an attempt to perform a more specific and relevant search. Later systems, including the Interactive Query Workstation (Cimino 1990) and the Internet Gopher (Hales 1993) allowed users to perform searches against a variety of resources besides Medline. The

DeSyGNER system supported the integration of books, tutorials, and simulation systems into a radiologist's clinical workstation (Greenes 1991). Researchers at LDS Hospital and the University of Utah, led by Homer Warner, integrated a diagnostic decision support system (Iliad) with a large clinical information system (HELP) (Wong 1994).

The advent of the World Wide Web has contributed a great deal to reducing the barriers to accessing health information (Hersh 1999). As a result, integrated systems became much easier to develop. Among the first was the MedWeaver system, which used a query formulator to translate a user's information request into a searchable form that could be passed to a retrieval manager. The retrieval manager, in turn, could access a variety of information resources and produce an integrated view of all information retrieved (Detmer 1997). Integration of clinical information systems with Web-based retrieval systems followed soon after. One system, at Duke University, integrated Web-based clinical practice guidelines into a system for documenting well-child visits to a pediatric clinic (Porcelli 1999). The ActiveGuidelines system at the Palo Alto Medical Clinic integrated Web-based guidelines with a clinical information system and allow users to invoke guidelines based on relevant topics in a patient's electronic medical record (problems, medications, etc.). Users could then import recommendations from the guidelines into the clinical information system, in the form of physicians' orders (Tang 2000). The same group later produced the PAMFOnline system, which linked online health resources to a system for allowing patients access to their medical records, in order to address patient information needs (Tang 2003).

16.4 INFOBUTTONS

Several integration efforts have taken the form of context-specific links to online resources, integrated into clinical systems. These links, called *infobuttons*, not only invoke relevant resources, but anticipate information needs and initiate retrieval strategies to help the user navigate resources (Cimino 1997). Infobutton research has included studies of information needs and their contexts. We focus here on infobutton implementations at Columbia University and Intermountain Health Care, and briefly mention various approaches (some called infobuttons, some not) by other research groups.

16.4.1 Understanding the Context of Information Needs

Studies in areas such as medical informatics, anthropology, knowledge management, and pervasive computing have highlighted the role of context in predicting the nature of workers' information needs. In an ethnographic study of physicians' information needs, Forsythe et al. (1992) stated that understanding a question correctly requires interpretation in the light of the context in which it was expressed. Khedr and Karmouch (2004) define context as "information about physical characteristics (such as location and network elements), the system (such as applications running and available services), and the user (such as privacy and presence)" and state "the environment becomes context-aware when it can capture, interpret, and reason about this information." Fischer and Ostwald propose that the context of the problem dictates the workers' information demands (Fischer 2001). More specifically in the health care information retrieval domain, Lomax and Lowe stated "in the effort to characterize information seeking, it is important to accurately define the types of information clinicians may use as well as states of information need which trigger information search and retrieval" (Lomax 1998).

Research has been conducted as an attempt to understand the context in which those information needs arise while users are interacting with a computer. Pratt and Sim showed, with their Physician Information Customizer, that formal representation of information about users could help information retrieval systems identify articles of greatest interest to those users (Pratt 1995). Cimino (2002) hypothesized that if computer systems are able to capture the context in which common information needs occur, such systems would be able to predict those information needs, automatically translating them into queries that can be executed by online resources. Del Fiol (2005) demonstrated the importance of context in an XML-based order set model. In this model, context defines the care settings where an order set (and individual orders within the order set) can be used, the patients (in terms of age, gender, and clinical condition) that are eligible for this order set, and the providers who can use this order set to write orders.

16.4.2 Infobutton Development at Columbia University

Infobutton development at Columbia can be traced back to an NLM-sponsored project to explore ways in which clinical data could be translated, using the UMLS, into MeSH terms to support automated bibliographic searches. Dubbed "the Medline Button," the system allowed users to select patient diagnoses and procedures, coded in ICD9-CM in a mainframe-based clinical information system, and use them to search a Medline database running on the same mainframe (see Figure 16-1) (Cimino 1992).

Despite the physical proximity of the systems, integration was extremely difficult, due to the disparate nature of the two systems. However, even with the technical problems solved, the Medline Button failed to garner much attention from clinician-users. We realized that this was because the system failed to address real information needs of real users.

The first step to rectifying this problem was to develop a representational scheme for capturing user information needs. We codified these needs into generic queries, which took the form "Is <disease 1> caused by <disease 2>?" (to take the example from Figure 16-1). These generic queries could then be generated to correspond to real users' questions and then the blanks (e.g., <disease 1>) could be filled in based on the context in which the user was asking the question (Cimino 1993).

While this work was in progress, the World Wide Web emerged as a major environment for clinical informatics research. With the creation of Web-based clinical and knowledge systems, the barriers to interfacing disparate systems, such as we experienced with our two mainframe applications, were largely removed. We began to explore ways to link our new clinical information system to online resources such as Dxplain (Elhanan 1996) and a variety of bibliographic, textual, and graphical resources (Zeng 1997).

These initial experiments led to the implementation of infobuttons in the New York Presbyterian Hospital's clinical information system, WebCIS (see Figure 16-2).

Admission Profile
Admission Date: 01/03/95 Discharge Date: 02/16/95 Location: M6GS
Doctor: CIMINO, JAMES J Discharge Summary: N
Primary Diagnosis: 410.71 ACUTE MI,SUBENDO INFARC, INITI
Select Terms You Are Interested in:
Diseases:
X 410.71 ACUTE MI, SUBENDO INFARC, INITI
X 780.3 CONVULSIONS
_ 507.0 FOOD/VOMIT PNEUMONITIS
426.0 ATRIOVENT BLOCK COMPLETE
415.1 PULMON EMBOLISM/INFARCT
453.8 VENOUS THROMBOSIS NEC
428.0 CONGESTIVE HEART FAILURE
F8 = for more information
Help=F1 Search MEDLINE=ENTER

MEDLINE Queries from Admission Profile
Select a question: _
1. Does Myocardial Infarction cause Convulsions?
2. Is Myocardial Infarction caused by Convulsions?
3. Does Myocardial Infarction occur with Convulsions?
Help=F1 Search MEDLINE=ENTER

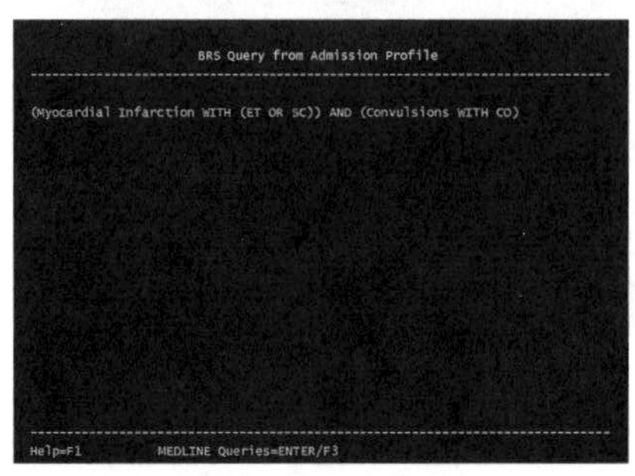

FIGURE 16-1 Screen shots from the Medline Button. The top left screen shows patient diagnoses, coded in ICD9-CM, in the clinical information system. When the user selects two ICD9-CM diagnoses (in this case, "ACUTE MI, SUBENDOC INFARC, INITI" and "CONVULSIONS") and presses the F8 key, the Medline Button translates the diagnoses into MeSH terms ("Myocardial Infarction" and "Convulsions" and presses the F8 key, the Medline Button translates the diagnoses into MeSH terms ("Myocardial Infarction" and "Convulsions," in this case) and presents several possible questions of interest to the user (shown in the top right screen). When the user selects a question (in this case, question 2 "Is Myocardial Infarction caused by Convulsions?"), the system generates the Medline search strategy shown in the bottom left screen that, in turn, produces the search results shown in the bottom right screen (in this case, one article was returned). The user can then go on to review the citation and abstract (not shown).

The infobuttons were inserted into WebCIS in a variety of places, including applications for viewing laboratory results, microbiology culture results, microbiology antibiotic sensitivity results, and pharmacy orders (shown in Figure 16-2). Analysis of log files showed that, depending on the context, users preferred infobuttons as much as nine to one over other available information resources (Cimino 2003).

16.4.3 Infobutton Development at Intermountain Healthcare

In September 2001, Intermountain Healthcare (IHC) integrated infobuttons with the medication list, problem list, and laboratory results modules of IHC's clinical information system, known as HELP2 (see Figure 16-3) (Reichert 2002). Infobuttons are placed next to each clinical concept (e.g., medication, problem) in these modules. When an infobutton is clicked, the user is presented with a list of questions about the concept of interest. The user can also select from a list of resources that cover the domain of the questions under consideration. When the user selects one of the questions, a search request is sent to the target resource, which then returns the search results.

The HELP2 infobuttons use coded clinical data from the IHC clinical data repository (CDR) to dynamically generate and send search requests to

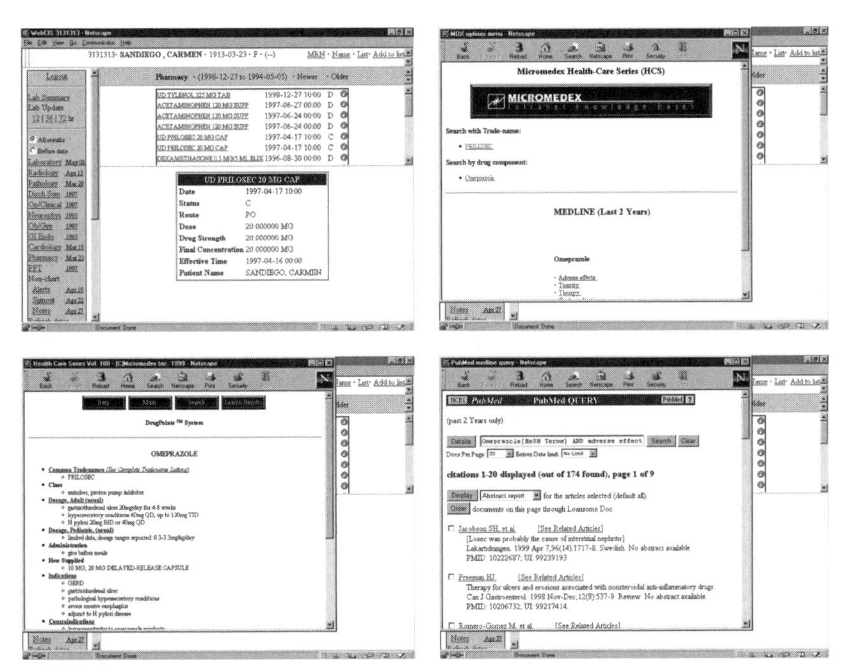

FIGURE 16-2 Screen Shots of the Initial Infobutton Implementation at New York Presbyterian Hospital. The top left image shows a typical WebCIS screen, in this case a display of pharmacy orders; the infobuttons are the white-"i"-in-blue-circle icons, to the right of each medication. The top right image shows the result of clicking on an infobutton (in this case, the one to the right of UD PRILOSEC 20 MG CAP): a screen pops up with links to two resources, Micromedex and Medline (PubMed). Note that the infobutton has extracted the trade name PRILOSEC for use in searching Micromedex and has also used a terminology knowledge base to recognize that the drug has the ingredient Omeprazole, which is suitable for use in searching both resources. The bottom left image shows the result of clicking on the Micromedex Omeprazole link, and the bottom right image shows the result of clicking on the Medline Adverse effects link.

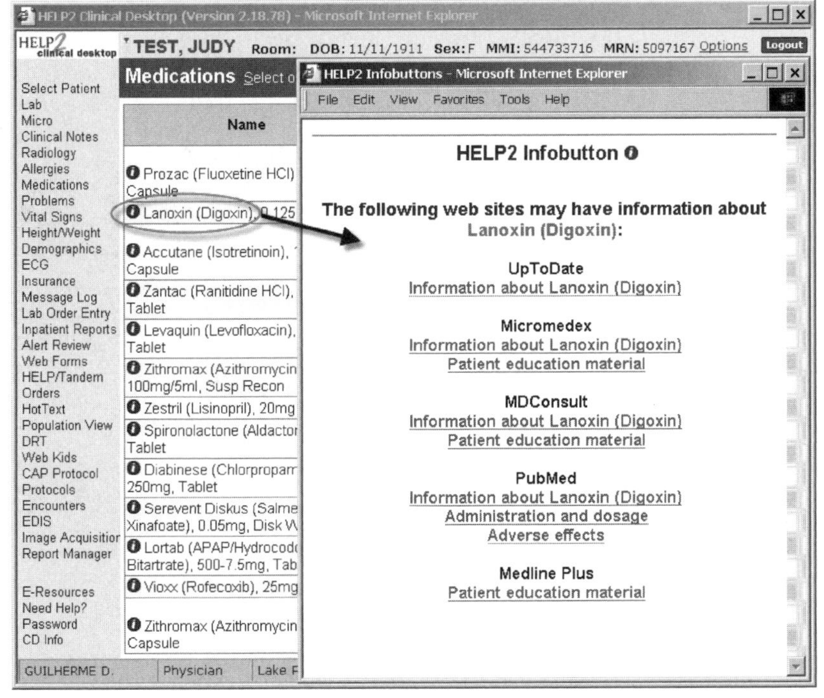

FIGURE 16-3 A medications list screen from IHC's HELP2 system, showing infobuttons (left) and the result of selecting the infobutton next to the medication Lanoxin (Digoxin) (right).

different resources available at IHC. The main resources currently in use at IHC include Elsevier's MDConsult (MD Consult, St. Louis, MO), Clin-eguide (Wolters Kluwer Health, Inc., Amsterdam), Thomson's Micromedex (Thomson Micromedex, Greenwood Village, Colorado), UpToDate (UpToDate, Waltham, MA), and NLM's PubMed. All CDR coded data values represent clinical concepts maintained within the IHC terminology server. Using the terminology server, each coded concept is translated into a suitable standard terminology, such as ICD-9-CM,[1] LOINC (Huff 1998), and the National Drug Codes (NDC)[2] (see Table 16-1). However, some resources currently do not support coded concepts; in these cases the terminology server is used to translate the CDR coded values into English expressions ("free-text"). Besides the main clinical concepts retrieved from the CDR, the HELP2 infobuttons also take into account context information expressed in terms of patient age and gender and the particular HELP2 module in which the infobutton is located. Moreover, a modifier (e.g., diagnosis, treatment, prognosis, patient education) can be added to the query based on the question that the user selected.

The utilization of infobuttons has been constantly increasing since their initial release. For instance, in 2004, infobuttons were used 17,656 times (58% higher than the same period in 2003) by 1,035 users (see Figures 16-4 and 16-5). The infobuttons in the medications module were the ones with highest use (64.9%), followed by the modules for viewing lab test results (21.5%), and managing problems lists (13.6%). Although only 27.8 percent of the infobutton users were physicians, these users accounted for the majority of the infobutton sessions (63.2%).

16.4.4 Other Infobutton Development Work

Several research groups are interested in linking clinical systems with knowledge resources. For example, MINDscape, at the University of Washington, integrates a digital library and electronic medical record. That system uses an "i" icon next to each term in the system's problem list to provide a link to a term-specific template that, in turn, provided links to a variety of resources. However, the links in MINDscape were all hard-coded, creating problems with maintenance and with passing details about the user's context (Fuller 1999).

TABLE 16-1 List of HELP2 modules that have infobuttons, the resources used by these infobuttons, and the code systems that are used to create the queries.

HELP2 Module	Resource	Code systems
Problem list	Clin-eguide, MDConsult, PubMed, UpToDate, MedlinePlus	ICD9-CM (and free text)
Medications	Micromedex, Clin-eguide, UpToDate, MedlinePlus	NDC codes (and free text)
Lab results	Clin-eguide, MDConsult	LOINC (and free text)

[1] http://www.cdc.gov/nchs/icd9.htm
[2] http://www.fda.gov/cder/ndc/index.htm

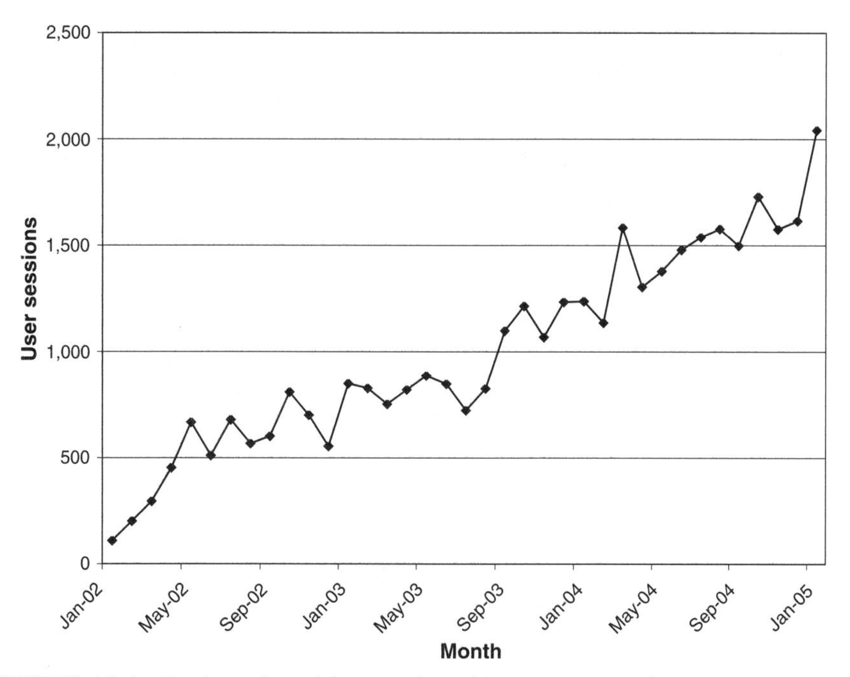

FIGURE 16-4 Total number of hits on the infobuttons at IHC from January 2002 to January 2005.

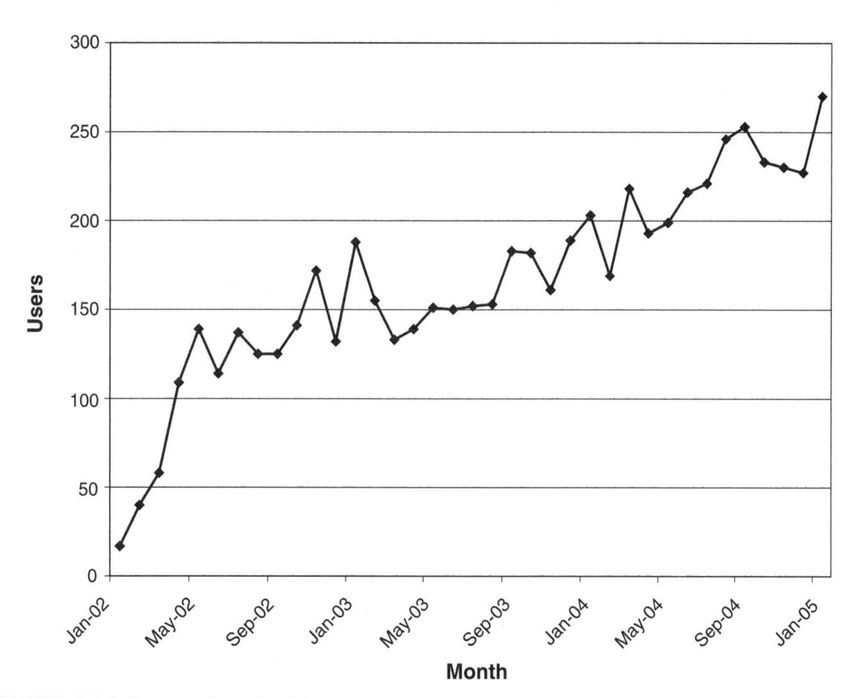

FIGURE 16-5 Number of infobutton users at IHC from January 2002 to January 2005.

Ruan and colleagues have inserted infobuttons into the clinical information system at the Geissen University Hospital. Their approach to information passing included the use of a data dictionary to provide additional information about the data from the clinical record. The dictionary provided not only links to specific relevant resources, but also links to related concepts that, in turn, could provide additional links (Ruan 2000).

SmartQuery, at Oregon Health and Science University, collects a variety of terms from the patient record: ICD9 diagnoses, laboratory tests, and terms extracted from dictated reports. These terms are then translated into MeSH terms and used to search five different resources. Preliminary evaluation showed that the system was clinically useful (Price 2002).

KnowledgeLink, at Partners HealthCare, embeds "look-up" buttons within the electronic medical record wherever a medication is displayed to the user. In a two-month evaluation after the application was launched in January 2003, the users were randomly prompted with a quick survey after using the application. The survey data indicated that medication queries were most often about dosing (33%), side effects (21%), and drug interactions (12%). In addition, answers were found in 74 percent of the queries, supporting a previously made medical decision in 59 percent of the cases and changing a medical decision in 25 percent of the cases (Maviglia 2003).

Recently, a group at Vanderbilt University described a study of the Patient Care Provider Order Entry with the Integrated Tactical Support (PC-POETS) component of their WizOrder system. PC-POETS provides links from order entry screens to online resources with searches for the item being ordered. The study showed that deployment of these links increased frequency of access to online resources by a ratio of over nine to one (Rosenbloom 2005).

16.5 MANAGING INFOBUTTONS

The experience with linking clinical systems to online resources has been consistently positive, but technical issues have constrained their deployment. For example, integrating infobuttons into the clinical system at New York Presbyterian Hospital required customized programming for each link. Each link, in turn, was associated with a customized program to provide the connections to the resources. The special programming required inhibited experimentation with additional resources and queries. Researchers at the University of Washington had a similar experience with MINDscape (Fuller 1999). The next logical step in developing infobuttons was to decouple the clinical systems from the knowledge resources to allow for more flexible connections. At Columbia, this has resulted in an Infobutton Manager, and Intermountain Health Care has developed an E-Resources Manager.

16.5.1 Columbia University's Infobutton Manager

The Infobutton Manager (IM) approach at Columbia involved three design components (Cimino 2002). The first component was the standardization of the set of context information that would be passed to the IM, including user information (user ID, profession, and institution), patient information (patient ID, age and gender), clinical task being performed, and clinical data being reviewed. A CGI program has been developed that accepts these items as input parameters. Integrating the clinical system to the IM thus required system

```
<a href="https://flux.cpmc.columbia.edu/webcisdev13/wc_infomanage.cgi? MRN=3131313&
info_institute=CPMC&info_med=1600&info_context=LabDetail& info_usertype=MD&info_age=
22&info_sex=F"><img src="info.gif"></a>
```

FIGURE 16-6 Example of a link to the Infobutton Manager. Most of the HTML code is the same for each link. The clinical system needs to provide specific parameter values, such as info_med, which, in this case, contains the code for the laboratory test being reviewed ("1600" is the code for a serum glucose test).

developers simply to insert a hyperlink to the CGI call that included the values of all the data items as parameters. Adding new IM links at different points in the program can thus be accomplished by reusing programming code to create the CGI calls where desired. Figure 16-6 shows an example of a link to the IM.

The second design component was a table of user questions (the Infobutton Table). Each question has been determined through Columbia's empirical studies of clinician information needs. For each question, the developers identify a resource that can answer the question and establish a method for transferring (and, where necessary, translating) clinical data to the resource to direct the retrieval process (Cimino 2004). The Infobutton Table therefore contains a unique question ID, a natural language version of the question (to display to users), and the URL for carrying out the search. Table 16-2 shows a part of the IM Infobutton Table.

The third design component was a table, called the Context Table, a sample of which is shown in Table 16-3. When the IM is called (because a user has clicked on an infobutton icon in the clinical information system), the IM receives the context parameters and matches them against rows in the Context Table. For each row that matches, the Infobutton ID is selected and the corresponding query in the Infobutton Table is retrieved. Each query is assembled into a URL that displays the natural language question and contains the link to the resource. The set of question-resource-links is assembled into a Web page that is passed back to the user, as shown in Figure 16-7.

The advantages of the IM over the Infobutton approach have been several. First, the integration into the clinical information system of the link to the IM is simplified and standardized. System developers need only be told where to insert the link; the link itself is essentially the same no matter where it is inserted.

A second advantage of the IM approach is the flexibility of adding questions and resources. In one case, the chief medical officer of the hospital requested that a heparin administration guideline be added as an infobutton related to the laboratory display for partial thromboplastin time (PTT) results. Within five minutes, links were established for PTT results, as well as for heparin orders, with one guideline invoked for adult patients and a second guideline invoked for pediatric patients. These links were available immediately to the 4,000 users of the system.

A third advantage of the IM approach is that it is not necessarily institution-specific. Some of the resources are only available within the NYPH Intranet, but others (such as PubMed) are available to all. The IM uses the "Institution" parameter to determine which questions will be appropriate for a user at a particular institution. A default institution of "Generic" has been created to allow outside users to obtain questions that use publicly available resources.

A number of other institutions have begun to take advantage of the IM by including links into their own systems. In one case, the New York Office of Mental Health has used the IM to provide drug information about the items in their patients' computerized medication lists. In another case, the developers of the Regenstrief Medical Record System, in Indianapolis, have added links to the IM for laboratory items, using LOINC codes as their controlled terminology, as well as

TABLE 16-2 A sample of rows from the Infobutton Table in Columbia's Infobutton Manager. The Infobutton ID is the unique identifier for the query, the Question is the natural language version of the query (for display in the user's Web browser), and the URL is the link to the target information resource.

Infobutton_ID	Question	URL
9	National Guidelines Clearinghouse	http://www.guideline.gov/search/searchresults.aspx?Type=3&txtSearch=<>&num=10
10	PubMed	http://www.ncbi.nlm.nih.gov/entrez/query.fcgi?dispmax=20&db=PubMed&cmd=Search&term=<> +&doptcmdl=DocSum
11	What is the differential diagnosis of <> (PubMed)?	http://www.ncbi.nlm.nih.gov/entrez/query.fcgi?dispmax=20&db=PubMed&cmd=Search&term=<>[MeSH+Terms]+AND+diagnosis[MeSH+Subheading]&doptcmdl=DocSum
55	OneLook (defintion)?	http://www.onelook.com/?w=<>&ls=a
70	What does the CPMC Lab Manual say about this test?	http://cpmclabinfo.cpmc.columbia.edu/<>
120	What are NYPH guidelines for managing adult patients with elevated INR due to warfarin?	http://infonet.nyp.org/Pharmacy/Forms/INR-policy-final-adult.pdf

TABLE 16-3 A sample of rows from the Context Table in Columbia's Infobutton Manager. Contexts refer to functions in the clinical information systems, such as review of detailed laboratory, pathology, and radiology results (LabDetail, PathDetail, and RadiologyReport, respectively), as well as inpatient and outpatient drug orders. Age Groups are Newborn, Infant, Adolescent, Young adult, Middle aged, and Elderly. Concept and Concept Name refer to the code and name of the class of clinical data that evoke the context. Institution identifies the various organizations that use infobuttons in their systems. ID refers to the Infobutton ID in Table 16-2.

Context	User Type	Age Group	Sex	Concept	Concept Name	Institution	ID
LabDetail	MD, Others	I,C,A,Y, M,E,N	M,F	50	MEASURABLE ENTITY	CPMTEST, CPMC	9
LabDetail	MD, Others	I,C,A,Y, M,E,N	M,F	30007	PATIENT PROBLEM	RMRS2, LDS, RMRS, CPMTEST, CPMC, GENERIC	9
InPatientDrugs, OutPatientDrugs, Sensitivity, OutPatDrugOrd	MD, Others	I,C,A,Y, M,E,N	M,F	30	PHARMACOLOGIC SUBSTANCE	RMRS2, LDS, RMRS, CPMTEST, CPMC, NYOMH, GENERIC	9
LabDetail	MD, Others	I,C,A,Y, M,E,N	M,F	50	MEASURABLE ENTITY	CPMTEST, CPMC	10
LabDetail, Microbiology	MD, Others	I,C,A,Y, M,E,N	M,F	30007	PATIENT PROBLEM	RMRS2, LDS, RMRS, CPMTEST, CPMC, GENERIC	10
InPatientDrugs, OutPatientDrugs, OutPat DrugOrd	MD, Others	I,C,A,Y, M,E,N	M,F	30	PHARMACOLOGIC SUBSTANCE	RMRS2, LDS, RMRS, CPMTEST, CPMC, NYOMH, GENERIC	10
LabDetail	MD, Others	I,C,A,Y, M,E,N	M,F	30007	PATIENT PROBLEM	RMRS2, LDS, RMRS, CPMTEST, CPMC, GENERIC	11
Radiology Report, PathDetail	MD, Others	I,C,A,Y, M,E,N	M,F	1	MEDICAL ENTITY	NYSPI, LDS, RMRS, CPMTEST, CPMC, NYOMH, GENERIC	55
LabDetail, LabOrder	MD, Others	I,C,A,Y, M,E,N	M,F	93	LABORATORY DIAGNOSTIC PROCEDURE	CPMTEST, CPMC	70
LabOrder	MD, Others	Y,M,E	M,F	33888	COAGULATION STUDIES	CPMTEST, CPMC	120
LabDetail	MD, Others	Y,M,E	M,F	32863	INTERNATIONAL NORMALIZED RATIO (INR) CALCULATIONS	CPMTEST, CPMC	120
InPatientDrugs, OutPatientDrugs, OutPat DrugOrd	MD, Others	Y,M,E	M,F	31433	WARFARIN PREPARATIONS	CPMTEST, CPMC	120
LabDetail	MD, Others	Y,M,E	M,F	32163	PLASMA PROTHROMBIN TESTS	CPMTEST, CPMC	120

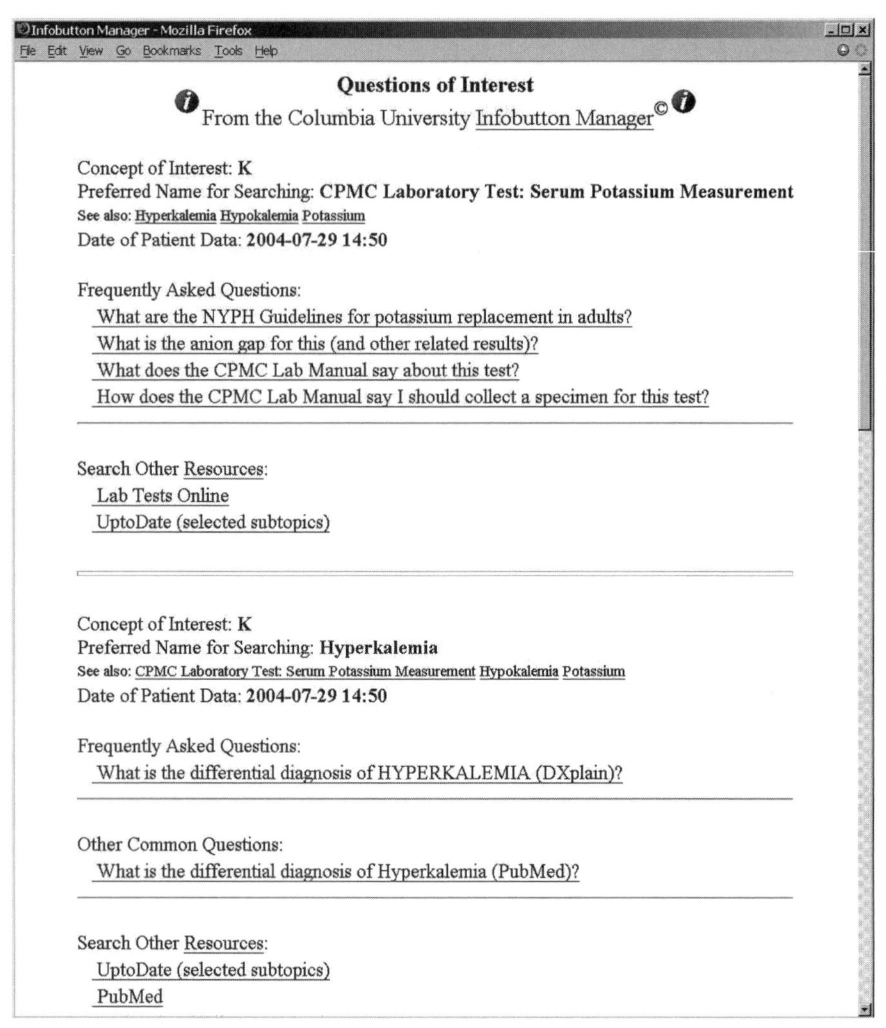

FIGURE 16-7 An example of output from Columbia's Infobutton Manager. This was evoked when the user clicked on the infobutton icon next to a serum potassium ("K") result.

links for medication-related items (McGowan 2004). Others interested in using the IM are invited to do so.[3] Although Columbia University has applied for a patent on the IM, researchers and other noncommercial users (including health care providers) are welcome to obtain free licenses for use of the IM technology.

16.5.2 IHC's E-Resources Manager

The initial version of the HELP2 infobuttons had several limitations similar to the initial Columbia infobuttons implementation. For instance, the basic infobutton routines were duplicated and manually customized for each HELP2 module for which infobuttons were enabled. Moreover, the addition, modification, or removal of resources required changes to each infobutton routine associated with every affected HELP2 module. These limitations imposed restrictions in terms of the number and variety of resources that could be efficiently enabled

[3] http://www.dbmi.columbia.edu/cimino/Infobuttons.html

and maintained for the HELP2 infobuttons. Also, partially because of these limitations, the HELP2 infobuttons have been configured to access only four "general purpose" resources that provide content in a wide variety of clinical domains, and have not taken advantage of very specific external and internal content collections that can provide more focused, domain-specific answers.

In 2004, a new software component called E-resources Manager (ERM) was developed to handle all infobutton requests originated from the HELP2 modules. The ERM is composed of four core components: e-resource profiles, e-resource selection, question builder, and query translator (see Figure 16-8).

An e-resource profile uses an XML-based file to characterize the relevant contexts for the resource, the context parameters the resource is able to handle, the proprietary query syntax used to express these parameters, and the code systems used to represent the parameter values. The main context parameters include the clinical information system (CIS) module, the main clinical concept of interest to the user, the age and gender of the patient, and the user role (e.g., physician, nurse, patient).

The XML-based profiles can describe two main types of resources: generic and domain-specific. Generic resources usually contain information regarding multiple aspects of a large number of medical diagnoses, including common clinical manifestations and laboratory findings, differential diagnosis list, recommended therapies, prognosis, and so on. PubMed is the most obvious example of this type of resource. Using PubMed, one can obtain information about virtually any diagnosis and all its relevant aspects directly from the scientific papers indexed by MEDLINE. Many other examples of generic resources are currently available, including some that provide synthesized information derived from the best available evidence. Generic resources with

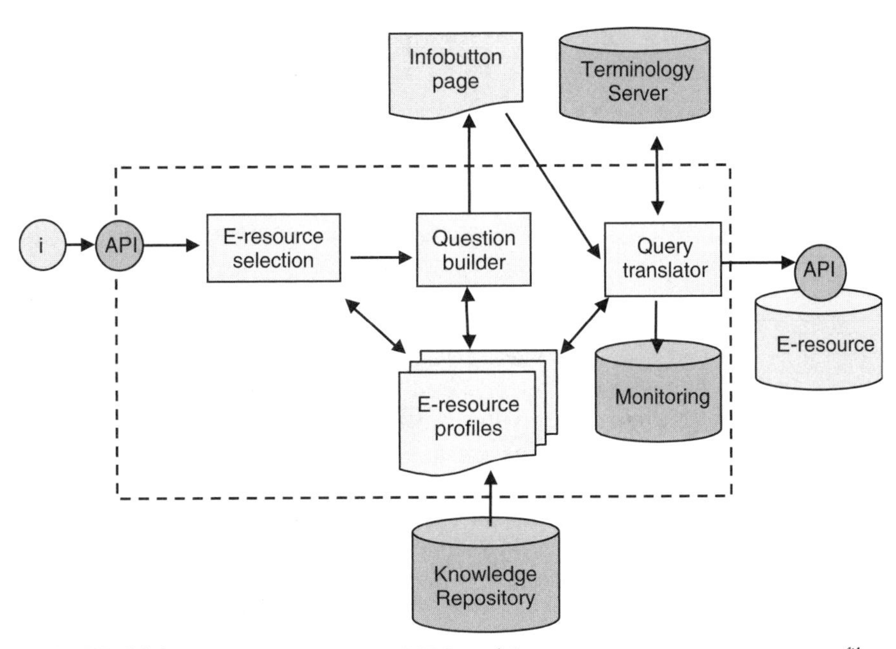

FIGURE 16-8 E-resources Manager (ERM) and its core components: e-resource profiles, e-resource selection, question builder, and query translator. The i represents an infobutton call, and the API represents the programmatic interfaces between the Clinical Information System and the ERM, and between the ERM and the resource.

preappraised and synthesized information can arguably provide evidence-based answers to busy clinicians in a timely manner (Magrabi 2005; Westbrook 2005). IHC currently licenses four commercial products that can be considered to be generic resources with such preappraised and synthesized information: MD Consult, Clin-eguide, Micromedex, and UpToDate.

Domain-specific resources cover subdomains of knowledge applicable to certain clinical conditions, medical subspecialties, patient age groups, and so on. An example is the American Diabetes Association Web site,[4] which has a large amount of content specific to the management of diabetes. Other examples are content collections developed internally by health care organizations and academic institutions, especially those that include relevant regional or local information (Del Fiol 2004a).

The e-resource profiles are stored in the IHC Clinical Knowledge Repository (CKR), an XML database (Oracle 10g) coupled with a set of services for searching and retrieving content. Content is stored in the CKR as XML documents. Each XML document is assigned a unique ID and a version number. Documents can easily be retrieved using a simple HTTP GET or POST request with the document ID specified as a parameter. The CKR infrastructure also contains an XML schema directory with reusable complex types and data types. One of the important reusable complex types defined in the directory is the context type, which allows any piece of knowledge in the CKR to be context-aware. The CKR search and retrieval services are also context-aware, enabling queries that will return content that matches a specific context. The e-resource profiles rely on the CKR context-awareness to represent the context covered by each resource. In addition to playing the role of an infobutton manager, the CKR is also a content provider for infobuttons.

Since the end of 2002, the CKR has been used at IHC to store, search, and retrieve various types of clinical documents, from reference information (e.g., patient education handouts, discharge instructions, practice guidelines) to "executable" content (e.g., order sets, resource profiles). More detail on the CKR and its context representation model can be found elsewhere (Del Fiol 2005). The CKR content, including the resource profiles, is created using a generic knowledge authoring tool that was developed to help clinical experts create and maintain knowledge content without the intervention of information technology personnel (Hulse 2005).

When the ERM is called by a HELP2 module, the e-resource selection component makes use of the profiles to identify the resources that can provide the best possible answers. The e-resource selection component uses the query context in which the HELP2 infobutton was activated to identify the resources that best match that context, as expressed in their profiles. Next, using the profiles of the selected resources, the question builder creates the questions that each of the matching resources is able to handle, and presents these questions as hyperlinks in a new HTML page (infobutton page). The resulting infobutton page will contain generic and/or domain-specific questions, depending on the types of matching resources. When the user clicks on one of the questions, a new request to the ERM is made. The query translator then transforms all the parameters and concept codes from the selected question into the specific syntax of the target resource. As with the previous IHC

[4] http://www.diabetes.org

infobutton implementation described earlier, the terminology server is used to translate nonstandard coded concepts to standard code systems, such as LOINC, ICD9-CM, and NDC, or to textual representations. Finally, an HTTP request is submitted to the target resource and the complete request is logged by the ERM monitoring infrastructure.

The ERM allows knowledge engineers and medical librarians to include new resources by simply adding new profiles to the CKR. Each profile can be easily created and maintained using the CKR authoring tool. If a resource is considered obsolete, or if it should no longer be used, the knowledge engineer and medical librarian responsible for that profile can simply change its status to inactive, using the CKR authoring tool. Changes and additions to the collection of ERM profiles are made instantly available to the HELP2 infobuttons, without requiring any changes or recompilations of either the ERM or HELP2.

16.6 INFOBUTTON STANDARDIZATION

One of the major problems complicating the integration between health information systems (HIS) and health information resources is that HIS vendors may wish to implement infobuttons in various modules of their applications (e.g., order entry, clinical notes, nurse charting, lab results). As previously explained, infobuttons are typically implemented by integrating the HIS module with an Infobutton Manager (IM). Since a different vendor may provide the latter, such integration is complicated by the lack of a standard application program interface (API) (see Figure 16-9).

In order to address this issue, the HL7 Clinical Decision Support Technical Committee has been developing a standard for infobutton APIs (Del Fiol 2004b). The goal of the proposed standard is to address the two problems previously described by defining a standard set of messages to support the communication between HISs and IMs, and between IMs and resources (see Figure 16-10).

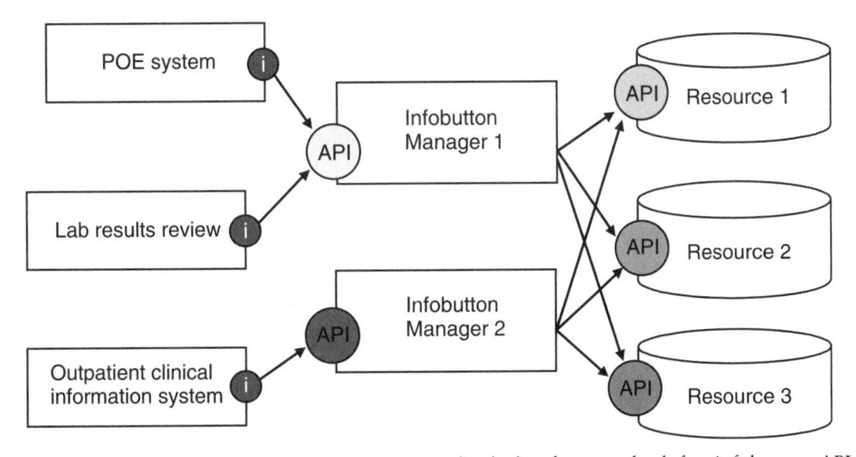

FIGURE 16-9 Current integration scenario: the lack of a standard for infobutton APIs requires the custom development of multiple interfaces among clinical information systems and infobutton managers and among infobutton managers and content resources.

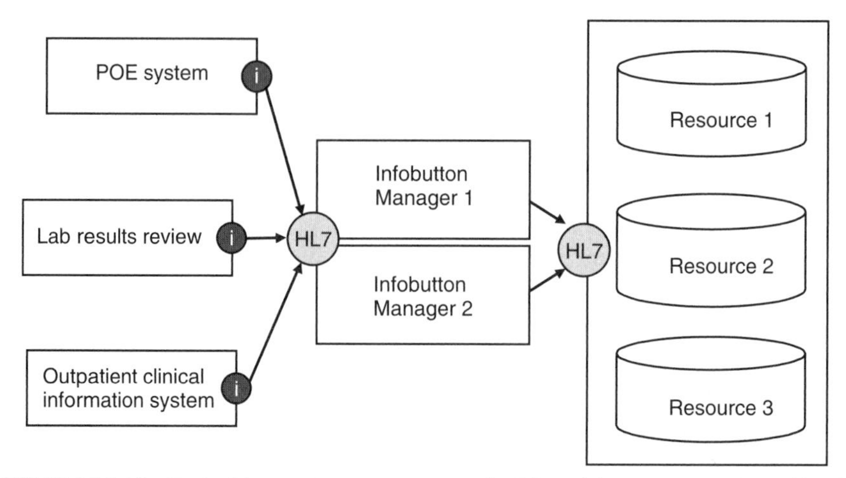

FIGURE 16-10 Desired integration scenario: a standard for infobutton APIs is adopted by the various parties involved in an infobutton transaction, simplifying the development and maintenance of infobuttons.

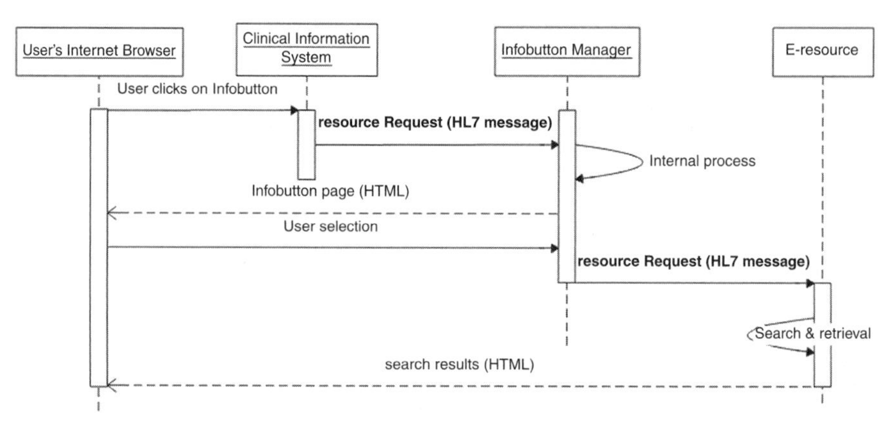

FIGURE 16-11 Sequence diagram depicting a typical HL7-compliant infobutton transaction involving a clinical information system, and infobutton manager, and a resource. Arrows indicate steps where HL7 messages are exchanged. The meanings of the arrows are described in the text.

In a typical HL7-compliant infobutton transaction (which may vary from implementation to implementation), the following steps are performed (see Figure 16-11):

1. A clinician using a clinical information system (e.g., order entry) clicks on an infobutton.
2. The clinical information system sends an HL7-compliant message to an infobutton manager, passing the main concept that the user is interested in (typically the concept next to the infobutton) and a set of parameters representing the context of the interaction between the user and the computer (e.g., patient age and gender, user role, care setting, module of the information system where the infobutton is located).
3. The infobutton manager receives the message, processes it, and returns an infobutton HTML page with a list of questions that are applicable to the context captured by the clinical information system and the clinical concept that the user is interested in.

4. The user selects one of the questions and a new request is submitted back to the infobutton manager, which in turn builds and sends an HL7-compliant message to a target resource.
5. The resource receives the HL7 message and runs a query against its content collection.
6. The resource returns an HTML page with the search results to the user's Internet browser. Although the current version of the HL7 proposal does not cover a standard response message from resources back to infobutton managers, such a message is expected to be part of a future version of the specification.

Implementations in which an infobutton manager is built into the clinical information system would not require HL7-compliant communication between those two components. A similar scenario applies when a resource provides an infobutton manager component.

The current version of the infobutton proposal defines a list of parameters (see Table 16-4) that will eventually become part of an HL7 version 3 message specification (Del Fiol 2004b). As part of this process, parameters will be associated with a vocabulary domain consisting of a value set or a controlled terminology.

With the aim of providing a stepwise transition and wider adoption of the standard by resources and clinical information systems, the following directives have been proposed: 1) human-readable representations of codes are required attributes; 2) clinical information systems and infobutton managers

TABLE 16-4 Brief description of the main parameters defined in the HL7 proposal for infobutton APIs.

Search parameter	
mainSearchConcept	The main clinical concept (e.g., medication, lab test result, disease) of the infobutton request. Typically, this is the concept displayed by the clinical information system next to an infobutton.
Modifier	Restricts the search criteria by specifying a subdomain of interest (e.g., indications, contraindications, dose) related to the mainSearchConcept. The modifier typically is selected by the user from an infobutton page that presents a list of questions (e.g., "what are the contraindications of digoxin").
Context	
TaskContext	The type of application the user is using, or the task that the user is trying to perform (e.g., order entry, lab results, problem list, medications list).
Age	Patient's age.
Gender	Patient's gender.
AssociatedCondition	Associated conditions that a given patient presents with in addition to the main search concept. This parameter can be used to further restrict the search criteria.
UserRole	Role of the user who initiated the infobutton request (e.g., MD, RN).
UserLanguage	Language of the user of the clinical information system.
CareSetting	Care setting that the patient is being cared for (e.g., inpatient, outpatient).
Content recipient	
RecipientLanguage	Language of the person who will be the final recipient of the content.
RecipientRole	Role of the person who will be the final recipient of the content (e.g., the patient).

should formulate requests as completely or fully-specified as possible to capture an optimal representation of the underlying context; and 3) resources should process the parameters that they can handle, ignoring the ones that cannot be processed by their search engines. Likewise, human-readable representations should be used whenever the resource has not implemented a particular code system conveyed in the HL7 message.

These directives allow applications to create fully specified requests that will execute on any HL7-compliant resource, without requiring major modifications to the internal indexing and content structure of resources. Although such modifications are desired in the long term, they should not be an impediment for a wider adoption of the standard and consequently a higher level of interoperability.

16.7 STANDARDS FOR INFORMATION RESOURCES

Even with an established standard for evoking infobutton managers, there would remain several other integration issues. The first is that online resources provide a variety of methods for handling information requests. Consequently, links provided by the infobutton manager must resort to a variety of methods and tricks to automate the retrieval of answers to questions (Cimino 2004). A second problem is that information resources, by and large, do not recognize the controlled terminologies used in clinical information systems. As a result, clinical data must be translated into forms recognizable by resources which, in most cases, do not themselves even use controlled terminologies (Cimino 2005).

Researchers have long called for the conversion of information resources into forms that could be readily integrated into clinical information systems (Cimino 1991; Greenes 1991). The advent of the World Wide Web has provided a partial solution, by allowing disparate systems to communicate through the standard of Uniform Resource Locators (URLs) to integrate Web-based guidelines into Web-based clinical systems (Cimino 1995), but this has solved only the syntax issue, not the semantics or terminology issues.

In order to address the semantics issues, researchers are turning to XML. In particular, XML is proving to be a useful way to mark up guidelines in order to facilitate access to specific, relevant parts of large guidelines. For example, the ActiveGuidelines project, described earlier, was accomplished by marking up sections of Web-based guidelines with tags that identified specific tests and treatments discussed in the guidelines. A clinical information system could then retrieve the parts of the guideline that were relevant to a particular order being considered by a user by searching for the relevant tag (Tang 2000). The tags succeeded in representing the guideline semantics, but they were not standardized.

In a second project, researchers at Intermountain Health Care marked up guidelines in a manner similar to that of Tang. However, they adopted three different standard XML-based document models (one each from the United Kingdom, the United States, and Germany) to represent guidelines in formats that were not only semantically useful, but also standardized (Eduljee 2003). These guidelines in turn have been integrated with XML-based order sets in a clinician order entry system (Del Fiol 2005).

Another challenge encountered when dealing with disparate information resources is that each resource generally is geared not to providing specific answers to specific questions (e.g., "What is the treatment for X?") but rather to providing information "chunks" that may or may not contain the answer (e.g., "Here are some documents that discuss the treatment of X; the answer may be in here someplace"). Information resource providers are starting to recognize the need for interfaces to their products that provide answers, rather than documents about the answers. With the emergence of the HL7 Infobutton Manager standard, and the intent by HL7 to develop standards for these question-specific interfaces, there is some impetus for content providers to begin to develop solutions.

The foremost example of an Infobutton Manager-accessible, question-specific interface is the InfoButton Access, currently being made available in product form from Micromedex. InfoButton Access provides APIs to Micromedex products that are question-specific; they accept as parameters specific topics (such as drugs or diseases) and then answer the specific question about the specific topic.[5]

16.8 CONCLUSION

The belief that better informed decisions lead to better patient outcomes is one of the underlying tenets of evidence-based medicine and, indeed, health care education. The integration of information resources into clinical information systems appears to be a viable approach to automated support of clinician decision-making. By lowering barriers to information at the moment when it is needed, the hope is that clinicians will use the information to make better-informed decisions. Though the impact on, and quality of, decisions are difficult to measure, we at least are seeing the increased access to information resources that we believe is a necessary (although not sufficient) step in the right direction.

Additional work is needed to adapt existing clinical information systems such that they are able to export context information for use in information retrieval, just as work is needed to adapt information resources to standardize on methods for facilitating that retrieval. However, communal efforts, such as those of HL7, are beginning to bear fruit.

RECOMMENDED RESOURCES

Cimino, J. J. (1996). Linking patient information systems to bibliographic resources. *Methods Inf Med* 35(2): 122–126.

A review of various methods used to link clinical systems to online resources.

Stead, W. W., Miller, R. A., Musen, M. A., and Hersh, W. R. (2000). Integration and beyond: Linking information from disparate sources and into workflow. *J Am Med Inform Assoc* 7(2): 135–145. Review.

A more recent review of integration work, with some in-depth examination of technologies.

Cimino, J. J. (2000). From data to knowledge through concept-oriented terminologies: Experience with the Medical Entities Dictionary. *J Am Med Inform Assoc* 7(3): 288–297.

This paper provides details on methods for translating clinical terms for use in information retrieval.

[5] http://www.thomson.connectthe.com/home/

Cimino, J. J., Li, J., Bakken, S., and Patel, V. L. (2002). Theoretical, empirical and practical approaches to resolving the unmet information needs of clinical information system users. *Proc AMIA Annu Fall Symp*, 170–174.

This paper provides more details on the Infobutton Manager.

Del Fiol, G., Rocha, R., and Cimino, J. J. (2004). HL7 Infobutton Standard API Proposal. Draft, 24 Feb 2004, <http://www.hl7.org/Library/Committees/dss/HL7-Infobutton-API-v2.2-20040224.doc>, accessed June 10, 2005.

This is the current draft HL7 specification for Infobutton Managers.

Del Fiol, G., Rocha, R. A., Washburn, J., Rhodes, J., Hulse, N., Bradshaw, R., and Roemer, L. K. (2004). "On-demand" access to a multi-purpose collection of best practice standards. *Proc of the 26th Annual International Conference of the IEEE ECMS* 3342–3345.

This paper provides more details on structuring content resources for infobutton access.

Hersh, W. R. (2003). *Information Retrieval: A Health and Biomedical Perspective*. New York: Springer.

This book includes extensive background material on clinicians' information needs.

ACKNOWLEDGMENTS

The authors would like to acknowledge Dr. Paul D. Clayton for his guidance on the Infobutton projects at both IHC and Columbia. Both authors are supported by Infobutton grant 1R01LM07593 from the National Library of Medicine.

REFERENCES

Allen, M., Currie, L. M., Graham, M., Bakken, S., Patel, V., and Cimino, J. J. (2003). The Classification of Clinicians' Information Needs While Using a Clinical Information System. In Musen, M. A., ed., *Proc AMIA Fall Symp* 26–30.

Chen, E. S. and Cimino, J. J. (2003). Automated discovery of patient-specific clinician information needs using clinical information system log files. In Musen, M. A., ed., *Proc AMIA Fall Symp* 145–149.

Chueh, H. and Barnett, G. O. (1997). "Just-in-time" clinical information. *Acad Med* **72**(6): 512–517.

Cimino, C. and Barnett, G. O. (1990). Standardizing access to computer-based medical resources. In Miller, R. A., ed. *Proceedings of the Fourteenth Annual Symposium on Computer Applications in Medical Care* 33–37. Washington, D.C.

Cimino, J. J. and Sengupta, S. (1991). IAIMS and UMLS at Columbia-Presbyterian Medical Center. *Med Decis Making* **11**(4 Suppl): S89–93.

Cimino, J. J., Johnson, S. B., Aguirre, A., Roderer, N., and Clayton, P. D. (1992). The Medline Button. In Frisse, M. E., ed., *Proceedings of the Sixteenth Annual Symposium on Computer Applications in Medical Care,* Baltimore, MD, 81–85. New York: McGraw-Hill.

Cimino, J. J., Aguirre, A., Johnson, S. B., and Peng, P. (1993). Generic queries for meeting clinical information needs. *Bull Med Libr Assoc* **81**(2): 195–206.

Cimino, J. J., Socratous, S. A., and Clayton, P. D. (1995). Automated guidelines implemented via the world wide web (poster). In Gardner, R. M., ed., *Proceedings of the Nineteenth Annual Symposium on Computer Applications in Medical Care,* New Orleans, LA, 941. Philadelphia: Hanley & Belfus.

Cimino, J. J. (1996). Linking patient information systems to bibliographic resources. *Methods Inf Med* **35**(2): 122–126.

Cimino, J. J., Elhanan, G., and Zeng, Q. (1997). Supporting infobuttons with terminological knowledge. *Proc AMIA Annu Fall Symp* 528–532.

Cimino, J. J., Li, J., Bakken, S., and Patel, V. L. (2002). Theoretical, empirical and practical approaches to resolving the unmet information needs of clinical information system users. *Proc AMIA Annu Fall Symp* 170–174.

Cimino, J. J., Li, J., Graham, M., Currie, L. M., Allen, M., Bakken, S., and Patel, V. (2003). Use of online resources while using a clinical information system. In, Musen, M. A., ed., *Proc AMIA Fall Symp* 175–179.

Cimino, J. J., Li, J., Allen, M., Currie, L. M., Graham, M., Janetzki, V. et al. (2004). Practical considerations for exploiting the world wide web to create infobuttons. *Proc Medinfo* 277–281.

Covell, D. G., Uman, G. C., and Manning, P. R. (1985). Information needs in office practice: Are they being met? *Ann Intern Med* **103**(4): 596–599.

Cullen, R. J. (2002). In search of evidence: family practitioners' use of the Internet for clinical information. *J Med Libr Assoc* **90**(4): 370–379.

Currie, L. M., Graham, M., Allen, M., Bakken, S., Patel, V., and Cimino, J. J. (2003). Clinical information needs in context: An observational study of clinicians while using a clinical information system. In Musen, M. A., ed., *Proc AMIA Fall Symp* 190–194.

Del Fiol, G., Rocha, R. A., Washburn, J., Rhodes, J., Hulse, N., Bradshaw, R., and Roemer, L. K. (2004a). "On-demand" access to a multi-purpose collection of best practice standards. *Proc of the 26th Annual International Conference of the IEEE ECMS* 3342–3345.

Del Fiol, G., Rocha, R., and Cimino, J. J. (2004b). *HL7 Infobutton Standard API Proposal*. Draft, 24 Feb 2004, <http://www.hl7.org/Library/Committees/dss/HL7-Infobutton-API-v2.2-20040224.doc>, accessed June 10.

Del Fiol, G., Rocha, R. A., Bradshaw, R. L., Hulse, N. C., and Roemer, L. K. (2005). An XML model that enables the development of complex order sets by clinical experts. *IEEE Trans Inf Technol Biomed*, inpress.

Detmer, W. M., Barnett, G. O., and Hersh, W. R. (1997). MedWeaver: integrating decision support, literature searching, and Web exploration using the UMLS Metathesaurus. *Proc AMIA Annu Fall Symp* 490–494.

Eduljee, A. and Rocha, R. A. (2003). Practical evaluation of clinical guideline document models. *AMIA Annu Symp Proc* 836.

Elhanan, G., Socratous, S. A., and Cimino, J. J. (1996). Integrating DXplain into a clinical information system using the world wide web. In Cimino, J. J., ed., *Proc Am Med Inform Assoc Annual Fall Symp* (formerly SCAMC), Washington, DC, 348–352. Philadelphia: Hanley & Belfus.

Ely, J. W., Burch, R. J., and Vinson, D. C. (1992). The information needs of family physicians: Case-specific clinical questions. *J Fam Pract* **35**(3): 265–269.

Ely, J. W., Osheroff, J. A., Ebell, M. H., Bergus, G. R., Levy, B. T., Chambliss, M. L., and Evans, E. R. (1999). Analysis of questions asked by family doctors regarding patient care. *BMJ* **319**(7206): 358–361.

Ely, J. W., Osheroff, J. A., Gorman, P. N., Ebell, M. H., Chambliss, M. L., Pifer, E. A., and Stavri, P. Z. (2000). A taxonomy of generic clinical questions: Classification study. *BMJ* **321**: 429–432.

Dee, C. and Blazek, R. (1993). Information needs of the rural physician: A descriptive study. *Bull Med Libr Assoc* **81**(3): 259–264.

Ely, J. W., Osheroff, J. A., Chambliss, M. L., Ebell, M. H., and Rosenbaum, M. E. (2005). Answering physicians' clinical questions: obstacles and potential solutions. *J Am Med Inform Assoc* **12**(2): 217–224.

Fischer, G. and Ostwald, J. (2001). Knowledge management: Problems, promises, realities, and challenges. *IEEE Intell Syst* 60–72.

Forsythe, D. E., Buchanan, B. G., Osheroff, J. A., and Miller, R. A. (1992). Expanding the concept of medical information: An observational study of physicians' information needs. *Comput Biomed Res* **25**(2): 181–200.

Fuller, S. S., Ketchell, D. S., Tarczy-Hornoch, P., and Masuda, D. (1999). Integrating knowledge resources at the point of care: opportunities for librarians. *Bull Med Libr Assoc* 87(4): 393–403.

Gosling, A. S., Westbrook, J. I., and Spencer, R. (2004). Nurses' use of online clinical evidence. *J Adv Nurs* **47**(2): 201–211.

Green, M. L., Ciampi, M. A., and Ellis, P. J. (2000). Residents' medical information needs in clinic: Are they being met? *Am J Med* **109**(3): 218–223.

Greenes, R. A. (1991). A "building block" approach to application development for education and decision support in radiology: Implications for integrated clinical information systems environments. *J Digit Imaging* **4**(4): 213–225.

Hales, J. W., Low, R. C., and Fitzpatrick, K. T. (1993). Using the Internet Gopher protocol to link a computerized patient record and distributed electronic resources. In Safran, C., ed., *Proceedings of the Seventeenth Annual Symposium on Computer Applications in Medical Care*, Washington, DC, 621–625. New York: McGraw-Hill.

Hersh, W., Pentecost, J., and Hickam, D. (1996). A task-oriented approach to information retrieval evaluation. *J Am Soc Inf Sci* **47**: 50–56.

Hersh, W. R. and Hickam, D. H. (1998). How well do physicians use electronic information retrieval systems? *JAMA* **280**(15): 1347–1352.

Hersh, W. (1999). "A world of knowledge at your fingertips": The promise, reality, and future directions of on-line information retrieval. *Acad Med* **74**(3): 240–243.

Huff, S. M., Rocha, R. A., McDonald, C. J., De Moor, G. J., Fiers, T., Bidgood, W. D. Jr. et al. (1998). Development of the Logical Observation Identifier Names and Codes (LOINC) vocabulary. *J Am Med Inform Assoc* **5**(3): 276–292.

Hulse, N. C., Rocha, R. A., Del Fiol, G., Bradshaw, R. L., Hanna, T. P., and Roemer, L. K. (2005). KAT: A flexible XML-based knowledge authoring environment. *J Am Med Inform Assoc* [Epub ahead of print].

Khedr, M. and Karmouch, A. (2004). Negotiating context information in context-aware systems. *IEEE Intell Syst* 21–29.

Lomax, E. C. and Lowe, H. J. (1998). Information needs research in the era of the digital medical library. *Proc AMIA Symp* 658–662.

Loonsk, J. W., Lively, R., TinHan, E., and Litt, H. (1991). Implementing the Medical Desktop: tools for the integration of independent information resources. In Clayton, P. D., ed., *Proceedings of the Fifteenth Annual Symposium on Computer Applications in Medical Care*, Washington, DC, 574–577.

Lucas, B. P., Evans, A. T., Reilly, B. M., Khodakov, Y. V., Perumal, K., Rohr, L. G. et al. (2004). The impact of evidence on physicians' inpatient treatment decisions. *J Gen Intern Med* **19**: 402–409.

Magrabi, F., Coiera, E. W., Westbrook, J. I., Gosling, A. S., and Vickland, V. (2005). General practitioners' use of online evidence during consultations. *Int J Med Inform* **74**(1): 1–12.

McColl, A., Smith, H., White, P., and Field, J. (1998). General practitioners' perceptions of the route to evidence based medicine: a questionnaire survey. *BMJ* **316**: 361–365.

Maviglia, S. M., Strasberg, H. R., Bates, D. W., and Kuperman, G. J. (2003). KnowledgeLink update: Just-in-time context-sensitive information retrieval. *AMIA Annu Symp Proc* 928.

McGowan, J. J., Overhage, J. M., Barnes, M., and McDonald, C. J. (2004). Indianapolis I3: The third generation Integrated Advanced Information Management Systems. *Bull Med Libr Assoc* **92**(2): 179–187.

Miller, R. A., Gieszczykiewicz, F. M., Vries, J. K., and Cooper, G. F. (1992). CHARTLINE: Providing bibliographic references relevant to patient charts using the UMLS Metathesaurus Knowledge Sources. In Frisse, M. E., ed., *Proceedings of the Sixteenth Annual Symposium on Computer Applications in Medical Care*, Baltimore, MD, 86–90. New York: McGraw-Hill.

Osheroff, J. A., Forsythe, D. E., Buchanan, B. G., Bankowitz, R. A., Blumenfeld, B. H., and Miller, R. A. (1991). Physicians' information needs: Analysis of questions posed during clinical teaching. *Ann Intern Med* **114**(7): 576–581.

Osheroff, J. A. and Bankowitz, R. A. (1993). Physicians' use of computer software in answering clinical questions. *Bull Med Libr Assoc* **81**(1): 11–19.

Porcelli, P. J. and Lobach, D. F. (1999). Integration of clinical decision support with on-line encounter documentation for well child care at the point of care. *Proc AMIA Ann Symp* 599–603.

Powsner, S. M., Riely, C. A., Barwick, K. M., Morrow, J. S., and Miller, P. L. (1989). Automated bibliographic retrieval based on current topics in hepatology: Hepatopix. *Comput Biomed Res* **22**: 552–564.

Powsner, S. M. and Miller, P. L. (1991). From patient records to bibliographic retrieval: A Meta-1 front-end. In Clayton, P. D., ed., *Proc Fifteenth Ann Symp Comp Appl Med Care*, Washington, DC, 526–530.

Powsner, S. M. and Miller, P. L. (1992). Automated online transition from the medical record to the psychiatric literature. *Methods Inf Med* **31**(3): 169–174.

Pratt, W. and Sim, I. (1995). Physician's information customizer (PIC): Using a shareable user model to filter the medical literature. *Medinfo* **8**(Pt 2): 1447–1451.

Price, S. L., Hersh, W. R., Olson, D. D., and Embi, P. J. (2002). SmartQuery: Context-sensitive links to medical knowledge sources from the electronic patient record. *Proc AMIA Symp* 627–631.

Reichert, J. C., Glasgow, M., Narus, S. P., and Clayton, P. D. (2002). Using LOINC to link an EMR to the pertinent paragraph in a structured reference knowledge base. *Proc AMIA Symp* 652–656.

Rosenbloom, S. T., Geissbuhler, A. J., Dupont, W. D., Giuse, D. A., Talbert, D. A., Tierney, W. M. et al. (2005). Effect of CPOE user interface design on user-initiated access to educational and patient information during clinical care. *J Am Med Inform Assoc* **12**(4): 458–473.

Rousseau, N., McColl, E., Newton, J., Grimshaw, J., and Eccles, M. (2003). Practice based, longitudinal, qualitative interview study of computerised evidence based guidelines in primary care. *BMJ* **326**: 314.

Ruan, W., Burkle, T., and Dudeck, J. (2000). An object-oriented design for automated navigation of semantic networks inside a medical data dictionary. *Artif Intell Med* **18**(1): 83–103.

Stead, W. W., Miller, R. A., Musen, M. A., and Hersh, W. R. (2000). Integration and beyond: Linking information from disparate sources and into workflow. *J Am Med Inform Assoc* **7**(2): 135–145. Review.

Tang, P. C. and Young, C. Y. (2000). ActiveGuidelines: Integrating Web-based guidelines with computer-based patient records. *Proc AMIA Symp* 843–847.

Tang, P. C., Black, W., Buchanan, J., Young, C. Y., Hooper, D., Lane, S. R. et al. (2003). PAMFOnline: Integrating EHealth with an electronic medical record system. *Ann Symp Proc/AMIA Symp* 649–653.

Thompson, C., McCaughan, D., Cullum, N., Sheldon, T., Mulhall, A., and Thompson, D. (2001). Research information in nurses' clinical decision-making: what is useful? *J Adv Nurs* **36**: 376–388.

Westbrook, J. I., Gosling, S., and Coiera, E. (2004). Do clinicians use online evidence to support patient care? A study of 55,000 clinicians. *J Am Med Inform Assoc* **11**: 113–120.

Westbrook, J. I., Coiera, E. W., and Gosling, A. S. (2005). Do online information retrieval systems help experienced clinicians answer clinical questions? *J Am Med Inform Assoc* **12**(3): 315–321.

Wong, E. T., Pryor, T. A., Huff, S. M., Haug, P. J., and Warner, H. R. (1994). Interfacing a stand-alone diagnostic expert system with a hospital information system. *Comput Biomed Res* **27**(2): 116–129.

Wyatt, J. C. (2000). Clinical questions and information needs. *J R Soc Med* **93**(4): 168–171.

Zeng, Q. and Cimino, J. J. (1997). Linking a clinical system to heterogeneous information resources. *J Am Med Inform Assoc* **4**(Suppl): 553–557.

17 THE ROLE OF STANDARDS: WHAT WE CAN EXPECT AND WHEN

ROBERT A. GREENES

17.1 KEY STANDARDS AND THEIR BENEFITS

We have discussed a variety of standards initiatives related to clinical decision support in the preceding chapters of this section of the book. At this point, in order to put those initiatives in perspective, it is useful to reflect on what the principal reasons are for interest in such standards. In other words, we ask the question, "How will such standards help in achieving the goal of wide dissemination and adoption of CDS?"

17.1.1 CDS Development with and without Standards

The key advantage that standardization can provide is the *ability to share and re-use knowledge* once it is created. Let us consider first what happens in the absence of a standards-based approach. We will then reflect on how standards might help. The discussion to follow is summarized in Table 17-1.

The first task, before CDS development is actually undertaken, is the creation of knowledge to be ultimately used as a basis for the CDS. In Section III we discussed three main classes of methodology utilized for generation and validation of clinical knowledge for CDS. In order to discover useful knowledge through such methodologies, specialized expertise, not only with respect to the medical domains being studied, but also in the application of the methodologies, is needed. Studies aimed at discovering knowledge, furthermore, are generally difficult and time-consuming to carry out. Typically, we learn about such results through their publication in the literature.

Having accomplished the goal of discovering a valid, and possibly important clinical relationship—that is, a unit of knowledge that might be applicable for CDS—then what? Typically the technology transfer process characterized by Balas and Boren (2000) kicks in at this point—some of the knowledge derived as a result of these efforts makes its way into publication, although considerably less of it becomes adopted in practice (14% of the original findings), and the adoption process itself occurs over a protracted period of time (average of 17 years). For example, a decision to apply specific knowledge to practice in a particular setting may depend on the presence of a

Clinical Decision Support: The Road Ahead

TABLE 17-1 Tasks involved in deploying knowledge in operational settings.

Task	Usual practice	How standards can help
Knowledge generation and validation	Researchers publish findings	Findings accessible via external repositories, including both positive and negative results
Useful knowledge identified	Primary or literature research locally, based on need or driven by local champion	Identification by specialty bodies, or other authoritative groups; knowledge bases organized by domains, purposes, and other attributes to facilitate access; local efforts use external knowledge bases as starting point
Adoption in practice for some knowledge	Review to decide whether it can be utilized in local environment	Recommendation and prioritization by authoritative bodies
Encoding in computer-executable form	Knowledge engineering to make the knowledge unambiguous and interpretable	Standards-based representation
Local adaptation	Modification based on local practices and constraints	External knowledge as starting point for customization, tools for doing adaptation
Implementation, debugging, and operational use of CDS	Integration into an application	Ability to implement via external execution engine; ability to provide modular CDS services; standard interfaces of information model to host data sources; standard mapping of results to actions to be carried out in host; standards-based invocation of CDS
Knowledge update	Changes identified and encoded	Updates received locally from external communal resource, reviewed for applicability
Incorporation of updates in applications	Applications using knowledge found and updated	Isolation of knowledge use, because of implementation via external CDS services or modules, facilitates update

local subject expert or champion who believes that using that knowledge will offer benefits, and who is thus motivated to take on the work.

When it comes to CDS, it is likely, though undocumented, that even more attrition of potentially useful knowledge occurs, given that some of the discovered knowledge may not be amenable to expression in computer-executable form, or suitable applications environments do not exist for its use, or data necessary for its execution are not accessible or practical to acquire.

If the assessment of usefulness of a knowledge resource makes it to this point, what typically happens next is that the knowledge is encoded and incorporated into CDS applications in academic medical center clinical IT systems or into clinical IT vendor products. However, this is not straightforward either. It has been shown that the process of encoding knowledge in executable form is not performed reliably, in that knowledge in the form of published results or narratives or even guidelines is subject to many ambiguities and differences in interpretation, resulting in often quite different renderings from the original. See, for example, an analysis of the process of assembling knowledge to create clinical practice guidelines (Peleg, Gutnik et al. 2006), and a study of differences in the encoding of published guidelines into the Guideline Interchange Format version 2 (GLIF2) (Ohno-Machado, Gennari et al. 1998). Still further effort is required to adapt guidelines, or other knowledge, to account for local practices and policies, processes, workflow, available resources (e.g., laboratory tests, imaging technologies, or specialized surgical expertise), and other constraints (e.g., relating to financial, personnel, or time limitations).

Thus the general inertia in the technology transfer process is compounded in the case of CDS adoption by the burdens and costs of rendering the knowledge into executable form, and adapting it for use in various application settings and on various platforms—not to mention building and debugging the applications, and deploying them operationally. Additional burden is associated with maintaining the knowledge. The greater the degree to which the knowledge is integrated into clinical applications, also, the more effort is required to modify instances of its use when update becomes necessary.

The onus of these processes and the protracted time frame involved thus provide the major impetus for pursuing the development of standards. It would be highly advantageous if many of these processes could be done only once for a given item of knowledge, and if the knowledge would thenceforth be widely available, so that it could be accessed and used by anyone who wanted to implement or use CDS functionality. It would also be desirable if the whole sequence of processes could be accelerated so that the benefits of applying the knowledge could be realized more rapidly.

But let us now examine in detail what needs to occur for these benefits to be realized. We begin with the knowledge generation and validation process. To reduce the time lag and the multiple points at which potentially useful discoveries get left by the wayside during the technology transfer process, it would be helpful if knowledge, once discovered, could be maintained in widely accessible repositories, as a starting point for use by other researchers. This should include not only positive but negative results, and should include annotations regarding reasons for nonapplicability when that is determined. The source of the knowledge, its authors and their credentials, its provenance, and other ways of assessing its quality should be annotated as well.

However, were such repositories to exist, the knowledge contained in them would not yet be in directly usable form. The principal target for standardization in CDS is the specification of well-defined, unambiguous *representations of knowledge*, for example, for the logic expressions in alerts, reminders, and drug order validation and interaction checking rules, and for the Medical Logic Modules (MLMs) that incorporate such logic, as discussed in Chapter 12, or for the depiction of the sequences of steps and process flow in clinical guidelines, which was the subject of Chapter 13. Other secondary targets for standardization in support of CDS are the *information models* for referencing data used by CDS and the *specifications for results* of CDS. Chapters 14 and 15 have focused on efforts at standardization of the information model: Chapter 14 discussed use of well-defined vocabularies and taxonomies for naming the concepts corresponding to the data elements, and Chapter 15 described the use of archetypes to define the attributes of the data elements necessary in particular contexts to make them truly unambiguous. Methods for *accessing* knowledge or for *invoking* CDS are also potential targets for standardization. The former may be through ontologies defining attributes of knowledge that can be used to organize and retrieve from the repositories items of certain types or for particular purposes. In Chapter 15, the role of ontologies was explored. Chapter 16 further elaborated on the retrieval aspects of this theme by discussing an emerging standard for an infobutton manager that can use such attributes to assemble resources from external knowledge bases. Another aspect of invocation is the prospect of a standard for a service-oriented architecture (SOA) method to initiate the performance of CDS by an external Web service. Such a proposed standard is in draft form at the time of this writing, as a joint project of the Clinical Decision Support Technical Committee and SOA Special Interest Group of HL7 (http://www.hl7.org). A discussion of Web services as a way of delivering CDS is presented in Chapter 23.

Many standardization efforts that bear on CDS beyond those covered in the previous chapters are also under way. Our omission of them does not suggest that those covered are more definitive. As should be apparent from the discussions in the preceding chapters of this section, the state of the standardization process for even the most well-developed and mature among the clinically relevant standards is still in considerable flux. Also, in the preceding discussion, we mentioned the need for standardizing the specification of the result produced by CDS so that it can be acted upon by a host environment. This focus has not been pursued actively, except for a few efforts (Essaihi, Michel et al. 2003; Tu, Musen et al. 2004) to define possible actions that can be initiated as a result of CDS (particularly focusing on guidelines). The use of organizing attributes and relations is being explored, such as for management of and access to Knowledge Element Groups (KEGs), as discussed in Chapter 15 (see also (Sordo, Hongsermeier et al. 2006)), but this work has not reached a level ready for standardization. Thus, although a variety of standards are needed to robustly support CDS, some of which are now well defined, many are still primitive or nonexistent. We will return to this issue in Section VII as we consider the road ahead.

If the needed standards were to exist, the expectation is that the tasks of implementing CDS could be more easily carried out. Tasks that still needed to be done locally would also be aided by standardization, partly because efforts could be confined to those tasks, and partly because methods and tools for modifying standard representations could be made available to aid in the tasks. Local adaptation would be based on a well-defined starting point, so that dependencies could be tracked. Implementation, debugging, and operational use of CDS could be localized to an external execution engine and to well-defined interfaces to the host system for invocation of CDS, data access, and result communication and mapping to actions. Knowledge management/update, and incorporation of updates in applications could be done more easily by relying on a central source for authoritative knowledge, and updating based on tracked relationships to the knowledge sources.

17.1.2 Beyond the Standards

The premise underlying standardization efforts is that having standards such as those identified earlier would stimulate sharing and reuse *because* it would enable a number of developments and changes in procedures. The following are some of these potential capabilities and the benefits that could be derived from them.

- Collections of discovered knowledge of various types could be made widely available in the form of knowledge bases. Evidence-based medicine repositories such as from the Cochrane Collaboration (Herxheimer 1993) are examples of this. If the knowledge generation and validation have been done by authoritative, respected experts, this would obviate the need to rely entirely on local experts for carrying out or redoing such efforts in each institutional setting or for each vendor-based system. That is, sites could adopt a set of authoritative rules or guidelines instead of having to develop them locally.
- The management of the knowledge bases under the aegis of external content provider entities (commercial, professional society, government, consortial, or other) would relieve local sites or vendor systems from having to undertake this task.
- Knowledge could be flexibly provided in a variety of ways. For example, it could be made available for access by end users or systems when needed (e.g., through Web interfaces). Alternatively, knowledge content could be provided by downloading and importing it into local environments.
- If the knowledge has been encoded into executable form by knowledge engineers and software engineers supported via the external provider, this effort also would not need to be redone in each setting. The knowledge might still need to be translated or adapted to local platform-specific representations and interfaces, but even this could be reduced to the extent that standards-based interfaces to data and application services were supported in the local platforms.
- Beyond translation and interfacing to a host platform, local efforts could be confined to customization and adaptation to local requirements and constraints.
- Updates to knowledge could be coordinated by the provider of knowledge, communicated to users, and details of provenance and versioning maintained. Although the process of updating the instances in which the knowledge is used in local settings would still be difficult, at least part of the burden of creating and tracking updates would be borne by the provider of the knowledge.
- Given economies of scale of effort that could be devoted to knowledge update, external knowledge bases would be likely to be kept more up-to-date and reliable than those that are developed or maintained locally.
- If suitable mechanisms existed for curation and management of the external knowledge bases, knowledge developed and created by local experts could also be uploaded and incorporated into those knowledge bases. This might serve to create a collaborative community for continuous knowledge base development and improvement.

It is thus important to recognize that, in order to stimulate sharing and reuse of knowledge, it is insufficient to just define standards for knowledge representations, interfaces, and modes of access and invocation. What must also occur is the creation of artifacts that use them, as well as models for their integration into the life cycles of knowledge generation, knowledge management, and CDS method development/implementation (see Chapter 1, subsection 1.4.5.1). Three principal classes of artifacts are needed for standards-based CDS:

1. **Knowledge bases.** We have referred to external knowledge bases earlier, but have not discussed under what aegis they come about, what the business models are for their ongoing support, how they are structured, their mode of access or interface, or how they are maintained and updated, These tasks are beginning to be addressed by knowledge management development projects at certain large academic medical centers, two examples of which are presented in Chapters 21 and 22. But the prospect of external knowledge bases being available on a broad scale is still not generally realized. We will return to this topic later.
2. **Tools for authoring and update.** Once we have an external standardized knowledge base of sufficient scale and utility for widespread use, it becomes feasible to invest in the development of a robust set of tools for authoring, review, editing, and publishing of knowledge of the types in the knowledge base. Further, to the extent that collaborative authoring and update are desirable, it is possible to provide other content management and collaboration capabilities that aid this process. For example, tools can be created to facilitate identification of similar knowledge to that which is being authored or modified, to provide a starting point for the work, and to detect potential inconsistencies, contradictions, redundancies, and gaps in the knowledge relating to a specific topic. Efforts to build and use such tools at large academic medical centers seeking to do their own knowledge management are reviewed in Section VI as well, but again there is little effort to provide generic capabilities such as these on a broad scale.
3. **Tools for execution.** Once we have standardized knowledge in executable form, it is feasible to consider the development of execution tools that will operate on such knowledge, and that can be invoked by host environments. Certainly execution can be done in a host-specific way, in terms of its degree of integration with the host platform, its databases, and its applications. But independent "execution engines" can alternatively be developed. The use of an external execution engine, however, requires the adoption of a means for interacting with various host environments in standard form, 1) so that the engine can be invoked by the host when specific CDS functionality is needed, 2) so the engine can obtain data from host EHRs, and 3) so the engine can return the results of its evaluations in order that appropriate actions can be initiated by the host systems. The impetus for developing such external engines would be the existence of standards-based knowledge bases that are suitable for use in CDS and the perceived benefits by medical center IT systems management and clinical IT system vendors to provide interfaces to them in exchange for not needing to implement versions of the CDS internally in the host platform or to maintain it. As noted previously, an example of an approach to implementing interfaces to external execution engines for CDS using Web services-oriented architectures is described in Chapter 23, and a proposal for a standard services invocation method is currently being developed in HL7.

17.2 HOW IMPORTANT ARE STANDARDS?

As compelling as the benefits of standardization appear to be for stimulation of knowledge sharing and reuse, we now pause to consider what evidence there is that this will make a substantial difference in the rate of dissemination and adoption of CDS. Although Chapters 1 and 8 have identified a number of barriers and areas of inertia that have impeded adoption of CDS, it is not immediately clear that the potential for access to external resources and reuse through standardization will have significant benefit in overcoming these barriers. Perhaps more important barriers are those that relate to the difficulties in adapting successful demonstrations for use in settings with different operational environments, practice styles, organizational approaches, incentives, and constraints. Those issues can certainly be addressed by having repositories of shared experiences, and a body of collective wisdom about how to go about the introduction of various forms of CDS in new environments. We have commented earlier (see Chapter 1) about the value of having a better scientific understanding of the human engineering and organizational strategies that are most effective. But would the availability of shared or reusable knowledge bases themselves play a significant role in accelerating adoption?

In truth, the case for the benefits of sharing and reuse as a driver is weak. There are not many instances one can point to where such sharing and reuse in fact has occurred. Perhaps the best examples are "commodity" compendia such as those containing drug formulary or drug–drug interaction data, which can be obtained from commercial vendors as well as some nonproprietary sources. These usually are provided in the form of simple tables that can be incorporated in relational databases in a host system and used in CPOE or other applications. Bibliographic database search (e.g., via PubMed) and other online resources such as collections of clinical guidelines (http://www.guidelines.gov) or clinical trials (http://www.clinicaltrials.gov) are also valuable (e.g., for access via an infobutton manager), although they do not provide sources of executable knowledge.

Sharing of Arden Syntax Medical Logic Modules (MLMs) is possible. Approximately 240 MLMs have been made publicly available on the Arden Syntax Web site (http://cslxinfmtcs.csmc.edu/hl7/arden/) by Columbia-Presbyterian Medical Center, many if not all of which were provided more than 10 years ago and have not been updated. No contributions from others have been provided on this Web site, and there appear to be no other sites that currently share MLMs except within the user groups of particular vendors.

A consortium of vendors, academic groups, and professional specialty society representatives known as the Institute for Medical Knowledge Implementation (IMKI) was formed in 2001 with the goal of creating and jointly contributing to a shared pool of knowledge resources such as MLMs, but after two to three years, it foundered for lack of commitment by participants to making knowledge content available. OpenClinical, founded by the Advanced Computation Laboratory, Cancer Research UK, is an international organization intended, according to its Web site (http://openclinical.org/) "to promote awareness and use of decision support, clinical workflow and other advanced knowledge management technologies for patient care and clinical research." Goals include disseminating development tools and techniques for CDS, but the site offers no compendium of executable knowledge resources, at least at present.

The upshot of this is that even a convincing existence proof for the value of shared, standards-based knowledge resources is largely lacking. Nonetheless, it feels right to many experts that having such resources would be valuable, for reasons such as those listed earlier. We will further consider prospects for realizing those potential advantages in Section VII, as we consider the road ahead.

Note also that the main purpose for standards development to date in other areas has not been for sharing of external resources but for interoperability of systems, to facilitate the processing of transactions. This focus on transactions has made economic sense in that standards for messaging and transfer of data enable disparate systems and applications to communicate and cooperate as part of a value chain, whether for business or clinical purposes. When one considers sharing of external resources as a driver, the questions that come quickly to the top concern who owns, maintains, and takes responsibility for the resource, what the business model is for sharing and supporting of it, and how it adds value to participants. Such questions need to be addressed in pursuing the goal of standardization of knowledge resources.

A further concern is that the knowledge bases, tools for authoring and editing, and tools for execution rely on the formation of initiatives, either in the public or private realm, that will organize around the objectives of building and supporting these artifacts, in order to ensure that they are viable and robust. Although their desirability is clear, will this actually occur? There are two nonmutually exclusive scenarios under which this would be most likely to happen. First, efforts such as those by academic medical centers in knowledge management may engender confidence in the approach, leading them to scale up and partner to replicate such capabilities outside of the walls of those centers. Alternatively, new ventures or initiatives might form to do similar activities. Stakeholders in either of these scenarios might include academic medical centers, professional specialty organizations, government agencies, standards development organizations, payers, and IT vendors. The initiatives might be supported through government or foundation funding, contributions of consortia participant organizations, or business investment. Yet, of course, there is no way to tell how likely or when such initiatives will occur.

Thus be warned. What we have posited as advantages largely have been unproven, and the questions raised remain mostly unanswered. What follows is thus firmly in the category of conjecture.

Even though the role of standardization efforts in this realm is to define the representation of knowledge, interfaces to it, organizing schemas, and invocation methods—*not* to create the repositories, authoring/editing tools, and execution tools—nonetheless, I believe that by having such standards, efforts such as we have described earlier, devoted to developing the repositories and tools, are more likely to be undertaken. Entire ecosystems built around a standard, such as for knowledge representation, have generally not occurred to date, but the possibility is intriguing. This is increasingly possible, given the ability to create self-contained Web services or API-invoked units of functionality as modular components.

A consequence of modularity is that the barriers to development of a component or entry into a marketplace through a Web service are greatly reduced. Without standards, we have the usual lack of critical mass and focus that prevents forward movement. Having the repositories and tools, even in rudimentary form, could stimulate refinement of them, and could facilitate explorations of new opportunities for creation of value, in terms of means for carrying out knowledge generation and validation activities, managing knowledge resources, and delivering knowledge content to users or applications that need it. This may involve new organizational and business models devoted to dissemination and reuse of shared knowledge content, involving modes of collaboration and commercial development not previously possible.

Would creation of standards-based publicly available knowledge bases, seeded with significant initial content, and provision of open-source tools for authoring and editing actually stimulate use? Would it stimulate multiparty collaboration and refinement of the knowledge as well as of the tools? Would the existence of standards, knowledge bases, and tools prompt systems managers to provide interfaces and means of using such knowledge in their systems? Would clinical IT system vendors also provide such interfaces? Would demand stimulate the creation of a commercial marketplace of content, tool, and service providers that can provide added value to clinical systems? These are all intriguing possibilities that could provide sufficient motivation to proceed in exploring how to accelerate such activity. Whether that will or should occur is a decision that stakeholders will need to make. We will return to this in Section VII.

REFERENCES

Balas, E. A. and Boren, S. A. (2000). Managing clinical knowledge for health care improvement. *Yearbook of Medical Informatics*. Van Bemmel, J. and McCray, A. T. Stuttgart, Schattauer Verlagsgesellschaft mbH: 65–70.

Essaihi, A., Michel, G. and Shiffman, R. N. (2003). Comprehensive categorization of guideline recommendations: creating an action palette for implementers. *AMIA Annu Symp Proc* 220–224.

Herxheimer, A. (1993). The Cochrane Collaboration: making the results of controlled trials properly accessible. *Postgrad Med J* **69**(817): 867–868.

Ohno-Machado, L., Gennari, J. H., Murphy, S. N., Jain, N. L., Tu, S. W., Oliver, D. E., et al. (1998). The guideline interchange format: a model for representing guidelines. *J Am Med Inform Assoc* **5**(4): 357–372.

Peleg, M., Gutnik, L. A., Snow, V. and Patel, V. L. (2006). Interpreting procedures from descriptive guidelines. *J Biomed Inform* **39**(2): 184–195.

Sordo, M., Hongsermeier, T., Kashyap, V. and Greenes, R. A. I., *Germany, 2006* (2006). Partners Healthcare order set schema: An information model for management of clinical content. *Computational Intelligence in Healthcare*. Berlin, Springer-Verlag, in press.

Tu, S. W., Musen, M. A., Shankar, R., Campbell, J., Hrabak, K., McClay, J., et al. (2004). Modeling guidelines for integration into clinical workflow. *Medinfo* **11**(Pt 1): 174–178.

V

ORGANIZATIONAL, BUSINESS, AND SOCIAL CHALLENGES

18 ORGANIZATIONAL AND CULTURAL CHANGE CONSIDERATIONS

JOAN S. ASH

This chapter addresses organizational and cultural impediments to widespread clinical decision support adoption. It describes how the type of hospital or health care institution may influence acceptance and therefore implementation strategies concerning decision support, how issues of organizational and personal control and autonomy are associated with decision support, how different stakeholders view decision support, and how an analysis can be done to assist development of implementation and maintenance strategies.

18.1 INTRODUCTION

Clinical decision support (CDS) is a comprehensive term that could include anything from paper order sets and guidelines to computerized alerts, but in this chapter, use of the term will be limited to computer-based "passive and active referential information as well as reminders alerts, and guidelines" (Bates et al. 2003). The definition includes only computerized CDS because the organizational and individual behavior issues of computer-based CDS are different from paper-based decision support. The issues become more complex when information technology enters the picture: this is in part due to the fact that, in addition to the many governance and prioritization issues of traditional decision support, there are interface and access issues specific to information technology.

Clinicians are highly trained, intelligent individuals whose medical education emphasizes clinical decision-making. The practice of medicine *is* medical decision-making. Any implication that physicians need help with decision-making may be taken as a subtle suggestion that their professional skills are lacking. Efforts to provide CDS might be seen as a threat, generating a natural fight-or-flight response. On a rational level, clinicians use, value, and welcome information resources they feel they need, but on a more emotional level, they may resent it when information they may not feel they need is not only thrust upon them, but may force them to waste time better spent in other endeavors.

The implementation process by organizations for CDS programs might best be viewed as a delicate balancing act. Acceptance of CDS depends a great deal on the type of decision support, the reason for its use, how good it is, and the value the clinician places on it at any particular point in time. With all

Clinical Decision Support: The Road Ahead

these variables changing over time, and with technology further complicating them, the challenge of managing CDS systems is difficult and never-ending. Given the potential for better patient care, however, the effort must be made. The good news is that some organizations have now had considerable experience implementing these systems, and there are lessons we can learn from them. Unfortunately, organizational issues in decision support have not been systematically studied and little has been published about them. There are a few papers offering guidance based on experience (Bates et al. 2003; Feldstein et al. 2004), and there is an implementation manual outlining an ideal process (Osheroff et al. 2005). There are studies about the organizational aspects of implementing clinical guidelines (Davis and Taylor-Vaisey 1997; Cabana et al. 1999; Solberg et al. 2000; Trivedi et al. 2002) and issues related to electronic prescribing with decision support (Miller et al. 2005). These sources have been drawn upon for this chapter. In addition, qualitative research by the author's interdisciplinary research team over the past seven years, while focused on computerized physician order entry (CPOE) in particular, has uncovered numerous decision support themes. Through observations and interviews at ten sites, we have learned a good deal about how the balancing act has been conducted and what has worked best (Ash et al. 2003, 2005).

This chapter will outline the organizational issues related to CDS and suggest strategies for addressing them within a framework for transition management that has been used with success. Although the primary focus here is CDS in the inpatient environment, similar principles apply to ambulatory and office practice settings.

18.1.1 Framework for Addressing Organizational Change and Transitions

Changing physician behavior involves changing organizational behavior. Decision support, as part of a larger CPOE implementation effort, involves everyone in the organization, not just the clinicians. As the Institute of Medicine study on safety asserts, organizational culture must be imbued with a sense of safety, and the focus should be on the total systems approach rather than on individual blame (Institute of Medicine Committee 2003). Garside provides further description of the organizational change needed for safety (Garside 1998).

Changing an organization's culture is usually a gradual process. In the classic change theory outlined by Lewin (1951), change happens in three phases, when something that has been in place gets unfrozen, changes, and freezes again. This does not offer informatics professionals a useful framework, however, because information systems are in constant flux with continuous improvements and in an ongoing slushy, rather than frozen, state. A more useful framework is that outlined by Bridges (2003) and depicted in Figure 18-1. Also a three phase model, it progresses from 1) an ending, to 2) a neutral zone, to 3) a new beginning, and keeps recycling as people progress through a psychological readjustment. Although it may seem odd to begin with an ending, it makes sense when one thinks about major life events such as a death in the family or a job change. Bridges offers strategies for dealing with each phase, such as a definite marking of the ending (a goodbye) and identification of who is losing what. The neutral zone is a phase people go through at different rates while they are adjusting, and communication and education become important during this phase. Finally, the new beginning marks a new mental state, a point at which hands-on training, for example, might be most

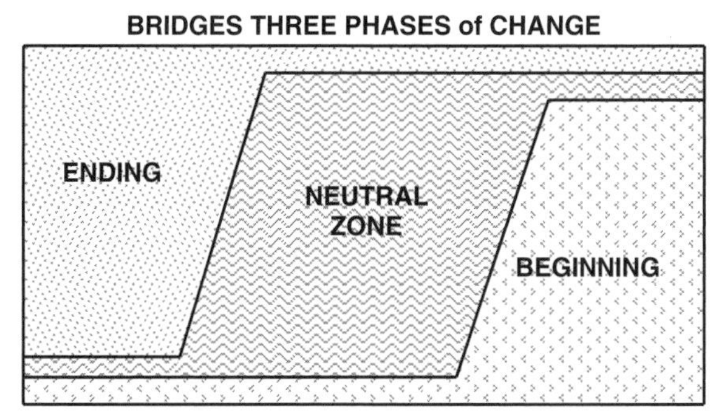

FIGURE 18-1 Bridges' three phases of transitions.

effective. The framework helps to explain why implementation of clinical systems is accepted at different rates by different people and why continuous communication, education, and training need to be planned.

18.1.2 Identifying the Barriers and Facilitators for Implementing CDS Systems

For our purposes, a barrier is anything that hinders acceptance of CDS system guidance.

Cabana et al. (1999) outlined seven barriers to paper-based clinical practice guideline adherence identified in reviewed studies: lack of awareness, lack of familiarity, lack of agreement, lack of self-efficacy (lack of confidence in one's own ability, to provide smoking cessation counseling, for example), lack of outcome expectancy, inertia of previous practice, and external barriers. External barriers include lack of time and support staff. These same barriers might very well apply to computer-based CDS systems, with the addition of numerous technology barriers. These might include confusion because of poor screen design, lack of computer skills, lack of direction due to unclear instructions about how to proceed, inaction because of mistimed interventions (at the end of an ordering session rather than at the time an order is entered), lack of trust in the quality of the data (about the patient, for example) or in the recommendation itself, or even emotional barriers such as annoyance or rage that may cloud thinking (Sittig et al. 2005).

Barriers differ across settings, so an analysis needs to be done prior to an organizational decision support effort. One study of expert opinion about (presumably paper) guideline implementation notes that variables related to organizational characteristics are more important than either guideline characteristics or the external environment (Solberg et al. 2000). Multiple strategies for addressing a host of factors, related to the guidelines themselves, the practice organization, and the external environment, are needed.

18.1.3 Stakeholder Analyses and Lewin's Force Field Analysis as Useful Techniques

An early step in the Bridges model is identifying who stands to gain or lose in any transition effort. A stakeholder analysis is also the first step recommended in the Implementers' Workbook (Osheroff et al. 2005). This includes identification not only of individual groups of clinicians, information technology personnel, quality assurance personnel, and administration, but also existing

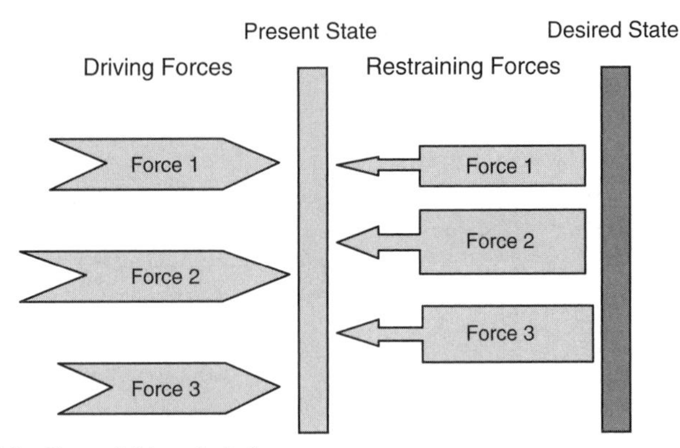

FIGURE 18-2 Force field analysis format.

committees. Such an analysis will uncover groups that can be targeted with change management efforts, and it will also help identify what the barriers are.

One aspect of the Lewin change model that can be applied to informatics implementation projects is Force Field Analysis, depicted in Figure 18-2. By first identifying the forces that facilitate the change or transition and those that provide barriers, and then by outlining strategies that can strengthen the facilitating forces and/or weaken the barriers, plans for reaching the desired state can be made. The following sections will describe barriers and facilitators to clinical systems implementation that can be addressed in a force-field analysis, issues that need to be considered in planning efforts, and those involved in developing and maintaining CDS systems.

18.2 ORGANIZATIONAL ISSUES RELATED TO CLINICAL DECISION SUPPORT

18.2.1 Differences among Kinds of Organizations and Cultures

Teaching hospitals differ from nonteaching (community) hospitals in a number of respects when considering decision support. In most teaching hospitals, physicians still in training, interns, and residents, enter the orders (Ash et al. 1999). In some, other individuals in the hierarchy such as higher-level house officers and attending/faculty physicians, in fact, are discouraged from entering orders. Therefore, the person entering the order and receiving decision support assistance is still at a learning stage of development. Interns are often quite grateful for any help that an information system can offer. As they gain in experience, they can become annoyed by alerts about things they already know, however. Teaching hospitals with high levels of decision support capability tend to provide everything they can at every "tier" of severity (mild, moderate, severe), independent of who the recipient is. However, it might be preferable for a system to decide what to provide depending on the level of training the clinician has. In nonteaching hospitals, clinicians are already trained and less tolerant of annoying interruptions. Most systems have the ability to alert only for the most critical reasons, so with a three-tiered system of alerts where the third tier includes only extremely urgent messages, only the third tier would be activated in a community hospital. For example, in teaching hospitals, all medication alerts might go directly to the physician. In a community hospital, the pharmacy might screen alerts first so that they are filtered and the physician gets fewer medication alerts to avoid fatigue and annoyance.

The rate of implementation of decision support functionality might be different for nonteaching hospitals. Community hospitals may want to progress slowly from reactive decision support (a severe allergy alert) to suggestive types of decision support (corollary orders the physician might consider in addition to the order just entered) to full clinical pathways and protocols. There appear to be psychological differences in receptivity to the three types of interventions.

Governance and control issues are somewhat different in the two types of hospitals, which has an impact on decision support. Whereas control issues arise in teaching hospitals with efforts at cutting costs that force physicians to behave in ways the administration or quality assurance committees want them to, in the community hospital, administrators who may have the same desires probably need to tread more gently. This implies that even greater attention needs to be paid to the organizational issues involved in planning and careful implementation and maintenance.

Clinical pathways that are developed in a multidisciplinary manner involving nurses and others and outlining treatment day by day may actually be easier to develop in a community hospital than an academic center, because of the stability of staff and because there are fewer clinicians and therefore fewer egos involved. In community hospitals, clinical pathways are appreciated by everyone on the health care team. When decision support can make larger numbers of staff grateful, this positive attitude can be infectious.

There is some debate about whether certain kinds of decision support such as order sets undermine the educational efforts in teaching hospitals. Most experts agree that in some ways decision support undermines education, because clinicians-in-training do not need to learn the same details they might in a paper environment, due to the ability to depend on the information system to provide help. On the other hand, residents seem to feel that these systems help them to do the right thing and that the systems will catch mistakes. By alerting the resident to potential mistakes, he or she learns without causing harm.

18.2.2 Issues of Control, Autonomy, and Trust

A stakeholder analysis and determination of who stands to lose or gain is useful for identifying threats to control, power, and autonomy related to decision support. Those highest in the organization see decision support as a way to modify physician behavior so that the organization can save money, be more efficient, or otherwise meet institutional goals. Those in the quality assurance arena see it as a way of gaining adherence to suggested guidelines, assuring regulatory compliance, changing clinician behavior, and tracking and monitoring adherence. The effort is required by clinicians, but the benefits may be realized in these other aspects of the organization, not by the clinicians.

An organization may want data for benchmarking, but clinicians may view such tracking as Big Brother-ish. Clinicians need to trust those who are instituting the quality initiative. If a physician is tapped to take on a quality assurance role, he may be then viewed by his colleagues as a "company man" who has gone over to the other side. Organizations may in truth be implementing systems so that the hospital overall can become more efficient: for example, cost alerts may be given to control physician medication or test ordering patterns. Although the goal may be to control physician behavior, more gentle ways of doing it that suggest and guide rather than mandate may be more successful.

Implementation is easiest for CDS modules that clinicians do not care about very much and do not require extra time. For example, clinical guideline acceptance has been hampered by a sense that developers are not local and may have special interests they are promoting (Wendt et al. 2000). As a result, customization at the local level appears to be important for success (Waitman and Miller 2004). CDS systems often interfere with the workflow of clinicians, a workflow that may be highly individual. It is because of this individuality that personal order sets are so acceptable. Even though it takes time for the clinician to build his own order set, it gives him a sense of autonomy and control. This is also a reason why it is not difficult getting clinicians involved in building or modifying decision support entities: they feel that, if they can have an impact on the system, they can have greater control. Personal order sets are extremely popular with community physicians because they save them time and can become a sense of pride when others want to borrow them. Personal order sets might be considered early on to encourage community physicians to use CPOE, even though they cause maintenance problems later. Eventually, they may be able to be replaced by more general order sets used by everyone. As long as the personal order sets migrate into more general order sets so that those that the clinicians use often remain available to them, there should be little resistance.

Making anything mandatory may be more difficult in a community hospital. Physicians, who bring patients, and therefore business, to the hospital, can take their business elsewhere. In designing decision support interventions, then, care must be taken that there be as much flexibility as possible. Also, it is important to give the users a sense of control or ability to influence the development and usage of the capabilities. Alerts developed by or at least endorsed by physicians have a higher degree of acceptability than those that are foisted upon them. In most cases, there should be mechanisms for the physician to acknowledge an alert, for example, and proceed to conduct care as he or she sees fit. In addition, according to the Ten Commandments outlined by Bates et al. (2003) (see also Chapter 5), the amount of time and effort required of clinicians to provide additional information should be minimized, so that workflow is not adversely affected.

Many of the control, autonomy, and trust issues are related to the culture of the organization. The cultures of teaching hospitals are different from those of community hospitals. For one thing, there is a strict hierarchy associated with levels of training. For another, there are strong department chairs and differences in power dependent on which areas bring in the most revenue. Clinicians are powerful, and those with power can exert it on house staff. Cultures within different teaching hospitals vary depending on the strength of the power and the nature of the fiefdoms. For decision support purposes, the power can be tapped by gaining the support of key department chairs. Community hospitals are less hierarchical, with little hierarchy within the ranks of the physicians. Informal champions and opinion leaders hold power only by virtue of their expertise or personalities, yet these are the people who need to be involved in decision support efforts. Of course, cultures vary within specialties in either type of hospital: these might be considered subcultures that may vary in their power levels.

18.2.3 Difference between Commercial and Locally Produced Decision Support

Locally produced decision support is hard to develop and maintain, but it is more easily accepted because most likely the users have helped to develop it. It can be tailored to fit the workflow and local interests (Miller et al. 2005). Most often it is fragmented, however, because it has been developed in a reactive way. When a need or gap is identified, a decision support module is developed, often because it captures the interest of an individual developer. On the other hand, commercial decision support modules or applications, although they may require extensive modification, do not need to be developed from scratch and they may cover the entire spectrum of needs in a less fragmented way. Some commercial vendors of CDS modules have responded actively to customer demand for more and more rules. Logic included in commercial systems may be at a lower level of sophistication and perhaps more rudimentary than that in place at advanced academic sites, but it is improving all the time. Some vendors encourage the sharing of rules among hospitals that have purchased their systems. More hospitals need to both share their locally developed rules this way and also take advantage of those developed by others. With some commercial systems and with the Veterans Affairs CPRS system, there may be a surfeit of riches, however, to the extent that difficult decisions need to be made about which available CDS modules to select.

18.2.4 Upsides and Downsides to Clinical Decision Support from the User Perspective

When doing a stakeholder analysis, the user perspective on the upsides and downsides of CDS systems must be considered most important (Ash et al. 2000). Behaviorally, such systems can aid decision-making and help workflow. From the user perspective, the greatest upside to computerized CDS might be its availability at times and in places when the clinician is off site. For example, if a physician decides to enter a medication order after she has left the patient's hospital unit, she might use CPOE in her office across the street and receive decision support help while sitting at her own comfortable workstation.

There are a number of downsides of CDS systems, alluded to earlier. Decision support modules can seem too controlling, and they can evoke strong emotions (Sittig et al. 2005). They can produce unintended consequences like alert fatigue (Ash et al. 2004). Some experts claim they can undermine education and learning, but others say they assist these endeavors. In an astute summary, Wendt et al. (2000) describe many reasons why decision support functions cause problems for users. These include:

- The fact that routine medical work needs broad knowledge support, yet most CDS modules provide advice on something very specific
- The idea that much decision support is not patient-specific
- The loss of interpersonal discussion
- The questionable validity of some of the suggestions
- The questionable quality of data that the clinician has given the system
- The additional time on the part of the clinician
- The problems with interface designs that are not intuitive
- The lack of integration between different parts of some systems such as results retrieval and CPOE (Wendt et al. 2000)

In other words, the decision support modules may not seem relevant to a user if they are not patient-specific, if they interfere with doctor-patient communication, and also if they interfere with doctor-doctor communication by eliminating the need for "curbside" discussions. The user must believe that the information, both the guidance offered and the data about the patient that has been entered into the system, is correct. Any time the clinician is asked to enter information in response to an alert, the time this takes is problematic. Any time the user must spend time to evaluate the appropriateness of a guideline, alert, or citations in a literature search, it is considered a downside.

18.2.5 Cognitive, Emotional, and Environmental Issues

Since physicians may want discussion and psychological support when making difficult decisions, this social aspect of decision-making can be turned into an advantage by those implementing decision support. The social aspect can be especially important during the development of the decision support module. It was actually this kind of collaborative effort that saved the day when residents threatened to strike because of implementation of the medication ordering system at the University of Virginia in 1989 (Massaro 1993a, b). There was no decision support, and the entry of orders was taking too long, so the system was discontinued. The house staff started a residents' organization, leaders of which took it upon themselves to help build order sets in collaboration with clinical leaders and information technology staff, and the order sets became widely accepted, once CPOE was reimplemented. This is also an example of how emotional these issues can become. Although decision support saved the Virginia implementation, it can also generate many negative emotions, including guilt, anger, sadness, hostility, and even disgust (Sittig et al. 2005). Care must be taken to design and implement systems in ways that avoid as much as possible these negative emotions and reactions to them.

Many of the Ten Commandments outlined by Bates (see Chapter 5) suggest ways to decrease barriers that such negative emotions might generate: be mindful of time constraints and demands, deliver information in real time at point of need, suggest and do not stop an action unless critical, take advantage of the easy and high impact interventions by doing them first, avoid asking for more information than is absolutely necessary, be responsive to feedback, and manage the knowledge base so that physicians trust the content offered them.

Alert fatigue, or overdoses of alerts, is perhaps the greatest barrier to decision support. For example, although the VA has an extensive library of rules available to all of its hospitals, when one hospital discovered that 90 percent of its alerts were being overridden, most were turned off. There is debate about the greatest acceptable override rate, but something around 50 percent seems reasonable to most experts. Strategies for avoiding overdoses of alerts include specifying levels of urgency and responses required for different alerts—avoiding the extra annoying step of an explanation and asking for it only if it is really needed; turning off alerts when monitoring discovers they are consistently ignored; fine-tuning the alerts—a low override rate shows good specificity of warnings; and considering having some alerts go only to pharmacy rather than physicians (e.g., for duplicate orders).

18.2.6 Addressing the Issues Judiciously

Now that the relevant issues have been described, an example of the use of the force field analysis technique will introduce a discussion of strategies that have proven useful in decreasing the barriers and increasing the facilitating forces.

Beginning at the level of the individual clinician who is to receive the decision support, the barriers that can be lowered include emotions, lack of awareness, lack of familiarity, lack of self-efficacy, inertia, time constraints, and lack of computer skills. Emotions can be mitigated by using kinder language in alerts and by improving screen designs and fine-tuning alert conditions. Lack of awareness, familiarity, self-efficacy, and computer skills can be addressed with education, training, and communication. Inertia can be addressed more globally through the culture and social system, including recruitment of nursing and pharmacy staff to provide support and encouragement. Time constraints need to be addressed through a combination of better technology, education about the value of decision support, and adequate technical support. Hard-to-overcome constraints include lack of physician belief in the content/recommendation provided by a decision support rule (lack of agreement) and the belief that the intervention will not be successful (e.g., smoking cessation recommendation to the patient).

Regarding technology, barriers that can be addressed are design problems, lack of clarity on how to proceed once given a decision support message, mistimed interventions, and lack of trust in the input data. Addressing all these involves improving systems design. Even the lack of trust in the data placed in the system by clinicians can be overcome. For example, physicians are well aware that it can be easy to enter orders on the wrong patient. If the screen design makes it evident who the patient is and safeguards are in place, like disallowing two patient records to be open at one time, wrong patient errors will be reduced and trust in the system will increase.

Finally, the higher level organizational issues of power, control, autonomy, and trust of administration need to be evaluated and addressed. They need to be viewed through the eyes of the clinicians who will be recipients of decision support, because these are all issues that are not in themselves barriers. Control of the quality measures in an organization should rest with quality assurance and administration, but whether the clinicians view that control negatively or positively is important.

A checklist of barriers and forces needed to overcome them can be developed by populating the Force Field Analysis diagram as shown in Figure 18-3. First, consider the driving forces that may be especially strong in your organization: these might include the existence of strong education and training programs, excellent communication channels with clinical staff, strong clinical systems committees, an innovative organizational culture, the ability to improve systems design, high levels of trust, a clear organizational vision, and outstanding quality and safety initiatives. Planners should also honestly recognize such restraining forces as a negative prior experience implementing a particular decision support program, lack of clinician time so that stress levels are high, and lack of sufficient numbers of information technology staff with clinical backgrounds. The relative strengths of the forces can be assessed and resources identified to strengthen the facilitating forces and decrease the restraining forces. Once that is done, planning can incorporate mechanisms for changing the forces.

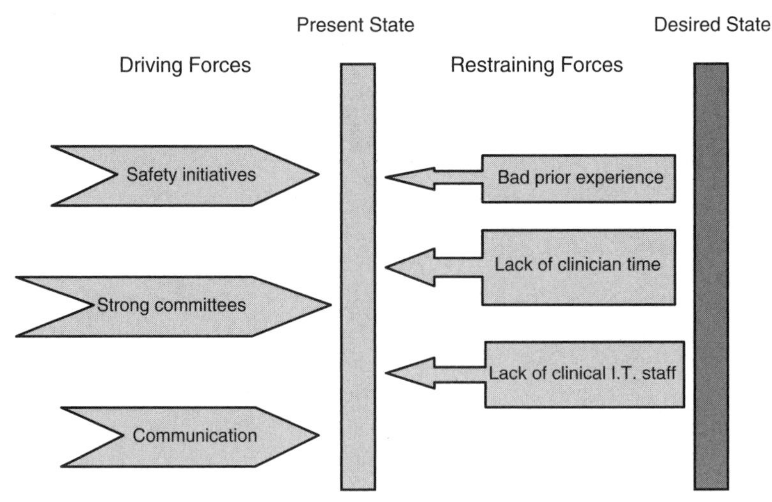

FIGURE 18-3 Force field analysis for a decision support initiative.

18.3 PLANNING WITH THESE ISSUES IN MIND

Planning needs to begin early, and a structure for continuous planning needs to be in place. Like any planning effort, it needs to begin with a vision. The vision needs to mesh with that of both a knowledge management strategy and strategic planning efforts at the organizational level. Then a situation analysis is needed. It should include an inventory of present technology that can provide the infrastructure for the decision support program and identification of clinical information systems committees already in existence. The situation analysis would include identification of the problems or weaknesses and the strengths, many of which would be identified in the force-field analysis. For the purpose of decision support, the quality gaps and needs for different components or types of decision support need to be described—these are the targets for the planning. The motivation for decision support, especially environmental trends, need to be made explicit.

Once the targets for the plan are identified, a financial overview can be done. As outlined in the next chapter, this is exceedingly hard to do, and at the beginning of any change effort, best guesses may need to be enough. What is important at this stage is the need for agreement that both intangible benefits as well as tangible ones be factored in. For example, costs of laboratory tests can be decreased with elimination of duplicate tests, which is a tangible benefit, but the ultimate goal of better patient care is in many respects intangible and needs to be considered.

Next, the plan should include goals and objectives and action plans. Possible interventions can be listed, prioritized, selected, and scheduled. Plans for monitoring the results must be built in, as well as a plan for planning. It is never too early to identify measures which will be used to define success. Baseline measurements should be done so that post-implementation changes can be documented. According to the Bridges model for transition management (Bridges 2003), there should be a parallel plan for monitoring the people and organizational aspects, including the effects on the clinician and nonclinician staff. The Implementer's Guide includes helpful worksheets to be completed during the planning process (Osheroff et al. 2005).

18.3.1 Vision and Philosophy

The plan needs to begin with the vision statement. Along with a clearly worded statement such as, "To be a leader in patient safety by implementing a broad range of computerized decision support entities," a philosophy, which must be part of the organizational culture or become part of it, must be behind the vision. For a CDS program to be successful, the Ten Commandments (see Chapter 5) should be followed. In addition, the following beliefs are useful.

The attitude should be that the clinician is usually right, and therefore the system must provide acceptable ways to avoid inflexibility, and to get around rules considered inappropriate in specific situations. A rule should not block or stop a clinician from doing what he feels is best for the patient, but rather lead him down the right path. This attitude can encourage physicians to accept the system, but it also is a recognition that no system is perfect. In fact, another belief must be that no decision support entity will be perfect, so you can only do your best and if it is minimally disruptive to a clinician's workflow, it can be accepted. A good rule of thumb that should be adopted by implementers is that one can anticipate only about half of the problems that will occur with each CDS module before it is perfected. Another helpful attitude is that one should not implement a new decision support module every time physician behavior change is desired. This will lead to alert fatigue and do more harm than good. Consider other ways to change behavior.

18.3.2 Organizing for Planning

The next chapter will describe various committee structures for providing a rational process for developing and reviewing plans. Every organization that already has a structure in place does it differently, depending on the history, structure, and culture of the organization. There is no recipe for a successful CDS committee structure. There are some general principles that can lead to success, however. First, physicians need to be heavily involved in the work of the committees that are charged with planning and selecting decision support. This may be obvious, but because pressures for quality enhancement are put primarily on administration and quality assurance staff, and because physicians are busy and may feel threatened by the focus on error reduction, they sometimes are not included. When physicians are appointed to committees, they sometimes are not truly involved or listened to. It is also important to bear in mind that clinicians may be called upon to do much of the work (data entry, response to alerts, etc.) yet may not be the primary recipients of the benefits of decision support. Getting clinicians involved in making decisions about decision support and in helping to develop decision support can be a selling point for CPOE in general, and for individual decision support entities in particular. Beyond the benefit of receiving expert clinical input, involvement of clinical opinion leaders can help to spread the word and encourage other clinicians to better accept the systems.

Another principle is to take advantage of the social needs of clinicians. Especially with the advent of information systems and the ability to enter orders and do clinical documentation remotely, they no longer have the opportunity to get together face to face as often as in the past. Decision support committees truly need clinician expertise, and clinicians can find this forum for interaction with their peers appealing, so well organized committees can serve both functions. The committees need to be multidisciplinary, including committed and respected clinicians, and they need to enjoy their work. The latter need can be met by having well-organized, on-time, well-scheduled, regular meetings with refreshments. In teaching hospitals, lunch meetings can be scheduled, but in community hospitals, early morning, before private physicians begin patient rounds, seems common. This may be unpopular with information services and operations staff, but it gives a signal to the physicians that their presence is important.

There is some debate about whether to have people paid to be part of the committee structure. As mentioned, sometimes physicians who are paid part time by the hospital for this work are considered by their colleagues to have gone over to the other side. Usually those who work on decision support committees must put in considerable amounts of time, so some kind of reward, even if it is not monetary, is important. This might include recognition in media, awards, and other incentives such as being the first to try out a new program.

A final underlying principle is that there must be a communication plan to provide backup for every decision support module that is implemented. Indeed, lack of training and communication can cause confusion, frustration, and risk (Horsky et al. 2005). There are many ways to communicate with clinical staff, and each organizational culture has its own most effective mechanisms. Sometimes the opening screens of the clinical information system are well read and effective, so these "hello" screens can alert clinicians about updates and changes made to the decision support systems. If these screens are ignored, however, more effective and complex communication media will be needed. Many organizations make sure that changes are mentioned at all important meetings that might be attended by clinicians. Some organizations have monthly "pizza meetings" that become popular gathering times for users to talk about both problems with the system and also new enhancements and changes. This may also be an opportunity to do training, and another incentive might be continuing medical education credit. It is especially important that clinicians be trained about what alerts mean and how to handle them. This improves the physician's sense of confidence, helping to overcome lack of self-efficacy, one of the major barriers to implementation.

18.4 DEVELOPMENT, IMPLEMENTATION, AND MODIFICATION

18.4.1 Preparing

Planning is of course the best preparation for development of an entire decision support program. For individual decision support module development, however, it is wise to look first at paper-based decision support. We sometimes forget that all health care organizations have had some kind of decision support in place prior to information system implementation (e.g., forms for order sets), and these can be gathered and inventoried. Many will no longer be in use, so the forms need to be carefully reviewed, but some can provide a starting point. They can be compared to any decision support rules that are available from a vendor, whether it be the vendor of the CPOE system or a third-party vendor of decision support software. A similar methodology might be used for clinical guidelines or pathway development.

Computer-based information systems are often blamed, when a standard policy that routinely has been ignored is suddenly exposed after it has been automated. A wise strategy is to identify for each decision support module whether it embodies a new enforcement of an old policy. If it does, plan to begin enforcing the policy with adequate notice prior to computerizing it. This will hopefully avert system blame.

When a new CDS module is being designed, clinician users must be included in the interface and screen design development. Many places have found that house officers can be paid to moonlight to do testing. After all, they have a vested interest in making sure the decision support is of high quality and is easy to use. Other types of clinicians and support personnel, including nurses, laboratory staff, and pharmacists, also need to be included at an early stage both for their clinical expertise and also their knowledge of the workflow processes that will be impacted by an intervention.

18.4.2 Committee Work

Early decision-making must include: 1) determination of what needs to be in place at both basic and more advanced levels over time, 2) establishment of a process and identification of responsible people for screening for new knowledge, and 3) identification of a mechanism for updating and revision of decision support modules. Someone needs to make decisions about priorities, for building from scratch or for selecting and modifying vendor-supplied rules. There are several processes that committees can use for identifying gaps that decision support can fill. For example, they can institute a regular process of reviewing new evidence, and for screening available guidelines, identifying problems that need addressing, and fielding suggestions from pharmacy and therapeutics committees and other chartered committees. Because of the nature of decision support systems and their relation to quality and safety, decisions by these committees can be viewed as policy and therefore need careful approval by the organization's decision makers. Recommendations from individuals, especially physicians, should be welcomed, but they need to go through the proper channels. They need to be vetted against the grand plan and prioritized. It is important that every suggestion that is received is not only acknowledged, but that its disposition is made clear to the person who made the suggestion. This is the only way that continuing involvement from rank-and-file clinicians can be motivated. For rules that are already in place, one person needs to be identified as clearly responsible for each, so that feedback can be given to that person and so that everyone knows who the responsible party is. That person should also be part of the regular review process, on a timeline, for keeping the rule up to date. If that person leaves the organization, someone else needs to be assigned to the role, to maintain continuity.

Aside from committees, getting groups of clinicians together to evaluate systems in the later phases of development is good from the point of view of both social integration and acceptance. Focus groups are immensely valuable at this point. Different decision support modules will generate different levels of anxiety among clinicians. The testing, piloting, and implementation strategies need to be designed with this in mind. For alerts that might spur resentment, clinical champions should be involved at each phase. Then when the messages are released, communicating about the change, the messages can mention that physician colleagues were included in the planning process.

18.4.3 Providing Resources for Support and Training

End user support and training go hand in hand. At teaching hospitals, new clinical staff usually are trained in the use of that organization's clinical system in a classroom setting, but should also be offered continuous ongoing training. This ongoing training can be in a group or individual setting; in community hospitals, both initial and ongoing training usually are done one-on-one. This is mentioned here because it is important to realize that decision support modules may be modified over time, as a result of which, users may need more than simple notification of the change: they may need some hands-on training if the update is substantial. This varies with the individual and with the type of change involved, but users have been known to walk away when they become frustrated or do not know how to respond to a suggestion. Monthly pizza sessions can provide an appropriate venue for reviewing changes with interested users and for giving instruction. Support staff, however, also need to be available to address problems as they occur before frustration sets in.

18.4.4 Strategies

Organizations have developed and used a number of general strategies successfully, but since each hospital or clinic has a different culture, and since each specialty and unit has its own culture, careful decisions must be made about whether each strategy fits the particular culture. Many feel that surgery is a good place to start because it is more protocol-driven than other specialties. As a result, order sets, for example, can save surgeons a great deal of time. On the other hand, decision support for general medicine is harder to implement and has a higher chance of not being appropriate, because the problems are so broad. Another strategy is to go for the low hanging fruit, for the most useful yet easy kinds of decision support. Many places have started with readily accepted general order sets with defaults such as dosages of medications at recommended levels, then have added alerts about allergies, and moved gradually toward drug/drug interactions and suggestions for corollary orders.

Provision of searchable information resources is another readily accepted and unobtrusive initial means of offering decision support. It is undesirable for the clinician to need to leave a patient record application in order to search elsewhere for reference information, however. Thus, it is preferable for the resources to be linked directly from or accessed from within the clinical system, e.g., via infobuttons (see Chapter 16). Eventually, organizations may find it useful to introduce clinical pathways, and some places find that having them available in lay language is useful for communicating with patients.

Vendor user groups of physicians share experiences, and it behooves an organization to send involved clinicians to these meetings. It is a valuable social networking opportunity, and often decision support modules are discussed, along with experiences in their use at other sites; this knowledge can be brought back to be discussed by relevant committees.

As mentioned, another strategy that has proven useful is to allow personal order sets in the beginning and then move away from them in favor of incorporating the most useful ones as general order sets. Even though it may seem cruel to take away a well-loved tool, implementers have found that by the time favorite order sets have been used for a period of time, the clinicians most likely will understand the difficulty of keeping them up to date and can be convinced to use general order sets that include their beloved features.

18.5 EVALUATION AND MAINTENANCE

There are personal and organizational issues related to both evaluation and maintenance of decision support. As discussed earlier, user involvement is needed for evaluating satisfaction levels with different CDS capabilities. However, it may also be useful to assess the cognitive aspects of decision support, such as the frequency with which alerts are being overridden, reasons they are being overridden, alternative ways they can be worded so that they are acted upon with higher frequency, and ways of increasing their appropriateness.

18.5.1 Have Data to Back You Up and Gain Involvement: Impact Assessment and Other Techniques

In the planning stage, metrics for success have been identified and baseline data gathered. It is likely that metrics tracked included usage, usability, and financial and clinical benefits. Numbers can be generated by the system itself for measures like how often an intervention was triggered (for alerts, for example) or sought (information search or review of passive guidelines). The acceptance or override rates and response time of the system also can be tracked automatically, and reports can be generated. Other aspects of the system such as acceptability to clinicians need to be studied in other ways, however, and require a concerted effort. Surveys of user satisfaction can be conducted. Observations of actual use can be done. Actual clinical outcomes are hard to measure and may call for data mining or manual chart reviews. If it is discovered that a particular module is not having an impact on care, an investigation is needed.

18.5.2 Soliciting Clinician Feedback

Each time a new decision support module is made available, clinician feedback should be encouraged. Otherwise, fine tuning, and therefore acceptance, will be difficult. Even if the alert has been tested in a laboratory setting and been reviewed by the appropriate committees, and the clinicians are aware of it and have been trained, it may not perform as hoped when in "production." Having a feedback button on the screen with the name of the responsible clinician owner clearly noted is very useful to capture the reasons for this. During introductions of new capabilities, the information services staff needs to be extra diligent and perhaps plan to put in extra time in order to offer assistance and also to make modifications when problems are discovered.

18.5.3 Knowledge Management

Section VI deals with knowledge management from a technical perspective, but there are a number of organizational and behavioral issues to consider here. First is the difference in emphasis on knowledge management between teaching and nonteaching hospitals. Second is the issue of local autonomy and control. Third is the importance of a systematic organizational approach.

The main difference between knowledge management in teaching and nonteaching settings is that in the former, a major focus is knowledge, as a result of which one would expect clinicians to be highly motivated to provide input into maintaining decision support content. This is indeed the case, although often there are disagreements and strong opinions about what improvements or other changes need to occur. Although expertise may be readily available in an academic organization, consensus may be hard to reach. On the other hand, the business of the community hospital is much less about knowledge, so there may be less interest in providing ongoing input, but gaining agreement may be easier.

One reason clinicians ignore alerts is that they may have found that they are out of date or do not pertain to the local setting. When a physician receives a cost alert that is no longer a realistic figure, he tends to pay less attention to all those he receives from that time onward. Given that even a community hospital may have 75 clinical pathways and 100 alerts, a regular review process is absolutely necessary. There must be a clinician responsible for each one, with support from quality assurance and information services staff. Changes should be documented and communicated with users.

Regardless of the setting, there needs to be a catalogue of decision support modules so that users, information systems staff, and involved clinical decision support committees know what is available. If this effort is begun early, when there is not much in the way of decision support, the catalogue can grow gradually. Someone must be put in charge of this ongoing effort so that it does not get out of control when more advanced and complex decision support tools are added, and so that continuity is maintained, even when local experts or champions of particular kinds of CDS depart.

18.5.4 The Importance of Ongoing Organizational Support

As with most initiatives, a concerted effort to build a decision support system may involve a good deal of energy in the beginning, but that energy may later wane, especially if high-level organizational management does not provide adequate resources. Unlike other system implementation initiatives, this one is ongoing and often accelerating. It seems that once decision support becomes part of the organizational culture, it propagates. There is actually a danger that CDS will be looked at too often as the method to be used whenever anyone wants to change clinician behavior. As a result, demand on clinician time and patience may increase, beyond what the end user clinicians can tolerate. To avoid this, a planning process must be in place for vetting possible initiatives before implementation.

18.6 CONCLUSION

The delicate balancing act referred to at the beginning of the chapter is an organizational management and change management challenge. The challenge can be met only if sufficient staffing resources are available. Planning, prioritization, training, implementation, and evaluation are activities that need talented, committed, and high-level staff members—and there must be enough of them. The organization must be willing to dedicate significant ongoing financial resources to this endeavor.

RESOURCES

www.CPOE.org has as its primary focus computerized provider order entry, but decision support resources of interest include an annotated bibliography and a list of considerations for implementing CPOE that is more comprehensive than what has been published in the literature.

www.himss.org/asp/davies_organizational.asp provides access to papers describing winners of the Davies Award and most describe organizational issues related to CDSS.

ACKNOWLEDGMENTS

The author's research was funded by grant LM06942 from the U.S. National Library of Medicine, National Institutes of Health. Special thanks are extended to Richard Dykstra, M.D. for assistance with figures and Dean Sittig, Ph.D. and Ken Guappone, M.D. for analysis and review.

REFERENCES

Ash, J. S., Gorman, P. G., Hersh, W. R., Lavelle, M., Poulsen, S. B. (1999). Perceptions of house officers who use physician order entry. *Journal of the American Medical Informatics Association Supplement, AMIA Proceedings* 471–475.

Ash, J. S., Gorman, P. N., Lavelle, M., Lyman, J. (2000). Multiple perspectives on physician order entry. *Journal of the American Medical Informatics Association Supplement, AMIA Proceedings*, 27–31.

Ash, J. S., Stavri, P. Z., Dykstra, R., Fournier, L. (2003). Implementing computerized physician order entry: The importance of special people. *International Journal of Medical Informatics* **69**: 235–250.

Ash, J. S., Stavri, P. Z., Kuperman, G. J. (2003). A consensus statement on considerations for a successful CPOE implementation. *Journal of the American Medical Informatics Association* **10**(3): 229–234.

Ash, J. S., Berg, M., Coiera, E. (2004). Some unintended consequences of information technology in health care: The nature of patient care information system related errors. *Journal of the American Medical Informatics Association* **11**(2): 104–112.

Ash, J. S., Sittig, D. F., Seshadri, V., Dykstra, R. H., Carpenter, J. D., Stavri, P. Z. (2005). Adding insight: A qualitative cross-site study of physician order entry. *International Journal of Medical Informatics* **74**: 623–628.

Bates, D. W. et al. (2003). Ten commandments for effective clinical decision support: Making the practice of evidence-based medicine a reality. *Journal of the American Medical Informatics Association* **10**(6): 523–530.

Bridges, W. (2003). *Managing Transitions: Making the Most of Change*, 2e. Cambridge, MA: Perseus Press.

Cabana, M. D., Rand, C. S., Powe, N. R., Wu, A. W., Wilson, M. H., Abboud, P-A. C., Rubin, H. R. (1999). Why don't physicians follow clinical practice guidelines? A framework for improvement. *JAMA* **282**(15): 1458–1465.

Davis, D. A., Taylor-Vaisey, A. (1997). Translating guidelines into practice: A systematic review of theoretic concepts, practical experience and research evidence in the adoption of clinical practice guidelines. *Canadian Medical Association Journal* **157**(4): 408–416.

Feldstein, A. et al. (2004). How to design computerized alerts to ensure safe prescribing practices. *Joint Commission Journal on Quality and Safety* **30**(11): 602–613.

Garside, P. (1998). Organizational context for quality: Lessons from the fields of organizational development and change management. *Quality in Health Care* 7(Supp): S8–15.

Horsky, J., Kuperman, G. J., Patel, V. L. (2005). Comprehensive analysis of a medication dosing error related to CPOE. *Journal of the American Medical Informatics Association* **12**(4): 377–382.

Institute of Medicine Committee on Quality of Health Care in America. (2003). Crossing the quality chasm: A new health system for the 21st century. Washington, DC: National Academy Press.

Lewin, K. (1951). *Field Theory in Social Science*. New York: Harper & Row.

Massaro, T. A. (1993a). Introducing physician order entry at a major academic medical center: I. Impact on organizational culture and behavior. *Academic Medicine* **68**: 20–25.

Massaro, T. A. (1993b). Introducing physician order entry at a major academic medical center: II. Impact on medical education. *Academic Medicine* **68**: 25–30.

Miller, R. A., Gardner, R. M., Johnson, K. B., Hripcsak, G. (2005). Clinical decision support and electronic prescribing systems: A time for responsible thought and action. *Journal of the American Medical Informatics Association* **12**(4): 403–409.

Osheroff, J. A., Pifer, E. A., Teich, J. M., Sittig, D. F., Jenders, R. A. (2005). Improving outcomes with clinical decision support: An implementer's guide. Chicago, IL: HIMSS.

Sittig, D. F., Krall, M., Kaalaas-Sittig, J., Ash, J. S. (2005). Emotional aspects of computer-based provider order entry: A qualitative study. *Journal of the American Medical Informatics Association* **12**(5): 561–567.

Solberg, L. I. et al. (2000). Lessons from experienced guideline implementers: Attend to many factors and use multiple strategies. *Journal of Quality Improvement* **26**(4): 171–188.

Trivedi, M. H. et al. (2002). Development and implementation of computerized clinical guidelines: Barriers and solutions. *Methods of Information in Medicine* **41**(5): 435–442.

Waitman, L. R., Miller, R. A. (2004). Pragmatics of implementing guidelines on the front lines. *Journal of the American Medical Informatics Association* **11**(5): 436–438.

Wendt, T., Knaup-Gregori, Winter, A. (2000). Decision support in medicine: A survey of problems of user acceptance. *Studies in Health Technology and Informatics* 77: 852–856.

19 MANAGING THE INVESTMENT IN CLINICAL DECISION SUPPORT

JOHN GLASER and TONYA HONGSERMEIER

The implementation of clinical decision support is a complex technical, medical, workflow, cultural, and support challenge. This challenge must be managed. This chapter will cover four areas of management challenges.

- Knowledge management will be discussed, including clinical decision support as a form of knowledge that must be managed, the boundaries of knowledge management, key functions of knowledge management, and the "business case" for knowledge management.
- The organization of the clinical decision support effort will be reviewed including objectives of the organization, examples of organizational structures and processes, a review of some of the organization at Partners HealthCare as an example, and observations on organization.
- Key IT strategies and considerations will be examined including legacy systems, knowledge management tools, and application foundations.
- The evaluation of the impact and value of knowledge management will be discussed.

19.1 INTRODUCTION

A health care provider organization's information technology (IT) strategy may have concluded that computer-based clinical decision support (CDS) is a critical contributor to efforts to improve the quality and efficiency of medical care and patient care operations. Computerized Provider Order entry (CPOE), with drug–drug interaction logic, may be seen as central to efforts to improve patient safety. An Electronic Health Record (EHR), with health maintenance reminders, can be an important approach to disease management efforts.

This conclusion may lead to the acquisition of new clinical information systems that possess necessary tools or the leveraging of the capabilities of existing information systems. Further, this conclusion will require that the organization establish management structures and processes that enable it to identify CDS priorities, develop the required content, manage the knowledge that is expressed through clinical information systems, and evaluate the impact of CDS. The organization will likely need to design changes in workflow and definitions of best clinical practice.

The implementation of CDS is a complex technical, medical, workflow, cultural, and support challenge. This challenge must be managed.

The preceding chapter dealt with cultural and personal challenges caused by the introduction of CDS and approaches to dealing with them. This chapter discusses the management strategies and tactics that are likely to be necessary to create and sustain effective computer-based CDS. Specifically, the chapter will provide an overview of knowledge management, the organization of the CDS effort, key IT strategies, and the evaluation of impact of CDS.

19.2 KNOWLEDGE MANAGEMENT

The development of any set of management structures and processes that surround a significant information technology can benefit from a discussion of the concepts that will guide and frame the development. For example, a discussion of the integration of an organization's applications should begin with attempts to answer the question, what does integration mean to us? The organization can develop very different strategies, for example, single vendor or interface engine, based on very different answers.

This section provides some concepts and context that should guide the organization's discussion of CDS. In this chapter we consider CDS as a form of knowledge that must be managed, the boundaries of knowledge management, key functions of knowledge management, and the business case for knowledge management.

19.2.1 Clinical Decision Support as a Form of Knowledge to Be Managed

With its newly formed IT strategy, the organization may quickly jump to discussions of decision support technologies and ontologies. These are important conversations, but in several ways they are premature and probably incomplete.

Clinical decision support is a class of knowledge application tactics. The class seeks to ensure that the clinician has the right information necessary to make the right decision about what to do for a particular patient.

A quick leap to a decision support conversation focuses on the application of knowledge but may fail to consider equally important aspects of knowledge discovery and knowledge asset management. This quick leap may also fail to consider other IT-based tactics for knowledge application. These tactics include Web conferencing, virtual collaboration, Web-based courses, and repositories of studies and practices.

The organization would be well served to step back and engage in an overall discussion of knowledge management. Such a discussion will force consideration and creation of processes designed to identify the "best" knowledge, ensure that knowledge currency is maintained, align the application of knowledge to organizational goals, and require that the organization consider the full range of IT-based and non-IT-based tactics.

A more holistic view of knowledge management is important, but it can fall prey to various "traps" such as fuzzy boundaries, incomplete understanding of the scope of knowledge management processes, and a complex business case. These issues are discussed in the following sections.

19.2.2 The Boundaries of Knowledge Management

Knowledge management can have diffuse boundaries—boundaries that encompass the entire organization. Translational research is knowledge management. Quality improvement is knowledge management. Training residents and allied health professionals is knowledge management. Training for managers on human resource issues is knowledge management.

If knowledge management is defined too broadly, it will be perceived (rightfully so) as too broad to be tractable and defying the ability to be managed by a common set of structures and processes. An organizational phenomenon that is too broad will be seen as unmanageable and hence dismissed from the management discussion. For example, no one in an organization proposes to be in charge of "decision-making."

Boundaries can be defined in several ways, with each way being based on a different core concept.

- **Clinical goals.** Knowledge management can focus on specific goals to improve clinical performance; for example, reduce medication errors or reduce inappropriate radiology procedure utilization.
- **Disease.** Knowledge management can focus on IT-based and non-IT-based knowledge that is applied to treatment of specific diseases. These diseases may be those for which there is a specific set of managed care financial incentives, a high prevalence in a hospital's service area, or organizational focus on developing clinical excellence.
- **Application.** Knowledge management can address the broad array of knowledge that is contained in or expressed through specific applications (e.g., CPOE or the EHR).
- **Knowledge implementation tactic.** Knowledge management can focus on a specific implementation tactic, such as health maintenance reminders or clinical pathways, which might cut across applications and diseases.

An organization may pursue more than one concept. All the concepts reflect "understandable" boundaries; that is, you can explain them to a room full of practicing physicians and they will "get it." And none of the concepts is inherently superior to any other concept.

These concepts also supply a context. Knowledge management or decision support that has no context has no value. Achieving a clinical goal or improving the care of the chronically ill provides a reason for pursuing knowledge management.

19.2.3 The Key Functions of Knowledge Management

Knowledge management, however an organization defines its boundaries, is essentially comprised of three key functions: knowledge application, knowledge discovery, and knowledge asset management. They are organized in a circle (see Figure 19-1) to emphasize that the knowledge management process is one of continuous learning and knowledge dissemination. (These functions correspond to the three life cycle processes introduced in Chapter 1 and discussed further in Chapters 21 and 24.)

Knowledge application is the art of leveraging knowledge at the right places in workflow to achieve a strategic objective. Knowledge discovery is the

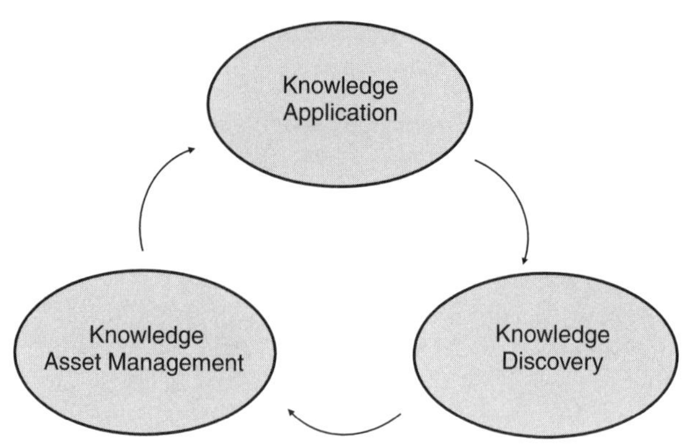

FIGURE 19-1 Knowledge management core processes.

process of analyzing data for the purpose of understanding performance, reporting, predicting, and/or harvesting new knowledge. Knowledge asset management is a set of processes for creating, validating, updating, and deploying knowledge.

Most clinical information systems typically are optimized to support knowledge application more effectively than discovery or asset management. Clinical decision support management mechanisms must encompass these three aspects of knowledge management.

19.2.4 The Business Case for Knowledge Management

The business case seeks to answer the question, "Why should we divert scarce resources to knowledge management?"

Provider organizations invariably are confronted with tight budgets; capital budgets are constrained and proposals to add expenses to operating budgets are subject to tough scrutiny. Knowledge management requires a budget, and obtaining this budget requires that knowledge management competes effectively with other budget proposals.

The knowledge management business case faces several challenges:

- The term "knowledge management" is often too abstract and intangible for concrete, action-oriented hospital managers. They may not fund it because the term "knowledge management" gets in the way; it doesn't mean anything to them.
- The knowledge management proponents may defend their case using terms such as "ontologies" or "semantics." These terms are incomprehensible to most hospital managers, and generally these managers will not support the funding of something that they don't understand.
- The organization may have no working experience with knowledge management, and hence it is not sure how to organize the function or what clinical value will be realized. Managers are often quite conservative and hesitant to launch undertakings that they are unsure of their ability to manage.

Successful business cases have several attributes. First, they link a proposal to an accepted organizational strategy or goal. For example, external business drivers for value-based purchasing, such as Leapfrog Group certification, are now emerging that "require" knowledge-enriched clinical information systems.

TABLE 19-1 Linkage of organizational goals to knowledge needs.

Organizational goal	Example knowledge need	Benefit
Patient safety	Drug–drug interaction checking in CPOE	Leapfrog compliance, reduced length of stay
Cost management	Radiology and medication order guidance in outpatient CPOE	Payer contract incentives
Health Employment Data Set (HEDIS)	Health maintenance reminders	Payer contract incentives
Joint Commission Core Measures (JCAHO)	Surgical site infection prevention protocols	Hospital accreditation, increased reimbursement
Disease management	Diabetes management protocols	Payer contract incentives

The creation or augmentation of knowledge management capabilities is often tightly linked to an overall investment in clinical information systems or medical care improvement. For example, clinical decision support is an aspect of an overall acquisition of a hospital information system and the CDS costs are not presented separately. In this case, the knowledge management resources piggyback on the overall resource request, with the overall request being considered in light of organizational goals.

Table 19-1 provides several examples of how knowledge management infrastructure can be explicitly aligned with business objectives to yield a tangible gain.

Second, the level of resources needed, such as staff and information systems, is deemed to be reasonable. Reasonableness is hard to empirically derive. Often organizations start with a small number of staff and gradually increase effort, as they understand the nature of the challenge. Other times, benchmark data from other organizations provide guidance on needed resources. Regardless, in order to proceed, the expense is deemed to be worth it.

Third, the business case describes the management structures and processes needed to manage this knowledge. For example, who should make sure that our health maintenance reminders are kept current? How do we determine if our guidance on radiology procedure ordering is leading to reduced radiology costs? Providing thoughtful answers to these questions helps to assure management that the invested resources are likely to result in the desired gains.

Fourth, the information technology infrastructure needed is defined. This infrastructure can include knowledge libraries, a rules engine, or collaboration tools. The tools proposed are no more and no less than is needed. And the tools and suppliers chosen offer an evolutionary technology path that is robust and enduring.

19.3 ORGANIZATION OF THE EFFORT

Organization refers to structures and processes needed to manage knowledge application, discovery, and asset management. This section will discuss objectives of organization, examples of organization structure and processes, a review of aspects of the organization at Partners HealthCare as an example, and observations on organization.

19.3.1 Objectives of Organization

Clinical decision support requires management structures and processes. These structures and processes are intended to accomplish several objectives:

- Identify new types of knowledge that need to be incorporated into the organization's clinical information systems (e.g., the addition of a new Black Box warning to a medication order entry application).
- Ensure that knowledge can be clinically defended through review of the literature or consensus-based decisions by appropriate clinical staff.
- Ensure that existing knowledge is reviewed at an appropriate frequency to determine if "old" knowledge needs to be revised.
- Recognizing finite information technology and clinical resources, provide direction on priorities for incorporating or modifying knowledge.
- Educate the clinical staff on the rationale for the knowledge.
- Assess the impact of existing knowledge application tactics on provider decisions and practices to determine if the desired outcomes are being achieved.
- Review strategies to improve the effectiveness of existing knowledge application tactics (e.g., Does a computer-based intervention impede workflow or does the application interface confuse rather than inform the user?).
- Guide the efforts of information technology staff and/or the application vendor to ensure that appropriate specifications have been developed and testing has been performed.

Invariably an organization will have several forums that pursue these objectives. The Pharmacy and Therapeutics Committee can be charged with managing all medication-centric knowledge for an inpatient clinical system. A Diabetes Advisory Council may be convened to develop decision support content to improve the health maintenance processes for a diabetic population. A committee formed to reduce the costs of care operations may decide to examine the possibility of reducing inappropriate radiology procedure utilization through CPOE. A committee that manages the evolution of an organization's clinical information systems may examine the systems to determine if there are "knowledge gaps," that is, areas where minimal knowledge exists in the system such as guidelines for the treatment of asthmatics.

The result of assigning knowledge management tasks to a range of forums can lead to a complex maze of decision-making. Although each individual assignment may be the right assignment, the maze needs to be coordinated, conflicts may require resolution, and the resulting demands on the information technology staff will require prioritization.

19.3.2 Examples of Organization

Several examples of approaches to organization are presented here. These examples are adapted from Pifer et al. (2005).

19.3.2.1 Example 1

A Medical Information Systems Committee (MISC) is charged with overseeing the design and implementation of clinical information systems for the organization. The MISC is also responsible for ensuring that the clinical information systems conform to all regulations, JCAHO requirements, and the organization's policies.

The MISC has multistakeholder representation and reports to an Executive Medical Committee. The MISC has a subcommittee that oversees the development of clinical decision support. This subcommittee receives requests from various task forces, committees, and user groups. The subcommittee requests IT assessment of the costs and time required to fulfill the request. The subcommittee recommends priorities and forwards its recommendations to the MISC for approval.

19.3.2.2 Example 2

The Information Technology Strategy and Policy Committee (ITSPC) is responsible for strategic, policy, and tactical decisions for all the organization's information systems and information management. The committee is composed of senior clinical, administrative, and IT leadership.

A Clinical Information Systems Committee reports to the ITSPC and is responsible for all patient care systems including clinical decision support. The Clinical Information Systems Committee is responsible for reviewing all requests for CDS, identifying required resources, prioritizing requests, and monitoring the effectiveness of existing CDS.

19.3.2.3 Example 3

The Clinical Systems Advisory Committee (CSAC) is responsible for providing direction and monitoring progress on the acquisition and implementation of clinical information systems. The CSAC members are senior leaders from across the organization.

Requests for decision support are sent to the CSAC for review, analysis of costs and effort, and prioritization. Decision support requests that are approved are sent to a Clinical Data and Documentation Committee, a committee of the Medical Staff organization, to ensure that the requests conform to organizational policy and are supportive of organizational efforts to improve patient safety and medical care.

19.3.3 Knowledge Management Organization at Partners HealthCare

The preceding examples center on the management of clinical decision support priorities and ensuring that CDS activity is linked into, and fits with, other supporting activities such as the implementation of a clinical information system or medical policies. In this section, a specific aspect of the management of clinical decision support at Partners HealthCare is examined.

At Partners, a series of strategic initiatives were launched in 2003 targeting physician order entry adoption, quality improvement, disease management, patient safety, and medical cost management. The patient safety team formed a subgroup on medication decision support, and this subgroup initially focused attention on physician order entry.

The Partners information systems organization has a group responsible for medication knowledge development and deployment at the enterprise level and multiple groups at the hospitals and physician practice levels, who are responsible for the various, site-specific physician order entry systems. A careful inventory of the current state revealed that there were nonuniform decision support practices across the sites and that a bottom-up approach to decision-making was impeding progress toward the sharing of bestpractices enterprisewide. For example, specialized pediatric, geriatric, and renal dosing decision support was maintained by a Partners-level pharmacy team but in production at only one hospital site.

To address the inconsistent decision support practices, the patient safety group first built a realigned model for governance (see Figure 19-2). The Medication Safety Steering Committee is composed of representative patient safety leadership and informatics expertise. It meets monthly to drive strategic priorities and is the final arbiter of disagreements that arise regarding application systems or medication content.

The Medication Knowledge Committee is composed of physicians, nurses, and pharmacists functioning in a variety of site-based drug safety roles, who gather monthly to guide and provide advice on the creation and maintenance of drug decision support content.

The Medication Systems Committee is composed of representatives of the various teams (largely information systems staff) supporting medication-related applications such as physician order entry, pharmacy systems, and medication administration systems across Partners. This team manages the planning and deployment of the IT portion of medication safety initiatives.

The Medication Knowledge Engineering team receives guidance from the Medication Knowledge Committee and site medication safety committees. This team of pharmacists and medical informatics staff creates and maintains the specialized medication knowledge bases such as physician-friendly drug–drug interaction checking, hematology–oncology dosing, pediatric dosing, renal dosing, and geriatric dosing.

After this alignment was formalized, the Medication Safety Steering Committee sanctioned and expanded the membership of the informal pediatric, renal, and geriatric subject matter expert teams to provide ongoing review and advice to the Medication Knowledge Engineering Team regarding

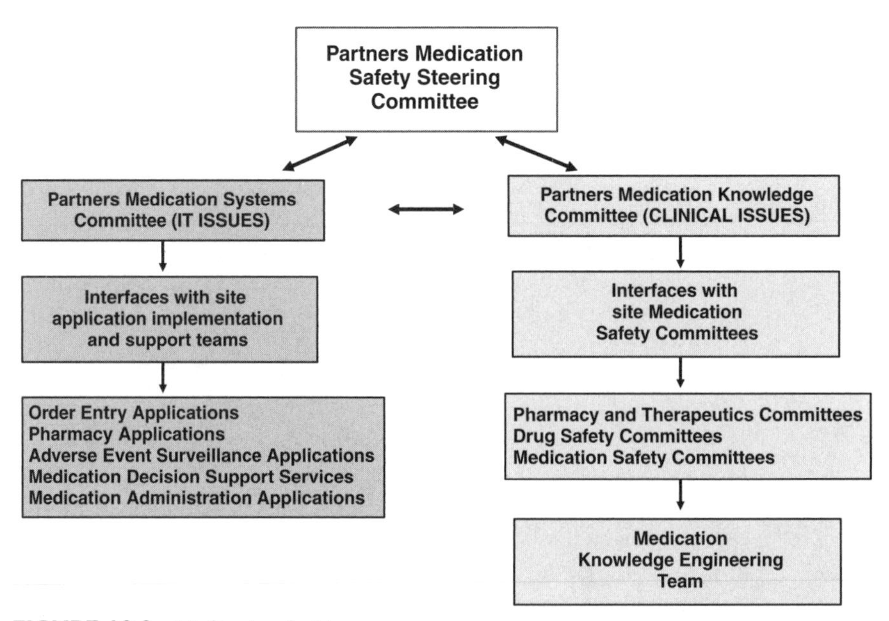

FIGURE 19-2 Medication decision support governance.

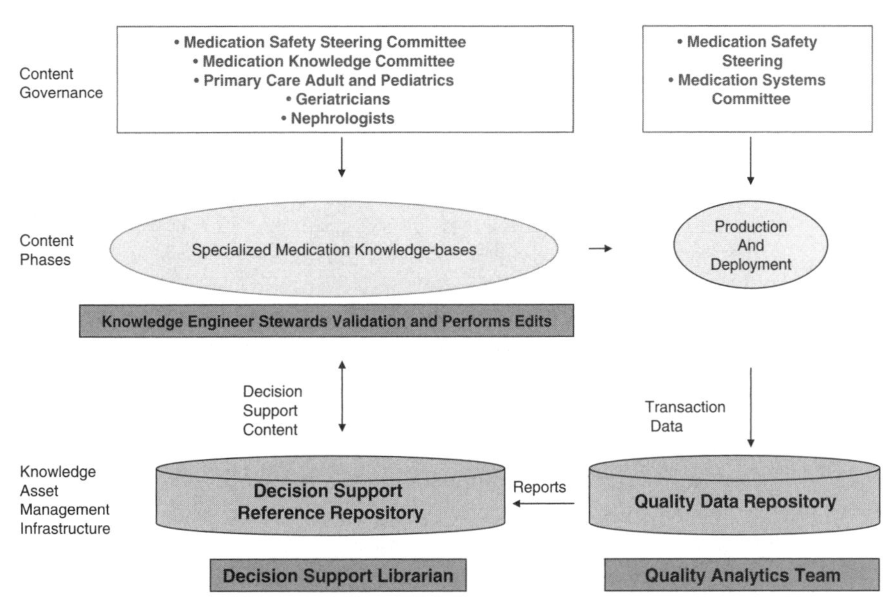

FIGURE 19-3 Process flow diagram for medication decision support knowledge maintenance.

knowledge maintenance. This step was a foundation to build acceptance for broader deployment of this kind of decision support in the inpatient and outpatient settings.

The Medication Knowledge Engineering team receives inputs from users, the Medication Safety Steering Committee, hospital pharmacists, and updates by the vendor drug information supplier. Changes and updates to the drug information that are not straightforward are brought to monthly meetings of the Medication Knowledge Committee. The subject matter expert teams for specialized dosing conduct reviews by therapeutic class annually.

Figure 19-3 provides a high-level diagram of the knowledge development and maintenance process. The figure illustrates the clinical content responsibilities of the Medication Knowledge Committee and the information technology responsibilities of the Medication Systems Committee. The figure notes one of the responsibilities of the Knowledge Engineering Team, which is the management of the Decision Support Reference Repository. The figure also illustrates the importance of ensuring that ongoing analysis of decision support impact on behavior and outcomes is performed, through a Quality Analytics Team, to provide feedback on the effectiveness of decision support.

19.3.4 Observations on Organization

As can be seen in the preceding examples, there is no single way to organize. However, there are several commonalities and guidelines that can guide the organizing of knowledge management.

19.3.4.1 Leverage Existing Committees

For a hospital, an existing Pharmacy and Therapeutics (P&T) Committee should be asked to manage medication-centric knowledge. An existing committee devoted to improving cardiac care should be asked to oversee knowledge related to hypertension and congestive heart failure guidelines.

Computer-based CDS is viewed as simply another tool available to the committee. This tool may be new to them, and they may need time and education to become comfortable with using the tool and understanding the strengths and weaknesses of the tool. Nonetheless one should direct knowledge management to existing forums where the necessary domain expertise exists.

The use of existing care-oriented committees helps to address several critical aspects of knowledge management and clinical decision support. First, the committees invariably possess the expertise necessary to determine the clinical utility of a specific decision support recommendation. Though "anyone" can propose a specific type of decision support, the experts must review and approve it. The use of an existing, appropriate committee can help silence squabbles about who is "the expert" on specific decision support content. Second, decision support must be maintained. Content may need to be updated and should be regularly reviewed. This knowledge maintenance should be a formal responsibility of the committee. Third, education of clinicians must often occur to explain why the decision support was implemented. The committee can be given this responsibility. Fourth, the committee is in the best position to prioritize requests. For example, a patient safety committee will have the best organizational perspective on the major patient safety issues. Fifth, these committees are usually in the best position to "discover" new knowledge or new opportunities to deploy existing knowledge. This discovery can be based on the experiences of the organization or the review of the discoveries of others.

19.3.4.2 Examine Committee Composition

Knowledge often spans domains. For example, there are obviously medication-centric rules that are of great interest to a committee focusing on cardiac care. To the degree that there is likely to be a significant set of knowledge that spans several committees, there should be cross-committee representation, such as a member of the P&T Committee on the Cardiac Care Committee. In general, this cross-committee representation is already in place; the boundary-spanning issues were present before the introduction of clinical information systems. Nonetheless, it can be useful to review committee composition and ensure that appropriate cross representation is in place.

Cross-representation should not only account for clinical discipline, but overall perspective. For example, it is important that clinicians representing the strategic concerns of the health system be balanced by those representing usability and efficiency concerns. Respected clinical champions can be those in management positions as well as the clinicians in a community practice who are greatly respected by their peers.

The addition of an information technology staff member (preferably with medical informatics training) to these committees is highly desirable. As an alternative to regular committee attendance by such individuals, some organizations appoint an IT staff member to act as a committee liaison, who is invited to committee meetings as needed. Regardless of approach, this person can help the committee members focus on the most feasible and effective informatics strategies to address a particular challenge (e.g., alerts at the time of ordering and the use of defaults and options for incorporating the knowledge into the workflow). Furthermore, this person can direct the transformation of clinical guideline decisions into proper design specifications.

19.3.4.3 Ensure IT Review and Assessment

Clinical decision support proposals must be examined from an IT perspective. The CDS technology will have limitations, some of which may mean that certain proposals cannot be practically implemented. The IT effort required to implement a new proposal must be understood. The IT staff that must "codify" and test the decision support will have a backlog that needs to be prioritized. Decision support can be a significant consumer of processing power, hence the machine performance of a specific CDS rule and the rules in aggregate must be monitored.

19.3.4.4 Define Oversight Group

The actions of individual committees will often conflict. The conflict can center on:

- The definition of appropriate knowledge, such as different opinions on best practices such as geriatric dosing or cholesterol management
- Tradeoffs between practicing best care and operational realities; for example, the primary care physicians are so harried that additional health maintenance reminders will fall on deaf ears
- Prioritization of scarce organizational resources; for example, budget limitations mean that some ideas can be implemented but not all ideas

In addition to the need for resolving conflicts, these individual committees must be coordinated. Coordination can be necessary for many reasons. For example, it may be the case that different committees independently embark on duplicative knowledge strategies (e.g., an inpatient Congestive Heart Failure (CHF) team and a CHF disease management team). Different groups may be considering investments in redundant tools (e.g., different teams independently investing in analytic infrastructure).

Decision support must conform to the organization's medical policy, hence policy assurance must be determined. At times, the decision support idea may lead to a need to alter policy. Decision support may also indicate the need to examine organizational roles; for example, who should respond to an asynchronous panic lab value alert? This oversight committee must have members who can bridge into other important organizational groups (e.g., compliance), and have processes that enable it to turf some issues to those other forums.

An existing committee can be assigned responsibility for oversight of knowledge management discussions and decisions. Many organizations have committees that have broad responsibility for care improvement; for example, an integrated delivery system may have a Chief Medical Officer's forum.

In several of the examples cited earlier, this oversight group is one that has been formed to provide overall direction for the implementation and management of the organization's clinical information systems. The placing of decision support oversight responsibility in such a committee is common. This orientation is usually a reflection of the need for such committees during the implementation of major clinical information systems. These implementations are massive and complex undertakings, and a committee of senior leaders is necessary to ensure that progress is being made. During implementation, clinical decision support efforts will begin, and it is natural that such efforts become the purview of the committee.

However, IT is a tool, and a natural evolution of tool oversight involves the transition from a tool-centric committee to a care-centric committee that has tools at its disposal, such as the Partners patient safety team.

As an example of this transition, many organizations had Internet Strategy Committees at the turn of the millennium. As understanding of the Internet increased, virtually all of these committees were disbanded, with responsibility for determining the best approaches to tool (the Internet) use being turned over to groups responsible for various aspects of business performance.

19.4 KEY IT STRATEGIES AND CONSIDERATIONS

Several chapters in this book have addressed specific aspects of the information technology and logic and data design of clinical decision support. This section will address three overall IT strategy considerations: legacy systems, tools, and applications as foundations. These considerations examine three critical aspects of defining and implementing the information technology infrastructure necessary for providing decision support.

19.4.1 Legacy Systems

How can an organization address the challenge of implementing robust, content-enriched, continually updated computer-based decision support, while working within the constraints of legacy information systems investments?

In pursuing the application of information technology to effect decision support, the organization will confront the reality of its legacy systems investments. In a large integrated delivery system, there may be several clinical information systems from multiple vendors. Each of these systems may have their own decision support technologies, and these technologies are likely to be of variable sophistication and utility. An organization need not be a large delivery system to face this challenge. A community hospital might find differing decision support capabilities in its laboratory, pharmacy, and hospital information systems.

Replacing these investments may not be practical. The organization may not have enough capital, or the replacement would consume an unacceptable amount of the capital budget. Replacement can take years to implement, but the organization needs care improvements in the near term. Moreover, some clinical information systems work well in large hospitals but not in the small physician's practice; hence in a large health system there may be little prospect of finding one system that effectively addresses the needs of all constituents.

There is no easy answer to this challenge. It is possible that advances in service-oriented architectures will enable an organization to have knowledge services (e.g., medication services) that effectively interoperate with heterogeneous applications. However, such a capability is not generally available in the market today. Service-oriented architectures for CDS are discussed further in Chapter 23.

Faced with this problem, the organization can take several steps to make the most of its legacy investments. Most of these steps are applicable in situations where there is a homogenous clinical information system across the organization.

1. Define the content areas that are important to drive the business. There are several content areas that can have a tangible effect on an organization's performance. For example, numerous studies have demonstrated the impact of antibiotic decision support in the inpatient setting on total antibiotic expenses, cost per case, length of stay, and adverse drug events (Evans et al. 1990; Pestotnik et al. 1996; Evans et al. 1998). In the outpatient setting, payer contracts are increasingly aligned with performance on HEDIS (Health Employment Data and Information Set) measures for asthma, diabetes, cardiovascular, and women's health management.
2. Define the applications that will be the focus of applying decision support. These applications are likely to include CPOE, the EHR, and pharmacy systems.
3. Evaluate the decision support capabilities of these applications; for example, what kind of drug decision support, order sets, templates, rules, and reporting do these applications support? This evaluation will lead to the identification of the "lowest common denominator" of tools, in effect establishing the limit to which decision support can be consistently implemented across the enterprise. If it appears that the limitations of the legacy infrastructure are woefully inadequate to meet strategic goals, decision support component suppliers are emerging on the market (such as Theradoc, Vigilanz, Cereplex) that can significantly augment the native capabilities of legacy systems at much less than the cost of a new infrastructure purchase.
4. Define types of knowledge that must be acquired. In most health care delivery organizations, formal structures and resources are often lacking to undertake the process of designing and maintaining decision support knowledge in clinical information systems. It is always cheaper to buy content than it is to bear the cost of developing content. There are several companies and associations that have developed excellent guideline content. New suppliers (such as Wolters-Kluwer, Zynx, and Micromedex) of prebuilt rules and order sets are also emerging. Some clinical information systems vendors now offer order sets, rules, and documentation templates through commercial partnerships with content publishers and/or academic health systems.
5. Define types of knowledge that must be "home-grown." There will be several areas where the organization will opt to use its own content or modify external content based on the organization's experiences. For example, at Partners, physician-friendly drug–drug interaction checking content is developed because commercial products have resulted in an intolerable rate of alert overrides.
6. Define strategies and resources needed to manage consistent knowledge across a heterogeneous set of applications and cultures, for example, applications across large academic health centers and small community hospitals. For example, if we have to implement a new health maintenance reminder across six different applications in four different organizations, how will we do that? How do we ensure that the logic is consistent across the organizations, and in different application environments with which it must be integrated? Ensuring consistency and currency might require that a person at each organization, or for each relevant application, be tasked with implementing content. These individuals can be managed by a corporate person who ensures coordination.

7. Develop/acquire an infrastructure for knowledge asset management. The organization must be able to keep a repository or library of the content that it has implemented across the enterprise. This library may be constrained to the subset of content that has been determined to have significant value and/or must be consistent across all care settings. The asset management tools should enable the searching of the library, support audit trails, and assist the organization in ongoing content management, for example, by identifying content that is due for a regular review. Some of the commercial content suppliers are beginning to offer tools to support inventory and subject matter review of content.

In the course of determining how to invest in knowledge management infrastructure, an organization must fully understand the comparative strengths and weaknesses of their legacy environment with respect to key functional capabilities. This assessment will lead to some form of the steps just outlined.

19.4.2 Knowledge Management Tools

Vendor systems often are designed with proprietary database design tools, typically called "knowledge editors," to build different content types such as rules, order sets, and documentation templates. Few vendor solutions offer functional support of other critical aspects of knowledge management such as governance, knowledge inventory, knowledge vetting, and design of complex cross-functional content such as disease management protocols. Hence, many clinical information systems are undernourished from a knowledge perspective. The advent of personalized medicine will bring a faster rate of change in commonly accepted standards of clinical practice. This will create an even higher burden on the part of health care systems to invest in updating the content embedded in their clinical systems.

An inventory and library of decision support design specifications is a critical component of any knowledge asset management strategy. At Partners, the knowledge management team performed an inventory and cataloguing of all decision support knowledge in production across the enterprise. A taxonomy was designed that enabled us to tag all the content specifications and publish them to a searchable portal. This portal has enabled clinical leadership to aggregate, compare, and analyze the robustness of content around strategic areas such as diabetes, cardiovascular disease, and adverse drug event surveillance.

Collaboration tools are useful to support subject matter expert review and validation of content. There are many tools on the market supplied by content management vendors and more recently, commercial clinical content suppliers. Such tools can facilitate virtual, asynchronous vetting of decision support design specifications among clinicians that are often too busy to attend meetings. Further, they enable capture of an audit trail for decisions made. At Partners, the collaboration portals are being deployed in a manner aligned with the strategic initiatives to support cross-disciplinary interaction (see Chapter 21). For example, the diabetes and the cardiovascular groups advise and consult one another regarding lipid management. The medication cost reduction group and the geriatric group collaborate on cost-effective pain management in the elderly.

Content management systems are useful to support the scheduled maintenance, versioning, and overall life cycle management of content. Typical clinical system vendor knowledge editors do not support easy capture of critical data such as authorship, purpose, subject matter expert validation, date of last update, schedule of next review, and the like. Further, innovative content management systems now support greater reuse and propagation of knowledge. For example, if an organization designs a set of rules and order sets for the use of beta-blockers in patients with coronary artery disease, it saves time and reduces errors if the addition of new beta-blockers to the formulary is automatically propagated for inclusion in these rules and order sets. Further, rules might suggest an alternative therapy if the patient has a contraindication to these drugs. With genomics and personalized medicine, the clinical guidelines for drug indications and contraindications will change rapidly; hence content management tools will become increasingly useful to synchronize ever-growing libraries of decision support content.

19.4.3 Foundations

The pursuit and progressive experience with knowledge-rich clinical information systems can lead the organization to begin to think of itself as implementing application foundations rather than strictly a set of clinical information system applications (Glaser et al. 2003). A foundation provides the broad ability to perform a never-ending series of application-leveraged small, medium, and occasionally large advances and improvements in organizational performance.

For example, a computerized provider order entry system can be viewed as a foundation to improve physician decision-making. Once the system is implemented, the organization can introduce an unending series of decision support rules and guides. These rules can address medication safety, the appropriateness of test and procedure orders, and the display to physicians of data relevant to a given order.

In effect, applications become the foundation necessary to achieve the core goals of enabling ongoing delivery of new CDS and improving workflow. This view of applications as foundations has several ramifications.

Clearly, there will be a flurry of intense effort as the foundation is laid. Introduction of CPOE and EHRs is difficult work that requires great skill and significant resources. But once the foundation is in place, implementation is a continual process—there is an ongoing implementation of decision support. In fact, implementation of a clinical information system never stops. Hence, organizational information system processes and management mechanisms must be adjusted to reflect this never-ending implementation strategy. This can imply that implementation teams do not disband and/or that there is a formal handoff of responsibility from the team that installed hardware and trained staff to the team that carries on ongoing implementation of decision support and workflow improvement.

The foundation must be able to evolve gracefully and support ongoing implementation. Tools are essential that enable rule development, the safe addition of local modifications, incorporation of new data types and coding conventions, and efficient interoperability with other systems. The foundation must be able to capitalize on new technologies with minimal disruption and to support growing organizational sophistication in applying the tools to improve care processes. In many ways, technologies and tools that enable ongoing implementation are more important than the present functionality of the application. This emphasis will affect the orientation of application requests for proposal (RFPs) and system selection criteria.

RFPs generally center on application functionality. The RFP process for a foundation must be changed to place a greater emphasis on tools and core technologies. In addition, an implementation that never stops implies that using the RFP in an effort to fully define all functionality that will ever be needed will be misguided. Experience will be the teacher.

Assessing the return on investment (ROI) of a foundation during the process of deciding capital budgets is more difficult than determining the ROI of an application. Although it is essential to continue to evaluate the ROI, it is difficult to do because the path of evolution is not always clear and implementation is never-ending. In acquiring and implementing a foundation, the organization is investing in "an ability." It is difficult to assign an ROI to an ability. In a similar fashion, it is difficult to assign an ROI to a well-educated workforce or having healthy capital reserves.

19.5 EVALUATION OF THE IMPACT AND VALUE OF KNOWLEDGE MANAGEMENT

If the organization has identified decision support as a critical strategic enabler and has, as a result, committed resources to acquiring and implementing needed information systems and support resources, it will ask, "Have our investments been effective? How much is it costing us to achieve our gains? Where must we focus our decision support resources next?"

The evaluation of the impact and value of knowledge management must address three areas:

- The strength of alignment of the content to business goals and strategies
- Organizational performance relative to key measures
- The efficiency and effectiveness of the knowledge management function to enable rapid-cycle learning

Evaluation does require that an organization have an approach to clinical data management and analysis. Assessing clinical performance and the impact of an intervention on that performance requires a set of well-defined data of known accuracy and timeliness. This approach must develop means to resolve issues that often plague the collection and management of necessary data.

Many health systems have poor access to clinical data for measurement and rely, instead, on billing and administrative data. The architecture of a typical transaction-oriented database is not optimized to support analysis. Further, the data that must be aggregated to enable deep analytics typically are located in many databases across an organization or in paper charts.

In the absence of a clinical data management and analysis strategy, those engaged in the process of understanding and reporting on clinical performance must often bear the cost and time delays of, for example, chart abstraction labor to collect clinical data, which consequently slows the translation of such insights into quality improvement.

19.5.1 Alignment

It is very useful for health care organizations to take a "begin with the end in mind" approach to decision support. In this way, business goals are linked to relevant measurement parameters and consequently, required decision support strategies.

Table 19-2 contains a sampling of the Acute Myocardial Infarction (AMI) JCAHO core measures to illustrate alignment between quality performance strategy and clinical decision support components. These measures are mapped to the necessary sources of discrete data for measurement and necessary knowledge components to achieve performance.

Specific goals, measures, and decision support "tuples" are the centerpiece of alignment. Those measures can be complemented by measures that provide a form of overall assessment of alignment. For example, measures that might serve as complements include:

- Degree of knowledge asset coverage for key business-impact measures such as HEDIS, JCAHO (Joint Commission on Accreditation of Healthcare Organizations), NQF (National Quality Forum), and Pay-For-Performance Contracts. For example, to meet HEDIS and JCAHO requirements for congestive heart failure, are the necessary clinical documentation elements, order sets, decision support rules, and measurement algorithms in production for inpatient, case management, and outpatient systems?
- Application end-user satisfaction with clinical decision support. Are clinicians satisfied with CDS content? Is the right balance achieved between quality improvement and workflow enhancement?

19.5.2 Performance

Table 19-2 also illustrates how decision support effectiveness can be measured in terms of direct impact on business performance. Effective knowledge management practices should result in better performance on key measures. Such measures can be translated into higher reimbursement on payer contracts or improved quality of care. Following are examples of the kinds of performance measures that can be used to assess decision support effectiveness. Clinician acceptance of decision support recommendations is also a barometer. An organization should anticipate and accept a minimum override rate because few decision support systems are so specific that recommendations are always clinically correct. Conversely, if an override rate is too high, the decision support is probably overly sensitive and task-interfering.

Examples of performance measures include:

- Quality Performance: HEDIS, JCAHO, CMS, NQF, and Pay-For-Performance contracts measures
- Adverse Event Rate: Drug events, bedsores, nosocomial infections, falls, confusion, and so on
- Compliance rate with decision support: Sensitivity and specificity analysis, override rates
- Malpractice: Insurance costs and trends in claims

TABLE 19-2 Acute myocardial infarction core measures.

Example core measure	Core measures description	Measurement data sources	Clinical knowledge for decision support
AMI-1	Aspirin at arrival for acute myocardial infarction (AMI)	1. For aspirin on arrival, electronic medication administration record for administration within 24 hours of arrival 2. For patient contraindications, clinical documentation, allergies, laboratory data, problem list 3. Time of admission is from Admission/Discharge/Transfer system	1. AMI admission order set with aspirin on arrival order 2. Documentation template for aspirin contraindication capture 3. Interactive alerts to notify physician if patient has contraindication to aspirin
AMI-2	Aspirin prescribed at discharge	Discharge orders from prescribing/ordering application	AMI discharge order set with aspirin
AMI-3	Angiotensin converting enzyme inhibitor (ACEI) for left ventricular systolic dysfunction (LVSD)	1. For ACE inhibitors, electronic medication administration record 2. For patient contraindications, clinical documentation, allergies, laboratory data, problem list 3. For LVSD, echo report has discrete field that indicates LVEF $< 40\%$	1. Discharge order set with ACEI on discharge order if LVSD present 2. Rules that indicate ACEI order is defaulted if echo report or problem list include LVSD 3. Documentation template in echo report with field for EF $< 40\%$ 4. Documentation template for ACEI contraindication capture
AMI-4	Smoking cessation advice/counseling for acute myocardial infarction (AMI) patients	1. Nursing and/or physician documentation template	1. AMI admission order set containing smoking cessation counseling order 2. Documentation template for smoking cessation counseling

19.5.3 Knowledge Management Function and Organizational Learning

Keeping an inventory of decision support knowledge current with commonly accepted standards of practice can be a costly business. It means investing in a team that conducts ongoing literature review, or obtains and adapts commercial content, to ensure that changes in the standard of practice are rapidly incorporated into the decision support content. As noted earlier, the advent of molecular medicine will increase the speed of change in clinical knowledge, presenting new challenges for decision support maintenance. In addition, the knowledge engineering team must work closely with the analytics team that evaluates performance data to determine how decision support must change to achieve strategic objectives. With each successive stage of decision support capability health care performance becomes increasingly transparent. Rapid-cycle learning is a critical goal of knowledge management.

Added to these costs of content management, the organization will need to bear the expense of license fees, tools, and the sunk cost of clinical time spent on CDS management.

Some illustrative measures of knowledge management effectiveness are:

- Cost of knowledge maintenance team plus content license and asset management infrastructure.
- Cycle time for content update. This cycle can be measured as duration of time it takes to convert an agreed-upon guideline into a decision support specification and then into production. This measure assumes there is a business cost to delayed alignment.
- Cycle time for content agreement. This measure evaluates broader organizational effectiveness in getting agreement on enterprise-wide guidelines. For many organizations, this can take longer than converting the guideline into decision support.

19.6 CONCLUSION

Clinical decision support represents a class of tactics for applying medical knowledge to achieve superior performance. An organization should devote strategic discussions to knowledge management overall to ensure that it has defined appropriate boundaries, understands the functions of knowledge management, and is able to prepare a business case that ensures necessary investments of organization resources.

Organizations need a set of management structures and processes to ensure that an investment achieves desired organizational goals. Clinical decision support management structures and processes must achieve goals that include linkage to organizational strategies, prioritization of resources, and determination of the impact of clinical decision support. Although there is some variation in the organizational approaches of different health care providers, common guidelines do emerge.

Clinical decision support implementation and management do require the consideration of key aspects of how an organization's clinical and business strategies drive the IT strategy. Specifically, this chapter discussed the application of clinical decision support across legacy systems, CDS tools, knowledge acquisition and maintenance approaches, and the view of application systems as foundations.

Clinical decision support has one overarching goal—improving organizational performance. Achieving this goal requires ensuring strategic alignment, measuring performance relative to goals, and continuous improvement of the efficiency and effectiveness of the knowledge management function.

REFERENCES

Davenport, T. and Glaser, J. (2002). Just-in-time delivery comes to knowledge management. *Harv Bus Rev 80(7): 107–111.*

Evans, R. S., Pestotnik, S. L., Classen, D. C. et al. (1998). A computer-assisted management program for antibiotics and other antiinfective agents. *NEJM* **338**: 232–260.

Evans, R. S., Pestotnik, S. L., Burke, J. P., Gardner, R. M., Larsen, R. A., Classen, D. C. (1990). Reducing the duration of prophylactic antibiotic use through computer monitoring of surgical patients. *DICP* **24**(4): 351–354.

Glaser, J. and Flammini, S. (2003). Taking the application foundation view. *Healthcare Inform* 26–28.

Pestotnik, S. L., Classen, D. C., Evans, R. S., Burke, J. P. (1996). Implementing antibiotic practice guidelines through computer-assisted decision support: clinical and financial outcomes. *Ann Intern Med* **124**(10): 884–890.

Pifer, E. A., Teick, J. M., Sittig, D. F. and Genders, R. A. (2005). *Improve Patient Care with Clinical Decision Support* (Chicago, IL: Healthcare Information and Management Systems Society (HIMSS).

Weingart, S. N., Toth, M., Sands, D. Z., Aronson, M. D., Davis, R. B., Phillips, R. S. (2003). Physicians' decisions to override computerized drug alerts in primary care. *Arch Intern Med* **163**(21): 2625–2631.

20 LEGAL AND REGULATORY ISSUES RELATED TO THE USE OF CLINICAL SOFTWARE IN HEALTH CARE DELIVERY

RANDOLPH A. MILLER and SARAH M. MILLER

20.1 INTRODUCTION

This chapter discusses current legal and regulatory issues related to the use of clinical software in medical practice. Its purpose is to illustrate how the legal system, through tort law, may impose responsibility on software vendors and their users (who may be health care institutions, individual practitioners, or even patients) for possible harm that medical software might cause to patients. By describing regulatory issues, we also seek to highlight ways that institutions' internal governance can minimize such harm. This commentary first discusses scenarios of legal liability that vendors, hospitals, and doctors might face for patients' injuries arising from two genres of clinical software—software embedded in medical devices, and decision support software provided to licensed, practicing clinicians that enhances clinicians' abilities to collect, manage, and draw inferences from patient-related data and from general biomedical information. This discussion excludes otherwise relevant and important privacy-related concerns such as the protection and management of patient information. The discussion also describes the responsibilities of individual practitioners, as well as institutions and governments, in optimizing patient safety whenever decision support software is employed in the clinical setting. Some of the material presented herein was first explored during earlier collaborations of author RAM with Kenneth Schaffner, MD, PhD, and Alan Meisel, JD, at the University of Pittsburgh in the 1980s [1] and with Reed Gardner, PhD, of the University of Utah and others in the 1990s [2, 3].

20.2 LEGAL ISSUES RELATED TO USING EMBEDDED AND FREE-STANDING DECISION SUPPORT SOFTWARE IN CLINICAL SETTINGS

Two decades ago, in an article in the *Annals of Internal Medicine*, Miller et al. discussed the question of legal liability for injuries resulting from use of

Clinical Decision Support: The Road Ahead

computer software in health care [1]. That discussion followed an earlier, more general series of articles addressing the broader issue of liability for software-related injuries [4–8]. In the ensuing years, hospitals have increased their reliance on automated patient record systems and automated medical devices, and clinicians have increasingly embraced clinical decision support software (CDS) to assist with diagnosis and treatment. To date, no American courts have elucidated the conditions under which vendors, care-providing institutions, or clinicians (e.g., physicians and nurses) might be liable for harm to patients arising from the use of computer software [9]. But given these trends, it seems likely that courts may soon address the issue.

The issue of liability arising from the medical use of software programs continues to attract widespread coverage in both the legal and biomedical literature. Most commentators have explored possible theories of liability, and have concluded that the tort system offers injured plaintiffs—in this case, patients—the best chance of remedies [1, 10, 11]. American tort law distinguishes between intentional and unintentional injuries. For intentional injuries, the plaintiff must show only that the defendant has intentionally caused the plaintiff's injuries in order to recover damages. If the defendant has unintentionally caused the plaintiff's injuries, tort law can apply several different standards to determine whether the defendant should be considered legally responsible, and hence liable for damages. American tort law at present relies on two major standards of liability: negligence and strict liability [12–14].

The principle of negligence holds that defendants are liable for unintentionally causing a plaintiff's injuries where the defendant caused such injuries due to wrongful or unreasonable conduct. In the medical context, most jurisdictions hold that a defendant's conduct is negligent when it diverges from the customary treatment or medical practice that would be followed by the profession. Most jurisdictions also consider national custom rather than the way most doctors in a particular locality would treat a particular condition. So, in most jurisdictions, a plaintiff in a medical malpractice case would need to show that the standard of care in the medical community was not being followed by the physicians involved [1].

A minority of U.S. jurisdictions rely on a different rule in the medical context. Following the case *Helling v. Carey*, these jurisdictions may consider physicians negligent even if they followed national custom, if a "reasonably prudent" physician would have followed another treatment that might have averted the patient's injuries [15]. Some jurisdictions thus may impose liability even if the physician's treatment adhered to general custom, on the grounds that "a whole calling may have unduly lagged behind" with respect to adopting new treatments and methods of care [16]. In some jurisdictions, the physician would be considered negligent if the provided care diverged from the care that a reasonably prudent physician would have exercised under the circumstances. In *Helling*, a court found ophthalmologists negligent when they failed to administer a glaucoma test to a patient under 40 despite a general medical custom to administer such tests only to patients over 40 [17]. However, many have criticized *Helling*'s holding as inefficient for the health care profession as a whole and argued that custom should prevail as the general standard [18]. Moreover, in a jurisdiction that followed *Helling*, the plaintiff generally would need expert testimony to establish that a reasonably prudent physician would have followed the noncustomary practice [19]. Thus, if the liability standard is negligence, a defendant would be held responsible for a plaintiff's unintentional injuries if the defendant's conduct caused such injuries and if the defendant's conduct either diverged from customary practice or fell below an objectively "reasonable" level of care.

The principle of strict liability, on the other hand, does not consider whether the defendant was exercising precautions or following customary practice. Instead, it merely requires proof that the defendant's conduct was the direct cause of the plaintiff's injuries. Originally, strict liability applied to situations where the defendant was engaged in an inherently hazardous activity. Even if the defendant was exceptionally cautious in conducting such activities, courts allowed anyone injured as a result to recover. The rationale was that such activities carried an inherent risk of harm and that the defendant, in choosing to conduct such an activity, should bear its ultimate costs [20]. Compensation of the injured is thus a primary goal of strict liability [1].

In the past several decades, the principle of strict liability has been extended to situations where a manufacturer's product ends up harming the consumer. The Restatement of Torts (Third) defines a product as follows:

> Tangible personal property distributed commercially for use or consumption. Other items, such as real property and electricity, are products when the context of their distribution and use is sufficiently analogous to the distribution and use of tangible personal property that it is appropriate to apply the rules stated in this Restatement [21].

Under the present doctrine of products liability, a plaintiff harmed by a seller or manufacturer's product could recover for such injuries if they were caused by manufacturing, design, or warning defects in the product, regardless of the care the seller or manufacturer used in manufacturing, designing, and marketing the product. As presently defined, a manufacturing defect means that the product did not comply with the manufacturer's own design standards. A design defect means that the product has been designed in such a way that it carries unreasonable risks for the consumer. A warning defect means that, absent warning labels on the product about its intended use and possible hazards that would not be obvious to the consumer, the product may be unreasonably dangerous [22].

Rationales for strict liability for harms caused by products include:

- That consumers have imperfect information and cannot adequately assess a product's safety on their own
- That manufacturers use their market power to rely on standard form contracts protecting themselves from liability
- That manufacturers are the most preventable parties, whereas preventing consumers from experiencing accidents would be difficult
- Where the accident is not preventable, manufacturers should be held liable because they can spread the risk of the product through pricing mechanisms

In recent decades, products liability has been extended to a range of situations on the grounds that it represents the only way to protect consumers from otherwise unaccountable manufacturers or sellers. It has been extended to apply not only to the product's original manufacturer, but to anyone involved in the stream of commerce who is selling or distributing the particular product [23]. Based on the extension of the doctrine, some commentators have argued that the strict liability standard used in products liability cases should govern courts' disposition of claims from software-related injuries. Based on this approach, software vendors would be held liable for all injuries caused by malfunctions, irrespective of whether the vendor exercised due care when developing the product. Commentators, however, have not addressed whether hospitals using such software should also be strictly liable for these harms [1, 14]. In the software context, products liability likely would center on a design defects theory, as opposed to manufacturing or warning defects. Software, as a more intangible product, is more easily and perfectly replicated than tangible goods, making it unlikely that the software a buyer received from a vendor would have any unique differences from the source code (unless, for some reason, the manufacturer inserted potentially faulty end-user-specific customizations prior to distribution of the software to the local customer, which would be unusual). Hence, manufacturing defects theories are inapposite. Similarly, problems arising from software are unlikely to be the sort where a warning would have allowed a user to avert potential harms, making warning defects theories inapplicable. Instead, injuries arising from the use of software are most likely to relate to design defects such as programming errors or miscalculations that relate to how the software was created and how it functions.

Whether negligence or products liability is the applicable liability standard has enormous implications in the context of clinical software. Some commentators have suggested that products liability should be the general liability standard applied in this context. Although courts appear unlikely to adopt such a sweeping rule, it could have significant effects on whether manufacturers of CDS systems would enter the marketplace [1]. Vendors, hospitals, and physicians are likely to alter their perceptions of the necessary institutional precautions, possible costs, and relative benefits of clinical software based on whether exercising care and/or following customary practice is enough to absolve them of responsibility for patients' unintentional injuries.

The authors of this chapter argue that the use of clinical software is too varied for either negligence or products liability to be the appropriate standard for all possible situations where the use of such software ultimately harms the patient. The authors contend, first, that products liability as a standard, if applicable, should apply only to vendors, and not to hospitals. Second, products liability should apply to vendors only in certain situations. Software-related injuries likely to arise in a medical context are diverse, and thus strict products liability, though appropriate in some situations, cannot be universally applied. Instead, the applicability of strict liability may be a function of the extent to which the software relied upon is automated as a "closed loop" that provides little opportunity for human intervention. For example, cardiologists (and their patients) now rely upon embedded computer software to perform arrhythmia detection within implantable cardiac pacemakers of various sorts. In such settings, when the software malfunctions, holding a vendor responsible is most likely to be appropriate, since the pacemaker represents a "closed loop" system not easily inspected or interrupted by clinicians in most situations. In such settings, embedded software in a device is considered to be an integral part of the medical device, and such devices can be considered products. In contrast, when physicians use computer software as a diagnostic aid, strict liability is less apposite, as the physicians are using the software to enhance what is ultimately their own judgment and professional responsibility regarding diagnosis. Moreover, vendors of clinical software and its users—hospitals and physicians—may face different liability standards based on their relative ability to prevent accidental injuries, and the desirability of distributing such costs.

In order to illustrate the range of legal issues associated with clinical software, this chapter therefore considers two broad scenarios: medical device software and CDS software. The authors conclude that although products liability may be an appropriate means of holding vendors responsible for defects in medical device software, it should not be applied to most decision support situations.

20.2.1 Software Used in Medical Devices

Hypothetical examples of cases in which flaws in software (or the operation of software) could produce catastrophic errors include 1) a defect in the embedded software that regulates cardiac pacemaker function, 2) software errors ("glitches" or "bugs") that misreport life-critical serum chemistry test results from a laboratory system to an electronic medical record system, and 3) a programming error in an electronic prescribing system that alters intended doses in prescriptions written by doctors in a manner not readily apparent to the physicians. The FDA (U.S. Food and Drug Administration) estimates that software flaws in medical devices were responsible for approximately 7 percent of all medical device recalls from the early to mid-1990s, and that the figure is likely to rise: of the 10,000 categories of medical devices available in the United States, approximately half rely on embedded software to function [24].

The most serious examples of medical device software-related injuries to patients to date have involved radiation treatment devices. In two separate cases, the software design features of these devices led to multiple severe patient injuries, including some deaths. In the late 1980s, programming errors in the Therac-25, a then leading-edge radiation therapy device, caused two patients' deaths and severely injured another when a software bug caused the device to ignore technicians' corrections to dosage inputs [24, 25]. The device administered up to 15 times the estimated fatal dosage of radiation. More recently, the FDA prohibited Missouri company Multidata Systems from manufacturing or distributing radiation treatment devices after numerous patients in a Panama facility died of radiation overexposure attributed to software malfunctions in Multidata's device [24, 26].

Such device-related software malfunctions create several possible bases for liability. First, the vendors of such devices could be held liable for producing defective software. Second, technicians or physicians using these devices to treat patients could be liable for their patients' resulting injuries. In this scenario, absent egregious behavior by the technicians or physicians, vendors are much more likely to bear responsibility. The authors conclude that products liability may be an appropriate standard for vendors in this situation.

20.2.1.1 Vendors' Responsibilities for Devices Containing Embedded Software

Patients could try to hold vendors responsible for injuries arising from software defects by claiming that the vendor was negligent in developing the device's software. However, vendors are unlikely to face substantial liability if the applicable tort standard is negligence, even in a seemingly egregious situation like the Therac-25 case. Although shortcomings in the vendor's software demonstrably caused serious harm to patients, plaintiffs would still have a difficult time showing that the vendor failed to take sufficient precautions or diverged from the industry's general safety practices. Even with extensive testing and software debugging, the precise software flaw in the Therac-25, which prevented modification of the technician's instructions if the instructions were entered at a particularly rapid rate and a particular modification was made, would not necessarily have been apparent. Absent evidence of very poor programming practices or a vendor's failure to test the software, plaintiffs may find it difficult to hold vendors responsible for their injuries. Moreover, if an institution's technicians have made even seemingly trivial modifications to the equipment or to software, vendors may be able to avoid or mitigate liability on the grounds that the institution contributed to the negligence or that it constituted an intervening cause of the accident that should break the chain of liability.

A hospital using a defective device containing embedded software might sue the vendor based on a contractual theory of breach of an implied or express warranty that the device would perform as intended. However, as several commentators have noted, vendors are unlikely to be held liable under contractual theories, since vendors generally include warranty disclaimers in their contracts [27, 28].

Given these challenges, commentators have turned to products liability as a possible theory that would hold vendors legally responsible for the harmful consequences of software malfunctions [29]. Although no court has ever applied products liability standards to computer software or online sites [9], products liability standards appear to be well suited to redress injuries arising from medical device software defects and to provide sufficient incentives for efficient vendor behavior. As outlined earlier, a plaintiff would need to show that the device was a product, that the product was defective for products liability purposes, and that the product caused the plaintiff's injuries [1].

Pursuing such a theory first would require plaintiffs to demonstrate that the computer software in a device-defect case should be considered a good rather than a service. The Restatement of Torts (Third), excerpted earlier, distinguishes between products and services and defines products as "tangible personal property distributed commercially for use or consumption [30]." Irrespective of whether the software is incorporated within the medical device or sold separately for the purpose of operating the device, such software would appear to meet the definition of a product for the purposes of products liability.

Indeed, the Ninth Circuit hypothesized as much in its decision in *Winter v. G.P. Putnam's Sons* [31]. There, the court opined that defective computer software that diverged from its intended functions, like a defective technical tool that a user relied upon, might be considered a product. Similarly, in *C.G. Bryant v. Tri-County Electric Membership Corporation*, electricity was considered a product for the purposes of products liability, because the irregularities in the utility company's electricity supply caused the plaintiff's sawmill to burn down [32].

Some have questioned classifying software as a product because its programming aspects are seen to represent more of a service [30], but medical device malfunctions caused by defective software would appear to fall squarely in the category of products. Software defects in medical devices should be encompassed within products liability, because the physical manifestations of the software program have caused harm. The software used to operate medical devices has no function outside of the device; its sole intended use is within a physical product for medical care. This distinction is significant because it is not true of all clinical software; for example, CDS software may be used in a less tangible way by a clinician to choose among alternative therapies.

Assuming the relevant software in a case of a defective device is considered a product, a products liability suit would most likely proceed under a theory of design defects, as discussed earlier. Currently, American jurisdictions are evenly split among three approaches to assessing whether there is a design defect. The first approach, based on the Restatement (Third) of Torts, allows a plaintiff to show a product is defective through three further theories. First, the plaintiff can present a reasonable alternative design (RAD) for the product, illustrating that the RAD would have reduced the foreseeable risks of harm and that the omission of the RAD prevented the product from being reasonably safe [33]. Alternately, a plaintiff can show the product is defective through §3 of the Restatement based on a *res ipsa loquitor* theory, or the idea that the product can be inferred to have caused the plaintiff's harm because the injury is of a kind that would not ordinarily result absent a defective product, and the specific harm is one that might come from such a defect. Finally, the plaintiff could show that the product in question has a manifestly unreasonable design [34]. Examples include exploding cigars—a product with such a low utility to consumers and such a high level of risk that the product, from a risk-utility standpoint, should never have been marketed.

The second approach, based on §402A of the Restatement (Second) of Torts, requires a plaintiff to show that a product is unreasonably dangerous based on one of two further theories: that the magnitude of risks in the product's design outweighs the product's utility, or that the product is manifestly defective because the plaintiff's injury is not the type of harm that could have occurred without a serious flaw in the product [35]. This is similar to the *res ipsa* theory in the Restatement (Third).

Finally, the third approach, or the consumer expectations test, allows plaintiffs to show a design defect through two scenarios (derived from [35–37]). If the product is of such type that an average consumer could have definite expectations regarding how a product was supposed to function, and the plaintiff's accident arose in that context, the issue of whether the product conformed to reasonable consumer expectations is left to the jury [38]. If, instead, the plaintiff's accident is of a type beyond the knowledge or expectation of an ordinary consumer, the plaintiff must rely upon expert testimony to provide a risk-benefit analysis similar to those relied upon under the approaches of the Second and Third Restatements [39, 40].

Whatever the jurisdiction, arguments in medical software defects cases seem likely to center on some version of a risk-benefit analysis and the issue of whether the plaintiff's injury manifestly suggests a product defect (*res ipsa*). These are the predominant aspects of the first two approaches, and medical software, as a highly technical product, is likely to fall outside an ordinary consumer's knowledge. The Therac-25 case, for example, might have been well suited to a *res ipsa* theory, since an excessive dose of radiation is not an injury patients are likely to suffer save device malfunctioning, and the evidence had ruled out technician error. Cases where the device has a less severe flaw—for example, if the software programming failed to adequately prevent known side effects of a particular device-related intervention—would likely require a risk-utility test or, in some jurisdictions, proof of a reasonable alternative design. Plaintiffs would find this route far more difficult. Beyond the expense of hiring expert witnesses capable of proposing an alternative product or sufficiently analyzing all the risks and relative benefits of the device, the likelihood of pinpointing the precise error or omission in the software code would be low as well as time-intensive. The code is generally proprietary information, although plaintiffs might acquire it in discovery. Given the complexity and magnitude of most software, however, and the fact that errors in only a few lines of software can be responsible for catastrophic malfunctions, a plaintiff might be required to review hundreds of thousands of lines of code in order to find a few lines that could have been altered to prevent the malfunction.

The use of software-based devices in clinical settings may also raise difficult questions about causation [1]. In many cases, technicians may not have followed the vendor's instructions precisely. For example, in Panama, the technicians operating Multidata's radiation device used five lead shields around patients rather than four; because Multidata's software had trouble identifying the five shields, the technicians identified the five shields as one large shield, causing the machine to drastically miscalculate the appropriate radiation dosage. Vendors may be able to point to this sort of behavior to limit their own liability. Multidata has relied on a misuse defense, arguing that the technicians' modifications could not have been reasonably foreseen by the vendor at the time the product was sold [41]. Depending on the type of modifications made by users in hospitals, and the foreseeability of such modifications at the time of sale, such a defense may prevail.

Plaintiffs realistically may recover only for injuries caused by the most blatant and severe software malfunctions; however, the normative case for applying products liability standards in this scenario remains strong. The scenario corresponds well with the general rationales for applying products liability. Because the software code is embedded in the device and opaque to the user, neither doctors nor patients can assess whether the device is safe. Indeed, doctors and patients may not even be able to examine the software's content, as such information is proprietary. Users generally depend upon the vendor's operating manual or technical support for assistance. Additionally, vendors of software-operated devices have almost universally relied upon contractual language that disclaims warranties, eliminating contractual liability. Furthermore, vendors of software-operated medical devices appear to be the most preventable parties, since they have the greatest knowledge of software content, can most easily test devices for problems, and have the ability to adjust the code, unlike users. Finally, although some software bugs are inevitable given the complexity and length of code required for most devices, vendors, rather than patients or doctors, should be liable because they can include the costs of nonpreventable accidents in the device price.

20.2.1.2 Technicians and Physicians' Responsibilities While Using Devices Containing Embedded Software

Patients could also potentially bring a case against the technicians or physicians who operated a malfunctioning device containing embedded software, as well as against the clinical facility where the device was used. The question of whether hospitals and clinics can be liable for defective devices implanted or used in surgery has been raised with increasing frequency in the context of medical recalls on defective pacemakers, breast implants, and Teflon jaw implants. Courts in this context have thus considered whether hospitals can be considered part of the distribution chain of the device in question for purposes of products liability. With a few exceptions, courts generally have found that hospitals are not subject to strict liability in such situations [42]. The main rationale has been that even where hospitals have charged patients markup for the devices, hospitals are health care service providers rather than product distributors. Even though a product may be used as part of patient care, it is being used because it is essential to providing a course of treatment [43]. Even courts that have considered hospitals as potential distributors have declined to apply products liability standards to them on public policy grounds, concluding that making hospitals accountable for thoroughly testing all medical products used in a clinical setting would be unreasonable and would detract from hospitals' primary mission of providing patient care [44]. Thus, though hospitals can be strictly liable for defective products sold in their gift shop, in general, they will not be liable for defects in products used as part of medical care. The more a particular device seems inseparable from the service of assisting in patient care, the less likely it is that any court would apply a strict liability standard to the hospital [23, 45]. Thus, it is particularly unlikely that a hospital would be strictly liable for harms arising from malfunctioning software in a radiation device, or similar software-based errors.

Hospitals, technicians, and physicians could still be held responsible for injuries arising from malfunctioning devices under negligence. As mentioned earlier, jurisdictions differ on the standard used to evaluate negligence. In most jurisdictions, the question would be whether the care provided fell below the standard of care customary in the medical profession. In a few jurisdictions—the ones that follow *Helling*—the question would be whether a "reasonable physician or technician" would have acted differently. In either case, if all relevant precautions were followed in using the device, it would be difficult to prove negligence. However, in the Multidata case, the technicians diverged from instructions without authorization in operating the device and failed to adequately observe patients receiving the treatment. A Panamanian court ultimately convicted the technicians of involuntary manslaughter for criminally negligent behavior [41].

In conclusion, in situations involving software-based malfunctions in a clinical context, it seems appropriate to hold vendors accountable through products liability. The application of such a standard in this context comports with the general purposes behind products liability, because in this type of situation the vendor is uniquely situated to control possible malfunctions. Hospitals should not be held liable under the same standard, because technicians and physicians have little to no control over how the product functions as a program. Moreover, hospitals are using the devices only as part of patient care; it is difficult to compare hospitals to product distributors. Finally, as current case law suggests, holding hospitals strictly liable for such devices would have extremely detrimental effects for patient care. It is appropriate to hold hospitals liable in situations where technicians or physicians fail to adequately supervise patients receiving treatment when device malfunctions occur.

20.2.2 CDS Software Used by Licensed Practitioners during Medical Practice

The previous analysis suggested that products liability is an appropriate standard to apply to vendors who sell malfunctioning software-containing medical devices when injury to patients occurs. In this section, we discuss whether CDS software should be governed by the same standard. CDS software enhances practitioners' abilities to collect, manage, and draw inferences from patient-related data and from general biomedical information.

We describe two possible scenarios for liability involving its use: reliance upon erroneous clinical advice provided by such software, and failure to use CDS software when its use would have prevented improper treatment of the patient. Again, two possible classes of defendants are considered: vendors who produce the CDS software, and physicians—and their hospitals—that rely on CDS software in treating the patient.

We argue that CDS software should not be governed by a products liability standard for either class of defendants, because the decision support context materially differs from defective software-operated devices. Scenarios arising from the use of CDS software appear more complex, and thus appear to require different sets of liability rules. Vendors should not, and are unlikely to be, held liable under a products liability standard. However, if vendors' CDS software provides erroneous advice to its users, vendors may be considered negligent. Questions arise as to whether a vendor would or should be held to the same standard of care as a physician might be in a situation where the CDS software dispensed erroneous advice. Additionally, if a doctor relied upon CDS software that provided erroneous information, it is likely that the doctor might be considered negligent.

20.2.2.1 Erroneous Information Provided by CDS Software

Vendors' Responsibilities for Erroneous Information Provided by CDS Software

Vendors are unlikely to face strict liability in the event that their software dispenses erroneous advice to licensed clinicians, ultimately causing harm to patients. CDS software is unlikely to be considered a product for purposes of products liability. Most courts have considered computer software to be a "good" within the meaning of the Uniform Commercial Code (UCC) [46, 47], and thus governed by the UCC. If, however, the vendor has provided institution-specific programming and tailored the software to the hospital's particular needs, courts are more likely to consider the contract to involve a service [48]. The UCC usually sustains warranty disclaimers provided that they are obvious from the contract [49, 50]. Patients are unlikely to prevail under third-party beneficiary theories (i.e., patients injured as "bystanders" to a contract between the vendor and the hospital), as there is no privity of contract (i.e., mutual or successive relationship to the same rights) between patients and the hospital on the one hand, and the hospital and vendor on the other [51]. Even if there were such a connection, vendors' disclaimers would still limit, if not eradicate, patients' ability to recover under breach of contract theories. Second, patients and/or hospitals might sue vendors for injuries arising from software defects through the torts system, by claiming that the vendor was negligent in developing the device's software. To prevail on this claim, plaintiffs would need to demonstrate that the vendor owed them a general duty of care, that the vendor breached this duty by failing to take adequate precautions to check the software code or by employing programming practices that were below the customary level of vendors' practices, that the vendor's negligence with the software proximately caused the plaintiff's injuries, and that the plaintiff's damages are recoverable within the tort system.

Decisions relating to whether something is a product for purposes of products liability have long distinguished between harms arising from the functionality of a thing and the ideas it contains. The former is generally classified as a product, the latter is not. For example, in *Winter v. G.P. Putnam's Sons*, the Ninth Circuit distinguished between things that graphically illustrated technical information, like a compass, and things like books identifying poisonous and edible mushrooms, which it considered more like instructions on how to use a technical device [31]. The latter was not considered a product in the case, which found that the ideas and expression in a text or other work could not be considered a product because of their intangible properties. Furthermore, in *James v. Meow Media*, a Kentucky court found that the violent images and ideas contained in video games and other media could not be considered products in a case where these ideas were alleged to have inspired a school shooting. That court suggested that the ideas and images had no tangible expression or physical manifestation in of themselves; the tangible actor was the school shooter, not the video games [52]. Finally, comments to §19(a) of the Third Restatement have approved of such decisions, again distinguishing between information and the tangible medium in which it appeared [53].

Thus, CDS software generally can be distinguished from software within medical devices. Medical devices can be classified as products because their tangible effects, such as targeted radiation, are the source of patients' injuries. CDS software, however, does not replace the judgment or the functions of a physician; instead, it augments the physician's existing knowledge by providing further information [1]. It is true that CDS software may also involve record entries or other data that might provide the physician with erroneous information if the software were to malfunction; however, unlike medical devices used for patient treatment, such information is easily verified by a physician, and should be verified in the course of treatment. Even patient record information remains closer to an idea than to something capable of tangible expression.

Moreover, to include CDS software within products liability would be inconsistent with the purposes of imposing strict liability. Products liability would find a vendor of CDS software liable for errors in information provided to doctors even if the doctor should have known that the program's information was blatantly false; for example, when the computer-suggested regimen included prescribing a medication wholly and obviously unrelated to the patient's illness. Doctors, as possessors of specialized expertise, should be incentivized to take the utmost care when treating patients. Unlike possible accidents involving the use of medical devices to perform treatment, CDS software represents what game theorists call a sequential care situation: even if the software provides an erroneous diagnosis, the doctor subsequently acting upon the information possesses specialized knowledge of the patient and of general medical conditions, and is therefore in the best position to evaluate and reject faulty or inapposite advice. This is, however, not to say that vendors of CDS software should be able to avoid liability on the grounds that the attending physician, rather than the software's erroneous information, was the proximate cause of the patient's injuries. Instead, both the vendor and physician should be held accountable.

Thus, negligence appears to be the appropriate standard with respect to vendor liability for nondevice-related CDS software. The question remains whether a vendor would be held to the same standard as a physician might be in such a situation. Vendors might face a different general standard given the nature and purposes of CDS software and the fact that vendors are not necessarily physicians. In jurisdictions that generally look to medical custom, vendors of CDS software might be considered negligent if their software failed to dispense the advice or information that a reasonably prudent physician would provide. However, such jurisdictions could also hold vendors to the standard of custom within the decision support field, looking to whether in general, competitors' software would have provided the appropriate advice or information. Alternately, jurisdictions might define custom in this area as available, disseminated knowledge in the medical field, on the grounds that CDS software compiles this knowledge and intends to provide physicians with a broader range of knowledge than the unaided physician might possess. It is difficult to anticipate how a court would resolve the matter. It seems clear, however, that difficulties will arise in defining the standard of care to which CDS software vendors should be held. At the very least, CDS software must at least meet the standard of care to which the unaided physician would be held. But to hold vendors merely to this standard may be at odds with the purpose of CDS software.

One possible exception to the proposed rule that vendors should be subject to a negligence standard may arise in situations where a patient (or family member, in the case of a child) relies on purchased medical software installed on or accessed by a home-based personal computer (PC) to diagnose the patient's symptoms and to provide advice as to whether additional medical treatment is necessary. In such a scenario, licensed practitioners would be bypassed. The patient might inappropriately rely upon the software to select appropriate therapy; for example, the PC-based software product might suggest taking two acetaminophen tablets for a headache that actually was caused by meningococcal meningitis, a treatable and rapidly progressive, often fatal infection. One might argue that products liability should apply in this situation, drawing an analogy to the case of the pilot relying upon a misleading aeronautical chart [54].

This scenario suggests the difficulties in making legal distinctions between products and services. Purchased medical software is being sold as a product, but it is intended to replace the service of qualified medical advice. Applying products liability would hold the software maker liable for any harms arising from advice given by the software. Applying a negligence standard would involve the same difficulties in defining the standard of care as described earlier; however, the negligence standard should then at least hold the software to the same standard of care as the medical profession as a whole. We argue that a negligence standard is more appropriate in this situation, and that the proper standard should be the same as whatever standard applies to CDS software in the previous example.

Whereas aeronautical charts rely on exact knowledge to provide crucial in-flight information, diagnostic software used by the consumer ideally is based on the best available knowledge prevailing in the medical profession. A negligence standard should be sufficient to incentivize vendors of such software to keep abreast of changing medical knowledge with respect to the intended CDS software function, for example, to incorporate progress in new approaches to diagnosis in new releases of a CDS software package.

Practitioners' and Healthcare Facilities' Responsibilities for Erroneous Information Provided by CDS Software

Given that products liability is unlikely to apply to hospitals whose physicians or technicians use medical devices with software malfunctions, it is even less likely to apply to hospitals using CDS software. Decision support software is even further integrated into the process of providing patients with a course of treatment. Even the small number of jurisdictions that hold hospitals strictly liable for defective implants or pacemakers would be unlikely to extend the rule to the use of CDS software, because such software is incidental to the service of providing medical advice, and is not passed on to the consumer the way an implanted device or treatment might be [20].

Instead, physicians (and the hospitals that employ them) may be considered negligent if they fail to question erroneous advice given by CDS software and proceed to provide improper care to the patient. Again, the standard for determining negligence would vary depending on the jurisdiction, and would either be custom or the care provided by a "reasonably prudent physician" [1, 15–19, 55]. Liability in such cases would depend on the care provider's actions in the particular case. In this respect, the use of CDS software would not expose the care provider to any additional grounds for liability. The care provider is ultimately accountable for the care given, and held to the same standard of care irrespective of whether CDS software was used. This is consistent with the aims of CDS software: when functioning properly, it can help enhance diagnostic abilities and prevent misdiagnosis or other errors. However, decisions in treatment are ultimately left to the care provider, and the care provider should be considered responsible for these decisions.

Finally, any eventual lawsuit involving erroneous advice dispensed by CDS software is likely to involve joint and several liability. Joint and several liability holds multiple defendants—in this scenario, the vendor, hospital or clinic, and physician—responsible for the ultimate injuries suffered by the patient. It would mean that any liability assigned to one of the defendants would be shared by all, and one defendant could compel the others to contribute to any damages awarded to the patient. Joint and several liability could apply in this situation because the ultimate harm to patients is an "indivisible harm"—meaning that without *both* the software vendor *and* the attending physician's negligence, harm to the patient would not have occurred [56]. In other words, the software vendor's erroneous advice was harmful only because the attending physician failed to correct it, or the attending physician may have recommended a course of treatment only because it was recommended by the CDS software.

20.2.2.2 Failure to Use CDS Software to Prevent Medical Errors

The previous section has highlighted possible areas of legal liability when physicians rely on CDS software. However, it is also possible that licensed practitioners could be considered negligent if they failed to use CDS software to avert medical errors. A clinician can be considered negligent due to omissions in care as well as for overt actions.

As mentioned earlier, most jurisdictions look to whether the physician followed the national standard of care when determining whether the physician's conduct was negligent. If the use of CDS software became the national norm in medical practice, a physician who failed to use such a program, to the detriment of a patient, might thus be liable for the patient's injuries. But at present, the use of CDS software does not appear to have reached the level of custom.

In those jurisdictions that follow *Helling* and look to a more general reasonableness standard (other than custom), clinicians could be found liable for failing to use CDS software to avert an error, if evidence were introduced that a reasonably prudent physician would have done so in the case at hand. This would require expert testimony and would likely be left to a jury to assess [53]. Thus, if reliable CDS software were available and its use might have prevented the patient from injuries caused by the chosen treatment, even if such software is not used by a majority of physicians, it is possible that a particular physician might be found liable for having failed to use that technology if this failure caused the patient's injury [1]. Some precedent exists for this particular scenario. A Washington court found that a physician's failure to consult available literature, such as Medline, went against general notions of "good basic medicine" and constituted negligence in a case where consulting such literature might have led to a proper diagnosis:

> Finally, we address the government's concern expressed at oral argument as to "how a doctor ought to know that he doesn't know" whether there is information that need be disclosed. To justify ignorance of this type of risk would insulate the medical profession beyond what is legally acceptable. Here, there is expert testimony . . . that it would be "just good basic medicine" to conduct a literature search or contact specialists in response to a direct question to a physician such as the one posed here. With the demands of their profession, no one can expect doctors to have all material information stored in their minds. We do not decide the extent to which a literature search must reach. Some limits are appropriate. This may best be defined by reference to what is material and reasonably available. As we have stated, a risk is not material unless expert testimony can establish its existence, nature, and likelihood of occurrence . . . A literature search will thus put a physician on notice of these risks [55].

As this quote demonstrates, even in jurisdictions that define medical malpractice based on a general reasonableness standard, courts faced with this issue will ultimately weigh difficult questions about what sort of knowledge doctors should be expected to possess. The court distinguished between the doctor's own knowledge and knowledge from other, readily available sources. In this case, CDS software could have been one of many possible sources of the type of knowledge that would have avoided the patient's injuries. Instances where the use of CDS software would have been the only means of preventing the patient's injuries are less common, but it has recently been demonstrated for certain intensive care unit protocols that computer-based advice/adjustments are superior to purely human-mediated attempts to follow carefully defined protocols for ventilator management and hemodynamic monitoring.

In conclusion, this section discussed how tort law might treat CDS software use during patient care, and how different levels of liability may affect the incentives of the vendors and health care providers who develop and use it. Whether products liability should apply to vendors who develop and sell such software should depend upon whether the software is used in an automated medical device (closed system context) or whether the licensed practitioner can make an independent decision regarding the adequacy of the program's advice before treating the patient (open system context). In any event, products liability should not apply to hospitals using any such software.

20.3 RESPONSIBILITY FOR CDS SOFTWARE AT THE INSTITUTIONAL LEVEL AND POTENTIAL GOVERNMENTAL REGULATION

Clinical software systems are defined as algorithmic programs, related knowledge bases, and embedded interfaces, that directly contribute to the delivery of health care, in contrast with inventory or accounting functions. Clinical software systems are ubiquitous in medium- to large-sized health care facilities, although CDS systems represent only a small minority of such systems. Health care practitioners, clinical facilities, industry, and regulatory agencies share an obligation to manage clinical software systems responsibly using a common framework [2]. The previous section of this paper reviewing legal issues related to CDS systems indicated that use of clinical software systems does not often cause substantial harm to patients. However, concerns for safety at both the individual practitioner and institution levels must be addressed. Portions of the following discussion are reproduced with permission from the *Annals of Internal Medicine* [2].

20.3.1 The Complexity of Institutional Clinical Software Environments

Because clinical software systems are diverse and complex, determination of their safety is difficult. In an ACMI Distinguished Lecture in the late 1990s, Dr. Clement McDonald of the University of Indiana estimated that every large academic medical center in the United States has at least three dozen major software systems installed. Such systems serve a variety of purposes, including (among others) billing, inventory control, scheduling, compliance, electronic mail, message handling/routing, ADT (admission, discharge, and transfer) patient census functions, various laboratory functions, radiology image capture and retrieval, pharmacy ordering and dispensing, electronic patient chart/electronic health record functions, care provider order entry (CPOE), electronic textbook/reference functions, and clinical decision support.

Thousands of clinical software products compete in the commercial marketplace. A large number of "home-grown" systems exist, and variable quality, biomedically oriented World Wide Web sites proliferate in an uncontrolled manner. Most overall institutional installations are one-of-a-kind. Significant functional changes occur when a software product is integrated locally into a clinical information management infrastructure. Upgrades and maintenance also increase the variety and complexity of clinical systems. If there are six possible vendors (including the institution itself for "home-grown" products) for each of the presumed 36 major systems at a large institution, and three possible versions of each software package (most recent release, recent past release, and institutionally customized older release), then the number of potential configurations at an institution would be 18th to the 36th power (six vendors times three configurations for each of 36 systems). This impossibly large number of system configurations in an institution's environment is further complicated by the observation that variability in interactions between clinical software programs and individual users may cause unpredictable outcomes not related to software malfunctions.

Because of high local variability in both system configurations and usage patterns, it is not possible for a centralized monitoring agency such as the FDA (Food and Drug Administration) to monitor local software environments for safety at all institutions in the United States. Just as monitoring of safety for human subjects research was delegated by Congress to be shared among the FDA, NIH, and local Institutional Review Boards (IRBs), monitoring of clinical software for patient safety is arguably only possible at the local level [2].

20.3.2 Past and Current FDA Regulation of Clinical Software Systems

Through its mandate from Congress to safeguard the public, the FDA has regulated marketing and use of medical devices [2]. Section 201(h) of the 1976 Federal Food, Drug, and Cosmetic Act defines a medical device as any "instrument, apparatus, implement, machine, contrivance, implant, in vitro reagent, or other similar or related article, including any component, part, or accessory, which is ... intended for use in the diagnosis of disease or other conditions, or in the cure, mitigation, treatment, or prevention of disease ... or intended to affect the structure or any function of the body" [57]. In the past, FDA representatives have stated that clinical software programs, whether associated with biomedical devices or stand-alone, are "contrivances," and therefore fall within the FDA's realm of responsibility [2].

The FDA regulates medical devices that are commercial products used in patient care, devices used in the preparation or distribution of clinical biological materials (such as blood products), experimental devices used in research involving human subjects [2]. Commercial vendors of specified types of medical devices must register as manufacturers with the FDA and list their devices as products with the FDA. Upon listing, FDA classifies medical devices by categories. In its regulation of classified medical devices, the FDA usually takes one of three courses of action [2]. First, the FDA can "exempt" specific devices, or categories of devices, deemed to pose minimal risk. Second, the FDA employs the so-called 510(k) process—premarket notification—for nonexempt systems. Through the 510(k) process, manufacturers attempt to demonstrate that their devices are equivalent, in purpose and function, to low-risk (FDA Class I or Class II) devices previously approved by FDA (or to devices marketed before 1976). Such devices can be cleared by FDA directly. Finally, the FDA requires premarket approval (PMA) for higher-risk (FDA Class III) products and for products with new (unclassified) designs invented after 1976. Through the premarket approval process, a manufacturer provides evidence to the FDA that a product performs its stated functions safely and effectively. Premarket approval is especially important for those products that pose significant potential clinical risk. The processes of 510(k) premarket notification and premarket approval typically take a few to many months to complete, and may involve numerous iterations [2].

Exemption can take place in two ways: a device can be exempt from registration, and thus not subject to 510(k) requirements at all; or, a category of listed (classified) devices may be specifically exempted from certain regulatory requirements [2]. Whenever a nonexempt product is modified substantially (as defined by FDA guidelines), the vendor must reapply to the FDA for new clearance through the 510(k) or premarket approval mechanisms.

In mid-1996, the U.S. Food and Drug Administration (FDA) called for new discussions on the regulation of software programs as medical devices [58]. Previously, a 1989 draft policy [59], never adopted formally, served as guidance for FDA conduct. The draft policy recommended that the FDA exempt from regulation "generic" software (e.g., content-free spreadsheet and database programs), educational systems merely providing information, and systems that generated advice for clinician-users in a manner that they could easily override. In response to the 1996 FDA announcement, a consortium of health information-related organizations developed and published in 1997 recommendations for public and private actions that were intended to accomplish responsible monitoring and regulation of clinical software systems [2, 3]. The list of 1997 consortium members included the American Medical Informatics Association (AMIA), the Center for Healthcare Information Management (CHIM); the Computer-based Patient Record Institute (CPRI), the Medical Library Association (MLA), the Association of Academic Health Sciences Libraries (AAHSL), the American Health Information Management Association (AHIMA), the American Nurses Association (ANA), and the American College of Physicians (ACP). Not all boards of directors of all consortium members formally endorsed the consortium recommendations [2]. Dr. Reed Gardner at the University of Utah subsequently obtained NIH grant funding to develop prototypic models of the Software Oversight Committees (SOCs) at each of four separate health care centers that had advanced information systems installed (see [60] for a discussion of early SOC activity at Brigham and Women's Hospital in Boston).

20.3.3 1997 Health Care and Informatics Consortium Recommendations

The consortium stated that users, vendors, and regulatory agencies (including the FDA) should (after testing in limited environments) adopt their 1997 recommendations, as described here (and detailed more fully in [2, 3]).

Recommendation #1: The consortium proposed four categories of clinical software system risks and four classes of measured regulatory actions as a template to use in determining how to monitor or regulate any given clinical software system (see Table 20-1, reprinted with permission from *Annals of Internal Medicine* [2]).

Decisions about whether to install, and how to monitor clinical software systems should take into account: (a) the clinical risks posed by software malfunction or misuse; (b) the extent of system autonomy from user oversight and control, that is, the lack of opportunity for qualified users, such as licensed practitioners, to recognize and easily override clinically inappropriate recommendations (or other forms of substandard software performance); (c) the pattern of distribution and degree of support for the software system, including local customization; (d) the complexity and variety of clinical software environments at installation sites; (e) the likelihood that systems and their environments will evolve and change over time; and (f) the ability of proposed monitors or regulators to detect and correct problems in a manner that protects patients [2, 3].

TABLE 20-1 Recommendations for monitoring and regulation of clinical software systems (reprinted with permission from *Annals of Internal Medicine* 1997 [2]).

Regulatory class	Class A	Class B	Class C	Class D
FDA oversight	Exempt from regulation	Excluded from regulation	Required: Simple registration and postmarket surveillance	Required: Premarket approval and postmarket surveillance
Local Software Oversight	SOC optional	SOC mandatory	SOC mandatory	SOC mandatory
SOC role	Monitor locally	Monitor locally instead of FDA	Monitor locally, report problems to FDA as appropriate	Assure adequate local monitoring without replicating FDA activity
Software risk category				
Category 0: Informational or generic systems provide factual content or simple, generic advice ("give flu vaccine to eligible patients in mid-autumn"); general programs, such as spreadsheets, databases	All software in category			
Category 1: Patient-specific systems providing low-risk assistance with clinical problems give simple advice (e.g., suggest alternative diagnoses or therapies without stating preferences) with ample opportunity for practitioners to ignore or override		All software in category		
Category 2: Patient-specific systems providing intermediate-risk support on clinical problems have higher clinical risk (e.g., generate diagnoses or therapies ranked by score) but easy opportunities for practitioners to ignore or override, so net risk is intermediate		Locally developed or locally modified systems	Commercially developed systems not modified locally	
Category 3: High-risk, patient-specific systems have great clinical risk, and little or no opportunity for practitioners to intervene (e.g., a "closed loop" system that automatically regulates ventilator settings)		Locally developed noncommercial systems		Commercial systems

Recommendation #2: The consortium recommended local oversight of clinical software systems whenever possible, through creation of SOCs. (see Table 20-1).

Local software installation sites have the greatest ability to detect software problems, analyze their impact, and develop timely solutions. The consortium thought that it would be advantageous to develop a program of institutional and vendor-level controls for the majority of clinical software products, rather than to mandate universal, national-level monitoring, which is likely to be cumbersome, inefficient, and costly [2, 3].

Institutional Review Boards (IRBs) represent a federally mandated local monitoring process for the conduct of clinical research. Like IRBs, SOCs are intended to serve as guardians to ascertain that institutional processes affecting patients are carried out properly. However, the analogy between SOCs and IRBs is not complete, since their responsibilities differ. SOCs will have to monitor processes far less discrete than formal research protocols. To protect patient safety and to insure that software programs do not disrupt the integrity of clinical practice, local SOCs should enlist members with expertise in health care informatics, health care delivery, data quality, biomedical ethics, patients' perspectives, and quality improvement. SOCs should work with system administrators, users, and vendors to make sure that the appropriate ongoing monitoring is in place to detect adverse events, address them, and insure that the overall system performs as designed. SOCs might apply pressure to correct vendor/product-related problems when the usual means fail. SOCs need not install software or monitor software function directly; however, they must insure that others do so in a manner that protects patient safety and institutional integrity. The consortium thought that it would be important to specify ethical guidelines for SOC conduct and rules for avoiding conflicts of interest between local employees (including SOC members) and commercial vendors (including employee-owned enterprises) [2, 3].

Recommendation #3: Budgetary, legal, and logistic constraints limit the type and number of systems that the FDA can regulate effectively. The consortium recommended that the FDA focus its regulatory efforts on those systems that pose the highest levels of clinical risk and that give limited opportunities for competent human intervention (see Table 20-1). This parallels the earlier discussion in this chapter regarding strict products liability standards for "closed loop" systems that had little chance for clinician intervention, and negligence standards for "open" systems providing advice to licensed practitioners directly.

The majority of clinical software systems should be exempt from FDA regulation (see Table 20-1). The FDA should require producers of certain intermediate-risk clinical software systems to list them as products with the FDA for simple monitoring purposes, requiring that such products undergo the 510(k) or PMA processes if more than a threshold number of validated adverse events are reported. The FDA and manufacturers should develop new, comprehensive, appropriate standards for clinical software product labeling (for example see [61]).

Recommendation #4: The consortium recommended that health care information system vendors and local software producers adopt a code of good business practices. The practices should include (but not be limited to) guidelines for quality manufacturing processes; standardized, detailed product labeling; responsible approaches to customer support; and utilization of industry-wide standards for electronic information handling and sharing—including standards for health care information format, content, and transport (see Table 20-1).

Recommendation #5: The consortium recommended that health information-related organizations work with other groups, including clinical professional associations, vendor organizations, regulatory agencies, and user communities, to advance our understanding and knowledge of approaches to regulating and monitoring clinical software systems (see Table 20-1).

20.4 CONCLUSION

This chapter has described both the individual's and institution's responsibilities and legal obligations regarding various forms of clinical software systems.

In summary, it is conceivable that both products liability and negligence may be applicable legal standards of liability for software used in a clinical context. Given the complexity of current programs and aspects of programs that resemble both goods and services, we conclude that the applicability of these differing standards may depend on the particulars of the relevant software. It is likely that clinical software embedded in medical devices in general will be held to strict products liability standards, whereas CDS software that provides advice to licensed practitioners will be judged by negligence standards.

The 1997 health care and informatics consortium provided recommendations on how to develop and realize processes for responsible monitoring and regulation of clinical software systems. The consortium's goal was to encourage a coordinated effort to safeguard patients, users, and institutions, as clinical systems are implemented to improve clinical care processes, but widespread formal adoption of those principles has not yet occurred. The need to develop innovative and more effective software systems while monitoring installed systems for patient safety remains a critical unmet need in the American health care system.

ACKNOWLEDGMENTS

Randolph A. Miller's efforts in preparing this manuscript have been supported in part by Grants 5-G08-LM-05443, 1-R01-LM-06226, 1-R01-LM06591-01, and 1-R01-LM007995-01 from the National Library of Medicine. We would like to thank Professor Douglas A. Kysar, 2005 Visiting Associate Professor, Harvard Law School, for his insightful comments on an earlier version of the manuscript. This chapter is based in part on earlier collaborations of author RAM with Kenneth Schaffner, MD, PhD, and Alan Meisel, JD, at the University of Pittsburgh in the 1980s [1] and with Reed Gardner, PhD, at the University of Utah and others in the 1990s [2, 3]. Permission has been granted by the *Annals of Internal Medicine* to reproduce herein portions of copyrighted materials previously published in that journal [2, 3].

REFERENCES

1. Miller, R. A., Schaffner, K. F., Meisel, A. (1985). Ethical and legal issues related to the use of computer programs in clinical medicine. *Annals of Internal Medicine* **102**: 529–536.
2. Miller, R. A., Gardner, R. M. (1997). Summary recommendations for responsible monitoring and regulation of clinical software systems. *Ann Intern Med* **127**(9): 842–845.
3. Miller, R. A., Gardner, R. M. (1997). Recommendations for responsible monitoring and regulation of clinical software systems. *J Am Med Inform Assoc* **4**(6): 442–457.
4. Gemignani, Products liability and software, 8 Rutgers Computer & Tech. L.J. 173, 196–199 (1981).
5. Note: Strict products liability and computer software: Caveat vendor, 4 Computer/L.J. 373 (1983).
6. Note: Negligence: Liability for defective software, 33 Okla. L. Rev. 848, 855 (1980).
7. Note: Computer software and strict products liability, 20 San Diego L. Rev. 439 (1983).
8. Note: Easing plaintiffs' burden of proving negligence for computer malfunction, 69 Iowa L. Rev. 241 (1983).
9. Rustad, M. L. (2004). Punitive damages in cyberspace: Where in the world is the consumer? *Chapman Law Review* 7(39).
10. American Law Institute. (1965). Restatement (Second) of Torts. St. Paul, Minnesota: American Law Institute Publishers.
11. Uniform Commercial Code. (1962). Chicago: National Conference of Commissioners of Uniform State Laws.
12. Keeton, W. P. (1984). Prosser and Keeton on Torts. 5th ed. St. Paul, Minnesota: West Publishing Co.
13. Epstein, R. A. (1980). Modern Products Liability Law. Westport, Connecticut: Quorum Books.
14. American Law Institute. (1998). Restatement (Third) of Torts: Products Liability. St. Paul, Minnesota: American Law Institute Publishers.
15. Peters, P. G. (2000). The quiet demise of deference to custom: malpractice law at the millennium, 57 Wash. & Lee L. Rev. 163.
16. Hooper, T. J. (1932). 60 F. 2d 737 (2d Cir. Wash.).
17. *Helling v. Carey* (1974). 83 Wn. 2d 514 (Wash.).
18. Epstein, R. A. (1992). The path to the T.J. Hooper: The theory and history of custom in the law of tort, 21 *J Legal Stud* 1.
19. Peters, 171–172, 185–187.
20. *Rylands v. Fletcher*, L. R. 3 E. & I. App. 330 (H. L. 1868).
21. Restatement (Third) of Torts, §19.
22. Restatement (Second) of Torts, §402A; Restatement (Third) of Torts, §2 and §3.
23. *Silverhart v. Mount Zion Hospital* (1971). 20 Cal. App. 3d 1022, 98 Cal. Rptr. 187.
24. Gage, D., McCormick, J. (2004). Can software kill? eWeek.com, Mar. 8, 2004; accessed at http://www.eweek.com/article2/0,1759,1544225,00.asp.
25. Leveson, N. J., Turner, C. S. (1993). An investigation of the Therac-25 accidents. *IEEE Computer* **26**(7): 18–41.
26. American Health Line. (2003). FDA blocks Missouri company from making radiation devices.
27. Dahm, L. L. (1995). Restatement (Second) of Torts Section 324A: An innovative theory of recovery for patients injured through use or misuse of health care information systems. *The John Marshall Journal of Computer & Information Law* 14(2): 91–92.
28. Gable, J. K. (2001). 5 Quinnipiac Health L.J. 127. An overview of the legal liabilities facing manufacturers of medical information, 140–141.
29. Dahm, Notes 3–6.
30. Gable, 146–147.
31. *Winter v. G. P. Putnam's Sons* (1991). 938 F.2d 1033, 1036, Ninth Circuit, July 12, 1991.
32. *Bryant v. Tri-County Elec. Membership Corp.* (1994). 844 F. Supp. 347, 349 (KY).
33. Restatement (Third) of Torts, §2(b).
34. Restatement (Third) of Torts, §2(e).
35. Restatement (Second) of Torts, §402A.
36. *Barker v. Lull Engineering Co.* (1978). 20 Cal. 3d 413 (CA).
37. *Green v. Smith & Nephew AHP, Inc.* (2001). WI 109 (WI).
38. *Heaton v. Ford Motor Co.* (1967). 248 Ore. 467 (OR).

39. *Soule v. General Motors Corp.* (1994). 8 Cal. 4th 548 (CA).
40. *Vautour v. Body Masters Sports Indus.* (2001). 147 N.H. 150 (NH).
41. McCormick, J., Steinert-Threlkeld, T. (2004). Panama technicians found guilty. *Baseline*; accessed at http://www.baselinemag.com/print_article2/0,2533,a=139604,00.asp.
42. The main exception was Missouri, whose courts permitted the application of products liability to hospitals in this situation. However, the Missouri legislature disagreed, and subsequent Missouri court decisions have concluded that the Missouri legislature's intent was to statutorily bar the extension of such liability. See, for example, *Budding v. SSM Healthcare Sys.*, 19 S.W.3d 678 (MO 2000).
43. *Ayyash v. Henry Ford Health Systems*, Mich. App., 210 Mich. App. 142, 533 N.W.2d 353 (1995) (strict liability not applied to hospital for defective jaw implants); *St. Mary Medical Center, Inc. v. Casko*, 639 N.E.2d 312, 315 (Ind. 1994) (strict liability not applied to hospital for defective pacemaker); *Hoff v. Zimmer*, 746 F. Supp. 872 (W.D. Wis. 1990) *(strict liability not applied to hospital for failure of prosthetic hip); Easterly v. HSP of Texas, Inc.*, 772 S.W.2d 211 (Tex. Ct. App. 1989) *(strict liability not applied to hospital supplying epidural kit with defective needle); Hector v. Cedars-Sinai Medical Center,* 180 Cal. App. 3d 493, 225 Cal. Rptr. 595 (1986).
44. *Parker v. St. Vincent Hosp.* (1996). NMCA 70, 122 N.M. 39, 46, 919 P.2d 1104, 1111 (Ct. App. 1996).
45. *Magrine v. Krasnica* (1967). 94 N.J. Super. 228 [227 A.2d 539].
46. *Advent Sys. v. Unisys Corp.* (1991). 925 F.2d 670, 675–676 (3d Cir.).
47. *RRX Indus. v. Lab-Con, Inc.* (1985). 772 F.2d 543, 546 (9th Cir.).
48. *Micro-Managers, Inc. v. Gregory.* (1988). 434 N.W.2d 97, 100 (Wis. Ct. App.).
49. Dahm, 91–92 and note 77.
50. Uniform Commercial Code 2-316(1).
51. Dahm, 93.
52. *James v. Meow Media* (2000). 90 F. Supp. 2d 798, 810-1 (United States District Court for the Western District of Kentucky).
53. Restatement (Third) of Torts, §19(a) Comment d.
54. *Aetna Casualty & Surety Co. v. Jeppesen & Co.* (1981). 642 F.2d 339 (9th Cir.).
55. *Harbeson v. Parke Davis, Inc.* (1984). 746 F.2d 517, 525 (9th Cir.).
56. *Kingston v. Chicago & N.W. R. Co.* (1927). 191 Wis. 610, 211 N.W. 913.
57. Public Law 94-295. (1976). Medical Device Amendments to the Federal Food, Drug, and Cosmetic Act (passed on May 28, 1976).
58. Federal Register. July 15, 1996. **61**(136): 36886–36887.
59. Young, F. E. (1987). Validation of medical software: Present policy of the Food and Drug Administration. *Ann Intern Med* **106**: 628–629.
60. Abookire, S., Martin, M. T., Teich, J. M., Kuperman, G. J., Bates, D. W. (1999). An institution-based process to ensure clinical software quality. *Proc AMIA Symp* 461–465.
61. Geissbuhler, A., Miller, R. A. (1997). Desiderata for product labeling of medical expert systems. *Internat J Med Informatics* **47**(3): 153–163.

VI

KNOWLEDGE MANAGEMENT APPROACHES

21 KNOWLEDGE MANAGEMENT INFRASTRUCTURE: EVOLUTION AT PARTNERS HEALTHCARE SYSTEM

TONYA HONGSERMEIER, VIPUL KASHYAP,
and MARGARITA SORDO

21.1 INTRODUCTION

Partners Healthcare Systems (PHS) is a heterogeneous integrated delivery system comprising numerous hospitals and thousands of physicians. The clinical information systems that support care across the enterprise are equally heterogeneous, and enriched with over a decade's worth of advanced decision support knowledge generated by research teams who have studied how to improve quality and safety through clinical informatics (Bates et al. 1999, 2003; Davenport and Glaser 2002; Boxwala et al. 2004). In recent years, the focus has expanded to investing in organizational alignment and systems that enable PHS to become an effective learning organization.

A Knowledge Management for Clinical Decision Support division was created within the Clinical Informatics Research and Development Group of Partners Healthcare Information Systems. The goal is to reduce the cost and speed the rate of acquisition of decision support knowledge that drives PHS care quality, safety, business performance, usability, and consumer loyalty. In addition, in collaboration with the PHS Enterprise Clinical Services group, this division seeks to architect an anticipatory, proactive decision support framework that will significantly reduce reliance on interruptive, task-interfering modes of decision support to promote quality and safety. The core objectives for the next three years are to:

1. Establish effective, efficient knowledge management practices across PHS through the implementation of knowledge management infrastructure and support of organizational alignment necessary for optimal knowledge asset management.
2. Decouple knowledge, decision support engines, and workflow engines for rapid agility customization and enhancements.
3. Implement context-aware, proactive decision support services through the implementation of clinical decision support infrastructure that enables knowledge delivery to be personalized in every encounter to

Clinical Decision Support: The Road Ahead

the context of the user, the patient, the setting, and commonly accepted best practices.

4. Support data-driven learning by leveraging the quality data warehouse to support a closed-loop learning cycle of enterprise-wide knowledge discovery and performance improvement.
5. Achieve personalized medicine and translational medicine by leveraging the integrated genomic and phenotypic data repositories, in collaboration with the Harvard Partners Center for Genetics and Genomics, for knowledge discovery and translational medicine.
6. Align knowledge assets with existing and emerging industry, state, and federal terminology standards (such as SNOMED); quality regulations (such as Joint Commission on Accreditation of Healthcare Organizations, Health Employment Data Information Set, National Quality Forum, Leapfrog); pay-for-performance contracts; and industry guidance wherever possible.

As a society, we are still early in our evolution from a medical culture that values reliance on memory to one that values effective distribution of effort between the information management that systems do well and the personalized caregiving only humans can deliver. Partners Healthcare Information Systems has been a trailblazer in the use of informatics to improve health care quality for well over a decade. Some of the most quoted published research on the impact of computerized physician order entry and rule-based decision support on patient safety was conducted here (Kuperman et al. 1995, 1996, 2001; Bates et al. 1999, 2003; Jha et al. 1999), some of which is reviewed in Chapter 5. A goal of Partners now is to promote an enterprise approach to unify and align, maintain, extend, and deepen those capabilities so that it can deliver only the highest quality care, and maintain its national leadership position. The intent is to extend the best of the clinical application functionality, knowledge sources, and decision support capabilities across all of Partners so that, regardless of where care is delivered, all patients receive uniformly superior quality care.

21.2 RAPID INNOVATION DISCOVERY AND ADOPTION: KEY INFRASTRUCTURE REQUIREMENTS

As biomedical research changes the standard of care at a rapid pace, the knowledge bases required for safe care are growing exponentially. Infrastructure for knowledge management, particularly the knowledge that is tightly embedded in clinical applications, is a quality imperative. Winnie Schmeling, RN, PhD is Senior Vice President of Organizational Improvement at Tallahassee Memorial Healthcare. This site is among 12 sites that were selected by the Robert Woods Johnson Foundation for the Pursuing Perfection grant program. She defines knowledge management as a "systematic process for making sure that everybody knows what the best of us knows" to deliver excellent care (Hongsermeier and Schmeling 2002).

Knowledge Management in practice is comprised of three key processes, as introduced in Chapter 19 (corresponding to the 3 life cycle processes described in Chapter 1 and further discussed in Chapter 24). These include knowledge application, knowledge discovery, and knowledge asset management. They are organized in a circle to emphasize that the

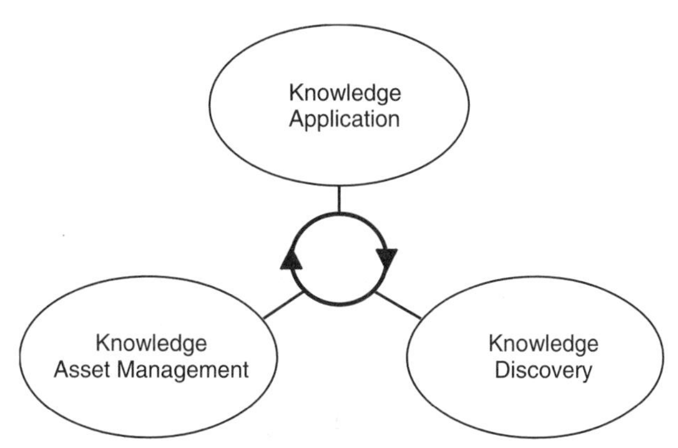

FIGURE 21-1 Knowledge management components for a learning organization.

knowledge management process is one of continuous learning and knowledge dissemination (see Figure 21-1).

Knowledge Application is the science of leveraging knowledge at the right places in clinical workflow to support enhanced caregiver effectiveness, work satisfaction, patient satisfaction and overall care quality. This view of Knowledge Application is broad and inclusive of the traditional definition of CDS. Knowledge can be delivered through clinical applications in a variety of forms such as templates for documentation, specialty data review flow sheets, order sets for protocol-driven care management, dashboards and reports for feedback, or alerts and reminders. Figure 21-2 (adapted from FCG Corp.) shows a model for the continuum of informatics approaches to Knowledge Application. This model ranges from passive access to knowledge to anticipatory CDS. In a high-performance learning organization, process and outcomes data are captured as a byproduct of workflow and rapidly analyzed to discover new knowledge or determine new models for process improvement. In essence, robust knowledge application evolves into and merges with knowledge discovery, whereby the knowledge repository "learns" from the clinical outcomes in the patient database.

Most major vendor systems on the market today are delivering CDS in safety-net mode. For example, when a clinician prescribes imipenem (a penicillin-related antibiotic) in a patient with renal failure and pneumonia, a vendor system might typically interactively warn of a potential adverse drug effect. The more

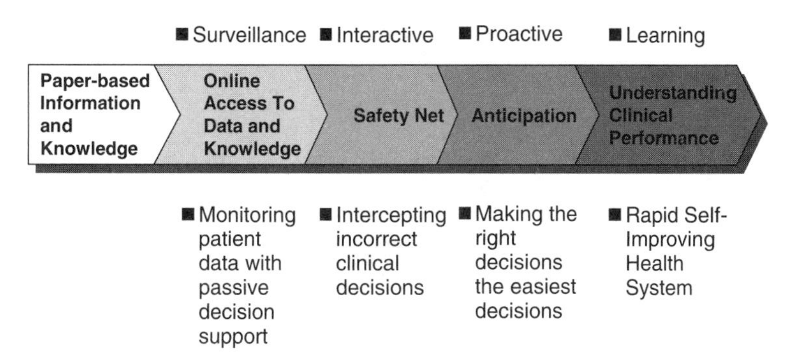

FIGURE 21-2 A continuum of knowledge application and discovery.

complex the patient, the more frequent the counteractive alerts, hence, the risk of alert fatigue. At Partners, the decision support system avoids alert fatigue by precalculating creatinine clearance and surfacing a safer default dose if needed, with information notifying the clinician of the suggested dose (see Figure 21-3). Clinicians have positively responded to this nontask-interfering design approach that makes the right care decisions the easiest care decisions.

The vendor community will be required to push development in the direction of anticipatory modes of decision support particularly in light of clinician concerns about speed and usability of information systems. This is illustrated by the recent events at Cedars-Sinai Medical Center in Los Angeles. The computerized physician order entry system was turned off in early 2003 after hundreds of physicians complained that it slowed down the process of filling their orders. Further, with the expected arrival of gene-diagnostics as additional data to be considered in therapeutic decision-making, exclusive reliance on the safety-net mode of decision support will become increasingly untenable.

Knowledge Discovery is the clinical research process of analyzing data for the purpose of understanding performance, reporting, predicting, and harvesting new knowledge on how to improve. With each successive stage of decision support in Figure 21-2, health care performance becomes increasingly transparent. As processes for caregiver communication, results access, ordering, documentation, task completion, and patient monitoring become supported by expert systems, more data are collected about clinical rationale, process, and outcomes. In a well-designed analytic data repository, these data can be mined to identify new knowledge that can, in turn, be disseminated through knowledge-enriched clinical applications. Rapid-cycle learning is one of the goals of knowledge management.

Currently, most hospitals have poor access to clinical data for measurement and rely, instead, on billing and administrative data. The architecture of a typical

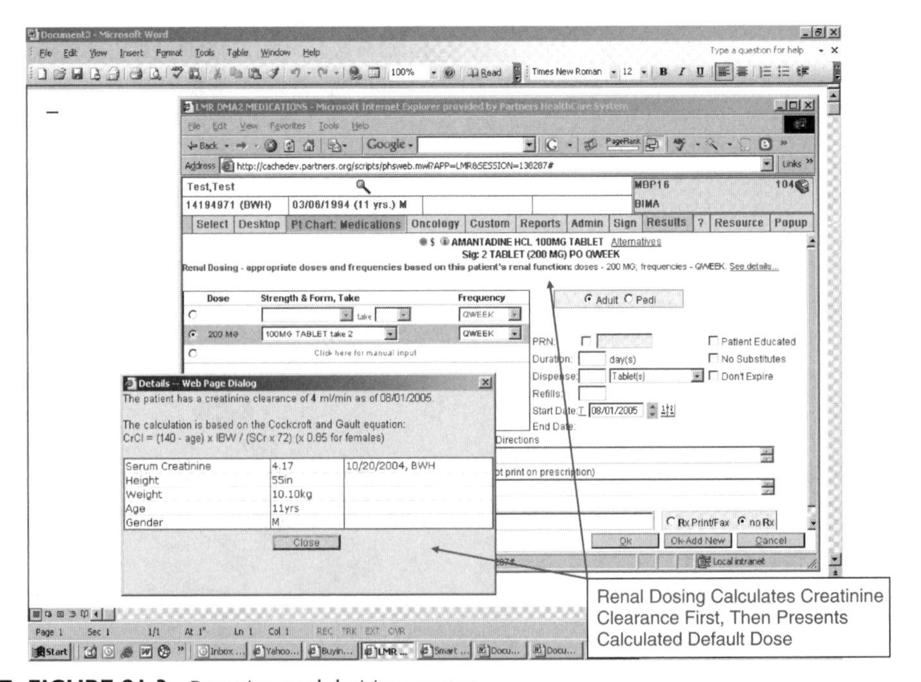

FIGURE 21-3 Proactive renal decision support.

transaction-oriented Clinical Data Repository (CDR) is not optimized to support measurement. Further, the data that must be aggregated to enable deep analytics often is located in many databases across an integrated delivery system. In the absence of an analytic data repository, those engaged in the process of research and performance measurement must bear the cost and time delays of chart abstraction and data acquisition to aggregate clinical data, which consequently slows the translation of such insights into quality improvement.

Knowledge Asset Management is one of the more challenging aspects of knowledge management. In nonhealth care industries, the focus of asset management is typically on free-text document management. In health care, we must also focus on acquisition and maintenance of knowledge encoded in workflow applications to support clinical care, administrative functions, and clinical research.

Developers of clinical systems in both academic and commercial settings typically have used nongeneralizable approaches to knowledge engineering and development of their products. By and large, these systems are proprietary in design and the knowledge within them is not easily extracted, exported, shared or disseminated (Greenes et al. 2004). Further, they often supply proprietary building blocks for knowledge editing but few road maps for how to leverage all the capabilities to optimize safety. Support for processes such as subject matter expert collaboration and vetting of computable guidelines is also weak, at best. It is not uncommon that, due to idiosyncratic and varied implementation approaches, even clients of a single vendor are challenged to share or pool their encoded knowledge so that they can progress more quickly in their implementations. The life cycle of knowledge engineering ranges from raw literature evidence to evidence summarization (e.g., that compiled and offered by companies such as Zynx, Clineguide, and MD Consult), to human-readable specifications for knowledge encoding, and then finally, knowledge encoding.

Much of the recent work in knowledge management research has focused primarily on the processes related to guideline modeling, knowledge encoding, and the sharing of encoded knowledge. This work is essential for progress with knowledge sharing and is receiving heightened attention with the publication of the recent IOM report, *Patient Safety: Achieving a New Standard of Care*. However, successful dissemination of standards for knowledge encoding will depend on vendor adoption of knowledge-based elements and information models into their proprietary application architectures and data structures. Most Electronic Health Record (EHR) knowledge editors are not able to utilize guidelines written in a guideline expression language or standard syntax. Therefore, until standards are agreed upon and adopted by the clinical systems vendors, the primary vehicle for sharing knowledge bases such as rules among health care providers utilizing disparate vendor systems will be human-readable specifications. Newly emerging suppliers of structured content such as order sets (e.g., Wolters-Kluwer, Micromedex, Zynx) are developing vendor relationships so that content can be uploaded via custom application programming interfaces.

21.3 KNOWLEDGE ASSET MANAGEMENT INFRASTRUCTURE

21.3.1 Internal Analysis of Requirements for Knowledge Management Infrastructure

The first step at Partners Healthcare Information Systems in the development of a knowledge management infrastructure was to evaluate the state of current knowledge management practices and technologies to determine how solutions could be crafted. The aim was to identify the requirements for an end-to-end management framework to ensure an optimal and consistent experience for knowledge engineers in a collaborative environment. In order to evaluate the state of current practices, a series of interviews was conducted with champions and multiple domain experts involved in knowledge engineering of key knowledge assets within the various clinical systems at Partners (Boxwala et al. 2004; Sordo et al. 2004). From these interviews many challenges were identified along the key stages of the content life cycle including:

Guideline selection:

- Lack of organizational alignment and clear stewardship of decision support
- Inconsistent policies for prioritizing proposed knowledge encoding projects
- Lack of interproject coordination leading to inconsistencies and duplication of effort
- Lack of policies, procedures, or systems support for development and maintenance of content
- Lack of organization standards for storing, searching information

Guideline vetting and validation:

- Lack of sanctioned subject matter expert teams to review content
- Poor communication among authors leading to inconsistencies and uncertainty in validity of enterprise content
- Lack of tools to support guideline vetting and validation by subject matter experts before encoding
- Lack of shared repositories of documentation about content in production
- In many instances, transactions must be generated in the system to generate screen shots to determine state of content in production

Guideline conversion into specifications for encoding:

- Lack of standard formats for structuring information
- Information that exists organized, structured, and displayed according to author's own criteria

Guideline encoding:

- Lack of propagation, inheritance or reuse of knowledge resulting in increased labor to encode or revise content
- Project competition between knowledge enhancements and functional enhancements to software because much knowledge encoding is engineer-dependent
- Delays and bottlenecks between when a decision is made to change content and when such changes can be activated in the decision support system
- Lack of definition management infrastructure

Guideline revision:

- Information kept in "silos," isolated repositories created by authors in isolation from other authors; therefore, information is missing, duplicated, inaccurate, recreated many times, and often not accessible to users

- Lack of tools for creating, storing, organizing, retrieving, reusing documents
- Lack of tools for versioning and auditing to keep track of changes to content
- Lack of reports to determine content impact on clinical outcomes or usability

Out of these challenges, a series of functional requirements were defined for a robust collaboration, content editing, and content management infrastructure. The primary benefits of integrating collaboration, content editing, and content management tools into enterprise-wide knowledge management platforms are to provide users with the means to aggregate and structure content that is currently widespread within various PHS institutions; to make such knowledge accessible to authorized users within the organization; to enable standardization of clinical best practices regarding clinical decision support; and to acknowledge and maintain content differences where applicable. Within this framework, collaborative processes would be transparent to clinical experts to understand how content decisions were defined and updated.

The philosophy that has been adopted is that a unifying content management environment will encourage processes that are repeatable and transparent across the organization, supporting reusable content stored in a definitive source accessible to all users. Such a strategy can help reduce costs of creating, managing, distributing and updating content information in a way that effectively supports PHS business performance by increasing quality and consistency of decision support assets. Ideally, such a knowledge management infrastructure must include some of the following capabilities:

Knowledge Reuse via a Centralized Repository:

- Storage and reuse of knowledge building blocks such as elements from terminologies, information models, and ontologies
- Storage and reuse of structured clinical knowledge objects such as order sets, documentation templates, clinical decision support rules
- Storage and reuse of unstructured knowledge documents

Support for Content Management and Collaboration Processes:

- Ability to define and enforce life cycles for different types of knowledge objects
- Staging area for rules creation and validation before publishing into production
- Provenance, versioning, and auditing of knowledge objects
- "Knowledge event management" such that changes in a content area such as drug information are propagated to all other related or dependent knowledge assets
- Access to and production of metadata as a byproduct of knowledge creation workflows

Ability to support collaboration:

- Well-defined workflow-based collaboration around structured knowledge objects
- Ad-hoc collaboration around unstructured knowledge objects
- Interproject coordination and collaboration across multifaceted domains (e.g., geriatric prescribing and stroke management)

Support for Knowledge Authoring and Creation Environments:

- Ability to define templates and forms
- Ability to support reuse by surfacing appropriate building blocks and operations for composing these objects
- Definition and modeling of workflows and life cycles
- Modeling of dependencies between knowledge objects

Figure 21-4 depicts the envisioned unified collaboration and content management environment where clinical and nonclinical users under different roles may collaborate among themselves to access, create, publish, and deploy content in multiple formats. A Web-based interface may provide searching, indexing capabilities to ease access to content.

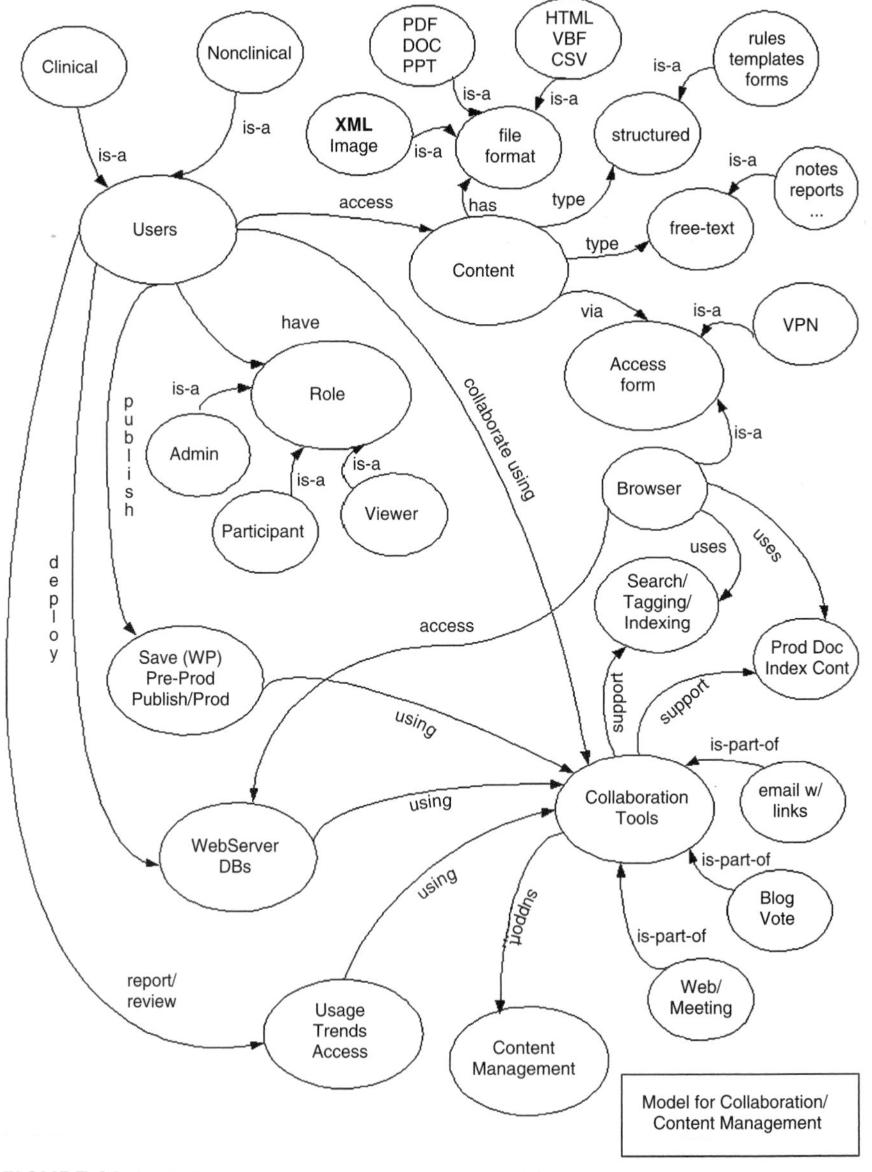

FIGURE 21-4 Collaboration/content management model.

Structured knowledge management processes require the monitoring of content through its various phases of development from creation, review, management, delivery, and maintenance.

The content creation phase includes planning, design, authoring, and revision:

- Planning activities such as analysis, collaboration, and decision-making are frequently not controlled by a content management system. Planning involves identifying the need for new content and determining what kind and type of information will most effectively satisfy such need. A good planning strategy supports the appropriate, predefined standards to ensure the quality repeatable processes, and consistency of information—a solid foundation for reusability.
- Design involves two aspects: Visual design and content/structural design. The former deals with the physical appearance of information, for example, templates. The latter involves identifying people involved in content development processes, standards already in place, limitations, and needs.
- Authoring means creating new content or accessing existing content in the form of text, graphics, and/or media. Authoring of domain-specific content such as structured templates for authoring rules should also be supported. It is essential to identify and fully understand all authoring processes; identify who creates/reviews content and their role(s) in the process; identify similarities in processes so they can be standardized and replicated.
- Revision may involve one or many reviewers. It may be carried out in single/multiple stages of refinement until it is ready for delivery. It is important to identify the tasks involved in the revision process: people responsible for reviewing content, the roles they play, and who gives the final approval.

Content Management involves several processes necessary to keep track of content at all stages of development:

- Version control ensures that each time content is saved, it is versioned. Changes can be tracked and comparisons among versions can be performed. It is important to have the proper versioning criteria in place (e.g., whole document vs. content-within-document versioning, product version, author, date, etc.). The ability to "roll back" to a prior version will be necessary in certain circumstances.
- Authoring access controls keep track of who accesses content, and when. Document life cycle management is also important to capture (a spectrum from read to read/write capabilities (based on role of user), as well as read-only access upon content publication.
- Version control should also be cognizant of the structure of the content, for instance, the ability to version the antecedent of a rule as opposed to the whole rule, rule version, author, date, and so on. The ability to roll back rules also involves deleting the instances of the rule being executed in the current production systems, and compensating for any of its effects.
- Knowledge change event monitoring and publishing to support propagation of content changes into the target decision support environments are essential. Such procedures must specify valid content formats, notification methods, location of repositories, and access requirements based on user roles and privileges.

- Definition management process and tools to ensure consistency and concept-based representation of structured enterprise knowledge assets.

With this current state analysis and functional requirements vision in hand, the following series of steps toward the development of a knowledge management infrastructure were defined:

1. Inventory and publish a document library portal of all knowledge in production across Partners Healthcare System.
2. Implement vendor tools to support collaborative content development and content life cycle management for unstructured content.
3. Implement tools to support authoring and maintenance of structured content leveraging the commercial content management infrastructure.

This road map that is now in its second stage requires a continuous balance of supporting the content needs of legacy systems while new services are implemented with new editing requirements and capabilities. The discussion that follows focused primarily on how the content maintenance infrastructure that publishes into transaction-side applications and services is being built.

21.3.2 Implementation of an Enterprise Clinical Decision Support Documentation Portal

The design and development of a first-generation knowledge management portal involved a painstaking inventory process that identified and catalogued clinical decision support content across Partners entities, including application-specific content such as order sets, documentation templates, reports, and rules, as well as decision support services content (shared among POE applications) such as physician-centric drug–drug interaction checking and expert geriatric and renal dosing. The inventory process included the interviewing of all identifiable content authors and knowledge engineers. Because there is no standard taxonomy for a document library, it was determined that it would be most pragmatic to derive a Partners-centric knowledge taxonomy to tag, aggregate, and organize information.

A knowledge management portal was developed to provide access to the knowledge assets. The portal supports three modes of access: keyword search, taxonomy navigation, and filter-based search. Taxonomy navigation (see Figure 21-5) enables the user to drill through content categories via four domains: Safety, Uniform Quality, Disease Management, and Trend Management. For example, a user can navigate to a lipid management reminder via *Uniform Quality → Clinical Disciplines → Cardiology (Medical)* or via *Disease Management → Cardiovascular Disease*.

Filter-based search (see Figure 21-6) enables comparative retrieval. Examples of filters include hospital entities, clinical disciplines (e.g., cardiac surgery or cardiology), applications, content type (e.g., order sets or interactive alerts), or populations (e.g., geriatric or pediatric). Comparative retrieval and evaluation, for example, might be done to inspect order sets created by different hospital entities, such as those for a cardiac surgery protocol.

Keyword-based search enables an exploratory and discovery mode of accessing content. For instance, one can retrieve all content that contains the words digoxin or heparin.

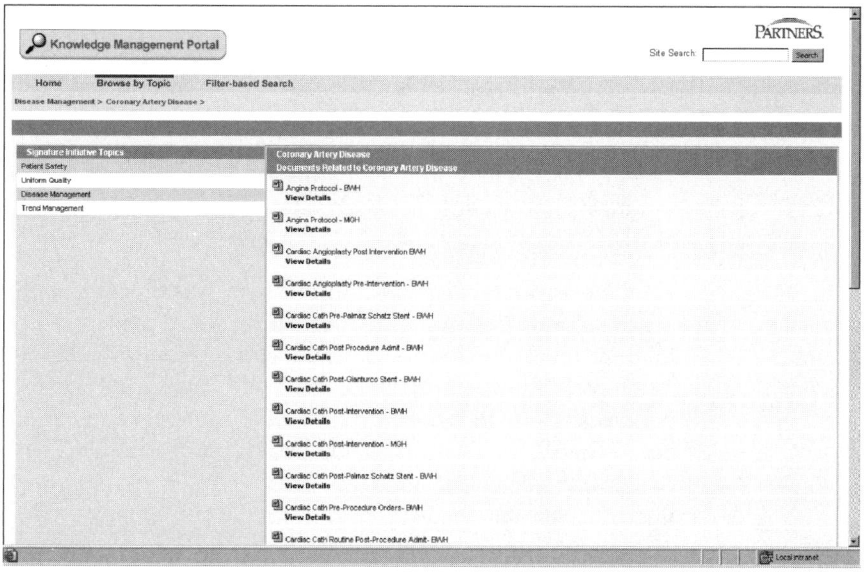

FIGURE 21-5 Taxonomy navigation.

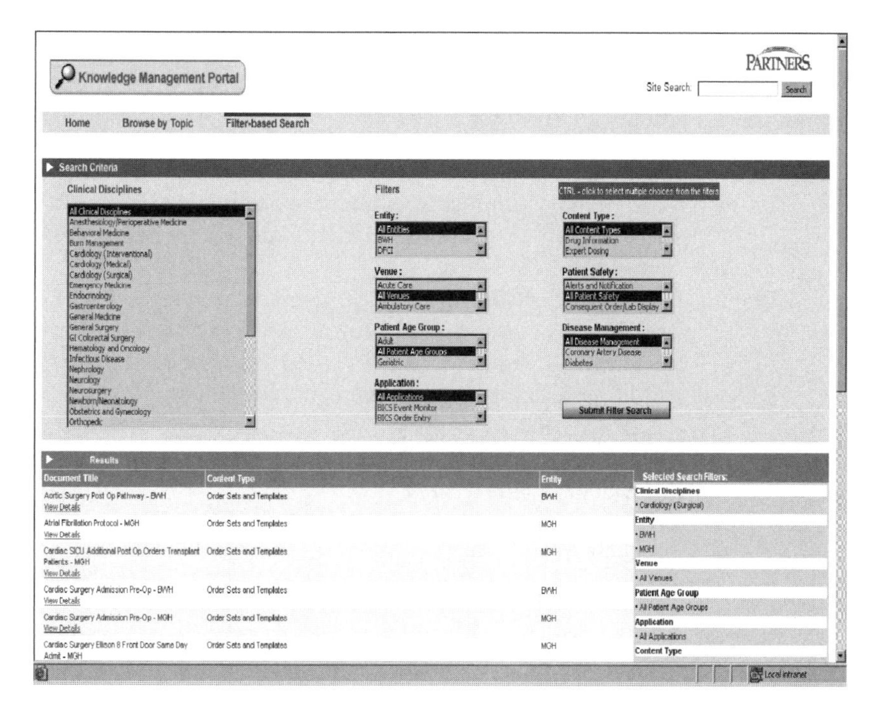

FIGURE 21-6 Filter-based search.

The administrative interface of the portal supports basic content management functions such as uploading content to the portal, annotation of the content with various filters and values, and assigning content to appropriate nodes in the knowledge taxonomy. The administrative interface also supports creation and management of the knowledge taxonomy and the various filters and values.

The portal has been deployed across the enterprise and has provided deeper insights into the organizational and technological requirements for a knowledge management infrastructure. Organizational insights included the need for processes for extending the taxonomy and enhancing the search interface. Further, in working with each knowledge engineering group to obtain content specifications, a clearer picture emerged of the deep and unmet need for improved workflow support for knowledge maintenance on the part of quality improvement leadership, subject matter experts, information systems personnel, knowledge stewards, and editors. Some examples of portal impact include:

- **Content Auditing**: Quality improvement stakeholders could now easily evaluate order sets across the enterprise for alignment with Joint Commission on Accreditation of Healthcare Organizations (JCAHO) core measures. However, further benefit will be realized when content relevant to JCAHO measures, Health Plan Employer Data and Information Set (HEDIS), and other value-based purchasing business drivers is tagged as such.
- **Content Sharing**: Project teams implementing physician order entry systems can evaluate content from other PHS sites with more mature implementations, enabling them to present a richer or more well-developed set of decision support design options to their respective clinical stakeholders. However, since PHS is a matrix organization, content that has been designated by enterprise leadership committees as having a high priority for common clinical specifications will need additional metadata tags in order to be recognized as such. Further, given the heterogeneous systems environment, some content specifications cannot be implemented in all systems. For example, advanced expert renal dosing is possible only with an interface to the medication services module, an interface that vendor-supplied systems cannot support today.
- **Content Acquisition**: Each project team engaged in knowledge engineering had unique processes and workflows for maintaining documentation of content. These relied on a wide variety of approaches including filing of documents related to content development, duplicate documentation of content published into applications, and/or generation of reports from editing tools, if such tools were available. The knowledge management team worked with each project team to develop unique processes for publishing production content updates to the portal.

Technological insights were also gleaned from this process. For example, some queries to the portal were unable to be expressed. These queries could be met by a combination of filter-based search and taxonomy-based browsing. Consider the query, "*Retrieve all content related to the clinical discipline cardiology and created by the entity Mass General Hospital.*" This can be done in two steps: 1) Browse to the node corresponding to "cardiology" in the taxonomy; and 2) apply the filter "Entity = MGH" to the content items classified under that concept. If mixed-mode access were not supported, we would need to create another filter with values corresponding to the subcategories of "Clinical Discipline" in order to address that query. Support for mixed mode access helps simplify the taxonomy and collections of filters and values. To further simplify access structures, if a node had just one child, the documents under that node were rolled up to its parent node. Also, the implementation supports the situation in which the same concept can have multiple parents, so the taxonomy is actually a directed acyclic graph and not a hierarchy.

21.3.3 Implementation of Commercial Infrastructure for Collaboration and Content Life Cycle Management

The deployment of an enterprise-wide portal for clinical knowledge was the first component of the Knowledge Management Infrastructure and denoted the first step in enabling knowledge reuse across the organization. This infrastructure is now being extended to support the activities of knowledge asset creation and maintenance that satisfy the broader design requirements outlined in subsection 21.3.1. The perspective has been adopted that knowledge authoring and creation editors must be tightly integrated with content management functionality such as versioning, life cycle support, and metadata management. To support collaborative editing and effective content maintenance, a logically centralized knowledge repository is highly desirable. It may be noted here that PHS hosts a highly federated environment of home-grown and vendor applications, each with a variety of knowledge editing capabilities. The vendor content management infrastructure being deployed can logically integrate these distributed knowledge editing environments, creating the illusion of a centralized knowledge repository.

An evaluation of the content management vendors in the market was performed, utilizing a variety of sources to develop selection criteria (Garvin 1993; Chatzkel; Dawson 2000; Detlor 2000; Lehto and Marttiin 2000; Soliman 2000; Fischer and Ostwald 2001; O'Leary 2001; Smith 2004; Sordo et al. 2004; Adams et al. 2005; Lausen et al. 2005; Puschmann and Alt 2005; Wing 2005) and the EMC/Documentum Content Management platform was selected. Some of the key considerations in the vendor selection were:

- Support for management of structured custom knowledge objects because of the desire to move away from unstructured document management to structured knowledge management.
- Support for configuration of a variety of content life cycles and workflows. As the KM solution is deployed throughout PHS, it is quite likely that different business units will have their own specific life cycles and workflows for specific pieces of knowledge. Given the likelihood of frequent changes to the life cycles and workflows as organizational practices evolve thus required the ability to configure these changes without frequent software updates.

Solutions for collaboration, content management, and content metadata design have been implemented. The knowledge portal taxonomy was moved to the Documentum platform such that this infrastructure is supporting the maintenance of the knowledge portal rather than the homegrown content upload tool.

Some knowledge creation and vetting processes are ad hoc and cannot be structured as well-defined workflows. This is typically the case when groups of subject matter experts collaborate with each other and generate a consensus around clinical knowledge such as guidelines, medication dosing, and the like. The approach taken to support this collaboration process has been by deploying eRooms implemented on the EMC/Documentum Collaboration solution platform. An example of a clinical knowledge artifact is an algorithm for glycemia management or a decision table, each column of which corresponds

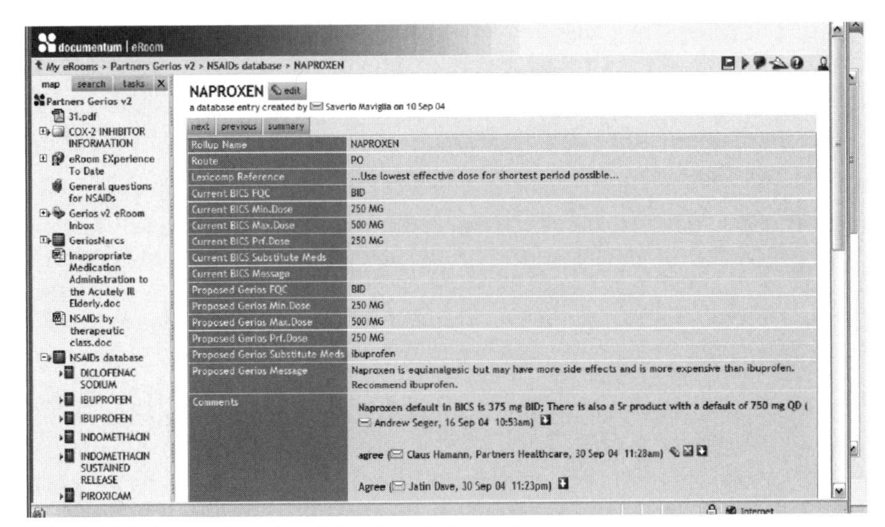

FIGURE 21-7 Collaborative clinical knowledge vetting.

to various parameters related to patient state and each row corresponds to dosing decisions that need be to taken conditioned on the patient state parameters associated with the row. In either case, collaboration tools enable vetting at the row level of a decision table. Vetting can be accomplished through votes and discussion threads (see Figure 21-7).

Pieces of knowledge on which consensus has been achieved are then moved to the content server where they can be maintained by using the associated life cycles and workflows. Content that is edited or engineered into production environments is tagged as such and published to the enterprise knowledge portal. Preproduction content is accessible only to relevant subject matter experts and knowledge engineering personnel.

21.3.4 Toward a Knowledge Management Platform

The next issue to consider is how to build on the currently implemented infrastructure to achieve two goals: 1) transition from unstructured document management to structured knowledge management, in the process, identifying and providing a set of building blocks to knowledge engineers for knowledge creation and authoring; and 2) enhancing the content management processes to support dependency propagation (i.e., identifying and propagating the impact of a change in one knowledge object to other related knowledge objects across the knowledge management infrastructure). The approach taken to incorporate various knowledge sources in the knowledge management infrastructure is illustrated in Figure 21-8.

Step 1: Logical Integration of Knowledge Source into a Centralized Knowledge Repository

As discussed earlier, a logical centralized knowledge repository is the key architectural component of the knowledge management infrastructure that has been designed. The individual knowledge sources contributing to this logical repository may be physically distributed. Knowledge objects in a given knowledge source need to be integrated into the Documentum Content Server. This can be achieved in the following ways:

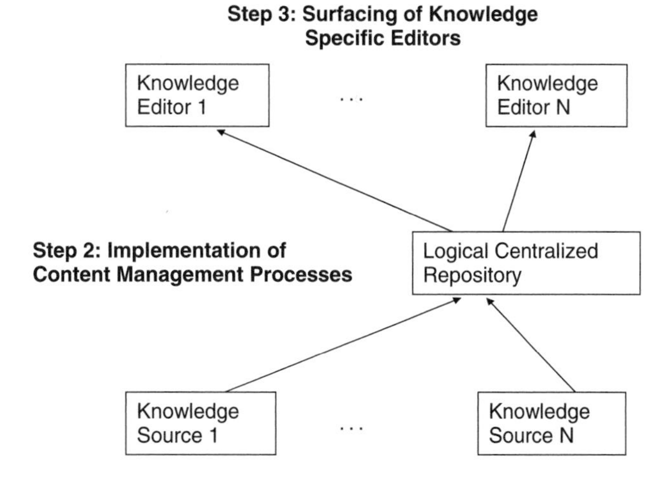

FIGURE 21-8 Building a knowledge management platform.

- If the knowledge is stored in relational databases, then these tables can be registered with the content server. The objects in these tables are then available natively within the content repository and can be then manipulated directly by the content server. This option is the preferred one and is planned for the incorporation of knowledge sources with structured objects such as terminological concepts managed by a terminology server such as HLi and a Business Rules Management Server (BRMS).
- If the knowledge is available as XML files on a file share, the Documentum infrastructure can accept XML feeds and load objects directly into the content repository.
- For knowledge that is currently hard-coded within legacy databases, the process being carried out involves externalizing this knowledge to a relational database structure with the intention of expanding the approaches to the maintenance and enrichment of this content. This will involve either remodeling and redeployment through vendor solutions such as a BRMS or the creation of new editors and services such as those for order entry or clinical documentation.
- In other cases, it may be necessary to implement custom code to extract and create knowledge objects directly through the Documentum Foundation Classes (DFC) API.

Step 2: Implementation of Content Management Processes

Once the knowledge objects have been loaded into the content server, content management processes will be implemented for creation and maintenance of these objects. This will be followed by dependency propagation across different types of knowledge objects.

- Each of these knowledge objects can now be created and managed via content life cycles and workflows that will be enforced by the Documentum content server. Furthermore, these processes are specified declaratively and are highly configurable. They can be easily and quickly changed in response to changing business conditions. Also, processes can be defined that will enable the content server to publish these knowledge objects to a transactional system (e.g., publishing of order sets to a CPOE implementation under various conditions, perhaps periodically or in response to an update operation). However content management tools are available for certain knowledge objects (such as the HLi terminology engine for terminological concepts) that have an efficient and robust implementation for typical content management operations (e.g., versioning) for those objects. Those content management operations will be delegated to the content management engine for that knowledge object, and other operations will be implemented by the content server. Enforcement of user access privileges will be implemented uniformly across all knowledge objects by the Documentum platform.

- Dependency propagation across knowledge objects can be viewed as being of two types: 1) dependency propagation across knowledge objects of the same type (e.g., the impact of deleting a concept in a terminology on other related concepts), and 2) dependency propagation across different knowledge types (e.g., the impact of deleting a concept on clinical decision support rules that reference that concept). Specialized content management tools have efficient and robust implementations of dependency propagation across knowledge objects of the same type. For instance, HLi has implemented content update functionality for terminological concepts. In that scenario, this functionality will be delegated to the specialized content management tool. Dependency propagation across different types of knowledge objects will be implemented by the Documentum platform. This is enabled by the integration of various knowledge sources into a logically centralized repository as discussed in Step 1.

Step 3: Surfacing of Knowledge-Specific Editors

The perspective is adopted that knowledge editors surface content management functionality supported by the KM Infrastructure. The implementation of a centralized knowledge repository and content management functionality enables the enhancement of editors with new functionality. There are knowledge editors currently in use at Partners for authoring and creating multiple types of knowledge content. At the same time certain knowledge objects such as terminological concepts come with their own specialized editing environments such as the Lexscape tool by HLi.

- Existing functionality of these editors will be reimplemented by invoking functionality from the content server. New functionality supported by the content server will be surfaced in these existing editors. This would require invoking methods on the APIs and leveraging the authoring integration services supported by Documentum. For knowledge objects that do not have well-defined editors, the third-party vanilla content editors available with the Documentum platform will be leveraged.
- The editors will reflect the appropriate editorial and governance policies. These policies can be implemented as life cycles/workflows in conjunction with appropriate group-based access control that will be

enforced by the content server. Enterprise Knowledge Management and Personal Knowledge Management both will be supported and distinguished. Depending on the access rights of a given user, only a subset of the knowledge available in the centralized repository may be surfaced in the knowledge editor.

21.3.4.1 Incorporation of Structured Knowledge Objects

Over time, a goal is to incrementally incorporate various types of structured knowledge objects. There are three categories of structured knowledge objects to consider:

- Basic knowledge objects such as terminological concepts, relationships and facets that will be used as building blocks for other knowledge objects.
- Complex knowledge objects such as rules and knowledge element groups (KEGs) that contain references to building blocks such as terminological concepts.
- Smart modules that combine a variety of complex knowledge objects such as rules, context-driven data review, context-driven documentation, and context-driven order sets. The first of these modules will be targeted at outpatient encounter management of patients with coronary artery disease, diabetes, and congestive heart failure.

Figure 21-9 depicts the layered approach to building out the Knowledge Repository being adopted. Terminological concepts provide the core building blocks of the knowledge infrastructure. The next layer of knowledge consists of knowledge objects (e.g., rules, KEGs) that have references to terminological

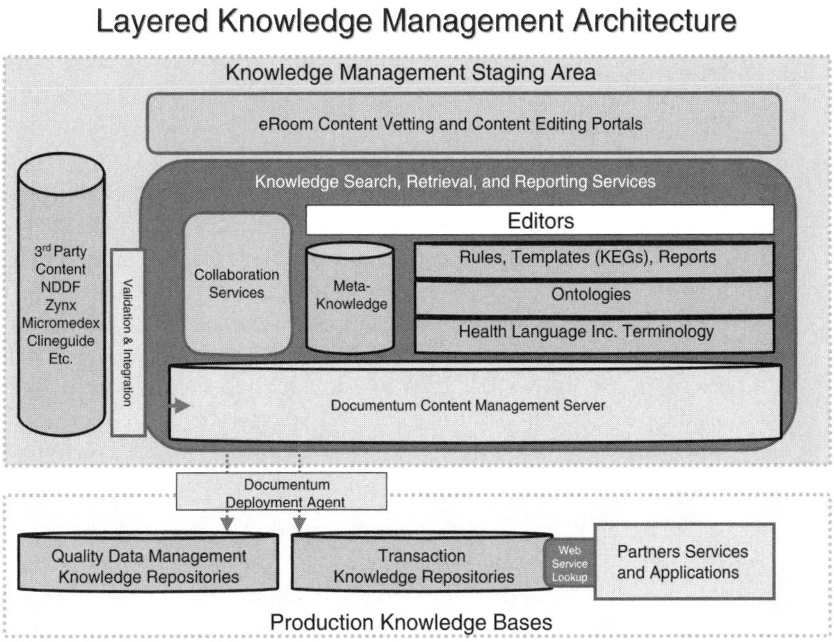

FIGURE 21-9 Layered knowledge repository.

concepts. Each of these knowledge objects comes from a different knowledge source and is integrated according to the approach described earlier. As discussed, structured knowledge objects (e.g., rules, terminological concepts) are associated with their own content management tools and are stored in well-defined repositories structured according to well-defined information models and schemas. The challenge here is to "map" these information models into object types in the content repository and to model and enforce dependency propagation across these structured knowledge objects.

Finally, most of the structured knowledge objects come with specialized editors. The intention is to surface these editors on top of the content repository on the Documentum platform and expose new functionality supported by the content server to these specialized editors. It is essential to provide a comprehensive editing environment, a knowledge creation dashboard, that contains plug-ins to support surfacing of appropriate knowledge object editors for any given knowledge creation task. In addition to enabling reuse, this dashboard will enable visualization and specification of dependencies between various knowledge objects and track the propagation of dependencies across those objects.

21.3.4.2 The Role of Ontology Management

Terminology servers such as HLi are very good for creation and maintenance of coded concepts. However, additional tools and techniques are required to maintain knowledge that typically is represented using a set of constraints on ontologies and information model elements. One example is that of defining a condition for a rule as all patients with diabetes except gestational diabetes. Another example, the definition of a drug contraindication to a statin or fibric acid for lipid management, is a complex definition that consists of a collection of constraints such as "liver function test 3 times the upper limit of normal" (see Figure 21-10). It is important for such a definition to be reused across a variety of knowledge objects such as clinical performance analysis queries, reporting to accreditation organizations, surfacing as a contraindication status in clinical data review, a documentation template to elicit contraindication observations, and lipid management decision support rules (Kashyap et al. 2005, 2006).

These complex definitions can be represented using constructs in an ontology management language such as OWL (Web Ontology Language 2006). The ability to support semantic constraints with ontology management tools will offer some notable advantages. It will be possible to express, manage, and reuse complex pieces of knowledge—something not well supported in terminology management or rule management environments. This in turn will provide a rich substrate for enabling sophisticated inferencing operations in the knowledge creation and management context such as bridging ontologies, merging ontologies, and inconsistency checking. This inference functionality is likely to be very useful in the context of knowledge editing and maintenance. Some interesting use cases are:

- It will be possible to retrieve knowledge objects based on descriptions of the patient state (e.g., to retrieve all documentation templates that deal with concomitant hyperlipidemia, diabetes, and coronary artery disease).
- Whenever a concept definition changes, it could render other related knowledge objects inconsistent or render it equivalent to other knowledge objects.

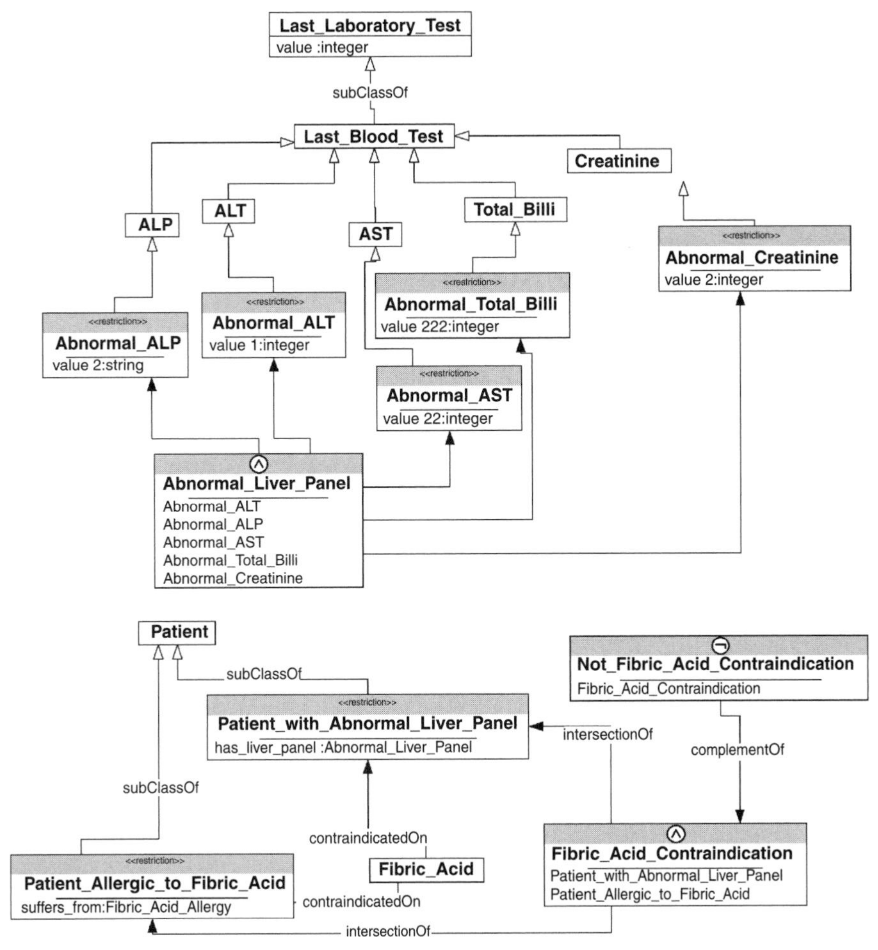

FIGURE 21-10 Definition of fibric acid contraindication.

- The ontology engine can surface definition versions and, with Documentum content management server support, relevant dependent knowledge objects to enable the knowledge engineer to ensure that knowledge integrity is maintained.

This capability will initially be deployed in the knowledge management context, to publish content into the existing transactional infrastructure. Quality data management personnel will need to return to the knowledge repository to interpret the richer semantic meaning of the transaction data. Over the long term, to externalize recognition and classification from the rules engine, the PHS enterprise clinical services team will introduce the ontology engine into the transactional infrastructure.

21.4 CONCLUSION

This chapter represents a snapshot of a multiyear undertaking to develop a knowledge management infrastructure that must, by necessity, serve the needs of a large, extremely heterogeneous application environment with an enormous inventory of homegrown content in production. The infrastructure needed to support knowledge transparency and content governance have been built, and the focus is now on tackling the deeper challenges of knowledge engineering that undermine the ability to expand or change content in production once it is determined that such changes are necessary. To accomplish this, new integration approaches are being pursued among vendor solutions for content management, business rules management, terminology management and ontology management. The authors believe this approach is essential to support managing knowledge at the speed of change anticipated, particularly with the advent of molecular medicine now occurring (Kashyap et al. 2004; Kashyap and Hongsermeier 2005). It is common in health care to focus on the computing power required for analytics or for meeting the event management requirements of patient data transactions. A knowledge management infrastructure such as described can be viewed as a knowledge-event management framework that will support structured knowledge discovery, acquisition, and maintenance for the era of personalized medicine.

REFERENCES

Adams, P., Nutter, D., Rank, S., Boldyreff, C. (2005). Using open source tools to support collaboration within CALIBRE. *Proceedings of the First International Conference on Open Source Systems*. Genova, 11th–15th July 2005. Marco Scotto and Giancarlo Succi, eds., 61–65.

Bates, D. W., Teich, J. M., Lee, J., Seger, D., Kuperman, G. J., Ma'Luf, N. et al. (1999). The impact of computerized physician order entry on medication error prevention. *J Am Med Inform Assoc* **6**(4): 313–321.

Bates, D. W., Kuperman, G. J., Wang, S., Gandhi, T., Kittler, A., Volk, L. et al. (2003). Ten commandments for effective clinical decision support: making the practice of evidence-based medicine a reality. *J Am Med Inform Assoc* **10**(6): 523–530.

Boxwala, A. A., Denekamp, Y., Greenes, R. A., Kuperman, G. J., Middleton, B. L. (2004). Survey and evaluation of knowledge bases for clinical decision support. *Medinfo*.

Chatzkel, J. Conversation with Alex Bennet, former deputy CIO for enterprise integration at the US Department of the Navy.

Chin, T. (2003). Doctors pull the plug on paperless system. *American Medical News,* Feb. 17.

Davenport, T. H., Glaser, J. (2002). Just-in-time delivery comes to knowledge management. *Harvard Business Review* **80**(7).

Dawson, R. (2000). Knowledge capabilities as the focus of organization development and strategy. *Journal of Knowledge Management* **4**(4): 320–327.

Detlor, B. (2000). The corporate portal as information infrastructure towards a framework for portal design. *International Journal of Information Management* **20**: 91–101.

Fischer, G., Ostwald, J. (2001). Knowledge management: Problems, promises, realities and challenges. *IEEE Intelligent Systems* 60–72.

Garvin, D. A. (1993). Building a learning organization. *Harvard Business Review* **71**(4): 78–91.

Greenes, R. A., Sordo, M., Zaccagnini, D., Meyer, M., Kuperman, G. J. (2004). Design of a standards-based external rules engine for decision support in a variety of application contexts: Report of a feasibility study at Partners HealthCare System. *Medinfo* **11**(Pt 1): 611–615.

Hongsermeier, T., Schmeling, W. (2002). Knowledge management: The use of knowledge strategies to transform healthcare. *HIMSS 2002 Proceedings*, Atlanta.

Hongsermeier, T., Kashyap, V., Hanrahan, J. (2005). Collaborative authoring of decision support knowledge: A demonstration. *AMIA 2005 Annual Symposium*, Washington, DC.

Jha, A. K., Kuperman, G. J., Teich, J. M., Leape, L., Shea, B., Ritternberg, E. et al. (1999). Identifying adverse drug events: Development of a computer-based monitor and comparison with chart review and stimulated voluntary report. *JAMA* **5**(3): 305–314.

Kashyap, V., Hongsermeier, T., Aronson, S. (2004). Can semantic web technologies enable translational medicine? (Or can translational medicine help enrich the semantic web?)

Tech Report No. CIRD-20041027-01, http://www.partners.org/cird/pdfs/Semantic_Web_Translational_Medicine.pdf.

Kashyap, V., Hongsermeier, T. (2005). Towards a national health knowledge infrastructure (NHKI): The role of semantics-based knowledge management Tech Report No. CIRD-20051123-01, http://www.partners.org/cird/pdfs/White_Paper.pdf.

Kashyap, V., Morales, A., Hongsermeier, T. (2005). Creation and maintenance of implementable clinical guideline specifications. Tech Report No. CIRD-20051027-01, http://www.partners.org/cird/pdfs/clinical_guidelines.pdf.

Kashyap, V., Morales, A., Hongsermeier, T., Li, Q. (2005). Definitions management: A semantics-based approach for clinical documentation in healthcare delivery, industrial track. *Proceedings of the 4th International Semantic Web Conference,* November 2005.

Kashyap, V., Morales, A., Hongsermeier, T. (2006). On implementing clinical decision support: Achieving scalability and maintainability by combining business rules and ontologies. Tech Report No. CIRD-20060322-01, http://www.partners.org/cird/pdfs/CIRD-20060322-01.pdf.

Kuperman, G. J., Teich, J. M., Bates, D. W., McLatchey, J., Hoff, T. (1995). Representing hospital events as complex conditionals. *AMIA* 137–141.

Kuperman, G. J., Teich, J. M., Bates, D. W., Hiltz, F. L., Hurley, J. M., Lee, R. Y. et al. (1996). Detecting alerts, notifying the physician, and offering action items: A comprehensive alerting system. *AMIA* 704–711.

Kuperman, G. J., Teich, J. M., Gandhi, T. K., Bates, D. W. (2001). Patient safety and computerized medication ordering at Brigham and Women's Hospital. *Journal on Quality Improvement* **27**(10): 509–521.

Lausen, H., Ding, Y., Stollberg, M., Fensel, D., Lara Hernandez, R., Han, S. K. (2005). Semantic web portals: State-of-the-art survey. *Journal of Knowledge Management* **9**(5): 40–49.

Lehto, J. A., Marttiin, P. (2000). Lessons Learnt in the Use of a Collaborative Design Environment. *Proceedings of the 33rd Hawaii International Conference on System Sciences.*

O'Leary, D. E. (2001). How knowledge reuse informs effective system design and implementation. *IEEE Intelligent Systems* 44–49.

Puschmann, T., Alt, R. (2005). Developing an integration architecture for process portals. *European Journal of Information Systems* **14**(2): 121–134.

Smith, M. A. (2004). Portals: Toward an application framework for interoperability. *Communications of the ACM* **47**(10): 93–97.

Soliman, F. (2000). Application of knowledge management for hazard analysis in the Australian dairy industry. *Journal of Knowledge Management* **4**(4): 287–294.

Sordo, M., Hongsermeier, T., Greenes, R. (2004). Knowledge management for clinical decision support. Requirement analysis of collaboration tools for content management. Tech Rep DSG-TR-2004-23.

Sordo, M., Hongsermeier, T., Kashyap, V., Greenes, R. (2005). Partners Healthcare Order Set Schema V1.3. Technical Report DSG-TR-2005-041.

Sordo, M., Hongsermeier, T., Kashyap, V., Greenes, R. (2006). A knowledge management approach for grouped elements: Applications in order sets, documents and flowsheets. Technical Report DSG-TR-2006-001.

Tang, P. (2003). Patient safety: Achieving a new standard of care. Washington, DC: National Academy Press.

Web Ontology Language (OWL), W3C, http://www.w3.org/2004/OWL/, 2006.

Wing, L. (2005). Investigating success factors in enterprise application integration: A case-driven analysis. *European Journal of Information Systems* **14**(2): 175–187.

22 THE CLINICAL KNOWLEDGE MANAGEMENT INFRASTRUCTURE OF INTERMOUNTAIN HEALTHCARE

ROBERTO A. ROCHA, RICHARD L. BRADSHAW,
NATHAN C. HULSE, and BEATRIZ H. S. C. ROCHA

22.1 CLINICAL KNOWLEDGE MANAGEMENT AT INTERMOUNTAIN HEALTHCARE

22.1.1 Intermountain Healthcare

Intermountain Healthcare (Intermountain) is a not-for-profit integrated delivery system of 21 hospitals, with over 150 service sites, an employed physician group with over 580 physicians, and an insurance plan located in Utah and southeastern Idaho (IHC Annual Report 2004). Intermountain's facilities range from major tertiary-level teaching and research facilities to small hospitals and clinics in rural communities, corresponding to 2,449 beds total. Intermountain provides over 50 percent of all care delivered in the region, with over 122,000 inpatient admissions and 5.2 million outpatient encounters annually. Intermountain has received the "Nation's Top Integrated Health Care System" award five times during the last six years (IHC named nation's top health system 2006). Intermountain has been named six times one of the nation's "100 Most Wired" health systems during the past seven years (IHC honored for high tech systems 2006). Intermountain's superior performance is partly attributed to state of the art clinical information systems that promote uniform quality of care within its facilities.

22.1.2 Clinical Information Systems Infrastructure

Since 1995, Intermountain has been building a new clinical information systems infrastructure, known as HELP2 (Clayton et al. 2003). The new clinical system infrastructure was created to progressively replace the HELP System, Intermountain's legacy hospital information system (Kuperman et al. 1991). In HELP2, Intermountain's clinical systems are being delivered to care providers through a web-based shell developed in-house, known as the "Clinical Desktop." Currently, the Clinical Desktop offers functionality such as laboratory results, radiology images and reports, surgery reports, clinical

Clinical Decision Support: The Road Ahead

notes, medication lists, problem lists, and order entry. Between June and August of 2005, an average of 5,809 clinicians (1,559 physicians) in diverse clinical settings used the Clinical Desktop at least once a month. Users can access the Clinical Desktop from Intermountain's Intranet or from the Internet, using either a Virtual Private Network (VPN) connection or a secure token device.

The main components of the HELP2 core infrastructure are the Clinical Data Repository (CDR) (3M Clinical Data Repository 2005), the Healthcare Data Dictionary (HDD) (3M Health Care Data Dictionary 2005), and the Enterprise Master Patient Index (EMPI) (3M Enterprise Master Patient Index 2005). The EMPI is a central register of all patients that are cared for in Intermountain's inpatient or outpatient facilities. The HDD is a vocabulary server that provides functions such as code mapping, support for data encoding, hierarchies, and semantic relationships. Presently, the vocabulary server at Intermountain contains 875,088 concepts, 5,066,366 surface forms (designations), and 4,690,470 relationships. The CDR is a longitudinal EMR that captures data from all clinical encounters, both outpatient and inpatient. Data are stored in the CDR directly from clinical applications or through HL7 interfaces (Health Level Seven 2005). As of September 2005, the CDR contained records of 5,302,450 patients.

Another important component of the overall information systems infrastructure of Intermountain is the Enterprise Data Warehouse (EDW). The EDW currently receives data feeds from almost all administrative and clinical databases used by Intermountain systems. One of the most relevant resources within the EDW is the CDR data mart, where researchers and data analysts with appropriate permissions can access all the clinical data from the original CDR.

Within the past five years, two new components have been added to the HELP2 core infrastructure: "Foresight" and the "Clinical Knowledge Repository" (CKR). Foresight is a homegrown decision logic execution engine coupled with a sophisticated clinical data monitor. Foresight implements and extends within HELP2 the decision support capabilities that made the original HELP System famous (Kuperman et al. 1991; Haug et al. 2003). The CKR is the central repository of all knowledge assets used by Intermountain clinical systems, especially HELP2, and it includes a wide variety of electronic knowledge resources that are produced by internal teams of clinical experts. The CKR and Foresight are the primary components of Intermountain's Clinical Knowledge Management (CKM) software infrastructure.

22.1.3 Management of Clinical Processes and Conditions

Intermountain's core clinical strategy is to provide high value care by effectively managing clinical conditions and processes, while improving medical outcomes and member satisfaction at the lowest necessary cost. Intermountain Clinical Programs (CPs) are the vehicles responsible for developing and implementing this strategy of best practice (Intermountain Clinical Programs 2006). A CP can be considered a specialized clinical advisory panel that develops tools and processes to help Intermountain clinicians consistently deliver high quality clinical care. Ten CPs have been created to date, covering the most common processes and conditions managed by Intermountain within the following broad clinical areas: Cardiovascular Medicine, Intensive Medicine, Intensive Pediatrics, Neuromusculoskeletal, Oncology, Pediatrics subspecialties, Preventive Care, Primary Care, Surgery, and Women & Newborns (Intermountain Clinical Programs 2006).

Each CP has a guidance council that coordinates its activities. The guidance council establishes the development priorities of the CP, taking into account the most prevalent or variable diagnostic conditions (James and Hammond 2000), and also key patient safety processes. Each council is responsible for a series of interdisciplinary development teams and specialized workgroups. Development teams and workgroups are staffed by a relatively small number of practicing clinicians that provide not only expert domain knowledge, but also vital local or regional representation. A senior physician recognized as a system-wide domain expert is commonly the leader of each one of these teams. In addition to practicing clinicians, each team is also staffed with outcomes analysts and data architects. Whenever necessary, these teams are also directly supported by knowledge engineers and clinical education professionals. Each team meets at least once every three months.

Development teams and specialized workgroups are directly involved with the following knowledge management activities:

- Creation of conventional knowledge assets that are used for training and education, including practice guidelines, referral and intervention indication guidelines, and patient educational materials
- Review and approval of clinical operations standards and protocols (a dedicated group of clinical writers has been developing these interdisciplinary practice standards (Hougaard 2004))
- Development and implementation of condition or processes-specific data collection forms, data marts, data extraction routines, and quality assurance reports
- Development and implementation of computable or interactive knowledge assets, including decision rules (e.g., alerts, reminders), protocols, interdisciplinary care plans, and order sets
- Selection and customization of external knowledge resources obtained from public domain sources, or through licensing from commercial vendors

The set of knowledge assets and processes developed for managing a particular clinical condition or clinical process is known as a "disease management system." Cannon describes in some detail the analysis that precedes the development of a disease management system (Cannon 2004).

The activities of each CP are expected to lead to measurable improvements in the following areas: clinical outcomes, patient safety, patient satisfaction, and also cost structure optimization. Every year, each CP formally defines a series of corporate-wide clinical goals that underscore these improvement areas. Once approved by the Intermountain Board of Trustees, these corporate clinical goals guide the development and implementation of each disease management system. Development teams are responsible for tracking the implementation progress of their respective disease management systems, and are also accountable for goal achievement. Larsen et al. describe the diabetes disease management system developed and implemented at Intermountain, and demonstrate its ability to reduce the risk of patients with diabetes developing diabetes-related complications (Larsen et al. 2003).

The development and implementation of a disease management system relies directly on the comprehensive clinical and information management infrastructure available at Intermountain. In other words, the success of each CP depends largely on direct and comprehensive support from: 1) regional and system-wide physician, administrative, and clinical operations leaders needed to implement best practice; 2) staff support personnel and systems necessary to measure clinical, financial, and satisfaction outcomes for key clinical processes; and 3) staff and systems necessary to develop, disseminate, support, and maintain the knowledge assets necessary to implement best practice. These infrastructure elements of the disease management system evidently require information technology, but the CKM infrastructure is mostly pertinent to the third element.

The subsequent sections will describe in some detail the CKM software infrastructure that has been implemented at Intermountain, along with some utilization data demonstrating how extensively it is being used. The data presented will reflect the knowledge content and software infrastructure that were in production use as of September 2005, unless stated otherwise.

Intermountain and GE Healthcare have recently signed an agreement establishing a collaboration to "create a best-practices based clinical software program that will enhance the patient care process in hospitals and clinics and accelerate the adoption of electronic health records (EHR)" (GE Healthcare & IHC establish new research, 2006; GE Healthcare & Intermountain Healthcare to provide wide-reaching IT system 2006). It is appropriate to conclude that the current CKM software infrastructure will continue to be used at Intermountain for as long as HELP2 remains in operation, but it is premature to indicate how much of the existing functionality will be incorporated as part of the envisioned "digital healthcare solution" (GE Healthcare & Intermountain Healthcare to provide wide-reaching IT system 2006).

22.2 KNOWLEDGE ASSETS

The CKM infrastructure is capable of handling multiple types of content, ranging from simple textual documents to large binary files. This diversity of assets reflects the broad definition of what Intermountain considers knowledge content. In essence, assets required to implement and support a disease management system are considered clinical knowledge content.

22.2.1 Learning Strategies

Clinical decision-making processes determine the information needs of clinicians and provide opportunities for knowledge dissemination and learning (Marriott et al. 2000; Fisher and Ostwald 2001). Taking into account the broad categories of clinician information needs (Wyatt 2000; Currie et al. 2003), as well as the approaches used for accessing and delivering the information (Fischer and Ostwald 2001), the knowledge assets managed by the CKM infrastructure can be grouped into the following learning strategies:

- **In-depth learning.** This mode corresponds to long periods of self-regulated learning, which support activities like general continuous education and clinical research. Self-regulated learning may be triggered by clinical practice, but is not frequently associated with a specific patient. Within the CKR, knowledge assets in this group are represented by sanctioned links to external (freely available or licensed) full-text biomedical databases (e.g., EBSCO Biomedical Libraries (Biomedical Libraries 2005), and MEDLINE/PubMed (MEDLINE/PubMed 2005)). Clinicians always initiate *in-depth learning* sessions, but these sessions normally do not coincide with clinical activities that require the use of a clinical information system.

- **Just-in-time learning.** This is a more focused form of self-regulated learning, where clinicians have better defined questions and limited time to find the appropriate answers. Focused learning activities usually are triggered by clinical practice and have clear associations with a specific patient or group of patients. Knowledge assets in this group correspond primarily to sanctioned links to summarized textual resources that are licensed from external entities (e.g., Clin-eguide (Clin-eguide 2005), MD Consult (MD Consult 2005), Micromedex (Micromedex 2005), and UpToDate (UpToDate 2005)). Just-in-time learning frequently coincides with clinical activities that require the use of a clinical information system, but the clinician is responsible for initiating the action to obtain the information.
- **Best-practice learning.** This is also a form of focused self-regulated learning, but emphasizes specific disease management systems and local or regional regulations and standards. These learning activities are almost always triggered by clinical practice and have very direct associations with care being delivered to a specific patient. Knowledge assets in this group are represented by guidelines, protocols, policies, procedures, and other care standards that are developed and maintained by Intermountain (e.g., Care Process Models (see subsection 22.2.6.1), Collaborative Practice Guidelines (see subsection 22.2.6.2)). Best-practice learning almost always coincides with clinical activities that require the use of a clinical information system and the clinician is again responsible for accessing the information.
- **Individualized learning.** This is a form of learning that results from the application of accepted best practices to remind or critique clinicians' decisions. The application of the knowledge takes into account the unique circumstances resulting from disease management processes being used for a specific patient within a specific clinical setting. Knowledge assets in this group are represented by executable or interactive assets, including computerized alerts and protocols (see subsection 22.2.6.4). Individualized learning always coincides with clinical activities that require the use of a clinical information system and are always patient-specific. In this group, the information is always delivered to the user via asynchronous or synchronous notification methods. The asynchronous methods normally are implemented using messages that are routed to the clinician's inbox, or are delivered using real-time communication channels (e.g., pagers, mobile phones). The synchronous methods are implemented using notifications that interrupt the interaction of the user with the clinical information system.

The learning strategies just described have direct influence on how the knowledge assets are created, clustered, and presented and on the mechanism or system used to disseminate them. Consequently, the CKM infrastructure has to accommodate different content models and content presentation requirements, but also be able to support multiple dissemination strategies.

Particularly in terms of dissemination, the CKM infrastructure has to support searching and retrieval mechanisms that range from simple hyperlinks that can be easily embedded into stand-alone documents or Web pages, to application programming interfaces that can be seamlessly integrated into clinical applications.

22.2.2 Asset Availability and Variation

Most knowledge assets are created with the expectation that they will be applicable to a relatively broad clinical audience. However, the need to support local variations and contextual restrictions that may be associated with knowledge assets is also very important. Presently, the CKM infrastructure has methods capable of handling the following kinds of contextual restrictions and local variations:

- **Conditional availability of an asset.** The whole knowledge asset is accessible only if predefined conditions are satisfied. Presently, these predefined conditions include patient demographics (e.g., age group and gender), provider roles and scope of practice (e.g., MD, RN, attending), care setting characteristics (e.g., inpatient, ICU, ER, LDS Hospital), and type of clinical application (e.g., CPOE, bedside documentation).
- **Conditional availability of content.** Specific portions of a knowledge asset are accessible only if predefined conditions are satisfied, using the same conditions just described. This particular method can be used to accommodate local content variations; that is, distinct portions of an asset will be exposed when it is accessed from different care settings or locations.
- **Dynamic substitution or generation of content.** Discrete portions of a knowledge asset are dynamically generated (predefined formulas or functions) or substituted (values and parameters provided by the requesting service). These methods can be used to maximize the reuse of assets that need to be dynamically configured to accommodate contextual settings (e.g., patient's age and weight, local cost of medication dose).

22.2.3 Content Frameworks

A *content framework* can be understood as an integrated set of software components, (meta-) content artifacts, and processes that are necessary for the development and maintenance of a specific type of knowledge asset. The wide diversity of knowledge assets used within Intermountain's disease management systems requires the utilization of multiple *content frameworks*. The degree of integration of the various *content frameworks* with the CKM infrastructure ranges from *basic* integration (e.g., the CKR is used for asset storage), to *full* integration, where all CKM software components and content artifacts are used. The most important *content frameworks* currently in use are described next. Subsection 22.3 describes the main software components of the CKM infrastructure.

22.2.3.1 XML-based Content Framework

The diverse nature of the knowledge assets supported by the CKM infrastructure, combined with the wide range of applications (e.g., clinical systems, browsers, search engines, portals) that make use of them, led to the implementation of an XML-based knowledge content framework (Tiwana and Ramesh 2001). The combination of an XML-based content framework and a Web-based services architecture (see subsection 22.3.1.3) was considered ideal for providing a flexible and extensible infrastructure for integrating, managing, and sharing knowledge content. The XML-based framework is the main content creation process used by the CKM infrastructure, responsible for 77 percent of all knowledge assets stored in the CKR.

The XML-based content framework supports three different classes of knowledge assets. Incremental levels of XML markup and content encoding characterize these classes. Assets that contain text with some well-defined XML tagging, but mostly to delimit their general internal sections (e.g., headings, subheadings, hyperlinks), are classified as *minimally structured* content. Minimally structured assets do not contain coded elements. Knowledge assets with textual content combined with extensive XML markup, including tags at the paragraph, sentence, and clause or term level, as well as some coded elements, are classified as *highly structured* assets. Highly structured assets are used by more sophisticated applications that require dynamic extraction and manipulation of content, such as context-aware retrieval tools (e.g., infobuttons, as discussed in Chapter 16) (Cimino et al. 1997; Del Fiol et al. 2004)). The third content class is known as *fully structured*, characterizing richly tagged knowledge assets with extensive encoded portions. Fully structured assets can be viewed as executable content, since they contain sufficient markup and encoding that can be interpreted and rendered by applications and services within or outside the CKM infrastructure.

22.2.3.2 Other Content Frameworks

Despite the advantages of the XML-based content framework, certain types of knowledge assets rely on other (external) content frameworks for their creation and initial review. In these cases, the CKM infrastructure is used basically for storage, including version control, and unified access and retrieval. Normally, these external content frameworks also use the CKM infrastructure to obtain direct user feedback and monitor asset utilization. Non-XML content frameworks are responsible for 23 percent of all knowledge assets presently stored in the CKR.

An example of an external content framework is that of documents that are created primarily for printing (paper products). Knowledge assets that also become paper products require a separate content framework because the XML-based framework does not easily support elaborate presentation and high-volume printing requirements. Several patient and provider educational materials produced by the various CPs utilize this framework. These assets are ultimately loaded into the CKM infrastructure as PDF documents, preventing any dynamic extraction and manipulation of the content. Recognizing this limitation, extensions to the CKM services are being explored to generate simple PDF documents (see subsection 22.3.1.3) with the aim of more direct integration with desktop publishing software packages with native XML support.

Another example of an external content framework is that of executable decision logic (e.g., rules and protocols). Executable rules and protocols are implemented in Java as independent modules. Each module contains a variable number of rules that pertain to a specific topic. For example, a set of rules used to identify critically altered laboratory results is an example of a module. Similarly, a set of rules that define the optimal management of intravenous insulin therapy, based on current and previous blood glucose levels, is another example of a decision logic module. Given the fact that the decision logic is implemented in Java, any open source or commercially available Java integrated development environment (IDE) can be used to author these rules. Rules can access data stored in multiple different databases through an abstract set of data retrieval services (Steiner et al. 2004). All the messages (e.g., alerts, reminders) resulting from the execution of these rules are constructed using coded concepts from Intermountain's vocabulary server.

The current framework enables the development of a wide range of executable rules and protocols, from simple to very complex, but it also requires specialized authors that have substantial programming experience using Java. In order to address this limitation, recently we have been exploring formalisms to represent simple decision rules in XML (e.g., RuleML (RuleML Homepage 2005)). We have also been following the efforts of the HL7 Clinical Decision Support Technical Committee (e.g., GELLO (Sordo et al. 2003), VMR (Johnson et al. 2001), Arden Syntax (HL7 Arden Syntax Special Interest Group 2002; Jadhav and Sailors 2003), as potential alternatives for more complex rules and protocols, with the understanding that these alternatives will require significant extensions to the existing Java-centric content framework.

22.2.4 Meta-Content Artifacts

As mentioned earlier, an XML-based content framework is used as the core content creation process supported by the CKM infrastructure. The framework supports multiple categories of knowledge assets and it is able to produce three classes of content (e.g., minimally structured, highly structured, and fully structured). Each asset category is associated with a set of XML documents that define and support the content framework. Examples include XML documents that define content structure and encoding, the authoring and review interfaces, and others that repurpose the content for different uses. These *meta-content artifacts* are described as follows, and also summarized in Figure 22-1.

- **Content schema.** Defines the markup and encoding of a given content category, including its overall structure, the cardinality of its tags, and the datatypes of the tag values. These structural properties are vital for the creation and maintenance of valid and consistent content category instances. The *content schema* is a required artifact and it is represented in XML Schema (XML Schema 2005).
- **Content authoring form.** Defines a Web form that is used to create and edit knowledge assets. The Web form contains fields and widgets that map directly to the structural elements defined by the *content schema*, including text fields, check boxes, and combo boxes (with or without coded values). The *content authoring form* is a required artifact and it is represented in XML, following a proprietary structure that is used by the CKM authoring tool (see subsection 22.3.2).
- **Content instance template.** Defines a default starting point from which new assets of a given category can be created easily, without requiring the author to start at the root element of the *content schema*. The *content*

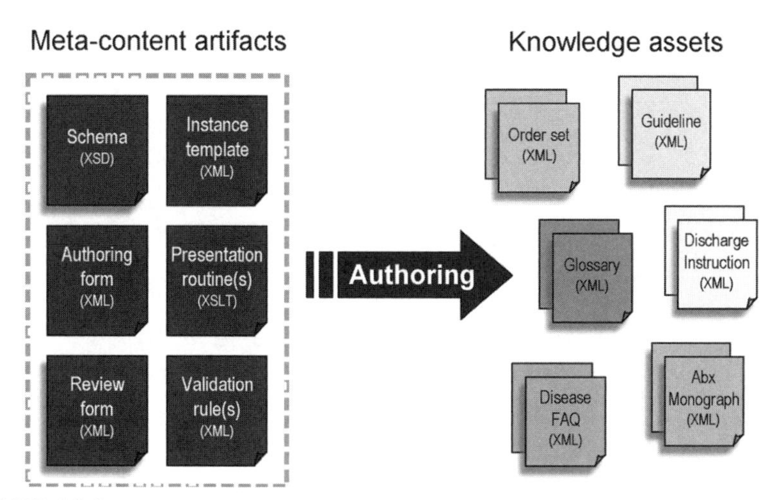

FIGURE 22-1 Overview of the XML-based content framework, demonstrating the meta-content artifacts required by the authoring process to create a wide variety of knowledge assets.

instance template is also a required artifact and is represented in XML, conforming to the structure and content defined by the *content schema*.

- **Presentation routine.** Defines a default transformation routine that is used to render the knowledge asset as an HTML, PDF, or XML document. This default transformation is used by the CKM authoring tool preview function (see subsection 22.3.2), and by the CKM asset retrieval service if a specific transformation is not defined. The *presentation routine* is a required artifact that is represented as an eXtensible Stylesheet Language Transformation (XSLT) (XSL Transformations 2005).
- **Content review form.** Defines a default Web form that is used to guide the review of knowledge assets. The Web form contains fields and widgets that map to elements defined by a proprietary schema used by the CKM content review tool (see subsection 22.3.3). Not every content category has a specific *content review form*, in which case a generic form can be used instead. The *content review form* is a required artifact represented in XML.
- **Content validation rules.** Defines one or many validation rules that extend the basic XML Schema validation process, including cross-attribute and cross-element validation, as well as validation that relies on data that are external to the knowledge asset (Hanna et al. 2005). A *content validation rule* is an optional artifact also represented in XML.

The *meta-content artifacts* usually are designed and implemented when a new asset category is created. Knowledge engineers are responsible for creating these artifacts. Figure 22-2 summarizes the relationships between *meta-content artifacts* and the various phases of the XML content framework.

Knowledge engineers are highly encouraged to use as a starting point a set of shared libraries and directories that contain basic artifact components. These basic components represent carefully defined XML datatypes, attributes, elements, and XML Schema fragments that ensure the consistency and integration of the knowledge management framework (Del Fiol et al. 2005). Most *content schemas* share (reuse) these basic components; a few asset

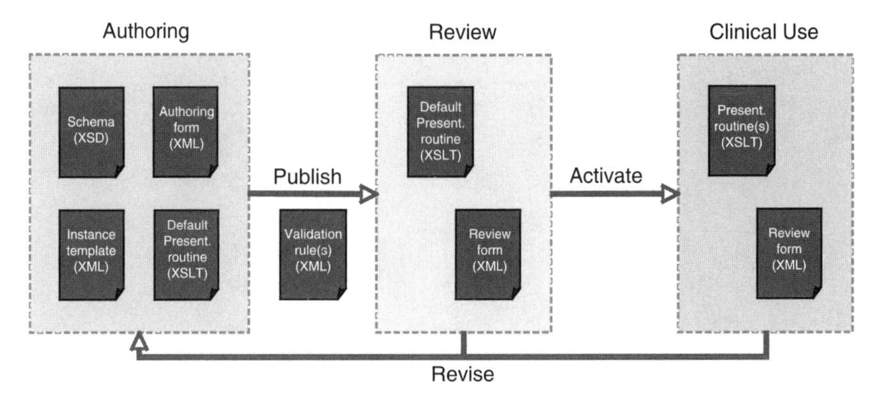

FIGURE 22-2 Overview of the relationships between meta-content artifacts and the various phases of the XML content framework.

categories do require focused extensions. All extensions or modifications to these basic artifacts require review and approval by all knowledge engineers.

Currently, most *meta-content artifacts* are created using a commercial XML editor (*Altova's® XML Suite* (XML Suite 2005)). Once created, *meta-content artifacts* are loaded into the CKR and can leverage most of the same XML document management processes that are available for traditional knowledge content instances. Table 22-1 summarizes the number of *meta-content artifacts* presently available, along with some examples.

22.2.5 Content Meta-Data

Another important component of the CKM infrastructure is a rich set of elements that represent common characteristics and properties of all knowledge assets. This set of elements is known as *content meta-data*, or simply as *header*. Every knowledge asset has a header, irrespective of the content framework used to create and maintain it. The header is a fully structured XML document that is created and maintained using the XML content framework. The header is based partially on meta-data models published in the literature, such as the Guideline Elements Model (GEM) (Shiffman et al. 2005), the Dublin core meta-data (Darmoni et al. 2001; Dublin core metadata initiative 2005), and guideline appraisal instruments (The AGREE Collaboration 2005).

Figure 22-3 presents the content schema with the primary elements of the header, and Figure 22-4 presents an example of a header. The most important header elements are:

- **Identification.** A set of elements that uniquely identify the knowledge asset. Identification includes elements like a unique numeric symbol (*ID*), a version number (*Version Number*), a preferred display name (*Name*), a unique preferred virtual name (*Virtual Name*), and the purpose of the asset. Assets can be retrieved using either the asset *ID* or the asset *Virtual Name*. Currently, the CKR has 30,195 assets total, out of which 8,759 are in use.
- **Version.** A set of elements that characterize each version of a knowledge asset. Version includes elements like the process used to develop it, a summary of the modifications that occurred since the previous version, the rational for these modifications, the release date, the status, the

TABLE 22-1 Meta-content artifacts currently used by the XML-based content framework.

Meta-content	Total	Active	Revisions average/ median	Examples
Schemas	587	48	10/7	Basic data types directory, Common meta-data (header) schema, Order set schema, Review form schema, Protocol schema, Antibiotic Monograph schema
Authoring forms	603	37	15/11	Common meta-data form, Order set form, Laboratory diagnostic findings form, Care plan form, Index form, Literature citation form, Emergency services discharger instructions form
Instance templates	318	43	5/2	Common meta-data template, Order set template, Calculation template, Clinical form instructions template, Patient education FAQ template
Presentation routines	2,873	76	29/13	Common meta-data presentation routine, Order set preview routine, CPOE order set presentation routine, CPOE order set with calculation presentation routine, Antibiotic monograph presentation routine
Review forms	12	6	2/2	Default review template, inter-author review template, order set review form
Validation rules	8	(N/A)	(N/A)	ASCII characters validation, linked asset validation, required field validation

custodian(s) (clinician(s) or knowledge engineer(s) responsible for the modifications), and the proposed next revision date. The development process element documents the authoring, review, and approval process that was used, including the domain experts that were involved and their respective roles (e.g., authors, reviewers, endorsers). The relevant references consulted during the development of the new version can also be represented. In terms of status, 68 percent of the CKR assets are *inactive* (obsolete versions of existing assets), 5 percent are *under development*, 3 percent are *under review*, and 25 percent are *active*. Each CKR asset has been revised 3.45 times on average.

- **Category**. An element that specifies the type of asset (e.g., order set, discharge instruction, care process model, antibiotic monograph). Each knowledge asset can belong to only a single category. In the case of assets produced with the XML content framework (see subsection 22.2.3.1), *category* corresponds to the underlying content schema. In the case of assets produced with other content frameworks (see subsection 22.2.3.2), *category* corresponds to the intended clinical purpose of the asset (e.g., patient education handout, care process model). The category names are made as specific as needed in order to distinguish types of assets that may have similar clinical purposes. The classification and refinement of the asset categories is a continuous process supported by the vocabulary

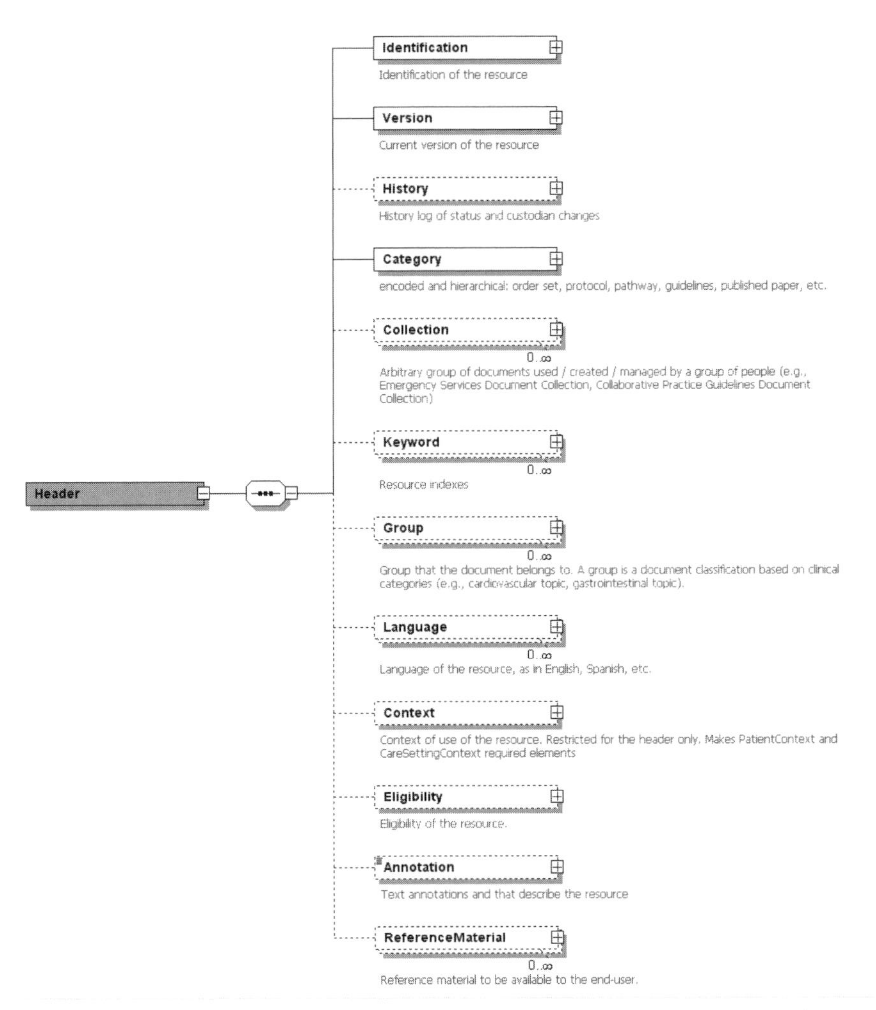

FIGURE 22-3 Diagram of the header showing the most important content schema elements.

server (see subsection 22.1.2 and Chapter 20). Presently, the CKR has 67 categories total. The average number of assets per category is 451 and the median is 151. Table 22-2 presents a list of the asset categories responsible for 80 percent of the active CKR assets.

- **Collection.** A repeating element that identifies the collection(s) to which the asset belongs. A collection is an arbitrary set of assets that is useful to a group of users. Collections are used to narrow the focus of browsing and searching efforts to highly relevant sets of assets. The CKR presently has ten collections defined.
- **Keyword.** A repeating element that specifies alternate names and related terms that can be used to locate an asset. Valid keywords include synonyms, acronyms, common abbreviations, and numeric or alphanumeric codes from controlled terminologies. Keywords can also be coded concepts defined in the vocabulary server, enabling documents to be indexed using the same controlled terminologies used by HELP2.
- **Context.** A set of elements that represents expressions used to define when an asset is available or not for retrieval. Context expressions act as retrieval filters that prevent assets from being inappropriately retrieved,

```
<?xml version="1.0" encoding="UTF-8" ?>
- <Header xmlns:xsi="http://www.w3.org/2001/XMLSchema-instance">
- <ResourceID>
    <ObjectIdentifier>51059854</ObjectIdentifier>
    <ObjectRelease>48</ObjectRelease>
    <ObjectVersion>48</ObjectVersion>
    <ObjectPreferredDisplayName>General Surgical / Post-Operative Orders</ObjectPreferredDisplayName>
  </ResourceID>
- <CurrentVersion>
  - <Date>
      <DateTimeValue>2005-08-09T14:35:50</DateTimeValue>
    </Date>
  - <Purpose>
      <StringValue>Set of post-operative orders commonly used by General Surgeons</StringValue>
    </Purpose>
  + <Modification>
  - <ResourceStatus>
    - <CodedValue ontologyID="50550871" ontologyName="IHC Healthcare Data Dictionary">
        <ConceptIdentifier>1024</ConceptIdentifier>
        <InstanceIdentifier>10990</InstanceIdentifier>
        <ResourceSurfaceForm>Active</ResourceSurfaceForm>
      </CodedValue>
    </ResourceStatus>
  + <Custodian>
  + <Publisher>
  + <DevelopmentProcess>
  </CurrentVersion>
+ <DocumentHistory>
- <DocumentCategory>
  - <CodedValue ontologyID="50550871" ontologyName="IHC Healthcare Data Dictionary">
      <ConceptIdentifier>50349744</ConceptIdentifier>
      <InstanceIdentifier>50342500</InstanceIdentifier>
      <ResourceSurfaceForm>Order Set</ResourceSurfaceForm>
    </CodedValue>
  </DocumentCategory>
- <Keyword>
  - <SearchTerm>
    - <CodedValue ontologyID="50550871" ontologyName="IHC Healthcare Data Dictionary">
        <ResourceSurfaceForm>Surgery</ResourceSurfaceForm>
      </CodedValue>
    </SearchTerm>
  </Keyword>
- <Keyword>
  - <SearchTerm>
    - <CodedValue ontologyID="50550871" ontologyName="IHC Healthcare Data Dictionary">
        <ResourceSurfaceForm>Operative</ResourceSurfaceForm>
      </CodedValue>
    </SearchTerm>
  </Keyword>
- <Keyword>
  - <SearchTerm>
    - <CodedValue ontologyID="50550871" ontologyName="IHC Healthcare Data Dictionary">
        <ResourceSurfaceForm>Procedure</ResourceSurfaceForm>
      </CodedValue>
    </SearchTerm>
  </Keyword>
+ <Context>
</Header>
```


FIGURE 22-4 Example of a header instance ("General Surgical Post-Operative" order set).

whenever the context of use of the asset is known to the requesting application (see subsection 22.2.2). For example, knowledge assets for specific age groups will be retrieved only when the patient's age falls within appropriate intervals. *Context* defines 15 possible dimensions that can be used to create conditional expressions, but currently only five are being actively used: patient gender (e.g., *male, female*), patient age group (e.g., *12 months to 36 months, 17 years old and under, 18 years old and above*), facility (e.g., *LDS Hospital, Primary Children's Medical Center*), unit (e.g., *Shock-Trauma ICU at LDS Hospital, Neonatal ICU at Dixie Medical Center*), and type of unit (e.g., *intensive care unit, emergency room*, or *coronary care unit*).

TABLE 22-2 Asset categories (excluding meta-content artifacts) responsible for 80% of the active assets presently available in the CKR, as of September 2005.

Category	Collection	Active assets	Total assets	Versions (average)
Discharge Instruction	Emergency Department	1,438	6,697	4.7
Literature Citation	Collaborative Practice Guidelines	1,025	1,764	1.7
Image	*(Multiple)*	664	893	1.3
Table or Tool	Collaborative Practice Guidelines	576	2,510	4.4
Footnotes	Collaborative Practice Guidelines	426	698	1.6
Order Set	CPOE	352	1,909	5.4
Symptoms	Collaborative Practice Guidelines	350	873	2.5
Discharge Instruction	*(Multiple)*	325	435	1.3
Risk-factor Causes	Collaborative Practice Guidelines	300	511	1.7
Nested Order Set	CPOE	291	1,154	4.0
Problem	Collaborative Practice Guidelines	270	1,784	6.6
Folio Document	*(Multiple)*	250	264	1.1
Protocol	Collaborative Practice Guidelines	204	1,520	7.5
Lab Diagnostic Findings	Collaborative Practice Guidelines	203	465	2.3
Clinical Form	Clinical Forms	191	374	2.0
Order Set	Collaborative Practice Guidelines	177	570	3.2
Index	*(Multiple)*	158	182	1.2
Total		*7,200*	*22,603*	*3.1*

22.2.6 Knowledge Assets

The number and diversity of knowledge assets have been constantly increasing since the CKR was first released in June 2003. The CKR presently has 30,195 knowledge asset instances, subdivided into 67 asset categories. Thirty-eight asset categories are currently fully supported by the XML-based framework. Described here are some details of four representative examples of knowledge assets.

22.2.6.1 Care Process Models

A *Care Process Model* is the most important type of clinician educational material produced by a CP. Each Care Process Model provides a detailed overview of the disease management system associated with a given clinical condition or process, in an effort to help providers deliver the best possible care in a consistent and integrated way. Typically, a Care Process Model establishes key treatment goals, provides a series of algorithms to guide the medical management of the disease, and presents detailed information regarding indicated and contraindicated medications, as well as details about the management of related conditions. Examples of Care Process Models include "Acute Low Back Pain," "Attention-Deficit/Hyperactivity Disorder (ADHD)," "Asthma," "Depression," "Diabetes," "Heart Failure," "Hypertension," and "Pneumonia." A complete list of Care Process Models is available at Intermountain's main Internet site (Intermountain Clinical Programs 2006).

All Care Process Models exist today in the CKR only as PDF documents, since they normally are printed and distributed as paper booklets, but their conversion to XML is planned for 2006. Access to the documents is primarily through static HTML links available in Intranet and Internet pages. However, taking advantage of the rich meta-data available in the CKR, Care Process Models recently have also been made available through disease-specific infobuttons within the Problem List of Intermountain's Clinical Desktop.

22.2.6.2 Interdisciplinary Collaborative Practice Guidelines

In 1996, Intermountain created the Clinical Consistency Project to develop and implement interdisciplinary standards, aiming at reducing variability of bedside care (Hougaard 2004). These interdisciplinary standards are known as the Collaborative Practice Guidelines (CPG) collection. The collection has been produced and released electronically since its early stages, initially using a proprietary SGML-based format. The CPG collection now uses a set of 16 highly structured content schemas and it is fully supported by the CKM infrastructure (Del Fiol et al. 2004).

The CPG collection is organized into five document groups: problem, risk for problem, protocol, procedure, and teaching plan. Each document group follows a specific *content schema*. The main sections of these *content schema* are tasks, goals, documentation needs, risk factors, symptoms, laboratory and diagnostic findings, and literature references. One of the main characteristics of the CPG collection is that documents are modular and reusable (Del Fiol et al. 2004).

Clinicians can search and browse the CPG collection at the bedside using a Web-based tool called the *CPG Viewer* (see subsection 22.3.4.1). The CPG collection presently has 3,690 documents in active use. Since April 2005, the *CPG Viewer* has been used 23,963 times, including 39,924 searching events that retrieved 122,266 documents, corresponding to an average of 1.69 searches and 5.01 documents retrieved per session.

22.2.6.3 Order Sets

Order sets are predefined groups of orders relevant to the management of specific clinical conditions or diagnoses (Ash et al. 2003; Cowden et al. 2003; Payne et al. 2003). Within Intermountain, order sets not only are considered key factors in physicians' acceptance of the CPOE system, but also as disease management system intervention instruments that can advance the implementation of best practices. Successful development and implementation efforts have demonstrated that order set developers and practicing clinicians (order set users) have to be in continuous communication, leading effectively to the sense of shared ownership of the resulting knowledge content. Intermountain's inpatient CPOE system is a module of HELP2. All CPOE order sets are created, maintained, and accessed using the CKM infrastructure.

The order sets are based on six highly structured content schemas. These content schemas define all structural and functional properties of the order sets, including the ability to reuse frequent sets of orders across multiple order sets, and the ability to dynamically display or hide segments of an order set according to predefined context parameters. The reuse of order set fragments simplifies content maintenance and also helps to promote widespread compliance with protocols and practice standards. The feature to dynamically display or hide content based on context parameters (e.g., facility, inpatient unit, patient age group) has enabled the development of order sets that can accommodate local or regional content variations. A detailed description of the order set content schemas can be found elsewhere (Del Fiol et al. 2005).

The CKR currently has 82 order sets that are in active use. Only 21 distinct order sets are responsible for 80 percent of all CPOE sessions. Since January 2005, an average of 35.6 unique physicians have used the HELP2 CPOE every month, accumulating 2,714 sessions total.

22.2.6.4 Computerized Alerts and Protocols

Six protocols, three alerts, and two critiquing services have been implemented in the HELP2 environment using the Foresight infrastructure since June 2003. Since its release, Foresight has generated 93,245 alerts total.

The first protocol developed was the Pediatric Ventilator Weaning Protocol. This protocol helps manage the extubation of pediatric patients. A recent evaluation of the protocol, not yet published, showed a decrease in 20 hours in the time required for extubation when the protocol was used. The second protocol developed manages patients that are chronically anticoagulated with the drug warfarin. The rules interpret the INR results of the patient and give suggestions of what should be done next about the drug dosage and when the patient should be tested again. The protocol also generates alerts when the patient fails to do a new test.

Four other protocols (management of hyperglycemic patients, community acquired pneumonia, post-liver transplant management, and early shock identification for pediatric patients), three different sets of alerts (adverse drug events alerts for creatinine doubling, newborn bilirubin, and critically altered laboratory results alerts), and two services for detecting drug–drug interactions and drug–allergies have been added. Table 22-3 presents detailed information about all these assets.

Rules and protocols are thoroughly tested before being released for routine use. There are four phases in the validation process. The first phase happens during rule creation, when the rules developer is required to continuously test the behavior of the rules with test data in the development environment. Once the rules are considered ready for initial testing, the second validation phase begins, requiring the latest version of the rules be copied to the test environment. The test environment contains a recent copy of the production patient data (CDR), enabling the execution of the rules against large volumes of "real" data. This second testing phase continues until the rules are considered correct and ready for formal testing.

The third validation phase is considered the formal testing phase. An independent software quality assurance (QA) team is responsible for this third validation phase. The QA team, with support from knowledge engineers, Foresight software engineers, and rule developers, initiates the formal testing by creating *test cases* that aim at executing all the branches of the logic represented by the rules. Once the test cases are created, the QA team uses them for testing the rules and confirming that the results obtained are the expected ones. These same test cases are also used for regression testing of the rules when upgrades to Foresight are deployed, in an effort to guarantee that the rules continue to produce the expected results.

The fourth and final validating phase, known as the acceptance-testing phase, is initiated after the QA team certifies that the rules produce the expected results in the test environment. The rules are copied to the production environment and the rules developers, in collaboration with the QA team, verify if the released rules produce the correct results, again using the same test case developed during the formal testing phase.

The actual deployment of the new rules and protocols normally is not perceived by HELP2 users. Rule and protocol deployments do not require system downtime. Once a new set of rules or protocol is uploaded to the production CKR environment, Foresight is immediately notified. Foresight

TABLE 22-3 Details about the alerts, critiquing services, and protocols currently implemented using the Foresight infrastructure.

Title	Type	Initial use	Counts (until September of 2005)							Current status
			Enrollees	Alerts	Users	Rules	Date-driven events	Data gets	Revisions	
Critically altered laboratory results	Alert	Mar-2005	All Emergency Department patients (8 sites)	3,003	All Emergency Department clinicians	56	46	1	9	Routine use
ADE creatinine doubling	Alert	June-2004	N/A	445	0	1	1	2	5	Under review
Bilirubin	Alert	June-2004	All newborns (2 sites)	8,024	4	5	4	2	10	Routine use
Allergies	Critiquing service	Mar-2005	N/A	N/A	All outpatient CPOE users	N/A	All prescribed drugs	2	2	Routine use
Drug–drug interactions	Critiquing service	Sep-2003	N/A	N/A	All outpatient CPOE users	N/A	All prescribed drugs	1	1	Routine use
Early shock identification	Protocol	Nov-2005	Pediatric patients (1 site)	N/A	N/A	20	10	2	1	Pilot use
Community-acquired pneumonia	Protocol	Nov-2004	N/A	33	5	22	12	0	9	Pilot use
Glucose management	Protocol	Sep-2004	ICU patients (2 sites)	9,917	All ICU clinicians	28	3	1	12	Routine use
Post-liver transplant	Protocol	May-2004	348	18,162	5	22	8	3	9	Routine use
Chronic anticoagulation	Protocol	June-2003	881 (2 sites)	31,076	20	32	3	4	11	Routine use
Pediatric ventilator weaning	Protocol	June-2003	0 (1 site)	1,550	5	61	34	1	5	Not in use

retrieves the new rules or protocol from the CKR and immediately starts making use of the new logic.

22.3 SOFTWARE INFRASTRUCTURE

In a similar manner to that in other vertical domains and industries, the effective management of knowledge in health care requires at least three interrelated processes: creation, integration, and dissemination (Alavi and Leidner 2001; Fischer and Ostwald 2001; Stefanelli 2004; Bali 2005; Guptill 2005). The three processes form a cycle that can be described as the *knowledge life cycle*. The implementation of a *knowledge life cycle* depends on clearly defined processes, appropriate resources, a well-prepared interdisciplinary team, and an integrated and flexible software infrastructure. Ultimately, the result of a successful knowledge life cycle is to "make it easy to do it right," that is, provide "the right information, in the right format, at the right time, without requiring special effort" (James 2001). (Note that the knowledge life cycle described here corresponds to the knowledge management life cycle described in Chapter 1.)

The CKM software infrastructure includes a series of software components designed to support the complete life cycle of knowledge assets. Figure 22-5 presents an overview of the CKM software infrastructure and its interactions with other Intermountain systems and tools.

22.3.1 Clinical Knowledge Repository

The Clinical Knowledge Repository (CKR) is a document-centric database that stores knowledge assets as XML documents, as well as other common file types (or MIME types) such as PDF, MS Word, Bitmap, JPEG, and so on. The primary goal of the CKR is to provide a permanent and convenient *storehouse*

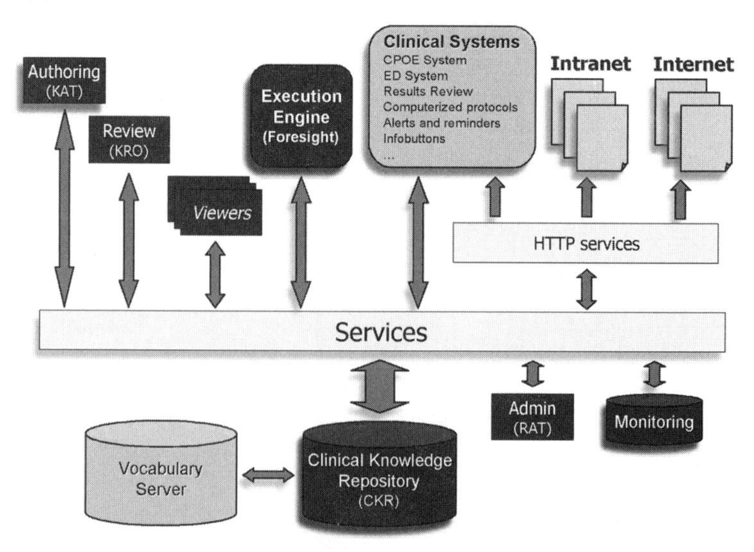

FIGURE 22-5 Overview of the CKR software infrastructure and its relations to clinical systems and Intranet or Internet resources. Bidirectional arrows imply read and write services, and unidirectional arrows imply read-only services.

where all categories of knowledge assets can be centrally managed and consistently retrieved. The CKR is the cornerstone component of the CKM software infrastructure.

22.3.1.1 Implementation

In 2001, a group of informaticians and knowledge engineers spent approximately nine months defining and refining the CKR requirements. During this period, commercially available content management solutions were also reviewed. The review process included the exchange of detailed requirements, in-house demonstrations, and, in a few cases, the implementation of prototypes. The decision to proceed with an in-house development of the CKR was made in early 2002. The main reasons for this decision included:

- Lack of support of external products for XML Schema, which was due to the fact that the XML Schema W3C recommendation had been approved only recently (*XML Schema* 2005)
- Lack of support for leading XML-based authoring tools, despite the fact that most products were able to store and retrieve XML documents
- Absence in external products of flexible application programming interfaces (APIs) that would enable an adequate level of integration with HELP2 applications and the implementation of optimal knowledge dissemination strategies
- Impossibility of achieving the required integration of external products with controlled terminology content (e.g., hierarchies, value sets) and services provided by Intermountain's vocabulary server
- Search and retrieval limitations of external products imposed by a restricted set of metadata elements, which would preclude the implementation of context-enabled searching and retrieval, as well as dynamic content configuration and presentation

The CKR was built upon a standard J2EE architecture with *best-of-breed* software on each tier. The persistence layer was implemented with Oracle 9i (Oracle Database 2005). Oracle 9i provided out-of-the-box indexing and searching of XML data, textual data (Clob), and binary data (Blob), as well as access to conventional relational data. BEA WebLogic was selected as the application server to host middle-tier Java-based services (BEA WebLogic Platform 2005). Services are accessible via URL requests that generally respond with XML unless otherwise specified. For example, CKR clients access a search service via a URL that returns asset metadata in XML, including asset URLs. Asset URLs are structured to request raw XML, transformed XML (typically HTML), or PDF output (using XSL-FO). XSLT also is executed on the middle-tier, enabling efficient asset linking and reuse. The CKR clients include Intranet and Internet browsers (MS Internet Explorer is the only browser presently supported), as well as other middle-tier applications.

22.3.1.2 Database

The CKR is implemented using a commercial relational database management system (Oracle RDBMS (Oracle Database 2005)), with extensive use of features supporting native XML data: XMLType datatype, XML indexing using Oracle Text, XPath-based querying, and XPath-based data extraction.

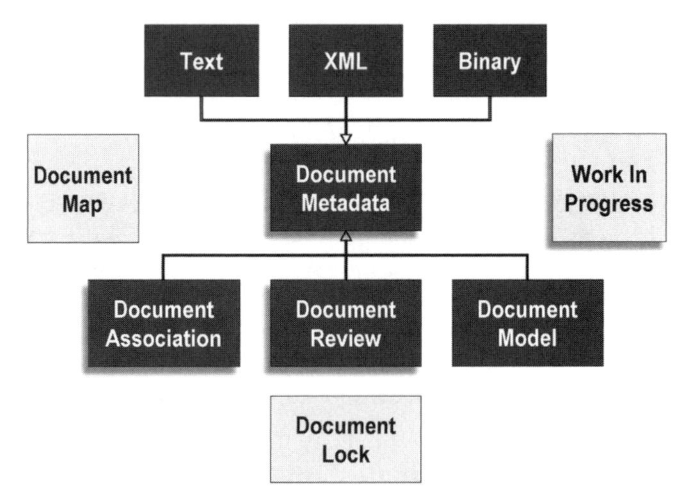

FIGURE 22-6 Diagram of the CKR database with the most important tables.

Figure 22-6 shows a diagram of the most important CKR tables. The *Document Metadata* is the core CKR table, storing all the necessary asset metadata (*header*—see Figure 22-3). Each of the surrounding tables, *Text, Binary*, and *XML*, store the asset itself (*body*), and each table uses a different data type that is compatible with the MIME type of the asset: text, binary (e.g., PDF, MS Office files), and XML, respectively. The table *Document Model* stores the *schemas* of the XML assets.

The *Document Review* table is used to store all the feedback (reviews) received from reviewers regarding any of the knowledge assets. The *Document Association* table is used to store explicit associations between knowledge assets. The *Document Map* table is used to represent all the *meta-content artifacts* required by each asset category (see subsection 22.2.4). The *Document Lock* table is used to keep track of all the assets that are locked for editing, and the *Work In Progress* table is used to store all the XML-based assets that are still being created or revised. The authoring tool is the primary user of these last two tables (see subsection 22.3.2).

Taking into account performance requirements, the CKR database model enables direct asset retrieval without requiring combinations (joins) of data across multiple tables. The *Document Model* table is separate from the *XML* table also for performance reasons, since the *Document Model* table gets the highest volume of reads.

Currently, the CKR database is implemented in four distinct environments: software development, knowledge engineering, testing and quality assurance, and production. The software development environment, used primarily by the software engineers, is used for the development of new CKR services and CKM applications. The knowledge engineering environment is where more complex knowledge assets initially are created, including, for instance, the meta-content artifacts required by a new asset category and the rules required by a new computerized clinical protocol. Knowledge engineers are the principal users of the knowledge engineering environment. The testing and quality assurance environment is where the overall software and content testing and validation happen before a production release. Finally, the production environment represents the CKR that is used by deployed

HELP2 applications, as well as all the Intranet and Extranet sites and portals. The production environment also is used by the CKM applications that are available to end users, including the tools for asset authoring and review. Therefore, reviews and new assets produced by clinicians (end users) are created in the production environment. The CKR administration tool (see subsection 22.3.1.4) provides the necessary functions for managing the four environments.

22.3.1.3 Services

A service layer mediates all forms of access to the CKR, including searching, retrieval, and creation of new assets. A straightforward set of services enables the retrieval of knowledge assets via HTTP requests and responses. Table 22-4 contains a list of the generic HTTP services presently implemented.

The most fundamental service, the asset retrieval service, requires only the document's *ID* parameter. A URL such as *https://kr.ihc.com/kr/Dcmnt?id=51059854* retrieves the document 51059854 (a commonly used order set) in its native MIME type form (XML in this case). Retrieving a specific version of an asset can be accomplished by adding the *Version Number*. The URL *https://kr.ihc.com/kr/Dcmnt?id=51059854&vrsn=44* retrieves version 44 of document 51059854. When the *Version Number* is not specified, the asset retrieval service does not necessarily retrieve the highest *Version Number*. Each asset version also has a status, with a certain order of precedence. From lowest to highest, the status precedence order is: *inactive, under development, under review*, and *active*. When the *Version Number* is not specified, the asset retrieval service will return the asset with the highest status. For example, if there are five *Version Numbers* (0 to 4) for a given asset, where *Version Numbers* 0 to 2 are *Inactive, Version Number* 3 is *Active*, and *Version Number* 4 is *Under Review*, the asset that will be retrieved is *Version Number* 3. If, for whatever reason, the maximum numeric version is desired, the parameter *maxVrsn* can be used to request the highest noninactive version of an asset.

Retrieving the *header* of an asset requires a very similar URL. Simply by changing the *id* parameter to *hid* indicates to the asset retrieval service that the *header* should be returned instead. For example, *https://kr.ihc.com/kr/*

TABLE 22-4 List of CKR services that can be accessed using HTTP clients.

Service name	Responsibilities
Asset Retrieval	Retrieve requested asset or its header (by *ID*) and, if specified, execute requested XSLT
Asset Search	Search for knowledge asset meta-data based on simple input parameters and return *IDs* of matching assets
Advanced Asset Search	Search for knowledge asset meta-data based on one or multiple input parameters and return meta-data values based on input parameters (see Table 22-5 for more details)
Asset Work-In-Progress (WIP) Retrieval	Retrieve requested asset or its header (by *ID*) from the work-in-progress table, if specified, execute requested XSLT
Asset Retrieval & PDF Generation	Retrieve requested asset (by *ID*), execute the requested XSL-FO, and return generated PDF
Monitor Event	Record event parameters received into the monitoring infrastructure

Dcmnt?hid=51059854&maxVrsn=Y retrieves the *header* of the latest version of the document 51059854. Frequently, XSLT assets contain URL links that utilize the *hid* parameter to obtain meta-data details needed during asset transformation.

Assets can also be transformed using the asset retrieval service by adding another parameter to the URL with the associated asset identifier, as in *tfrm=50358866* (in this case an order set preview transformation). Using the *tfrm* parameter, any XML asset can be transformed by an XSLT asset stored in the CKR. Setting the *tfrm* parameter value to *default* indicates to the asset retrieval service that the standard XSLT assigned to the asset category should be used (see subsection 22.2.4). XSLT versions can be directly specified using the *tfrmver* parameter. When the parameter *tfrmver* is not specified, the XSLT *Version Number* with the highest status is retrieved (as described earlier). In summary, if the intent was to retrieve the latest active version of the order set 51059854 and transform it using the latest active version of the transformation 50358866, the URL would be *https://kr.ihc.com/kr/Dcmnt?id=51059854&tfrm=50358866.*

The second fundamental service is the advanced asset search service. Table 22-5 shows the most important parameters accepted by the advanced search service. There are several implementations of the asset search service for various business-specific needs, but they are all variations of the same advanced search service. The different implementations of the advanced asset search service include: *CPOE search, CPG search, KAT search, KRO search, Simple Search*, and *Emergency Department discharge instructions search.*

TABLE 22-5 Most important parameters used by the CKR advanced asset search service.

Parameter	Multiplicity	Default	Description
searchPhrase	0..n	(none)	The text string to search for. Multiple values are and'ed together.
xPath	0..n	(none)	The *Xpath* fields to search within. *Xpath*[i] associates with *searchPhrase*[i]. If no *XPath* is specified and a *searchPhrase* is specified, the *searchPhrase* will be used to search against the complete *header*. Multiple *XPath* values are associated with *searchPhrases* are and'ed together.
facilityId	0..n	(none)	The encoded *facility(ies)*—LDS Hospital, Cottonwood Hospital, etc. *Facilities* are or'ed together and then and'ed with the total phrase.
unitId	0..n	(none)	The encoded *unit*—LDS Hospital Critical Care ICU, Cottonwood Hospital Emergency Department, etc. *Units* are or'ed together and then and'ed with the total phrase.
genderId	0..n	(none)	The encoded *gender(s)* is/are and'ed with the total phrase.
ageId	0..n	(none)	The encoded *ageId(s)* is/are and'ed with the total phrase.
statusId	0..n	(none)	The encoded *status(es)* of the assets being searched for is/are and'ed with the total phrase.
tfrm	0,1	(none)	The asset *ID* of the XSLT to transform the results of the search.

TABLE 22-6 Examples of CKR services that support the CKM software tools.

Service name	Responsibilities	Access
Document Service	1. Insert new asset, new version of asset 2. Change asset status 3. Transfer ownership of asset 4. Validate asset 5. Fetch review template/questionnaire 6. Get review consolidation 7. Add/remove custodians	CKR tools
Work-in-progress (Wip) Service	1. Create new Wip 2. Save/update Wip 3. Publish Wip (move from *Wip* to *KRDocument*) 4. Edit KRDocument (copy *KRDocument* to *Wip*) 5. Add/remove custodians	CKR tools
Login Service	1. Authentication 2. Profile management	CKR tools
Email Service	1. Send e-mail 2. Send review notifications	Intranet clients
Http Review Service	1. Fetches/stores asset-specific review 2. Review consolidation 3. Notify custodians when a review has been submitted	Intranet clients
User Service	1. Insert user and user parameters 2. Update user and user parameters	CKR tools
Notification	1. Add/remove users from asset specific notification list 2. Upon asset update, e-mail users on notification list that update occurred 3. Send notifications to client applications of asset change	CKR tools
Validation Service	1. Perform XSD-based validation when asset is published to CKR (overrides XSD validation messages with user-friendly messages) 2. Conditional validation—based on XML values 3. Validation against external entities (terminology-based validation) 4. Validation uses on-the-fly configuration (abstract of chain-of-responsibility pattern)	CKR tools

Besides the core search and retrieval services, an additional set of services has been implemented to support the CKM software tools. Table 22-6 contains examples of the services used by the CKM software tools.

22.3.1.4 CKR Administration

A suite of software tools and procedures has been created to perform administrative functions against the CKR database. One example of these administration tools is a set of database-driven PL/SQL procedures developed for transferring assets between the three CKR environments. These procedures are needed particularly during a software release cycle, enabling the deployment of new asset categories and their corresponding *meta-content* artifacts into the production environment.

Another example is the *Repository Administration Tool* (RAT). RAT was created to enable reliable and efficient maintenance functions against large sets of CKR assets. RAT performs advanced content validation and XSLT updates against predefined sets of assets. These sets of assets are defined using SQL select statements. Running these batch processes within RAT allows the CKR administrator to execute the defined job and review the results before committing any changes. Changed assets are backed up and changes are logged when assets are updated. RAT is used extensively during software updates, releases, and also for revisions and minor enhancements of the CKR content.

Other operational and maintenance functions are served through custom-built SQL and PL/SQL procedures, enabling simple operations such as batch asset status changes and deletion of temporary assets.

22.3.2 Authoring

The primary authoring tool of the CKM infrastructure is the *Knowledge Authoring Tool* (KAT). KAT is a Web-based knowledge editor that interacts with the CKR through a service layer. KAT allows knowledge authors to create and update XML documents using Microsoft's Internet Explorer Web browser (see Figure 22-7). The XML editor used by KAT is the browser edition of Authentic®, produced by Altova (Authentic Browser Edition 2005). Summarized here are some of the main features of KAT. Detailed information about KAT, including extensive information about requirements and implementation, can be found elsewhere (Hulse 2005; Hulse et al. 2005).

Within KAT's main page (see Figure 22-8), two separate tables are presented to the knowledge author, enabling quick access to all assets of interest. The *Work In Progress* table contains assets (documents) that are currently under development, but not yet published for others to access or review. The *Published Documents* table contains all the published assets that have been created by the knowledge author. Both tables can be sorted and filtered using client-side scripting, which enables effective access to a desired asset. Hyperlinks in each row of the two tables represent shortcuts to all main functions of KAT, including editing, duplicating (saving as), previewing, publishing, changing status, and transferring (or sharing) ownership to other author(s).

Using the CKR infrastructure, KAT is capable of supporting modular content production, where any given knowledge asset can be linked to other relevant assets that complete its content. These content links, known as nested content, are implemented by referencing the unique identifiers (*IDs*) of the other relevant assets. At run-time, the retrieval service assembles the complete knowledge asset by exhaustively traversing all the nested content links. This function promotes content reuse among knowledge authors.

KAT can be used to author practically any type of XML document, provided that the appropriate *meta-content artifacts* have been created and are explicitly declared in the *Document Map* table (see Figure 22-6). KAT loads the data from the *Document Map* table into memory upon start-up, using these data to access the necessary *meta-content artifacts* required during the authoring process of a given asset category. A new row in the *Document Map* table is essentially all that is needed to bring about a new asset category; that is, it does not require a redeployment of KAT, or a recompilation of its source code. Consequently, KAT remains independent of the content it is able to create, which ensures excellent flexibility to the CKM infrastructure and much desired autonomy to knowledge engineers. The increasing number of asset categories being created using the XML content framework, as presented in Table 22-2, confirms the benefits obtained with this feature.

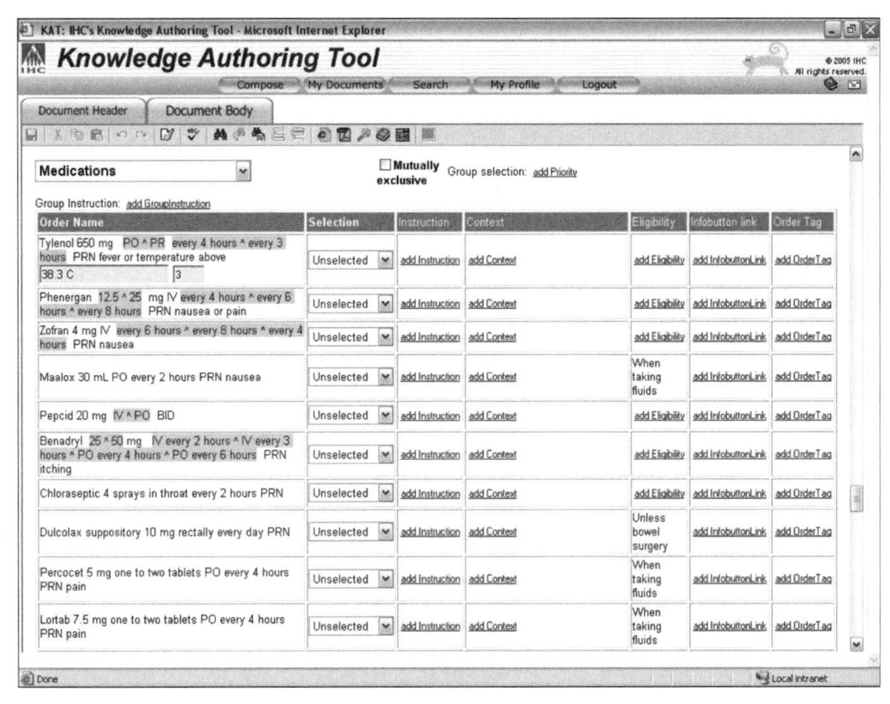

FIGURE 22-7 An example of KAT as it is used to edit an XML-based order set. The Web forms include text fields, combo boxes, tables, and other objects typical in HTML-based Web forms. Toolbar buttons across the top allow the user to save, validate, check spelling, and search for content to be nested.

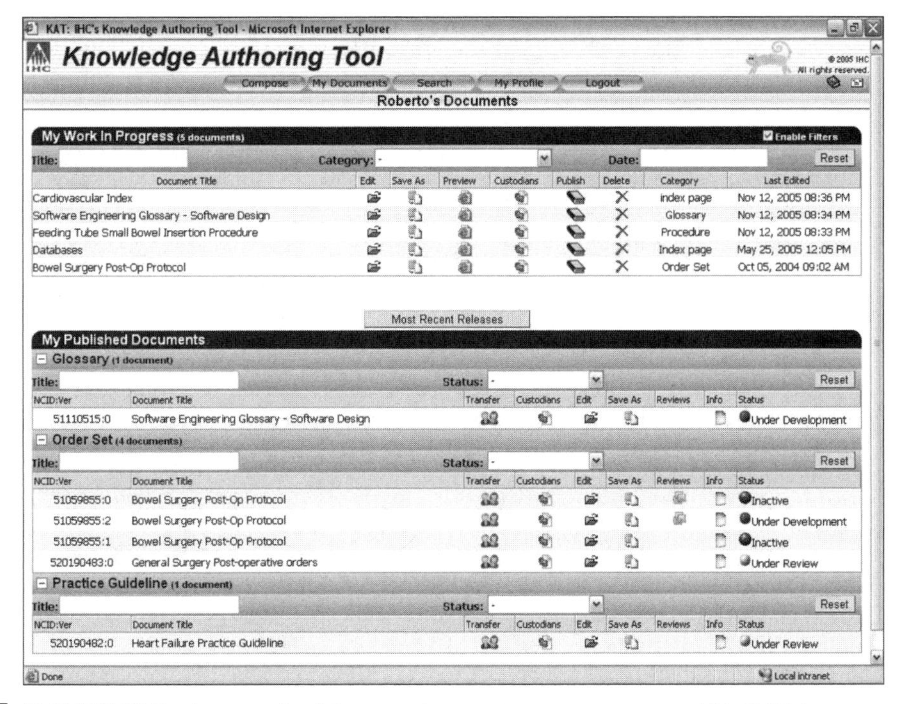

FIGURE 22-8 An example of the asset (document) management page within KAT, known as *My Documents*. Notice the two main areas: *My Work In Progress* and *My Published Documents*. From this page, authors can edit, publish, and manage the knowledge assets they own.

The KAT monitoring infrastructure was activated in May 2004. Since this date, 170 unique users have used KAT, and 38 different authors use it on a monthly basis. Since January 2005, a total of 516 new documents have been created using KAT.

22.3.3 Review

The primary review tool of the CKM infrastructure is the *Knowledge Repository Online* (KRO). KRO is a Web-based knowledge browser capable of searching and retrieving CKR assets using a service layer (see Figure 22-9). Described next are some of the most relevant features of KRO. Detailed information about an initial prototype of KRO, including its core requirements, can be found elsewhere (Wilkinson 2003).

KRO was designed primarily to enable an open and distributed review process. KRO makes it possible for practicing clinicians (i.e., end-users of the knowledge assets) to provide direct feedback to the authors about these assets. KRO exposes all the published assets stored in the CKR to nearly all Intermountain clinicians through Intermountain's Intranet. In order to create a review, clinicians can use the search page available in KRO to find the desired document (see Figure 22-10). Another option, which is certainly more convenient for clinicians, is to gain access to KRO directly from the HELP2 application or intranet Web page that is presenting the desired knowledge asset. In this case, the clinician simply clicks at the *Knowledge-button* and is taken directly to the appropriate review form within KRO (see Figure 22-11).

Whenever a review is submitted through KRO, the author is promptly notified by e-mail. The actual reviews are also stored in the CKR, using the

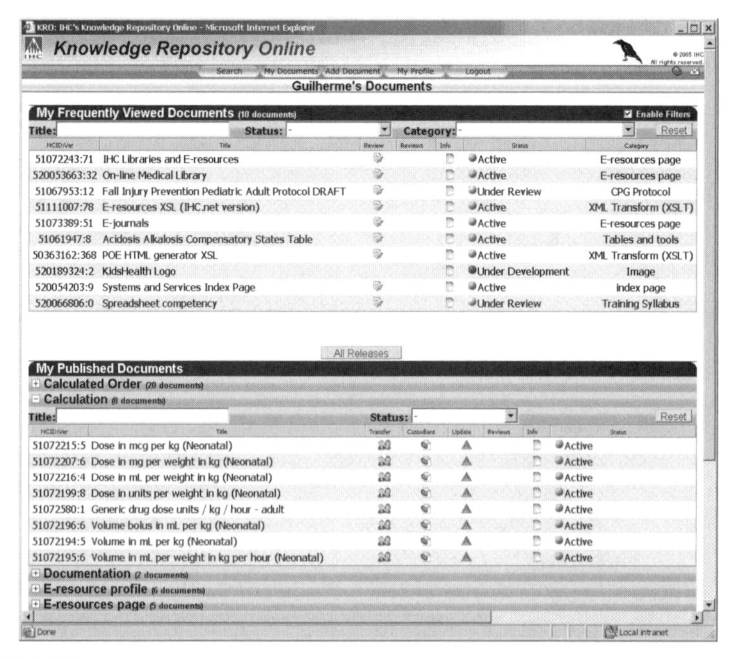

FIGURE 22-9 An example of the asset management page within KRO. Notice the two main areas: *My Frequently Viewed Documents* and *My Published Documents*. From this page, authors can add and read reviews, change asset status, and also update assets that have been previously uploaded.

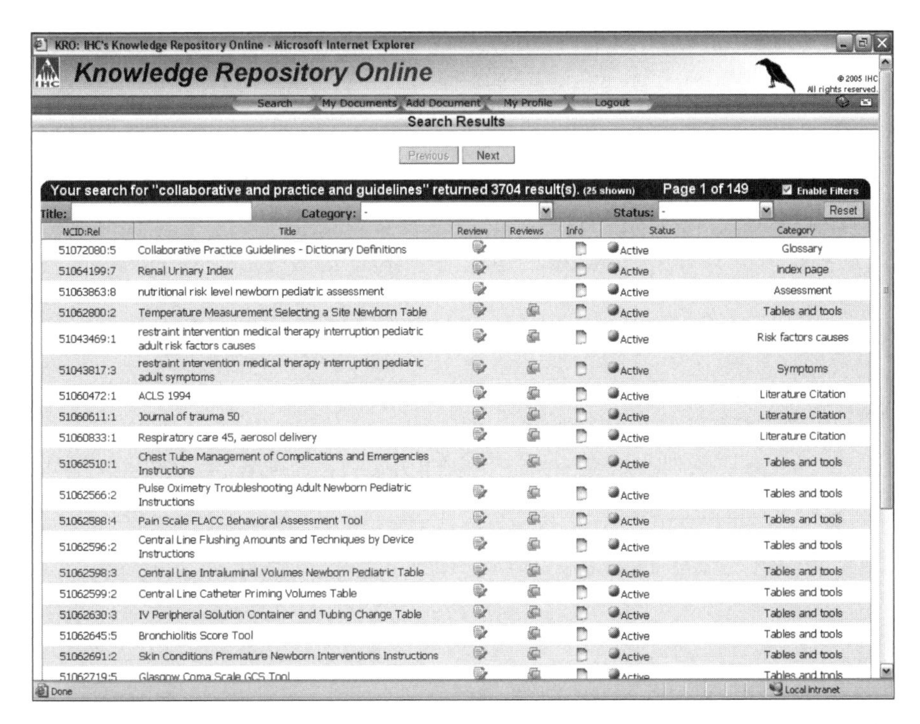

FIGURE 22-10 An example of the search results page within KRO. The user searched for *collaborative practice guidelines* and 3,690 documents were found, out of which 25 are displayed. In this screen the user can create a review (third column) and also read reviews created by others (fourth column).

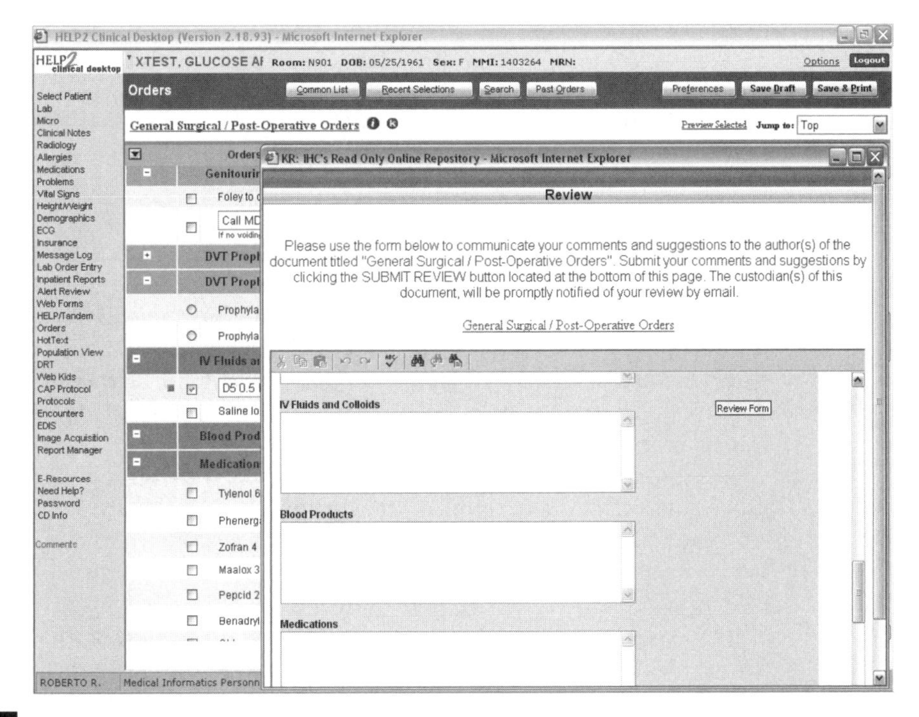

FIGURE 22-11 Example of a knowledge asset (order set) review form being requested directly from with the CPOE module of HELP2.

Document Review table (see Figure 22-6). KRO users can read all the available reviews, but the names of the reviewers are not disclosed. Also through KRO, clinicians can examine the meta-data of an asset (any version), and subscribe to e-mail alerts that keep them informed about new versions of the assets in which they are interested.

The KRO monitoring infrastructure was activated in February 2005. Since this date, 312 unique users have used KRO, and 70 different reviewers use it on a monthly basis. Presently 70 users are registered for the e-mail notification service offered by KRO, corresponding to a total of 137 unique documents. Since January 2005, 410 reviews have been created.

22.3.4 Searching and Browsing

As previously described, the CKR services provide a reasonably complete set of APIs to search and retrieve knowledge assets. Many different clinical applications, Web portals, and static Intranet and Extranet pages use these APIs (Del Fiol et al. 2004). Alternatively, KRO also provides a fairly complete set of searching and browsing functions, enabling Intranet users to find and preview any asset stored in the CKR (see subsection 22.3.3).

22.3.4.1 Knowledge Portals

Within the CKM infrastructure, *knowledge portals* are specialized tools for searching and browsing CKR document collections. A knowledge portal combines traditional hypertext navigation with flexible searching functions. Examples of methods used to support content navigation are lists of asset names (titles) in alphabetical order, manually created (static) indexes that group assets by topic, and dynamic indexes (queries) that arrange assets using different meta-data values. The searching functions used by a knowledge portal include simple keyword search, as well as advanced searching options that enable users to constrain search results using meta-data elements, or select keywords from a list of known words generated from the asset collection in question.

The first knowledge portal implemented is used for searching and browsing the CPG asset collection (see subsection 22.2.6.2). This knowledge portal, known as the *CPG Viewer*, offers to clinicians a detailed table of contents, several custom-built indexes, and all other searching and navigation methods just mentioned (see Figure 2-12). The CPG Viewer also leverages some of the KRO functions, including the subscription to e-mail alerts communicating that CPG documents have been updated, and the *knowledge-buttons* that enable users to offer feedback to the authors of the CPG documents. The CPG Viewer also makes extensive use of the CKM monitoring infrastructure, generating detailed searching and navigation trails for every user session.

Since its release (June 2004), the CPG Viewer monitoring infrastructure has recorded 30,682 sessions total, resulting from 6,215 unique users. Since March of 2005, an average of 1,868 unique users has accessed the CPG Viewer every month, with an average of 10.9 documents retrieved per user session. Users access the CPG Viewer an average of 2.16 times per month. Additional details about the CPG Viewer can be found elsewhere (Xu et al. 2004; Xu et al. 2005).

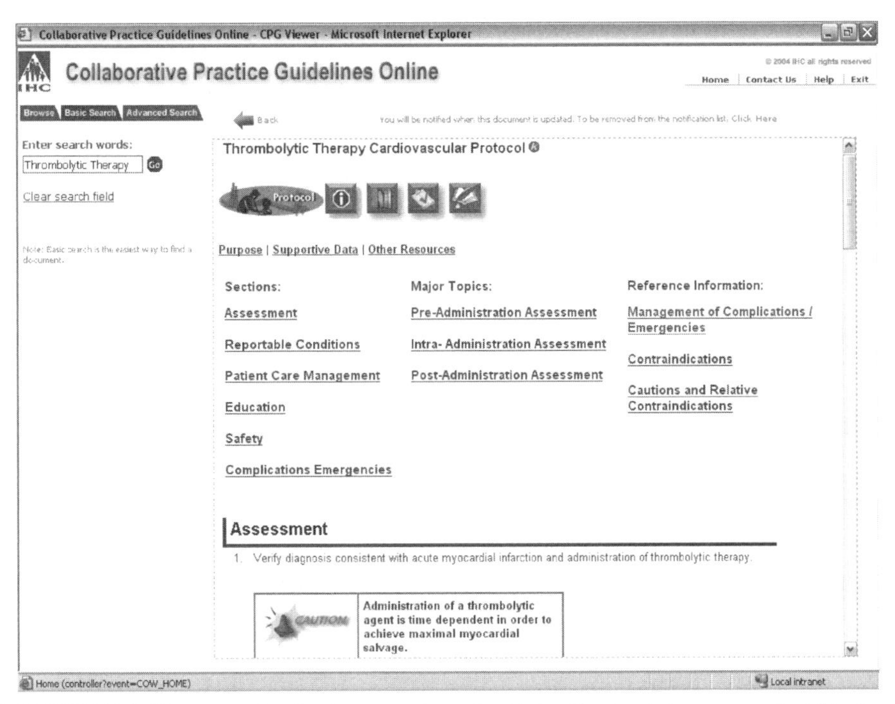

FIGURE 22-12 Example of a document from the CPG collection being presented within the *CPG Viewer* application.

A new searching and browsing tool currently is undergoing the final stages of testing. The new tool implements the same searching and browsing options available in the CPG Viewer, with a few enhancements, including the ability to be launched from within a Web page or application. Most importantly, the new searching and browsing tool can be configured to work with essentially any CKR asset collection. The configuration corresponds to a fully structured XML document (knowledge portal profile) that is created and maintained using the XML content framework. With this new tool, knowledge engineers will be able to configure and deploy knowledge portals within minutes.

22.3.4.2 Infobuttons

Infobuttons are highly specialized knowledge retrieval tools (Cimino et al. 1997; Reichert et al. 2002; Del Fiol, Rocha and Cimino 2005). Within the CKM infrastructure, infobuttons are used to generate queries against the CKR collections, as well as external knowledge resources, using context-dependent information obtained from HELP2 and patient data extracted from the CDR. The goal of an infobutton is to anticipate the information needs of a clinician at the point-of-care, enabling the user to find the appropriate answers with just a hyperlinks. More details about infobuttons and their implementation using the CKR infrastructure can be found in Chapter 17 of this book.

22.3.5 Execution Engine

Foresight is the principal knowledge execution engine of the CKM infrastructure. Foresight is a J2EE compliant application deployed in a BEA WebLogic application server (BEA WebLogic Platform 2005). Foresight is composed of five integrated modules (see Figure 22-13):

1. **Data-drive**. Module responsible for gathering data and performing the initial filtering. Data-drive continually receives copies of all the data transactions performed against the CDR. Using a configuration file, the data of interest to the protocols known to Foresight are selected and forwarded to the process manager. Every time a new clinical protocol (set of rules) is uploaded, the rule developer registers all the rules and data dependencies in this data-drive configuration file. Before forwarding the data, data-drive also translates the data according to the internal data model adopted by Foresight (Steiner et al. 2004). Approximately 49 percent of all clinical data received by data-drive are forwarded to the process manager.
2. **Time-drive**. Module responsible for holding data for predetermined periods of time. Time-drive enables the activation or reactivation of rules at certain times of the day, or after a specified period of time, where the holding time can be specified in seconds to years. Time-drive releases the data to the process manager.
3. **Data gateway**. Module that implements the synchronous (immediate) activation of Foresight. The data gateway essentially bypasses data-drive and time-drive and sends the data directly to the process manager.
4. **Process manager**. Module responsible for coordinating and monitoring the distribution of the data and the activation of rules and other Foresight modules.
5. **Rules manager**. Module that receives the data from the process manager and verifies which rules are to be executed. The rules manager can activate four types of rules: alert manager rules, alert routing rules, data access services rules, and clinical rules. The alert manager rules are used to create the output messages (e.g., alerts, reminders) generated by Foresight. The alert routing rules are responsible for dispatching the output messages to previously specified locations or

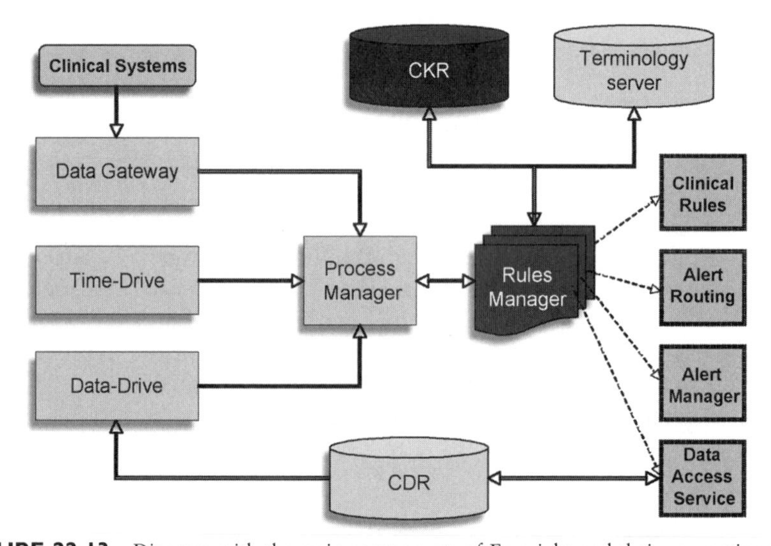

FIGURE 22-13 Diagram with the main components of Foresight and their connections with other components like the CKR, the terminology server, the CDR, and the clinical systems.

devices. These devices and locations are defined in a separate configuration file, and can be edited at any time. The data access services rules specify the constraints and functions that need to be applied against the patient data retrieved from the CDR, as required by the clinical rules. Finally, the clinical rules correspond to the decision logic specified by the clinical protocol (e.g., *If serum potassium greater than 6.0, then serum potassium is elevated.*).

All new clinical rules are uploaded to the CKR, where they are permanently stored and continuously versioned. In real-time, the CKR notifies Foresight that new rules are available. Foresight immediately retrieves the new rules and starts executing them. In order to update a set of clinical rules, a knowledge engineer or rule developer downloads it from the CKR, makes the necessary changes using the appropriate framework (see subsection 22.2.3.2), and uploads it back to the CKR as a new version (using KRO).

Foresight is controlled and monitored through a Web-based application called Runtime Administration & Deployment (RAD). RAD is used for managing all Foresight modules, including data-drive and time-drive, as well as all rules, triggers, and configuration resources. Using RAD an authorized user can verify which rules are running, how frequently each rule is being executed, and how long it is taking to execute each rule. RAD is also used to activate and deactivate, in real-time, any Foresight module or rule, either individually or as a set. All Foresight configuration files are represented in XML and stored in a local database managed by RAD.

22.3.6 Monitoring Infrastructure

Within the context of the CKR, the monitoring infrastructure is used to track utilization of the services and applications, and also to analyze the performance and availability of the database. A generic Java API is used to persist generic events and their associated parameters. Knowledge engineers and application developers are responsible for defining the monitoring events. Monitoring events can have one or multiple parameters that are relevant to the event. These monitoring events can range from simple results of performance and uptime queries to detailed content navigation from within an application. An event pattern-matching algorithm has also been implemented, enabling the identification of events that develop predefined characteristics. Patterns, users to be notified, as well as user-specific notification messages, are defined in configuration files ensuring flexibility.

Within the context of the CKM applications (e.g., KAT, KRO, CPG Viewer, infobuttons), the monitoring infrastructure is used for tracking utilization patterns, as well as frequency and duration of user sessions. Using the monitoring infrastructure, virtually any event within a session can be recorded and time-stamped, along with its parameters. Examples of recorded events include login, logout, traversal of hyperlinks, searching (including keywords and constraints used), and use of toolbar functions.

Within the context of Web pages and portals, where users are not authenticated and the boundaries of a session are not well defined, the monitoring infrastructure is used primarily for monitoring HTTP events that search or retrieve assets from the CKR.

Since the beginning of 2005, the monitoring infrastructure has logged 41,972 sessions and 4,723,510 events. The monitoring infrastructure presently tracks 82 distinct events with 77 parameters total. Six applications are using the monitoring infrastructure.

Detailed and continuous utilization monitoring is vital for the effective management of any large collection of knowledge assets. Monitoring logs are valuable resources for knowledge engineers and domain experts, providing important clues to understand which knowledge assets are used more frequently, in which context these assets are being used, and sometimes also some basic information regarding user behavior. Of particular interest are detailed monitoring records of users' interactions with knowledge assets. These interactions are not limited to asset authoring and review sessions, and should also include situations where clinical users are able to customize the knowledge content to fulfill the requirements of a particular patient or to reflect their personal preferences.

ACKNOWLEDGMENTS

We would like to thank all those that have contributed to the development of the CKM infrastructure, in particular, Bryan Austad, Sharon Bigelow, Guilherme Del Fiol, Timothy Hanna, Lorrie Roemer, Venkatesh Rudrapatna, Randy Secrist, David Steiner, Steven Wilkinson, and Xiaomin Xu. We also want to thank David A. Burton, Paul D. Clayton, Terry P. Clemmer, and Brent C. James for their continuous mentoring during the development and implementation of the CKM infrastructure.

REFERENCES

3M Clinical Data Repository, 3M Health Information Systems [Online]. (2005). Available at http://www.3m.com/us/healthcare/his/products/records/data_repository.jhtml. Accessed November 5, 2005.

3M Enterprise Master Patient Index, 3M Health Information Systems [Online]. (2005). Available at http://www.3m.com/us/healthcare/his/products/records/master_index.jhtml. Accessed November 5, 2005.

3M Health Care Data Dictionary, 3M Health Information Systems [Online]. (2005). Available at http://www.3m.com/us/healthcare/his/products/records/data_dictionary.jhtml. Accessed November 5, 2005.

Alavi, M., Leidner, D. E. (2001). Knowledge management and knowledge management systems: Conceptual foundations and research issues. *MIS Quarterly* **25**(1): 107–136.

The AGREE Collaboration [Online]. (2005). *Appraisal of guidelines for research & evaluation (AGREE) instrument.* Available at http://www.agreecollaboration.org/. Accessed November 5, 2005.

Ash, J. S., Stavri, P. Z., Kuperman, G. J. (2003). A consensus statement on considerations for a successful CPOE implementation. *J Am Med Inform Assoc* May–June, **10**(3): 229–234.

Authentic Browser Edition, Altova Inc. [Online]. (2005). Available at http://www.altova.com/features_browser_a.html. Accessed November 5, 2005.

Bali, R. K., editor. (2005). *Clinical Knowledge Management: Opportunities and Challenges.* Idea Group Publishing.

BEA WebLogic Platform, BEA Systems, Inc. [Online]. (2005). Available at http://www.bea.com/framework.jsp?CNT=index.htm&FP=/content/products/weblogic/platform/. Accessed November 5, 2005.

Biomedical Libraries, EBSCO Publishing Inc. [Online]. (2005). Available at http://www.epnet.com/biomedical/default.asp. Accessed November 5, 2005.

Cannon, H. E. (2004). Assessing the need for a clinical COPD care program in a managed care organization. *J Manag Care Pharm* **10**(4 Suppl): S17–S21.

Cimino, J. J., Elhanan, G., Zeng, Q. (1997). Supporting *infobuttons* with terminological knowledge. *Proc AMIA Annu Fall Symp* 528–532.

Clayton, P. D., Narus, S. P., Huff, S. M., Pryor, T. A., Haug, P. J., Larkin, T. et al. (2003). Building a comprehensive clinical information system from components. The approach at Intermountain Health Care. *Meth Inf Med* **42**: 1–7.

Clin-eguide, Wolters Kluwer Health Inc. [Online]. (2005). Available at http://www.clineguide.com/products/what_is_clineguide/body.html. Accessed November 5, 2005.

Cowden, D., Barbacioru, C., Kahwash, E., Saltz, J. (2003). Order sets utilization in a clinical order entry system. *AMIA Annu Symp Proc* 819.

Currie, L. M., Graham, M., Allen, M., Bakken, S., Patel, V., Cimino, J. J. (2003). Clinical information needs in context: An observational study of clinicians while using a clinical information system. *AMIA Annu Symp Proc* 190–194.

Darmoni, S. J., Thirion, B., Leroy, J. P., Douyere, M. (2001). The use of Dublin Core metadata in a structured health resource guide on the internet. *Bull Med Libr Assoc* **89**(3): 297–301.

Del Fiol, G., Rocha, R. A., Bradshaw, R. L., Hulse, N. C., Roemer, L. K. (2005). An XML model that enables the development of complex order sets by clinical experts. *IEEE Trans Inf Technol Biomed* **9**(2): 216–228.

Del Fiol, G., Rocha, R. A., Cimino, J. J. (2005). *HL7 Infobutton Standard API Proposal* [Online]. Available at http://www.hl7.org/Library/Committees/dss/HL7-*Infobutton*-API-v2.2-20040224.doc. Accessed November 5, 2005.

Del Fiol, G., Rocha, R. A., Washburn, J., Rhodes, J., Hulse, N., Bradshaw, R., Roemer, L. K. (2004). "On-demand" access to a multi-purpose collection of best practice standards. *Proc of the 26th Annual International Conference of the IEEE ECMS* 3342–3345.

Dublin core metadata initiative. (2005). *Dublin core metadata element set* [Online]. Available at http://www.dublincore.org/documents/dces/. Accessed November 5, 2005.

Fischer, G. and Ostwald, J. (2001). Knowledge management: Problems, promises, realities, and challenges. *IEEE Intell Syst* 60–72.

GE Healthcare & IHC establish new research center to develop electronic health record technologies, Intermountain Healthcare [Online]. (2006). Available at http://intermountainhealthcare.org/xp/public/aboutihc/news/article26.xml. Accessed April 9, 2006.

GE Healthcare & Intermountain Healthcare to provide wide-reaching IT system, GE Healthcare [Online]. (2006). Available at http://www.gehealthcare.com/company/pressroom/releases/pr_release_10225.html. Accessed April 9, 2006.

Guptill, J. (2005). Knowledge management in health care. *J Health Care Finance* **31**(3): 10–14.

Hanna, T. P., Del Fiol, G., Bradshaw, R., Roemer, L. K., Hulse, N., Rocha, R. A. (2005). Customized document validation to support a flexible XML-based knowledge management framework. *AMIA Annu Symp Proc* 291–295.

Haug, P. J., Rocha, B. H., Evans, R. S. (2003). Decision support in medicine: Lessons from the HELP System. *Int J Med Inf* **69**: 273–284.

Health Level Seven (HL7) [Online]. (2005). Available at http://www.hl7.org/about/. Accessed November 5, 2005.

HL7 Arden Syntax Special Interest Group & Clinical Decision Support Technical Committee. (2002). *Arden Syntax for Medical Logic Systems*, version 2.1. *Health Level Seven.*

Hougaard, J. (2004). Developing evidence-based interdisciplinary care standards and implications for improving patient safety. *Int J Med Inf* **73**(7–8): 615–624.

Hulse, N. C., Rocha, R. A., Del Fiol, G., Bradshaw, R. L., Hanna, T. P., Roemer, L. K. (2005). KAT: A flexible XML-based knowledge authoring environment. *J Am Med Inform Assoc* **12**(4): 418–430.

Hulse, N. C. (2005). *Development and evaluation of an authoring environment for a centralized knowledge repository*. Ph.D. Dissertation. Department of Medical Informatics, University of Utah.

IHC Annual Report [Online]. (2004). Available at http://intermountainhealthcare.org/xp/public/documents/corp/annualreport.pdf. Accessed April 9, 2006.

Intermountain Clinical Programs, Intermountain Healthcare [Online]. (2006). Available at http://intermountainhealthcare.org/xp/public/physician/clinicalprograms/. Accessed April 9, 2006.

IHC honored for high tech systems, Intermountain Healthcare [Online]. (2006). Available at http://intermountainhealthcare.org/xp/public/aboutihc/news/article35.xml. Accessed April 9, 2006.

IHC named nation's top health system, Intermountain Healthcare [Online]. (2006). Available at http://intermountainhealthcare.org/xp/public/aboutihc/news/article25.xml. Accessed April 9, 2006.

Jadhav, A., Sailors, M. (2003). Structuring healthcare knowledge bases: An analysis of explicit and implicit structures in Arden Syntax and an XML schema representation of Arden Syntax. *AMIA Annu Symp Proc* 875.

James, B. C., Hammond, M.E. (2000). The challenge of variation in medical practice. *Arch Pathol Lab Med* **124**(7): 1001–1003.

James, B. C. (2001). Make it easy to do it right (editorial). *N Engl J Med* **345**(13): 991–992.

Johnson, P. D., Tu, S.W., Musen, M. A., Purves, I. (2001). A virtual medical record for guideline-based decision support. *Proc AMIA Symp* 294–298.

Kuperman, G. J., Gardner, R. M., Pryor, T. A. (1991). *HELP: A dynamic hospital information system*. New York: Springer-Verlag.

Larsen, D. L., Cannon, W., Towner, S. (2003). Longitudinal assessment of a diabetes care management system in an integrated health network. *J Managed Care Pharm* **9**(6): 552–558.

Marriott, S., Palmer, C., Lelliott, P. (2000). Disseminating healthcare information: Getting the message across. *Qual Health Care* **9**(1): 58–62.

MD Consult, Elsevier Inc. [Online]. (2005). Available at http://www.mdconsult.com/offers/AboutMDC.html/standard.html. Accessed November 5, 2005.

MEDLINE/PubMed, National Library of Medicine [Online]. (2005). Available at http://www.ncbi.nlm.nih.gov/entrez/query.fcgi?DB=pubmed. Accessed November 5, 2005.

Micromedex, Thomson Healthcare, Inc. [Online]. (2005). Available at http://www.micromedex.com/index.html. Accessed November 5, 2005.

Oracle Database, Oracle, Inc. [Online]. (2005). Available at http://www.oracle.com/database/index.html. Accessed November 5, 2005.

Payne, T. H., Hoey, P. J., Nichol, P., Lovis, C. (2003). Preparation and use of preconstructed orders, order sets, and order menus in a computerized provider order entry system. *J Am Med Inform Assoc* **10**(4): 322–329.

Reichert, J. C., Glasgow, M., Narus, S. P., Clayton, P. D. (2002). Using LOINC to link an EMR to the pertinent paragraph in a structured reference knowledge base. *Proc AMIA Symp* 652–656.

RuleML Homepage [Online]. (2005). Available at http://www.ruleml.org. Accessed November 5, 2005.

Shiffman, R. N., Karras, B. T., Agrawal, A., Chen, R., Marenco, L., Nath, S. (2000). GEM: A proposal for a more comprehensive guideline document model using XML. *J Am Med Inform Assoc* **7**: 488–498.

Sordo, M., Ogunyemi, O., Boxwala, A. A., Greenes, R. A. (2003). Software specifications for GELLO: An object-oriented query and expression language for clinical decision support [Online]. Technical Report DSG-TR-2003-02. Decision Systems Group, Brigham & Women's Hospital, Harvard Medical School, Boston, MA. Available at http://www.hl7.org/Library/Committees/dss/GELLOspecsJan041%2EPDF. Accessed November 5, 2005.

Stefanelli, M. (2004). Knowledge and process management in health care organizations. *Methods Inf Med* **43**(5): 525–535.

Steiner, D. J., Coyle, J. F., Rocha, B. H., Haug, P., Huff, S. M. (2004). Medical data abstractionism: Fitting an EMR to radically evolving medical information systems. *Medinfo* **11**(Pt 1): 550–554.

Tiwana, A., Ramesh, B. (2001). Integrating knowledge on the Web. *IEEE Internet Comput* **5**(3): 32–39.

UpToDate [Online]. (2005). Available at http://www.uptodate.com/about/index.asp. Accessed November 5, 2005.

Wilkinson, S. (2003). *Knowledge repository online: An electronic knowledge development and maintenance process*. Master's thesis. Department of Medical Informatics, University of Utah.

Wyatt, J. C. (2000). Clinical questions and information needs. *J R Soc Med* **93**(4): 168–171.

XML Schema, World Wide Web Consortium (W3C) [Online]. (2005). Available at http://www.w3c.org/XML/Schema. Accessed November 5, 2005.

XML Suite, Altova Inc. [Online]. (2005). Available at http://www.altova.com/suite.html. Accessed November 5, 2005.

XSL Transformations, World Wide Web Consortium (W3C) (XSLT) [Online]. (2005). Available at http://www.w3.org/TR/xslt. Accessed November 5, 2005.

Xu, X., Del Fiol, G., Rocha, R. A. (2004). Towards a flexible Web-based framework to navigate clinical reference documents. *Medinfo* 1915.

Xu, X., Rocha, R. A., Bigelow, S. M., Wallace, C. J., Hanna, T., Roemer, L. K. (2005). Understanding nurses' information needs and searching behavior in acute care settings. *AMIA Annu Symp Proc* 839–843.

23 INTEGRATION OF KNOWLEDGE RESOURCES INTO APPLICATIONS TO ENABLE CLINICAL DECISION SUPPORT: ARCHITECTURAL CONSIDERATIONS

KENSAKU KAWAMOTO

The previous two chapters have described approaches to knowledge management in which the knowledge base required for clinical decision support (CDS) is maintained in a repository that is independent of the clinical applications that utilize the knowledge base to provide decision support. This separation between the knowledge base and individual applications is desirable, as it facilitates knowledge management and the reuse of knowledge resources across multiple applications. At the same time, such knowledge management processes must be coupled with strategies for deploying the knowledge resources into operational settings, so that the resources can be used to deliver proactive decision support at the point of care.

The objective of this chapter is to provide an overview of how knowledge resources can be integrated into clinical applications in order to enable decision support. To meet this objective, this chapter first defines four high-level tasks associated with this knowledge integration process. Second, this chapter discusses important architectural issues that must be considered when devising a strategy for accomplishing these tasks. Third, this chapter provides four case studies in order to provide concrete examples of how CDS knowledge resources can be integrated into clinical applications. Fourth, this chapter discusses lessons learned from the way in which the health care marketplace has reacted to different knowledge integration approaches. Finally, this chapter concludes by examining how many of the challenges associated with knowledge integration can be addressed through the use of standard services being specified by Health Level 7 and the Object Management Group in their joint Healthcare Services Specification Project (HSSP). In particular, focus is placed on the emerging HSSP Decision Support Service specification, which seeks to standardize the input/output interface for services that use patient data to generate patient-specific conclusions.

Clinical Decision Support: The Road Ahead

23.1 INTRODUCTION

When managing a clinical decision support (CDS) knowledge base, significant benefits can be achieved by keeping the knowledge base separate from the clinical applications that make use of the knowledge base to provide decision support. These benefits include the facilitation of knowledge maintenance tasks such as periodic content review, content updating, and version management, as well as the enabling of knowledge reuse across multiple applications. Given these benefits, this approach to knowledge management has been adopted by health care institutions such as Partners HealthCare (see Chapter 21), Intermountain Healthcare (see Chapter 22), and Vanderbilt University Medical Center (Geissbuhler et al. 1999), as well as by commercial knowledge vendors such as First DataBank and Cerner Multum.

The knowledge resources maintained in these repositories can vary widely in terms of scope and focus. These resources could include medication knowledge (e.g., minimum and maximum doses, drug–drug interactions, drug–allergy contraindications), decision rules, grouped knowledge elements (e.g., order sets, documentation templates), and clinical practice guidelines encoded in a machine-interpretable format.

In order for the medical knowledge contained in these repositories to impact a patient's care optimally, the knowledge must be integrated into clinical applications so that patient-specific decision support can be provided to care providers in clinical settings. Moreover, the research literature indicates that such CDS interventions should be 1) delivered automatically as part of routine clinical workflow, 2) presented as actionable recommendations, and 3) provided at the time and location of clinical decision-making (Kawamoto et al. 2005a). Indeed, computer-based CDS interventions possessing these features have led to significant improvements in clinical practice in over 90 percent of published randomized controlled trials (Kawamoto et al. 2005a).

The objective of this chapter is to provide an overview of how knowledge resources can be leveraged to deliver CDS effectively (i.e., proactively, at the appropriate point in clinical workflow, and as actionable recommendations). To meet this objective, the chapter begins by defining several terms central to the ensuing discussion.

23.2 TERM DEFINITIONS

For the purposes of this chapter, the following term definitions will be used.

Software architecture. The high-level blueprint of a software system, which describes the functional components of the system as well as the interactions between these components (Krafzig et al. 2004).

Clinical application. A software program designed to support patient care.

Clinical application with CDS capabilities. A clinical application that is capable of using patient data to provide patient-specific assessments or recommendations to clinicians or to other health care stakeholders (e.g., patients).

Clinical decision support system (CDSS). An application with CDS capabilities.

CDS application. A clinical application with CDS capabilities.

CDS module of a clinical application. The software component of a clinical application that enables CDS capabilities within the application.

Knowledge resource execution engine. A software program that is designed to interpret the contents of a specific type of knowledge resource in order to facilitate the provision of decision support using resources of that type. A clinical application's CDS module may interact with a given knowledge resource's execution engine in order to make use of the resource. Also, an execution engine may in itself constitute the primary component of a clinical application's CDS module.

Knowledge integration architecture. A software architecture in which a clinical application is designed to leverage medical knowledge stored in an independent knowledge base in order to provide clinicians and other health care stakeholders with CDS functionality.

Having completed these term definitions, this chapter will now outline four high-level tasks involved in the integration of medical knowledge resources into clinical applications.

23.3 KNOWLEDGE INTEGRATION TASKS

In order to provide proactive decision support using an external knowledge resource, a clinical application must complete a number of tasks. Although these tasks could be expressed in more granular terms, this chapter classifies the steps involved into four high-level tasks for the sake of simplicity. These four tasks are described in detail, next.

23.3.1 Task 1: CDS Module Invocation at Point of CDS Opportunity

In order to make use of a knowledge resource to provide proactive decision support, an application's CDS module must be integrated into the application in such a way that the module is invoked at points in the application's workflow where CDS is needed. A CDS module can be invoked through either (a) direct invocation of the module by the application or (b) the broadcasting of application events, which are evaluated by a CDS module and identified as opportunities for decision support. For (a), a software engineer designates points within the application code at which a particular CDS module should be invoked in response to a specific user action. For example, a computerized provider order entry (CPOE) system can be designed so that the system's CDS drug allergy screening module is invoked immediately following the entry of a new medication order by a clinician. In approach (b), messaging is used to notify a CDS module regarding events taking place within the clinical application, and the CDS module evaluates these events to identify opportunities for providing decision support.

For example, a CPOE system can be set up to send a Health Level 7 (HL7) message to the pharmacy system when a new medication order has been placed by a clinician, and the CDS module of the pharmacy system can be designed to detect this message as an opportunity to check for drug–drug interactions, drug–allergy contraindications, and drug dosage appropriateness using a medication knowledge base. As another example of this approach, an application designer can couple the provider alert module of an electronic health record (EHR) system with an event listener and configure the EHR

TABLE 23-1 Commonly leveraged CDS opportunities within the clinical workflow.

Setting	Target users	Workflow context	CDS content
Inpatient	Clinicians	Computerized order entry	Drug–allergy contraindication alert Drug–drug interaction alert Drug dosage recommendation or alert Corollary order recommendation Order set presentation
		Laboratory result entered into system	Page to responsible clinician regarding critical lab values
	Pharmacists	Medication order processing	Drug–allergy contraindication alert Drug–drug interaction alert Drug dosage alert
Outpatient	Clinicians	Review of patient information	Patient summary with disease management and health maintenance recommendations
		Encounter documentation	Documentation template containing data collection and action recommendations
		E-prescribing	Drug–allergy contraindication alert Drug–drug interaction alert Drug dosage recommendation or alert
	Nurses	Patient intake	Standing orders (e.g., for vaccination)
	Patients	Between visits	Letter or phone call to remind patient of overdue health maintenance procedures

system so that it notifies the event listener regarding the events of interest to the listener. Using this approach, a software engineer can set up the alert module to detect and respond to events within the EHR system (e.g., the storage of serum creatinine values into the clinical database) that represent opportunities for providing decision support using a relevant knowledge resource (e.g., a decision rule that evaluates whether a patient taking gentamicin may be developing renal insufficiency due to the antibiotic, as evidenced by increasing serum creatinine levels).

Table 23-1 summarizes some of the most common workflow contexts in which an application's CDS module can be invoked. Of note, the scenarios listed in the table represent only a subset of the many potential CDS opportunities that exist within the clinical workflow.

23.3.2 Task 2: Data Retrieval

In order to provide decision support using a knowledge resource, an application's CDS module must retrieve the data required by the resource for drawing patient-specific conclusions. This section describes notable aspects of this task.

23.3.2.1 Data Requirement Types

The data required for generating conclusions regarding a patient typically include demographic and "act" data, where an act refers to any act or service constituting health care services (Health Level 7, 2006). A knowledge resource may require data on such demographic data elements as age, birth date, and gender, as well as data on such health care acts as encounters, diagnoses, procedures, observations, treatment goals, and medication orders.

In addition to demographic and act data, some knowledge resources may require data regarding the context in which the resource is being utilized. Context data that may be required by a knowledge resource include: 1) the type of end-user (e.g., physician, nurse, patient); 2) the care setting (e.g., hospital, emergency department, primary care clinic, specialty clinic); 3) the current point in the clinical workflow (e.g., clinician entering admission orders in inpatient setting, nurse processing patient in outpatient setting); and 4) local preferences or constraints (e.g., formulary constraints, availability of specialized personnel or equipment, institution-specific antibiotic sensitivity patterns, miscellaneous disease management preferences entrenched in the local care setting). A knowledge resource may use such context information in order to make sure that the course of action recommended is appropriate for the local context.

23.3.2.2 Data Sources

The requisite data just described are primarily obtained from data repositories or from end-users. In addition, an application's CDS module may provide a knowledge resource with information on the relevant context, such as the role of the current user and the current workflow context.

Ideally, all or most of the structured patient data required for providing decision support would be obtained from preexisting data contained in one or more data repositories, given that clinicians dislike having to manually enter data in order to obtain advice from a clinical application (Bates et al. 2003). Data repositories that contain structured patient data relevant to decision support include the databases of department-specific systems such as laboratory systems and pharmacy systems, the databases of administrative and billing systems, back-end databases of EHR systems, enterprise-wide data warehouses, and the claims databases of health insurers.

Although it would be ideal if all required data could be retrieved from data already captured in one or more data repositories, this is not always possible. In these cases, the missing data must be obtained from end-users of the clinical application. In seeking data from a human user, it is important that the data request be limited, relevant, and minimally duplicative of other data entry tasks (Bates et al. 2003). Moreover, data entry requests should be presented within the application's native user interface when possible, so as to minimize disruptions to the clinical workflow (Johnson et al. 2001a).

23.3.3 Task 3: Generation of Patient-Specific Inferences

Following retrieval of the required patient data, an application's CDS module must leverage the appropriate knowledge resource to generate patient-specific conclusions. A CDS module typically obtains these conclusions from an execution engine that is designed to interpret the contents of a given knowledge resource. The types of conclusions that can be generated through the use of a knowledge resource can be quite variable. For example, a knowledge resource may be used to provide a differential diagnosis based on the patient's symptoms and signs; recommend a set of admission orders given a patient's primary diagnosis; identify the optimal dose of a medication given the patient's comorbidities, weight, and renal function; recommend an empiric antibiotic regimen for a patient with an unidentified infection; or determine whether a patient is due for health maintenance or chronic disease management procedures.

23.3.4 Task 4: Communication of Actionable, Context-Relevant Recommendations

Once patient-specific conclusions have been reached, an application's CDS module must use the conclusions to generate actionable, context-relevant recommendations that are communicated to the right people, in the right format, and at the right point in the clinical workflow. Of note, the same conclusion may need to be communicated in very different ways depending on the clinical context. For example, a conclusion that an infant is overdue for immunizations may need to be formulated into a reminder message that is presented to the patient's primary care physician when she opens the patient's record within an EHR system; into a standing order that is communicated to the patient's nurse at the beginning of an outpatient encounter; or into a letter that is sent to the patient's parents to remind them of their child's need for the immunizations. The language used would depend on the context as well.

23.4 ARCHITECTURAL CONSIDERATIONS

In designing a strategy for fulfilling the four knowledge integration tasks just described, a number of architectural issues must be considered (see Table 23-2). This section will examine these issues in detail and describe common ways in which these issues are addressed.

23.4.1 Role of Knowledge Resource

A fundamental issue to resolve when designing a knowledge integration architecture is the role that the knowledge resource should play in the four-step knowledge integration process just described. In many cases, knowledge integration architectures address this issue using one of two approaches.

In one common architectural pattern, which will henceforth be referred to as a Knowledge Resource-Centric Knowledge Integration Architecture, the knowledge resource is used to control steps 2 (data retrieval), 3 (inference generation), and 4 (generation and communication of actionable, context-relevant recommendations) of the knowledge integration process. In this architecture, the execution engine associated with a knowledge resource constitutes the primary component of a clinical application's CDS module. Knowledge integration approaches that use this pattern include the Arden Syntax (Karadimas et al. 2002), the GLIF3 Guideline Execution Engine (GLEE) (Wang et al. 2004), PRODIGY (Johnson et al. 2001a), and SAGE (Ram et al. 2004). In this approach (see Figure 23-1), instructions specified in the knowledge resource are used by the execution engine to retrieve required data; to generate patient-specific inferences; to formulate actionable, context-relevant recommendations based on the inferences; and to communicate the recommendations to the appropriate end-users. In addition, the

TABLE 23-2 Issues to consider when designing a CDS knowledge integration architecture.

Issue	Common approaches to issue
Role of knowledge resource	Knowledge resource controls knowledge integration tasks 2, 3, and 4; knowledge resource may also control task 1. Knowledge resource primarily controls task 3.
Scope of supported institutions	Knowledge integration architecture does not include within its scope the use of knowledge resources by outside institutions. Knowledge integration architecture includes within its scope the use of knowledge resources by outside institutions.
Scope of supported CDS application types	Knowledge integration architecture designed to support a specific type of CDS application. Knowledge integration architecture designed to support multiple CDS application types.
Support for different approaches to representing medical knowledge	Deployment of multiple knowledge integration architectures, each specific to a given knowledge representation approach. Common components reused across knowledge integration architectures. Generic service interface used to encapsulate knowledge represented using different formalisms.
Security	Standard protective measures put into place.
Need for fast execution	Various performance-optimizing approaches used.
Variability in information models and terminologies	Information must be provided to the knowledge resource using the resource's information model (e.g., as specified in a vMR). Terminology service used to reconcile differences in the level of specificity at which concepts are specified within a given vocabulary and to translate between vocabularies.
Adaptation of decision support to local context	Knowledge resource considers local context when generating patient-specific conclusions. CDS module of a clinical application considers local context when invoking knowledge resource and/or when formulating CDS communications based on conclusions generated using the knowledge resource.
Requisite infrastructure	Specific EHR or CPOE implementation required. Specific virtual medical record (vMR) implementation required. Highly trained personnel with specialized skills required.
Leveraging of existing infrastructure	Knowledge integration architecture designed around existing infrastructure. Existing capabilities leveraged via service interfaces.

knowledge resource may control step 1 of the knowledge integration process by providing instructions on when the resource should be used to take advantage of a CDS opportunity. For example, a decision rule that evaluates a patient for hyperkalemia may specify that it should be leveraged whenever a serum potassium value is entered into the laboratory database.

In a second common architectural pattern, which will henceforth be referred to as an Application-Centric Knowledge Integration Architecture, the knowledge resource is used primarily to control the third step of the

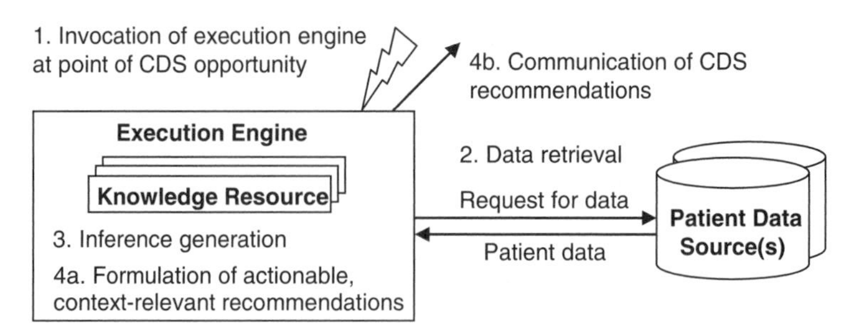

FIGURE 23-1 Knowledge Resource-Centric Knowledge Integration Architecture. The numbers refer to the four knowledge integration tasks specified in subsection 23.3.

knowledge integration process, that is, the step in which patient data are used to generate patient-specific conclusions. In this approach (see Figure 23-2), the knowledge resource does not specify when it should be used, how the data required by the resource should be retrieved, or how the inferences generated should be communicated to end-users. Instead, these knowledge integration tasks are controlled by the clinical application. Knowledge resources that utilize this pattern include commercial medication knowledge resources and SEBASTIAN (Kawamoto et al. 2005b).

In comparing the two approaches, an important advantage of a Knowledge Resource-Centric Knowledge Integration Architecture is that a given knowledge resource more completely specifies the information necessary for providing decision support in particular clinical contexts. On the other hand, an Application-Centric Knowledge Integration Architecture provides greater flexibility with regard to how and when a knowledge resource is used. This flexibility is important because a given knowledge resource may be useful in many different clinical contexts (see subsection 23.3.1) and because each context may require a markedly different approach to communicating the CDS results to end-users (see subsection 23.3.4).

23.4.2 Scope

As with any software initiative, the design and implementation of a knowledge integration architecture depends heavily on what is and is not included within the scope of the initiative. In particular, the scope of supported institutions and applications has a profound impact on how a knowledge integration architecture is designed.

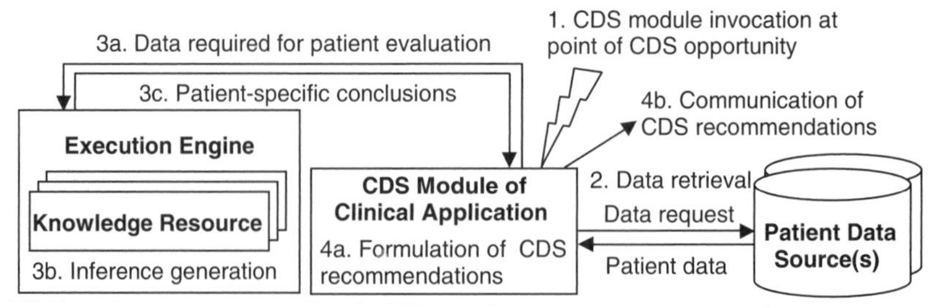

FIGURE 23-2 Application-Centric Knowledge Integration Architecture. The numbers refer to the four knowledge integration tasks specified in subsection 23.3.

23.4.2.1 Scope of Supported Institutions

A critical scope decision involves the determination of which institutions are to be supported by the knowledge integration strategy. A decision might be made, for example, to support the operational use of knowledge resources by a single hospital; by all hospitals that use a particular vendor's CPOE system; or by all clinics within a regional health information organization (RHIO) that utilize one of five supported EHR products. For a clinical information system (CIS) vendor, the decision is usually to include within its scope only those institutions using its products. On the other hand, knowledge vendors will usually attempt to support as many institution-specific and clinical IT vendor systems as possible to increase the potential market for their products.

The decision regarding the scope of supported institutions has both obvious and subtle implications for the knowledge integration approach as well as for the structure and content of the knowledge resources themselves. For example, if a collection of order sets is intended to primarily support institutions that use a particular vendor's CPOE system, it is likely that such order sets are encoded in a format that integrates easily with that vendor's CPOE system but not with another vendor's CPOE system. As another example, if knowledge repositories and accompanying knowledge integration architectures are being designed for environments with EHR and CPOE systems, the integration architecture may make the availability of such systems a prerequisite for using the knowledge resource. In a country such as the United States where the adoption of EHR and CPOE systems remains quite low, this type of a prerequisite can significantly limit the number of health care institutions that can make use of the knowledge resource.

The previous examples demonstrate how scope limitations often result in the placement of constraints on knowledge integration architectures. These constraints are desirable in that they reduce the cost and complexity involved in implementing a solution that meets the needs included within the project scope. However, if the scope of the initiative is increased at a later date, it may be difficult or impossible to adapt the existing knowledge integration strategy to meet the new project requirements. Thus, anticipated increases in scope requirements should be considered and addressed as early as possible in the design process, provided that adequate resources are available to do so.

23.4.2.2 Scope of Supported CDS Application Types

A second critical scope decision involves the identification of the CDS application types that are to be supported. Many existing knowledge integration architectures are designed to support specific types of CDS applications, such as event-based alerting systems, CPOE systems with CDS functionality, and interactive consultation systems. Although more difficult, it is also possible to design a generic knowledge integration architecture that is capable of supporting multiple CDS application types. This can be done, for example, through the use of an Application-Centric Knowledge Integration Architecture (see subsection 23.4.1) in which the knowledge resource and its execution engine are separated from the application-specific system components that identify CDS opportunities and deliver context-appropriate recommendations to end-users.

Limiting the scope of supported application types allows more constraints and assumptions to be placed on the knowledge integration architecture, thereby reducing the cost and complexity involved in designing and implementing the system. However, as a consequence, it can be difficult or impossible for a given knowledge integration architecture to be adapted to meet the needs of a new CDS application type that had not been included within the initial scope of the project.

23.4.3 Support for Different Knowledge Representation Approaches

The diversity of approaches used to represent knowledge presents a significant challenge when attempting to provide decision support using knowledge resources. More than a dozen named knowledge representation formalisms have been described in the literature, including the Arden Syntax (Karadimas et al. 2002), Asbru (Shahar et al. 1998), EON (Tu et al. 1999), GASTON (de Clercq et al. 2001), G-CARE (Overhage et al. 1995), GEM (Shiffman et al. 2000), GLARE (Terenziani et al. 2001), GLIF3 (Boxwala et al. 2004), GUIDE (Ciccarese et al. 2005), HGML (Hagerty et al. 2000), Prestige (Gordon et al. 1999), PRODIGY (Johnson et al. 2001a), PRO*forma* (Sutton et al. 2003), SAGE (Tu et al. 2004), Siegfried (Lobach et al. 1997), SEBASTIAN (Kawamoto et al. 2005b), and Stepper (Svatek et al. 2003). In addition, vendors and health care institutions often encode medical knowledge using proprietary or unique approaches that have not been described in the literature.

The challenge of designing a strategy for using knowledge resources to provide decision support would be greatly simplified if all resources were encoded using the same knowledge representation formalism or if resources encoded in one format could be unambiguously translated into another format. With regard to the possibility of a single knowledge representation formalism, there is active work within the HL7 CDS Technical Committee to attempt to define a standard, machine-executable representation format for clinical practice guidelines (see Chapter 13). However, the most appropriate knowledge representation approach may depend on the type of knowledge being modeled. For example, a complex clinical guideline may be best represented using Task-Network Models (Peleg et al. 2003), and knowledge on medication contraindications may be best represented using a relational database format specifically designed to capture this type of knowledge. Thus, it is likely that the health IT community will never be able to agree upon a single approach to modeling all knowledge relevant to clinical decision support. Furthermore, with regard to the possibility of translating unambiguously between knowledge representation formats, this approach generally has been found to be difficult, or even impossible, due to significant differences between various approaches to representing knowledge (Boxwala et al. 2001).

The reality, then, is that knowledge integration architectures will need to continue to deal with multiple knowledge representation formats for the forseeable future. Given this situation, institutions often develop and maintain separate knowledge integration architectures for each type of knowledge resource that is being leveraged. For example, a given institution may create an infrastructure for using the Arden Syntax for providing event-based alerts, while maintaining a separate mechanism for providing decision support during the order entry process using a commercial medication knowledge base. In implementing these resource-specific knowledge integration architectures, an institution may attempt to reduce duplicative effort by encapsulating commonly used components into shared modules or services that can be reused by multiple knowledge integration architectures.

As an alternative to resource-specific knowledge integration architectures, a generic service interface can be used to provide access to medical knowledge represented using different formalisms. This is the approach used by SEBASTIAN (Kawamoto et al. 2005b) and by the emerging HL7 Decision Support Service standard, both of which will be described in greater detail later in the chapter.

23.4.4 Security and Performance Considerations

23.4.4.1 Security Considerations

As with any application handling personally identifiable health information, a clinical application with CDS capabilities must incorporate appropriate security safeguards. These safeguards include standard protective measures such as user authentication, user authorization, and message encryption. Although this chapter will not discuss this issue in greater detail, it is important to note that security concerns must be explicitly addressed when designing a knowledge integration architecture.

23.4.4.2 Need for Fast Execution

In many types of CDS, especially those that are incorporated in interactive applications (e.g., in CPOE), it is critical that end-users are not made to wait while the CDS recommendations are being generated. Indeed, system speed has been found to be critical to clinicians' acceptance of real-time interactive CDS systems (Bates et al. 2003). Thus, clinical applications with CDS functionality typically include various performance-optimizing features. For example, terminology translations can be performed prior to run-time, and time-intensive data queries can be conducted prior to an anticipated user session (Overhage et al. 1995).

23.4.5 Need for Information Model and Terminology Reconciliation

The data sources available to the CDS module of a clinical application often store data using information models and terminologies that differ from the information model and terminologies used by a knowledge resource. As discussed in Chapter 14, reconciling these differences represents one of the greatest challenges of using external knowledge resources to provide decision support in an operational setting.

With regard to the information model, the problem arises from the fact that there are numerous valid ways for modeling clinical information in a structured, coded format. For example, one system might represent Streptococcal pneumonia of the upper right lobe as a diagnosis with a single identifying code (streptococcal pneumonia of the upper right lobe), whereas another system might represent the infection as a diagnosis with a primary code (pneumonia) qualified by a code for the causative pathogen (*Streptococcus pneumoniae*) and a code for the anatomical location (upper right lobe of the lung) (Parker et al. 2004). As another example, an EHR might represent a vaccination as a procedure, whereas many American billing systems represent a vaccination both as a procedure and as a diagnosis indicating the patient's need for a vaccination.

In addition to variability in the information model, significant variability exists in the terminologies used by knowledge resources and by information systems to identify clinical concepts. For example, diagnoses can be identified using terminologies such as SNOMED CT, ICD9, and ICD10; observations can be identified using terminologies such as LOINC and SNOMED CT; and medications can be identified using terminologies such as NDC, SNOMED CT, and RxNorm. In addition, many institutions and vendors use proprietary code sets to identify clinical concepts. This large variety of terminologies would not be a significant problem if one-to-one mapping between the codes in different terminologies was feasible. However, this mapping is often not possible due to significant differences in the scope and granularity of terminologies. For example, concepts contained in one vocabulary may not be included in a second vocabulary intended to cover the same clinical domain. Also, if two terminologies define codes at different levels of granularity (e.g., ICD9 versus SNOMED CT), it is often not possible to find semantically equivalent codes in both vocabularies. As a result, the original semantics of a clinical concept can be distorted or lost when a concept is translated to a different terminology.

Also of note, even when both a knowledge resource and its data sources utilize the same terminologies, mismatches may exist with regard to the level of granularity at which concepts are specified. For example, a knowledge resource may specify a clinical concept at a relatively abstract level (e.g., lung cancer), whereas its data source may record relevant data at a much more granular level (e.g., small cell carcinoma of the right lung).

In order to make use of a knowledge resource to obtain decision support, a strategy must be in place for reconciling any information model or terminology differences that exist between a knowledge resource and its data sources. With regard to differences in the information model, a knowledge resource typically mandates that required data be provided using the information model specified by the knowledge resource. For example, a knowledge resource may specify its information model requirements in terms of a virtual medical record (vMR), which is an abstract representation of an EHR that defines a patient information model and an interface for the knowledge resource's execution engine to interact with the underlying record system (e.g., to query for required patient data) (Johnson et al. 2001b). To use a knowledge resource with a vMR requirement, a clinical application must wrap its EHR with the specified vMR interface, so that the knowledge resource's execution engine can make query requests and retrieve query results using an information model understood by the knowledge resource.

With regard to differences in terminologies, a common approach taken by many knowledge integration architectures is to reconcile these differences using a terminology service. Terminology services can be implemented in-house using the mapping between terminologies provided by the U.S. National Library of Medicine's Unified Medical Language System (National Library of Medicine 2006). Also, terminology services can be obtained from a vendor such as Apelon, Inc. In either case, the terminology service typically is used to accomplish two tasks. One task of a terminology service is often that of identifying codes subsumed by a parent code within a given terminology (e.g., the ICD9 codes for insulin-dependent diabetes mellitus and for non-insulin-dependent diabetes mellitus, which are subsumed by the code for diabetes mellitus). This type of operation often is required to make sure that a general concept referenced by a knowledge resource (e.g., diabetes mellitus) is translated into all the specific codes that a data source might use to record relevant patient data. Another task for which a terminology service often is called upon is the translation of codes between vocabularies (e.g., the conversion of a set of ICD9 codes representing diabetes mellitus to a set of equivalent SNOMED CT codes).

23.4.6 Adaptation of Decision Support to Local Context

As noted earlier, CDS results must be packaged and delivered in different ways, depending on the local clinical context. In order for the CDS module of a clinical application to provide contextually relevant recommendations to the end-user, the knowledge resource itself can explicitly consider contextual factors when generating its conclusions. Alternatively, the CDS module can provide context-appropriate decision support to end-users without requiring the knowledge resource to take the context into account. For example, instead of having a decision rule regarding the need for Pap testing provide different recommendations based on the care setting, an application's CDS module can simply make sure that the decision rule is invoked only in an appropriate care setting (e.g., outpatient primary care). Similarly, instead of having a decision rule regarding beta-blocker therapy take the local formulary into account in order to recommend a specific beta-blocker, an application's CDS module can have the decision rule return a set of acceptable treatment options and then select the most appropriate option based on formulary preferences defined at the level of the local application.

23.4.7 Infrastructure Considerations

23.4.7.1 Infrastructure Requirements

The infrastructure requirements associated with a knowledge integration architecture should be kept to a minimum, so as to facilitate the use of the architecture by a wide range of health care institutions. However, the infrastructure requirements for many architectures can be quite substantial. For example, some knowledge integration architectures require a specific underlying EHR or CPOE system in order to function correctly, whereas other architectures including GUIDE (Ciccarese et al. 2005), PRODIGY (Johnson et al. 2001a), and SAGE (Ram et al. 2004) require a specific vMR implementation. Moreover, some knowledge representation formalisms are conceptually quite complex and can be difficult to understand. In these cases, it may be challenging to identify and recruit personnel who can understand the subtleties of the knowledge representation approach and design a robust knowledge integration architecture appropriate for that approach.

23.4.7.2 Leveraging of Existing Infrastructure

The cost of implementing a knowledge integration architecture can be significantly reduced by leveraging existing infrastructure. As a result, knowledge integration architectures frequently are designed around existing infrastructural components such as specific EHR or CPOE systems. Because of significant heterogeneity in the IT infrastructure available at different institutions, however, a knowledge integration architecture designed to work in one institution may not be redeployable in a different institution.

To mitigate this problem, a knowledge integration architecture can be designed so that key infrastructural components are accessed through standard service interfaces. This architectural approach allows components underlying the services to be changed without disrupting the overall system architecture. For example, an industry-standard vMR interface would allow the underlying EHR system to be changed without affecting the knowledge integration architecture, and a standard Decision Support Service interface would allow the underlying knowledge representation approach to be altered without disrupting the overall architecture. Standardization work currently being conducted to realize this vision is described in detail at the end of this chapter.

23.5 CASE STUDIES

The following four case studies provide concrete examples of how medical knowledge can be integrated into clinical applications. The first two case studies will examine the knowledge integration approaches used by the Arden Syntax and by SAGE. These approaches utilize a Knowledge Resource-Centric Knowledge Integration Architecture, in which the knowledge resource is used to control all four steps of the knowledge integration process (see Figure 23-1). This section will then examine the knowledge integration strategies used by commercial medication knowledge resources and by SEBASTIAN. These approaches represent examples of an Application-Centric Knowledge Integration Architecture, in which the knowledge resource does not dictate when the knowledge resource is used, how the required data are obtained, or how the conclusions generated by the resource are communicated to end-users (see Figure 23-2). This section will provide only a high-level overview of these approaches; a detailed review of these and other approaches is outside the scope of this chapter.

23.5.1 Case Study 1: Arden Syntax

23.5.1.1 Background

The Arden Syntax is a standard for representing medical knowledge in a machine-executable format (see Chapter 12) (de Clercq et al. 2004). The Arden Syntax was first introduced in 1989, adopted as an American Society of Testing and Machinery (ASTM) standard in 1992, and is currently being maintained within HL7 as an American National Standards Institute (ANSI) standard. A number of CIS vendors currently provide support for the Arden Syntax within their product lines.

23.5.1.2 Knowledge Representation Approach

In the Arden Syntax, medical knowledge is encapsulated into ASCII files known as Medical Logic Modules (MLMs). Each MLM contains sufficient logic to support a single medical decision. Structurally, a MLM consists of a *maintenance* section, a *library* section, and a *knowledge* section. The *maintenance* and *library* sections specify metadata regarding the modules, and the *knowledge* section contains the core content of a MLM. Within the *knowledge* section, the *data* subsection defines local variables used in the rest of the

MLM, such as patient data, clinical events, and the destination of CDS messages. In addition, the *knowledge* section contains an *evoke* subsection that specifies when the module should be used to evaluate a patient, a *logic* subsection that provides the instructional logic for drawing patient-specific conclusions, and an *action* subsection that specifies how conclusions regarding patients should be communicated to end-users.

23.5.1.3 Knowledge Integration Architecture

A typical architecture for providing decision support using Arden Syntax MLMs is shown in Figure 23-3 (Karadimas et al. 2002). In this architecture, an event listener associated with the MLM execution engine listens for CIS events such as the storage of a serum sodium level or the ordering of a specific medication. If an event specified in the *evoke* section of a MLM is identified, the MLM execution engine is invoked and begins to evaluate the relevant MLM (task 1 of the four knowledge integration tasks outlined earlier). The execution engine then retrieves the patient data specified in the MLM *data* section by querying patient data sources associated with the CIS (task 2).

Using these data, the processing instructions specified in the MLM *logic* section are used to determine whether a particular condition holds true for a patient (task 3). Examples of conditions evaluated by MLMs include whether a medication dose is inappropriate given the patient's renal function or whether a patient is in need of a particular health care intervention. If it is deemed that the evaluated condition does in fact hold true for the patient, instructions in the MLM *action* section are used to generate a CDS message (task 4a) and to deliver that message to the appropriate destination (task 4b). The message may be communicated to the CIS (e.g., to provide a warning within a CPOE system) or to a non-CIS destination, such as an e-mail address or a pager number.

23.5.1.4 Strengths and Limitations

An important strength of the Arden Syntax is that it has been accepted as a standard by ASTM, HL7, and ANSI. As a result, a number of vendors have incorporated support for the Arden Syntax into their CIS products. A second

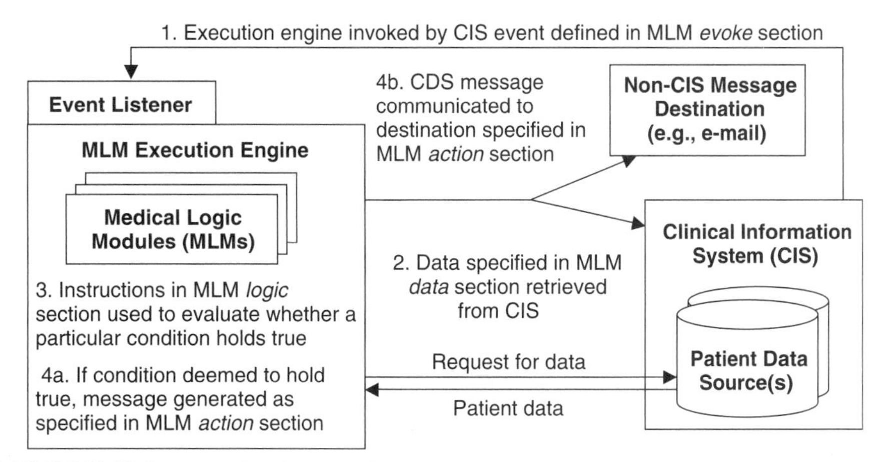

FIGURE 23-3 Typical architecture for providing decision support using Arden Syntax Medical Logic Modules (MLMs). The numbers refer to the four knowledge integration tasks specified in subsection 23.3.

strength, then, lies in the fact that the infrastructure necessary for authoring and executing MLMs is available within several commercial CIS products. A third strength of the Arden Syntax is the relative simplicity of its knowledge representation approach, as this simplicity facilitates understanding and use.

On the other hand, a primary limitation of the Arden Syntax is that all references to the local data environment are specified in a platform-specific (i.e., vendor- or institution-specific) manner within curly braces ({ }). This issue, which is commonly referred to as the "curly braces problem," limits the ease with which MLMs can be shared across institutions. Also, a second limitation is that the infrastructure required to use Arden Syntax MLMs is substantial. In particular, MLM execution engines require that MLMs first be compiled into a machine-executable format (e.g., a Java or C++ class), but writing such a compiler is a difficult and time-consuming task (Karadimas et al. 2002). Finally, a third limitation of the Arden Syntax is that each MLM explicitly specifies when it is to be used and how CDS messages are to be constructed and delivered. As discussed in subsection 23.4.1, this type of approach limits the flexibility with which a given knowledge resource can be leveraged for delivering decision support in various clinical contexts.

23.5.2 Case Study 2: SAGE

23.5.2.1 Background

The SAGE (Standards-Based Sharable Active Guideline Environment) project is a joint academic-commercial endeavor that seeks to develop a standards-based approach for encoding and deploying machine-executable clinical practice guidelines (see Chapter 13) (Ram et al. 2004). Started in 2001, the SAGE project was launched with a budget of approximately $18 million, and the initiative received approximately half of its funding from the U.S. National Institute of Standards and Technology's Advanced Technology Program. The academic partners in the project are University of Nebraska Medical Center, Mayo Clinic–Rochester, Intermountain HealthCare, and Stanford University; and the commercial partners are IDX Systems Corporation (now part of GE Healthcare) and Apelon, Inc. The key deliverables from the project are 1) a standards-based guideline representation model, 2) a guideline authoring environment, and 3) an interoperable guideline deployment system. As of 2005, guidelines for select immunizations and for diabetes management have been encoded in the SAGE format, and a prototype implementation of the guidelines has been tested successfully using IDX Carecast™ as the underlying CIS.

23.5.2.2 Knowledge Representation Approach

In SAGE, clinical practice guidelines are modeled as *activity graphs* encoded using the SAGE *recommendation-set* formalism (Tu et al. 2004). SAGE activity graphs consist of four types of nodes that are connected to each other in a flowchart-like fashion. These four types of nodes are

- *Context* nodes, which define the clinical contexts in which the guideline should be used to evaluate a patient
- *Decision* nodes, which list alternative downstream nodes and specify the criteria that must be met for reaching those nodes
- *Action* nodes, which specify the work items that should be performed by a computer system or by a health care provider

- *Routing* nodes, which are used for branching and synchronization of multiple concurrent processes

23.5.2.3 Knowledge Integration Architecture

The knowledge integration architecture for SAGE is shown in Figure 23-4 (Ram et al. 2004). In this architecture, the SAGE execution engine interacts with the CIS through vMR service interfaces. The SAGE vMR utilizes a patient information model that is based on the HL7 Reference Information Model (RIM) (Health Level 7 2006), and clinical concepts are specified using standard vocabularies such as SNOMED CT and LOINC.

Within this architecture, the SAGE execution engine is invoked when its event listener detects the occurrence of a CIS event specified in a guideline's *context* node, such as a patient checking into an outpatient pediatric clinic (task 1). Following this invocation, the execution engine traverses the nodes of the relevant guideline. When a *decision* node is encountered, the execution engine first uses a terminology server to obtain all codes that are subsumed by the code used in the guideline to identify a clinical concept. For example, if a *decision* node specifies the need for a patient's weight using the SNOMED CT code for *weight finding* (SNOMED CT *107647005*), the terminology server returns all codes subsumed by this code (e.g., the SNOMED CT code for *normal weight*, *43664005*) (Ram et al. 2004). The execution engine then retrieves the data through the CIS's vMR query service by specifying the set of subsumed codes and the vMR object type of interest (e.g., Observation) (task 2). If the terminology specified in the query request is different from the terminology used by the CIS, this difference must be reconciled through a translation process (see subsection 23.4.5).

Once the required data have been retrieved from the vMR, the SAGE execution engine uses the *decision* node to evaluate the patient and to identify which *action* nodes should be invoked (task 3). When an *action* node is reached, a request is made to the vMR's action service to perform select work items (task 4). Example work items that can be requested by an *action* node include the obtaining of informed consent, the communication of an alert, and the creation of a pending order within the CIS's order entry system.

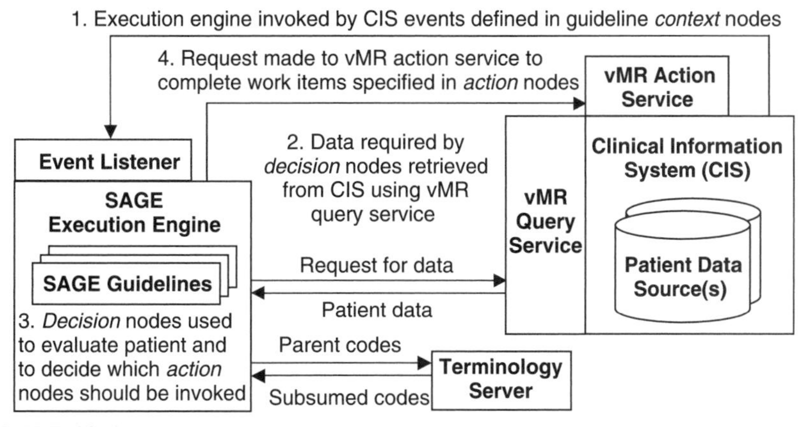

FIGURE 23-4 SAGE knowledge integration architecture. The numbers refer to the four knowledge integration tasks specified in subsection 23.3. vMR = virtual medical record.

23.5.2.4 Strengths and Limitations

An important strength of the SAGE approach is that the execution engine interacts with the host CIS through a vMR service interface. As a result, the execution engine can be reused in different clinical contexts as long as the CIS is wrapped by the vMR service interface specified by SAGE. Also, a second strength of SAGE is its ability to deliver decision support through the local CIS's native user interface. A third strength of the approach is that its guidelines specify all the information required for providing decision support in operational settings. Finally, a fourth strength of SAGE is its use of relevant standards. For example, the vMR information model is based on the HL7 RIM, and standard vocabularies are used to represent clinical concepts.

With regard to weaknesses, one limitation of SAGE is that it has not yet been used to support clinical decision-making in real clinical settings. A second limitation is that the infrastructure required for using the approach is quite significant. These infrastructure requirements include a SAGE execution engine, which is currently not available outside of the SAGE project; a CIS meeting the functional and information model requirements assumed by SAGE; and the implementation of a SAGE-compliant vMR service interface around the CIS. Because CISs can vary significantly with regard to their functional capabilities, and also because many institutions have not yet implemented a robust CIS, the CIS infrastructure requirement may prevent some health care institutions from making use of the SAGE framework. Finally, like Arden Syntax MLMs, SAGE guidelines explicitly specify when they are to be used and what actions should follow as a result of the conclusions reached. As discussed in subsection 23.4.1, this type of approach limits the flexibility with which a given knowledge resource can be leveraged to deliver different types of decision support in divergent clinical contexts.

23.5.3 Case Study 3: Commercial Medication Knowledge Resources

23.5.3.1 Background

Medication errors are prevalent and costly; as such, they represent an attractive target for CDS interventions designed to improve care quality and ensure patient safety (Kaushal et al. 2003). The knowledge base required for providing medication decision support can be developed in-house by well-resourced health care institutions. However, given the vast number of medications that must be considered, it is usually more feasible to use a commercial medication knowledge base "as is" or as the foundation of an institutionally tailored knowledge base (Reichley et al. 2005). In the United States, Cerner Multum and First DataBank offer commercial medication knowledge resources with significant market penetration. Core functionality supported by these knowledge resources include the identification of potentially dangerous medication doses, the identification of drug–allergy contraindications, and the identification of drug–drug interactions.

23.5.3.2 Knowledge Representation Approach

Both Cerner Multum and First DataBank offer their medication knowledge bases in proprietary relational database formats. This approach works well for this domain, as much of the knowledge required for medication decision support can be represented in a straightforward manner using relational databases. In addition, both vendors offer application programming interfaces (APIs) that can be used to access the knowledge contained in their databases.

23.5.3.3 Sample Knowledge Integration Architecture

Because the commercial medication knowledge resources allow the CDS implementer to determine when to use the resource, how to retrieve the data, and how to communicate the CDS results, these knowledge resources can be integrated into clinical applications in many different ways. Thus, the knowledge integration architecture presented next represents just one example of how these resources can be leveraged to provide medication decision support.

Figure 23-5 provides a sample architecture in which a commercial medication knowledge base is used by a pharmacy system to alert pharmacists regarding medication doses considered to be dangerously high or low for a particular patient. This architecture is based on published descriptions of how BJC HealthCare in St. Louis, Missouri, has used Cerner Multum's knowledge base to provide medication dosing alerts to pharmacists (Miller et al. 1999; Reichley et al. 2005). In this sample architecture, the CDS module of the pharmacy system is invoked when a new medication order has been placed by a clinician (task 1). The CDS module then retrieves relevant data from the clinical data repository, such as the patient's age, gender, height, weight, comorbidities, and serum creatinine level (task 2).

The execution engine associated with the medication knowledge base then uses this information to determine whether the dose of the ordered medication is within acceptable limits (task 3). In making this determination, the execution engine uses information in the commercial drug database that defines the maximum and minimum allowable doses for a medication given a particular patient profile (e.g., age > 18, creatinine clearance > 30 mL/min). The execution engine also takes into account local extensions to the database aimed at reducing false positive alerts, such as the degree to which the dosing limits should be relaxed for a particular drug.

If it is determined that a medication dosage in fact does fall outside of acceptable limits, the CDS module generates an alert that includes patient demographics, relevant laboratory results, and information regarding the drug order triggering the alert (task 4a). These alerts are then communicated to pharmacists via a secure Web page, fax, or network printer (task 4b). Of note, an alert may be suppressed if it is deemed to be duplicative, as might be the case if a similar alert had been generated for the same patient in the recent past.

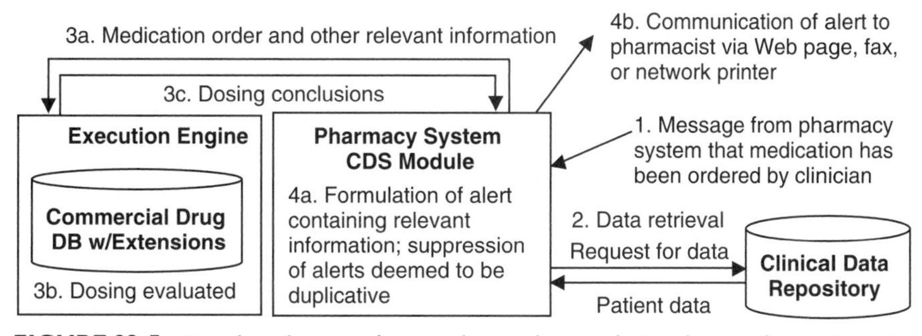

FIGURE 23-5 Sample architecture for providing medication dosing alerts to pharmacists using a commercial medication knowledge base. The numbers refer to the four knowledge integration tasks specified in section 23.3. DB = database.

23.5.3.4 Strengths and Limitations

An important strength of commercial drug databases is that they represent knowledge in a format that is easy to comprehend. As a result, it is relatively simple to make extensions to the knowledge base and to implement a custom execution engine that makes use of the local extensions. A second strength of commercial drug databases is that they maintain very large bodies of knowledge that are updated on a regular basis. This availability of comprehensive, up-to-date content has been critical to the market success of these knowledge bases. Finally, a third strength of commercial medication knowledge resources is that they place minimal restrictions on how they are used to meet end-user needs. As a result, these knowledge resources can be leveraged in a variety of clinical contexts, and CDS inferences can be communicated to end-users using the approach deemed most appropriate for a given clinical context.

Despite these strengths, one important limitation of commercial drug databases is that they tend to be conservative for medical/legal reasons (Reichley et al. 2005). Consequently, local adaptations may be required to reduce the incidence of false-positive alerts. Also, and very importantly, the knowledge representation formats and APIs used by these resources are specific to the pharmacy domain and cannot be directly leveraged for use in other medical domains.

23.5.4 Case Study 4: SEBASTIAN

23.5.4.1 Background

Initially described in the literature in 2005, SEBASTIAN is a decision support Web service developed at Duke University (Kawamoto et al. 2005b). As a Web service, SEBASTIAN communicates with client systems using extensible markup language (XML) messages transmitted over the Internet (Cermai 2002). The main service operation offered by SEBASTIAN is a patient evaluation operation, in which SEBASTIAN receives patient data as the input and returns patient-specific, machine-interpretable conclusions as the output. To date, SEBASTIAN has been used to implement four distinct CDS applications:

- A system that provides clinicians with diabetes care recommendations in the outpatient setting
- A system that generates individually tailored care reminder letters for patients
- A system that provides clinics with reports that list the patients most in need of services, along with identified care needs and recommended actions
- A system that e-mails alerts to health care providers regarding care issues requiring follow-up

23.5.4.2 Knowledge Representation Approach

SEBASTIAN encapsulates medical knowledge in XML documents known as Executable Knowledge Modules (EKMs). Each module consists of four sections: a *maintenance* section, a *library* section, a *knowledge* section, and a *logic* section. The *maintenance* section contains general maintenance information such as the title, identifier, and version number of the EKM, and the *library* section consists of bibliographic information such as keywords and references. The *knowledge* section then defines the data required for evaluating patients using the module. Data requirements for an EKM may include demographic data (e.g., gender, race, age) as well as data on health care acts (e.g., observations, diagnoses, procedures). The patient information model used by SEBASTIAN is based on the HL7 RIM, and health care acts are preferentially identified using standard terminologies included in the National Library of Medicine's Unified Medical Language System (UMLS). A clinical concept in an EKM can be specified in terms of multiple vocabularies, and a terminology service is used by the EKM authoring environment to facilitate the translation of concepts between vocabularies and the identification of codes subsumed by a parent code.

In addition to specifying the EKM data requirements, the *knowledge* section also specifies the format and meaning of the machine-interpretable results that will be returned to the client. For example, a knowledge module that evaluates whether a patient with diabetes is due for a glycated hemoglobin test may specify that it will return a result code indicating whether the patient is 1) ineligible because he does not have diabetes, 2) eligible but not in need of the test due to a test on record from the previous six months, or 3) eligible and in need of the test. Furthermore, the EKM may specify that it will return a result parameter that specifies when the last test was conducted and a parameter that specifies the value of the patient's last test.

Finally, an EKM contains a *logic* section, which specifies how the patient data provided by the client will be used to generate the CDS results promised in the *knowledge* section. To generate the results, SEBASTIAN creates Java classes corresponding to each EKM in its knowledge repository. Within these Java classes, required data elements are made available to the module author as native Java objects. For example, if a module requires data on glycated hemoglobin tests from the past year, the Java class will contain an array that is populated at runtime by Observation objects that represent glycated hemoglobin tests from the previous year. Given this setup, standard programming techniques can be used to manipulate the patient data and to generate the CDS results promised in the *knowledge* section.

The encoding of decision logic using a native programming language is associated with significant advantages, including:

- The widespread availability of robust programming environments designed for the language
- The ability to create utility classes to handle common operations
- The ease with which medical knowledge stored in external knowledge repositories (e.g., commercial drug databases) can be accessed and leveraged
- The ability to invoke external CDS execution engines using any interfacing mechanism supported by Java

To provide decision support using EKMs, SEBASTIAN offers four service operations to its clients. The primary service operation offered is a patient evaluation operation. In this operation, a client specifies the EKMs to use for evaluating a patient, and the client also submits the patient data required by the EKMs. In return, SEBASTIAN returns CDS results regarding the patient as specified in the EKMs' *knowledge* sections. SEBASTIAN also offers three supplemental operations to support the patient evaluation operation, to allow a client to identify the knowledge modules that meet client search criteria; obtain descriptions of selected modules, including descriptions of the results that will be returned following patient evaluation; and identify the data required for evaluating a patient using specified knowledge modules.

23.5.4.3 Sample Knowledge Integration Architecture

As noted earlier, SEBASTIAN has been used to implement four distinct CDS applications to date. One of these applications is the Diabetes Reminder System (DRS); this system uses EKMs related to diabetes management to provide clinicians with diabetes care recommendations in the outpatient setting (Kawamoto et al. 2005b). The high-level system architecture of the DRS is outlined in Figure 23-6. When a patient with diabetes checks into a clinic, the intake nurse requests a diabetes care reminder sheet for the patient through the DRS Web site (task 1). The DRS's CDS module then retrieves the patient data specified as data requirements in the *knowledge* sections of the diabetes EKMs (task 2a). The required data are retrieved from the local institution's clinical data repository and from a DRS-specific database that collects data not otherwise collected in a coded format in the clinical data repository (e.g., whether a microfilament foot exam was done). The CDS module then consolidates the patient data retrieved from the two data sources into a single representation of the patient (task 2b). Following this step, the CDS module makes a request to SEBASTIAN to evaluate the patient using its diabetes EKMs (task 3a). SEBASTIAN then uses the decision rules encoded in the EKM *logic* sections to generate CDS results regarding the patient (task 3b), and these evaluation results are communicated back to the CDS module (task 3c).

Once it has received the CDS results from SEBASTIAN, the DRS's CDS module generates an XML representation of a diabetes care reminder sheet (task 4a). This XML document is forwarded to an XML transformation Web service (task 4b), which uses an XSL style sheet to convert the XML content into a PDF document (task 4c). This PDF document is then streamed to the Web browser (task 4d), so that the nurse can print the sheet and attach it to the patient's chart for the clinician to review. The diabetes care reminder sheet consists of a section that lists relevant patient data, a section that provides diabetes care recommendations, and a section where clinicians can record data

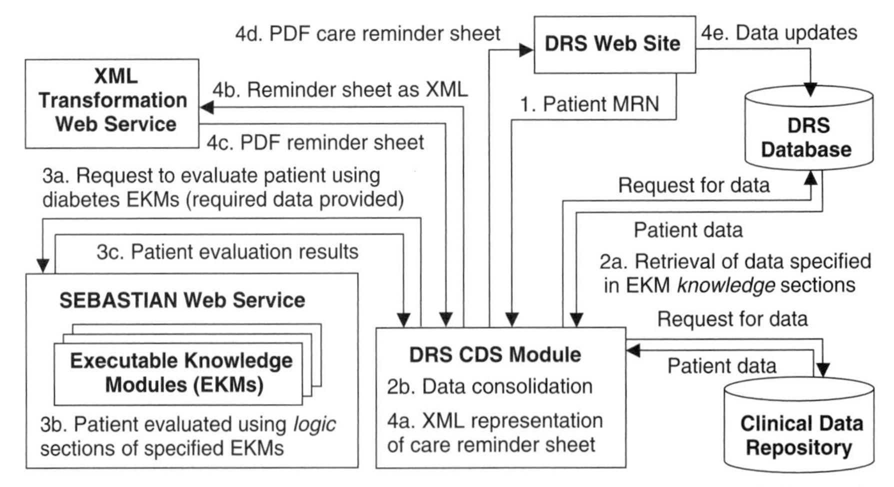

FIGURE 23-6 System architecture for Diabetes Reminder System (DRS), which provides diabetes care recommendations using SEBASTIAN. The numbers refer to the four knowledge integration tasks specified in subsection 23.3. MRN = medical record number.

not otherwise collected in a coded format. Any updates to the DRS database are entered by clinic support staff through the DRS Web site (task 4e).

23.5.4.4 Strengths and Limitations

An important strength of SEBASTIAN is that it does not specify how and when EKMs are to be used. As a result, the same EKMs can be leveraged in various clinical contexts to provide context-appropriate decision support. For example, the four CDS applications that have been implemented using SEBASTIAN provide diabetes care recommendations using the same set of EKMs (Kawamoto et al. 2005b). A second strength of SEBASTIAN is that its only infrastructure requirements are an Internet connection and the capacity to exchange data using XML. A third strength of SEBASTIAN is that it could be used to centrally manage executable medical knowledge on behalf of an institution or a region, since one SEBASTIAN instantiation can support multiple applications operating in diverse clinical environments. A fourth strength of SEBASTIAN is its use of relevant standards, including the HL7 RIM and standard vocabularies included in the UMLS. A fifth strength of this approach is that it is not limited to the representation of medical knowledge from a particular clinical domain. Finally, a sixth strength of SEBASTIAN is that external knowledge resources and decision engines can be invoked from within the EKM *logic* section. Thus, medical knowledge that has been represented using different formalisms could potentially be accessed through a common SEBASTIAN interface.

On the other hand, one important limitation of SEBASTIAN is that its usefulness has not yet been validated for several important types of CDS applications, such as CPOE systems with CDS capabilities. A second important limitation of SEBASTIAN is that less than a hundred EKMs have been implemented to date, with a focus primarily on preventive care, inappropriate resource utilization, and diabetes management. The SEBASTIAN knowledge base will need to be expanded significantly in order to more comprehensively meet the CDS needs faced by health care professionals in various clinical contexts. Finally, a third limitation of the SEBASTIAN approach is that the local application's CDS module must specify when to use SEBASTIAN (task 1), how to retrieve the required data (task 2), and how to communicate the CDS results to end-users (task 4).

23.6 LESSONS LEARNED FROM MARKET ADOPTION PATTERNS

As exemplified by the case studies just presented, knowledge resources can be integrated into applications using a variety of approaches. In practice, however, only a handful of these approaches have found relatively widespread acceptance within the health care marketplace. These relatively successful approaches include the Arden Syntax and commercial medication knowledge resources. However, most CDS capabilities are still implemented using approaches that are specific to a given application or institution. Furthermore, the availability of CDS capabilities remains the exception rather than the norm in most health care settings. Through examination of these market trends, this section will attempt to draw generalizable conclusions regarding what is required for a knowledge integration approach to be adopted in operational clinical settings.

23.6.1 Lessons Learned from Market Adoption of Arden Syntax

As discussed in the last section, the Arden Syntax is one of the more widely adopted approaches to providing decision support using centrally managed knowledge resources. This relatively widespread adoption arises from the support for the Arden Syntax provided within a number of CIS products (Karadimas et al. 2002) and also from the relative ease with which MLMs can be understood and authored. However, the Arden Syntax is not commonly implemented by individual health care institutions that have not purchased a CIS product with built-in support for the approach, due to the cost and difficulty associated with implementing an execution environment for Arden Syntax MLMs (Karadimas et al. 2002). Based on these observations, the following lessons can be gleaned regarding the requirements for the market's acceptance of a CDS knowledge integration architecture.

Standardization can greatly facilitate market adoption, but endorsement by a standards body is not in itself sufficient. The acceptance of the Arden Syntax as a standard by ASTM and by HL7 played an important role in several vendors' decisions to incorporate support for the standard within their CIS products. However, the Arden Syntax is still far from representing a dominant approach to decision support within the health care community. Thus, standardization is important, but not sufficient, for the adoption of a CDS approach by the health care industry. Moreover, despite the standardization, there has been minimal reuse of knowledge resources through cross-vendor sharing of MLMs.

Implementation costs should be minimized. An important factor limiting the more widespread adoption of the Arden Syntax is the cost and difficulty associated with implementing the execution environment required for using MLMs. To facilitate widespread adoption of a knowledge integration approach, the costs associated with implementation and maintenance should be minimized.

Simplicity is critical. The Arden Syntax is frequently criticized as being too simplistic. For example, it is often noted in the academic literature that it is cumbersome to represent complex clinical practice guidelines using the Arden Syntax (Peleg et al. 2001). However, the relative simplicity of the Arden Syntax has been critical to its success to date, as it facilitates all aspects of the CDS delivery process, from the authoring of MLMs to the design and implementation of an execution environment. Thus, an important lesson learned from the success of the Arden Syntax is that simplicity should be a core goal when designing a knowledge integration architecture.

23.6.2 Lessons Learned from Market Adoption of Commercial Medication Knowledge Resources

Commercial medication knowledge resources, such as those offered by Cerner Multum and by First DataBank, are widely used in hospital settings to provide medication decision support (see subsection 23.5.3). Because these knowledge resources do not dictate when they are to be used or how CDS results are to be communicated to end-users, they can be leveraged in a variety of clinical contexts to meet user needs. Based on the success of these resources, the following insights can be derived.

The availability of a large underlying CDS content base greatly facilitates the adoption of a knowledge integration architecture. The success of the commercial medication knowledge resources can be attributed in large part to the fact that these resources maintain vast knowledge bases that are updated on a regular basis. Clearly, these products would have been much less successful if they had been marketed as content-free frameworks for knowledge representation and integration, rather than as knowledge integration frameworks coupled with extensive knowledge bases.

Simplicity is critical. The approach used by commercial drug databases to represent and deliver knowledge is quite straightforward (see subsection 23.5.3). This simplicity makes it easy for a system engineer to understand the resource and to leverage it to provide decision support. Also, these resources can be adapted relatively easily to meet local requirements (see subsection 23.5.3).

Flexibility facilitates adoption. Because commercial medication knowledge resources do not specify when they are to be used or how their CDS results are to be communicated, they can be adapted to support various types of CDS applications operating in diverse clinical and technical environments. This flexibility greatly increases the contexts in which these knowledge resources can be leveraged to provide CDS capabilities.

23.6.3 Lessons Learned from Prevalence of Application- and Institution-Specific Approaches

Given the high cost involved in maintaining a machine-executable medical knowledge base, it would be highly desirable if a knowledge integration architecture supported the reuse of medical knowledge across applications and institutions. However, knowledge integration approaches are frequently designed in an application-specific or institution-specific manner. For example, computer-interpretable order sets may be formulated in a format that works only within a specific vendor's CPOE product, or an architecture for deploying decision rules may work only within a specific institution's CIS environment. Based on the prevalence of such application- and institution-specific approaches, the following observations can be made.

Clinical applications and health care institutions differ significantly in terms of their CDS functional requirements. The prevalence of application- and institution-specific approaches to representing and integrating medical knowledge indicates that the CDS needs of a clinical application depends heavily on the local context. As a result, a knowledge integration approach must be very flexible if it is intended to support diverse CDS functional requirements within heterogeneous clinical settings.

To fully leverage existing knowledge resources, a knowledge integration architecture must support diverse knowledge representation approaches. The prevalence of application- and institution-specific knowledge integration approaches is accompanied by a prevalence of application- and institution-specific approaches to knowledge representation.

In order to leverage these existing knowledge resources, a knowledge integration architecture must be capable of supporting various approaches to representing and encoding medical knowledge.

23.6.4 Lessons Learned from Limited Adoption of CDS Capabilities in General

One of the most striking observations regarding the market's adoption of knowledge integration approaches is that so few health care institutions have adopted a robust strategy for integrating knowledge into their applications. Based on this observation, the following lessons can be inferred.

Implementation costs should be minimized. Given the demonstrated ability of CDS applications to improve clinical practice (Kawamoto et al. 2005a), the limited adoption of CDS capabilities indicates that the cost of implementing effective decision support often is perceived to be too high relative to the anticipated benefits. To facilitate more widespread adoption of CDS capabilities, the costs associated with knowledge integration approaches should be minimized.

Knowledge integration approaches should be as simple as possible to understand and to use. An important barrier to the implementation of CDS capabilities is the complexity associated with many approaches to knowledge representation and integration. To enable widespread adoption, knowledge integration approaches should be made as simple as possible to understand and to use.

23.6.5 Summary of Lessons Learned

As just discussed, many lessons can be drawn from how the health care marketplace has reacted to different knowledge integration approaches. These lessons can be summarized as follows:

- Simplicity is critical.
- Implementation costs should be minimized.
- The CDS content available for a knowledge integration architecture should be made as comprehensive as possible.
- Support should be provided for knowledge resources encoded using different knowledge representation approaches.
- The architecture should be flexible, so as to support diverse CDS needs in differing IT environments.
- Endorsement of an approach by a standards body can accelerate market adoption.

These lessons form the core motivation for the service-based approach to decision support proposed in the next section of the chapter.

23.7 PROPOSAL FOR A SOA APPROACH TO CLINICAL DECISION SUPPORT

In recent years, enterprises in various industries have begun to reorganize their IT capabilities within the framework of a Service-Oriented Architecture (SOA). This section describes how a SOA approach to decision support can overcome many of the challenges associated with knowledge integration. Also, this section discusses how standard services being specified by the HL7

Healthcare Services Specification Project could serve as the foundation of a SOA approach to decision support. Even though these service specifications are still in proposal form and not yet implemented, we describe here a CDS approach based on these services for four reasons. First, based on the lessons learned earlier, a SOA approach to decision support appears to be the most promising approach available for overcoming the challenges encountered when attempting to integrate medical knowledge into applications. Second, there is increasing evidence from other industries that a SOA approach to system design and implementation is associated with significant, tangible benefits. Third, there is an established and maturing SOA implementation technology in the form of Web services. Fourth, the service specifications are being developed as part of an active HL7 initiative, and these specifications are expected to move toward standardization close to the time of the publication of this book.

23.7.1 SOA Overview

In a SOA, core business capabilities are encapsulated within independent software services, and these services are leveraged by various front-end applications to fulfill identified business requirements (He 2003; Booth et al. 2004; Krafzig et al. 2004). Key properties of a SOA include the following.

Business-oriented services. A service in a SOA typically encapsulates functionality in terms that are meaningful from a business perspective. As such, services are typically relatively broad in scope, and each service typically provides relatively few operations in conjunction with relatively large, complex service inputs and/or outputs.

Message-based interactions with "black-box" implementations. Each service in a SOA is defined in terms of the messages exchanged between the service and its clients. Service implementation details, such as the programming language used and the structure of any underlying databases, are deliberately abstracted away in a SOA. This "black-box" approach to implementation allows a client to interact with a service without understanding the complexities of how the service is implemented. Also, a given service interface can be implemented using various underlying approaches, and a legacy software system can be used to meet the requirements of a service as long as the legacy system can be wrapped with the appropriate service interface.

Communication over a network. Although not required, SOA messages typically are exchanged across a network, such as an intranet or the Internet.

Platform neutrality. Messages in a SOA are communicated using a platform-neutral, standardized format. XML messages generally are used to fulfill this requirement for platform neutrality. SOA services typically are implemented as Web services, in which services communicate with their clients using XML messages transmitted over the Internet (Cermai 2002). Moreover, Web service messages often are encoded using a specific XML-based communication protocol known as the Simple Object Access Protocol (SOAP). It is important to note, however, that the use of a specific implementation technology (e.g., SOAP Web services) is not mandated by the SOA.

Service description and discovery. Service interfaces are described using a platform-neutral description language, such as the Web Service Definition Language (WSDL). Also, SOA services should be associated with a mechanism for discovering their existence. Typically, services are registered in a service repository, such as an online registry conforming to the Universal Description, Discovery, and Integration (UDDI) specification.

Loose coupling. In a SOA, an individual service is designed to be as independent as possible from other services as well as from front-end applications that invoke the service. This limited interdependence between services and other software components within a SOA often is referred to as loose coupling. Because individual services have limited external dependencies, they can be orchestrated together in various ways to meet application-specific functional requirements. This ability to efficiently reuse services to fulfill new application requirements is a key strength of the SOA.

23.7.2 SOA Benefits and Alignment with Knowledge Integration Needs

The use of a SOA is associated with several drawbacks, including a performance penalty incurred from the use of verbose messages to communicate between system components and occasional message transmission failures arising from network connectivity problems. These limitations, however, are overshadowed by the many important benefits associated with the use of a SOA (Krafzig et al. 2004). These benefits are summarized here.

Simplicity. In a SOA, complex problems can be decomposed into smaller, more manageable problems that are addressed by individual services. This significantly reduces the complexity involved in designing and implementing systems. Furthermore, the mechanism by which a specific service meets its functional obligations is hidden from the service consumer. This "black-box" nature of services means that service consumers are shielded from potentially complex implementation details that exist underneath a service interface.

Technology independence. The SOA itself is independent of any specific implementation technology, such as SOAP Web services. Moreover, even when a specific SOA technology has been chosen (e.g., SOAP Web services using WSDL for service description and UDDI for service registration), the services can be implemented using any available technology as long as the interface requirements are fulfilled. Thus, enterprises are able to make use of services regardless of the technologies used for implementation (e.g., Java, .NET, object-oriented databases, Microsoft Windows, Linux, etc.).

Reusability. In a SOA, capabilities that already exist within legacy systems can be reused by exposing the functionality through platform-neutral service interfaces. Furthermore, a given service can be reused by multiple applications and by other services in order to meet various business requirements.

Flexibility. One of the most important benefits of a SOA is that it is able to adapt to different situations in a flexible manner. Systems that lie underneath of service interfaces can be changed as needed, and new business requirements can be rapidly fulfilled by leveraging existing services and by creating new services as needed.

Market stimulation. A SOA allows a service consumer to obtain software functionality easily from a remote service provider. This ease of transaction can stimulate both the demand for and supply of SOA services, especially if the interface used by a service is standardized across vendors. For example, a standard Decision Support Service interface specification could stimulate the creation of a robust knowledge resource marketplace wherein various knowledge vendors offer a diverse array of CDS content through a common service interface.

Cost savings. Finally, a critical benefit of a SOA lies in the potential for significant cost savings. Potential sources of cost savings include the ability to reuse existing IT assets to meet new business requirements; the ability to more easily accommodate changes in business processes or business relationships aimed at improving efficiency; and the simplification and modularization of the IT landscape, which can reduce the time and cost involved in designing, implementing, testing, and maintaining individual IT systems.

With regard to clinical decision support, the SOA benefits just described address many of the key knowledge integration needs identified in subsection 23.4 and summarized in subsection 23.6.5. As outlined in Table 23-3, a SOA approach to decision support can

- Facilitate adoption of a knowledge integration approach by making it easier for institutions to understand and to use the approach
- Allow knowledge resources encoded using different approaches to be delivered through a common, platform-neutral service interface
- Facilitate the creation of a large CDS knowledge base through the reuse of knowledge already encoded using various approaches and through the stimulation of a market for CDS content delivered as a service
- Provide the flexibility required for dealing with diverse CDS needs encountered in various IT environments
- Reduce implementation costs, thereby increasing the pool of health care institutions that can make use of the approach to provide clinical decision support

TABLE 23-3 SOA strength and corresponding CDS knowledge integration need addressed by strength.

SOA strength	CDS knowledge integration need addressed
Simplicity	Need for simplicity to facilitate adoption of knowledge integration approach
Technology independence	Need to provide support for knowledge resources encoded using different knowledge representation approaches
Reusability	Need for CDS knowledge base to be as comprehensive as possible
Market stimulation	Need for CDS knowledge base to be as comprehensive as possible
Flexibility	Need to support diverse CDS needs in diverse IT environments
Cost savings	Need to minimize implementation costs

23.7.3 Services Useful for Clinical Decision Support

In considering a SOA approach to decision support, it is instructive to identify services that could serve as the building blocks for CDS applications. Services that may be useful for the implementation of a CDS application include:

- A decision support service (DSS), which uses patient data to make machine-interpretable inferences regarding the patient
- A common terminology service (CTS), which provides access to various terminology operations
- An entity identification service (EIS), which enables the identification of entities (e.g., patients) across systems
- A record location and access service (RLAS), which facilitates the retrieval of patient records across systems, and which also allows for fine-grained queries for patient data
- A patient record updating service (PRUS), which allows the service consumer to update the patient record
- A CIS action brokering service (CABS), which permits the service consumer to invoke various actions within a CIS. Of note, the patient data query functionality of the RLAS, the PRUS, and the EABS comprise the primary service capabilities that a CIS would need to implement in order to meet the requirements of a vMR service (Johnson et al. 2001b).

Health care institutions could use these and other services in order to implement CDS capabilities more rapidly and more efficiently than is currently possible. Figure 23-7 provides an example architecture of an outpatient care reminder module of a CIS that could be implemented using the services just described. In this sample architecture, the care reminder module in Health System A is invoked by a message from the CIS that a patient has checked into an outpatient clinic (task 1 of the four knowledge integration tasks outlined earlier). The CDS module then retrieves relevant patient data from Health

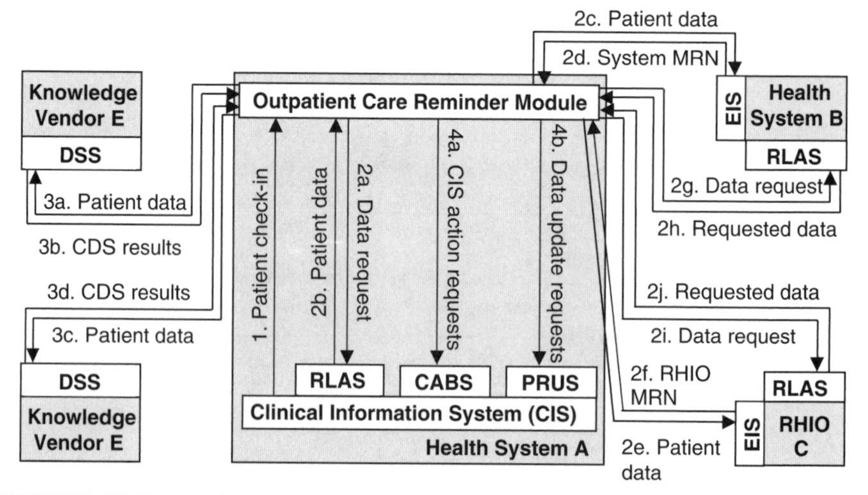

FIGURE 23-7 Sample Service-Oriented Architecture (SOA) for outpatient care reminder module of a CIS. The numbers refer to the four knowledge integration tasks specified in subsection 23.3. CABS = CIS Action Brokering Service; DSS = Decision Support Service; EIS = Entity Identification Service; PRUS = Patient Record Updating Service; RHIO = Regional Health Information Organization; RLAS = Record Location and Access Service.

System A through its RLAS (tasks 2a and 2b). Also, the care reminder module provides identifying demographic data (e.g., name, birth date, address, phone number) to the EISs of Health System B and Regional Health Information Organization (RHIO) C, learns that the patient is registered in those systems, and receives the patient's medical record numbers (MRNs) within those systems (tasks 2c, 2d, 2e, and 2f). The CDS module then uses these system-specific MRNs to retrieve relevant patient data from Health System B and from RHIO C (tasks 2g, 2h, 2i, and 2j). Then, after consolidating all the patient data that it has collected, the care reminder module provides the patient data to the DSSs offered by two knowledge vendors and retrieves CDS inferences regarding the patient, such as whether the patient is overdue for any preventive health procedures or is in need of a medication change to better control a chronic illness (tasks 3a, 3b, 3c, and 3d).

Based on the CDS results obtained, the CDS module makes a request to the CABS to perform appropriate actions (e.g., place pending orders within the order entry system for required procedures and generate an alert that is visible when the clinician opens the patient's record) (task 4a). The care reminder module also makes a request to the PRUS to update the patient's record to note relevant data regarding the CDS communications provided (e.g., when, why, and to whom any alerts were sent) (task 4b). Of note, although not shown in the figure, a CTS could be used to provide terminology support during many of the steps involved in this CDS process. For example, a RLAS could use a CTS to translate coded data request parameters into codes that are understood by the local clinical data repository.

23.7.4 Need for Standardization of Services to Facilitate Semantic Interoperability

A health care institution could reap some of the benefits associated with a SOA even if vendors and institutions used different service interfaces to provide the same type of service. However, such heterogeneity in service interfaces would make it more difficult to achieve semantic interoperability, because service clients would have to deal with multiple interfaces when requesting similar services from different providers. In the sample knowledge integration architecture described in Figure 23-7, for example, the cost of implementing the CDS capability would increase substantially if Health System A, Health System B, and RHIO C used different EIS and RLAS service interfaces and the knowledge vendors provided their CDS capabilities through incompatible service interfaces. Thus, in order to facilitate semantic interoperability, health care service interfaces should be standardized where feasible and appropriate. This is precisely the objective of the standardization effort described in the following section.

23.7.5 HL7-OMG Healthcare Services Specification Project (HSSP)

The Healthcare Services Specification Project (HSSP) is a project that aims to standardize the interfaces for software services important to the health care industry. Initiated in 2005, the HSSP is being pursued as a joint initiative between HL7 and the Object Management Group (OMG). HL7 is the premier standards development organization within health care, whereas OMG is an open-membership, not-for-profit consortium that produces and maintains computer industry specifications for interoperable enterprise applications.

OMG is perhaps best known for its specification of the Unified Modeling Language (UML) and the Common Object Request Broker Architecture (CORBA).

Within this partnership, HL7 identifies and prioritizes candidate services. Then, for each service designated for standardization, HL7 specifies the functional requirements of the service, the information model for the service payloads, and functional conformance criteria. The end result of this HL7 process is a computationally independent functional specification of a service, which is referred to as a Service Functional Model (SFM). Once a SFM is balloted on and adopted as a HL7 standard, it is refined within OMG to develop computationally dependent service specifications (e.g., a SOAP Web service specification) as well as at least one reference implementation.

As of January 2006, the HSSP is actively working on the specification of service interfaces for the Entity Identification Service (EIS), Common Terminology Service (CTS), and Decision Support Service (DSS), which were briefly described in subsection 23.7.3. Also, the HSSP is actively working to specify the service interface for the Record Location and Updating Service (RLUS), which combines the functionality of the Record Location and Access Service (RLAS) and the Patient Record Updating Service (PRUS). Given its direct relevance to the focus of this chapter, the DSS is described in greater detail in the next section. It is anticipated that HL7 Service Functional Models will be available as balloted standards by late 2006 for the EIS and RLUS and by early 2007 for the CTS and DSS. In addition, it is anticipated that the HSSP will begin standardization work on additional health care services during 2006.

23.7.6 HSSP Decision Support Service

The purpose of the HSSP DSS project is to define a common service interface for fulfilling a core functional requirement shared by all decision support systems—that is, the need to draw conclusions regarding individual patients. The DSS interface is based on the service interface used by SEBASTIAN (see subsection 23.5.4). However, it is important to note that the DSS interface is being specified in a generic manner, so that it can serve as a common platform for delivering medical knowledge that has been encoded using different knowledge representation approaches.

A DSS provider can be conceptually understood as the guardian of one or more modules of medical knowledge, wherein each DSS knowledge module is capable of utilizing coded patient data to arrive at machine-interpretable conclusions regarding the patient under evaluation. The scope of a DSS knowledge module is the assessment of a single patient in a specified topic area. The topic area may be narrow (e.g., the need for a glycated hemoglobin test for a patient with diabetes) or broad (e.g., the existence of contraindications to any medications prescribed or about to be prescribed for a patient).

When requesting a patient evaluation, a DSS client specifies the knowledge modules to use for the evaluation, and the client also submits the patient data required by the knowledge modules. In return, the DSS returns inferences regarding the patient in a format that has been predefined for that knowledge module. For example, an online immunization registry might submit a patient's age, comorbidities, allergies, and past immunizations to a DSS and request that the patient be evaluated using the service's immunization knowledge module. In return, the DSS might return a list of the vaccines for which the patient is ineligible due to contraindications, a list of the vaccines for which the patient is up-to-date, and a list of the vaccines for which the patient is due.

In order to acquire patient evaluations in this manner, a client must be able to obtain several supplemental pieces of information from a DSS. These supplemental information needs include the need to identify the knowledge modules that could be used to meet client needs; the need to know what patient data must be submitted to the DSS in order to obtain an accurate evaluation; and the need to know the meaning and format of the CDS results that will be returned by the DSS following a patient evaluation. Supplemental service operations are provided to meet these information needs.

Because the specification and updating of machine-executable decision logic represents one of the most expensive aspects of implementing CDS capabilities within clinical applications, the use of a DSS should significantly reduce the effort required for providing robust decision support to clinicians and other health care stakeholders. In conjunction with other service specifications being developed by the HSSP, it is hoped that the DSS specification will facilitate the more widespread implementation of CDS capabilities, which in turn should result in higher quality care and improved patient safety.

23.7.7 Need for Constraints on HSSP Services to Ensure Interoperability

Like most standards designed to be useful in heterogeneous environments, HSSP service specifications allow for a significant degree of implementation flexibility. For example, the RLUS specification currently places relatively few restrictions on the types of patient data that can be retrieved by the service. Similarly, the DSS specification currently places minimal restrictions on the types of data that can be required for evaluating a patient or on the types of CDS results that can be returned by the service. As a third, more generic example, HSSP services do not mandate the use of a specific SOA implementation technology such as SOAP Web services.

Although this type of flexibility is useful, it does introduce the potential for divergent implementations that are not interoperable. Thus, in order to ensure interoperability, it is important that mechanisms exist for specifying and publicizing appropriate constraints. For example, the HSSP may specify at the OMG level that all service specifications must include an implementation profile for SOAP Web services, and a given service specification may define conformance profiles that are associated with specific constraints and requirements. Also, implementation guides could be developed that define how various standards, including HSSP service specifications, should be constrained and coordinated in order to fulfill specific functional requirements. A logical choice for an organization that could specify such implementation profiles would be Integrating the Healthcare Enterprise (IHE) (Siegel et al. 2001). Started in 1998 as a joint effort of the Radiological Society of North America and the Healthcare Information and Management Systems Society, IHE seeks to precisely define how existing standards should be used to complete a particular task. IHE focused initially on radiology, but it has already started to expand into other clinical domains. Thus, IHE implementation profiles could potentially be developed for transactions that involve the use of HSSP services, including the HSSP Decision Support Service.

23.8 CONCLUSION

The provision of proactive, actionable clinical decision support represents one of the most promising strategies available for improving care quality and promoting patient safety. At the same time, however, the availability of robust CDS capabilities remains the exception rather than the norm in most health care settings in the United States and elsewhere.

Multiple factors have contributed to the limited adoption of CDS technologies, including reimbursement models that fail to reward the delivery of higher quality care, limited research regarding return on investment, and the widely publicized problems associated with several large-scale health IT initiatives. An additional important barrier, which has served as the focal point of this chapter, is the cost and complexity associated with encoding medical knowledge in a machine-executable format and then integrating that knowledge into heterogeneous environments to deliver effective decision support. As discussed in subsection 23.3, the high-level tasks involved in integrating knowledge into applications is fairly straightforward, involving CDS module invocation, data retrieval, patient evaluation, and decision support communication. However, as discussed in subsection 23.4, many issues complicate the implementation of these tasks, such as heterogeneity in the information models and terminologies used by different IT systems, significant variability in the underlying knowledge representation approaches used, and the diversity of clinical contexts in which CDS capabilities are to be integrated.

Based on the way in which the health care marketplace has reacted to existing approaches to knowledge integration (see subsection 23.6), it appears clear that more widespread adoption of CDS capabilities will require a knowledge integration approach that is simple, flexible, content-rich, relatively inexpensive, and standardized. It is my belief that the SOA approach to decision support outlined in subsection 23.7 in fact could fulfill these criteria and lead to more widespread availability of CDS capabilities, especially if the approach is coupled with the standardized services that are currently under development by the joint HL7-OMG Healthcare Services Specification Project. The history of medical informatics is replete with predictions of rapid progress that failed to materialize; however, it appears feasible that clinical decision support will play a much larger role in routine health care within 10 to 15 years. The key to the realization of this vision will be the development and refinement of a knowledge integration approach that is flexible and content-rich yet easy to understand and to implement.

RECOMMENDED RESOURCES

Health Level 7. (2006). Technical committees and special interest groups. http://www.hl7.org/Special/committees.

This page within the HL7 Web site provides links to committees within HL7, including the HL7 CDS Technical Committee and the HL7 Services Oriented Architecture (SOA) Special Interest Group (SIG). The SOA SIG is coordinating the efforts of the Healthcare Services Specification Project, and the CDS TC is leading the DSS specification project.

Krafzig, D., Banke, K., and Slama, D. (2004). Enterprise SOA: Service-Oriented Architecture Best Practices. Indianapolis: Prentice Hall PTR.
This book provides numerous insights into how enterprises can more efficiently fulfill complex business requirements through the use of a Service-Oriented Architecture.

OpenClinical. (2006). OpenClinical home: knowledge management technologies and applications for health care. http://www.openclinical.org/home.html.
This Web site provides a comprehensive overview of commercial and academic initiatives related to knowledge management and decision support.

ACKNOWLEDGMENTS

I would like to thank Dr. David F. Lobach for his critical review of the chapter.

REFERENCES

Bates, D. W., Kuperman, G. J., Wang, S., Gandhi, T., Kittler, A., Volk, L. et al. (2003). Ten commandments for effective clinical decision support: Making the practice of evidence-based medicine a reality. *J Am Med Inform Assoc* **10**: 523–530.

Booth, D., Haas, H., McCabe, F., Newcomer, E., Champion, M., Ferris, C., and Orchard, D. (2004). Web Services Architecture. W3C Working Group Note 11 February 2004. http://www.w3.org/TR/2004/NOTE-ws-arch-20040211.

Boxwala, A. A., Peleg, M., Tu, S., Ogunyemi, O., Zeng, Q. T., Wang, D. et al. (2004). GLIF3: A representation format for sharable computer-interpretable clinical practice guidelines. *J Biomed Inform* **37**: 147–161.

Boxwala, A. A., Tu, S., Peleg, M., Zeng, Q., Ogunyemi, O., Greenes, R. A. (2001). Toward a representation format for sharable clinical guidelines. *J Biomed Inform* **34**: 157–169.

Cermai, E. (2002). Top ten FAQs for Web services. http://webservices.xml.com/pub/a/ws/2002/02/12/webservicefaqs.html.

Ciccarese, P., Caffi, E., Quaglini, S., and Stefanelli, M. (2005). Architectures and tools for innovative Health Information Systems: the Guide Project. *Int J Med Inform* **74**: 553–562.

de Clercq, P. A., Blom, J. A., Korsten, H. H., and Hasman, A. (2004). Approaches for creating computer-interpretable guidelines that facilitate decision support. *Artif Intell Med* **31**: 1–27.

de Clercq, P. A., Hasman, A., Blom, J. A., and Korsten, H. H. (2001). Design and implementation of a framework to support the development of clinical guidelines. *Int J Med Inform* **64**: 285–318.

Geissbuhler, A. and Miller, R. A. (1999). Distributing knowledge maintenance for clinical decision-support systems: the "knowledge library" model. *Proc AMIA Symp* 770–774.

Gordon, C. and Veloso, M. (1999). Guidelines in healthcare: The experience of the Prestige project. *Stud Health Technol Inform* **68**: 733–738.

Hagerty, C. G., Pickens, D., Kulikowski, C., and Sonnenberg, F. (2000). HGML: A hypertext guideline markup language. *Proc AMIA Symp* 325–329.

He, H. (2003). What is Service-Oriented Architecture? http://webservices.xml.com/pub/a/ws/2003/09/30/soa.html.

Health Level 7. (2006). HL7 Data Model Development. http://www.hl7.org/library/data-model.

Johnson, P., Tu, S., and Jones, N. (2001a). Achieving reuse of computable guideline systems. *Medinfo* **10**: 99–103.

Johnson, P. D., Tu, S. W., Musen, M. A., and Purves, I. (2001b). A virtual medical record for guideline-based decision support. *Proc AMIA Symp* 294–298.

Karadimas, H. C., Chailloleau, C., Hemery, F., Simonnet, J., and Lepage, E. (2002). Arden/J: An architecture for MLM execution on the Java platform. *J Am Med Inform Assoc* **9**: 359–368.

Kaushal, R., Shojania, K. G., and Bates, D. W. (2003). Effects of computerized physician order entry and clinical decision support systems on medication safety: A systematic review. *Arch Intern Med* **163**: 1409–1416.

Kawamoto, K., Houlihan, C. A., Balas, E. A., and Lobach, D. F. (2005a). Improving clinical practice using clinical decision support systems: A systematic review of trials to identify features critical to success. *BMJ* **330**: 765–768.

Kawamoto, K. and Lobach, D. F. (2005b). Design, implementation, use, and preliminary evaluation of SEBASTIAN, a standards-based Web service for clinical decision support. *Proc AMIA Symp* 380–384.

Krafzig, D., Banke, K., and Slama, D. (2004). *Enterprise SOA: Service-Oriented Architecture Best Practices*. Indianapolis: Prentice Hall PTR.

Lobach, D. F., Gadd, C. S., and Hales, J. W. (1997). Structuring clinical practice guidelines in a relational database model for decision support on the Internet. *Proc AMIA Symp* 158–162.

Miller, J. E., Reichley, R. M., McNamee, L. A., Steib, S. A., and Bailey, T. C. (1999). Notification of real-time clinical alerts generated by pharmacy expert systems. *Proc AMIA Symp* 325–329.

National Library of Medicine. (2006). Unified Medical Language System. http://www.nlm.nih.gov/research/umls.

Overhage, J. M., Mamlin, B., Warvel, J., Warvel, J., Tierney, W., and McDonald, C. J. (1995). A tool for provider interaction during patient care: G-CARE. *Proceedings—The Annual Symposium on Computer Applications in Medical Care* 178–182.

Parker, C. G., Rocha, R. A., Campbell, J. R., Tu, S. W., and Huff, S. M. (2004). Detailed clinical models for sharable, executable guidelines. *Medinfo* **11**: 145–148.

Peleg, M., Boxwala, A. A., Bernstam, E., Tu, S., Greenes, R. A., and Shortliffe, E. H. (2001). Sharable representation of clinical guidelines in GLIF: Relationship to the Arden Syntax. *J Biomed Inform* **34**: 170–181.

Peleg, M., Tu, S., Bury, J., Ciccarese, P., Fox, J., Greenes, R. A. et al. (2003). Comparing computer-interpretable guideline models: A case-study approach. *J Am Med Inform Assoc* **10**: 52–68.

Ram, P., Berg, D., Tu, S., Mansfield, G., Ye, Q., Abarbanel, R., and Beard, N. (2004). Executing clinical practice guidelines using the SAGE execution engine. *Medinfo* **11**: 251–255.

Reichley, R. M., Seaton, T. L., Resetar, E., Micek, S. T., Scott, K. L., Fraser, V. J. et al. (2005). Implementing a commercial rule base as a medication order safety net. *J Am Med Inform Assoc* **12**: 383–389.

Shahar, Y., Miksch, S., and Johnson, P. (1998). The Asgaard project: A task-specific framework for the application and critiquing of time-oriented clinical guidelines. *Artif Intell Med* **14**: 29–51.

Shiffman, R. N., Karras, B. T., Agrawal, A., Chen, R., Marenco, L., and Nath, S. (2000). GEM: A proposal for a more comprehensive guideline document model using XML. *J Am Med Inform Assoc* 7: 488–498.

Siegel, E. L. and Channin, D. S. (2001). Integrating the Healthcare Enterprise: A primer. Part 1. Introduction. *Radiographics* **21**: 1339–1341.

Sutton, D. R. and Fox, J. (2003). The syntax and semantics of the PROforma guideline modeling language. *J Am Med Inform Assoc* **10**: 433–443.

Svatek, V. and Ruzicka, M. (2003). Step-by-step mark-up of medical guideline documents. *Int J Med Inform* **70**: 329–335.

Terenziani, P., Molino, G., and Torchio, M. (2001). A modular approach for representing and executing clinical guidelines. *Artif Intell Med* **23**: 249–276.

Tu, S. W. and Musen, M. A. (1999). A flexible approach to guideline modeling. *Proc AMIA Symp* 420–424.

Tu, S. W., Musen, M. A., Shankar, R., Campbell, J., Hrabak, K., McClay, J. et al. (2004). Modeling guidelines for integration into clinical workflow. *Medinfo* 174–178.

Wang, D., Peleg, M., Tu, S. W., Boxwala, A. A., Ogunyemi, O., Zeng, Q. et al. (2004). Design and implementation of the GLIF3 guideline execution engine. *J Biomed Inform* **37**: 305–318.

VII

THE ROAD AHEAD

24

A PROPOSED STRATEGY FOR OVERCOMING INERTIA

ROBERT A. GREENES

As we reach this final chapter, having explored the barriers to, the requirements of, and the initiatives for encouraging adoption of clinical decision support, one conclusion appears inescapable—that the pace of adoption will not significantly accelerate on its own. Yet the exigencies of the present are manifold and pressing, and we can't afford to continue to abide the slow progress. Therefore I believe that we need a fresh strategy. In this chapter I venture into uncharted territory, by positing that new communally developed and supported mechanisms are required and by elaborating on a proposed strategy to bring this about. My hope is that this will stimulate discussion and action and that the suggestions contained herein will be helpful in accomplishing the goal of broad dissemination and wide use of high-quality CDS.

24.1 EXISTING APPROACHES NOT WORKING

Over the past four and a half decades, the pursuit of CDS has been stimulated by three main kinds of interests, as we have reviewed in Chapter 2: the intellectual challenge of understanding and improving on the cognitive processes of the human; the desire to address important issues in patient safety, health care quality, and access to health care; and business and policy reasons relating to allocation of limited resources and control of costs of an increasingly expensive health care system. Those approaches to CDS based on motivations of error prevention and quality improvement have tended to be carried out largely in academic settings and have been ad hoc, as we discussed in Section II. Business and policy reasons for implementation of CDS, like those for promotion of computer-based clinical systems and the EHR, have frequently been tied to changes in health care financing and reimbursement models, efforts to shift care from hospitals to office or home, introduction of managed care, and approaches to curbing overutilization by requiring preapproval/prior authorizations for high-cost procedures, referrals, or medications. CDS has been introduced in those situations as a means of addressing government or payer regulations and restrictions or as a defensive measure by health care organizations and providers to ward off such intrusions. As a result, business-oriented uses of CDS have tended to be implemented in institution-specific

Clinical Decision Support: The Road Ahead

fashion. In fact, it has been argued by some that CDS—in such forms as logic rules, order sets, and documentation templates—is so dependent on local needs, constraints, and preferences that there is little benefit to sharing, given the need for local adaptation and customization.

We shall return to this point later. In any case, technical advances that we have reviewed in previous chapters, including computer technologies and systems architectures, and development of some of the important standards needed for data and knowledge representation and communication, as well as increased understanding of organizational strategies to encourage CDS use, should make the process easier. Nonetheless, these developments have not translated into significant initiatives to accelerate the pace of adoption, perhaps because of perceptions of the high degree of dependence on local adaptation and customization noted earlier.

24.2 NEED FOR NEW MECHANISMS

I believe that at least part of the inertia is due to the fact that, in order to systematically address the problem of accelerating adoption on a broad scale, there is a need for considerable infrastructure, tools, and resources that is both daunting to individual efforts and requires concerted action that has not yet become organized. I also believe, and will argue later, that a considerable body of shared knowledge content is needed in addition to or as a basis for local adaptation and customization. Addressing the inertia thus requires that a much more coordinated, communal approach be developed to overcome barriers, align motivations, obtain support, and establish the mechanisms that will be needed.

Specifically, to move ahead, I believe that a set of three activities should be pursued, discussed in more detail in the three Sections 24.6, 24.7, and 24.8.

24.2.1 Identifying Key Societal Drivers and Setting Priorities

Ideally the impetus for a concerted effort needs to derive from its alignment with the most important societal drivers for CDS. In Chapter 2, we described many factors that have contributed to the growing recognition of the need for CDS. In this chapter, we focus on five of the most important current drivers. Aligning with societal drivers can help to garner political and financial commitment to the effort, as opposed to a piecemeal strategy that has no real constituency. The proposed strategy is aimed at prioritizing these drivers, and selecting as an initial focus the forms of CDS that are most feasible to deliver in order to address those priorities. In the words of Alexandre Dumas, "Nothing succeeds like success." By successfully addressing a limited but important target and then moving on systematically to other targets, rather than expending huge efforts on assaulting the entire range of possible foci, it will be much easier to gain and sustain support, commitment, and enthusiasm from various stakeholders crucial to achieving the long-range goal.

The choosing of priorities should most naturally be organized and orchestrated at a national level, with participation of relevant stakeholders, as we will discuss further, although the process may be modified or influenced by local or regional considerations. In some cases priorities may also align with international interests. Thus coordination of the process of selecting priorities will need to be cognizant of local/regional constraints and international collaboration opportunities.

24.2.2 Formalizing the Three Life Cycle Processes

In addition to collectively establishing priorities, it is necessary to put in place communal mechanisms and infrastructure for formalizing the three interrelated lifecycle processes introduced in Chapter 1 for 1) knowledge generation (KG), that is, creating and validating the knowledge required for CDS; 2) knowledge management (KM); and 3) CDS implementation and evaluation (IE). In most development efforts to date, these processes generally have been carried out informally, if at all. This may be satisfactory as long as the amount and variety of kinds of knowledge being managed are relatively small, and the kinds of uses of the knowledge reasonably controlled, within an institutional or vendor-based system environment. But, as pointed out in Chapters 21 and 22 and we will further discuss, any sizeable effort to implement, deliver, and maintain CDS becomes overwhelmed if it does not address the tasks of the three life cycle processes more systematically. The fundamental need I will argue for is for a set of infrastructure, resources, and tools, *external* to proprietary and one-of-a-kind implementations that support these life cycle processes.

I believe that this activity needs to be carried out and supported on a broad scale. Permanent operational entities made up of the necessary skill sets to carry out these tasks need to be established and adequately funded. Their work must be transparent, independent of any vested interests, balanced, respected as being of the highest quality, authoritative, and broadly accessible. These permanent entities should also likely be at a national level, but many of the formal methods and activities could benefit from international cooperation and collaboration. Therefore, as various nations move ahead with this approach, at differing paces, opportunities should be made available for them to join in and leverage specific efforts already established and then to contribute to those efforts.

The set of activities that must be carried out to support the three life cycle processes includes a number of tasks that are costly and time-consuming and require large-scale effort. This is discussed in more detail later. The data and knowledge bases, tools, and infrastructure to be developed should be aimed at both making the work of the responsible community-scale entities manageable, and also providing resources for the constituency that will need to deploy and manage CDS in their own sites, and for businesses that will offer or extend those capabilities in their products. Entities addressing rudiments of some of these processes already exist, and could be built upon. The responsible entities for each of the three life cycle processes need to identify and advocate for their funding requirements, champion the adoption of particular standards they need, and coordinate the participation of many individuals and groups.

24.2.3 Getting Specific: End-to-End Implementation Starting with High-Priority Focus Area(s)

The final element of the proposed strategy is to validate the approach by doing an end-to-end implementation aimed at widespread adoption of CDS but limited to one or more selected high priority areas, and then to use that experience as a basis for tackling further areas. It is essential that the steps taken be deliberate, well thought out, and iterative. By focusing on high priorities for CDS initially, the communal infrastructure, resources, and tools required can be developed to support these, and deployment of CDS can occur relatively quickly. Assuming these efforts are successful, the positive results can be expected to further solidify support and commitment for continuing and refining the approach for those forms of CDS, and also to provide a model that can be adapted and built upon for addressing others.

Another iterative aspect relates to the impact this approach could have on stimulating further development and refinement of needed standards. Successful deployment using even an imperfect formalization if a standard does not yet exist could provide the impetus for its adoption as a standard or its refinement. Any changes could then be reflected in updates to deployed systems, which the infrastructure would facilitate. In addition, the approach is almost certain to alter the commercial landscape, by providing new business opportunities for knowledge content, software, and services, as well as affecting the way in which integrated clinical information system solutions are positioned and sold. Ideally, the communal activities should not be in competition with the marketplace but should stimulate it, both in terms of priming the demand for the capabilities, and also supporting the entry of vendors into the marketplace. Last, the experience gained will also provide important lessons and input as the strategy is further refined and applied to additional opportunities for delivery of CDS.

This process should again likely be overseen and coordinated on a national basis. First steps might well occur in a pilot project of limited scope, perhaps regionally based or among selected participants, and would expand as the methods are refined. Such a process would provide important experience with the approaches needed to formalize the life cycle processes, and the opportunity to see how such communal initiatives are accepted, how readily they are adopted, the effect they have on the marketplace, and their ability to become relatively self-sustaining. Subsection 24.3 describes some of the considerations involved in adopting two possible candidates for initial focus, to illustrate how this might occur.

24.3 RATIONALE FOR COMMUNAL INFRASTRUCTURE, RESOURCES, AND TOOLS

A key element of the preceding strategy outline is the posited requirement for the creation of communal standards-based shared infrastructure, resources, and tools external to local or proprietary systems. This is no doubt the most speculative part of the proposal. Many would argue that the history of development of such shared capabilities has not been encouraging. Why would this now work? How would it come about? What entities would support it? How would such capabilities relate to commercial and local capabilities? As we also noted earlier, some would also maintain that the requirement for customization and local adaptation is so great that there is little benefit to be gained from a communal initiative. Reasons why I have come to the conclusion that such an effort is needed and feasible are the following.

24.3.1 The Core of Biomedical Knowledge Is Not Subject to Local Interpretation

Although knowledge must, of course, be interpreted in terms of local population and patient characteristics, and its application adapted to local resources, constraints, and clinical and business practices, nonetheless the underlying relationships of findings to outcomes, and effectiveness of procedures for diagnosis and treatment, or for estimation of prognosis, are independent of these factors. We should all be starting from a common base of well-researched, evidence-based, authoritative principles. Where does that exist now? How does one go about assembling pertinent knowledge (both positive and negative), organizing it, representing it in a formal manner, ideally standards-based, and making it available in order to use it or build on it? How does one update it?

The biomedical literature is clearly one source of such knowledge, in the form of the MEDLINE database made available worldwide via the PubMed service of the U.S. National Center for Biotechnology Information (NCBI). Besides the MEDLINE database, NCBI also maintains and provides access through PubMed to a wide variety of genetic and molecular data bases (Wheeler, Barrett et al. 2006), as do many other agencies and public and private institutions and consortia. There is a growing movement to improve access to the published literature (e.g., through the open access policy of the U.S. National Institutes of Health, which makes full text articles available rapidly through NCBI's PubMed Central). A clinical trials registry is maintained by the U.S. government via a Web site, http://www.guidelines.gov http://McCray 2000), and a recent editorial jointly published by an international group of the leading medical journals called for mandatory registration of clinical trials in a registry such as this that meets specific criteria, as a condition for publication (DeAngelis, Drazen et al. 2004).

Although these repositories of literature, research results, and clinical trials often have not included access to primary research data and details of analytic findings, particularly in the clinical domain, that added step is being taken by a number of consortia and other activities aimed at promoting sharing of research data and analyses. With respect to clinical knowledge in particular, "Evidence-based Practice Centers" (EPCs) in the United States and Canada, commissioned since 1997 by the U.S. Agency for Healthcare Research and Quality (AHRQ) (see http://www.ahrq.gov/clinic/epc/), carry out work to assemble knowledge on particular topics and produce evidence-based medicine reports. Similar efforts are carried out on a broader international scale by the well-known Cochrane Collaboration (Herxheimer 1993), which publishes quarterly the Cochrane Database of Systematic Reviews, as part of the Cochrane Library. These are early steps, and clearly much needs to be worked out regarding the proper treatment of intellectual property, but the paradigm of communal access and review is already established, and momentum for it is growing.

24.3.2 Useful Clinical Knowledge Is Too Vast to Be Rediscovered and Assimilated by Local or Proprietary Efforts Alone

Already onerous and contributing to daily unease by practitioners as well as the public, the need to deal with a barrage of information confronting us on a daily basis is poised to dramatically expand. We all need help in filtering this information, organizing it, and ranking it. Just considering peer-reviewed research alone, much of this stored knowledge is not relevant to health care. Other knowledge may not be reliable or ready for application. Still other knowledge is not organized in a way that it is useful.

As one example of an important category of clinical knowledge, consider the continued growth in the arsenal of diagnostic and therapeutic options. The prospect of overuse of expensive alternatives has resulted in efforts by payers to monitor and to curb utilization through a variety of approaches, one of which involves use of guidelines and processes for obtaining preapproval of expensive procedures, referrals, and medication orders. Although such efforts are regarded by physicians as intrusive, some sort of rational basis for choosing among the myriad available health care resources is clearly needed, and the role of CDS can be expected to be essential. Even if we assume that it is feasible to cope with the current array of options—a position that is not supported by the variability in use of proedures and in patient outcomes—consider the impact of needing to understand and appropriately utilize all the new biomarkers and genomic tests for risk prediction, disease detection, and choice of therapy that are coming on the scene.

We will soon have a situation much more confounding and potentially detrimental (as well as costly) than the dilemma when multichannel laboratory analyzers were first introduced (e.g., the Technicon SMA12® and SMAC20®, in 1966 and 1975, respectively) for performing multiple tests on a single blood serum sample. At that time, the concern was expressed (Weinstein and Pearlman 1981) about the probability, even in a completely normal individual, of having at least one false positive on a panel—which on a SMAC20, assuming each test has only a 5 percent false positive rate, is as high as 0.64 (that is, $1-(0.95)^{20}$), or slightly less than two out of three patients. Once a positive test is reported, the clinician is obligated to do further testing to verify or rule out the disease(s) suggested by the test.

Consider the implications in our current genomic (or "post-genomic" as some call it) era, when one can run a microarray analysis of a specimen providing 250 (or 250,000) results. As Kohane et al. (2006) point out, even with much more stringent criteria for positivity (e.g., false positive rates of only 1% or 0.1%), the likelihood of false positives is huge. How then should we use these tests wisely? Further, how should we interpret their results? And what should we do about the results, in terms of next steps? Clearly, CDS is needed to provide guidance about the ordering of tests, the interpreting of results, and the choice of therapy in these settings. Nonetheless, it can be argued that the difficulties in assimilating such knowledge in practice has already contributed (along with many other factors) to the manifest unevenness in the quality of health care.

Initiatives to assemble information such as that relating to problems like the preceding, distill knowledge from it and manage such knowledge require huge, ongoing, expensive efforts and could not and should not be undertaken separately by individual institutions or even moderate-sized organizational entities. It may be possible to introduce a manageably small set of CDS rules in local or proprietary fashion, and, as we have noted, adaptation of more generic knowledge sources in specific settings will continue to be needed. Yet the potential for CDS already far outstrips the ability of individual entities, even moderately large ones, to assemble and manage all the needed underlying knowledge for providing it, and to do regular updates of it.

24.3.3 The Lack of Collective Effort Inhibits the Growth of and Push for Shared Resources

This becomes a self-fulfilling prophecy. Without a set of communal infrastructure resources, if the pace for CDS adoption were to accelerate in response to the growing pressure for it, we would be unprepared and likely overwhelmed. CDS remains a cottage industry. Many of the same mistakes of the past 40+ years get repeated. I believe that much of the demand for CDS capabilities is latent, and can be expected to be unleashed once mechanisms are available to implement and support CDS robustly. In a sense, although on a much smaller scale, this may be analogous to what has happened with respect to the advent of the World Wide Web in many areas of our lives. Its existence has enabled people to perform queries that they would not even have thought of or considered doing before it came into being. It in fact has fostered a mindset that has now generated its own demand. We see this in health care with the ease with which we can do a Google® search for the answer to a question, which has prompted both physicians and patients to instantly turn to it when questions arise. Imagine the stimulus that a well-researched, evidence-based repository of knowledge, in standardized, computable form, and tools for delivery of it in local settings in a patient-specific manner at the time of need would have on the ability to implement and demand for CDS capabilities.

To achieve this robust means for providing CDS, collective efforts are needed for the establishment of infrastructure and methods to ensure that appropriate topics are studied to generate needed knowledge, to manage the knowledge, to provide tools for its dissemination, and to create a means for sharing of experiences with and methods for delivery of CDS. As we have seen in the first two chapters of Section VI, experiences with knowledge management are uncovering how large a task this is. When it comes to implementation, methods and approaches to delivery of CDS should build on tested methods that have been shown to be effective. We should not have to continue to rediscover the wheel, making many of the same errors over again. Although there is a growing call for CDS, it is not so easy to step up the process of implementing and supporting it. We have discussed in Chapters 1 and 8 how just accelerating the process of implementing advanced capabilities such as CPOE have backfired in several high-profile installations. Adding CDS to them or implementing it elsewhere in clinical applications without appropriate planning can lead to further negative experiences that set progress back. We need to lower the cost of implementing successful methods and approaches, and to build on the experiences, positive and negative, of those who have done this before. We also need to reduce the cycle time of experimenting, modifying the approach based on feedback, and replicating successful approaches. New architectures for modular delivery of CDS relying on invocation through application programming interfaces (APIs), messaging, or Web service calls (e.g., see Chapter 23) promise to facilitate this process.

24.4 ORGANIZATION OF PROCESS

To summarize the preceding, I believe that for the deployment of CDS to progress at other than the glacial speed that has occurred to date, the communities of interest—the stakeholders invested in delivering safe, high-quality, cost-effective care—need to proactively organize themselves to provide a guiding role in the evolution of CDS capabilities and the knowledge resources that they require. Robust external mechanisms such as we have briefly described already for formalizing the three life cycle processes will not come about quickly on their own.

Accelerating progress thus depends on organization and guidance by an oversight body (OB) that is in a position to influence how health care is organized and delivered, and how it is paid for.

Key responsibilities of the OB would be:

- To determine priorities for communal efforts to facilitate CDS adoption
- To establish and oversee permanent entities to carry out the formalization of infrastructure, resources, and tools to support the three life cycles
- To oversee the implementation of end-to-end processes to facilitate adoption of CDS for the selected priorities and their subsequent refinement and iteration in expanded or additional areas

The OB should include representatives of the health care professions, health services research, economics, and policy experts, payers, and the public. A reasonable way to accomplish this would be for the OB to be comprised of high-visibility, respected, and knowledgeable individuals representing these stakeholder categories. As noted earlier, such an OB would most naturally function at a national level, so as to be responsive to the overall needs of the country and to be able to garner the necessary support to carry out the work, but there may be related efforts that could occur on regional levels, or also internationally. Ideally such efforts should dovetail with and leverage the work of national OBs.

The permanent entities responsible for carrying out the communal development of infrastructure, resources, and tools to formalize the three life cycle processes would report to the OB. The OB would be responsible for ensuring that their composition is appropriate, that their functions are transparent and of high quality, that access to their products and services are broadly accessible, and that they are adequately funded. It would also oversee their interrelationship and coordination.

For the process of refinement of the overall strategy through iterative cycles of end-to-end implementation, one mechanism the OB could adopt would be to initiate and/or fund projects by institutions or consortia that would serve as appropriate test beds. It would probably be best for these projects to be of limited duration. If they are successful, they will provide feedback for improvement of the permanent infrastructure, resources, and tools available to all. However, it may be necessary to provide additional funds aimed specifically at technology transfer, in order to get successful projects to the point where they are self-sustaining at their local or consortial sites, and for refining the process of adoption of the approach at other sites. Ultimately the goal will be for further replication and adoption of established approaches to be supported through the commercial marketplace. This might also need to be stimulated through a series of small grants to business.

24.5 OVERVIEW OF STRATEGY

The strategy proposed is an iterative one. A key challenge is to get it started. The interplay of the three activities described earlier is depicted in Figure 24-1. In the rest of this chapter, we consider in more detail the problems of selection of priorities, supporting the life cycle process, and carrying out first steps. I believe that if we can set these in motion, we would be solidly on "The Road Ahead."

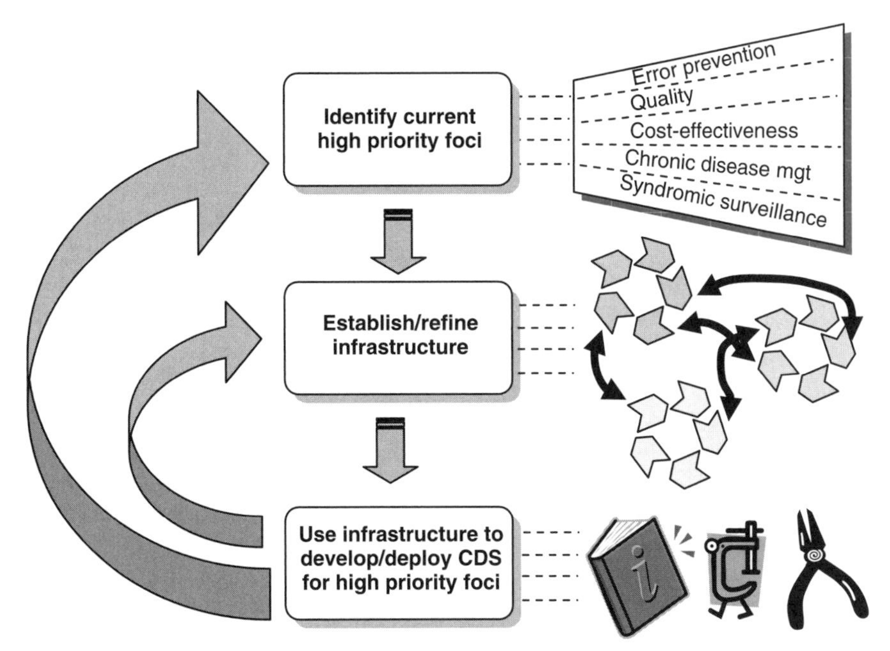

FIGURE 24-1 Priorities for CDS are likely to fall into five main areas (top). The three interrelated life cycle processes involved in generation of knowledge, knowledge management, and incorporation into functional CDS require infrastructure for supporting them (middle). The result of applying these life cycle processes to the priority areas will be knowledge bases and authoring and implementation tools (lower). The whole process iterates as we learn more about how to create infrastructure to support it, and as priorities change.

This discussion is intended to be complementary to a June 2006 white paper produced on contract from the U.S. Office of the National Coordinator for Health Information Technology to the American Medical Informatics Association (Osheroff, Teich et al. 2006), which outlines a proposed *Roadmap for National Action on Clinical Decision Support for the U.S.* In that white paper, a number of steps are proposed to create an environment conducive to the general goal we address. In this book, however, we take a somewhat broader view than those specific steps, tied as they are to the social and political environment and agenda of the United States at the time of the report.

24.6 IDENTIFYING KEY SOCIETAL DRIVERS AND SETTING PRIORITIES

In Section I of this book, we reviewed the vast array of possible ways to deliver CDS, ranging from simple to complex. They relate to a broad range of problem-solving and decision-making tasks of health care providers and patients that have been explored over the past 45 years or more. They can be used to augment the functionality of a variety of clinical information technology applications, and are most effective if integrated with those applications, able to use the EHR to obtain patient data, and able to offer patient-specific advice.

We now consider five drivers that represent concerns that appear to be of highest priority at the current time. In this section, we will examine the ways in which CDS can help to address them. Four of the drivers are selected from among the many influences on the growing recognition of importance of CDS cited in Chapter 2: the goals of 1) decreasing errors and improving patient safety, 2) improving quality through adoption of best practices, 3) increasing cost-effectiveness, and 4) optimizing the management of chronic disease. We also consider another driver not discussed in Chapter 2, that of 5) supporting public health initiatives.

The rationales for broad endorsement of the first three drivers were presented in Chapter 2 and won't be reviewed again here. The fourth driver, optimizing chronic disease management, is also motivated in large part by the same rationales. But its particular importance is by virtue of the extent of its impact, which has by all projections not yet been fully realized. As noted in Chapter 2, it is estimated that almost 80 percent of U.S. health care expenditures are already devoted to chronic disease (e.g., see (Ray, Collin et al. 2000)). Other developed nations are also facing this phenomenon, although with differing rates of progression. The fifth driver, arising from the public health sphere, is included because of its overall importance in terms of need for CDS, even though its impact on stimulating the delivery of CDS capabilities will be more indirect and long-term. Biohazards, manmade and natural, have resulted in a significant wake-up call for the need for public health systems for two kinds of capabilities: early detection of problems in the form of syndromic surveillance, and disaster response in the form of rapid mobilization and coordination of health care resources for the care of victims.

These drivers represent major contemporary social, political, and economic concerns regarding the health care systems in developed nations. In addition, as we have noted, these drivers are most likely to have the backing of key stakeholders and thus be able to marshal the financial resources needed for their pursuit. For each of the five drivers, we indicate the most effective roles for CDS, many of which have been evaluated by studies such as those reported in the first three chapters of Section II.

24.6.1 Error Prevention/Patient Safety

How CDS can help: The interventions that appear to work best are those aimed at ensuring that drug doses are within appropriate limits, checking for harmful drug–laboratory and drug–drug interactions, and identifying allergies that may be contraindications. By and large, the most successful interventions, in terms of impact, have been those that relate to medication ordering, usually through direct interaction via a CPOE application.

Another class of interventions involves generation of alerts notifying providers about abnormal laboratory results. In some cases the alerts are modified or qualified by logic concerning certain medications that the patient is receiving, or the relation of the date and time of the result to that of a previous abnormal result for the same test, or the interpretation of the test in the light of results of other tests or related clinical problems.

A final set of approaches that has been shown to be useful involves detecting possible adverse events (AEs), through monitoring of selected indicators such as laboratory results, emergency department visits, hospitalizations, and lengths of stay greater than the durations expected based on admitting diagnoses. This capability is not intended to provide direct CDS, but is valuable as a means for judging effectiveness of CDS, by determining baseline AE rates and comparing them with rates after various CDS interventions have been introduced.

24.6.2 Quality/Best Practices

How CDS can help: The most direct way to focus on quality is to adopt methods that help to ensure that identified target goals are met. One example is a variant of CPOE drug dose checking described in the previous section, in which dose modifications are suggested in specific circumstances, as in the elderly patient or the patient with renal failure.

Another major approach in this category is the use of reminders and suggestions to encourage optimal actions (e.g., to order HbA1c tests and do periodic ophthalmology, peripheral vascular, and podiatry referrals in diabetic patients; to perform approved screening examinations such as mammography or colonoscopy at appropriate intervals for patients who qualify; or to do immunizations at recommended intervals for patients at risk). CDS of this type may be of increasing importance in the near future, particularly in the United States, given the growing movement in the direction of utilizing quality measures as a basis for pay for performance (P4P)-based health care reimbursement. Approaches based on P4P for aligning incentives to achieve best practices, of course, can be misused if they rely on simplistic rules that require uncritical routine compliance with guidelines even if not appropriate in some circumstances (Boyd, Darer et al. 2005). Nonetheless it can be expected that some forms of P4P will be implemented broadly, as a result of which there will be considerable need for CDS to support physicians in achieving optimal compliance.

Order sets represent one of those rare classes of CDS interventions that not only provide useful advice but save time. Order sets are now widely used as a way of encouraging problem-specific and setting-specific best practices; at the same time, their use is noteworthy in that they provide efficiency when adopted, by obviating the need for time-consuming entry of individual orders in clinical settings in which sets of orders can be anticipated. The bundling of orders also provides a reminder function and has education value, especially to new physicians. This is an example of the CDS methodology of using knowledge to create groupings of items associated by a common theme or purpose; other approaches to applying this approach to CDS in creating structured data entry and reporting applications are less far along in demonstrating acceptance and effectiveness.

Clinical practice guidelines are another approach to encouraging best practices. However, the problem with guidelines, as discussed in detail in Chapter 13, has largely been that it is difficult to interface them directly with clinical systems, because of the many points in clinical processes in which they can interact. Guidelines as a whole are thus rarely implemented in executable form, but the logic in various decision steps nonetheless does find its way into CPOE recommendations, alerts, and reminders that are incorporated in various applications, in the forms discussed earlier. Also, through infobutton approaches, guidelines in human-readable form can be expected to be made available in patient-specific, context-specific form within a variety of clinical applications.

A subset of guideline knowledge is logic that helps to interpret the meaning of abnormal laboratory tests, particularly more esoteric tests or those for which in-depth interpretation involves considering many other clinical and laboratory findings. Infobuttons could be helpful here as well.

24.6.3 Cost-Effectiveness

How CDS can help: The key interventions for cost-effectiveness appear to be in the form of recommendations for use of specific medications (e.g., generics) within a class when alternatives are available, and automation of preapproval/prior authorization for specific clinical actions.

Displaying the charge for a test procedure has been equivocal in its effectiveness, with work at Brigham and Women's Hospital reporting limited effect (see Chapter 5) and that at Regenstrief Institute (see Chapter 4) reporting greater effect. Doing this for both tests and medications is an easy intervention that can be readily provided, so its use may gain some traction, but it is not expected to be among the highest priority approaches.

Preapproval/prior authorization represents one of the more intrusive kinds of CDS, largely imposed by payers, but can be expected to increase in use. For many years, health care insurers and payers in the United States have required preapproval for certain high-cost surgical procedures or specialty referrals. More recently, high cost imaging procedures have become a particular target for preapproval, since such procedures represent a growing proportion of health care expenditures. There is now gathering momentum in the United States also toward introducing a prior authorization process for prescribing of high cost medications. None of these preapproval and prior authorization applications have been broadly automated to date, although implementation of logic rules in the context of particular systems or payment plans have occasionally been done. Appropriateness criteria for radiological procedures have been published for many years by the American College of Radiology (ACR 2006), and at least two projects have aimed to make these available in computer-interpretable format (Kahn, Pingree et al. 1998; ACR 2000; Sistrom and Honeyman 2002).

Some academic medical centers have implemented programs for automating feedback about appropriateness of indications for imaging procedure orders as part of their CPOE process. For example, a commercial product is being used at Brigham and Women's Hospital to implement automated evaluation of insurer-endorsed criteria as a basis for imaging procedure authorization during CPOE in lieu of requiring telephone authorization by the insurer. Regarding prior authorization for pharmaceutical prescribing, first steps are being taken in the United States by the federal government in conjunction with the pharmaceutical industry by formalizing a Structured Product Label (SPL) containing parameters that could be used in processing of decision logic regarding appropriateness of a medication for specific indications. In early work underway by participants of the National Council of Prescription Drug Programs (NCPDP) and other collaborators, the use of the HL7-endorsed standard expression language GELLO (Sordo, Boxwala et al. 2004) is being evaluated as the formalism for creating a repository of logic rules based on the SPL (Sordo, Dunlop et al. 2006). The intent is that such a shared collection of rules would serve as a common starting point for implementation by payers of prior authorization programs and services, and user interfaces to CPOE and eRx applications that provide access to them by payers, pharmacy benefit management systems, and clinical information systems. In view of these activities, we can expect to see such uses of CDS increasingly being deployed.

24.6.4 Chronic Disease Management

How CDS can help: Because of the huge proportion of encounters of elderly patients with the health care system, it is advantageous in terms of sheer numbers to deploy CDS interventions that can achieve the three kinds of benefits cited earlier (promoting error prevention, quality, and cost-effectiveness), specifically with respect to the management of chronic disease. In addition, efforts to optimize care, in terms of recommended pathways, prevention, and home-based care rather than hospitalization may all play important roles (Bodenheimer 2003; Seroussi, Bouaud et al. 2004; Dorr, Wilcox et al. 2006). Improvement in data collection on patients with chronic disease may also lead to more refined management of such patients based on prediction from databases of similar patients.

24.6.5 Public Health Initiatives

How CDS can help: As noted earlier, this includes both syndromic surveillance and disaster response. In syndromic surveillance, the needs for CDS are less oriented toward the individual provider or patient, but rather are aimed at identifying suspicious syndromes in patients having encounters with the health care system, for reporting to public health agencies. In addition to employing various modeling approaches (heuristic/rule-based, or pattern recognition/data-based) for identifying candidate syndromes in individual patients, syndromic surveillance relies also on the compilation, refinement, and updating of epidemiologic databases, and the use of Geographic Information Systems (GISs) to analyze the data to detect temporal and spatial clustering of syndromes (Mandl, Overhage et al. 2004; Wagner, Espino et al. 2004). These latter analyses are beyond the scope of this book, but the success of such public health systems depends first and foremost on the detection and reporting of suspicious findings in individuals. Thus, environmental and biohazard concerns are becoming major reasons for seeking ways to identify key findings in individuals, and are driving both more complete capture of structured clinical data, increased access to specified data elements in the EHR through interoperable data exchange, and event-triggering of evaluation mechanisms (e.g., as a result of an emergency department visit or an elevated WBC) (Bourgeois, Olson et al. 2006). Other forms of CDS requiring such data access capabilities and event architectures can be expected to benefit from this stimulus.

Disaster response is aimed at individual patients but in the context of ad hoc or makeshift settings and with a need for coordination of logistics for large numbers of patients. The need for CDS focuses mostly on monitoring and tracking the status of victims, so that problems can be detected rapidly, appropriate health care personnel alerted, and needed equipment located. Portable wireless methods for monitoring pulse, pO_2, and ECG of patients, and technologies for sensing patient, provider, and equipment locations are being developed and tested (Chan, Killeen et al. 2004; Buono, Chan et al. 2005; Pino, Ohno-Machado et al. 2005; Waterman, Curtis et al. 2005). Beyond these applications, other CDS uses are expected eventually to be for wireless telemetry of patients in hospitals, as well as patients at risk for various conditions while at home, at work, or on the road. These applications are somewhat long-range, but disaster management concerns can be expected to stimulate their development.

Where are we now?

Commercial adoption of CDS to date has mostly been in the form of drug dose safety and interaction checks, the provision of alert and reminder mechanisms, and the use of order sets (see Chapter 7). These have largely depended on the roll out of EHRs and CPOE, but some of the drug dose and interaction check capabilities can also be expected to find their way into e-prescribing (eRx). Personal health records (PHRs) can make use of reminders and alerts to notify patients about the need for appointments or procedures, or provide advice about suggested educational materials.

Access to context-specific knowledge is becoming simpler to automate through the use of infobuttons (see Chapter 16). A standard for an infobutton manager currently being considered by HL7 is likely to accelerate use of the approach, because the standard will make it easier to interface to commercial and noncommercial clinical systems and encourage information provider vendors and public knowledge resource providers to make their information accessible in compatible form. This particular form of CDS bears on all these categories, by providing access to pertinent knowledge.

I don't believe that other approaches to CDS, although shown to be effective in some circumstances, are as likely to be considered to be of highest priority. We have already mentioned the difficulty in deploying clinical practice guidelines. Differential diagnosis, although readily available, is not widely used, owing largely to the difficulties in interfacing it to clinical applications and to patient data sources. As a consequence, it requires excessive manual data entry or is not sufficiently patient/context-specific to be useful except in occasional highly perplexing clinical circumstances. The use of documentation templates to improve data collection, and thereby provide a base for enhanced CDS, continues to be an important goal, but it is likely to continue to be addressed only incrementally, because its payoff is less direct and it demands increased effort and time by users.

We summarize the likely highest priority candidates for CDS, based on this discussion, in Table 24-1. It should be recognized, of course, that this list of priorities is arbitrary. An actual set of priorities adopted in a particular nation should ideally come out of a consensus process such as described in the introduction to this chapter, convened by an organizing body and representing appropriate stakeholders with both authority, knowledge, and the means for causing recommended priorities to be implemented, as well as for obtaining the necessary financial support for doing so.

24.7 FORMALIZING THE THREE LIFE CYCLE PROCESSES

A primary reason for formalizing the three life cycle processes underlying CDS development and dissemination is that such formalization could make it possible to create certain capabilities that would otherwise be impractical or less likely to come about through individual efforts. For example, formalization could identify repetitive tasks that might benefit from development

TABLE 24-1 Drivers for and likely high priority types of CDS.

Driver	Types of CDS	Priority/importance
1. Error prevention/ patient safety	• CPOE drug checks: dose, interactions, allergies	• High
	• Alerts for abnormal lab results	• High
	• Adverse event monitoring	• More important as background method for assessing effectiveness of preceding interventions
2. Quality/best practices	• Dose modifications in specific circumstances	• High
	• Reminders and suggestions, for screening, immunization, timed follow-ups	• High
	• Order sets	• High
	• Clinical guidelines	• Limited direct use • Mostly deployed by decomposition into the preceding types of interventions
3. Cost-effectiveness	• Medication substitutions	• High
	• Medication or procedure charge display	• Equivocal, but easy to do
	• Preapproval/prior authorization for procedures, specialty referrals, medications	• High
4. Chronic disease management	• Primarily approaches in the preceding categories, but aimed specifically at chronic diseases	• High
	• Other approaches aimed at home care, self-management, and disease prevention	• Probably effective, but less well-defined
5. Public health initiatives	• Event-triggered logic for identification of reportable conditions, e.g., presenting complaint or ICD9 code-based, or abnormal findings	• Primary impact on CDS likely to be indirect, in terms of increased emphasis on structured data capture, standard encoding and access to data, and provision of event trigger capabilities in systems
	• Personal monitoring and location tracking in disaster settings, to identify patients at risk and to notify providers	• Primary impact ad hoc and hopefully rare, but vital; other longer-term impact on development of wireless monitoring-based alerts for patients in nonhospital settings

of tools to make those tasks easier. Also, it could foster the collection, organization, and management of knowledge content of broad usefulness. In addition, it could facilitate interoperability and exchange of knowledge content and use of tools that operate on it, by advocating and promoting adoption of needed standards. Finally, the effort at formalization and standardization could help move activities aimed at dissemination and adoption into a communal, shared initiative that has more of an ability than individual efforts to attract a critical mass, to pool resources, and to gain economies of scale.

We consider the three life cycle processes of knowledge generation, knowledge management, and CDS implementation and evaluation in more detail. As we pointed out in Chapter 8, much of the CDS of the types identified in Table 24-1 arose as the result of efforts by investigators in academic medical centers, interested in establishing whether a particular approach to CDS was effective. Because of that focus in such studies, the investigators generally devoted little thought to the long-term need to maintain the capabilities being implemented, to update them as needs change, or to formalize them so they could be replicated and utilized elsewhere. When approaches have been adopted or implemented by vendors and incorporated in commercial systems, this has usually been done via proprietary implementations. The knowledge bases involved also usually have been maintained on a proprietary basis, with little external sharing, even when based on a standard formalism such as Arden Syntax (see Chapter 12). Thus each vendor, user group, or local site generally has had to start from scratch in identifying appropriate knowledge, determining how to represent it, and developing and testing a means for using it to provide CDS in a clinical IT application.

We seek to establish mechanisms *external* to proprietary and one-of-a-kind implementations to facilitate the collection and management of knowledge, and its application in a variety of different host environments and application settings. For each of these life cycles a permanent entity needs to oversee its processes, coordinate with related efforts, maintain repositories of needed information, lead or push for the development of required standards, and create the infrastructure to support collaboration, dissemination, and use of CDS.

As we discuss the three life cycles, it should become apparent that the groundwork for some of these efforts already exists. Only modest initial funding may be needed to get them started. But the need for their existence should be broadly recognized and subscribed to in order for the kind of change we want to actually come about.

24.7.1 Knowledge Generation (KG) Life Cycle

Regarding this first life cycle, authoritative knowledge bases are essential as a foundation for coping with the knowledge explosion and establishing a robust mechanism for CDS. As we noted in the beginning of this chapter, the prospect of biomarkers and genomic testing for diagnostic, therapeutic, and prognostic decision-making raises the stakes considerably higher. The decision of when it is appropriate to apply research findings to practice is a complex process. For new diagnostic tests or treatments, for example, this involves careful analyses of not only their efficacy but other implications in terms of side effects, costs, risks, and relative benefit in comparison to other alternatives. Such knowledge is often derived by analysis of the published literature, particularly by meta-analysis of clinical trials.

The assembly of such knowledge will be eased by databases, clinical trial registries, increased access to clinical trial data and results, and full open access to the public literature—encouraging trends we cited earlier in this chapter. We also noted organized efforts to create authoritative repositories of research knowledge, through government-funded initiatives like the U.S.-sponsored EPCs and the international efforts of the Cochrane Collaboration. These efforts are based on the principle that the sheer burden of doing all the evidence-based research needed to keep such knowledge bases current requires that the research be done by knowledgeable, well-staffed, sufficiently funded investigative groups, so that the results can be relied upon by clinicians and patients.

These efforts can serve as models or initial forms for what I will call Centers for Knowledge Generation (CKGs). Indeed the mission of a CKG could be based on an expansion of that of an EPC. Like EPCs, CKGs would oversee the assembly and analysis of existing knowledge. CKGs would focus on particular domains or problems in health care (e.g., cardiac care, or infectious disease, or on particular kinds of studies such as therapeutic trials). CKGs would also be in a position to function as advocates for needed research, since they could identify gaps in knowledge, and work with funding agencies to encourage funding to address those gaps. They would systematically identify reports of findings pertaining to their domains or problem foci, both positive and negative, and carry out assessments aimed at determining the quality, robustness, and specific features of the studies that would allow them to be compared with other studies. An example of the task of a CKG would be that of analyzing putative genetic markers and tests to determine optimal conditions for their use and interpretations of their results. Researchers would be encouraged to submit their results to the CKGs, perhaps even making it a precondition for publication, much as is already being required for registration of clinical trials by biomedical journals (DeAngelis, Drazen et al. 2004), as noted earlier.

The function of a CKG would extend beyond that of an EPC, in that a CKG would also ensure that its data and knowledge be represented using formal approaches and standards, wherever possible, and would help to develop and formulate such standards where lacking. Knowledge in particular areas may not exist in the form of clinical trials or meta-analyses that are the typical foci of EPCs (see Chapter 11). In those cases, efforts could be undertaken to assemble best available knowledge by other means such as human-derived and data-derived methodologies (see Chapters 9 and 10), or to commission or advocate for funding for studies for these purposes.

The formal knowledge thus encoded would be made accessible in knowledge bases, and would be searchable through a rich ontology of categories that would need to be developed. Thus CKGs could provide a means to access related knowledge (but see next section on Centers for Knowledge Management (CKMs), with which the CKGs would need to interact). Access to this knowledge through CKGs and CKMs hopefully would serve as an incentive to investigators to contribute their findings, much as is true for those contributing to public databases in genomics and molecular biology. Once results have been obtained, they would be made available for use by all. To the extent that individual patient data were included in accessible data sets, this of course would require robust mechanisms for deidentification that do not defeat the goals of the research, and models, templates, and procedures for appropriate conduct of research that can be approved by Institutional Review Boards.

The role of CKGs in formalizing and supporting the KG life cycle should thus be regarded as central. The specific tasks in the life cycle, and the approaches that need to be taken by a CKG, are indicated in Table 24-2. Most of the primary research will be carried out and published as always, by individual investigators and teams of researchers. The formalization of the life cycle aims to make the data and results more broadly accessible, and focuses

TABLE 24-2 Knowledge generation (KG) life cycle.

Life cycle task	Responsible party (or parties)	Approach
Generation	Researchers, CKGs	Human-based, data-derived, or meta-analysis-based methods (see Section III)
Validation	Researchers, CKGs	As part of a study, or in follow-up studies Explicit delineation of method used, source, and other indicators of quality and robustness
Refinement	Researchers, CKGs	Follow-up or when results are inconclusive or otherwise motivated by validation results
Representation	CKGs	Storing of the data and results of analyses in defined formats
Update	Researchers, CKGs	Expiration dates for planned re-review, or triggered by conflicting results
Infrastructure	CKGs	Tools, methods, repositories, and standards to support preceding tasks

on explicit delineation of the methods of validation, refinement if necessary, formal representation, and triggering of updates of the research at specified time intervals. The roles of a CKG would be to take ownership of the responsibility for making knowledge of a specific type available, by collecting and assembling it, by generating it itself or commissioning or advocating for studies, and ensuring that the appropriate steps are taken, either by the reporting investigators or by other groups that it commissions, for validating it, refining it, representing it, and updating it.

Besides overseeing the processes of the life cycle, CKGs would identify and facilitate or create a number of infrastructure capabilities, e.g., those aimed at creating the databases, developing ontologies for organizing their contents, determining formal representation requirements for specific kinds of data and results, and providing repositories of methodological resources such as tools for doing meta-analysis, tools for data mining and prediction, and tools for choosing appropriate methods for specific kinds of studies. Some of this work could be outsourced by contracts or grants to investigators, or by adaptation of publicly available or commercial tools.

24.7.2 Knowledge Management (KM) Life Cycle

New knowledge is not generated in a vacuum. The knowledge may amplify on or refine existing knowledge, or it may refute it. It may fill in gaps by addressing areas not previously addressed. Or its relation to existing knowledge may be unclear; for example, if it is contradictory but insufficiently well established to alter accepted practices. Thus as knowledge is generated and validated (via the KG life cycle), it needs to be assimilated into knowledge bases through a process of curation. It needs to be organized (e.g., via ontology tags) to facilitate its association with other knowledge pertaining to similar problems, and to enable its retrieval based on intended user, purpose, setting, or other attributes. These efforts would occur ideally under the auspices of Centers for Knowledge Management (CKMs). Since knowledge collections may be organized around different themes, such as knowledge bases relating to a disease like diabetes or to a clinical application such as CPOE, or to a kind of knowledge like alert messages, the responsible parties for KM should be those that have a particular commitment to seeing knowledge of a particular type or focus utilized in practice. Thus CKMs, analogous to CKGs, might focus on particular types of knowledge, kinds of uses, or problem domains. The focus of a CKM could be the same as that of a CKG, and the same organization could perhaps oversee both life cycle processes. But more likely would be a setting where a CKM focusing on something specific, such as CPOE rules for drug prescribing, might draw on knowledge from a variety of CKGs that deal with knowledge pertaining to drug efficacy in various circumstances.

The specific tasks in the KM life cycle, and the approaches that need to be taken are indicated in Table 24-3. Note that the knowledge bases that are assembled may have different rationales for their organization and maintenance than the knowledge repositories developed through the KG life cycle. The KG life cycle is concerned with creating and validating clinical precepts. The KM life cycle is concerned with means for facilitating the application to practice of those clinical precepts considered to be useful. The goal of KM is not to determine validity of an item of knowledge but to manage the corpus of knowledge in a domain to support its use in CDS. This process, of course, interacts strongly with the KG life cycle. For example, if an item of knowledge contradicts existing knowledge, the KM process could identify those contradictions, providing feedback to the CKG(s) responsible for that type of knowledge, suggesting additional studies or refinements that may need to be done.

TABLE 24-3 Knowledge management (KM) life cycle.

Life cycle task	Responsible party (or parties)	Approach
Curation and content management	CKM	Use of content management tools, ontologies for tagging of content to create and maintain knowledge bases.
Collaborative authoring and editing	CKM	Use of collaborative authoring, editing, and review tools to add and revise content.
Versioning and tracking of changes	CKM	Using these authoring tools, versioning to be automatically tracked as updates occur.
Standards-based dissemination	CKM	Use of existing standards where available. Otherwise, adoption of best possible approach. Work with standards bodies to encourage adoption. Revise to incorporate standards updates.
Localization and update	Individual implementations	Use of jointly developed KM and IE (see Table 24-4) tools to track where knowledge has been modified or incorporated locally, so that updates can be facilitated (still would usually need to be updated manually, however).
Infrastructure	CKM	Tools, methods, repositories to support preceding tasks.

Assimilation of the knowledge into a knowledge base needs to include a means for enabling users to obtain information about the origin and method of generation and validation of the knowledge (from the KG process), and to assess its reliability, relation to existing knowledge, and other factors. As the knowledge is updated over time, therefore, its update history, versions, and provenance need to be tracked.

For the knowledge to be sharable, it needs to be represented in an unambiguous form. For it to be computable and platform-independent, formal standards need to be used. The state of standards definition is unfortunately quite limited (see Chapter 17), with the most well-defined standards relating to rules and logical expressions (see Chapter 12). Given that state of affairs, it may turn out to be the case that KM processes need to be established for formal representation of specific kinds of knowledge, even if not fully worked out as standards. Depending on the stature of the CKM overseeing this activity and the quality and impact of its work, such efforts may serve to catalyze adoption of those representations as first versions of standards.

For CDS, standardization is important not only for the representation of the decision model itself but for the information model defining the data needed and the results produced. The information model needs to specify the vocabulary/taxonomy to be used and the terms within that vocabulary/ taxonomy for naming the concepts representing the data items (see Chapter 14). It also must specify the attributes that ground those data items in terms of particular archetypes (see Chapter 15).

As with the KG life cycle, a number of infrastructure tools and methods need to be developed to support the processes of the KM life cycle, including tools for management of the knowledge bases themselves, methods for ontology management and update, standards-based representation of content, and tools to facilitate collaborative authoring, review, and editing. To facilitate the dissemination and use of a particular type of CDS, tools need to be made available for providing access to knowledge of that form, and for adapting and utilizing the knowledge in building local versions of the CDS capabilities; therefore, they need to work with tools and resources that support the IE life cycle (see next subsection). The latter also need to include the ability to track such local adaptations and uses to facilitate their update when necessary. As with the KG life cycle, development of capabilities to support KM processes does not need to start from scratch. Research and development underway at various institutions such as described in Chapters 21 and 22 could serve as models for the infrastructure and tools to be provided on a more global basis. Open source and commercial tools such as for ontology management, content management, and collaborative authoring and editing could be adapted for use for these purposes as discussed in Chapter 21.

Another issue to be addressed, though, is where the content would come from. Experience with sharing of knowledge resources among institutions or across vendors has not been encouraging, as we discussed earlier. However, I believe this is a matter of achieving the appropriate critical mass. The experience with GenBank and other genomic and molecular databases shows that it becomes worthwhile to individuals to contribute to a communal resource when the resource is considered to be sufficiently valuable. Thus, this appears to be a matter of priming the resource with high-quality, needed knowledge, in order to attract contributors. This priming could either come from well-respected sources that are willing to contribute their knowledge for communal benefit, the establishment of consortia to jointly develop and collate shared knowledge, or the deliberate creation of resources of a particular type that will be broadly needed (e.g., under government aegis). In the United States, a number of academic medical centers have stated in public forums that they would be willing to share their clinical knowledge content, and government-based health care entities such as the Veterans Administration might be expected to do the same. We mentioned earlier the NCPDP work with the U.S. Food and Drug Administration to develop a standards-based specification for a structured product label (SPL) for medications and a shared repository of model rules based on parameters in the SPL for prior authorization approval of prescriptions, should prior authorization requirements be established by payers, as is currently considered likely. CKMs could work with EPCs and entities like the Cochrane Collaboration (or their evolution into CKGs, if that were to occur) to incorporate the results of their analyses into knowledge bases addressing particular clinical problems.

24.7.3 CDS Implementation and Evaluation (IE) Life Cycle

The final life cycle to consider is that which is aimed at determining the most effective way to provide CDS of specific types, or for specific classes of knowledge. Although it is desirable to encourage experimentation in the use of CDS, if we do so without formalizing the process, we run the risk of creating ever more one-of-a-kind implementations that can't be replicated easily. Given the momentum that is building for adoption of CDS, how do we facilitate an orderly transition of experiences and results at single sites to either commercialization, or ideally, to a broader multiplatform adoption process? How do we do the latter without imposing so many additional requirements on the experimentation and evaluation process that they serve as barriers to experimentation? Indeed, can we create a mechanism that both facilitates experimentation and also enables more rapid dissemination and adoption of useful results?

A key to answering these questions appears to be the creation of platforms on which CDS capabilities can be developed, as well as tools to facilitate the development process, where both the platforms and the tools rely on standards-based approaches. In other words, when it becomes desirable to disseminate a particular kind of CDS capability, rather than needing to reimplement it for a whole new systems environment, or having to extract the essence of a successful experiment from an environment in which it is embedded, the experiment itself is carried out in a way that is inherently portable. This can be done by building CDS capabilities as separate components or modules that can be invoked by applications rather than being embedded in the applications. There are several ways to accomplish this modularity; for example, modules can be designed to be invoked by Application Programming Interfaces (APIs), by messaging interfaces, or via Web service calls (the so-called "Service-Oriented Architecture" or SOA).

With methods such as these, the work that is necessary in order to incorporate a CDS approach in a new environment can be confined to the tasks of developing interfaces for (a) triggering the CDS module, (b) passing to the module the necessary host clinical data, and (c) accepting from the module the results of its evaluation. The host environment must also implement a means for responding to those results in the user interface and process flow of the particular application in which the CDS is used; certain paradigms for doing this effectively are likely to emerge from evaluation studies, and should be reported along with those that have been found to be ineffective. It may be that part of this could also be automated in terms of skeletal application tools and templates. The principal task to ensure portability is to require that the interfaces for functions (a) through (c) are standards-based. Further, the CDS module itself can be implemented in such a way that its decision method expects the knowledge that it uses to be in a standard representation, and maintained in an external knowledge base. Table 24-4 describes the IE life cycle in terms of tasks, responsible parties, and approaches involved, as well as infrastructure resources that would be useful.

Messaging standards are being developed for various kinds of CDS such as for the infobutton manager, as we discussed in Chapter 16. Chapter 23 describes a Web services-based approach to invocation of CDS, and at the time of this writing a proposed standard is being developed by HL7 for SOA-based invocation of such Web services. One way to stimulate adoption of these approaches would be to create open source prototypes or skeletal modules for various types of CDS using invocation methods such as these.

TABLE 24-4 CDS implementation and evaluation (IE) life cycle.

Life cycle task	Responsible party (or parties)	Approach
Decision model to be supported	CIE	Decision method designed to use knowledge in a standard representation and maintained in an external knowledge base; decision method implemented in a self-contained module
Application environment and interface	CIE, local sites	CDS module triggered by API, message, or Web service call, data required from host is supplied in accord with standards-based information model, and results received from module conform to the same information model; user interface and process for managing results host-specific, but may be able to be generalized as skeletal applications and templates
Evaluation of effectiveness	Local sites	Experience with implementation is evaluated, to determine how effective the model is in its interaction with the application, and with the user, and its impact on the care process; effectiveness of user interface and process flow adopted, both positive and negative, reported
Feedback	Local sites, CIE	Local site obtains feedback, and collective experiences are assembled at IEC
Modification and update	Local sites, CIE	Local sites modify approach, IEC updates its models and interfaces
Infrastructure	CIE	Prototypes/skeletal modules for CDS and skeletal applications and templates for host user interface and processing; tools for host-standard data mapping; repositories for reporting successes and failures, discussion boards, FAQs and other support for experimentation and evaluation; refinement of standards for messaging and SOA invocation for CDS

The open source community could then refine, modify, and extend these prototypes and skeletal modules, contributing improved versions and instantiations for various specific purposes.

To the extent that academic medical centers, with homegrown systems, and vendors wanted to invoke particular forms of CDS through such modular interfaces, and if they provided the necessary hooks in their applications for doing so, such open-source modules could be utilized directly. Having such hooks in applications would also probably stimulate the commercial marketplace to develop enhanced versions of such applications as well as to create other innovations that would provide added value.

As with the other two life cycles, we postulate the creation of one or more Centers for Implementation and Evaluation (CIEs) to systematically address these needs and to make resources broadly available to facilitate adoption. The CIEs could be focused on particular kinds of uses of CDS or could be more broadly based. There is currently only a limited marketplace of knowledge products, mainly in the form of collections of databases relating to drug formularies, drug interaction tables, and other compilations of tabular data. Once it becomes easier to introduce CDS into a variety of applications, and because of activities such as those of the CKGs and CKMs in support of the other two life cycles, a whole range of opportunities could become available for incorporating CDS in clinical IT systems, particularly in CPOE, EHRs, eRx, and PHR products. To the extent that the interfaces to the modules were generic or standards-based, CDS would be able to be invoked from any proprietary system that incorporates those interfaces, thus encouraging adoption and use. In addition, a set of tools could be made available through CIEs to enable proprietary systems and local sites to review and adapt knowledge used in CDS, when installing a resource as well as when updates occur. To do this, as noted in the previous section, the CIEs need to interact with the CKMs to provide tools that bridge the gap between update of knowledge bases and propagation of those updates in local implementations of CDS.

In summary, the best chance for the three life cycle processes to be formalized would be if entities were specifically created with the responsibilities for specifying and managing the tasks involved in those processes, championing their causes, fostering collaborations, promoting standards, and building infrastructure and tools. The three types of organizations would need to interact with each other where their respective foci overlap (see Figure 24-1), to ensure that their efforts are coordinated. As we discuss in the next section, the mechanisms for establishing and formalizing the needed processes would need to evolve over time. A good place to start, therefore, would be with one or more areas of CDS that are considered both high priority and amenable to formalization given the present state of the art.

24.8 GETTING SPECIFIC: END-TO-END IMPLEMENTATION STARTING WITH HIGH-PRIORITY FOCUS AREA(s)

I suggested in the introduction of this chapter that the oversight body (OB) should put into place an operational unit that is responsible for carrying out projects aimed at deploying the preceding resources for the purposes of actually stimulating the adoption of CDS, evaluating their effectiveness, and refining the process. The goal, as proposed in the discussion on organization of the effort in the introduction to this chapter, would be to start with a somewhat limited focus that has a high likelihood of success, learn from it, and continue to expand capabilities and breadth of focus over time. To begin and build on such iterative cycles of end-to-end implementation, I suggested mechanisms that included initiation and/or funding of projects of limited duration by institutions or consortia, additional funding for some to get them to a state in which they are self-sustaining, and other funds aimed at refining the process of adoption of the approach at other sites, with the ultimate goal that the process be supported through the commercial marketplace.

In order to begin to translate the general capabilities enunciated in the previous sections to practical steps for creating a robust means for dissemination and adoption of CDS, we consider two possible initial foci: rules-based knowledge, in the form of *if...then* rules, such as in CPOE medication checking, lab alerts, and procedure reminders; and order sets. We consider for each the reasons for importance, the status of support for CDS of that type in terms of standardization, existing or potential sources of shareable content, ease of implementation, and opportunity to be self-sustaining in terms of likelihood of stimulating commercial activity. These are only examples of the kinds of factors that should be considered. Actual prioritization should be done under the leadership of OBs with appropriate input from all relevant stakeholders, as we have previously discussed.

24.8.1 Rules-based Knowledge

24.8.1.1 Reasons for Importance

If...then rules are, as can be noted in reviewing Table 24-1, central to all of the five areas we posited as likely to be considered high priority targets for CDS. Error prevention/patient safety depends on rules that can be used to warn about potentially harmful actions before they occur, such as a dangerous medication dose or a contraindication due to allergy, and to alert providers by calling attention to situations that need action such as an abnormal lab result. Quality/best practices depend on rules that are used in reminders or in actionable parts of guidelines that are translated into real-time medication recommendations or alerts. Cost-effectiveness interventions make use of rules for medication substitution, or for preauthorization/prior approval for certain procedures, referrals, or prescriptions. Chronic disease management depends on rules used in reminders, suggestions for home care and prevention, and other recommendations. With regard to public health initiatives, syndromic surveillance depends on event detection triggers of rules that assess presence of possible reportable conditions, and disaster response depends on the ability to monitor patient conditions and alert providers in appropriate circumstances. Thus, a variety of clinical problems and purposes for CDS can be served simply by considering a knowledge base of rules.

24.8.1.2 Status of Standardization

As noted in Chapter 12, a standard for representing knowledge in the form of Medical Logic Modules, Arden Syntax, is already in wide use, albeit with little current sharing and with considerable need for customization for each platform. The GELLO expression language is an approved standard for encoding logical expressions used in rules, and supporting the HL7 Version 3.0 Reference Information Model, although not yet used except in prototype systems. Work would still be needed for any rule formalism, and on defining standard approaches to mapping the data elements used to vocabularies/taxonomies and specifying the particular archetypes needed.

24.8.1.3 Existing or Potential Sources of Shareable Content

A benefit of pursuing rules knowledge as an initial focus is that, as we have mentioned earlier, various current initiatives and actions are poised to stimulate the compilation of rules-based knowledge content by contributing their own rules knowledge to a common repository or by establishing consortia for creating rules knowledge repositories. How successful those efforts will be cannot yet be judged. The extent to which such knowledge would be used by others as a starting point for their own applications remains to be seen, given the limited experience with this occurring to date.

We have also mentioned the skeptics who believe that so much customization will be needed that sharing will be of limited value. But by focusing on this kind of knowledge and compiling well-researched, authoritative repositories, we would be in a position to determine how useful it is. For reasons discussed in the Rationale section of the introduction to this chapter, the need for such shared content could become self-fulfilling once utility is demonstrated.

24.8.1.4 Ease of Implementation

A study by Greenes, Sordo et al. (2004) at Partners HealthCare analyzed computer-based rules knowledge in use at the Brigham and Women's Hospital and Massachusetts General Hospital in 2002. This study examined 2,972 items of rules-based knowledge used in six main applications. Despite the large numbers of individual items of rules-based knowledge, there was a limited range of forms of such knowledge, in terms of approximately 250 rule types. A surprising further result was that logical expressions for this collection of rule types were constructed by Boolean combination of a very limited set of primitive types of relations (41 unique types) referencing a small set of data types (corresponding to 13 HL7 RIM data types).

A conclusion of that work was that it would be quite feasible to implement an external rules engine to support that existing set of rules knowledge. Such an implementation could be homegrown, open source, or based on a proprietary tool, and could operate on a standards-based representation of rules knowledge. Support for the limited range of primitive operations needed would not be difficult. Further, the host EHR standards-based data mapping effort could be confined to the small number of data types used. Last, the limited number of primitive types and data types used for constructing rules would lend itself to development of a user-friendly authoring tool that, for example, would enable creation of expressions by a wizard that permitted the user to select a supported primitive type, and fill in the attributes (specific data elements and target values) from choices limited to that type, and then combine these with Boolean logic. Further, because specification of SOA interfaces is expected to be standardized, this would make the incorporation of rules evaluation capability by an external rules engine easy to achieve in many different types of applications. The interfaces for host triggering of the rule evaluation CDS service, host-data and host-result mapping would be supported as part of the SOA.

24.8.1.5 Opportunity to Be Self-Sustaining in Terms of Likelihood of Stimulating Commercial Activity

Finally, there are considerable opportunities for business. Knowledge bases of authoritative rules can be provided. Rules engines can be provided in the form of implementation toolkits, or rules evaluation services can be offered. Customization/adaptation of existing applications or provision of skeletal applications and templates to incorporate rules is another prospect. Rules editing/authoring and collaboration tools for knowledge management based on public resources could be offered through the CKMs or enhanced versions for use in local KM and adaptation efforts could be made available as commercial products or through Web services.

24.8.2 Order Sets

24.8.2.1 Reasons for Importance

Focusing on deployment of order set capability as an initial effort would primarily address the goals of the first four of the five drivers for CDS. Predefined orders for particular circumstances can reduce the chance of introducing errors. Including orders that are considered to represent best practices, and making it easier to request such orders through their packaging as part of order sets should have a beneficial effect on quality. Similarly, by incorporating preselected orders for particular circumstances, order sets should encourage more cost-effective practices. For chronic disease encounters, the large variety of possible disease presentations (singly or in combination), treatment regimens, and complications almost cry out for order sets that incorporate optimal recommendations for management in particular circumstances.

We have cited two other benefits of order sets not shared by most other means of offering CDS—that their use saves physician time and that they can be readily customized to support individual preferences.

24.8.2.2 Status of Standardization

At the time of this writing, a proposed order set standard is close to adoption by HL7. This includes specification of the composition of individual orders.

Efforts such as those described in Chapter 15 (see also (Sordo, Hongsermeier et al. 2006)) are being developed for organizing orders and order sets into knowledge element groups (KEGs), and developing ontologies that enable them to be selected based on purpose or intention, disease process, setting, or other attributes that will facilitate cataloging of them, sharing, and adaptation. However, these efforts have not yet led to proposals for standardization.

24.8.2.3 Existing or Potential Sources of Shareable Content

With regard to content, CKMs might compile order sets for indications of particular importance such as for settings where there is high frequency of errors, undesirable outcomes, or noncost-effective practices, or a large variation in practice patterns. Repositories of order sets already exist in many institutions and vendor-based systems that address these needs. There are also businesses that offer order sets specifically. So a question to be determined is whether the adoption of best practices through order sets is sufficiently progressing through those available sources. If it were determined that achievement of CDS goals could be accelerated by efforts to specifically develop an authoritative communal order set knowledge base, such efforts could be undertaken as a targeted activity.

24.8.2.4 Ease of Implementation

The use of order sets largely impacts on one primary kind of application, CPOE, and to a lesser extent on e-prescribing. As a result, implementation of the capability to use order sets is a relatively circumscribed task. Models for how to do so through external services or modules have not yet been devised, however, so it is unclear how much effort would be needed to support the various tasks of incorporating or adapting an external standards-based collection of order sets in differing host environments. This would need to include the abilities to customize the order sets for individual practitioners, institutions, or environments, automatically select and suggest order sets in particular patient care settings, and modify or override the orders for a particular patient.

24.8.2.5 Opportunity to Be Self-Sustaining in Terms of Likelihood of Stimulating Commercial Activity

We have noted that there is already commercial activity in providing order sets, both in terms of capabilities embedded in clinical IT systems that provide CPOE and in the form offerings of purchasable/licensable collections of order sets. However, these are largely nonstandard and highly customized. A question to be addressed with respect to the role of a communal collection of order set knowledge is whether its existence would lead to requests for it by health care organizations, and stimulation of adoption of it by commercial IT systems and content vendors as a new starting point for an enhanced level of care.

Selecting the initial focus

Choosing between these two options for initial focus of CDS end-to-end implementation, or selecting another option that might surface from the priority development process, should involve considerations such as we have just outlined. By whatever criteria the selection is made, the important aspect is that the effort should seek to formalize all the tasks of the three life cycle processes that pertain to the deployment of this capability. The experience and feedback from doing this will itself be important to increase understanding of how to build the needed communal resources and how the overseeing organization and its component operational activities should function.

24.9 LOOKING AHEAD: EPILOGUE AS PROLOGUE

We are poised at a point where the need to accelerate efforts for CDS adoption is great, but where ill-conceived or inadequately founded efforts could contribute more to chaos than to benefit. We are already overwhelmed by knowledge, so just having many varieties of it deployed in the form of CDS is no guarantee that patient safety, quality, cost-effectiveness, or other objectives will be achieved. In fact, sorting through and reconciling conflicting knowledge may be particularly frustrating.

As we seek to accomplish approaches to sharing the results of knowledge generation and knowledge management required for the preceding, we also need to continue an active process of experimentation to learn how to best deploy CDS for maximum benefit and acceptability by users. Thus we need to lower the barriers for this process. By considering CDS as an external capability, we are also shifting the paradigm from a built-in functionality of a clinical system to an added value that can be incorporated into clinical applications in a variety of ways. This opens up the process not only to initiative and experimentation but also to business opportunities, by creating niches for content, software, and services that would otherwise not be there.

Thus there are many reasons for moving in the general direction outlined. The road up to this point has been a bumpy one that has been largely unpaved, so it is desirable to shift onto a paved road that will allow our speed to accelerate. However, we can see only part of the distance along the road ahead. Further along, the road may wind in one direction or another, and its path is indistinct so that the direction in which it ultimately goes is not clearly discernible. The action we need to do now is to begin to pave the road. However, we don't want to invest a lot in the paving process until we are sure where the road will take us. Therefore it is prudent to do so for only a short stretch of the road at a time, re-evaluate, and continue.

The process I have outlined appears to be a feasible one for getting us going after a long period of relative dormancy. It is bold, but it is also somewhat cautious and iterative, and requires thoughtful and deliberative effort to organize it and build initial infrastructure. It will require a concerted focus on the problem and a collective willingness to move ahead. It is encouraging to see efforts to do this mounting in various nations, both in standards efforts, national health care infrastructure development, EHR adoption, and professional and public calls to action.

So I do hope that this Epilogue will indeed be a Prologue. The road ahead beckons. The direction we take will shape health care quality, safety, and cost-effectiveness for generations to come.

REFERENCES

ACR (2000). American College of Radiology ACR Appropriateness Criteria 2000. *Radiology* **215**(Suppl): 1-1511.

ACR. (2006). ACR Appropriateness Criteria.American College of Radiology. from http://acr.org/s_acr/sec.asp?CID=1845&DID=16050.

Bodenheimer, T. (2003). Interventions to improve chronic illness care: evaluating their effectiveness. *Dis Manag* **6**(2): 63–71.

Bourgeois, F. T., Olson, K. L., Brownstein, J. S., McAdam, A. J. and Mandl, K. D. (2006). Validation of syndromic surveillance for respiratory infections. *Ann Emerg Med* **47**(3): 265 e261.

Boyd, C. M., Darer, J., Boult, C., Fried, L. P., Boult, L. and Wu, A. W. (2005). Clinical practice guidelines and quality of care for older patients with multiple comorbid diseases: implications for pay for performance. *JAMA* **294**(6): 716–724.

Buono, C. J., Chan, T. C., Brown, S. and Lenert, L. (2005). Role-tailored software systems for medical response to disasters: enhancing the capabilities of "mid-tier" responders. *AMIA Annu Symp Proc* 908.

Chan, T. C., Killeen, J., Griswold, W. and Lenert, L. (2004). Information technology and emergency medical care during disasters. *Acad Emerg Med* **11**(11): 1229–1236.

DeAngelis, C. D., Drazen, J. M., Frizelle, F. A., Haug, C., Hoey, J., Horton, R., et al. (2004). Clinical trial registration: a statement from the International Committee of Medical Journal Editors. *JAMA* **292**(11): 1363–1364.

Dorr, D. A., Wilcox, A., Burns, L., Brunker, C. P., Narus, S. P. and Clayton, P. D. (2006). Implementing a multidisease chronic care model in primary care using people and technology. *Dis Manag* **9**(1): 1–15.

Greenes, R. A., Sordo, M., Zaccagnini, D., Meyer, M. and Kuperman, G. J. (2004). Design of a standards-based external rules engine for decision support in a variety of application contexts: report of a feasibility study at Partners HealthCare System. *Medinfo* **11**(Pt 1): 611–615.

Herxheimer, A. (1993). The Cochrane Collaboration: making the results of controlled trials properly accessible. *Postgrad Med J* **69**(817): 867–868.

Kahn, C. E., Jr., Pingree, M. J. and Longworth, N. J. (1998). A multipurpose model of radiology appropriateness criteria. *Acad Radiol* **5**(3): 188–197.

Kohane, I. S., Masys, D. R. and Altman, R. B. (2006). The Incidentalome: A threat to genomic medicine. *JAMA* **296**(2): 212–215.

Mandl, K. D., Overhage, J. M., Wagner, M. M., Lober, W. B., Sebastiani, P., Mostashari, F., et al. (2004). Implementing syndromic surveillance: a practical guide informed by the early experience. *J Am Med Inform Assoc* **11**(2): 141–150.

McCray, A. T. (2000). Better access to information about clinical trials. *Ann Intern Med* **133**(8): 609–614.

Osheroff, J., Teich, J., Middleton, B., Steen, E., Wright, A. and Detmer, D. (2006). A Roadmap for National Action on Clinical Decision Support. Report. Bethesda, MD, American Medical Informatics Association.

Pino, E., Ohno-Machado, L., Wiechmann, E. and Curtis, D. (2005). Real-Time ECG Algorithms for Ambulatory Patient Monitoring. *AMIA Annu Symp Proc*: 604–608.

Ray, G. T., Collin, F., Lieu, T., Fireman, B., Colby, C. J., Quesenberry, C. P., et al. (2000). The cost of health conditions in a health maintenance organization. *Med Care Res Rev* **57**(1): 92–109.

Seroussi, B., Bouaud, J., Chatellier, G. and Venot, A. (2004). Development of computerized guidelines for the management of chronic diseases allowing to position any patient within recommended therapeutic strategies. *Medinfo* **11**(Pt 1): 154–158.

Sistrom, C. L. and Honeyman, J. C. (2002). Relational data model for the American College of Radiology Appropriateness Criteria. *J Digit Imaging* **15**(4): 216–225.

Sordo, M., Boxwala, A. A., Ogunyemi, O. and Greenes, R. A. (2004). Description and status update on GELLO: a proposed standardized object-oriented expression language for clinical decision support. *Medinfo* **11**(Pt 1): 164–168.

Sordo, M., Dunlop, R., Martin, R. D., McKinnon, B. M., Schueth, A. J. and Greenes, R. A. (2006). GELLO and ePrescribing: Exploring the Use of a Standard for Prior Authorization in Electronic Prescribing. *DSG Technical Report, DSG_TR_2006_002*. Report. Boston, Decision Systems Group, Brigham and Women's Hospital.

Sordo, M., Hongsermeier, T., Kashyap, V. and Greenes, R. A. I., Germany, 2006. (2006). Partners Healthcare order set schema: An information model for management of clinical content. *Computational Intelligence in Healthcare*. Berlin, Springer-Verlag, in press.

Wagner, M. M., Espino, J., Tsui, F. C., Gesteland, P., Chapman, W., Ivanov, O., et al. (2004). Syndrome and outbreak detection using chief-complaint data–experience of the Real-Time Outbreak and Disease Surveillance project. *MMWR Morb Mortal Wkly Rep* **53**(Suppl): 28–31.

Waterman, J., Curtis, D., Goraczko, M., Shih, E., Sarin, P., Pino, E., et al. (2005). Demonstration of SMART (Scalable Medical Alert Response Technology). *AMIA Annu Sympos Proc.*, Washington, D.C.

Weinstein, M. C. and Pearlman, L. A. (1981). Case Study #4: Cost effectiveness of automated multichannel chemistry analyzers. Report, Office of Technology Assessment, U.S. Congress: 1–36.

Wheeler, D. L., Barrett, T., Benson, D. A., Bryant, S. H., Canese, K., Chetvernin, V., et al. (2006). Database resources of the National Center for Biotechnology Information. *Nucleic Acids Res* **34**(Database issue): D173–180.

INDEX

Page numbers with "t" denote tables; those with "f" denote figures